Trace Metals in the Environment 5

Metals, Metalloids and Radionuclides in the Baltic Sea Ecosystem

Trace Metals in the Environment 5

Series Editor: Jerome O. Nriagu
Department of Environmental and Industrial Health
School of Public Health
University of Michigan
Ann Arbor, Michigan 48109-2029
USA

Other volumes in this series:

Volume 1: Heavy Metals in the Environment, edited by J.-P. Vernet (Out of Print)
Volume 2: Impact of Heavy Metals on the Environment, edited by J.-P. Vernet (Out of Print)
Volume 3: Photocatalytic Purification and Treatment of Water and Air, edited by D.F. Ollis and H. Al-Ekabi (Out of Print)
Volume 4: Trace Elements – Their Distribution and Effects in the Environment, edited by B. Markert and K. Friese

Trace Metals in the Environment 5

Metals, Metalloids and Radionuclides in the Baltic Sea Ecosystem

Piotr Szefer

Department of Food Sciences
Medical University of Gdańsk
80-416 Gdańsk, Poland

2002

ELSEVIER

Amsterdam · London · New York · Oxford · Paris · Shannon · Tokyo

ELSEVIER SCIENCE B.V.
Sara Burgerhartstraat 25
P.O. Box 211, 1000 AE Amsterdam, The Netherlands

First edition 2002

Library of Congress Cataloging in Publication Data
A catalog record from the Library of Congress has been applied for.

ISBN: 0 444 50352 8

∞ The paper used in this publication meets the requirements of ANSI/NISO Z39.48-1992 (Permanence of Paper).
Printed in The Netherlands.

To memory of my Parents

Acknowledgements

I particularly wish to express my special appreciation to Professor Jerome Nriagu, the Editor of *the Science of the Total Environment*, for encouraging me to write this book. I would like to thank Mrs. Mary Malin and Mr. Peter Henn, the Senior Publishing Editors, Mrs. Conny Kreinz, the Production Editor, as well as Mr. Simon Richert from Elsevier, for their co-operation, understanding and great patience. I particularly wish to thank Elsevier for their willingness to add extra material, even at a late date, to ensure that the book is up to date. I am also very grateful to Dr. Eric I. Hamilton, the Editor-in-Chief of *the Science of the Total Environment*, for his critical and constructive remarks concerning all my manuscripts published in the journal; scientific content of these papers constitutes important part of the book.

My most sincere thanks are extended to Dr. Geoffrey P. Glasby, Marine and Environmental Consultant from Sheffield, for many stimulating discussions during his visits to my laboratory. I am also especially indebted to Professor Philip S. Rainbow from the Natural History Museum in London for much helpful discussion which undoubtedly contributed to improvement of the book quality. My wife Krystyna and daughter Magdalena are heartily thanked for their patience and support. I would like to thank Dr. A. Latała and Dr. J. Warzocha for their help in the collection of literature data concerning geographical distribution of phyto- and zoobenthos in the marine environments.

I am grateful to various publishers and authors for permission to use figures, tables and photographs from previously published papers which are their copyright.

Many thanks to Urszula Wawrzyńska and Maksymilian Biniakiewicz from Printing-house of the Foundation for the Development of Gdańsk University who have contributed to the text typesetting of the manuscript.

Piotr Szefer

Gdańsk
Spring 2001

Preface

"The external world has proved to be surprisingly obedient to logic".

Bertrand Russel

The Baltic Sea is a unique basin, being productive with intensive fishing potential and has therefore been the object of many studies. It is a brackish, non-tidal, relatively shallow and semi-enclosed sea. The Baltic is located at a high latitude, hence one of its characteristic features is ice. Another unique geographical pattern are the archipelagos located off the coast of Stockholm which consist of more than 25 000 islands. The relative ionic concentration of toxic substances e.g. chemical elements is generally higher in the low-saline Baltic Sea than compared to the North Sea. The drainage area is densely populated, heavily industrialised and is characterized by intensive agriculture. Therefore this sea is thought to be extremely polluted and, with a wide range of contributing factors to its level of pollution, there are obvious implications for the people, flora and fauna in the surrounding Baltic states.

Although the Baltic Sea is divided into natural basins by bottom topography and into economic sectors by man it represents an integrated system, highly sensitive to what happens in its contact zones with the adjacent North Sea, the land and the atmosphere. Areas suffering from pollution are unevenly distributed within the sea. Among the key factors influencing this distribution are: distance from the transition zone between the North Sea and the Baltic Sea; local hydrologic and hydrographic conditions; the catchment area of the adjacent rivers and the extent of conservation measures in the adjacent areas. At the end of the 1960s great attention was paid to the marked deterioration of water and biota in the Baltic Sea, resulting in the preparation and signing of the Convention on the Protection of the Marine Environment of the Baltic Sea Area (i.e. the Helsinki

Convention) by all riparian countries. Considering the geopolitical situation in this region, the Helsinki Convention of 1974 should be regarded as a unique international agreement, covering all sources of pollution of the open sea areas of the Baltic. However, until 1992 the coastal zones were not included in the Helsinki Convention.

Since the beginning of the 1980's, a series of assessments covering the wide range of ecological problems has been published by the Helsinki Commission (HELCOM). These assessments, prepared by numerous expert groups, summarise scientific results from the beginning of the century and reflect the present status of knowledge resulting from the research and monitoring programmes. The achievements of these collective studies are utilised in this book as valuable background information and are cited under the name HELCOM. Also since the 1980's, our knowledge of the biogeochemistry of the Baltic Sea has improved remarkably with results being published at first mostly in national journals and later also in international journals with a biogeochemical and environmental pollution orientation. This book has partly synthesised the wide-ranging research done, and it is envisaged that it will prove to be a valuable addition to the literature.

The book discusses the distribution and cycling of metals, metalloids and radionuclides in the Baltic Sea and, where needed, in adjacent northern or other seas. The main aim of the book is to acquaint the reader with the distribution, bioavailability, fate and sources of chemical pollutants in the Baltic environment (seawater, suspended matter, bottom sediments, ferromanganese concretions, seaweed, plankton, molluscs, crustaceans, nereids, fish, waterfowls, marine mammals). The distribution of pollutants in the atmosphere (aerosol, wet and dry fall-out) as well as in the rivers of the Baltic catchment have also been considered. Justification for such an approach is that the atmosphere and most seas do not have borders, even in the case of such a basin as the semi-enclosed Baltic Sea which is connected with the North Sea via the Danish Straits. Therefore chemical elements and radionuclides are often transported long distances from their emission sources via atmospheric circulation, sea currents and rivers. Since the marine cycle of bioelements such as C, N, P and Si is often strictly related to the fate of metals and metalloids, some aspects concerning these nutrients have also been included in the book.

Because some organisms e.g. marine mammals, waterfowls and fish can be effective carriers of pollutants from even remote areas, concentration data for Baltic migrants were compared together (where needed) with those corresponding to non temperate zones e.g. sub-Arctic waters of the Northern Hemisphere. In the case of sedentary organisms, such as phyto- and zoobenthos, worldwide data were cited in the book because of the universal biomonitoring significance and utilisation of the sedentary bottom animals (e.g. Mytilidae) having a similar affinity to most trace elements irrespective of their geographical habitation. Knowledge of the chemical composition of Baltic benthal organisms and those from other geographical areas allows us to estimate the pollution status of compared marine environments, although it should be borne in mind that some environmental parameters e.g. salinity can influence bioaccumulation of several trace elements in biota.

In order to set the data in context, characteristics of the main features of both the abiotic (general characteristics, distribution, hydrological and geochemical features), and biotic (taxonomy – classification to particular categories, habitat, food habits) compartments of the Baltic Sea are presented. Particular components of the Baltic ecosystem are considered as potential monitors of pollutants. Budgets of chemical elements and the ecological status of the Baltic Sea in the past, present and future are presented. Estimates of health risks to man in respect to some toxic metals and radionuclides in fish and seafood are briefly discussed. The book is mainly directed to marine chemists, geochemists, environmentalists, biologists, ecologists, ecotoxicologists, educators in marine sciences as well as to students of oceanography. Although the Baltic Sea has been widely studied it is hoped that the book makes possible the identification of gaps in our environmental knowledge with certain sections establishing possible priorities, key areas or strategies for future research.

Piotr Szefer
Gdańsk, Poland
Spring 2001

Contents

Chapter 1

Introduction

A. CHARACTERISTICS OF THE BALTIC SEA BASIN

Regional setting

The general characteristics (meteorology and chemical oceanography; fishes and fisheries, pollution, geology, international management and co-operation) of the Baltic Sea including environmental state of its particular subareas have been well and detailed described in a number of major text books, monographs, reports and articles (see for example: Manheim, 1961; Hartmann, 1964; Fonselius, 1969; Magaard and Rheinheimer, 1974; Łomniewski et al., 1975; Gudelis and Emelyanov, 1976; Millero, 1978; Dybern and Fonselius, 1981; Ehlin, 1981; Blazhchishin and Lukashev, 1981; Grasshoff and Voipio, 1981; Hällfors et al., 1981; Kullenberg, 1981; Lisitzyn and Emelyanov, 1981; Ojaveer et al., 1981; Sjöblom and Voipio, 1981; Winterhalter et al., 1981; Blazhchishin, 1982a, 1982b, 1982c; Emelyanov and Pustelnikov, 1982; Elmgren, 1984; Fonselius et al., 1984; Falkenmark, 1986; HELCOM, 1986, 1998a; Augustowski, 1987; Franck et al., 1987; Ambio, 1990a, 1990b; Anon, 1990; Granéli et al., 1990; Mikulski, 1991; Emeis et al., 1992; Matthäus, 1992, 1993a, 1993b; Matthäus and Francke, 1992; Winterhalter, 1992; Bergström and Carlson, 1993; Hägerhäll, 1994; Majewski and Lauer, 1994; Emelyanov, 1995; Harff et al., 1995; HELCOM, 1996; Huckriede et al., 1996; Trzosińska and Łysiak-Pastuszak, 1996; Gingele and Leipe, 1997; Jensen et al., 1997, 1999; Lemke et al., 1997, 1998; Rheinheimer, 1998; Jansson and Dahlberg, 1999; Łysiak-Pastuszak, 1999; Sokolov and Wulff, 1999; Falandysz et al., 2000; Kautsky and Kautsky, 2000; Blomqvist and Heiskanen, 2001; Lemke et al., 2001) and therefore it is not the intention to repeat this published information.

Rather attention will be directed forward the presentation of these basic environmental problems shortly which are linked with the fate of selected chemical elements in the Baltic Sea.

The Baltic Sea is a young postglacial inland sea, with its drainage basin over four times its sea area (Fig. 1.1). The drainage basin – densely inhabited and urbanised is used mainly for agricultural and industrial purposes (Falandysz et al., 2000). The Baltic Sea is connected to the North Sea (Atlantic Ocean) via the

Fig. 1.1. Map of the Baltic Sea showing its large drainage basin. After Bergström and Carlsson (1993); modified.

Kattegat and narrow inlets of the Belt Sea and Sound – the transition zone. The Baltic Proper is the largest subdivision of the Baltic Sea. It has a surface area of 211 069 km^2 (51% of the whole sea) and the volume of 13 045 km^3 (60 % of the total) (Melvasalo et al., 1981; HELCOM, 1990, 1996). It covers the area between the Darss Sill (18 m depth) in the transition zone and the Gulfs of Bothnia, Finland and Riga. Several regions are distinguished based on the bottom topography: the Arkona Basin, the Bornholm Basin and the Gotland Basin (Fig. 1.2). The Gotland Basin in subdivided into its eastern and western parts. The Gdańsk Basin is a southward extension of the Eastern Gotland Basin; it is frequently treated as a separate natural region because the Gdańsk Deep (max. depth 118 m) acts as a sink for the suspended matter carried by the Vistula River, which is the largest river draining the Baltic Proper (Falandysz et al., 2000).

Continuous inflow of more saline water from the North Sea into the Baltic Sea is hampered by shallow sills. Only major inflows, approximately 100 km^3 in volume, reach the Bornholm Basin. To renew the deep or intermediate water lay-

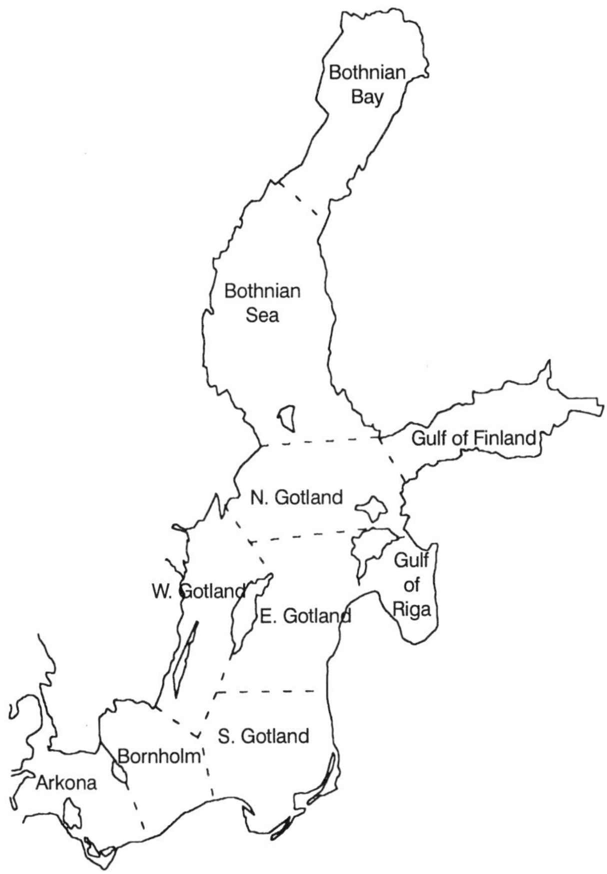

Fig. 1.2. Map of the Baltic Sea showing its subareas. After Danielsson (1998).

ers in the Gotland and Gdańsk Basins, even greater volumes of dense oceanic water of high salinity, low temperature and high oxygen concentration are required. These proceed in cascades eastward and northward through the Słupsk Furrow which has a sill depth of approximately 60 m. Major inflows occur at irregular intervals, mostly in winter. Their impact depends not only on the volume but also on its salinity and the duration of the event. The causes of these inflowing water are not well understood but meteorological and hydrological conditions play a great role (Falandysz et al., 2000).

Due to an extensive river run-off, there are pronounced horizontal salinity gradients in the surface layers of the Baltic Sea (Fig. 1.3). Moreover, rivers flowing into the Baltic Sea carry various types of pollutants that could negatively affect the ecological balance of the sea (Falkenmark, 1986). The salinity of surface water is highly variable within each region. In the Baltic Proper, it ranges from about 1 psu in estuarine areas up to 9 psu in the western region (HELCOM, 1986).

Cyberski (1995) reported statistically significant long-term trends in the seasonal outflows of the rivers draining into the Baltic whereas the mean annual flow rates of most rivers displayed only some fluctuations with time. These seasonal changes began in the 1920s and have accelerated since the 1970s. They coincide with the energy crisis and the resulting attempts to improve water storage facilities for electricity generating stations. Seasonal variations in the river outflow to the Baltic Sea as well as recent climatic changes may also affect different ele-

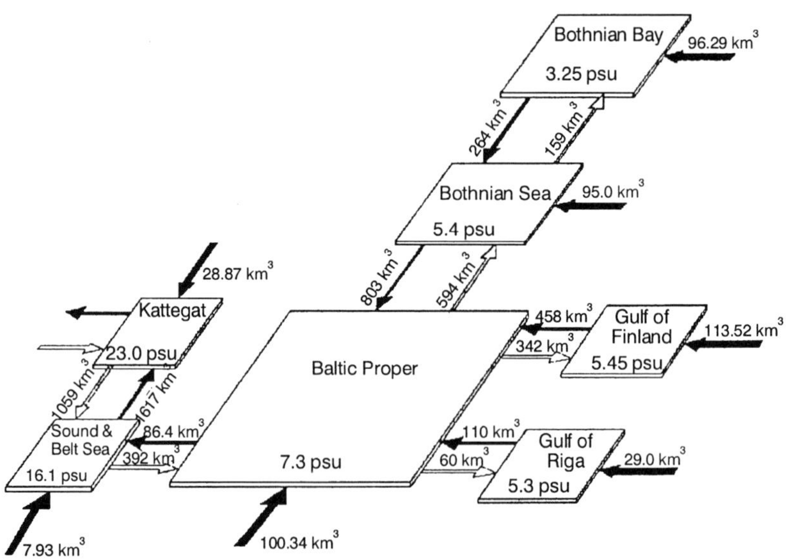

Fig. 1.3. Annual water exchange between the Baltic regions (km³), mean long-term salinity of surface water (psu) and regional riverine inflow (km³, thick arrows). After Falandysz et al. (2000); modified.

ments in the water balance. As an example, they may influence the salinity, one of the fundamental factors controlling environmental conditions and the distribution of biological species within the Baltic Sea (Falandysz et al., 2000).

A horizontal salinity gradient also exists in the deep waters of the Baltic Proper. Fonselius et al. (1984) studied 100-year series of salinity data. They found that salinity varied from over 14 psu to about 21 psu in the near-bottom layer of the Bornholm Deep, whereas in the southern and northern basins these variations were less, e.g. from over 11 to 14 psu in the Gotland Deep.

Changes in the surface water temperature in the Baltic Sea are governed by the increased continental influence in the east and the considerable north-south extent of the Baltic Sea (Melvasalo et al., 1981). In the Baltic Proper, the average winter sea surface temperatures are around 2°C. The extent of ice cover is very variable, depending on the severity of winter and the region (Majewski and Lauer, 1994). The mean sea surface temperature is 16–18°C in the southern part, about 16°C in the central part and 15–16°C in the northern part of the Baltic Proper during August. During 1989–1993, the mild winters caused positive water temperature anomalies (HELCOM, 1996). The deep waters have more or less stable temperatures (5–8°C), which are influenced by the frequency and season of the major inflows.

The relationships between separate elements of water budget and seasonal variations in water temperature result in marked vertical gradients in water density of the Baltic Sea. In summer, warm surface water is separated from the cold winter water by the thermocline at a water depth of approximately 20 m. The main barrier between the low salinity upper (isohaline) layers and higher salinity (heterohaline) deep layers occurs at 40–70 m, on the average, depending on the region and the period under consideration. Major inflows of water from the North Sea significantly change the location of the permanent halocline within the water column and the relative volumes of the isohaline and heterohaline layers (Falandysz et al., 2000).

The residence time of Baltic Sea water, estimated from the salinity distribution, to be in the range of 20–35 years, varies spatially. Those elements which take part in the biogeochemical processes spend much shorter time in the Baltic. Wulff et al. (1990) calculated that the average residence times for silicate, phosphorus and nitrogen compounds are 13, 11 and 5 years, respectively.

Flora and fauna in the Baltic Sea

The main natural factor determining the occurrence of species in the Baltic is low salinity, which limits the occurrence of many marine species as well as fresh water species resulting in a relatively low biodiversity (Falandysz et al., 2000). Most of the typically marine species (e.g. Echinodermata, Porifera, Anthozoa) do not occur in this region or occur on the edge of their distribution range, therefore even small changes in environmental conditions may influence their spatial distribution. A decreasing number of marine species along with diminishing salinity (due to increasing distance from the Danish Straits) is a characteristic feature of the Baltic Sea. The least number of species occur in waters with salinity ranging from 5 to 8 psu, that is, salinity of the northern part of the Baltic. Baltic Proper is thus a region intermediary between Kattegat and transition zone, reach in marine species, and Bothnian Sea, where only a few marine species occur. The low temperature is also important factor limiting immigration of marine organisms into the Baltic (Dahl, 1956; Segerstråle, 1957, 1972; Remane, 1958). In addition, the relatively young age of the Baltic having been a brackish sea for only 6000 years, should be taken into account. There are therefore not many species which can be regarded as typical Baltic, brackish-water species. Most species have immigrated to the Baltic Sea from near-by seas and freshwater bodies during different periods up its evolution, beginning with the last glacial period (about 12,000 years ago). There are four groups of natural immigrants in the Baltic flora and fauna. The first group consists of Northwest European euryhaline marine and brackish-water species, e.g. *Macoma balthica* – Bivalvia and *Clupea harengus* – Pisces, and the second are freshwater species, e.g. *Theodoxus fluviatilis* – Gastropoda and *Perca fluviatilis* – Pisces (Falandysz et al., 2000). The third and fourth groups include glacial relikts which reached the Baltic either through ice-dammed lakes from the Syberia, e.g. *Saduria entomon* – Isopoda, *Mysis relicta* – Mysidaecea, or by a westerly route through the sea, e.g. *Astarte borealis* – Bivalvia, *Pontoporeia femorata* – Amphipoda. This migration process still continues (Dahl 1956; Segerstråle 1957; Jansson, 1972; Magaard and Rheinheimer, 1974; Elmgren, 1984; Lozan et al., 1996).

The main coastal and marine biotopes

Sandy coasts (moraine landscape formed by glacial and postglacial processes) dominate the shores of Germany, Poland, Lithuania, Russia, Latvia as well as southern Sweden. Sandy coasts often have an accumulative-abrasive character; sandy beaches and dunes in various stages of succession (from white, green, grey dunes to brown dunes covered by forests – e.g. Łeba in Poland) are typical elements of such coasts. High active cliffs, so-called moraine cliffs built of clays and sands are also present. In the western part (e.g. Rügen Island) cliff and rocky coasts (bedrock on Bornholm) are found (Falandysz et al., 2000).

In the southern part of the Baltic Proper the characteristic elements are lagoons: Szczecin Lagoon (Oder Haff), Vistula Lagoon and Curonian Lagoon. The coastal lakes are also typical elements of the southern coasts.

They are a few types of coastal salty meadows as well as coastal bogs which are a typical element of the coastal marshes. These are pit bogs of two types – "high" fed by rain waters and "low" – fed by ground and surface waters. Large pit bog complexes are located along the southern coasts (e.g. along Łebsko Lake in Poland). "Low" pit bogs do not form large complexes, but are dispersed as small patches along the entire coast in meadow and pasture complexes.

The pelagic coastal biotopes are found within depths down to 15–25 m where interactions between waves and the see floor usually occur. Pelagic offshore biotopes are the water body of the open Baltic Sea area deeper than 15–25 m usually without interaction between wave orbits and the sea floor. The offshore biotopes can be divided into water body above and below the halocline (Falandysz et al., 2000).

The sea floor of the coastal zone is dominated by sandy sediments mixed with gravel deposits. In the deep water zone, silty sediments prevail (Lozán et al., 1996; HELCOM, 1998a).

Eutrophication

Seasonal and annual variations in the concentrations of nutrients in the Baltic Sea have been widely studied and extensively described in the scientific literature. Because of the differences in climate and bathymetry within the Baltic Sea, they are usually referred to particular regions and/or water bodies (Melvasalo et al., 1981; HELCOM, 1987, 1990, 1993, 1996).

Seasonal fluctuations in the nutrient concentrations in surface waters of the Bornholm and Gdańsk Deeps and the southern part of the Gotland Basin, averaged over 20 years, show distinct temporal and spatial differences in the accumulation pattern during the winter as well as the uptake by autotrophic organisms during spring. There is a time-lag of about 2–4 weeks in the accumulation and assimilation peaks, when moving from the Arkona Basin toward the northern Baltic. Another time-lag, of about 1–2 weeks, occurs between the coastal zone and the off-shore areas (Falandysz et al., 2000).

In the 1990s, the winter nutrient concentrations in the photic layer become much more equal throughout the off-shore area of the Baltic Proper. However, exceptions were found in the northern Baltic (the Landsort Deep with much elevated phosphate and nitrate content), as well as in the southern Baltic (the Gdańsk Deep with much elevated nitrate content). Comparing with the 1960s, an overall concentration increase took place: 1.5–5 times for nitrate and 2–3.5 times for phosphate, depending on the region. During the vernal phytoplankton blooms the pool of assimilable nitrogen and phosphorus compounds was already consumed by June–July in all areas except the estuaries. Nitrate depletion in warm water creates conditions promoting the growth of blue-green algae, which are able to make use of N_2 and add several hundred thousand tons of nitrogen to the waters of the Baltic Proper. From summer until December nitrogen is a limiting nutrient in the Baltic ecosystem, and the nitrogen content appears to be almost balanced in most regions, with respect to input versus uptake. However, some exceptions were recognised, *viz.* the Pomeranian Bay and the most inner part of the Gulf of Gdańsk, where phosphorus has becomes a temporary limiting nutrient at the beginning of summer since the 1980s (Trzosińska, 1992; Falandysz et al., 2000).

In contrast to nitrate and phosphate, silicate has never been the limiting factor for productivity of the Baltic Proper. However, since the 1980s, almost complete silicate consumption has occasionally occurred following vast phytoplankton blooms. In spite of some decline found in the 1990s in the silicate uptake, amplitudes in silicate concentrations were high, 5–7 mmol m^{-3} annually. Seasonal fluctuations of silicate display evident changes as a consequence of the autumnal species development. Such fluctuations were previously observed for the phosphate and nitrate, as well. Recently they flattened in the southern Baltic, where extremely low concentrations of nitrate and phosphate and the supersaturation of surface water with oxygen cover the whole summer and autumn, until December. This situation can be partly attributed to mild winters and variations in the riverine run-off. The accumulation of nutrients starts in January–February. At the peak of nutrient concentration during winter, the mean molar ratio of nitrate to phosphate is approximately 7 in the Bornholm Deep and the Gotland Basin, but as high as 10 in the Gdańsk Deep. When compared with the 1960s, this means an increase in the N/P ratio by few percent for the off-shore regions, and by 50 % for the Gdańsk Basin (Falandysz et al., 2000).

Before the eutrophication accelerated in the 1970s, the N/P ratios in the trophic zone of the Baltic Proper were significantly lower than the Redfield ratio (16:1), which reflected the steady state relations between the environment and the biota in the ocean. Even so, nitrate and phosphate have been taken up in proportions approximating the Redfield ratio. HELCOM (1987) investigated the uptake of nitrogen and phosphorus during the vernal phytoplankton bloom in the Bornholm Basin and found the relation to be about 15:1. A somewhat lower mean value (14:1) was found for the spring/summer species in the southern Baltic, including the off-shore and coastal areas (HELCOM 1996). Interregional defferences were, however, considerable. The mean uptake ratio of silicate versus phosphate was close to the Redfield ratio; it ranged from 13:1 in the Gotland Basin to 18:1 in the Bornholm Deep.

Variations observed in saturation with oxygen in the near-bottom water layer reflect a seasonality in the oxygen utilized in respiration and remineralisation processes, though they are to a certain extend overwhelmed by the hydrographic occurrences, such as occasional oceanic inflows, relatively slow water advection, vertical density gradient weakening northwards and the long stagnation period. Substantial fluctuations in the phosphate concentrations are connected with their resuspention or remobilization from the bottom sediments in accordance with alternating oxygen conditions. Silicate also accumulates in the deep waters whenever dissolved oxygen concentrations decline. On the other hand, decreasing redox potential promotes the denitrification activity. It has been calculated that denitrification is responsible for the overall nitrogen loss of 470 000 tons annually (HELCOM, 1990).

A variety of the input and sink mechanisms, as well as temporal and spatial differences in their efficiency, do not permit any realistic mass balance calculations. Nevertheless, nutrient budgets calculated by Wulff and Stigebrandt (Ambio, 1990) for phosphorus, nitrogen and silicate in particular parts of the Baltic Sea in 1971–1981 are very impressive and contain some management implications regarding the desired reduction in the pollution loads.

The first signs of the increasing fertility were reported in the mid-1970s (Melvasalo et al., 1981, HELCOM, 1987). The long-term trends, calculated by means of approximately 20 year data series, were in most cases highly significant and positive from the statistical point of view. In surface water of the Baltic Proper, the mean annual accumulation rates of phosphate during the winter seasons ranged from 0.015 to 0.26 mmol m^{-3} and of nitrate from 0.17 do 0.34 mmol m^{-3}, depending on the region. Even a higher rate, exceeding 2–4 times that of the surface water, was found for phosphate in the deep water layers. In spite of anoxic conditions, nitrate accumulated in some water layers of the Baltic deep basins (Nehring, 1989).

In the 1980s, when loads from external sources were still high, the rate of eutrophication slowed down. The most characteristic feature of that period was the long-lasting stagnation in the Baltic deep waters, the longest ever been observed during the Twentieth century. As a result of the diminishing salinity and increasing temperature of the deep waters, the weakening vertical density gradient supported downward transport of oxygen and upward transport of nutrients over a vast area of bottom at the intermediate water depths (HELCOM, 1990). The long-term increase in the phosphate and nitrate concentrations continued, but was, interrupted by periods with decreasing concentrations. It has been found almost cyclic behaviour in the phosphate and nitrate accumulation in the Gdańsk Deep of 3 and 6–7 years (HELCOM, 1990). This was probably caused by variations in the atmospheric circulation affecting both the riverine run-off and the oceanic inflows.

At present, the concentrations of assimilable compounds of phosphorus, nitrogen and silicates in the photic zone of the Baltic Proper are at a stable level, though sufficiently high to support intensive primary production. During the last few decades the phytoplankton primary production has almost doubled in some areas, with a resultant doubling of phytoplankton biomass and its subsequent sedimentation (Ambio, 1990).

Biological effects of eutrophication

Eutrophication is considered to be the main anthropogenic factor influencing life in the Baltic. The most important effects of eutrophication are such as increasing primary production, decrease in water transparency and increased organic matter sedimentation resulting in oxygen depletion occurrence. There is not much evidence of primary production increase, mainly due to large natural annual phytoplankton variability, relatively infrequent sampling, influence of local factors and, finally, changes in measurement techniques. However, intensity of phytoplankton blooms may be a general indicator of primary production increase. More frequent blooms of toxic algae may also be related to eutrophication. In the Baltic Proper, no major negative effects related to harmful algae have been observed during phytoplankton blooms, although blue green algae, toxic to mammals, have been found, e.g. *Nodularia spumigena, Anabaena lemmermanii, Micro-*

cystis aeruginosa, Aphanizomenon flos-aquae, and also *Dinophysis acuminata, D. norvegica* and *Prorocentrum minimum*. It has proved difficult to establish trends in the abundance and biomass of zooplankton, mainly due to lack of long-term measurements and to changes in sampling methodology. Distinctive, often drastic, changes, which might be an indirect indication of the influence of euthrophication on Baltic marine life, were observed in benthic macroalgae and vascular plant composition and distribution, during the 1970s. A decrease in water transparency may explain the decrease in depth range of bottom plants. Such changes were observed along the coasts of Latvia, Lithuania, Russia, Poland, Germany and the southern coast of Sweden. *Fucus vesiculosus* communities underwent the most drastic changes, and the community has vanished in some regions. In the shallow littoral zone, many species of red and brown algae have become extinct, e.g. *Fucus vesiculosus, Furcellaria lumbricalis*. Others, e.g. vascular plants such as sea grass – *Zostera marina* occur within more limited areas. In their place, opportunistic green algae (*Enteromorpha intestinalis, Cladophora* sp.) and filamentous red algae from the Ectocarpaceae genus *(Ectocarpus* and *Pilayella*) have become dominant (Falandysz et al., 2000).

Long-living bottom fauna also reflect the adverse effects of excessive nutrient discharges to the marine environment. Bottom organisms depend on food of pelagic origin. Increased sedimentation results in both positive and negative changes in benthos. Positive effects include an increase in biomass and abundance of macrozoobenthos observed in some regions above the halocline. Negative effects include a decrease in species diversity through elimination of species less resistant to environmental changes and a concomitant increase in opportunistic species. The most drastic, adverse changes are noted below the halocline. Long-term oxygen deficits, resulting from increased sedimentation, caused changes in species composition, domination structure, including, in some cases, even the total disappearance of the macroscopic life on the bottom. In the first half of the twentieth century, Bornholm, Gdansk and Gotland Basins were inhabited by numerous bottom fauna species. The total extinction of macrozoobenthos on the Bornholm Basin bottom was observed for the first time in the early 1950. Presently, the bottom of deeps below 70–80 m depth, shows no signs of macroscopic life, and sediments are covered by anaerobic bacteria. There is a lot to suggest that oxygen deficiency in the deep water has contributed to low effectiveness of cod spawning. Cod may hatch only in waters of 10–11 psu minimum salinity, which allows spawn to float in pelagic zone. In less saline waters the cod eggs fall down to the bottom and die. In the Bornholm Basin, where waters are sufficiently saline for effective spawning, oxygen deficits occurring lately as a result of lack of inflows and eutrophication, became a limiting factor in deep water zone ($<$ 70 m). Also, observed recently, decrease in salinity causing halocline uplift, which in turn, widens the water layer not influenced by convection mixing, diminishes effectiveness of cod spawning. In the shallow littoral zone, increasing sedimentation of organic matter together with a lack of water mixing contribute to summer oxygen deficiencies, which in turn adversaly influence primarily bottom perennial species (e.g. Pomeranian Bay, Gulf of Gdańsk) (Magaard and Rheinheimer, 1974; Jansson, 1972; Jarvekulg, 1979; Kautsky et al., 1986; Cederwall and Elmgren, 1990; Andell et al., 1994; Lozán et al., 1996; HELCOM, 1996, 1998a).

Industrial production in the drainage area

Several authors (Bruneau, 1980; Elmgren, 1989; Lithner et al., 1990; Backlund et al, 1992; Jonsson et al., 1996; Rheinheimer, 1998; Jansson and Dahlberg, 1999) reported on man's impact on the Baltic ecosystem as well as the past and recent pollution sources in its drainage area. Riverine and direct loads of pollutants (heavy metals and nutrients) into the Baltic Sea are an important environmental problem (HELCOM, 1993, 1998a). Therefore, the monitoring survey of trace elements and radionuclides is necessary to control the anthropogenic input of pollutants and contaminants to the Baltic Sea (HELCOM, 1991, 1993, 1997a, 1997b, 1998a, 1998b). The industries in the Baltic countries are largely based on locally available row materials, e.g. the deposits of Fe, Cu, Pb and Zn ores which support numerous steel mills and stainless steel works, copper and zinc smelters and aluminium refineries. Some major industrial regions located along the coasts of the Baltic Sea are presented in Fig. 1.4. This is reported that riverine heavy metals load is the largest source of total pollution load amounting to ca. 90%. The municipal and industrial wastewater discharges as well as diffuse discharges are probably the predominant anthropogenic sources in the riverine load (HELCOM, 1998a). According to Lithner et al. (1990) the anthropogenic loads of Cd, Pb and Hg to the Baltic Proper were from 5 to 7 times higher than the background loads. This pollutant input has been reflected by increasing concentrations of Cd, Cu and Zn in fish during 1980s. However, Pb showed a decreasing temporal trends possibly owing to the significantly reduced air emissions from car traffic in Finland, Sweden, Denmark and Germany (HELCOM, 1996; Jansson and Dahlberg, 1999). The Bothnian Bay catchment area comprises 260,675 km^2 of which 56% belongs to Finland, 44% to Sweden and < 1% to Norway (HELCOM, 1998a). According to Bruneau (1980) both Finland and Sweden have had steel mills on the Bothnian Bay. Finnish stainless steel plants possibly have discharged Ni and Cr from the pickling operations. The Finnish fertiliser plant located the most northern in the drainage area and Swedish forest industries – on the coast as well as pulp mills in this regions are suspected to be emitters of pollutants to the Bothnian Sea (Bruneau, 1980). The Bothnian Sea catchment area comprises 220,765 km^2, of which 80% belongs to Sweden, 18% to Finland and 2% to Norway (HELCOM, 1998a). Finland has copper smelters, Sweden – aluminium plant; a still mill and several stainless steel plants are located in the area. The chemical industry is predominantly located in Finland, i.e. refinery, fertiliser and chlorine plants and in Sweden – chlorine and PCV plants. It is important to note that chlorine plants are based on the mercury method but discharge of this element is very low owing to extensive measures to its reduce. Recently a non-mercury type

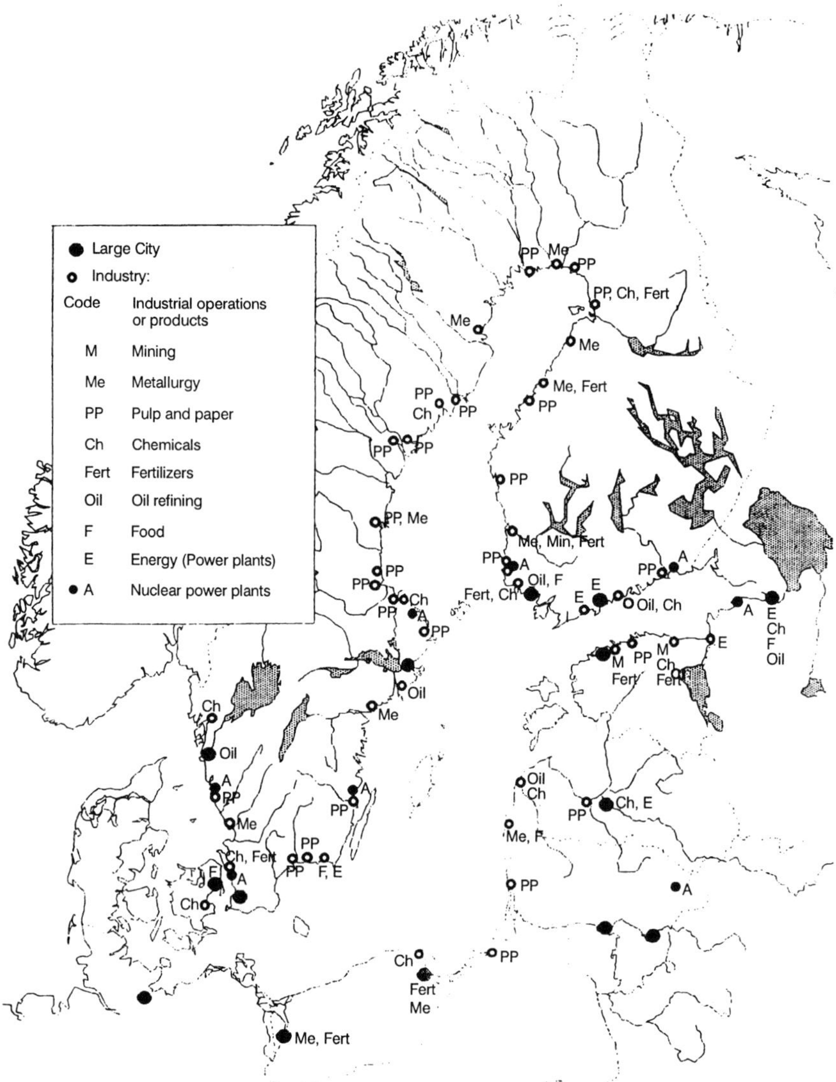

Fig. 1.4. Some major industrial regions and individual plants on the coast in the Baltic region. After Backlund et al. (1993); modified.

plants are applied. According to Bruneau (1980) the Finnish textile industry is mainly located either on the Baltic coast or on rivers flowing into the Baltic Sea; it could be responsible for entering many stable chemicals the Baltic Sea. The major pollutants load in this area is suspected to be from the pulp and paper industry, i.e. part of the Finnish plants are located inland, but whole the Swedish industry is concentrated on the coast (Bruneau, 1980). The catchment area of the

Gulf of Finland comprises 412,900 km^2, of which 67% belongs to Russia, 26% to Finland, 7% to Estonia and < 0.1% to Latvia (HELCOM, 1998a). In the Russia are located aluminium works and fertilise industry; the latter is also located on the Estonian coast. The Russia has also a highly developed chemical industry, petrochemical plants and the pulp and paper industry. Most of pollutants are transported through the Neva River to the Gulf of Finland, and thus affect the Baltic Sea. Scandinavian petrochemical installation is located on the Baltic coast; the Finnish steel mills, pulp and paper industry and manufacturing industry are located in this area and on the lakes and streams where pollutants are discharged either directly or through Lake Ladoga (Bruneau, 1980). The catchment area of the Gulf of Riga comprises 128,340 km^2, of which 39% belongs to Latvia, 20% to Belarus, 18% to Russia, 14% to Estonia and 9% to Lithuania (HELCOM, 1998a). According to Bruneau (1980) the drainage area of the Gulf of Riga seems to be rather poorly industrialised. In this region, however, were situated a very large refinery, petrochemical plant and some paper mills, mostly of small size. The catchment area of the Baltic Proper comprises 574,245 km^2, to which all Contracting Parties except Finland belong as well as Non-Contracting Parties of Belarus, the Czech Republic, Ukraine and Slovakia with the total area of 78,360 km^2. The catchment area of the Contracting Parties is divided into particular subareas as follows: 54% to Poland, 15% to Sweden, 9% to Lithuania, 3% to Russia, 2.6% to Germany, 2% to Latvia, 0.2% to Denmark and 0.2% to Estonia. The Polish rivers, the Vistula and Oder, enter the Baltic Proper transporting all pollutants even from industry located in such remote southern area as borders of the Czech Republic and Slovakia (HELCOM, 1998a). Poland is much industrialised country with most of the smelters and steel mills located in the south. In Poland is also produced copper, zinc and aluminium. There are also refineries, petrochemical centre, fertiliser production, mines, textile industry and pulp and paper mills. On the Swedish side, pollutants entering the Baltic Sea from Lake Mälaren originate partly from the central industrial district located in the vicinity of numerous old mines. There are also ammonia plants, fertiliser plants, small steel mills and stainless steel mills including the Oxelösund steel mill on the coast. Among other industries on the Swedish coast and on rivers discharging to the Baltic Proper are mostly pulp and paper mills. In the Russia fertilisers plants are located on the southeast side of the Baltic while in Germany the most important sources of pollutants are petrochemical plant and steel mills. However, most industrial pollutants are transported via rivers into the North Sea (Bruneau, 1980). The collective deposition of trace elements such as Zn, Cu, Cd, Pb, As, Hg, Cr and Ni to the Baltic Sea, i.e. to the Baltic Proper, Gulf of Finland and Gulf of Riga has been estimated (Lithner et al., 1990). The Chernobyl-derived ^{137}Cs has been evaluated as significant (65 TBq) for all Finnish rivers discharging into the Baltic Sea during 1986–1996 as well as for five Russian rivers (14 TBq) discharging from the former USSR during only 1986–1988 and for Polish Vistula River (18 TBq) during 1986–1996 (Gavrilov et al., 1990; Ilus and Ilus, 2000; Saxén and Ilus, 2000; Smith et al., 2000). Other rivers from the former USSR, i.e. Neva, Luga, Narva, Daugava and Neman provided ca. 2.6 TBq of ^{137}Cs to the Baltic Sea. It is reported that Swedish rivers have discharged ca. 150 TBq of ^{137}Cs into the Baltic Sea during 1986–1996 while contribution of the Oder River and smaller German rivers to the total radiocaesium activity has been evaluated at level of 10 TBq. Totally, ca. 300 TBq of ^{137}Cs has been discharged by these rivers into the Baltic Sea. The next sources of radionuclides, e.g. ^{137}Cs and ^{90}Sr are reprocessing plants in Western Europe providing since 1970s into the Baltic ca. 150 TBq of ^{137}Cs (Ilus and Ilus, 2000).

B. CHEMICAL ELEMENTS AND RADIONUCLIDES

(i) Classification of Chemical Elements and Radionuclides

Trace, minor and major elements

There are different classifications concerning the terminology used for groups of pollutants in various environmental compartments. According to Hopkin (1989) terminology of chemical elements as trace metals and heavy metals is not adequately defined. The metals and metalloids (semi-metals) include all of the elements except the noble gases (Group 0) and H, B, C, N, O, F, P, S, Cl, Br, I and At. To metalloids belong Si, Ge, As, Se, Sb and Te. The Periodic Table is presented in Table 1.1 (Morgan and Stumm, 1991). Groups Ia and IIa, i.e. the "s block" metals, form monovalent cations (alkali metal cations) and divalent cations (earth alkali cations), respectively. Groups IIIb through VIb belong to "p block" metal ions. The classification of elements into A- and B-type metal cations is based on the number of electrons in the outer shell. As can be seen in Table 1.2 type-A metal cations (hard acids) having the inert gas type (d^0) electron configuration form complexes mainly with F and O as donor atoms contained in ligands. Molecules of H_2O are more strongly attracted to these metals than are molecules of NH_3 and CN^-; no reaction occurs also with S^{2-} in aqueous solution. Addition of NH_3, alkali CN^- and alkali S^{2-} produced difficulty soluble precipitates. The hard Lewis acids (Al, Ti, Sn, Mn, Co, Cr, V and Ni) are termed lithophile because their mass excess in stream transport in respect to their atmospheric transport to the oceans. The type-B metal ions (soft Lewis acids) have tendency to coordinate preferentially with bases having I, S or N as donor atoms (Morgan and Stumm, 1991). These metals (Hg, As, Se, Sn, Pb) can be methylated and/or released to the atmosphere as vapours. In contrast, the soft Lewis acids are remarkably accumulated in the natural environments as well as potentially hazardous to ecology and human health because of their tendency to react with soft bases (SH- and NH-groups in enzymes). According to Williams (1981) chemical elements are distributed in the biosphere depending on various acid-base affini-

TABLE 1.1.

Periodic Table of the Elements. After Morgan and Stumm (1991); modified

Period	Group Ia	Group IIa	Group IIIa	Group IVa	Group Va	Group VIa	Group VIIa	Group VIII			Group Ib	Group IIb	Group IIIb	Group IVb	Group Vb	Group VIb	Group VIIb	Group 0
1 1s	1 H																1 H	2 He
2 2s2p	3 Li	4 Be											5 B	6 C	7 N	8 O	9 F	10 Ne
3 3s3p	11 Na	12 Mg											13 Al	14 Si	15 P	16 S	17 Cl	18 Ar
4 4s3d 4p	19 K	20 Ca	21 Sc	22 Ti	23 V	24 Cr	25 Mn	26 Fe	27 Co	28 Ni	29 Cu	30 Zn	31 Ga	32 Ge	33 As	34 Se	35 Br	36 Kr
5 5s4d 5p	37 Rb	38 Sr	39 Y	40 Zr	41 Nb	42 Mo	43 Tc	44 Ru	45 Rh	46 Pd	47 Ag	48 Cd	49 In	50 Sn	51 Sb	52 Te	53 I	54 Xe
6 6s (4f) 5d 6p	55 Cs	56 Ba	57* La	72 Hf	73 Ta	74 W	75 Re	76 Os	77 Ir	78 Pt	79 Au	80 Hg	81 Tl	82 Pb	83 Bi	84 Po	85 At	86 Rn
7 7s (5f) 6d	87 Fr	88 Ra	89* Ac															

*Lanthanide series 4f	58 Ce	59 Pr	60 Nd	61 Pm	62 Sm	63 Eu	64 Gd	65 Tb	66 Dy	67 Ho	68 Er	69 Tm	70 Yb	71 Lu
**Actinide series 5f	90 Th	91 Pa	92 U	93 Np	94 Pu	95 Am	96 Cm	97 Bk	98 Cf	99 Es	100 Fm	101 Md	102 No	103 Lr

Heavy boundary divides metals and metalloids (dashed boundary) from non-metals.
Type-B metals are marked by ×, borderline metals are marked diagonal lines (those closer to type A by \ and those closer to type B by /).

TABLE 1.2.

Classification of metal ions. After Morgan and Stumm (1991); modified

Type-A Metal Cations	Transition-Metal Cations	Type-B Metal Cations
Electron configuration of inert gas; low polarizability; "hard spheres"; (H^+), Li^+, Na^+, K^+, Be^{2+}, Mg^{2+}, Ca^{2+}, Sr^{2+}, Al^{3+}, Sc^{3+}, La^{3+}, Si^{4+}, Ti^{4+}, Zr^{4+}, Th^{4+}	One to nine outer shell electrons; not spherically symmetric; V^{2+}, Cr^{2+}, Mn^{2+}, Fe^{2+}, Co^{2+}, Ni^{2+}, Cu^{2+}, Ti^{3+}, V^{3+}, Cr^{3+}, Mn^{3+}, Fe^{3+}, Co^{3+}	Electron number corresponds to Ni^0, Pd^0 and Pt^0 (10 or 12 outer shell electrons); low elec- tronegativity; high polarizabili- ty; "soft spheres"; Cu^+, Ag^+, Au^+, Tl^+, Ga^+, Zn^{2+}, Cd^{2+}, Hg^{2+}, Pb^{2+}, Sn^{2+}, $T1^{3+}$, Au^{3+}, In^{3+}, Bi^{3+}
	According to PEARSON'S Hard and Soft Acids	
Hard Acids	**Borderline**	**Soft Acids**
All type-A metal cations plus Cr^{3+}, Mn^{3+}, Fe^{3+}, Co^{3+}, UO^{2+}, VO^{2+}	All bivalent transitions metal cations plus Zn^{2+}, Pb^{2+}, Bi^{3+}	All type-B metal cations minus Zn^{2+}, Pb^{2+}, Bi^{3+}
Also species such as BF_3, BCl_3, SO_3, RSO_2^+, RPO_2^+ CO_2, RCO^+, R_3C^+	SO_2, NO^+, $B(CH_3)_3$	All metal atoms, bulk metals I_2, Br_2, ICN, I^+, Br^+
Preference for ligand atom:		
$N \gg P$		$P \gg N$
$O \gg S$		$S \gg O$
$F \gg Cl$		$I \gg F$
Qualitative generalizations on stability sequence:		
Cations: Stability (charge/radius)	Cations: Irving-Williams order: $Mn^{2+} < Fe^{2+} < Co^{2} + < Ni^{2+} < Cu^{2+} > Zn^{2+}$	
Ligands: $F>O>N = Cl>Br>I>S$		Ligands: $S>I>Br>Cl = N>O>F$
$OH^- > RO^- > RCO_2^-$		
$CO_3^{2-} \gg NO_3^-$		
$PO_4^{3-} \gg SO_4^{2-} \gg CIO_4^-$		

ties, their kinetics and spatial partitioning, e.g. they are partitioned by membranes and compartments as well as by temporal partitioning. Metal toxicity is resultant of the chemical combination of metals and ligands, i.e. the Lewis acids and bases in organisms. The cellular bases are predominantly S, N and O as donor atoms in molecules of H_2O and solute bases, e.g. OH^-, HCO_3^-, HPO_4^{2-}. Among the acids are H^+, cations of the essential metals such as Na, K, Mg, Cr, Mn, Fe, Co, Ni, Cu, Zn, Mo as well as potentially toxic metals such as divalent Hg, CH_3Hg^+, Pb, Cd, Cr, etc. Some of the type-B metals belonging to the sulphur-seeking ("soft-soft") metals are the toxicants such as Hg, Pb, Tl, as well as the essential proteine and enzyme metals, e.g. Fe, Cu and Zn. Cations of the metals such as Mg, Ca, Be, Al, Sn, Ge and the lanthanides (showing tendency to exhibit "hard sphere" or A-type behaviour in their coordination compounds) are oxygen-seeking ('hard-

-hard'). Since H^+ shows a high affinity for donor atoms such as S, N and O; hence fundamental significance plays value of pH in metal binding in the biota (Morgan and Stumm, 1991). The lanthanides with Sc (21) and Y (39) are classified as the rare earth elements (REE). These elements with 3+ ions and decreasing radii, indicate strong ionic bonding and weaker covalent bonding characteristics. The lanthanides show tendency to exhibit "hard sphere" or A-type properties in their coordination compounds (Morgan and Stumm, 1991).

As can be seen in Table 1.3 trace elements are released into the atmosphere from natural and anthropogenic sources (fossil fuel combustion, cement production, extractive metallurgy etc.). It is found that Ag, As, Cu, Hg, Pb, Sb, Sn and Zn are the most potentially hazardous elements on a global or regional scale (Table 1.4). In this Table the geochemical scale is defined as 'global' when the effect of perturbation can be demonstrated at least in large part of the Northern Hemi-

TABLE 1.3.

Natural and anthropogenic sources of atmospheric emissions[a]. After Morgan and Stumm (1991); modified

Element	Continental Dust Flux	Volcanic Dust Flux	Volcanic Gas Flux	Industrial Particulate Emissions	Fossil Fuel Flux	Total Emissions, Industrial Plus Fossil Fuel	Atmospheric Interference Factor (%)[b]
Al	356 500	132 750	8.4	40 000	32 000	72 000	15
Ti	23 000	12 000	–	3600	1600	5200	15
Sm	32	9	–	7	5	12	29
Fe	190 000	87 750	3.7	75 000	32 000	107 000	39
Mn	4250	1800	2.1	3000	160	3160	52
Co	40	30	0.04	24	20	44	63
Cr	500	84	0.005	650	290	940	161
V	500	150	0.05	1000	1100	2100	323
Ni	200	83	0.0009	600	380	980	346
Sn	50	2.4	0.005	400	30	430	821
Cu	100	93	0.012	2200	430	2630	1363
Cd	2.5	0.4	0.001	40	15	55	1897
Zn	250	108	0.14	7000	1400	8400	2346
As	25	3	0.1	620	160	780	2786
Se	3	1	0.13	50	90	140	3390
Sb	9.5	0.3	0.013	200	180	380	3878
Mo	10	1.4	0.02	100	410	510	4474
Ag	0.5	0.1	0.0006	40	10	50	8333
Hg	0.3	0.1	0.001	50	60	110	27 500
Pb	50	8.7	0.012	16 000	4300	20 300	34 583

[a] All fluxes are in units of 10^8 g per year

[b] Atmospheric interference factor = [total emissions ÷ (continental + volcanic fluxes)] x 100

sphere (Morgan and Stumm, 1991). Chemical elements are partitioned in the biota by different acid-base affinities, i.e. by their kinetics, spatial partitioning and by temporal partitioning (Williams, 1981; Morgan and Stumm, 1991). One of important aspects of metal toxicity is the chemical combination of metal ions and ligands (Lewis acids and bases) in organisms. The cellular bases are predominantly S, N and O donor groups, i.e. H_2O and solute bases. The acids are as follows: H^+, the essential metal cations and potentially hazardous metals. Among the essential metals are Ca, Co, Cr, Cu, Fe, K, Mg, Mn, Mo, Na, Ni and Zn while to hazardous elements belong Ag, As, Bi, Cd, Cr, Cu, Hg, In, Pb, Sb, Se, Sn, Tl and Zn (Morgan and Stumm, 1991). The toxic elements (Hg, Pb, Tl) and the essential protein and enzyme metals (Cu, Fe, Zn) are classified to the sulfur-seeking ("soft-soft") type-B metals. To oxygen-seeking ("hard-hard") elements belong H^+ (having strong affinity for all donors, S, N and O) as well as Al, Be, Ca, Ge, Mg, Sb and the lanthanides (Morgan and Stumm, 1991).

Radionuclides

Radionuclides (isotopes, nuclides) present in the aquatic and terrestrial environment are classified as either of natural or anthropogenic in origin. Naturally-occurring radionuclides occur in the different ecosystems with primordial and cosmogenic provenience. To the most abundant radionuclides belong ^{40}K, members of the U and Th chains (^{234}U, ^{235}U, ^{238}U ^{226}Ra, ^{210}Pb and ^{210}Po) and the cosmogenic species, i.e. ^{3}H and ^{14}C (Ilus and Ilus, 2000).

Primary radionuclides

Primordial radionuclides are long-lived species have been ubiquites on the Earth since its formation, i.e. ca. 4.5 x 10^9 years ago (MacKenzie, 2000). The radionuclides ^{238}U, ^{232}Th and ^{235}U are the parent members of the uranium, thorium and actino-uranium radioactive decay series, respectively (Riley, 1971):

Uranium series

$$\underset{(99.27\%\text{ of U})}{^{238}\text{U}} \xrightarrow[\alpha]{4.51\times10^{9}y} {}^{234}\text{Th} \xrightarrow[\beta^-]{24.1\,d} {}^{234}\text{Pa} \xrightarrow[\beta^-]{6.66\,h} \underset{(0.0056\%\text{ of U})}{^{234}\text{U}} \xrightarrow[\alpha]{2.48\times10^{5}y} {}^{230}\text{Th} \xrightarrow[\alpha]{8.\times10^{4}y} {}^{226}\text{Ra}$$

$$\xleftarrow[\beta^-]{22.0\,y} {}^{210}\text{Pb} \leftarrow \text{short-lived daughters} \leftarrow {}^{218}\text{Po} \xleftarrow[\alpha]{3.83\,d} {}^{222}\text{Rn} \xleftarrow[\alpha]{1622\,y} {}^{226}\text{Ra}$$

Thorium series

$$\underset{(100\%\text{ of Th})}{^{232}\text{Th}} \xrightarrow[\alpha]{1.39\times10^{10}y} {}^{228}\text{Ra} \xrightarrow[\beta^-]{6.7\,y} {}^{228}\text{Ac} \xrightarrow[\beta^-]{6.13\,h} {}^{228}\text{Th} \xrightarrow[\alpha]{1.91\,y}$$

Actino-uranium series

$$\underset{(0.72\%\ \text{of U})}{^{235}\text{U}} \xrightarrow[\alpha]{7.1\times10^{8}y} {}^{231}\text{Th} \xrightarrow[\beta^-]{25.6h} {}^{231}\text{Pa} \xrightarrow[\alpha]{3.43\times10^{4}y} {}^{227}\text{Ac} \xrightarrow[\beta^-]{21.8y} {}^{227}\text{Th} \xrightarrow[\alpha]{18.4d}$$

These series are consisted of nuclides with very different geochemical fate and hence the radionuclides in the marine environments are not in secular equilibrium. In particular, Th and Pa are rapidly removed from the sea to the sediments resulting in significant deviation from the equilibrium value. The ratios of the concentrations of these both isotopes to that of U are much lower than the equilibrium values (0.05–0.2%) (Riley, 1971).

Cosmogenic radionuclides

A significant part of the cosmic rays reaching the earth has energies exceeding that binding the nuclei of atoms. Cosmic ray-produced radionuclides, e.g. ^{3}H, ^{10}Be, ^{14}C, ^{26}Al, ^{32}Si, ^{36}Cl and ^{41}Ca, are generated in the upper atmosphere due to absorption of cosmic ray energies by the nuclei of the atoms of atmospheric gases, e.g. O_2, N_2 and Ar. In consequence, they are fragmented to stable or unstable nuclei lighter than the parent nuclide. These cosmic ray-produced radionuclides are transported to the lower atmosphere and next to the oceans and to the continents. All of these radionuclides, e.g. ^{32}Si and ^{14}C, have been very useful as tracers of environmental processes (Riley, 1971; MacKenzie, 2000). An extensive overview of five decades of studies of cosmic ray produced nuclides in oceans has been presented by Lal (1999). It should be emphasised that human radiation exposure from man-made radionuclides constitutes an additional contribution to the natural dose from the background radiation, i.e. cosmic rays as well as primordial and cosmic ray-produced radionuclides, resulting in a world average individual exposure of ca. 2.5 mS yr^{-1} (MacKenzie, 2000). The presently established limit by the ICRP (International Commission for Radiological Protection) for the allowable maximum radiation dose from industrial releases of radionuclides is 1 mS yr^{-1} (MacKenzie, 2000).

Anthropogenic radionuclides

Anthropogenic-derived radionuclides are mainly released from several sources since the 1940s. Major their sources in the environment are nuclear weapons, nuclear power production, accidents, radioactive waste disposal, solid radioactive waste disposal and man-made radionuclides as tracers of environmental processes (Ilus and Ilus, 2000; MacKenzie, 2000). Fallout from nuclear weapons explosions represents the largest contribution of anthropogenic-derived radionuclides in the ecosystem. In the detonation of a nuclear bomb, radioactive fission products are generated from primary fission of ^{235}U or ^{239}Pu. The main radionuclides produced in nuclear weapons explosions and released to the atmosphere are as follows: ^{3}H, ^{14}C, ^{54}Mn, ^{55}Fe, ^{85}Kr, ^{89}Sr, ^{90}Sr, ^{91}Y, ^{95}Zr, ^{103}Ru, ^{106}Ru, ^{125}Sb, ^{131}I, ^{133}Xe, ^{134}Cs, ^{137}Cs, ^{140}Ba, ^{141}Ce, ^{239}Np, ^{238}Pu, ^{239}Pu, ^{240}Pu, ^{241}Pu and ^{241}Am. These radionuclides are deposited from the atmosphere to the surface of the earth as fallout comprising

TABLE 1.4.

Perturbations of the geochemical cycles of metals by society. The elements are grouped according to the scale for which such perturbations can be documented. After Morgan and Stumm (1991); modified

Element	1			2	3	4	5
	Scale of Perturbation			Most Diagnostic Environments	Mobility	Health Concern	Critical Pathway
	Global	Regional	Local				
Pb	+	+	+	A, Sd, I, W, H, So	v, a	+	F, A[c]
V	+	+, c	+	A	[d]	(+)	A?
As	(+)	+	+	A, Sd, So, W	v, s, a	+	A, W
Sn	(+)	+	+	A, Sd, W	v, a	+ [a]	F
Zn	(−)	+	+	A, Sd, W, So	v, s	E	F
Cd	(−)	+	+	A, Sd, So, W	v, s	+	F
Hg	(−)	+	+	A, Sd, Fish, So	v, a	+ [a]	F, (A)
Sb	(−)	+	+	A, Sd	v, s	(+)	F, W, A?
Cu	(−)	+	+	A, Sd, W, So	v, s	E	F?
Ag	(−)	+	+	A, Sd, W	(v)	(+)	?
Se	(−)	(+)	+	A	v, s, a	E	F
Ge	?	+	+	A, So, W?	v, s, a	(+)[a]	?
Ni	−	+	+	A, Sd	−	E	F, W, A?
Cr	−	+	+	A, Sd, W, Gw	s, v[b]	E	W, F
B	−	(+)	+	A, Sd, Gw	v, s	E	W
K	−	(+)	+	A	s	E	F
Pt	?	?	+	A, Sd	s	(+)	?
Pd	?	?	+	Sd	s	(+)	?
Mo	?	?	+	A, W, So, Sd	s	E	F, W
Tl	?	?	+	Em, So	v, s	(+)	A, F?
In	?	?	+	A, So, Em	v	(+)	?
Bi	?	?	+	A, So, Em	v	(+)	?
Be	?	?	+	A, So, Em	−	(+)	A
Ga	?	?	+	Em	v	(+)	?
Te	?	?	(+)	So	v, a?	(+)	?
Mn	−	c, +	+	A, Sd, W	r	E	A
Fe	−	c	+	A, Sd, W	r	E	F, A, W
Al	−	c	+	A, Sd	−	(+)	W?
Si	−	c	+	A	−	(E)	−
Ti	−	c	+	A, Sd, So	−		−
Co	−	c	(+)	Sd	r	E	F?
Na	−	c, +	+	W, A, So	s	E	F, W
Mg, Ca	−	c	+	A, Sd, Em	(s)	E	
Ba	−	c	+	A, Sd, Em		(+)	F, W
U	−	c	+	A, So, Gw	s	(+)	A, W
Th	−	c	+	A, So, Gw	−	(+)	A
Zr	−	c	+	A	−		−

Element	1			2	3	4	5
Y	–	c	(+)	Sd	–		–
Au	(–)	(–)	+	Sd	–	+	–
W	(–)	(–)	+	A	s?		–
Li, Rb, Cs	–	c	c	Sd	s	+	–
Rare Earths	–	c	c	Sd	–		–
Ta, Hf, Sc, Sr	–	c	c	Sd	–		–
Os, Ir, Ru, Rh	?	?	?	–	–		–

(1) + significant perturbation; (+) possible perturbation; (–) enriched relative to crustal abundances, but the enrichment may not by anthropogenic; – no perturbation; ? not enough information; c enhanced due to mobilization of crustal materials (soil, dust). (2) A air; Sd sediments (coastal, lake); So soils; I ice cores; W surface waters; Gw groundwaters; H humans; Em emission studies (only listed when little geochemical information is available). (3) v volatile; s soluble; r soluble only under reducing conditions; a mobile as alkylated organometallic species; – not mobile. (4) + toxic in excess; (+) toxic, but little data available; E essential, but toxic in excess. (5) F food; W water; A air; – no significant exposure likely.
[a] organometallic forms only
[b] hexavalent form volatile and toxic, trivalent form essential
[c] exposure through hand-to-mouth activity is critical for lead in children
[d] enriched relative to crustal abundance from fuel oil combustion (vanadium porphyrins)

components such as stratospheric (78%), local (12%) and tropospheric (10%) (MacKenzie, 2000). Since stratospheric fallout was globally dispersed and tropospheric fallout was mainly dispersed in the latitude of the nuclear test it is resulted in low contaminant level on a global scale with concentration greater in the Northern Hemisphere than in the Southern Hemisphere (Cambray et al., 1982; MacKenzie, 2000). Large quantities of radionuclides were produced in underground nuclear weapons tests but incomplete report is available to estimate the long-term environmental consequences of such tests (MacKenzie, 2000). The total radiological impact of atmospheric nuclear weapons tests is estimated at level of 3×10^7 manSv (70% of ^{14}C and other main contributors such as ^{137}Cs, ^{90}Sr, ^{95}Zr and ^{106}Ru). The bioaccumulative abilities of radionuclides depend on their biochemical properties as well as on the individual accumulation strategies of given organism for each nuclide. Generally, radioactive isotopes of metals follow the metabolic pathways their stable counterparts, if they exist, although differences in atomic weight may be responsible for slight differences in relative rates of reaction resulting in metabolic kinetics. These differences show tendency to be more significant for the isotopes with a low atomic weight, e.g. H and C. However, radioactive isotopes are represented in organisms in the expected amounts corresponding to proper proportion to their stable isotopes. The representation of artificial radionuclides not having their stable counterparts in aquatic biota is dependent on the extent and pattern of their release into the outer medium, their both radioactive and the biological half-lives and on other biological features of the organism considered (Phillips and Rainbow, 1993). The heavier transuranic radioisotopes are, as rule, strongly adsorbed by external tissues which are directly exposed to the ambient waters, i.e. the gills and exoskeleton. Similar route is observed for algae which are able to accumulate transuranic radioisotopes predominantly by passive adsorption (Phillips and Rainbow, 1993). Nuclear power production is next source of radionuclides; their most releases to the environment from the nuclear fuel cycle occur in the uranium mining and the fuel reprocessing stages resulting in the emission of ^{222}Rn to the atmosphere. It makes a local potential health hazard. Spent nuclear fuel contains significant amounts of fissile ^{235}U and ^{239}Pu and different countries have adopted various policies with respect to the fuel reprocessing (UNSCEAR, 1993). There have been numerous accidents concerning reactors and other nuclear facilities, certain satellites, nuclear weapons and large radiation sources. The most significant incidents are described in the below section (iv).

(ii) Chemical Elements as Environmental Pollutants

Relationship between man and ecosystem health has been explored, especially in respect to perturbed ecosystems (Table 1.4). This includes the pollution status of regions harmed by some catastrophes, large-scale pollution, environmental accidents and episodes etc. High-risk groups consuming extremely high quantities of trace metals present in specific assortment of seafood or offal concern seriously sea-side populations. Marine fish and shellfish may by the dominant dietary sources of Hg for local populations (US EPA, 1984; Mance, 1987; Von Burg and Greenwood, 1991; dos Santos et al., 2000; Gray et al., 2000; Maurice-Bourgoin et al., 2000). A spectacular example of aquatic pollution by a toxic metal is the Minamata incident, commencing in 1953. Fish and bird mortalities in waters of the partially landlocked Minamata Bay were observed in early 1950s. Dogs, pigs and especially cats were also suffered from this incident. Effluents from the Chisso factory contained significant amounts of different chemical elements including both methyl-Hg and inorganic Hg; hence the principal pathway of Hg exposition of animals and humans at Minamata area was postulated to be polluted seafood, i.e. fish and shellfish (Tsubaki and Irukayama, 1977; Bertram et al., 1985; Von Burg and Greenwood, 1991; Phillips and Rainbow, 1993; Harada, 1995; Ninomiya et al., 1995; Akagi et al., 1998). By the end of 1974, 107 of 798 officially verified patients had died. Other cases of Minamata disease (Harada, 1978, 1995) were noted in Niigata, Japan, caused by the discharge of Hg in the effluents from electric industrial plant. The consuming of polluted fish from local river caused this poisoning resulted in 6 deaths (Takizawa, 1979; Phillips and Rainbow, 1993). According to Tomiyasu et al. (2000) the sediments from the Minamata Bay were consistently found to contain Hg at level that highly exceeded background level. It is supposed that the Hg was mostly derived from the effluent from the chemical plant. The surficial sediments enriched in Hg are not stable and apparently still moving even though 30 years have passed since the last discharge of polluted effluent (Tomiyasu et al., 2000). Amongst other incidents and accidents resulting in release of Hg compounds to the environment, the most significant had place in the 1960s and early 1970s in Sweden, Canada and the United States, northern Iraq, Guatemala, Pakistan, Ghana (Förstner and Wittmann, 1983; Phillips and Rainbow, 1993; Akagi et al., 1995; Harada, 1996; Harada et al., 1999). For instance, in 1960 was noted epidemic outbreak when 221 patients were hospitalised (Damluji, 1962) as a result of the use of ethyl-Hg fungicide in 1956 (Jalili and Abbasi, 1961). However, the most dramatic epidemic ever has been recorded took place in northern Iraq in 1971–1972; the people have been affected there by massive poisoning due to the ingestion of homemade bread prepared from wheat seed that had been treated with a methyl-Hg fungicide. In consequence, the dressed seed was dumped in local rivers and lakes resulting in severe pollution of large area. These combined events strongly affected a huge part of local population; it is thought that 100,000–500,000 inhabitants have been suffering permanent disabilities (Förstner and Wittmann, 1983; Phillips and Rainbow, 1993). According to Bakir et al. (1973) 6530 patients were admitted to hospitals where 459 died. Based on epidemiological data it is indicated that over 2000 deaths occurred and more than 60,000 people were exposed (Greenwood, 1985; Von Burg and Greenwood, 1991).

Methylmercury in aquatic ecosystems is accumulated especially in fish constituting a major public health problem (Wheatley and Wheatley, 2000; Pilgrim et al., 2000). Its levels in the hair of fishermen are described anticipating that they represent the critical group for dietary exposure. For instance, the concentrations of Hg (total and MeHg) in hair of fishermen from Kuwait were twice the WHO 'normal' level (2.0 $\mu g\ g^{-1}$) (Al-Majed and Preston, 2000). The unlimited use of Hg in a gold mining process has resulted in the serious pollution of many aquatic and terrestrial ecosystems. Such anthropogenic emissions occur in almost gold mining operations in developing countries such as Brazil, Ecuador, Peru, Columbia, the Philippines and Tanzania (Akagi et al., 1995, 2000; Harada, 1996; Ikingura and Akagi, 1996; Harada et al., 1999; Rosa et al., 2000; van Straaten, 2000a, 2000b). Problem of large scale pollution has been and is still key topic and some of published papers (Pfeiffer and Lacerda, 1988; Nriagu et al., 1992; Akagi et al., 1995; Malm et al. 1995a, 1995b; Artaxo et al., 2000) presented abnormal levels of total mercury and methylmercury in environmental compartments such as air, soil, bottom sediments, fish and plants as well as in human hair and urine in order to evaluate the extent of environmental mercury pollution due to goldmining activities in the Amazon. An average of 63% of the Hg concentrations was associated with the gold mining activities (Artaxo et al., 2000). Biomass burning in tropical forests also seems to have contributed significantly to the Hg release to the atmosphere. Apptoximately 31% of the Hg concentrations was associated with the vegetation fire component (Artaxo et al., 2000). Long-range air mass trajectory analyses indicate the possibility that Hg occurs in the Amazon basin over two main routes: to the South Atlantic, and to the Tropical Pacific, over the Andes (Artaxo et al., 2000).

Chemical composition of aerosol particles from direct emissions of vegetation fires in the Amazon Basin has been estimated by Yamasoe et al. (2000). Global emission flux estimates exhibited that biomass burning could be important source of heavy metals and black carbon to the atmosphere. It is estimated that savanna and tropical forest biomass burning could emit ca. 1 Gg Cu yr^{-1}, 3 Gg Zn yr^{-1} and 2.2 Tg black carbon to the atmosphere, i.e. these values correspond to 2, 3 and 12%, respectively, of the global budget of these elements (Yamasoe et al., 2000).

Episode involving the poisoning of local population by other trace elements than Hg was noted in 1947 in the Jintsu River basin in Japan (Yamagata and Shigematsu, 1970; Phillips and Rainbow, 1993). It is believed that principal cause of the disease, named the Itai-itai syndrome were effluents enriched in Cd from a zinc mine operated in this area. As a result of this episode approximately 200 patients died prior 1966 (Phillips and Rainbow, 1993). In Zhejiang province, China, representing a highly exposed area, concentrations of Cd in rice were 50 times greater than those from control area (Nordberg et al., 1997). There was a significant dose-response relationship between Cd in urine and β_2-microglobulin excretion in urine, as an indicator of renal dysfunction. This report as first one concerns a dose-response association in the Chinese population group in Zhejiang province.

The nature of current anthropogenic sources of Hg is different than it was several decades ago. Many of the most significant emitters in the past, e.g. chlor-alkali industry, paint containing mercury additives and pharmaceuticals have been largely phased out with fossil fuel combustion and waste disposal remaining as the most significant recent sources (Sunderland and Chmura, 2000). Other example of toxic effect of chemical element to man is SO_4^{2-} which has been used recently as an air pollution indicator in the epidemiological studies in Beijing, China (Zhang et al., 2000). Main sources of this pollutant in the Beijing atmosphere are coal combustion, number of households using gas fuel, counts of motor vehicles and population density. Epidemiological studies have demonstrated that the air pollution in Beijing is associated with reduced immune function of children, chronic obstructive pulmonary disease, a total mortality and mortality due to cardiovascular disease, pulmonary heart disease, malignant tumor and lung cancer (Xu et al., 1994, 1995, 1995b; Zhang et al., 1995, 2000). Another environmentally-derived healthy problem named Keshan disease has been identified in Jilin province, China. This endemic cardiovascular disease is mainly caused by low Se levels in the environment (Ma and Zhang, 2000). Jilin province is one of the most seriously affected Chinese area by Keshan disease. The annual average incident rate of the disease is 90/100 000 and the rate of morbidity amounted to 10.2 % from 1959 to 1984.

It should be emphasised that Russia is one of the major sources of Pb air pollution in Europe (Snakin, 1997; Snakin and Prisyazhnaya, 2000). More than 40% of children in Russian cities may develop behaviour problems and learning difficulties caused by exposure to Pb (Bykov and Revich, 1997; Snakin and Prisyazhnaya, 2000). Pb in Russia is mainly provided to human organism via foodstuffs (ca. 50%); air (< 1%) and drinking water (2–3%). Direct Pb contamination with soil and dust is estimated at 10–12%. The greatest concentration of Pb has been detected in tin can packaging, wheat bran, gelatine and sea fruit products, i.e. fresh and frozen fish, molluscs and shellfish. Pb poisoning holds first place among professional intoxication reflected by rising from 9.4% in 1991 to 11.6% in 1995 (Snakin and Prisyazhnaya, 2000). In a residential area of Greater Calcutta ca. 50 000 people inhabit in the vicinity of lead factory in which are produced lead ingots and lead alloys. Many people, especially children, are effected by Pb toxicity (Chatterjee and Banerjee, 1999). Nriagu et al. (1996a, 1996b, 1997a, 1997b) provided first data pointing to childhood Pb poisoning as a growing public health problem in urban area of Africa. The strong relationships were found between blood Pb levels in children and whether the family owned a car or lived in a house on a tarred road. The studies documented the silent epidemic of childhood Pb poisoning in African cities and towns (Nriagu et al., 1996a, 1996b, 1997a, 1997b; Liggans and Nriagu, 1998). According to Shen et al. (1996) in China, childhood Pb poisoning might be widely pervasive as a result of rapid industrialisation and the use of leaded gasoline. The harmful health effects of childhood Pb poisoning provide evidence that this absolutely preventable disease warrants considerable public attention in China (Shen et al., 1996). Increased risk estimates for lung cancer in Pb exposed smelter workers have been demonstrated by Englyst et al. (2001). However, considerable As exposure also had place in most of the lung cancer cases. In this multifactorial exposure it has, however, not possible to distinguish the carcinogenic effects caused by Pb and As but a possible interaction between these metals may be involved in explaining the carcinogenic risks.

The deleterious effects of tributyltin (TBT), representing the most toxic form among Sn compounds, were the first indicated in Arcachon Bay, France, as the 'TBT problem' at the end of the 1970s (Alzieu, 1986, 2000). As a result of TBT releasing by antifouling paints to the area was shell abnormalities and reduced growth and settlement in oysters, *Crassostrea gigas*, cultured in the vicinity of marinas. In much polluted water, the production of the oysters was severely affected by a complete lack of reproduction resulted in a strong decline in the marketable value of the remaining stock (Alzieu, 2000). Imposex, i.e. the development of male sexual characteristics in female marine mesogastropods and neogastropods caused by TBT pollution, is a widespread phenomenon which has concerned several coastal species and more recently also offshore species (Evans et al., 1995, 1996, 2000c; Minchin et al., 1995, 1997; Tester and Ellis, 1995; Huet et al., 1996; Skarphédinsdóttir et al., 1996; Smith, 1996; Morgan et al., 1998; Poloczanska and Ansell, 1999; Santos et al., 2000; Shim et al., 2000; Hung et al., 2001). Later regulations prohibiting the use of TBT-based antifoulants on vessels less than 25 m in length have been highly effective in reducing TBT levels in coastal waters. However, larger vessels are still responsible for releasing of TBT and major harbours continue to be hot-spots of pollution (Evans and Nicholson, 2000). Therefore, environmental impact of TBT in coastal waters, microbial interaction, detoxification, accumulation as well as TBT regulatory strategies, pending actions, related costs and benefits are still the important subject of considerable debate (Bryan and Langston, 1992; Readman, 1996; Evans, 1999, 2000; Michel and Averty, 1999; Abbott et al., 2000; Champ, 2000; Evans et al., 2000a, 2000b; Gadd, 2000; Jacobson and Willingham, 2000; Strandenes, 2000). Unfortunately, yet the expected effects of a worldwide ban concerning TBT have not been observed at all (Abel, 2000).

The extensive flooding, especially occurred in river area of former or operating metalliferrous mining can be responsible for wide-spreading of heavy metals and metalloids far distance from pollution source. Example of such environmental events is flooding of the Severn catchment (United Kingdom) in January 1998 (Zhao et al., 1999). The exceptional Oder flood in summer 1997 has led to considerable additional metal pollution of the Szczecin Lagoon and the Pomeranian Bay, Baltic Sea (Siegel et al., 1998; Müller and Wessels, 1999). Large quantities of industrial and municipal waste were taken up by the Oder River and transported to the Baltic. According to Müller and Wessels (1999) compared with the mean concentrations of heavy metals such as Cu, Pb and Zn before flood, their maximum levels in $< 20\ \mu m$ fraction of suspended solid during the flood were from two- to four times higher.

(iii) Nutrients as Environmental Pollutants

Although the distribution of nutrients and their impacts to the Baltic ecosystem are not the principal topic of this book, different pollutants are mostly controlled by the nutrient dynamics (Wulff et al., 1990). Eutrophication resulting from an increase of the primary production often leads to an increase in the sequestering of trace elements in the sediments. Heavy metals being anthropogenic in origin are deposited in more recent sediments and often bioaccumulated in organic matter. In consequence, changes in bioavailability of both heavy metals and nutrients take place (HELCOM, 1996, Danielsson, 1998). Therefore, these issues are briefly discussed here.

The blooms of blue-green algae likewise *Nodularia* may be toxic to animals producing a peptide hepatoxin under particular conditions, which can pass through the food web to affect top consumers, e.g. man. The toxin degenerates liver cells promoting tumours and causing death from hepatic haemorrhage (Forsberg, 1993); such poisoning is defined as, e.g. paralytic shellfish poisoning, diarrhetic shellfish poisoning, venerupin poisoning or ciguatera (Phillips and Rainbow, 1993). Paralytic shelf poisoning (PSP) and/or ciguatera have been identified predominantly in the subtropical and tropical zones such as Australia (Gillespie, 1984; Hallegraeff and Sumner, 1986), Hong Kong (Holmes and Lam, 1985; Phillips, 1985; Morton, 1989) and especially in other the Indo-Pacific regions; e.g. India, Thailand, Indonesia, Philippines, Papua New Guinea (Maclean, 1989; Maclean and White, 1985). Fish and shellfish from these regions were killed mainly by principal toxic dinoflagellate species *Pyrodinium bahamense* var. *compressa*.

The consumption of seafood in the Indo-Pacific area constituted considerable public health problem (Phillips and Rainbow, 1993). The significant PSP incidences were reported also from temperate zones. For instance, in May 1968 the poisoning episode affected 78 persons inhabited Britain after consumption of soft tissue of blue mussel *Mytilus edulis* (Ayres, 1975). Further poisoning dinoflagellate-derived events appeared again in the north-east England in the summer of 1990, possibly attributing to a specific combination of elevated nutrient inputs from rivers and exceptionally warm and sunshine weather conditions, which may be favourable for algae growing (Phillips and Rainbow, 1993).

It is well documented that antropogenically-derived atmospheric N deposition to the North Atlantic Ocean is strictly linked with harmful algal bloom expansion (Paerl and Whitall, 1999). This phenomenon concerns especially such geographical areas as Eastern Gulf of Mexico, US Atlantic coastal waters, the North and Baltic Seas (Paerl, 1985, 1995; Anderson, 1989; Aksnes et al., 1989; Tester et al., 1991; Buskey and Stockwell, 1993; Hallegraeff, 1993; Anderson et al., 1994; ECOHAB, 1995; Howarth et al., 1996; Prospero et al., 1996; Paerl and Whitall, 1999). Along the Northeast US Atlantic coastline, expanding blooms of the noxious dinoflagellate *Alexandrium tamarense* have been identified (Anderson et al., 1994; Paerl and Whitall, 1999). The Western European coastal areas have been increasingly impacted by harmful algal bloom outbreaks in the post World II period resulting from agricultural, industrial and urban expansion in western and central Europe (Ambio, 1990; Paerl and Whitall, 1999). There are numerous examples of specific harmful algal bloom expansions in coastal and off-shore waters where atmospheric N deposition is significant, e.g. in the North Sea, Adriatic Sea, Western Mediterranean and Baltic Seas (Paerl, 1997; Paerl and Whitall, 1999). A great attention has been paid to toxic hypoxia-inducing dinoflagellate blooms in the North Sea and the Western Baltic (Paerl and Whitall, 1999). The Baltic Sea has been affected by increasing anthropogenic nutrients such as N, P, as well as by trace metals (Ambio, 1990). Extensive summer algae blooms have been reported for the Baltic Sea since the mid of 19th century. In the late 1800's and early 1900's was observed a mass occurrence of the blue-green alga *Nodularia spumigena* and *Aphanizomenon flos-aquae*. This event was attributed to an effect of nutrient inputs from rivers and coastal waters (Forsberg, 1993; Paerl and Whitall, 1999). In the summer of 1991 a very extensive bloom of *N. spumigena* was observed in the open Baltic Sea as well as along the southern and south-eastern Swedish coasts. Dogs mortalities caused by toxic *Nodularia* blooms have been observed in Denmark, Gotland and the Swedish coastal waters. In other Baltic areas, horses, cows, sheep, pigs, cats, birds and fish were also suffered from this event. Human problems concerning stomach complaints, headaches, eczema and eye inflammation have been caused by *Nodularia* blooms (Forsberg, 1993). In more saline regions of the Baltic Sea, e.g. the Skagerrak and Kattegat harmful algal bloom expansion has taken place (Asknes et al., 1989). For instance, during 1980's several dramatic blooms of toxic algae species such as *Prorocentrum*, *Dino-*

physis, *Dichtyocha*, *Prymnesium* and *Chrysochromulina* have been noted in the Kattegat. The recent blooms mostly killed pelagic organisms and the phyto- and zoobenthal organisms.

It has been reported (Billen et al., 1999) that by the end of the last century, large scale use of traditional technologies in some branches of industry (textile and paper industries, tanneries, candles factories etc.) accounted for a dominant part of the nutrient load carried by rivers in Western Europe and possibly was responsible for nutrient inputs to coastal zones similar to present one. In some areas such as the Southern Bight of the North Sea, algal proliferation as intense as presently detected was regularly occurring at the end of the 19th century (Billen et al., 1999).

Other consequence of eutrophication is the inadequate O_2 supply into marine environments. An example of the harmful deoxygenation of water giving rise to fish kills was producing nutrient-derived large mats of macroalgae in the Peel-Harvey Estuary, Western Australia (Birch et al., 1986). Similar event took place in the northern Adriatic Sea where diatom blooming in summer resulted in the production of mucilage affected tourism in north-eastern Italy and reducing fish catches (Justic, 1987; Degobbis, 1989; Phillips and Rainbow, 1993). Insufficient water exchange and increasing production of organic matter provide the next example, here, it is documented that during this century O_2 depletion in all deep waters of the Baltic Proper has taken place. It has devastating consequences for marine flora and fauna leading during the last decades to replacing of O_2 by H_2S in these bottom waters (Forsberg, 1993). Although the eutrophication generally increased fish productivity, however it can also cause negative environmental changes in fish populations. Threatened by O_2 depletion in Baltic deep basins are cod and plaice resulting in decreasing catches in Köge Bay in the Sound (Forsberg, 1993).

(iv) Radionuclides as Environmental Contaminants

It is reported that radioactive isotopes of trace elements generally show similar environmental behaviour and cycling to their stable counterparts but they are distinguished themselves by enhanced potential impacts on organism owing to their radioactivity emitted as alpha, beta and gamma radiation. Considering their abilities to bioaccumulate in biota, alpha emitters, e.g. ^{239}Pu, ^{238}U, are thought to be much effective in respect to their radiation-related impacts as compared to beta and gamma isotopes, e.g. ^{137}Cs, ^{58}Co (Förstner, 1980; Phillips and Rainbow, 1993). There are numerous examples of the contamination of aquatic and terrestrial ecosystems by radionuclides which excite justified great public and political interest. One of the first low-level emissions of radioactivity took place from the Hanford reactors in the Columbia River, Washington State (USA) released significant amounts of particular radionuclides, mainly ^{60}Co, ^{51}Cr and ^{65}Zn, to environs during 1940–1971 (Förstner, 1980; Phillips and Rainbow, 1993). Other example of discharging quantities of radioisotopes, i.e. ^{144}Ce, ^{137}Cs, ^{95}Nb, ^{106}Ru and ^{95}Zr to the marine environment are the nuclear reactors in Cumbria, the north-west England. Although such emissions have diminished recently, discharges from nuclear power station such as Sellafield (formerly named Windscale) could be tracked for far distance from their source (ISSG, 1990; Kershaw et al., 1999). Significant quantities of artificial radionuclides (^{137}Cs, ^{134}Cs, ^{90}Sr, ^{99}Tc) have been transported to the North Atlantic and Arctic from Sellafield, together with measurable amounts of Pu and Am (Aarkrog et al., 1985; Kershaw and Baxter, 1995; Kershaw et al., 1999). Discharges from other, although relatively less significant, radionuclide emissions of ^{137}Cs and $^{239+240}Pu$ derive from the nuclear reprocessing plant at La Hague in France (Förstner, 1980; Phillips and Rainbow, 1993). This plant, however, has dominated the supply of ^{129}I and ^{125}Sb (Kershaw and Baxter, 1995). Besides expected emission of radionuclides from nuclear and reprocessing facilities, significant quantities of radioisotopes contaminate aquatic and terrestrial environments from nuclear weapons testing and from accidents in nuclear reactors. The first man-made nuclear explosions took place at the end of the II World War, with the testing of a 19-kt device in the USA on 16 July 1945 as well as extremely tragic in consequences – the detonation of nuclear weapons over the Japanese cities of Hiroshima and Nagasaki on 6 and 9 August 1945, respectively (MacKenzie, 2000). A great attention was paid to the thermonuclear detonation in 1954 at Bikini Atoll resulting in contamination of large areas of the Marshall Islands. During 1952–1958 and 1961 to 1962, two main series of atmospheric nuclear testing were carried out resulting in signing the partial atmospheric test ban treaty by the former USSR, USA and UK in 1963. Thereafter, only a limited number of atmospheric tests was carried out by France, China and India, with the last one having been in 1980 (MacKenzie, 2000). A number of atmospheric tests mainly carried out in the Northern Hemisphere has been estimated to be 520, including 8 underwater, with a total explosive yield of 542 Mt. It is also reported that that there have been totally 1352 underground tests with total yield of 90 Mt (UNSCEAR, 1993).

A numerous series of nuclear incidents were noted anxiously, for instance those affecting the crew of the Japanese fishing vessel 'Fukuru Maru' (Phillips and Rainbow, 1993). The two major nuclear incidents were followed in 1957 by the next two at Kyshtym and Windscale reactors; from the latter were released into the atmosphere highly significant amounts of radionuclides, especially ^{137}Cs, ^{131}I, ^{89}Sr and ^{90}Sr. It is recently postulated that plutonium from the Kyshtym accident in the Urals has been much probably detected in deep basins of the Arctic Ocean (Beasley et al., 1995). In 1968 an aircraft from the US Strategic Air Command crashed near the Thule Airbase in NW Greenland releasing to the marine environment ca. 1 TBq $^{239+240}Pu$ (Aarkrog et al., 1984). After the accident, the marine sediments as well as benthic organisms, i.e. bivalves, shrimps and seastars, have been contaminated by Pu. However, the levels of Pu have been rapidly decreasing (Smith et al., 1994) A similar accident had place two years earlier at

Palomares in SE Spain (NEA, 1981). A number of American and Russian nuclear submarines have been lost in the world ocean. For instance, the Soviet Komsomolets submarine sank at a depth of 1700 m at Bear Island in the eastern part of the Norwegian Sea. The radioactivity in the wreck is estimated at 2.8 PBq ^{90}Sr and 3 PBq ^{137}Cs; the nuclear warheads probably contain 16 TBq $^{239+240}Pu$ (Joint Russian–Norwegian Expert Group, 1994). Some satellites, nuclear powered, can be incidentally sources of radioactivity. Sometimes they have burned up in the upper atmosphere resulting in the contamination of the ocean. Such accident took place in 1964 when a SNAP-9A nuclear power generator contained 0.6 PBq ^{238}Pu aboard a U.S. satellite re-entered the atmosphere of the southern. As a result, in seawater from the Southern Hemisphere has been detected an enhanced $^{238}Pu/^{239+240}Pu$ ratio than in ocean water from the Northern Hemisphere (National Academy of Sciences, 1971; Aarkrog, 1998).

Sea dumping was carried out since the late 1940s to the middle of the 1960s mainly by the U.S. in the Atlantic and Pacific Oceans as well as by the U.K. in the NE Atlantic (Aarkrog, 1998). In 1967 an international operation was initiated by the former European Nuclear Energy Agency which contributed to the deposition of ca. 0.3 PBq solid waste at a depth of 5 km in the eastern Atlantic Ocean. Other international operations were continued until 1982 when ca. 0.7 α PBq activity, 42 PBq β activity and 15 PBq tritium have been dumped in the North Atlantic (Commission of the European Communities, 1989). It has been assessed the radiological impact of the NEA (former European Nuclear Energy Agency) dumping activities resulting in some releases of Pu from the dumped waste. This source would be responsible for only a part of the total body-burden radioactivity in local benthal organisms, e.g. sea cucumbers; the remainder has been attributed to fallout (NEA, 1996). According to CRESP evaluation, the individual dose to a critical group consuming seafood components such as molluscs from the Antarctic Ocean was estimated to level of 0.1 μSv y^{-1} corresponding to ^{239}Pu and ^{241}Am as critical radionuclides. The indefinite collective dose to the world's population coming from sea dumping was estimated at 40,000 manSv with predominance of ^{14}C and ^{239}Pu (NEA, 1996; Aarkrog, 1998).

Radiocaesium discharges during the 1950s are responsible for a large fraction of the historical releases from US weapons production facilities According to Garten et al. (2000) the total quantities of radiocaesium released to the environment are ca. 24 PBq (640 kCi) at Oak Ridge, 3.0 PBq (80 kCi) at Hanford and 0.16 PBq (4.4 kCi) at the Savannah River Site. Among other accidents, being of significant importance for healthy status of human population was chemical explosion which took place in 1959 in radiochemical processing pilot plant at Oak Ridge National Laboratory (USA). As a result, ^{239}Pu was released to surrounding area (Eisenbud, 1963; Phillips and Rainbow, 1993). Among further series of incidents involving the nuclear industry most significant was event at Three Mile Island (USA). However, the most dramatic incident occurred at Chernobyl in former USSR where an explosion of reactor core of the nuclear plant took place in April 1986. The Baltic countries and greater part of central and western Europe have been contaminated principally by ^{131}I, ^{134}Cs and ^{137}Cs (INSAG, 1986; Phillips and Rainbow, 1993). Approximately two thirds of ca. 100 PBq ^{137}Cs corresponding to the Chernobyl-derived radiocaesium were deposited outside the former USSR. It is found that a significant part of the activity fell over the European marginal seas from which the Baltic Sea was the most affected by contamination (WHO, 1989; Aarkrog, 1998). ^{137}Cs and ^{90}Sr behavioural regularities in the southeastern part of the Baltic Sea have been reported by Styro et al. (2001). Within a couple of months after the Chernobyl accident mean concentration of ^{137}Cs in Baltic surficial water was, on the average, ca. 100 Bq m^{-3} and total inventory from this dramatic event was estimated to 4.5 PBq ^{137}Cs (HELCOM, 1995). Before the accident, the concentration of ^{137}Cs in Baltic water has been directly proportional to the salinity because this Sellafield-derived radionuclide crossed the Danish Straits with the inflow high salinity water from the North Sea. This distribution pattern changed after the Chernobyl accident, i.e. the concentration of ^{137}Cs in Baltic waters became inversely proportional to the salinity because the major source of the radionuclide was then identified in the low saline waters of the Baltic Sea. Since the Chernobyl accident, the Baltic Sea has been a mainly responsible for additional inflow of the contaminants to the North East Atlantic Ocean (Aarkrog, 1998). It has been computed that the Black Sea received 2–3 PBq ^{137}Cs due to the Chernobyl accident (European Commission, 1995) resulting in an increase of mean its concentration in water to ca. 50 Bq ^{137}Cs m^{-3}. The outflow from the Black Sea is the major source of 'additional' ^{137}Cs in the Mediterranean Sea (Aarkrog, 1998). The North Sea received 1.2 PBq ^{137}Cs from the Chernobyl event. It has been reported that the inventory from the accident in the NE Atlantic was estimated at about 5 PBq ^{137}Cs. In summer of 1987, the Chernobyl-derived ^{137}Cs amounting, on the average, to 5 PBq was also detected in surficial waters of the Greenland, Norwegian and Barents Seas as well as at the west coast of Norway and from the Faroe Islands. According to Aarkrog (1998) the total Chernobyl ^{137}Cs input to the World Ocean is estimated at level of 15–20 PBq; it is relatively significantly less than that estimated for nuclear weapons fallout because of tropospheric nature of this accident which as first of all has contaminated the surrounding European continental areas.

It is reported (Ilus and Ilus, 2000) that radionuclides in the Baltic Sea are from the fallout associated with nuclear weapons testing carried out in the Northern Hemisphere during 1950–1960; artificial radioactivity source in the Baltic area was also fallout from the most serious accident in nuclear facilities at the Chernobyl Nuclear Power Plant in 1986 and discharges from nuclear fuel reprocessing plants in Western Europe at Sellafield, United Kingdom and in La Hague, France, which are spread by sea currents through the Danish Straits into the Baltic Sea. To other sources of artificial radionuclides in this area belong authorised routine discharges from civilian nuclear installations including nuclear power plants (NPPs) and reasearch reactors (RRs) in the Baltic Sea region, discharges from other industrial facilities, hospitals, laboratories, etc. and small-scale dumping of radioactive waste in the past (Ilus and Ilus, 2000).

The Persian Gulf War resulted in injuries of US Coalition personnel by fragments of depleted U (DU). Biological effects of embedded DU have been estimated by Armed Forces Radiobiology Research. According to McClain et al. (2001) levels of U in the kidney were nephrotoxic after acute exposure. It is recommended to study of long-term health impact, since DU was found to be mutagenic and it transformed human osteoblast cells to a tumorigenic phenotype (McClain et al., 2001).

References

Aarkrog, A., 1998. A retrospect od anthropogenic redioactivity in the global marine environment. Radiat Protect Dosimetry **75**, 23–31.

Aarkrog, A., H. Dahlgaard, K. Nilsson and E. Holm, 1984. Studies of plutonium and americum at Thule, Greenland. Health Phys. **46**, 29–44.

Aarkrog, A., H. Dahlgaard, H. Hansen, E. Holm, L. Hallstadius, J. Rioseco and G. Christensen, 1985. Radioactive tracer studies in the surface waters of the northern North Atlantic including the Greenland, Norwegian and Barents Seas. Rit. Fiskideildar **9**, 37–42.

Abbott, A., P.D. Abel, D.W. Arnold and A. Milne, 2000. Cost-benefit analysis of the use of TBT: Case for a treatment approach. Total Environ. **258**, 5–19.

Abel, P.D., 2000. TBT – Towards a better way to regulate pollutants. Editorial. Sci. Total Environ. **258**, 1–4.

Akagi, H., O. Malm, Y. Kinjo, M. Harada, F.J.P. Branches, W.C. Pfeiffer and H. Kato, 1995. Methylmercury pollution in the Amazon, Brazil. Sci. Total Environ. **175**, 85–95.

Akagi, H., P. Grandjean, Y. Takizawa and P. Weihe, 1998. Mercury dose estimation from umbilical cord concentrations in patients with Minamata disease. Environ. Res. **77**, 98–103.

Akagi, H., E.S. Castillo, N. Cortes-Maramba, A.T. Francisco-Rivera and T.D. Timbang, 2000. Health assessment for mercury exposure among schoolchildren residing near a gold processing and refining plant in Apokon, Tagum, Davao del Norte, Philippines. Sci. Total Environ. **259**, 31–43.

Aksnes, D.L., J. Aure, G.K. Furnes, H.R. Skjoldal and R. Saetre, 1989. Analysis of the *Chrysochromulina polyepsis* bloom in the Skagerrak, May, 1988. Environmental conditions and possible causes. Bergen Scientific Centre Publication **No. BSC 89/1**.

Al-Majed, N.B., and M.R. Preston, 2000. Factors influencing the total mercury and methyl mercury in the hair of the fishermen of Kuwait. Environ. Pollut. **109**, 239–250.

Alzieu, C., 1986. TBT detrimental effects on oyster culture in France – evolution since antifouling paint regulation, in: Proc. of Oceans 86 Confererence Record. Organotin Symposium. 4 (Institute of Electrical and Electronics Engineers, New York), pp. 1130–1134.

Alzieu, C., 2000. Environmental impact of TBT: the French experience. Sci. Total Environ. **258**, 99–102.

Ambio, 1990a. Marine Eutrophication, Ambio **19**, 101–176.

Ambio, 1990b. Special Report: Current status of the Baltic Sea. No. **7**, 1–10.

Andell, P., J. Dürinck and H. Skov, 1994. Baltic marine areas of outstanding importance for wintering seabirds. WWF Bulletin **5**, 1–8.

Anderson, D.M., 1989. Toxic algal blooms and red tides. A global perspective, in: Red Tides: Biology, Environmental Science and Toxicology, eds. T. Okaichi, D.M. Anderson and T. Nemoto (Elsevier Science Publishing Co., Inc., New York, NY), 11–21.

Anderson, D.M., D.M. Kulis, G.J. Doucette, J.C. Gallagher and E. Balech, 1994. Biogeography and toxic dinoflagellates in the genus *Alexandrium* from the northeastern United States and Canada. Mar. Biol. **120**, 467–478.

Anon, 1990. Status of the Baltic Sea – A sea in transition. Ambio Special Reports **7**.

Artaxo, P., R. Calixto de Campos, E.T. Fernandes, J.V. Martins, Z. Xiao, O. Lindquist, M.T. Fernández-Jiménez and W. Maenhaut, 2000. Large scale mercury and trace element measurements in the Amazon basin. Atmos. Environ. **34**, 4085–4096.

Augustowski, B., 1987. Southern Baltic (Polish Academy of Sciences, Wrocław), (in Polish).

Ayres, P.A., 1975. Mussel poisoning in Britain with special reference to Paralytic Shellfish Poisoning. Environ. Health, July 261–265.

Backlund, P., B. Holmbom and E. Leppäkoski, 1992. Industrial pollution in the Baltic region, in: Industrial Emissions and Toxic Pollutants, ed. L. Rydén (Uppsala University, Sweden), **Session 5**, pp. 8–15.

Bakir, F., S. Damluji, L. Amin-Zaki, L. Murtadha, A, Khalidi, N. AlRawi, S. Tikriri, H. Dhakhir, T.W. Clarkson, J. Smith and R. Doherty, 1973. Methylmercury poisoning in Iraq. An interuniversity report. Science **181**, 230–241.

Beasley, T.M., L.W. Cooper, J.M. Grebmeier, K. Orlandini, and J.M. Kelley, 1995. Fuel reprocessing Pu in the Arctic Ocean Basin: evidence from mass spectrometry measurements, in: Proc. Conf. on Environmental Radioactivity in the Arctic. Oslo, August 1995.

Bergström S., and B. Carlsson, 1993. Hydrology of the Baltic Basin. SMHI Reports Hydrology **No. 7**, 21 pp.

Bertram, H., F. Kemper and C. Zenzen, 1985. Man – a target of ecotoxicological influences, in: Pollutants and Their Ecotoxicological Significance (John Wiley & Sons, New York), **Chapt. 26**, 429–431.

Billen, G., J. Garnier, C. Delinge and C. Billen, 1999. Estimate of early-industrial inputs of nutrients to river systems: implication for coastal eutrophication. Sci. Total Environ. **243/244**, 43–52.

Birch, P.B., G.G. Forbes and N.J. Schofield, 1986. Monitoring effects of catchment management practices on phosphorus loads into the eutrophic Peel-Harvey Estuary, Western Australia. Water Sci. Technol. **18**, 53–61.

Blazhchishin, A.I., 1982a. Bottom sediments of the Baltic Sea, in: Geology of the Baltic Sea, eds. V.K. Gudelis and E.M. Emelyanov (Wydawnictwo Geologiczne, Warsaw), 188–214 (in Polish).

Blazhchishin, A.I., 1982b. Main chemical constituents of the sediments of the Baltic Sea, in: Geology of the Baltic Sea, eds. V.K. Gudelis and E.M. Emelyanov (Wydawnictwo Geologiczne, Warsaw), 257–289 (in Polish).

Blazhchishin, A.I., 1982c. Sedimentary environment of the Baltic Sea, in: Geology of the Baltic Sea, eds. V.K. Gudelis and E.M. Emelyanov (Wydawnictwo Geologiczne, Warsaw), 131–141 (in Polish).

Blazhchishin, A.I., and V.K. Lukashev, 1981. Litho-stratigraphic characteristics of the late Quaternary deposits, in: Formation of sediments in the Baltic Sea, eds. A.P. Lisitsyn and E.M. Emelyanov (Izdatelstvo Nauka, Moscow), 232–246 (in Russian).

Blomqvist, S. and A.-S. Heiskanen, 2001. The challenge of sedimentation in the Baltic Sea, in: A System Analysis of the Baltic Sea, eds. F. Wulff et al. (Springer-Verlag Berlin Heidelberg) Ecological Studies Vol. **148**, 211-227.

Bruneau, L., 1980. Pollution from industries in the drainage area of the Baltic. Ambio **9**, 145–152.

Bryan, G.W., and W.J. Langston, 1992. Bioavailability, accumulation and effects of heavy metals in sediments with special reference to United Kingdom estuaries: a review. Environ. Pollut. **76**, 89–131.

Buskey, E.J., and D.A. Stockwell, 1993. Effects of a persistent "brown tide" on zooplankton populations in the Laguna Madre of South Texas, in: Toxic Phytoplankton Blooms in the Sea. Proc. 5th Intern. Conf. on Toxic Marine Phytoplankton (Elsevier), 659–665.

Bykov, A.A., and B.A. Revich, 1997. Children's health on the territories contaminated with lead. Issues of biological monitoring: risk assessment: prevention and treatment, in: Materials of a Russian-American Seminar, Moscow (in Russian).

Cambray, R.S., K. Playford and G.N.J. Lewis, 1982. Radioactive fallout in air and rain: results to the end of 1981. United Kingdom Atomic Energy Report AERE-R 10485.

Cederwall, H., and R. Elmgren. (1990) Eutrophication of the Baltic Sea. Ambio **19**, 109–112.

Champ, M.A., 2000. A review of organotin regulatory strategies, pending actions, related costs and benefits. Total Environ. **258**, 21–71.

Chatterjee, A., and R.N. Banerjee, 1999. Determination of lead and other metals in a residential area of greater Calcutta. Sci. Total Environ. **227**, 175–185.

Commission of the European Communities, 1989. The Radiological Exposure of the Population of the European Community from Radioactivity in North European Marine Waters – Project MARINA; Report EUR 12483 (EU, 1989).

Cyberski J., 1995. Recently observed and prognostic changes in water balance and their impact on the salinity in the Baltic Sea. Wydawnictwo Uniwersytetu Gdańskiego. Rozprawy i monografie Nr **206**, 1–210 (in Polish).

Dahl, E., 1956. Ecological salinity boundries in pikilohaline waters. Oikos **7**, 1–21.

Damluji, S., 1962. Mercurial poisoning with the fungicide granosan M. J. Fac. Med. Baghdad **4**, 83.

Danielsson, Å., 1998. Spatial Modelling in Sediments (Linköping Studies in Arts and Science, Sweden). 89 pp. + Appendices.

Degobbis, D., 1989. Increased eutrophication of the Northern Adriatic Sea. Second act. Mar. Pollut. Bull. **20**, 452–457.

Dos Santos, L. do. S.N., R.C.S. Müller, J.E. de S. Sarkis, C.N. Alves, E. da S. Brabo, E. de O. Santos and M.H. da S. Bentes, 2000. Evaluation of total mercury concentrations in fish consumed in the municipality of Itaituba, Tapajós River Basin, Pará, Brazil. Sci. Total Environ. **261**, 1–8.

Dybern, B.I. and S.H. Fonselius, 1981. Pollution, in: The Baltic Sea, ed. A. Voipio (Elsevier Scientific Publishing Company, Amsterdam), pp. 351–382.

ECOHAB, 1995. The Ecology and Oceanography of Harmful Algal Blooms. A National Research Agenda, US N.S.F./N.O.O.A. Publication (Woods Hole Oceanographic Inst., Mass., USA).

Ehlin, U., 1981. Hydrology of the Baltic Sea, in: The Baltic Sea, ed. A. Voipio (Elsevier Scientific Publishing Company, Amsterdam), pp. 123–134.

Eisenbud, M., 1963. Environmental Radioactivity (McGraw-Hill Book Company, Inc., New York).

Elmgren, R., 1984. Trophic dynamics in the enclosed, brackish Baltic Sea. Rapp. P.-v. Réun. Const. Int. Explorer. Mer. **183**, 149–166.

Elmgren, R., 1989. Man's impact on the ecosystem of the Baltic Sea: energy flows today and at the turn of the century. Ambio **18**, 326–332.

Emeis, K.-C., B. Larsen and B. Seibold, 1992. What is the environmental capacity of enclosed marginal seas? Approaches to the problem in the Baltic, North, and Mediterranean Seas, eds. K.J. Hs and J. Thiede, Use and Misuse of the Seafloor (John Wiley, New York), pp. 181–211.

Emelyanov, E.M., 1995. Baltic Sea, Geochemistry, Paleoceanography, Pollution (P.P. Shirshov Institute, Kaliningrad).

Emelyanov, E.M., and Pustenikov O.S., 1982. Suspended matter, its composition and the balance of sediments in the Baltic, in: Geology of the Baltic Sea, eds. V.K. Gudelis and E.M. Emelyanov (Wydawnictwo Geologiczne, Warsaw), 161–187 (in Polish).

Englyst, V., N.-G. Lundström, L. Gerhardsson, L. Rylander and G. Nordberg, 2001. Lung cancer risks among lead smelter workers also exposed to arsenic. Sci. Total Environ. **273**, 77-87.

European Commission, 1995. The Radiological Exposure of the Population of the European Community to Radioactivity in the Mediterranean Sea. Marina-Med. Project. Report EUR 15564, in: Proc. Seminar on Marina-Med. Rome, (EC, Luxembourg) May 1994.

Evans, S.M., 1999. Tributyltin pollution: the catastrophe that never happened. Mar. Pollut. Bull. **38**, 629–636.

Evans, S.M., 2000. Response to the letter from Nichols. Mar. Pollut. Bull. **40**, 713–715.

Evans, S.M., and G.J. Nicholson, 2000. The use of imposex to assess tributyltin contamination in coastal waters and open seas. Sci. Total Environ. **258**, 73–80.

Evans, S.M., T. Leksono and P.D. McKinnel, 1995. Tributyltin pollution: a diminishing problem following legislation limiting the use of TBT-based anti-fouling paints. Mar. Pollut. Bull. **30**, 14–21.

Evans, S.M., P.M. Evans and T. Leksono, 1996. Widespread recovery of dogwhelks, *Nucella lapillus* (L.), from tributyltin contamination in the North Sea and Clyde Sea. Mar. Pollut. Bull. **32**, 263–269.

Evans, S.M., A.C. Birchenough and M.S. Brancato, 2000a. The TBT ban: out of the frying pan into the fire? Mar. Pollut. Bull. **40**, 204–211.

Evans, S.M., A.C. Birchenough and H. Fletcher, 2000b. The value and validity of community-based research: TBT contamination of the North Sea. Mar. Pollut. Bull. **40**, 220–225.

Evans, S.M., E. Kerrigan and N. Palmer, 2000c. Causes of imposex in the dogwhelk *Nucella lapillus* (L.) and its use as a biological indicator of tributyltin contamination. Mar. Pollut. Bull. **40**, 212–219.

Falandysz, J., A. Trzosińska, P. Szefer, J. Warzocha and B. Draganik, 2000. The Baltic Sea, especially southern and eastern regions in: Seas at the Millennium: An Environmental Evaluation, Vol. I: Europe, The Americas and West Africa, ed. C.R.C. Sheppard (Pergamon, Elsevier), **Chapt. 7**, pp. 99–120.

Falkenmark, M., 1986. Hydrology of the Baltic Sea area: temporal fluctuations in water balance. Ambio **15**, 97–102.

Fonselius S., 1969. Hydrography of the Baltic deep basins III, Fishery Board of Sweden, Ser. Hydrogr., No. **23**, 1–97.

Fonselius S., J. Szaron and B. Öström, 1984. Long-term salinity variations in the Baltic Sea deep water. Rapp. P.-v. Reun. Cons. int. Explor. Mer, **185**, 140–149.

Forsberg, C, 1993. Eutrophication of the Baltic Sea (The Baltic Sea Environment, Uppsala, Sweden), pp. 32.

Förstner, U., 1980. Inorganic pollutants, particularly heavy metals in estuaries, in: Chemistry and Biogeochemistry of Estuaries, eds. E. Olausson and I. Cato (John Wiley & Sons, New York), pp. 307–348.

Förstner, U., and G.T.W. Wittmann, 1983. Metal Pollution in the Aquatic Environment. Second Edition (Springer-Verlag, Berlin).

Franck, H., W. Matthäus and R. Sammler, 1987. Major inflows of saline water into the Baltic Sea during the present century. Gerlands Beitr. Geophys. **96**, 517–531.

Gadd, G.M., 2000. Microbial interactions with tributyltin compounds: detoxication, accumulation, and environmental fate. Sci. Total Environ. **258**, 119–127.

Garten Jr., C.T., D.M. Hamby and R.G. Schreckhise, 2000. Radiocesium discharges and subsequent environmental transport at the major US weapons production facilities. Sci. Total Environ. **255**, 55–73.

Gavrilov, V.M., Z.G. Gritchenko, L.M. Ivanova, T.E. Orlova, V.P. Tishkov and N.A. Tishkova, 1990. Strontium-90, caesium-134 and caesium-137 in water reservoirs of the Soviet Union's Baltic region (1986–1988). Radiochemistry **No. 3**, 171–179 (in Russian).

Gillespie, N., 1984. Ecological and epidemiological aspects of ciguatera fish poisoning, in: Proc. of the Red Tide Workshop, Cronulla, (Australian Department of Science, Canberra) 18–20 June 1984.

Gingele, F.X., and T. Leipe, 1997. Clay mineral assemblages in the western Baltic Sea: recent distribution and relation to sedimentary units. Mar. Geol. **140**, 97-115.

Granéli, E., K. Wallstrom, U. Larsson, W. Graneli and R. Elmgren, 1990. Nutrient limitation of primary production in the Baltic Sea area. Ambio **19**, 142–151.

Grasshoff, K., and A. Voipio, 1981. Chemical Oceanography, in: The Baltic Sea, ed. A. Voipio (Elsevier Scientific Publishing Company, Amsterdam) pp. 183–218.

Gray, J.E., P.M. Theodorakos, E.A. Bailey and R.R. Turner, 2000. Distribution, speciation, and transport of mercury in stream-sediment, stream-water, and fish collected near abandoned mercury mines in southwestern Alaska, USA. Sci. Total Environ. **260**, 21–33.

Greenwood, M.R., 1985. Methylmercury poisoning on Iraq: An epidemiological study of the 1971–1972 outbreak. J. Appl. Toxicol. **5**, 148–159.

Gudelis, V., and E.M. Emelyanov (eds.), 1976. Geology of the Baltic Sea (Mokslas Publishers, Vilnius), (in Russian).

Hallegraeff, G.M., 1993. A review of harmful algal blooms and their apparent global increase. Phycologia **32**, 79–99.

Hallegraeff, G.M., and C. Sumner, 1986. Toxic plankton blooms affect shellfish farms. Aust. Fish. **45**, 15–18.

Harada, M., 1978. Congenital Minamata disease: intrauteric methylmercury poisoning. Teratology **18**, 285–288.

Harada, M., 1995. Minamata disease: methylmercury poisoning in Japan caused by environmental pollution. Crit. Rev. Toxicol. **25**, 1–24.

Harada, M., 1996. Characteristics of industrial poisoning and environmental contamination in developing countries. Environ. Sci. **4**, 157–169.

Harada, M., S. Nakachi, T. Cheu, H. Hamada, Y. Ohno, T. Tsuda, K. Yanagida, T. Kizaki and H. Ohno, 1999. Monitoring of mercury pollution in Tanzania: relation between head hair and health. Sci. Total Environ. **227**, 249–256.

Harff, J., W. Lemke, F. Tauber and E.M. Emelyanov, 1995. Geological mapping in the Baltic Sea. Geowissenschaften **13**, 442–447 (in German).

Hartmann, M., 1964. On the geochemistry of manganese and iron in the Baltic Sea. Meyniana **14**, 3–20 (in German).

Hägerhäll, A.B., 1994. Alien species in the Baltic marine environment. WWF Bulletin **4**, 10–14.

Hällfors, G., Å. Niemi, H. Ackefors, J. Lassig and E. Leppäkoski, 1981. Biological Oceanography, in: The Baltic Sea, ed. A. Voipio (Elsevier Scientific Publishing Company, Amsterdam) pp. 219–274.

HELCOM, 1986. Water balance of the Baltic Sea, M. Falkenmark (ed.), Baltic Sea Enviroment Proceedings No. **16**, 1–174.

HELCOM, 1987. First periodic assessment of the state of the marine environment of the Baltic Sea area, 1980–1985. Baltic Sea Environment Proceedings No. **17 B**.

HELCOM, 1990. Second periodic assessment of the state of the marine environment of the Baltic Sea, 1984–1988. Baltic Sea Environment Proceedings No. **35 B**.

HELCOM, 1991. Airborne pollution load to the Baltic Sea 1986–1990. Baltic Sea Environment Proceedings No. **39**.

HELCOM, 1993. First assessment of the state of the coastal waters of the Baltic Sea. Baltic Sea Environment Proceedings No. **54**.

HELCOM, 1995. Radioactivity in the Baltic Sea 1984–1991. Baltic Sea Environment Proceedings (Baltic Marine Environment Protection Commission, HELCOM) No. **61**.

HELCOM, 1996. Third Periodic Assessment of the State of the Marine Environment of the Baltic Sea, 1989–93; Background Document. Baltic Sea Environment Proceedings No. **64B**.

HELCOM, 1997a. Radioactivity in the Baltic Sea 1984–1991. Baltic Sea Environment Proceedings No. **61**.

HELCOM, 1997b. Airborne pollution load to the Baltic Sea 1991–1995. Baltic Sea Environment Proceedings No. **69**.

HELCOM, 1998a. The third Baltic Sea pollution load compilation. Baltic Sea Environment Proceedings No. **70**.

HELCOM, 1998b. Red list of marine and coastal biotopes and biotope complexes of the Baltic Sea, Belt Sea and Kattegat. Baltic Sea Environment Proceedings No. **75**.

Holmes, P.R., and C.W.Y. Lam, 1985. Red tides in Hong Kong waters – response to a growing problem. Asian Mar. Biol. **2**, 1–10.

Hopkin, S.P., 1989. Ecophysiology of Metals in Terrestrial Invertebrates (Elsevier Applied Science, London).

Howarth, R.W., G. Billen, D. Swaney, A. Townsend, N. Jaworski, K. Lajtha, J.A. Downing, R. Elmgren, N. Caraco, T. Jordan, F. Berendse, J. Freney, V. Kudeyarov, P. Murdoch and Z.-L. Zhu, 1996. Regional nitrogen budgets and riverine N and P fluxes for the drainages to the North Atlantic Ocean: Natural and human influences. Biogeochemistry **35**, 75–139.

Huckriede, H., S. Clasen and D. Meischner, 1996. Hydrographic and climatic changes recorded in Holocene sediments of the Central Baltic Sea. Baltica **9**, 76-91.

Huet, M., Y.M. Paulet and M. Glémarec, 1996. Tributyltin (TBT) pollution in the coastal waters of west Brittany as indicated by imposex in *Nucella lapillus*. Mar. Environ. Res. **41**, 157–167.

Hung, T.-C., W.-K. Hsu, P.-J. Mang and A. Chuang, 2001. Organotins and imposex in the rock shell, *Thais clavigera*, from oyster mariculture areas in Taiwan. Environ. Pollut. **112**, 145–152.

Ikingura, J.R., and H. Akagi, 1996. Monitoring of fish and human exposure to mercury due to gold mining in the Lake Victoria goldfields, Tanzania. Sci. Total Environ. **191**, 59–68.

Ilus, E., and T. Ilus, 2000. Sources of radioactivity, in: The Radiological Exposure of the Population of the European Community to Radioactivity in the Baltic Sea. Marina-Balt Project, ed. S.P. Nielsen.

Proceedings of a Seminar held at Hasseludden Conference Centre, Stockholm, 9–11 June 1998, European Commission, Directorate-General Environment, EUR 19200 EN (European Communities, 2000, Belgium), pp. 9–76.

INSAG, 1986. Post accident Review Meeting on the Chernobyl Accident. Summary Report (International Atomic Energy Agency, Vienna).

ISSG, 1990. The Irish Sea: An Environmental Review. Rep. of the Irish Sea Study Group (Liverpool University Press, Liverpool).

Jacobson, A.H., and G.L. Willingham, 2000. Sea-nine antifoulant: an environmentally acceptable alternative to organotin antifoulants. Sci. Total Environ. **258**, 103–110.

Jalili, M., and A. Abbasi, 1961. Poisoning by ethyl mercury toluene sulphonanilide. Brit. J. Ind. Med. **18**, 303.

Jansson, B.O., 1972. Ecosystem approach to the Baltic problem. Bull. Ecol. Res. Comm. NFR **16**, 82 pp.

Jansson, B.-O., and K. Dahlberg, 1999. The environmental status of the Baltic Sea in the 1940s, today, and in the future. Ambio **28**, 312–319.

Jarvekulg, A., 1979. Donnaja fauna vostočnoj časti Baltijskogo Morjia. Sostav i ekologija raspredelenija (Bottom fauna of eastern part of the Baltic Sea. Composition and ecology of distribution). Tallin.

Jensen, J.B., O. Bennike, A. Witkowski, W. Lemke and A. Kuijpers, 1997. The Baltic Ice Lake in the southwestern Baltic: sequence-, chrono- and biostratigraphy. Boreas **26**, 217–236.

Jensen, J.B., O. Bennike, A. Witkowski, W. Lemke and A. Kuijpers, 1999. Early Holocene history the southwestern Baltic: the Ancylus Lake stage. Boreas **28**, 437–453.

Joint Russian-Norwegian Expert Group, 1994. Radioactive Contamination of Dumping Sites for Nuclear Wastes in the Kara Sea. Results from the 1993 Expedition (Norwegian Radiation Protection Authority, Østerås).

Jonsson, P., A. Grimvall, M. Cederlöf and M. Hildén, 1996. Pollution threats to the Gulf of Bothnia. Ambio Spec. Rep. **8**, 22-27.

Justic, D., 1987. Long-term eutrophication of the Northern Adriatic Sea. Mar. Pollut. Bull. **18**, 281–284.

Kautsky, L., and N. Kautsky, 2000. The Baltic Sea, including Bothnian Sea and Bothnian Bay, in: Seas at the Millennium: An Environmental Evaluation, Vol. I: Europe, The Americas and West Africa, ed. C.R.C. Sheppard (Pergamon, Elsevier), **Chapt. 8**, pp. 121-133.

Kautsky, N., H. Kautsky, U. Kautsky and M. Waern, 1986. Decreased depth penetration of *Fucus vesiculosus* (L.) since the 1940s indicates eutrophication of the Baltic Sea. Mar. Ecol. Prog. Ser. **28**, 1–8.

Kershaw, P.J., and A.J. Baxter, 1995. The transfer of reprocessing wastes from north-west Europe to the Arctic. Deep Sea Res. **42**, 1413–1448.

Kershaw, P.J., D. McCubbin and K.S. Leonard, 1999. Continuing contamination of North Atlantic and Arctic waters by Sellafield radionuclides. Sci. Total Environ. **237/238**, 119–132.

Kullenberg, G., 1981. Physical oceanography, in: The Baltic Sea, ed. A. Voipio (Elsevier Scientific Publishing Company, Amsterdam), pp. 135–182.

Lal, D., 1999. An overview of five decades of studies of cosmic ray produced nuclides in oceans. Sci. Total Environ. **237/238**, 3–13.

Lemke, W., J.B. Jensen, O. Bennike and A. Witkowski, 1997. Sequence stratigraphy of Late Pleistocene and Holocene deposits in the Mecklenburg Bay, southwestern Baltic Sea, in: Proc. of the Fourth Mar. Geol. Confer. – the Baltic, eds. I. Cato and F. Klingberg, (Sveriges Geologiska Undersökning, Uppsala, 1995), Ser. Ca **86**, pp.117–122.

Lemke, W., R. Endler, F. Tauber, J.B. Jensen and O. Bennike, 1998. Late- and postglacial sedimentation in the Tromper Wiek northeast of Rügen (western Baltic). Meyniana **50**, 155–173.

Lemke, W., J.B. Jensen, O. Bennike, R. Endler, A. Witkowski and A. Kuijpers, 2001. Hydrographic thresholds in the western Baltic Sea: Late Quaternary geology and the Dana River concept. Mar. Geol. **176**, 191-201.

Liggans, G.L., and J.O. Nriagu, 1998. Lead poisoning of children in Africa, IV: exposure to dust lead in primary schools in south-central Durban, South Africa. Sci. Total Environ. **221**, 117–126.

Lisitzyn, A.P., and E.M. Emelyanov, 1981. Formatiom of sediments in the Baltic Sea (Akademia Nauk SSSR, Moscow) in Russian.

Lithner, G., H. Borg, U. Grimas, A. Göthberg, G. Neumann and H. Wrádhe, 1990. Metals – trends observed 1984–1988. Ambio Special Report **7,** 7–9.

Lozán, J. L., R. Lampe, W., Matthäus, E., Rachor, H. Rumohr and H. Westernhagen (eds.), 1996. Warnsignale aus der Ostsee: wissenschaftliche Fakten. (Parey, Berlin), 385 pp.

Łomniewski, K., W. Mańkowski and J. Zaleski, 1975. The Baltic Sea (Pańistwowe Wydawnictwo Naukowe, Warsaw), 508 pp. (in Polish).

Łysiak-Pastuszak, E., 1999. Ph.D. Thesis: An Impact of the Climatic and Hydrological Factors on the Variability in the Physical and Chemical Parameters in the Marine Environment of the Southern Baltic (University of Gdańsk), (in Polish).

Ma, B., and X. Zhang, 2000. Regional ecological risk assessment of selenium in Jilin province, China. Sci. Total Environ. **262**, 103–110.

MacKenzie, A.B., 2000. Environmental radioactivity: experience from the 20th century – trends and issues for 21st century. Sci. Total Environ. **249**, 313–329.

Maclean, J.L., 1989. Indo-Pacific red tides, 1985–1988. Mar. Pollut. Bull. **20**, 304–310.

Maclean, J.L., and A.W. White, 1985. Toxic dinoflagellate blooms in Asia: a growing concern, in: Toxic Dinoflagellates, eds. D.M. Anderson, A.W. White and D.G. Baden (Elsevier, New York), pp. 517–520.

Magaard, L., and G. Rheinheimer, (eds.), 1974. Meereskunde der Ostsee (Springer Verlag, Berlin).

Majewski, A., and Z. Lauer, (eds.), 1994. Atlas of the Baltic Sea (Institute of Meteorology and Water Management, Warszawa), (in Polish).

Malm, O., M.B. Castro, W.R. Bastos, F.J.P. Branches, J.R.D. Guimarães, C.E. Zuffo and W.C. Pfeiffer, 1995a. An assessment of Hg pollution in different goldmining areas, Amazon Brazil. Sci. Total Environ. **175**, 127–140.

Malm O., F.J.P. Branches, H. Akagi, M.B. Castro, W.C. Pfeiffer, M. Harada, W.R. Bastos and H. Kato, 1995b. Mercury and methylmercury in fish and human hair from Tapajos river basin, Brazil. Sci. Total Environ.**175**, 141–150.

Mance, G., 1987. Pollution Threat of Heavy Metals in Aquatic Environments Elsevier Applied Science), 372 pp.

Manheim, F.T., 1961. A geochemical profile in the Baltic Sea. Geochim. Cosmochim. Acta **25**, 52–70.

Matthäus, W., 1992. Water exchange between the North Sea and Baltic Sea. Geograph. Rundschau **44**, 626–631 (in German).

Matthäus, W., 1993a. Major inflows of highly saline water into the Baltic Sea. Paper presented at the ICES Statutory Meeting.

Matthäus, W., 1993b. Salt water influx and its significance for marine environment in the Baltic Sea. Wasser + Boden **12**, 922–928 (in German).

Matthäus W., and H. Francke, 1992. Characteristics of major Baltic inflows – a statistical analysis. Contin. Shelf Res. **12**, 1375–1400.

Maurice-Bourgoin, L., I. Quiroga, J. Chincheros and P. Courau, 2000. Mercury distribution in waters and fishes of the upper Madeira rivers and mercury exposure in riparian Amazonian populations. Sci. Total Environ. **260**, 73–86.

McClain, D.E., K.A. Benson, T.K. Dalton, J. Ejnik, C.A. Emond, S.J. Hodge, J.F. Kalinich, M.A. Landauer, A.C. Miller, T.C. Pellmar, M.D. Stewart, V. Villa and J. Xu, 2001. Biological effects of embedded depleted uranium (DU): summary of Armed Forces Radiobiology Research Institute research. Sci. Total Environ. **274**, 115-118.

Melvasalo T., J. Pawlak, K. Grasshoff, L. Thorell and A. Tsiban (eds.), 1981. Assessment of the effects of pollution on the natural resources of the Baltic Sea, 1980. Baltic Sea Environment Proceedings No. **5 B**, 426 pp.

Michel, P., and B. Averty, 1999. Contamination of French coastal waters by organotin compounds: 1997 update. Mar. Pollut. Bull. **38**, 268–275.

Mikulski, A., 1991. Comparative investigations on the water balance in the Baltic regions. Czasopismo Geograficzne **63**, 177–185 (in Polish).

Millero, F.J., 1978. The physical chemistry of Baltic Sea water. Thalassia Jugoslavica **14**, 1–46.

Minchin, D., J. Oehlmann, C.B. Duggan, E. Stroben and M. Keatinge, 1995. Marine TBT antifouling contamination in Ireland, following legislation in 1987. Mar. Pollut. Bull. **30**, 633–639.

Minchin, D., B. Bauer, J. Oehlmann, U. Schulte-Oehlmann and C.B. Duggan, 1997. Biological indicators used to map organotin contamination from a fishing port, Killybegs, Ireland. Mar. Pollut. Bull. **34**, 235–243.

Morgan, J.J., and W. Stumm, 1991. Chemical processes in the environment, relevance and chemical speciation, in: Metals and Their Compounds in the Environment – Occurrence, Analysis, and Biological Relevance, ed. E. Merian (VCH, Weinheim, New York, Basel, Cambridge), pp. 67–103.

Morgan, E., J. Murphy and R. Lyons, 1998. Imposex in *Nucella lapillus* from TBT contamination in south and south-west Wales: a continuing problem around ports. Mar. Pollut. Bull. **36**, 840–843.

Morton, B.S., 1989. Pollution of the coastal waters of Hong Kong. Mar. Pollut. Bull. **20**, 310–318.

Müller, A., and M. Wessels, 1999. The flood in the Odra River 1997 – Impact of suspended solids on water quality. Acta Hydrochim. Hydrobiol. **27**, 316–320.

National Academy of Sciences, 1971. Radioactivity in the Marine Environment (National Academy of Sciences, Washington, DC 20418).

NEA, 1981. The Environmental and Biological Behaviour of Plutonium and some other Transuranium Elements (OECD, Paris).

NEA, 1996. Co-ordinated Research and Environmental Surveillance Programme Related to Sea Disposal of Radioactive Waste. CRESP Final Report 1981–1995 (OECD, Paris).

Nehring D., 1989. Phosphate and nitrate trends and the ratio oxygen consumption to phosphate accumulation in central Baltic deep waters with alternating oxic and anoxic conditions. Beitr. Meeresk. Berlin, **59**, 47–58

Ninomiya, T., H. Ohmori, K. Hashimoto, K. Tsuruka and S. Ekino, 1995. Expansion of methylmercury poisoning outside of Minamata: an epidemiological study on chronic methylmercury poisoning outside of Minamata. Environ. Res. **70**, 47–50.

Nordberg, G.F., T. Jin, Q. Kong, T. Ye, S. Cai, Z. Wang, F. Zhuang and X. Wu, 1997. Biological monitoring of cadmium exposure and renal effects in a population group residing in polluted area in China. Sci. Total Environ. **199**, 111–114.

Nriagu, J.O., W.C. Pfeiffer, O. Malm, C.M.M. Souza and G. Mierle, 1992. Mercury pollution in Brazil. Nature (London) pp. 356– 389.

Nriagu, J.O., M.L. Blankson and K. Ocran, 1996a. Childhood lead poisoning in Africa: a growing public health problem. Sci. Total Environ. **181**, 93–100.

Nriagu, J.O., Ch. Jinabhai, R. Naidoo and A. Coutsoudis, 1996b. Atmospheric lead pollution in KwaZulu/Natal, South Africa. Sci. Total Environ. **191**, 69–76.

Nriagu, J.O., Ch.C. Jinabhai, R. Naidoo and A. Coutsoudis, 1997a. Lead poisoning of children in Africa, II. Kwazulu/Natal, South Africa. Sci. Total Environ. **197**, 1–11.

Nriagu, J.O., N.T. Oleru, Ch. Cudjoe and A. Chine, 1997b. Lead poisoning of children in Africa, II. Kaduna, Nigeria. Sci. Total Environ. **197**, 13–19.

Ojaveer, E., A. Lindroth, O. Bagge, H. Lehtonen and J. Toivonen, 1981. Fishes and fisheries, in: The Baltic Sea, ed. A. Voipio (Elsevier Scientific Publishing Company, Amsterdam), pp. 275–350.

Paerl, H.W., 1985. Enhancement of marine primary productivity by nitrogen enriched rain. Nature **315**, 747–749.

Paerl, H.W., 1995. Coastal eutrophication in relation to atmospheric nitrogen deposition: Current perspectives. Ophelia **41**, 237–259.

Paerl, H.W., 1997. Coastal eutrophication and harmful algal blooms: Importance of atmospheric deposition and groundwater as "new" nitrogen and other nutrient sources. Limnol. Oceanogr. **42**, 1154–1165.

Paerl, H.W., and D.R. Whitall, 1999. Anthropogenically-derived atmospheric nitrogen deposition, marine eutrophication and harmful algal bloom expansion: is there a link. Ambio **28**, 307–311.

Pfeiffer, W.C., and L.D. Lacerda, 1988. Mercury input into the Amazon region, Brazil. Environ. Tech Lett. **9**, 325–330.

Phillips, D.J.H., 1985. Monitoring and control of coastal water quality, in: Pollution in the Urban Environment, POLMET 85, ed. M.W.H. Chan, R.W.M. Hoare, P.R. Holmes, R.J.S. Law and S.B. Reed (Elsevier Applied Science Publishers, London), 559–565.

Phillips, D.J.H., and P.S. Rainbow, 1993. Biomonitoring of Trace Aquatic Contaminants. Elsevier Science Publishers Ltd., London, 371 pp.

Pilgrim, W., W. Schroeder, D.B. Porcella, C. Santos-Burgoa, S. Montgomery, A. Hamilton and L. Trip, 2000. Developing consensus: mercury science and policy in NAFTA countries (Canada, the United States and Mexico). Sci. Total Environ. **261**, 185–193.

Poloczanska, E.S., and A.D. Ansell, 1999. Imposex in the whelks *Buccinum undatum* and *Neptunea antiqua* from the west coast of Scotland. Mar. Environ. Res. **47**, 203–212.

Prospero, J.M., K. Barret, T. Church, F. Detener, R.A. Duce, J.N. Galloway, H. Levy, J. Moody and P. Quinn, 1996. Atmospheric deposition of nutrients to the North Atlantic basin. Biogeochemistry **35**, 27–73.

Readman, J.W., 1996. Antifouling herbicides – a threat to the marine environment. Mar. Pollut. Bull. **32**, 320–321.

Remane, A., 1958. Okologie der Brackwassers, in: Die Biologie des Brackerwassers, eds. A. Remane and Schlipper. Binnengewasser **12**, 126 pp.

Rheinheimer, G., 1998. Pollution in the Baltic Sea. Naturwissenschaften **85**, 318–329.

Ribbe, J., S.H. Müller-Navarra and H.A. Nies, 1991. One-dimensional dispersion model for radionuclides in the marine environment applied to the Chernobyl fallout over the northern Baltic Sea. J. Environ. Radioactivity **14**, 55–74.

Riley, J.P., 1971. The major and minor elements in sea water, in: Introduction to Marine Chemistry, eds. J.P. Riley, R. Chester (Academic Press London and New York) **Chapt. 4**, pp. 60–104.

Rosa, R.M.S.S., R.C.S. Müller, C.N. Alves, J.E. de S. Sarkis, M.H., da S. Bentes, E. Brabo and E.S. de Oliveira, 2000. Determination of total mercury in workers' urine in gold shops of Itaituba, Pará State, Brazil. Sci. Total Environ. **261**, 169–176.

Santos, M.M., N. Vieira and A.M. Santos, 2000. Imposex in the dogwhelk *Nucella lapillus* (L.) along the Portuguese coast. Mar. Pollut. Bull. **40**, 643–646.

Saxén R., and E. Illus, 2000. Discharge of ^{137}Cs by Finnish rivers to the Baltic Sea in 1986–1996, in: The Radiological Exposure of the Population of the European Community to Radioactivity in the Baltic Sea. Marina-Balt Project, ed. S.P. Nielsen. Proceedings of a Seminar held at Hasseludden Conference Centre, Stockholm, 9–11 June 1998, European Commission, Directorate-General Environment, EUR 19200 EN (European Communities, 2000, Belgium), pp. 333–347.

Segerstråle, S.G., 1957. Baltic Sea. Mern. Geol. Soc. America **67**, 751–800.

Segerstråle, S.G., 1972. The distribution of some malacostraca crustaceans in the Baltic sea in relation to the temperature factor. Merentutkimuslait. Julk. **237**, 13–26.

Shen X.-M., J.F. Rosen, D. Guo and S.-M. Wu, 1996. Childhood lead poisoning in China. Sci. Total Environ. **181**, 101–109.

Shim, W.J., S.H. Kahng, S.H. Hong, N.S. Kim, S.K. Kim and J.H. Shim, 2000. Imposex in the rock shell, *Thais clavigera*, as evidence of organotin contamination in the marine environment of Korea. Mar. Environ. Res. **49**, 435–451.

Siegel, H., W. Matthäus, R. Bruhn, M. Gerth, G. Nausch, T. Neumann and C. Pohl, 1998. The exceptional Oder flood in summer 1997 – Distribution patterns of the Oder discharge in the Pomeranian Bight. Dtsch Hydrogr. Z. **50**, 1–23.

Sjöblom, V, and A. Voipio, 1981. International management and cooperation, in: The Baltic Sea, ed. A. Voipio (Elsevier Scientific Publishing Company, Amsterdam), pp. 383–390.

Skarphédinsdóttir, H., K. Ólafsdóttir, J. Svavarsson and T. Jóhannesson, 1996. Seasonal fluctuations of tributyltin (TBT) and dibutyltin (DBT) in the dogwhelk, *Nucella lapillus* (L.), and the blue mussel, *Mytilus edulis* L., in Icelandic waters. Mar. Pollut. Bull. **32**, 358–361.

Smith, J.N., K.M. Ellis, A. Aarkrog, H. Dahlgaard and E. Holm, 1994. Sediment mixing and burial of the ^{239,240}Pu pulse from the 1968 Thule, Greenland nuclear weapons accident. J. Environ. Radioactivity **25**, 135–159.

Smith, P.J., 1996. Selective decline in imposex levels in the dogwhelk *Lepsiella scobina* following a ban on the use of TBT antifoulants in New Zealand. Mar. Pollut. Bull. **32**, 362–365.

Smith, J.T., R.T. Clarke and R. Saxén, 2000. Time-dependent behaviour of radiocaesium: a new method to compare the mobility of weapons test and Chernobyl derived fallout. J. Environ. Radioactivity **49**, 65–83.

Snakin, V.V., 1997. Report on Lead Contamination of the Environment of the Russian Federation and its Effect on Human Health (REFIA, Moscow) (in Russian), 233 pp.

Snakin, V.V., and A.A. Prisyazhnaya, 2000. Lead contamination of the environment in Russia. Sci. Total Environ. **256**, 95–101.

Sokolov, A., and F. Wulff, 1999. SwingStations: a web-based client tool for the Baltic environmental database. Comput. Geosci. **25**, 863–871.

Strandenes, S.P., 2000. The second order effects on commercial shipping of restrictions on the use of TBT. Sci. Total Environ. **258**, 111–117.

Styro, D., Zh. Bumeliene, M. Lukinskiene and R. Morkuniene, 2001. ^{137}Cs and ^{90}Sr behavioural regularities in the southeastern part of the Baltic Sea. J. Environ. Radioact. **53**, 27–39.

Sunderland, E.M., and G.L. Chmura, 2000. An inventory of historical mercury emissions in Maritime Canada: implications for present and future contamination. Sci. Total Environ. **256**, 39–57.

Takizawa, Y., 1979. Epidemiology of mercury poisoning, in: The Biogeochemistry of Mercury in the Environment, ed. J.O. Nriagu (Elsevier/North Holland Biomedical Press, Amsterdam) pp. 325–365.

Tester, M., and D. Ellis, 1995. TBT controls and the recovery of whelks from imposex. Mar. Pollut. Bull. **30**, 90–91.

Tester, P.A., R.P. Stumpf, F.M. Vukovich, P.K. Fowler and J.T. Turner, 1991. An expatriate red tide bloom: transport, distribution, and persistence. Limnol. Oceanogr. **36**, 1053–1061.

Tomiyasu, T., A. Nagano, N. Yonehara, H. Sakamoto, Rifardi, K. Ōki and H. Akagi, 2000. Mercury contamination in the Yatsushiro Sea, south-western Japan: spatial variations of mercury in sediment. Sci. Total Environ. **257**, 121–132.

Trzosińska, A., 1992. Nutrients, in: An assessment of the effects of pollution in the Polish coastal area of the Baltic Sea, 1984–1989. Stud. Mater. Oceanol. **61**, Mar. Pollut. (2), 107–130.

Trzosińska, A., and E. Łysiak-Pastuszak, 1996. Ecological status of the recent Baltic Sea. Wiadomości Instytutu Meteorologii i Gospodarki Wodnej **XIX**, 27–62 (in Polish).

Tsubaki, T., and K. Irukayama, 1977. Minamata Disease (Elsevier Scientific Publishing Co., Amsterdam).

UNSCEAR, 1993. Sources and effects of ionizing radiation. Report to the General Assembly by the United Nations Scientific Committee on the Effects of Atomic Radiation (United Nations, New York).

US EPA (US Environmental Protection Agency), 1984. Health effects assessment of mercury (Environmental Criteria and Assessment Office, Cincinnati, Ohio).

Van Straaten, P., 2000a. Human exposure to mercury due to small scale gold mining in northern Tanzania. Sci. Total Environ. **259**, 45–53.

Van Straaten, P., 2000b. Mercury contamination associated with small-scale gold mining in Tanzania and Zimbabwe. Sci. Total Environ. **259**, 105–113.

Von Burg, R, and M.R. Greenwood, 1991. Mercury, in: Metals and Their Compounds in the Environment: Occurrence, Analysis and Biological Relevance, ed. E. Merian (VCH, Weinheim, New York, Basel, Cambridge) 1045–1088.

Wheatley, B., and M.A. Wheatley, 2000. Methylmercury and the health of indigenous peoples: a risk management challenge for physical and social sciences and for public health policy. Sci. Total Environ. **259**, 23–29.

WHO, 1989. Health Hazards from Radiocaesium Following the Chernobyl Nuclear Accident. Report on a WHO Working Group. J. Environ. Radioactivity 10, 257–259.

Williams, R.J.P., 1981. Physico-chemical aspects of inorganic element transfer through membranes. Philos. Trans. R. Soc. London **B 294**, 57–74.

Winterhalter, B., 1992. Late-Quaternary stratigraphy of Baltic Sea basins – A review. Bull. Geol. Surv. Finland **64**, 189–194.

Winterhalter, B., T. Floden, H. Ignatius, S. Axberg and L. Niemistö, 1981. Geology of the Baltic Sea, in: The Baltic Sea, ed. A. Voipio (Elsevier Scientific Publishing Company, Amsterdam) pp. 1–121.

Wulff, F., A. Stigebrandt and L. Rahm, 1990. Nutrient dynamics of the Baltic Sea. Ambio **29**, 126–133.

Xu, X., J. Gao, D.W. Dockery and Y. Chen, 1994. Air pollution and daily mortality in residential areas of Beijing, China. Arch. Environ. Health **49**, 216–212.

Xu, X, H. Ding and X Wang, 1995a. Acute effects of total suspended particles and sulfur dioxides on preterm delivery: a community-based cohort study. Arch. Environ. Health **50**, 407–415.

Xu, X, D.W. Deckery, D.C. Christian and B. Li, 1995b. Association of air pollution with hospital outpatient visits in Beijing. Arch. Environ. Health **50**, 214–220.

Yamagata, N., and I. Shigematsu, 1970. Cadmium pollution in perspective. Bull. Inst. Public Health, Tokyo **19**, 1–24.

Yamasoe, M.A., P. Artaxo, A.H. Miguel and A.G. Allen, 2000. Chemical composition of aerosol particles from direct emissions of vegetation fires in the Amazon Basin: water-soluble species and trace elements. Atmos. Environ. **34**, 1641–1653.

Zhang, J., L. Wang, X. Pan, Y. Zhou, D. Meng and L. Zhang, 1995. The effects of air pollution on pupil's pulmonary function in Beijing (in Chinese). J. Environ. Health **12**, 206–209.

Zhang, J., H. Song, S. Tong, L. Li, B. Liu and L. Wang, 2000. Ambient sulfate concentration and chronic disease mortality in Beijing. Sci. Total Environ. **262**, 63–71.

Zhao, Y., S. Marriott, J. Rogers and K. Iwugo, 1999. A preliminary study of heavy metal distribution on the floodplain of the River Severn, UK by a single flood event. Sci. Total Environ. **243/244**, 219–231.

Chapter 2

Air and Water as a Medium for Chemical Elements

A. AEROSOLS AND ATMOSPHERIC FALLOUT

(i) Introduction

Knowledge of trace and minor element concentrations in marine aerosols, dry and wet atmospheric fallout is an important basis for pollution control. This information permits an understanding of transport processes, and also makes it possible to estimate the capacity of the atmosphere to take up pollutants (Bogen, 1974). In many environmental studies mosses (Schaug et al., 1990; Grodzinska and Godzik, 1991; Carpi et al., 1994; Steinnes, 1995; Äyräs et al., 1997; Berg and Steinnes, 1997; Fernández et al., 2000) and lichens (Hawksworth, 1971; Ferry et al., 1973; Puckett and Finnegan, 1980; Garty, 1993; Nimis et al., 1993, 2000; Riget et al., 2000) have been used in biomonitoring of atmospheric pollution; lichens are considered to reflect 'air pollution', 'air quality' or 'air purity'. According to Nimis et al. (2000) these terms, however, are not synonyms; 'pollution' means concentration lying above thresholds established by low (i.e. 'pollution' must be determined instrumentally, and that organism can not be considered as 'cheap recording gauges'). Term "air purity/quality' is actually defined non-operationally by biologists as indicators of air purity/quality. In their opinion, biomonitoring techniques just 'assess the deviations from normal conditions of pollution-reactive organisms' (Nimis et al., 2000). The biomonitoring data depend on several factors other than anthropogenic and therefore it is difficult to distinguish the effect of pollution from those of climate, substrate ecology and other anthropogenic disturbances (Seaward, 1995, Nimis et al., 2000).

The atmospheric load of trace elements mostly constitutes a significant proportion of their total input to the marine ecosystems. The concentration of some metals in the atmosphere may be much higher than that in terrestrial debris and the marine aerosols suggesting their anthropogenic origin (Buat-Menard and Chesselet, 1979). Nriagu (1979, 1989) and Nriagu and Pacyna (1988) have reported a global assessment of natural and anthropogenic emissions of trace metals to the atmosphere. Concentration data are available for the chemical composition of marine aerosol and atmospheric dry and wet fallout from all over the world (Bogen, 1973, 1974; Peirson et al., 1974; Boutron and Lorius, 1975, 1979; Duce et al., 1975, 1976, 1980, 1983; Ayling and Bloom, 1976; Goodman et al., 1976; Hopke et al., 1976; Annegarn et al., 1978; Cawse, 1978; Kowalczyk et al., 1978; Lawson and Winchester, 1978; Sadasivan, 1978, 1980; Boutron, 1979a, 1979b, 1980; Cambray et al., 1979; Chester et al., 1979, 1984, 1991, 1999, 2000; Maenhaut et al., 1979, 1981a, 1981b, 1983; Adams et al., 1980; Tanaka et al., 1980; Heidam, 1981, 1984; Andreae, 1982; Tomza et al., 1982; Lindberg and Harris, 1983; Weisel et al., 1984; Arimoto et al., 1985a, 1985b; Ashawa et al., 1985; Blank et al., 1985; Essien et al., 1985; Khemani et al., 1985; Dulac et al., 1987, 1989; Bergametti et al., 1989; Maring and Duce, 1989; Ecker et al., 1990; Sahu, 1990; Cheng et al., 1991; Dick, 1991; Chester and Bradshow, 1991; Infante and Acosta, 1991; Remoudaki et al., 1991; Injuk and van Grieken, 1995; Jickells, 1995; Matschullat and Bozau, 1996; Reimann et al., 1997a, 1997b; Jambers et al., 1999, 2000; Kerminen et al., 2000; Matschullat et al., 2000; Piña et al., 2000). Extensive reviews on the atmospheric concentrations and deposition of micropollutants over the North Sea have been reported by Struyf and Grieken (1993) and Injuk and Van Grieken (1995).

Heavy metals such as Pb, Cu and Zn have been routinely measured in Arctic air, reflecting the greater influence of Eurasian emission of Pb compared to North American emissions (Macdonald et al., 2000). A comprehensive Eulerian modelling has been developed to study the geographical transport of atmospheric airborne Hg species (Petersen et al., 1998).

According to Macdonald et al. (2000), the principal inputs of artificially-derived radionuclides to the Canadian Arctic have been fallout from atmospheric nuclear weapons testing and from the nuclear power plant accident at Chernobyl in 1986. Uranium and plutonium isotopes in the atmosphere have been studied by several authors, e.g. Sakuragi et al. (1983).

(ii) Chemical Elements in Atmosphere

The total emission of selected elements from various anthropogenic sources in Europe has been presented by several authors (Pacyna, 1983, 1984; Pacyna and Tørseth, 1997; Pacyna et al. 1984; Petersen et al., 1995, 1998; Petersen, 1996, 1999). The current state and future direction of numerical models in simulating atmospheric long-range transport of trace elements over Europe have been reviewed by Petersen (1996). Airborne heavy metals over European countries including Baltic States (Denmark, Finland, Germany, Lithuania, Poland, Russia, Sweden) with emphasise on their emission, long-range transport and deposition fluxes to natural ecosystems have been also extensively presented (Pacyna, 1984; Pacyna et al., 1984; Petersen et al., 1995). The Baltic Sea is surrounded by most industrialised nations where various pollutants, including heavy metals, are emitted to various environmental compartments (Kubin and Lippo, 1996; Matschullat and Bozau, 1996; Matschullat et al., 2000; Polkowska et al., 2001). They can enter the Baltic Sea, e.g. via the atmosphere which is dominant trajectory for some trace and minor elements. Atmospheric particles can be transported over thousands of kilometers and airborne particulate trace elements originating from anthropogenic sources can be detected even in remote areas (Stahlschmidt et al., 1997). Because the input of metallic pollutants changes with time, several authors have undertaken extended studies to provide a baseline for future monitoring. Some information is available for distribution of heavy metals or macro-elements in atmospheric component over the Baltic Sea (Brzezińska and Garbalewski, 1980; Rodhe et al., 1980; Andreae and Froelich, 1984; Schneider, 1984, 1987, 1993; Bolałek, 1985; Szefer and Szefer, 1986; Boström et al., 1989; Petersen et al., 1989; Hasanen et al., 1990; Suszkin, 1990; Szefer, 1990; Kersten et al., 1991a; Pacyna, 1992, 1993; Wrembel, 1993, 1994; Matschullat, 1997; Brügmann and Hennings, 2000; Gustafsson and Franzén, 2000; Jalkanen et al., 2000; Schneider et al., 2000; Sofiev et al., 2000; Urba et al., 2000). Modelling the atmospheric transport of trace metals from Europe to the Baltic Sea has been developed by Petersen et al. (1989). Several authors (Leppäranta et al., 1998; Granskog, 1999) have investigated chemical composition of particulate matter entrained into the ice cover of the Baltic Sea.

The concentration data of chemical elements in atmospheric precipitation near the Baltic Sea are presented in Table 2.1.

Based on climatic models a heterogeneous distribution of precipitation over the Baltic Sea area within an annual cycle is postulated. The atmospheric input consists of both components, i.e. wet and dry depositions (aerosol, dust). According to Brügmann and Matschullat (1997) the trace elements emitted from anthropogenic sources are mainly bound to small particles; their prime deposition route is wet fallout amounting to ca. 90% of the total deposition. The atmospheric deposition of trace metals in the Baltic area significantly changes from one to two orders of magnitude, depending mainly on meteorological conditions (Matschullat, 1997). These differences may also occur between different localities (e.g., monitoring stations).

In 1977, the atmospheric element input on the Baltic Sea was estimated as negligible. Hallberg (1991) calculated the atmospheric input of elements based on concentration data corresponding to 575 sediment samples from the Baltic Sea. According to Matschullat (1997) an approach of direct recalculation of emissions to input is not satisfactory, even if trajectories and distances are considered. It is recommended to use both measured and extrapolated data. Data from different groups and relating to at least full annual cycle of measurements have been used to calculate the probable input (Matschullat, 1997).

TABLE 2.1.

Concentrations of traace elements ($\mu g\ dm^{-3}$) and K, Na, Ca and Mg ($mg\ g^{-1}$) in atmospheric precipitation over the Baltic Sea and other northern areas

Region	Sampling date	Precipitation	Filter (μm)	N	Ag	Al	As	Ba	B	Be	Br	Ca	Cd	Co	References
Gdynia	1976	Rain (bulk precipitation)	0.45	12								4.75[b] 1.8–7.7	4.1 0.9–11.1	5.8 2.5–10.4	Szefer and Szefer, 1986
Swinoujscie	1976	Rain (bulk precipitation)	0.45	8								5.1[b] 0.9–9.9	4.1 0.5–12.0	5.1 0.8–14.1	
Hel Peninsula	1977–80	Rain (bulk precipitation)	0.45	23								2.72[b] 0.98–8.4	0.63 0.10–1.48	3.4 0.68–12.5	
Gdynia	1976	Rain (filtered)	0.45	12								4.29[b] 1.5–7.3	1.59 0.7–3.0	4.73 1.9–7.4	Szefer and Szefer, 1986
Swinoujscie	1976	Rain (filtered)	0.45	8								4.64[b] 0.7–8.9	3.04 0.5–11.5	3.73 0.8–8.8	
Hel Peninsula	1977–80	Rain (filtered)	0.45	23								2.18[b] 0.71–6.06	0.18 0.02–0.50	2.83 0.30–12.3	
	1997–98	Bulk deposition											16.2[a]		Schneider et al., 2000
Schleswig-Holstein	1976	Rain water		8	0.03–0.67		1.0–6.4	8.1–1300			10.0–76		1.1–32.0	0.08–0.72	Suszkin, 1990
	1987	Melted snow		1	0.32	2000	9	10	8.4				1.8	0.5	
Kiel, Western Baltic	Pre–1984	Aerosol					3.0*								Andreae and Froelich, 1984
Southern Baltic		Aerosol											0.5–6.5		Bolalck, 1985
Archipelago of Stockholm	1984–87	Rain (bulk precipitation)				16.1–66.1		1.31–3.84							Boström et al., 1989
Gulf of Finland															
Ähtäri, Utö, Virolahti	1991	Aerosol	2.5	21		71* 38–124	0.57* 0.44–0.77				2.71* 1.59–3.85				Jalkanen et al., 2000
Virolahti	1991	Aerosol	2.5	8		158* 10–461	1.43* 0.54–2.76				4.9* 0.83–11.9				Jalkanen et al., 2000
Helgoland, German Bight	1986	Aerosol				205*	2.1*	3.1*	1.2*	0.015*			1.9*	0.25*	Kersten et al., 1991
Kiel Bight	1981–83	Bulk deposition				394* 68–720	2.8* 0.2–5.4	7.9* 2.1–14		300* 124–476		435* 83–787			Schneider, 1987
Arkona, Island of Ruegen	1997–98	Bulk deposition											0.69[a]		Schneider et al., 2000
Gotland-Hoburg	1997–98	Bulk deposition											0.89[a]		Schneider et al., 2000
Southern Gotland-Preila	1997–98	Bulk deposition											0.72[a]		Schneider et al., 2000
North Sea	1972–73	Rain				2400	45				2300			3.2	Peirson et al., 1974
	1988–89	Aerosol				294* 21–887								0.25* 0.01–0.79	Chester and Bradshaw, 1991

* – ng m^{-3}
[a] – nmol dm^{-3}
[b] – $\mu g\ dm^{-3}$

TABLE 2.1. – continued

Region	Sampling date	Precipitation	Filter (μm)	N	Cr	Cs	Cu	Fe	Ge	Hg	K	Mg	Mn	Mo	Na	Ni	References
Gdynia	1976	Rain (bulk precipitation)	0.45	12			11.3 2.3–30.8					2.1^b 0.7–3.2	76 11.0–150			10.8 1.8–23.6	Szefer and Szefer, 1986
Swinoujscie	1976	Rain (bulk precipitation)	0.45	8			4.6 1.6–7.4					1.3^b 0.4–2.1	33 1.0–29.0			6.1 2.9–19.7	
Hel Peninsula	1977–80	Rain (bulk precipitation)	0.45	23			8.13 1.2–23.9	1650 200–2470			0.51^b 0.15–1.05	0.39^b 0.11–1.1	40.1 4.3–95.2		1.27^b 0.24–4.5	3.03 0.90–8.14	
Gdynia	1976	Rain (filtered)	0.45	12			2.53 0.1–7.2					1.57^b 0.6–2.6	45.1 5.3–77			4.97 0.7–9.6	Szefer and Szefer, 1986
Swinoujscie	1976	Rain (filtered)	0.45	8			0.7 0.03–3.7					1.0^b 0.1–2.0	16.6 1.0–35			2.57 0.4–4.7	
Hel Peninsula	1977–80	Rain (filtered)	0.45	23			0.91 0.01–3.6	0.03 0.01–0.08			0.30^b 0.13–0.84	0.36^b 0.14–0.96	19.3 2.5–64.4		1.69^b 0.28–11.2	1.84 0.60–5.71	
Schleswig-Holstein	1976	Rain water		8	1.6–13.0	0.06–0.53		110–7400		0.001–0.12						1.6–9.5	Suszkin, 1990
	1987	Melted snow		1	10		16	170		< 0.5			20	0.27		17	
Kiel, Western Baltic	Pre-1984	Aerosol							0.6*								Andreae and Froelich, 1984
Southern Baltic		Aerosol					4.5–25.2	27–860					1.1–26.1				Bolalek, 1985
Archipelago of Stockholm	1984–87	Rain (bulk precipitation)						20.1–59.4					7.21–15.6				Boström et al., 1989
Gulf of Finland																	
Ähtäri, Utö, Virolahti	1991	Aerosol	2.5	21			0.79* 0.66–0.95							0.11* 0.06–0.15		0.94* 0.68–1.30	Jalkanen et al., 2000
Virolahti	1991	Aerosol	2.5	8			1.41* 0.51–2.70							0.19* 0.08–0.38		1.89* 0.78–3.16	Jalkanen et al., 2000
Helgoland, German Bight	1986	Aerosol			1.9*		4.7*	249*					9.6*			2.9*	Kersten et al., 1991
Kiel Bight	1981–83	Bulk deposition			2.9* 0.2–5.6		7.7* 1.3–14	369* 90–648			300* 124–476		15* 4.0–26		1404* 440–2370	4.0* 1.7–6.3	Schneider, 1987
North Sea	1972–73	Rain			20	1.7	82	3.1*								23	Peirson et al., 1974
	1988–89	Aerosol			4.7* 0.1–25.0		6.3* 0.4–37.5	353* 2.0–1565					14.5* 0.20–84.5			3.8* 0.04–13.0	Chester and Bradshaw, 1991

* – ng m^{-3}
^{b} – μg dm^{-3}

TABLE 2.1. – continued

Region	Sampling date	Precipitation	Filter (μm)	N	P	Pb	Rb	S	Sb	Sc	Se	Si	References
Gdynia	1976	Rain (bulk precipitation)	0.45	12		64.6 22.2–141							Szefer and Szefer, 1986
Swinoujscie	1976	Rain (bulk precipitation)	0.45	8		52.5 17.1–99.5							
Hel Peninsula	1977–80	Rain (bulk precipitation)	0.45	23		47.7 10.7–119							
Gdynia	1976	Rain (filtered)	0.45	12		5.38 0.2–33.6							Szefer and Szefer, 1986
Swinoujscie	1976	Rain (filtered)	0.45	8		2.45 0.3–8.6							
Hel Peninsula	1977–80	Rain (filtered)	0.45	23		9.16 2.0–38.2							
	1997–98	Bulk deposition				17[a]							Schneider et al., 2000
Schleswig-Holstein	1976	Rain water		8			0.69–6.7		0.14–52.0	0.004–0.28	0.027–0.85		Suszkin, 1990
	1987	Melted snow		1	180	11		2200	1.7		< 1.0	190	
Kiel, Western Baltic	Pre-1984	Aerosol							1.4*				Andreae and Froelich, 1984
Southern Baltic		Aerosol				1.5–20.9							Bolalek, 1985
Archipelago of Stockholm	1984–87	Rain (bulk precipitation)			5.01–157			923–2400				27.8–110	Boström et al., 1989
Northern Sweden Kalix River catchment	1990–92	Rain		14				16.5[b] 9.4–31.5					Ingri et al., 1997
	1991	Snow		9				7.24 5.6–9.4					
Gulf of Finland													
Ähtäri, Utö, Virolahti	1991	Aerosol	2.5	21		6.53* 5.28–7.60							Jalkanen et al., 2000
Virolahti	1991	Aerosol	2.5	8		14.4* 4.52–24.6							Jalkanen et al., 2000
Helgoland, German Bight	1986	Aerosol				52.6*			1.9*	1.5*			Kersten et al., 1991
Kiel Bight	1981–83	Bulk deposition				53* 14–92	1.4* 0.3–2.5				1.6* 0.6–2.6		Schneider, 1987
Arkona, Island of Ruegen	1997–98	Bulk deposition				25[a]							Schneider et al., 2000
Gotland-Hoburg	1997–98	Bulk deposition				11[a]							Schneider et al., 2000
Southern Gotland-Preila	1997–98	Bulk deposition				14[a]							Schneider et al., 2000
North Sea	1972–73	Rain				190	< 60		5.9	0.49	1.75		Peirson et al., 1974
	1988–89	Aerosol				34.5* 0.6–190							Chester and Bradshaw, 1991

* – ng m^{-3}
[a] – nmol dm^{-3}
[b] – μmol dm^{-3}

TABLE 2.1. – continued

Region	Sampling date	Precipitation	Filter (μm)	N	Sn	Sr	Ti	V	Zn	Zr	References
Schleswig-Holstein	1976	Rain water		8	0.55–1.70	6.4–91.0			43–290	1.2–16.0	Suszkin, 1990
	1987	Melted snow		1	< 0.5	9.4	3.3	3.3	61	0.1	
Archipelago of Stockholm	1984–87	Rain (bulk precipitation)				1.94–5.28	0.25–1.16		14.3–32.9		Boström et al., 1989
Gulf of Finland											
Ähtäri, Utö, Virolahti	1991	Aerosol	2.5	21				2.11* 1.25–3.06			Jalkanen et al., 2000
Virolahti	1991	Aerosol	2.5	8				5.38* 2.46–8.18			Jalkanen et al., 2000
Helgoland, German Bight	1986	Aerosol				6.5*	13.7*	8.1*	46.1*		Kersten et al., 1991
Kiel Bight	1981–83	Bulk deposition				4.7* 2.0–7.4	27* 5.0–49	9.7* 4.9–15	57* 8–106		Schneider, 1987
North Sea	1972–73	Rain						< 40	2000		Peirson et al., 1974
	1988–89	Aerosol							41.0* 0.7–250		Chester and Bradshaw, 1991

* – ng m^{-3}

Spatial trends

Under an ECE-monitoring program, terrestrial mosses were used as biomonitors, to quantify the atmospheric inputs around the Baltic Sea. The data obtained indicated that the highest atmospheric loads of metallic pollutants comprise mainly the southern part of the Baltic catchment (the Black Triangle), e.g. Poland, Czech, Slovak Republic and Germany (Markert et al., 1996). The distribution of heavy metals in a transect of three countries, i.e. the Netherlands, Germany and Poland has been presented by Herpin et al. (1996). It is estimated, that in 1995 total depositions of Pb to the southern Baltic Proper from Germany and Poland were 73.2 and 34.1 tons yr^{-1}, respectively. For Cd the annual loads amounted to 3.42 and 1.57 tons (HELCOM, 1997). Table 2.2 presents the total annual deposition of the metals to the south Baltic Proper in respect to their total deposition in the Baltic Sea in 1991–1995 (HELCOM, 1997). Moderate loads have been detected in the near vicinity of the Baltic Sea, except some anomalies observed, e.g. along the eastern Swedish coast and Gulf of Bothnia, which decrease even further towards the Polar Circle. Hasanen et al. (1990) reported concentration data expressed as geometric means (ng m^{-3}) for sulphate S (2110), ammonium N (1950), nitrate N (750), Na (267), Al (218), Fe (215), Zn (22.0), Pb (11.7), Mn (6.7), V (5.9), As (0.82), Sb (0.28), La (0.19), Cd (0.17) and Co (0.11) in aerosol samples collected over the Baltic Sea. The levels of all the elements studied varied widely, by a factor of 5 to 10 depending on whether the air masses were coming from the heavily industrialized regions of Central Europe or from less polluted areas of Northern Europe. A strong positive correlation found between some elements suggested a common their provenience, e.g. from industry, energy production, traffic or natural environment sources (Hasanen et al., 1990).

TABLE 2.2.

Total annual deposition of metals to the south Baltic Proper in total deposition in respect to their the Baltic Sea in 1991–1995 (HELCOM, 1997)

	Pb (100 kg yr^{-1})				Cd (10 kg yr^{-1})			
	South Baltic Proper	Baltic Sea Deposition			South Baltic Proper	Baltic Sea Deposition		
		Wet	Dry	Total		Wet	Dry	Total
1991	1626			6280	651			2605
1992	1692			6024	698			2572
1993	1907			7026	749			2880
1994	2366			6774	968			2875
1995	1888			5769	779			2516
Average	1896	5858	517	6375	769	2354	336	2690
CV (%)	14				14			

The results from moss monitoring survey and other data were used to recalculate of input for the Baltic Sea surface area. Matschullat (1997) owing to use of the calibration procedure obtained the very good agreement between the sediment-originated data reported by Hallberg (1991) and those taken from the measurements and transport models of Rodhe et al. (1980). The estimation of atmospheric Pb inputs on the Baltic Sea was surprisingly close to the Pb input, calculated from total deposition data for open continental area in the eastern Erzgebirge (Matschullat and Bozau, 1996). In the formerly highly polluted area, fallout can now be comparable to rural areas without any distinct spatial trends; however individual events may be resulting in elevated concentrations and loads. An example of such situation is remarkable shift from S dominated system to N dominated system (Matschullat et al., 2000). According to Schneider et al. (2000) there is a weak gradient with decreasing deposition rates of Pb from the southwest towards the east and north. The spatial distribution pattern of Cd was characterised by an extreme deposition maximum at the Polish zone on the Hel Peninsula. Szefer and Szefer (1986) assayed the distribution of selected elements in rain water at three sampling sites of the southern Baltic coast. The mean concentrations of Cu, Mn and Ni were ca. twice as high in the samples from Gdynia than in those from Świnoujście. Atmospheric dispersion and physicochemical behaviour of Cd and Pb in rainwater after emission from lead works in Germany have been reported (Struck et al., 1996).

Temporal trends

Temporal trends in some metal concentrations in rain fallout at the Hel Peninsula were observed. Samples collected in May 1979 and at the beginning of the winter contained the greatest concentrations of Cd, Co, Cu, Mn, Ni and Pb (Szefer and Szefer, 1986). Scheider et al. (2000) documented a decreasing trend of the atmospheric deposition of Pb and Cd into the Baltic Sea during the past 10–15 years. This depletion is consistent with the decreasing trends for these two metals in Baltic surficial waters (HELCOM, 1996; Kremling and Streu, 2000).

Sources of chemical elements over Baltic coast

According to Schneider (1987) atmospheric sea salt contributed significantly to Na and Sr. Elements such as Al, Ba, Ca, Cr, Fe, K, Mn, Sr, Rb and Ti were mainly derived from mineral dust while As, Cu, Ni, Pb, V and Zn were predominantly anthropogenic in origin and for most of other elements was identified an anthropogenic component with nearly their constant composition. It suggests that trace elements over the Kiel Bight were mostly derived from the same source. It should be emphasised that a 40° wind sector directed to the south of the Bight was found as the major pathway for the trajectory of anthropogenic trace elements to the Kiel Bight, Western Baltic (Schneider, 1987).

(iii) Radionuclides in Atmosphere

Radiological monitoring of coastal Polish area of the Baltic Sea was performed by Isajenko et al. (2000). The concentrations (annual averages) of ^{7}Be, ^{137}Cs, ^{131}I, ^{40}K and ^{210}Pb in 52 aerosol samples were 3110±200, 0.9±0.1, 0.8±0.2,

18.5±1.6 and 407±36 μBq m^{-3}, respectively. These values are comparable to those reported for Finnish coastal area (Helsinki and Kotka) (Isajenko et al., 2000).

According to Macdonald et al. (2000) in addition to the direct atmospheric inputs of Chernobyl-derived radionuclides to the Arctic Sea, there would be contributions due to initial deposition in the North Sea and the Baltic Sea and subsequent advective transport into the Arctic Ocean with Atlantic water masses. Several authors (Salo et al., 1984; Pöllänen and Toivonen, 1994; Pöllänen et al., 1997) characterised transport of radioactive particles before and after the Chernobyl accident. The activity concentrations of radionuclides in ground-level air at Lovissa were reported by Ilus et al. (1992). In 1989–90 the major artificial radionuclide in surface air was detected ^{137}Cs originating from the Chernobyl fallout; its concentrations ranged from 2.4 to 28 μBq m^{-3} at Loviisa, the Gulf of Finland, and from 2.6 to 37 μBq m^{-3} at Olkiluoto, Bothnian Sea. ^{134}Cs was also frequently registered in 1989; however in 1990 its concentration was generally very low, i.e. below the detection limit. Besides radiocaesium other artificial radionuclides were detected only once at Loviisa. For instance, in September a low levels of ^{110m}Ag occurred at Loviisa; no gamma-emitting radionuclides, except of radiocaesium, were determined at Olkiluoto (Ilus et al., 1992). The concentrations of radionuclides in atmospheric precipitation originated from the Chernobyl fallout, i.e. ^{137}Cs and ^{134}Cs, have continued to decrease since 1988. Short-lived Chernobyl nuclides were no longer existed in precipitation samples at the end of 1990. The last records for ^{106}Ru and ^{125}Sb were made in the early 1989s and 1990s, respectively (Ilus et al., 1992). Long-term variation (1986–1998) of post-Chernobyl ^{90}Sr, ^{137}Cs, ^{238}Pu and $^{239,240}Pu$ concentrations in air in South Germany has been reported by Rosner and Winkler (2001).

The concentrations of U and Th in rain water at coastal regions of the southern Baltic during 1976–1980 ranged from 0.10 to 0.9 μg dm^{-3} and from 0.02 to 1.1 μg dm^{-3} (Szefer and Szefer, 1986).

B. TRIBUTARIES IN THE BALTIC CATCHMENT

(i) Introduction

General Characteristics of the Baltic Catchment Area

Among 51 large rivers, the hydrologically most important in the Baltic region are the following rivers (discharge in km^3 yr^{-1}): the Vistula (33.6) and Odra (Oder) (18.1) in Poland; the Nemunas (19.9) in Lithuania; the Daugava (20.8) in Latvia, the Luga and Neva (77.6) in Russia; the Kemijoki (17.7) in Finland; the Ångermanälven (15.4), Lule älv (15.3), Ume älv (14.2), Indalsälven (14), Dalälven (11.4) and Torne älv (11.3) in Sweden. It constitutes together 60.4% of the total average discharge of all tributaries of 446 km^3 yr^{-1} (HELCOM, 1993). The contribution of German discharge is smaller than 0.5%. The main rivers in the Bothnian Bay catchment area are Finnish Kemijoki, being the seventh largest river in the Baltic Sea area, and the Swedish river Lule älv (HELCOM, 1998a). The main rivers in the Bothnian Sea catchment area are the Ångermanälven and Indalsälven in Sweden and the Oulujoki in Finland (HELCOM, 1998a). The largest river flowing into the Baltic Sea is the Neva which drains from the Russian territory directly to the Gulf of Finland. Dominant percentage of the pollution load in this subregion is transported to the Baltic Sea via two large rivers, i.e. the Neva and the Narva (HELCOM, 1998a). The main river flowing into the Gulf of Riga is the Daugava, the fourth largest river in the Baltic catchment, which drains from the Latvian territory (HELCOM, 1998a). Three of the seven largest rivers flow in the Baltic Proper catchment, i.e. two of them named the Vistula (Wisła) and the Oder (Odra) drain from the Polish territory directly to the Baltic Sea. The Nemunas being the third largest river flows from the Lithuanian territory through the Curonian Lagoon into the Baltic Sea. The main river flowing into the Gulf of Riga is the Daugava, the fourth largest river in the Baltic catchment, which drains from the Latvian territory (HELCOM, 1998a). The total catchment of the Vistula comprises 194,420 km^2, of which 87% belongs to Poland; the remaining area belongs to Belarus, Ukraine and Slovakia. The total catchment area of the Oder comprises 118,840 km^2, of which 89% belongs to Poland; this area also includes the Czech Republic and Germany. The Neman (Nemunas) River belonging to the Lithuanian territory (46,700 km^2) drains areas of Belarus, Poland, Russia and Latvia. The 7,459 km^2 area of the Lithuanian territory is part of the catchment area of the river Venta and the River Bartuva, flowing to the Baltic Sea subregion from the Latvian territory. It is important to note that the majority of pollutants enter the Baltic Sea as dissolved or suspended forms in river waters. The Baltic Proper receives 21% of the total run-off into the Baltic Sea. Three main rivers: the Vistula, the Neman and the Oder introduce 72% of this riverine outflow. Their mean long-term flow rates amount to 1081, 664 and 574 m^3 s^{-1}, respectively. The fresh water from these big rivers and from several dozen of small ones enters the sea along the southern coast and is incorporated into the surface layer characterised by a prevailing counter-clockwise current system. Some rivers such as the Neman and Oder (and a considerable part of waste waters) enter the sea through lagoons and coastal lakes which have retention times of several weeks to several months. These reservoirs serve as natural purification basins and are, as a consequence, seriously degraded. However, a substantial reduction of the pollution load to the Baltic takes place there, although this is usually disregarded in input compilations.

Overview of Worldwide Literature

Continental material is transported from land to marine ecosystems via river system. According to Martin and Meybeck (1979) the total flux of dissolved and suspended matter transported by rivers is estimated to be 20×10^{15} g yr^{-1}, i.e.

15.5×10^{15} and 4.5×10^{15} g yr^{-1} for the solid and dissolved loads, respectively. The chemical composition of worldwide rivers is less recognised as compared to that of seawater. Several authors documented distribution of trace, minor- and major elements in world major rivers, e.g. Amazon, Orinoco, Parana, Mississippi, Colorado, Mackenzie, Saint Lawrence, Magdalena, Congo, Nile, Ganges, Mekong, Yellow River, Yangtze River, Yenissei, Amour, Ob, Elbe and UK Rivers (Spalding and Sackett, 1972; Spalding and Exner, 1976; Trefry and Presley, 1976; Eisma et al., 1978; Martin and Meybeck, 1978, 1979; Salomons and Eysink, 1979; Whitfield and Turner, 1979; Sholkovitz and Price, 1980; Li, 1981; Li et al., 1984a; Martin and Whitfield, 1983; Tanizaki and Nagatsuka, 1983; Tanizaki et al., 1984, 1985; Abaychi and DouAbal, 1985; Mart et al., 1985; Brügmann, 1990; Windom, 1990; Sholkovitz, 1993, 1995; Brügmann, 1995; Neal et al., 2000; Pettersson and Ingri, 2001). Water of some salt rivers on the northern Andes of Antofagasta (Chile) was also analysed for concentrations of selected heavy metals and metalloids (Queirolo et al., 2000). Baseline studies of the Slave River, NWT, in respect to its chemical quality have been performed by McCarthy et al. (1997); Pb concentration in water frequently exceeded national water quality guidelines in spite of unknown local anthropogenic sources of this metal in the immediate area of Forth Smith (Macdonald et al., 2000). Riverine sediments have been also analysed for concentrations of chemical elements (Winkels et al., 1998; Douglas and Adeney, 2000; Vital and Stattegger, 2000).

Some rivers have been assayed in respect to their abundance in some artificially-derived radionuclides; for instance Johnson-Pyrtle et al. (2000) have reported data of ^{137}Cs concentration in the Lena River, the second largest Siberian river discharging into the Arctic Ocean. This study suggested that the Lena River drainage basin is an important source of ^{137}Cs in the adjacent Arctic Ocean. Earlier investigations of ^{137}Cs in this Polar region were focused on assessing its influx to the Ob River-Yenisey River systems (Sayles et al., 1997; Stepanets et al., 1999). Concentrations of naturally-derived radionuclides such as ^{234}U, ^{238}U, ^{230}Th and ^{232}Th, ^{228}Th and ^{226}Ra were measured in water and sediments of the Amazon and Mississippi Rivers (Moore, 1967). Borole et al. (1982) reported the U isotope concentrations in rivers and estuaries of western India. Data for U concentrations in 29 rivers and eight estuaries have been presented by Windom et al. (2000). Carpenter et al. (1984) have reported activity profiles of excess ^{234}Th and ^{210}Pb, ^{232}Th, ^{230}Th, ^{234}U, ^{238}U and $^{228/232}Th$ activity ratios in the sedimentary column from central Puget Sound, USA. The literature data pertained to the distribution and behaviour of U in inland and estuarine waters have been reviewed by Szefer (1987).

(ii) Trace Elements in Riverine and Estuarine Systems

One of the important sources of trace elements in the Baltic Sea are the rivers among which the most significant contribution in pollution load entering have the Oder and Vistula. The total pollution load of trace elements in the Baltic ecosystem varies among the several subregions and depends on the population density, location of industry centres and the abundance as well as intensity of the exploitation of natural resources. The anthropogenic sources of riverine pollutants in the Baltic Sea are mainly industrial wastewater, leakage from products in use and those removed from service, "natural" degradation of pro-products as well as pollution from different types of land-use, e.g. fertilising, and mining (HELCOM, 1998a).

Several authors have reported the distribution and concentrations of chemical elements in rivers of the Baltic catchment basin (Ostrowski, 1963; Bojanowski and Koszałka, 1975; Ahl, 1977; Burman, 1983; Szefer, 1989, 1990; Löfvendahl, 1990; Löfvendahl et al., 1990; Pontér et al., 1990, 1992; Andersson et al., 1992, 1994, 1998a, 1998b; Anbar et al., 1996; Heybowicz and Borkowski, 1997; Matschullat, 1997; Mierzwiński and Niemirycz, 1997a, 1997b; Pettersson et al., 1997; Porcelli et al., 1997; Ingri et al., 1998; Land et al., 1999a; Müller and Heininger, 1999; Müller and Wessels, 1999; Niemirycz, 1999; Niemirycz and Bogacka, 1997; Gustafsson et al., 2000; Ingri et al., 2000). Ground or soil waters of the river watershed in Sweden, have been also analysed for concentration of major and trace elements (Land and Öhlander, 1997; Land et al., 1999b; Laaksoharju et al., 1999). A numerous Finnish stream water, ground water or lake water samples have been collected in the Baltic catchment and also analysed hydrochemically (Lahermo et al., 1995; Åström and Åström, 1997; Åström, 2001). There are numerously reported data on chemical elements in riverine sediments and soils of the Baltic Sea drainage basin (Helios Rybicka, 1991, 1992, 1993, 1996a, 1996b, 1996c; Helios Rybicka et al., 1994; Gustaffson and Jacks, 1995; Widerlund and Ingri, 1995, 1996; Edén and Björklund, 1996; Öhlander et al., 1996, 2000; Ingri et al., 1997, 2000; Aström, 1998; Land et al., 1999a, 1999b; Åström and Nylund, 2000; Kļaviņš et al., 2000). Geochemistry of till weathering in the Kalix River watershed has been characterised by Õhlander et al. (1991, 1996).

Due to incomplete river-derived matrix data, an overall pattern of the trace metals' load entering the Baltic Sea could not be given. However, the results reported (Bojanowski and Koszałka, 1975; Pontér et al., 1990; Szefer, 1990; Helios Rybicka, 1993, 1996a, 1996b, 1996c; Helios Rybicka et al., 1994; Pettersson et al., 1997; Müller and Heininger, 1999; Müller and Wessels, 1999) suggest that riverine heavy metals load has significant contribution to the total pollution flux, especially in estuarine areas. Besides particulate matter also alluvial sediments in the Baltic catchment and adjacent areas have been analysed for concentrations of some elements taking into account their early diagenesis and redox cycling (Müller and Förstner, 1975; Widerlund and Ingri, 1995, 1996; Nagaitsev, 1996). The municipal and industrial wastewater discharges including diffuse discharges within the river catchment zone are supposedly the most important compartment of the riverine load (HELCOM, 1998a). Reimann et al. (2000) have performed extensive Baltic soil survey resulting in recognising of the distribution of 41 major and trace elements (Al, As, Ba, Bi, Ca, Ce, Cl, Co, Cr, Cs, Cu, F, Fe, Ga, Hf, K, La, Mg, Mn, Mo, Na, Nb, Ni, P, Pb, Rb, S, Sb, Sc, Si, Sn, Sr, Ta, Th, Ti, U, V, W, Y,

Zn and Zr) in arable soils from 10 countries (Belarus, Estonia, Finland, Germany, Latvia, Lithuania, Norway, Poland, Russia, Sweden) around the Baltic Sea. The Nordic countries show considerably higher levels and variations for a number of elements, i.e. Al, Fe, (Mg, P), Ti, Ba, Sc, Sr and V in their agricultural soils. This is most probably a result of the relatively young age of the soils and of the climatic conditions, i.e. reduced weathering rates. There is remarkable relative enrichment of the topsoils from Germany, followed by Poland, in Pb while such enrichment is practically insignificant in Sweden. It is demonstrated that geology overwhelmingly dominates the chemical composition of the agricultural soils. High levels of some elements, e.g. K, Pb, Rb and Th reflect the granitic intrusions in the explored area. Industrial sources in Poland are mirrored only over very short distances while the high traffic density in Germany is responsible for elevated levels of Pb in the surficial agricultural soils throughout the country. Polluted soils are exposed to atmospheric and water erosion in the Baltic drainage basement (Reimann et al., 2000).

According to Matschullat (1997) of immediate importance for a calculation of the input are areas close to the river mouths. The river load over its course can only be estimated if the stream fluxes are not changed by locks, weirs or reservoirs. Further problems arise when coastal lakes or estuaries have to be taken into account because alteration of the residence times may lead to a considerable loss of the river loads. As it has been reported for several large catchments of Scandinavia (Löfvendahl, 1990), to recent man-made changes of natural element flows belongs the hydrochemical alteration caused by acid deposition and by soil and water acidification. The tributaries show very different transport capacities and discharge variations in subsequent years (Matschullat, 1997). It is assumed that tributaries spread from the southern to the eastern parts of the catchment will still prevail over the current trace-element input to the Baltic Sea because of anthropogenically released elements in this region. Other important point sources from single industries along the Finnish and Swedish coasts should not be neglected which affect, e.g. the Bothnian Bay with metal discharges (Bruneau, 1980). The part of catchment area in the south and southeast is more densely populated with higher traffic density. Their hinterland hosts mining operations, heavy industries and chemical plants which are still largely equipped with rather outdated technology (Matschullat, 1997).

The concentration data on chemical elements including REE and radionuclides in river water and riverine suspended matter of the Baltic drainage basin are given in Tables 2.3–2.5. Table 2.6 lists concentration data for riverine-estuarine sediments catchment.

Geographical distribution of elements

Southern catchment

Hydrologically the rivers Neva, Vistula and Odra (Oder) are the most important tributaries to the Baltic Sea. While available data are scarce for the Neva;

TABLE 2.3.

Concentrations of chemical elements ($\mu g\ dm^{-3}$) in river water of the Baltic drainage basin and other northern areas

Region	Sampling date	Fraction (μm)	N	Al	Ca	Cd	Co	Cu	References
Swedish rivers									
Dalälven	1991	< 0.45	1	67	5300				Andersson et al., 1994
Indalsälven		< 0.45	1	60	8100				
Kalixälven		< 0.45	1	42	2600				
Kalix-Kamlunge	1995	< 0.45	1		0.018*				
	1995	< 0.45	1		2100				Porcelli et al., 1997
		< 10 kD ultrafiltrate[a]	1		1400				
		> 10 kD colloid conc.[b]	1		200				
		10 kD filter rinse[c]	1		500				
		< 3 kD ultrafiltrate[a]	1		1600				
		> 3 kD colloid conc.[b]	1		100				
		3 kD filter rinse[c]	1		200				
Rautas	1995	< 0.45	1		0.015*				Andersson et al., 1998a
Kemijoki	1991	< 0.45	1	99	2900				Andersson et al., 1994
Kokenmäenijoki	1991	< 0.45	1	110	7000				
Narkån	1995	< 0.45	1		1480				Porcelli et al., 1997
Polish rivers									
Vistula									
Swibno	1973–74		1		59.7**	0.65	0.1	3.8	Szefer, 1989
	1975–77								
	1987	< 0.45	1		46.6**	1.8		7.4	
Malbork	1987	< 0.45	1		52.8**	2.3		5.8	
Vistula	1993	< 0.45	1	< 60	89.4**				Andersson et al., 1994
					97.0**				Löfvendahl et al., 1990

Region	Sampling date	Fraction (μm)	N	Al	Ca	Cd	Co	Cu	References
Oder			1			0.13	0.5	3	Pohl et al., 1998
Latvian rivers									
Daugava	1993–97		9		43.3**	0.06		1.2	Kļaviņš et al., 2000
Lielupe			4		96.2**	0.09		2.5	
Venta			7		52.5**	0.04		0.9	
Gauja			5		52.6**	0.03		1.9	
Salaca			3		54.5**	0.02		2.3	
Ciecere			1		40.8**	0.05		1	
Abava			2		83.8**	0.05		1.1	
36 rivers of the Baltic basin		.45	36		9.90** 0.04–90				Löfvendahl et al., 1990
11 Rivers of N. Sweden	1984–85		1909–1925		4.6±1.8**				Löfvendahl, 1990
11 Rivers of S. Sweden					16.3±17.2**				

[a] – Measured concentrations are ±5%, except for K, which are ±10%, and where noted otherwise.
[b] – The measured concentrations in the colloid concentrates have been corrected for the concentrations of < 3 kD / < 10 kD solutes and normalised to the total sample weight.
[c] – The measured concentrations in the acid rinse have been normalized to the total sample weights. Errors are ca. 7% of the reported concentration.
* – mmol kg^{-1}
** – mg dm^{-3}

TABLE 2.3. – continued

Region	Sampling date	Fraction (μm)	N	Fe	Hg	Ir	K	Mg	Mn	Na	Ni	References
Swedish rivers												
Dalälven	Pre-1994	< 0.45	1	70			640	920	5.4	2200		Andersson et al., 1994
Indalsälven		< 0.45	1	34			360	770	1	1400		
Kalixälven		< 0.45	1	1010			760	760	8.7	1100		
Kalixälv-Kamlunge	1990–91	< 0.45	2						293[b]			Ingri et al., 1997
	1992	< 0.45				23.5±1.4*						Anbar et al., 1996
	1995	< 0.45	1			17.4±0.9*						Anbar et al., 1996
Kalix-Kamlunge	1995	< 0.45	1	135			700	600	19	1440		Porcelli et al., 1997
		< 10 kD ultrafiltrate[c]	1	48			600	400	11	1160		
		>10 kD colloid conc.[d]	1	142			0	0	1.4	10		
		10 kD filter rinse[e]	1	51			30	100	4.3	50		
		< 3 kD ultrafiltrate[c]	1	43			400	500	13	1100		
		> 3 kD colloid conc.[d]	1	76			0	0	0.7	0		
		3 kD filter rinse[e]	1	31			30	0	1.4	30		
Kalixälv-Tärendö	1990–91	< 0.45	2						97.5[b]			Ingri et al., 1997
Rautas	1995	< 0.45	1	0.70^			0.015**	0.023**	0.09^	0.072**		Andersson et al., 1998a
Kemijoki		< 0.45	1	225			610	960	13.5	1300		Andersson et al., 1994
Kokenmäenijoki		< 0.45	1	151			2000	2400	17.3	7100		
Narkan		< 0.45	1	601			630	550	14	2520		Porcelli et al., 1997
5 Swedish woodland rivers	1990–91	< 0.45	10						410[b] 96–735			Ingri et al., 1997
5 Swedish mountain rivers	1990–91	< 0.45	10						136[b] 5–451			Ingri et al., 1997
Russian river												
Neva	1993	< 0.45	1			49.7±1.3*						Anbar et al., 1996
Polish rivers												
Vistula												
Swibno	1958–59	< 0.45	1	107			6000	7500	30	1500		Szefer, 1989
	1973–74	< 0.45							36			
	1987	< 0.45	1	61			4100	7000	11	24200		
Malbork	1987	< 0.45	1	52			4400	8600	5	23700		

Region	Sampling date	Fraction (μm)	N	Fe	Hg	Ir	K	Mg	Mn	Na	Ni	References
Vistula	1993	< 0.45	1			92.9±2.2*						Anbar et al., 1996
		< 0.45						13700				Löfvendahl et al., 1990
		< 0.45	1	42			4200	11800	21	48000		Andersson et al., 1994
Oder			1	1300[a]	0.025[a]						3.5	Pohl et al., 1998
Latvian rivers												
Daugava	1993–97		9					12800				Kļaviņš et al., 2000
Lielupe			4					24000				
Venta			7					17400				
Gauja			5					13600				
Salaca			3					13200				
Ciecere			1					26000				
Abava			2					24400				
36 rivers of the Baltic basin		< 0.45	36					1980 480–7200				Löfvendahl et al., 1990
11 Rivers of N. Sweden	1984–85		1909-					800±200		1700±300		Löfvendahl, 1990
11 Rivers of S. Sweden			1925					2900±1200		9400±3300		

* – 10^8 atoms kg^{-1}
** – mmol kg^{-1}
^ – μmol kg^{-1}
[a] – total concentration (dissolved and particulate bounded)
[b] – nmol dm^{-3}
[c] – Measured concentrations are ±5%, except for K, which are ±10%, and where noted otherwise.
[d] – The measured concentrations in the colloid concentrates have been corrected for the concentrations of < 3 kD / < 10 kD solutes and normalised to the total sample weight.
[e] – The measured concentrations in the acid rinse have been normalized to the total sample weights. Errors are ca. 7% of the reported concentration.

TABLE 2.3. – continued

Region	Sampling date	Fraction (μm)	N	Pb	S	Si	Sr	Zn	References
Swedish rivers									
Dalälven	1991	< 0.45	1			1200	16.9		Andersson et al., 1994
Indalsälven		< 0.45	1			1100	21.3		
Kalixälven		< 0.45	1			2200	10.8		
Kalixälv-Kamlunge	1990–91	< 0.45	2		31.5**				Ingri et al., 1997
Kalix-Kamlunge	1995	< 0.45	1			2600			Porcelli et al., 1997
		< 10 kD ultrafiltrate[a]	1			1900			
		> 10 kD colloid conc.[b]	1			100			
		10 kD filter rinse[c]	1			100			
		< 3 kD ultrafiltrate[a]	1			1800			
		> 3 kD colloid conc.[b]	1			ND			
		3 kD filter rinse[c]				100			
Kalixälv-Tärendö	1990–91	< 0.45	2		34.5**				Ingri et al., 1997
Rautas	1995	< 0.45	1			27.2^			Andersson et al., 1998a
Kemijoki	1991	< 0.45	1			2900	10.8		Andersson et al., 1994
Kokenmäenijoki		< 0.45	1			610	35.1		
Narkån		< 0.45	1			2570			Porcelli et al., 1997
5 Swedish woodland rivers	1990–91	< 0.45	10		19.3**				Ingri et al., 1997
5 Swedish mountain rivers	1990–91	< 0.45	10		34** 25–50				Ingri et al., 1997
Polish rivers									
Vistula									
Swibno	1973–74	< 0.45	1	0.47				80	Szefer, 1989
	1987	< 0.45	1					55	
Malbork	1987	< 0.45	1					38	
Vistula	1993	< 0.45	1			4700	395		Andersson et al., 1994
		< 0.45	1				504		Löfvendahl et al., 1990

Region	Sampling date	Fraction (μm)	N	Pb	S	Si	Sr	Zn	References
Oder			1	5.5			399	30	Pohl et al., 1998
Latvian rivers									
Daugava	1993–97		9	0.1	28.8*			4.4	Kļaviņš et al., 2000
Lielupe			4	0.3	119.2*			4.1	
Venta			7	0.3	44.8*			2.7	
Gauja			5	0.7	34.0*			4.8	
Salaca			3	0.6	31.9*			3.3	
Ciecere			1	0.1	42.0*			2.9	
Abava			2	0.1	50.8*			3.2	
36 rivers of the Baltic basin		< 0.45	36				44.5 15–234		Löfvendahl et al., 1990

^ – μmol kg^{-1}
* – mg SO_4^{2-} dm^{-3}
** – μmol dm^{-3}
[a] – Measured concentrations are ±5%, except for K, which are ±10%, and where noted otherwise.
[b] – The measured concentrations in the colloid concentrates have been corrected for the concentrations of < 3 kD / < 10 kD solutes and normalised to the total sample weight.
[c] – The measured concentrations in the acid rinse have been normalized to the total sample weights. Errors are ca. 7% of the reported concentration.

TABLE 2.3. – continued

River	Sampling date	Fraction (μm)	N	230-Th (ng kg^{-1} (× 10^5))	232-Th (ng dm^{-3})	U (t) (μg dm^{-3})	δ234-U (d)[a]	235-U (d) (mBq dm^{-3})	238-U (d) (μg kg^{-1})	References
Swedish rivers										
Kalixälven	1991	< 0.45	4	13.2	13.95		659		0.169	Andersson et al., 1995
				8.1–21.0	9.2–19.3		559–761		0.16–0.18	
Kalix - Kamlunge	1995	< 0.45					896±6		771±1**	Andersson et al., 1998a
		< 10 kD ultrafiltrate[b]							167±0.4**	
		< 0.45					896±7		184±0.3*	Porcelli et al., 1997
		< 10 kD ultrafiltrate[b]					883±17		40±0.1*	
		> 10 kD colloid conc.[c]					878±7		61*	
		10 kD filter rinse[d]					851±56		71*	
		< 3 kD ultrafiltrate[b]					847±16		45±0.1*	
		> 3 kD colloid conc.[c]					880±9		34*	
		3 kD filter rinse[d]					846±10		58*	
Narkän		< 0.45					770±8		187.2±0.4*	
Rautas	1995	< 0.45					1005±7		231±0.4**	Andersson et al., 1998a
		< 10 kD							94±0.2**	
Russian rivers										
Neva						0.3				Baturin and Kočenov, 1969
Luga						0.5				
Narva/Pljussa						0.4				
Polish rivers										
Vistula	1976–77				66 (N=4)	0.58 (N=10)				Szefer, 1977; Bojanowski and Szefer, 1979
						0.29–0.98				
	Pre-1983					0.34±0.02				Gellermann et al., 1983
	1985					0.90±0.12[e]	14.3±1.6[e]	0.5±0.1[e]	11.1±1.5[e]	Skwarzec, 1995
	1993	< 0.45		2.5±0.1	2.50±0.01		259±3		0.724±0.001	Andersson et al., 1995
7 German rivers	1982–83	< 0.45				11.7[e]				Gellermann and Stolz, 1997
9 rivers entering Bothnian basin	1953		9			0.43				Koczy et al., 1957
						0.2–0.7				
Kymmeneälv River entering Finnish basin	1953		1			0.4				
Kävlingea River entering Öresund	1953		1			1.4				
2 rivers entering Skagerrak	1953		2			0.5				

* – pg g^{-1}
** – pmol kg^{-1}
a – $\delta^{234}U = [(^{234}U/^{238}U)_s/(^{234}U/^{238}U)_{eq} - 1] \times 10^3$, where $(^{234}U/^{238}U)_{eq}$ is the secular equilibrium ratio of 5.472×10^{-5}
b – Measured concentrations are ±5%, except for K, which are ±10%, and where noted otherwise.
c – The measured concentrations in the colloid concentrates have been corrected for the concentrations of < 3 kD / < 10 kD solutes and normalised to the total sample weight.
d – The measured concentrations in the acid rinse have been normalized to the total sample weights. Errors are ca. 7% of the reported concentration.
e – mBq kg^{-1}

TABLE 2.4.

Concentrations of Fe and rare earth elements (pM) in river water from the Baltic drainage basin

Region	Sam pling date	Fraction (μm)	Fe	La	Ce	Pr	Nd	Sm	References
Swedish river									
Klix-Kamlunge	1997	< 0.2	5.51 2.14-9.36	757 374-1296	1123 437-1970	171 86.6-295	682 361-1213	107 58.1-190	Ingri et al., 2000
		Solution > 3 kD	0.12 0.07-0.20	31.9 15.3-64.9	56.5 18.6-106	10.2 5.11-18.9	42.6 21.1-68.1	11.2 5.12-17.0	
		Colloidal	3.63 1.55-5.70	503 238-790	713 271-1158	114 56.1-182	435 221-706	70 35.8-116	

Region	Sam pling date	Fraction (μm)	Eu	Dy	Ho	Er	Tm	Yb	References
Swedish river									
Kalix--Kamlunge	1997	< 0.2	25 16.5-41.0	81.1 48.0-140	16.6 10.0-34.0	49.8 34.1-82.5	8.46 6.05-12.9	55 40.3-84.4	Ingri et al., 2000
		Solution > 3 kD	4.87 < 3.3-7.57	8.45 5.48-11.8	2.95 < 3.0-4.79	6.28 < 3.0-11.1	< 3.0	13.1 6.99-16.0	
		Colloidal	14.4 7.1-21.2	45.7 26.2-75.9	15.9	31.6 19.7-46.3		28.2 18.5-48.2	

TABLE 2.5.

Concentrations of chemical elements (μg dm^{-3}) in particulate matter of river water of the Baltic drainage basin and other northern areas

Region	Sampling date	Fraction (μm)	N	Al	Ca	Cd	Co	Cu	References
Swedish rivers									
Dalälven	1991	> 0.45	1	400	80				Andersson et al., 1994
Indalsälven	1991	> 0.45	1	110	20				
Kalixälven	1991	> 0.45	1	127	70				
Kalix-Kamlunge	1995	> 0.45	1	6.4*	2.7*				Andersson et al., 1998a
Rautas	1995	> 0.45	1	7.2*	2.2*				
Kemijoki	1991	> 0.45	1	32	11				Andersson et al., 1994
Kokenmäenijoki	1991	> 0.45	1	310	52				
Polish rivers									
Vistula									
Swibno	1973–74	> 0.45	1				0.25	1.6	Szefer, 1989
	1987	> 0.45	1		800	0.5		2.8	
Malbork	1987	> 0.45	1		1800	0.4		1.5	
Vistula	1993	> 0.45	1	520	450				Andersson et al., 1994

* – %

TABLE 2.5. – continued

Region	Sampling date	Fraction (μm)	N	Fe	K	Mg	Mn	Na	Ni	References
Swedish rivers										
Dalälven	1991	> 0.45	1	330	150	53	52	70		Andersson et al., 1994
Indalsälven	1991	> 0.45	1	89	42	19	7.5	17		
Kalixälven	1991	> 0.45	1	520	47	32	14	40		
Kalix-Kamlunge	1995	> 0.45	1	15.4*	1.4*	1.5*	0.61*	1.9*		Andersson et al., 1998a
Rautas	1995	> 0.45	1	5.6*	2.4*	1.6*	0.22*	1.8*		
Kemijoki	1991	> 0.45	1	110	8	8	2.6	7		Andersson et al., 1994
Kokenmäenijoki	1991	> 0.45	1	220	110	60	20	62		
Polish rivers										
Vistula										
Swibno	1973–74	> 0.45	1				55		1	Szefer, 1989
	1987	> 0.45	1	879	500	300	93			
Malbork	1987	> 0.45	1	848	200	300	49			
Vistula	1993	> 0.45	1	640	140	89	31	81		Andersson al., 1994

* – %

TABLE 2.5. - continued

Region	Sampling date	Fraction (μm)	N	P	Pb	Si	Sr	Ti	Zn	References
Swedish rivers										
Dalälven	1991	> 0.45	1	15		2200	0.74	21		Anderson et al., 1994
Indalsälven	1991	> 0.45	1	4.7		400	0.15	7.4		
Kalixälven	1991	> 0.45	1	17		500	0.46	8.2		
Kalix-Kamlunge	1995	> 0.45	1	0.42*		22.3*				Anderson et al., 1998a
Rautas	1995	> 0.45	1	0.21*		27.6*				
Kemijoki	1995	> 0.45	1	2.8		90	0.08	1.8		Anderson et al., 1994
Kokenmäenijoki	1991	> 0.45	1	8.9		1100	0.61	19		
Polish rivers										
Vistula										
Swibno	1973–74	> 0.45	1		1.83				12	Szefer, 1989
	1987	> 0.45	1						37	
Malbork	1987	> 0.45	1						13	
Vistula	1993	> 0.45	1	89		1900	2.37	29		Anderson et al., 1994

* – %

new data from a Polish–Swedish and Polish–German joint projects give insight into current trace-element fluxes (HELCOM, 1998a). A major problem is the pollution of both the bottom and flood-plain sediments of the main rivers, the Vistula and the Oder, with heavy metals (Figs. 2.1 and 2.2). These pollutants are not only derived from mine waters but also are released by Zn, Pb and Cu ore

TABLE 2.6.

Concentrations of Al, Ca, Fe, K, Mg, Mn, Na, P and S (%) and other elements ($\mu g\ g^{-1}$ dry wt) in riverine-estuarine sediments of the Baltic catchment and till in the Kalix River watershed. The concentrations of Fe, Mn and S ($mg\ dm^{-3}$) and As ($\mu g\ dm^{-3}$) in pore water are also given

River	Sampling date	Segment (mm)	N	Al	As	Ba	Ca	Cd	Co	Cr	Cu	References
Swedish river												
Kalix River estuary	1991–92	0–20	4	5.65	41							Widerlund and Ingri,
				5.4–5.8	38–44							1995, 1996
		0–20	4**		1.65							
					1.32–1.95							
		20–320	24	6.26	69.9							
				5.8–6.7	10–171							
		20–320	24**		72.3							
					6.83–166							
		320–360	2	6.4	22.7							
				6.0–6.8	9.4–36							
		320–360	2**		47.8							
					27.8–67.7							
Kalix		to 80		12.9*		734	2.39*		< 5	45	8	Öhlander et al., 2000
				12.6–13.1		693–774	2.31–2.47			34–56	5.0–11	
		300–400		15.9*		528	3.37*		9.5	86	25.5	
				15.5–16.2		491–564	2.77–3.96		6.0–13	67–105	22–29	
		1000		14*		624	4.10*		12	87	36	
				13.9–14.1		571–676	3.50–4.69		7.0–17	62–112	17–55	
Polish rivers												
Vistula		0–100						3–140		4–420	25–470	Helios-Rybicka, 1993
Przemsza		Surficial						200[a]				Helios-Rybicka, 1993
Oder		Surficial						12[a]			1800[a]	Helios-Rybicka, 1996
Latvian rivers												
Daugava	1993–97		9					1.03	2.5		6.48	Kļavņš et al., 2000
Lielupe			4					1.67	4.49		13.4	
Venta			6					0.99	4.07		2.58	
Gauja			5					0.71	2.63		4.74	
Salaca			3					0.72	1.84		2.53	
Ciecere			1					0.65	1.72		7.05	
Abava			1					0.45	2.16		5.02	

* – Concentration expressed as oxides (%).
** – Concentration in pore water.
[a] – Maximum value.

TABLE 2.6. – continued

River	Sampling date	Segment (mm)	N	Fe	K	La	Mg	Mn	Na	Nd	Ni	P	References
Swedish river													
Kalix River estuary	1991–92		4	9.48				1.53					Widerlund and Ingri, 1995, 1996
				8.2–10.8				1.04–1.94					
			4**	0.27				2.46					
				0.03–0.57				0.003–5.87					
			24	8.45				0.34					
				7.0–9.3				0.217–0.465					
			24**	22.9				18.3					
				8.18–31.2				10.2–23.9					
			2	8.55				0.28					
				8.0–9.1				0.262–0.300					
			2**	25.8				18.8					
				21.9–29.6				18.3–19.3					
Kalix		to 80		4.3*	3.1*	10–< 11	0.77*	0.07*	3.55*	13.7	7.5	0.05*	Öhlander et al., 2000
				2.41–6.19	2.78–3.41		0.74–0.79	0.05–0.08	3.45–3.65	10.9–16.4	7.0–8.0	0.01–0.08	
		300–400		7.21*	2.28*	22.5	1.93*	0.10*	3.30*	27.7	30	0.24*	
				5.72–8.70	2.05–2.51	19–26	1.37–2.49	0.08–0.11	3.19–3.41	20.0–35.3	29–31	0.18–0.29	
		1000		6.64*	2.67*	37.5	2.28*	0.10*	3.80*	34.3	21	0.23*	
				4.99–8.29	2.30–3.04	31–44	1.67–2.89	0.07–0.12	3.79–3.81	28.6–39.9	15–27	0.14–0.31	
Polish rivers													
Vistula		0–100									10–150		Helios Rybicka, 1993
Oder		Surficial									350[a]		Helios Rybicka, 1996
Latvian rivers													
Daugava	1993–97		9								8.59		Kļaviņš et al., 2000
Lielupe			4								8.86		
Venta			6								4.49		
Gauja			5								4.65		
Salaca			3								3.86		
Ciecere			1								2.1		
Abava			1								3.27		

* – Concentration expressed ad oxides (%).
** – Concentration in pore water.
[a] – Maximum value.

TABLE 2.6. – continued

Region	Sampling date	Segment (mm)	N	Pb	S	Sc	Si	Sm	Sr	Ti	V	Y	Yb	Zn	Zr	References
Swedish river																
Kalix River estuary	1991–92		4		0.15											Widerlund and Ingri, 1995, 1996
					0.118–0.162											
			4**		66.5											
					60.3–76.0											
			24		0.58											
					0.124–1.38											
			24**		24											
					0.942–66.0											
			2		0.86											
					0.584–1.14											
			2**		1.51											
					1.39–1.63											
Kalix		to 80				9.5	69.4*	2.95	250	1.25*	85.5	18	2.5	22.5	645	Öhlander et al., 2000
						8.0–11.0	67.2–71.5	2.32–3.58	226–274	0.98–1.51	53–118	13–23	1.9–3.1	19–26	607–682	
		300–400				11.5	59.1*	5.56	229	0.95*	112	27.5	3.5	54.5	383	
						11.0–12.0	57.0–61.2	4.08–7.03	215–243	0.75–1.15	78–145	18–37	2.6–4.4	45–64	369–397	
		1000				15.5	63.9*	6.77	262	1.10*	117	35.5	4.2	53.5	456	
						12.0–19.0	60.6–67.1	5.62–7.92	240–284	0.85–1.35	80–153	24–47	3.1–5.3	37–70	393–519	
Polish rivers																
Vistula		0–100		57–760										250–5300		Helios Rybicka, 1993
Przemsza		Surficial		1000[a]										7000[a]		Helios Rybicka, 1993
Oder		Surficial		4000[a]										6700[a]		Helios Rybicka, 1996
Latvian rivers																
Daugava	1993–97		9	22.3												Kļaviņš et al., 2000
				9.5–65.5												
Lielupe			4	23												
				19.9–26.7												
Venta			6	24												
				12.4–55.3												
Gauja			5	18.4												
				14.1–28.1												
Salaca			3	12.1												
				9.21–16.7												
Ciecere			1	11.1												
Abava			1	12.3												

* – Concentration expressed as oxides.
** – Concentration in pore water.
[a] – Maximum value.

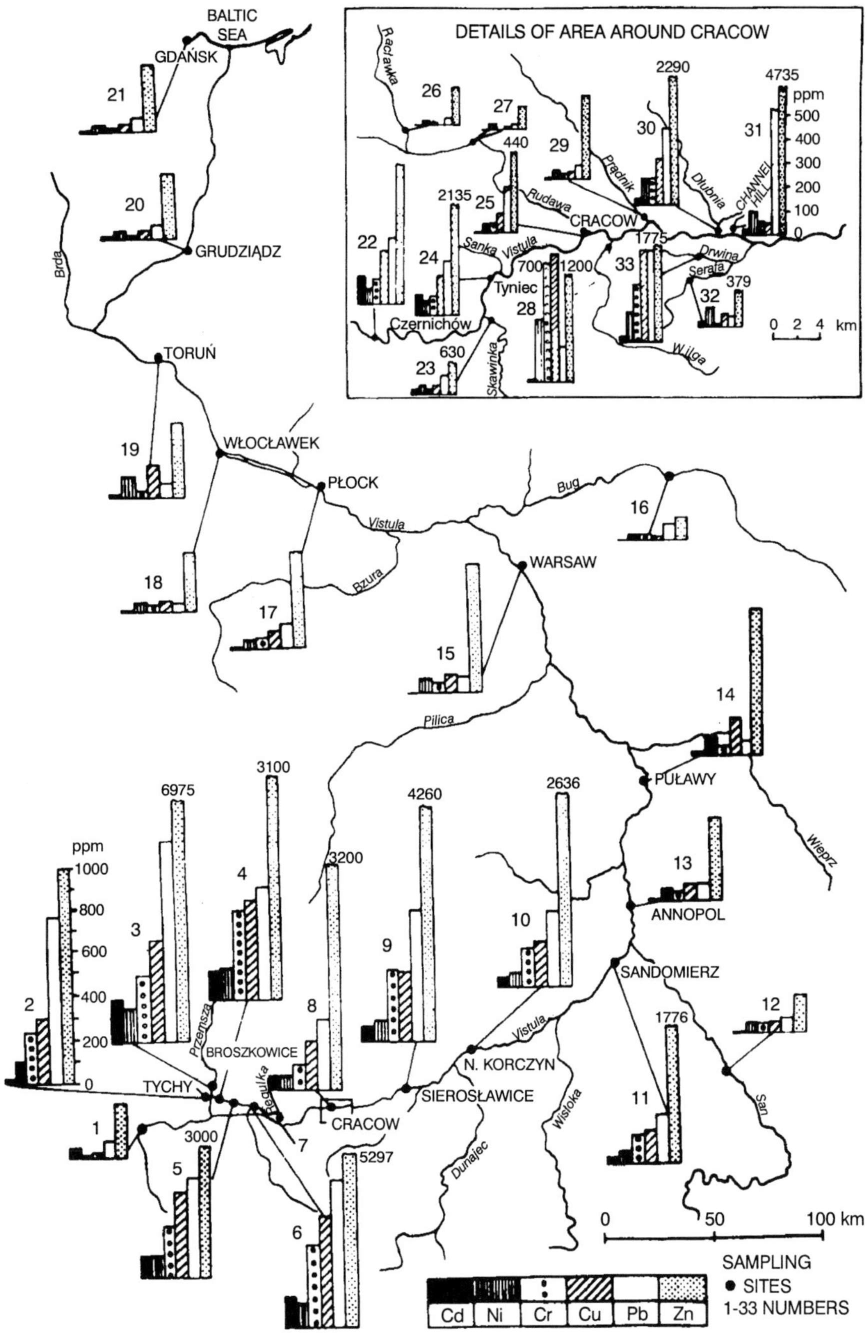

Fig. 2.1. Concentration (ppm) of heavy metals in sediments of the Vistula River and its tributaries, Poland (< 63 μm fraction). After Helios Rybicka (1992, 1993); modified.

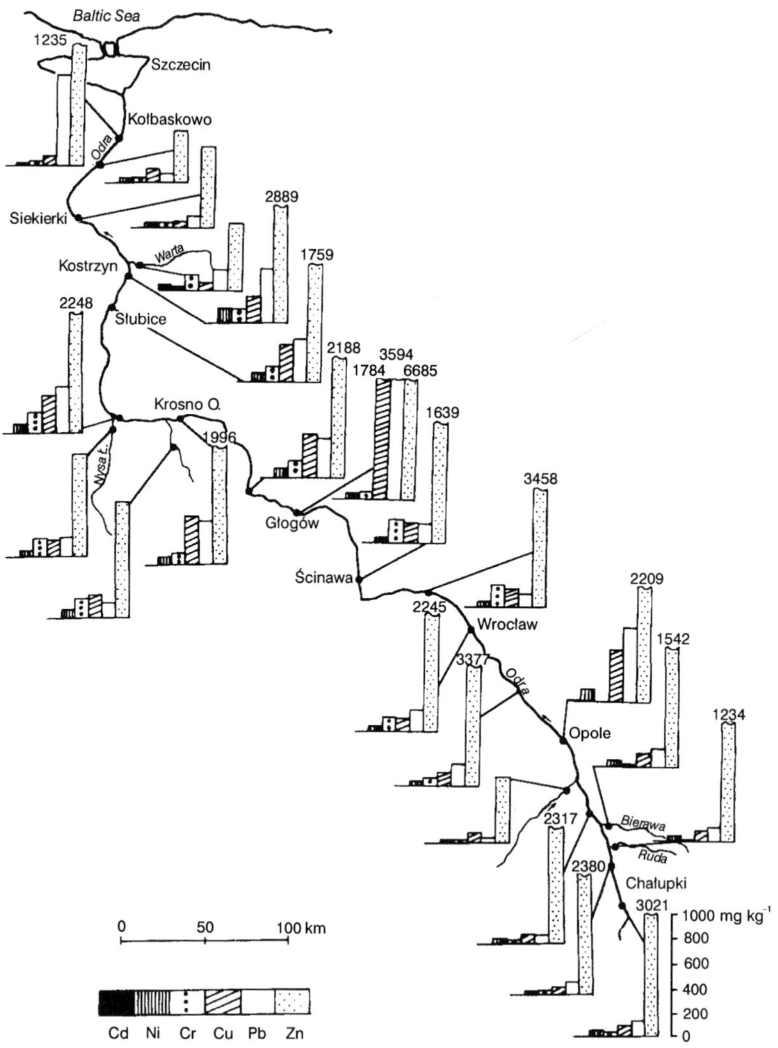

Fig. 2.2. Concentrations of heavy metals in the < 63 μm sediment fraction of the Odra River. After Helios Rybicka (1996); modified.

processing and smelting plants. It is assumed that tributaries from the southern to the eastern parts of the catchment (Poland and the Baltic countries) have the greatest contribution to the current trace-metal input (Matschullat, 1997). Poland is a rich mining area with abundant Zn–Pb and Cu deposits which are situated mainly in Silesia in the southern part of the country; Ag and Cd are associated with the Zn–Pb ores (Osika 1986). In fact, Poland has the highest abundance of Ag per unit area of any country (Singer, 1995). In 1991, 5.3 $\times$ 10^6 tonnes of Zn–Pb ores and 31 $\times$ 10^6 tonnes of Cu ores were mined (Helios Rybicka, 1996a). These mining operations have caused very significant environmental contamination. For example, Helios Rybicka (1996a) reported that, on average, 3331 tonnes of Zn, 448 tonnes of Pb, 515 tonnes of Cu, 40 tonnes of Cd, 443 tonnes of Cr and 284 tonnes of Ni are discharged into the Baltic Sea each year by rivers of the Polish drainage basin.

The Vistula River is 1092 km long and enter the Gulf of Gdańsk; it has a drainage area of 194,000 km^2 and introduces about 34 km^3 of freshwater annually. The Vistula and its tributaries traverse highly industrialised areas which include manufacturing (iron and steel, electrochemical and chemical, petroleum refining and light industry) and mining (coal, building stone, aggregate and petroleum) industries. Agricultural activity also contributes nutrients and eroded soil material to the river waters. In 1989, the Vistula was estimated to have carried 2,930 tonnes Zn yr^{-1}, 12.8 tonnes Cd yr^{-1}, 196 tonnes Pb yr^{-1}, 233 tonnes Cu yr^{-1} and 15.5 tonnes Hg yr^{-1}. Until recently, municipal systems in Poland discharged 900,000 m^3 of untreated sewage and 1,400,000 m^3 of partially treated sewage, much of which would have ended up in the Vistula; it is responsible for ca 10 % of the input of Pb, Cd, Cr, Cu and Zn. The Gdańsk region has been designated one of the ecologically endangered areas in Poland and one of the pollution "hot spots" in the Baltic.

The concentrations of heavy metals in Vistula River sediments are very high, mainly in the upper reaches of the river (Fig. 2.1). In the < 63 μm fraction the amount of metals ranges are as follows (in $\mu g\ g^{-1}$): 3–140 Cd, 57–760 Pb, 250–5,300 Zn, 25–470 Cu, 4–420 Cr and 10–150 Ni. The highest amount of metals was found in the sediment of the Przemsza River (Helios Rybicka, 1996a). There is a decrease of metal contents in the Vistula River sediments downstream of Cracow (Fig. 2.1).

The Vistula River drains the Upper Silesian Mining and Metallurgical Industrial District, hence there is a particularly strong accumulation of metals in the river sediments of this region. In the sediment of Upper Vistula River along a distance of about 200 km > 99% of both Cd and Pb and > 90% of Zn are of anthropogenic-industrial origin. The sediment sample from Gdańsk showed decreasing metal concentrations by factors of about 10 for Zn, 20 for Cd and 5 for Pb and Cu. Recent investigations (Macklin and Klimek 1992) have revealed very high concentrations of Zn (up to 11,000 $\mu g\ g^{-1}$), Pb (over 1,700 $\mu g\ g^{-1}$) and Cd (up to 150 $\mu g\ g^{-1}$) in the overbank alluvial sediments of Upper Vistula and Przemsza Rivers. Industrial waters discharged from Zn–Pb ore processing plants and also mine waters and meteoric waters infiltrating the waste dumps, are principal pollutants of the Przemsza River, a tributary of the Vistula River. The levels of heavy metals in the bottom sediments of these two rivers are extremely high, reaching values up to 7,000 μg Zn g^{-1}, more than 900 μg Pb g^{-1} and 200 μg Cd g^{-1} in the Przemsza River, and up to 6,000 μg Zn g^{-1}, 800 μg Pb g^{-1} and about 140 μg Cd g^{-1} in the Vistula River sediments The concentrations of Cd were particularly high and reached some of the highest values detected in river sediments anywhere in Europe (Helios Rybicka, 1992, 1993).

The Szczecin Lagoon and the Pomeranian Bay (southern Baltic) are supplied by the Oder (Odra) River, the second largest river in Poland after the Vistula, with a length of 854 km and an outflow of 14.5 km^3 yr^{-1}. This river drains the heavily industrialised heartland of Silesia and transports 90 tonnes Cu yr^{-1}, 792 tonnes Zn yr^{-1}, 15.5 tonnes Cd yr^{-1} and 104 tonnes Pb yr^{-1} to the Baltic (Neumann et al., 1996). Near the mouth of the Odra (Oder) River, the < 2 μm size fraction of the riverine sediments contains 426 μg Cu g^{-1}, 3114 μg Zn g^{-1}, 20 μg Cd g^{-1} and 1132 μg Pb g^{-1} (Helios Rybicka, 1996b). The concentrations of heavy metals in the Odra River sediments are also high with concentrations up to 6,700 μg Zn g^{-1}, 4,000 μg Pb g^{-1}, 1,800 μg Cu g^{-1}, 350 μg Ni g^{-1} and 12 μg Cd g^{-1} (Fig. 2.2).

Concentration gradients between the river mouth of the Peenestrom and the Oder Lagoon (German sector) showed, that only a very small percentage of river transported trace elements entered the Pomeranian Bay in August 1996. Most of heavy metals such as Zn (94.9%), Fe (93.6%), Cd (90.6%), Pb (90.4%), Co (83.4%), Cu (79.7%), Ni (68.3%) was retained in the inner coastal part of lagoon (Pohl et al., 1998). The exceptional Oder flood in summer 1997 has led to considerable additional metal pollution of the Szczecin Lagoon and the Pomeranian Bay, Baltic Sea (Siegel et al., 1998; Helios Rybicka and Strzebońska, 1999; Lehmann et al., 1999; Müller and Wessels, 1999; Protasowicki et al., 1999). On the other side, however, the high dilution effect of the flood water reduced the concentration of pollutants and therefore prevented a direct negative impact of trace elements on the Baltic ecosystem (Siegel et al., 1998). According to Müller and Wessels (1999) compared with the mean concentrations of heavy metals such as Cu, Pb and Zn before flood, their maximum levels in < 20 μm fraction of suspended solid during the flood were from two to four times higher. For instance, differently remarkable increase of Cu and Pb concentrations occurred in the river section from Scinawa to Nietkow due to the impact of mining and smelting industry. The river rich between Glogow and Nietkow was selected as a reference area with an industrial background and a typical pollutant pattern. Under the typical conditions (1996/97) a slight dilution can be noted for almost all heavy metals, except Cd, along Frankfurt/Oder and Schwedt sector. The mean concentration of Cd increased from 6.2 μg g^{-1} at Frankfurt to 7.7 μg g^{-1} at Schwedt, suggesting additional Cd source located downstream of Frankfurt (Müller and Wessels, 1999). Analyses of unsieved sediments collected after the flood from the eastern and the western Odra showed increased concentrations of Cr and Mn and decreased concentrations of Pb. The levels of Cd, Ni and Zn have remained unchanged (Protasowicki et al., 1999).

Northern catchment

Rivers of the northern Baltic catchment have been analysed for heavy metal levels less frequently than rivers of the southern catchment. However, these northern rivers, especially the Kalix River and its estuary (Northern Sweden) have been intensively assayed in respect to the distribution and concentration of dissolved and particulate forms of Mn, Fe, Al, P, S, Si, Ti, Sr as well as alkali and alkaline-earth elements Ca, Mg, K and Na (Pontér et al., 1990, 1992; Andersson et al., 1994; Ingri and Widerlund, 1994; Widerlund and Ingri, 1996; Porcelli et al., 1997; Gustafsson et al., 2000). It is pointed out that the mean particulate Fe/Al ratio (6.5) in the Kalix River is more than ten times the ratio in mean world river. Particulate Fe levels reached maximum values during early snowmelt and decreased during maximum discharge similar to its dissolved form. A sharp increase in non-detrital particulate Mn with simultaneous decrease in its dissolved form in early July is caused probably by two processes, i.e. addition of non-detrital particulate Mn to the river from lakes, and transformation of dissolved Mn to a particulate phase in the river (Pontér et al., 1990; Andersson et al., 1994). According to Widerlund and Ingri (1996) the sediments of the Kalix River estuary are an efficient trap for settling particulate Fe and Mn, although non-detrital species of the two metals, presumably reactive Fe- and Mn-oxides, form substantial fractions of the total input of Fe and Mn to the estuarine area. The results of the 3 kD ultrafiltration experiment being in a good agreement with those obtained from < 10 kD experiment have been discussed in respect to the distribution of K, Na, Ca, Mg, Fe, Mn and Si in the Kalix River (Kamlunge). From 62 to 76% of the total Na and K were in < 3 kD fraction. A significant < 3 kD fraction of the elements such as Mg, Ca and Si was comparable to < 10 kD fraction suggesting that these elements are mainly associated with colloids. Fe and Mn also appeared to be associated with colloids, likely either as oxyhydroxide phases or incorporated to organic colloids (Porcelli et al., 1997). Rare earth elements (REE) have been determined in filtered and suspended particulate phase of the Kalix River (Ingri et al., 2000). Approximately 10% of the REE were transported in detrital particles during winter while in spring-flood ca. 30% of the light REE (LREE) and 70% of the heavy REE were present in the detrital phase. It is found that the REE are associated with two colloidal (and particulate) phases, i.e. an organic-rich phase with associated Al–Fe and Fe-rich, as Fe-oxyhydroxide, inorganic phase. The temporal and seasonal variations of the Ce-anomaly in the suspended particulate phase are attributed to rapid transport of colloids enriched in REE, C, Al and Fe from the upper section of the till to the river at storm events. There was a fractionation of Ce from the other LREE during weathering and transport to the river with relative its enrichment at storm events caused by C-rich colloids (Ingri et al., 2000).

Comparison between chemical composition of northern and southern rivers

Intercomparison studies of northern rivers (Dalälven, Indalsälven, Kalixälven, Kemijoki and Kokenmäenjoki) and a southern river (Vistula River) have been carried out for the concentrations of dissolved and particulate Al, Ca, Fe, K, Mg, Mn, Na, P, Si, Sr and Ti (Andersson et al., 1994; Porcelli et al., 1997). Moreover, Anbar et al. (1996) have compared distribution of Ir as a member of PGEs (platinum group elements) in filtered (< 45 μm) waters of northern river (Kalixälven) and two southern rivers (Neva, Vistula). The Neva and Vistula Rivers drain weathered Phanerozoic sedimentary rocks and are polluted by metallurgical industries and coal combustion. It should be emphasised that these both rivers are characterised by a higher abundance of the dissolved elements in respect to the average world river composition (Andersson et al., 1994; Anbar et al., 1996). The concentrations of Ir in the Neva and Vistula Rivers are from two to five times greater those in the Kalixälven. The behaviour of Ir during estuarine mixing, i.e. in oxic waters from above halocline in the Baltic Sea (Fig. 2.3) has been studied by Anbar et al. (1996). The Ir was distributed uniformly in these waters and fell below the conservative mixing line defined by the North Sea and the mean Baltic river input (Fig 2.4). It indicates that ca. 75% of the dissolved Ir provided by rivers is removed rapidly from Baltic water supposedly into sediments enriched in Fe-and Mn-oxyhydroxides or organic matter. As it is well known, floculation of this ferromanganese phase takes place at the river mouths and in the Gulfs of the Baltic Sea influencing other trace elements.

Fig. 2.3. Map of the Baltic Sea showing the sampling stations and the geographic distribution of freshwater inputs. Dashed lines: 40-m depth contour. Dotted lines: 100-m depth contour. After Anbar et al. (1996).

The Vistula River displayed a substantially larger dissolved load of all the elements except Al and Fe as compared to the northern Precambrian rivers. Significantly higher levels of dissolved Ca and Sr in the Vistula as compared to those in the Precambrian rivers are an example of the importance of carbonate weathering in controlling the riverine transport of Sr in the vicinity of carbonate bedrock. The highest abundance of P in the Vistula, not correlated with Fe in contrast to the Precambrian rivers, can be explained by this that the draining area is densely

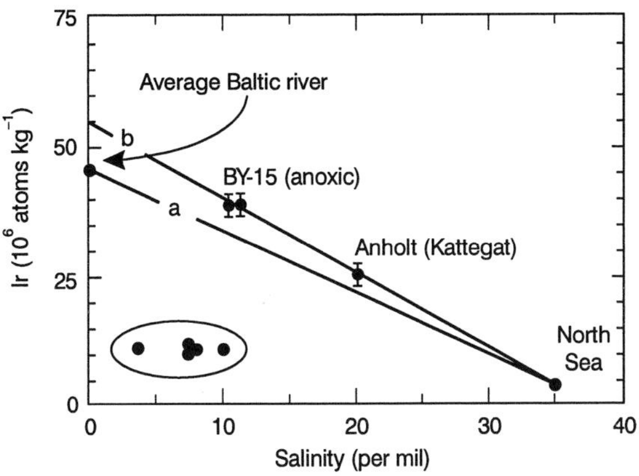

Fig. 2.4. Iridium in the Baltic Sea. The circled points are samples from BY-15 (oxic), BY-28, and F-2 waters. Curve a is the conservative mixing line between the North Sea and average Baltic river water. The flux weighted average river concentration is estimated with the assumption that the Kalixälven is representative of the rivers draining Precambrian terrain (226 $km^3\ year^{-1}$) and that the Neva and Vistula are representative of average river water in their drainage basins (112 $km^3\ year^{-1}$ and 114 $km^3\ year^{-1}$). The remaining drainage from Phanerozoic terrain (32 $km^3\ year^{-1}$) is assumed to be similar to the Neva. The result, $\sim 46 \times 10^8$ atoms kg^{-1}, is similar to the concentration in the Neva $(49.7 \pm 1.3) \times 10^8$ atoms kg^{-1}. The actual average concentration may be higher because anthropogenic inputs from southern Sweden are neglected. Curve b is the conservative mixing line defined by the samples from the North Sea, Kattegat, and BY-15 anoxic waters. After Anbar et al. (1996); modified.

populated and extensive agricultural and hence fertilised. Fe is substantially accumulated in particulate fraction of the Kalixälven and Kemijoki Rivers while particulate Mn concentrations reached the highest values in all the Precambrian Rivers. From comparison of the overall pattern of the riverine composition of the Baltic catchment clearly results that the particulate elements in river water draining the Precambrian basement are closely similar to the world mean particulate load except Fe, Mn and P (Porcelli et al., 1997).

Elevated levels of As in the sediments of the Bothnian Bay (Table 2.6) are attributed to emission from a sulphide ore smelter in the vicinity of Skellefteå, Sweden (Widerlund and Ingri, 1995; Borg and Jonsson, 1996). The sediment and pore water concentrations of Al, As, Fe, Mn and S have been determined in two sediment cores from the Kalix River estuary (Widerlund and Ingri, 1995, 1996). Excess of Fe most probably is dominated by Fe(III)-oxides, suggesting that the diagenetic remobilization of reactive Fe is incomplete. It means that dissolution of Fe (III)-oxides in anoxic sediments can be a slow process probably being under kinetic control of Fe transformation. Fe(III)-oxides persist in the anoxic zone of the Kalix estuary sediments and play role of a carrier for As down to segment depths of 10–15 cm. The following release of As into the pore water is controlled by reduction/dissolution of the Fe(III)-oxides resulting in internal cycling of As in the uppermost 10–15 cm, i.e. its upward diffusion in pore water, re-adsorption onto Fe(III)-oxides in the oxidised surface layer and reburial.

Chemical speciations of metals in riverine and estuarine sediments

Chemical speciation analyses of riverine and estuarine sediments (Fig. 2.5) have been performed by Helios Rybicka (1992). In Vistula River sediments, Cd, Cu, Mn and Zn occurred in less stable phases, i.e. exchangeable, carbonate and easily reducible, while Cr, Fe, Ni and Pb were more concentrated in residual and Fe-oxyhydrates phases. In the highly polluted Vistula River sediments the metals are combined with the more mobile forms, i.e. ion exchangeable, carbonates and Mn-Fe-oxyhydroxides; Cd, Cu and Zn are more mobile than the other metals.

In the estuarine system of Vistula River the clay minerals can play an important role in partitioning of trace metals. They accumulate metals, however, under specific conditions (salinity, pH, Eh) act as sources of metals, which at the presence of S^{-2} ions can form appropriate sulphides (Helios Rybicka, 1991). The direct uptake from the seawater solution of Cu and Cd by sulphides forming in the anoxic sediment should also be considered (Davies-Colley et al., 1985).

Generally smectite-illite group minerals besides chlorites and kaolinite are found in the Vistula River and Baltic Sea sediments. Usually these minerals are coated by Fe- and Mn-oxyhydrates and organic material – living and dead (Helios Rybicka, 1983). Solubility and mobility of sediment bound metals can be increased by many factors: i.e., lowering of pH and changing of redox conditions. For many metals a linear relationship has been found to exist between pH values and dissolved metal concentrations, hence the buffer capacity of the aquatic sediment is of prime importance (Förstner et al., 1986, Helios Rybicka, 1993). Under reducing conditions at the presence of sulphur ions some metals (Cd, Cu, Pb, Zn) can precipitate as sulphides (Helios Rybicka, 1991).

Based on the geochemical behaviour of the metals, they could be subdivided into two groups (Helios Rybicka, 1993):

- highly mobile elements – Cd, Zn, Cu;
- metals more strongly combined with the mineral phases of the sediments – Cr, Pb, Ni.

In the highly polluted sediments the metals are combined with the more mobile forms, i.e. ion exchangeable, carbonate and Mn-Fe-oxyhydroxide fractions. The mobility of metals in the aquatic system depends very strongly on the buffering capacity of the sediments, which is affected by the presence of calcite.

Trace element chronologies in Baltic coastal wetlands vs other Northern European areas

Besides needs for monitoring of metallic pollution, there is interest in estimating historical additions and accumulation rates in coastal wetland areas. It is well known that sediments are appropriate tool for this purpose (Bricker, 1993; Zwolsman et al., 1993). The consistent finding of many of these studies is registration of the recent decrease in heavy metal concentrations in sediments (Trefry et al., 1985; Dörr et al., 1991; Macdonald et al., 1991). Since salt marshes in coastal and estuarine systems are excellent areas for sediment record of pollution chronology;

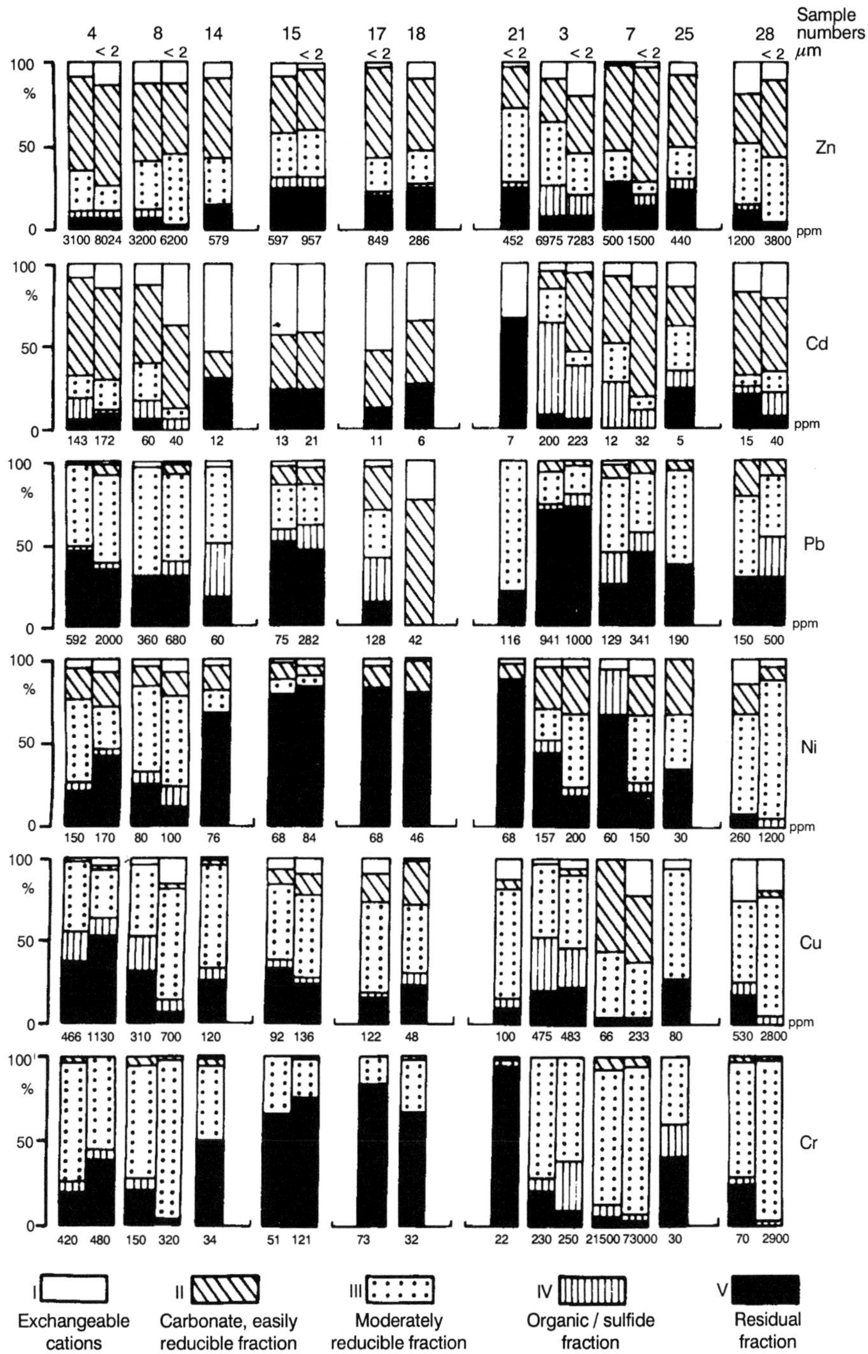

Fig. 2.5. The chemical speciation of heavy metals in the fractions < 63 µm and < 2 µm of selected river sediments, Poland. After Helios Rybicka (1993); modified.

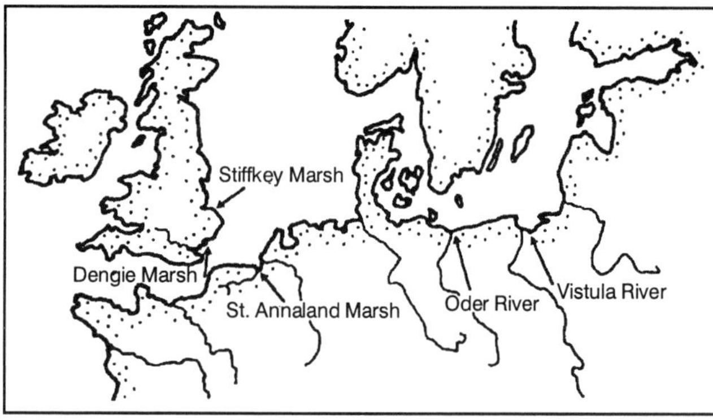

Fig. 2.6. Location of sampling sites along the northern European coastline. After Callaway et al. (1998); modified.

coastal wetlands of northern Europe, including Baltic southern Baltic estuaries (Fig. 2.6) have been studied in this respect. Sediment cores from the high and low marsh of Oder and Vistula Rivers, Poland; Stiffkey and Dengie Marses, UK; and St Annaland Marsh (Eastern Scheldt), The Netherlands, have been analysed for evaluation of heavy metal (Cd, Cu, Cr, Fe, Mn, Ni, Pb and Zn) chronologies (Callaway et al., 1998). Two cores were taken at each sampling site to evaluate the mobility of trace elements within the sediments. It is unlikely that diagenetic processes would result in similar chronologies for the two cores, because the cores were collected from different relative elevations and flooding regimes. Callaway et al. (1998) demonstrated graphical chronologies for metals in sediment cores from the above mentioned geographical areas. As can be seen in Figure 2.7 sediments from southern Baltic estuary, namely from the Oder River (dated back ca. 80 years, i.e. 1910s) showed the highest levels of the elements in any of the cores studied. This marsh is located within the Szczecin Lagoon and affected drastically by industries within and outside the lagoon (Helios Rybicka, 1996b; Szefer et al., 2000). Most of metals reached relatively low values in the oldest (deepest) segments corresponding to the high marsh (1910–1920). It is interesting to note that Pb levels reached a maximum value in sample related to ca. 1960 and have remained relative high and constant over the last three decades (Callaway et al., 1998). Metals such as Cd, Zn and Cr, Cu and Ni showed very similar distribution pattern, i.e.; Cd, Cr and Zn contents exhibited, like in the case of Pb, remarkable increases in 1960s. Furthermore, surface segments were also enriched in Cd and Zn, similarly to profiles of Cu and Ni (Fig 2.7). It suggested that the estuary of the Oder River is recently much polluted with these metals. This finding is confirmed by several authors (Neumann et al., 1996; 1998; Pohl et al., 1998; Siegel et al., 1998; Lampe, 1999; Müller and Wessels, 1999; Szefer et al., 2000; Glasby et al., 2001) who reported elevated levels of Cd, Cu, Pb and Zn in surface sediments from the Szczecin Lagoon.

The samples from the Vistula River were characterised by shortest chronology (dated back to the early 1950s) because of the high sediment accretion rate in

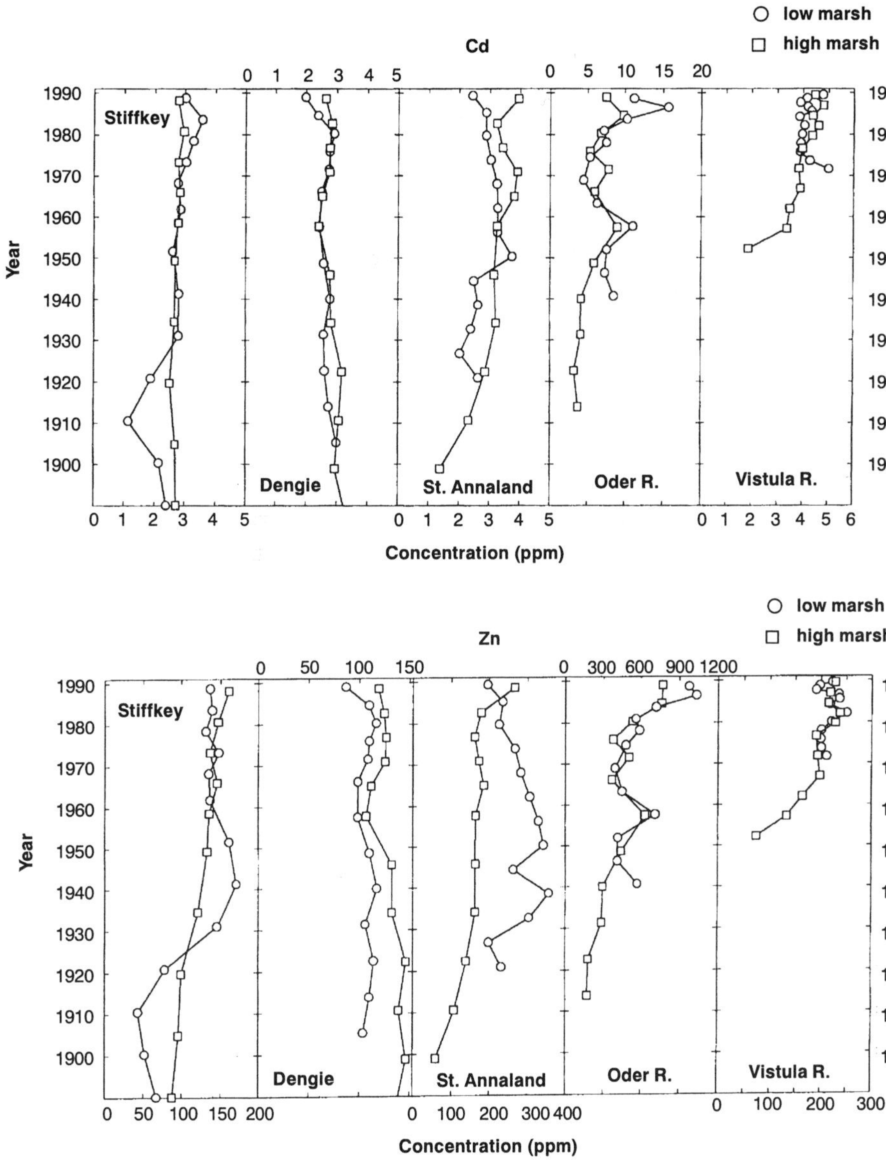

Fig. 2.7. Chronology of sediment Cd, Zn, Cr, Cu, Ni and Pb concentrations for high and low-marsh samples from all five sampling sites; depth profiles of Fe and Mn concentrations from the five sampling sites. Each panel includes data for high- and low-marsh samples. After Callaway et al. (1998); modified.

this area (Callaway et al., 1998). The concentrations of several metals in the samples studied were relatively smaller than those in the samples from the Oder River (Fig. 2.7) and corresponded to background values detected in this latter area. The high marsh cores showed sharp concentration increase for all metals

Fig. 2.7. – continued.

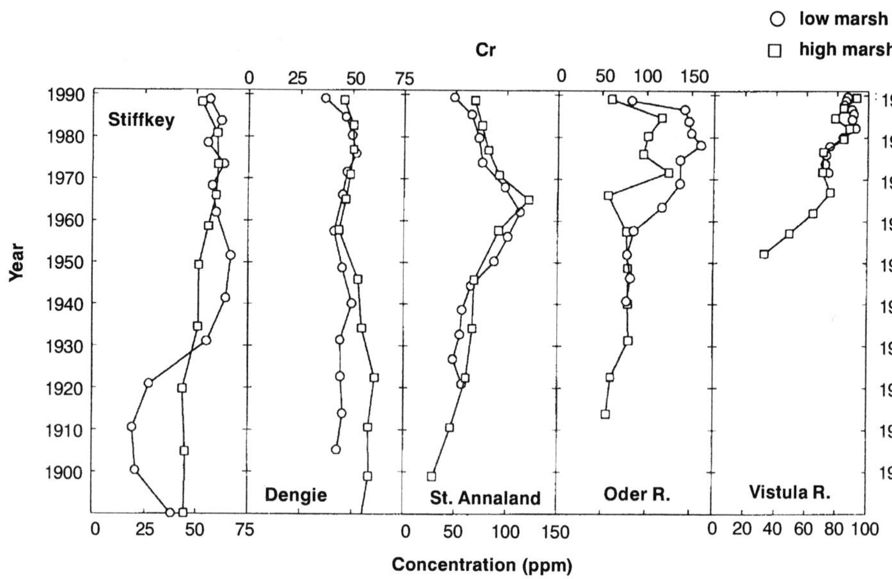

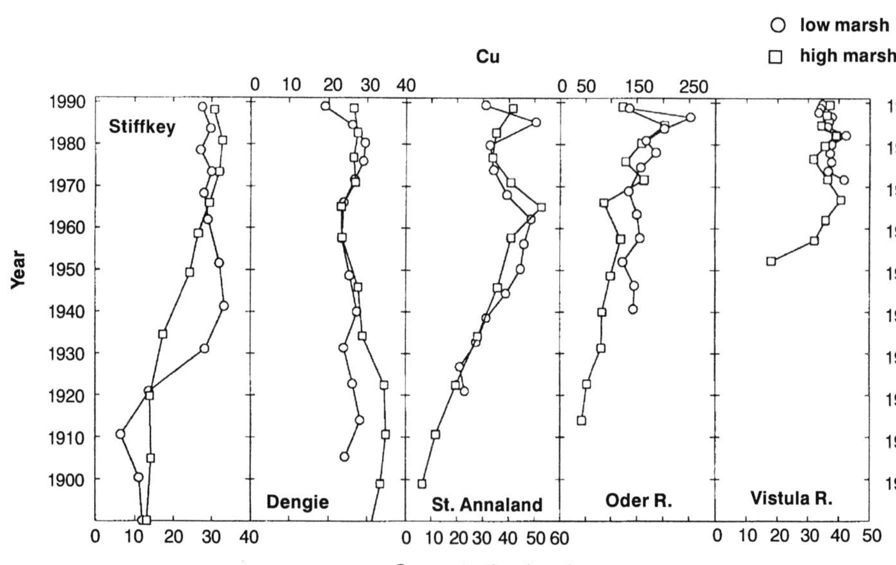

from 1950 to 1970 and after this date the levels have remained relatively constant. Similar (Cr, Cu, Ni and Pb) or lower (Cd, Zn) metal levels of in the Vistula sediments (collected in 1995) than those in the Oder sediments have been reported

Fig. 2.7. – continued.

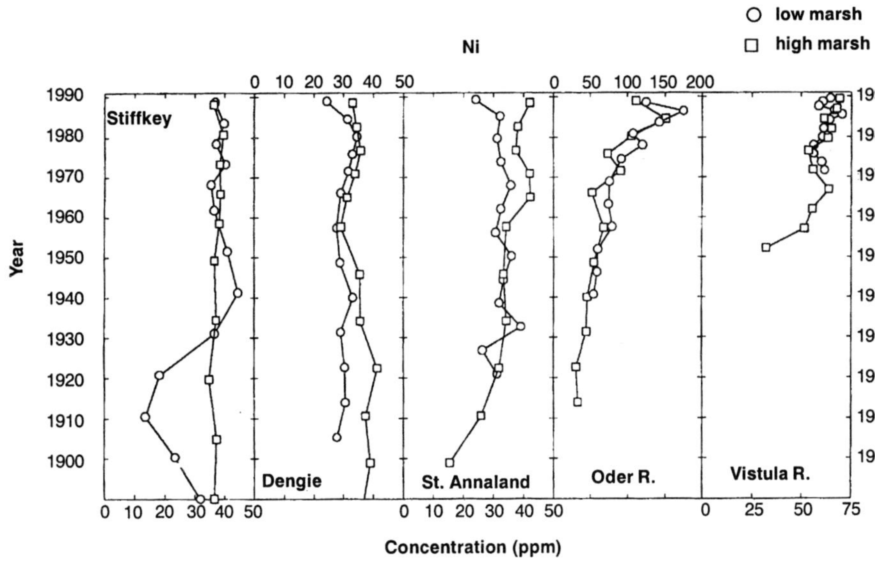

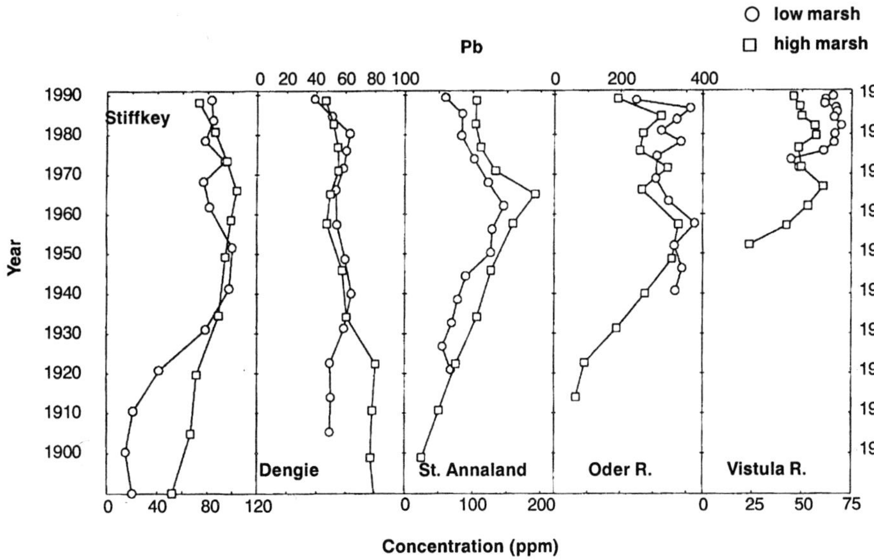

by Szefer et al. (2000) and Glasby and Szefer (1998). The difference in chronologies between these two Polish coastal wetlands may be explained by their different hydrological pattern which is one of the most important factors governing over metal concentrations in salt marshes (Williams et al., 1994; Callaway et al., 1998). Other important factor affecting chemical composition of sediments from the Oder and Vistula estuaries appears to be anthropogenic input of metals there.

Fig. 2.7. – continued.

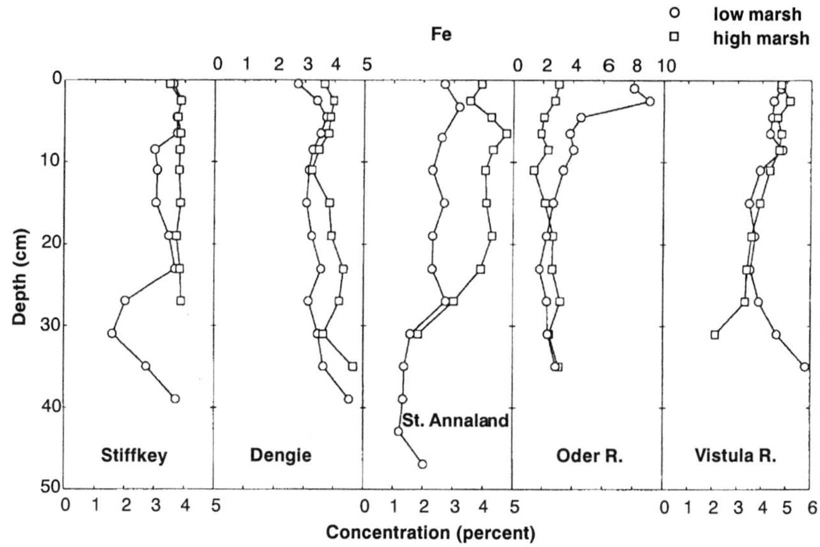

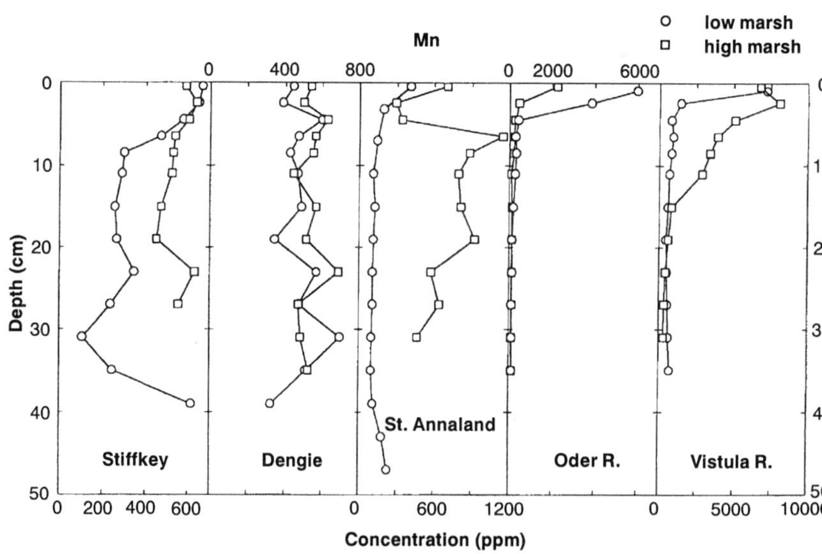

There were significant correlations between Cd, Cr, Cu, Ni, Pb and Zn in all the samples from the five coastal wetlands reflecting the resemblance of the different profiles for the various metals from a particular location (Fig. 2.7). This similar chronological pattern indicates similar sources of metals and their similar geochemical fate being in an agreement with data reported for intertidal and subtidal sediments (Bricker, 1993). It should be emphasized that most metals anthropogenic in origin showed similar profiles in spite of their association with various and specific industries (Valette-Silver, 1993; Callaway et al., 1998). However, these metals showed no significant co-associations with Fe and Mn (except Ni–Fe) suggesting that behavior of Cd, Cr, Cu, Pb and Zn was different from Fe and Mn; the latter elements are strongly dependent on local redox conditions (Callaway et al., 1998). Comparing distribution patterns of metals in Polish coastal areas with those from English and Dutch areas (Fig. 2.7) it is evidently marked that metal concentrations were the greatest in the cores from the Oder estuary (Poland) and have remained constantly high until to 1990. Chronological trends for the Vistula estuary showed slightly enhanced levels of metals in recent sediments. As can be seen in Figure 2.7 concentrations of metals have recently decreased in the vertical profile of sediments from Stiffkey Marsh (UK) and St. Annaland Marsh (the Netherlands), similarly to other sediment cores from western Europe, although the chronological trends for Dengie Marsh (UK) showed no significant metal contents changes from past to 1990.

(iii) Radionuclides in Riverine and Estuarine Systems

The concentrations of radionuclides in river water of the Baltic drainage basin are presented in Table 2.3. Changes in ^{137}Cs activity concentrations in Finnish rivers after the Chernobyl accident have been shown to decline as a result of slow sorption to clay minerals of soil in the basin of the five largest Finnish rivers. Removal of activity from the catchment had no significant effect on the long-term decline in ^{137}Cs activity levels in the rivers (Saxén and Ilus, 2000; Smith et al., 2000). Schimmack et al. (2001) considered possibility of replacement of ^{137}Cs by $^{239+240}Pu$ in agricultural areas contaminated with Chernobyl fallout.

The $^{87}Sr/^{86}Sr$ and $^{143}Nd/^{144}Nd$ ratios in river water reflect the age and geochemistry of the bedrock in the drainage area with typically higher values of $^{87}Sr/^{86}Sr$ and lower values of $^{143}Nd/^{144}Nd$ in rivers draining older rocks as compared to rivers draining younger rocks (Fisher and Stueber, 1976; Goldstein and Jacobsen, 1987). According to Andersson et al. (2001a) the Nd in the Kalix River is mainly transported in a colloidal phase dominated by organic C and Fe. The isotopic composition of Nd exported from a large boreal drainage basin does not directly reflect that of the bedrock in the catchment.

Some studies were performed of isotopic compositions of Sr in river and stream inputs from the Baltic Sea drainage basin (Åberg and Wickman, 1987; Löfvendahl et al., 1990; Andersson et al., 1992, 1994; Åberg, 1995; Land et al., 2000). The data demonstrate that rivers draining the Precambrian shield have small Sr concentration (0.02–0.08 $\mu g\ g^{-1}$) and high $^{87}Sr/^{86}Sr$ ratio (0.718–0,745). In contrast, rivers draining the Phanerozoic basin in the south and southwest have greater Sr concentration (0.2—0.5 μg g^{-1}) and lower $^{87}Sr/^{86}Sr$ ratio (0.710) (Åberg and Wickman, 1987; Löfvendahl et al., 1990; Andersson et al., 1992).

Baturin and Kočenov (1969) and Szefer (1987) reported data on migration of uranium in rivers and its residence time in World Ocean. The transport of uranium and thorium as well as their isotopes ^{234}U, ^{238}U, ^{232}Th and ^{230}Th through the rivers of Baltic drainage area have been investigated by several authors (Szefer, 1977; Bojanowski and Szefer, 1979; Gellermann and Fröhlich, 1984; Andersson et al., 1992, 1995, 1998a, 1998b; Skwarzec, 1995; Gellermann and Stolz, 1997; Porcelli et al., 1997; Andersson et al., 2001b). It has been shown that U in the Kalix River exists dominantly in the < 0.45 μm fraction; i.e. these particles are strongly enriched in U due to scavenging from the surrounding water. Particulate U is characterised by similar $^{234}U/^{238}U$ ratios to those in the < 0.45 μm fractions. This suggests that U on particles maintains isotopic equilibrium with colloidal and dissolved forms of U as the isotopic ratio changes along transport of this radionuclide along the river. Suspended particles in the Kalix River mouth are mainly consisted of Fe- and Mn(OOH) indicating strong correlation between U and Fe, but not with Mn (Porcelli et al., 1997). Both the low sediment flux and low total flux of U in the Kalix are caused by the low rate of chemical and mechanical weathering in this River watershed associated with low relief and local climatic conditions (Porcelli et al., 1997).

The greater concentration of U in waters of the Vistula River as compared to its natural concentration (deducing from U-salinity regression equation: Bojanowski and Szefer, 1979; Duniec et al., 1984; Gellermann et al., 1983; Löfvendahl, 1987, or the steady-state model advanced by Ku et al., 1977) much probably resulted from the substantial utilisation of phosphate fertilisers in the Vistula drainage area (Bojanowski and Szefer, 1979). Such positive deviations in the U concentrations have been also observed in world rivers flowing through regions where phosphate fertilisers are used (Spalding and Sackett, 1972; Sackett et al., 1973; Spalding and Exner, 1976).

The $^{230}Th/^{232}Th$ in the river waters of the Baltic drainage area was ca. twice the equilibrium value for $^{232}Th/^{238}U$ (3.8). The former ratio for the brackish waters was higher by a factor of 10–100 as compared to that for both river and seawater (Andersson et al., 1995). The significant increase in $^{230}Th/^{232}Th$ ratio in the Baltic waters over the riverine input shows that the Th isotopes enter the estuary as a mixture of two carrier phases. Andersson et al. (1995) inferred that ca. 96% of ^{232}Th in river water is carried by detrital particles while dissolved and colloidal phases were characterised by a much higher $^{230}Th/^{232}Th$ value. Almost all of the riverine ^{230}Th is removed in the low-salinity areas of the estuary, whereas the non-detrital phase is settled more slowly resulting in marked increase in $^{230}Th/^{232}Th$ in the brackish waters. The ^{230}Th excess in Baltic rivers is produced in U-rich, ^{232}Th-poor peatlands and trapped in authigenic particles and transported with the particulate form (Andersson et al., 1995).

(iv) Nutrients in Riverine and Estuarine Systems

Large rivers mostly contain a large amounts of groundwater leakage than small rivers. Concentration of nutrients in large rivers generally is reduced by nutrient retention and turnover in lakes, wet meadows and periodically in flooded riparian areas (HELCOM, 1998a). Such self-purification processes cause significant reduction a nutrient transport via rivers to the Baltic Sea. In contrast, these processes are markedly reduced where the rivers are regulated by straightening or surrounded by dikes and where the riparian areas are drained (HELCOM, 1998a). Humborg et al. (2000) presented a long-term assessment of estuarine eutrophication in Szczecin Lagoon, southern Baltic.

Concentrations of nutrients such as N, organic C, P and Si in the Baltic catchment area have been reported by Taylor (1984), Grimvall et al. (1991), Pastuszak (1995a, 1995b), Pastuszak and Nagel (1996), Rahm et al. (1996), Voß and Struck (1997) and HELCOM (1998a). Lagoons and estuaries, possessing a high self-purification potential, are specific links between rivers and the seas. An example of such water system, characterised by intensive suspended matter turnover, is the Oder Estuary (Mohrholz et al., 1998; Grelowski et al., 2000) – one of the most polluted coastal waters of the southern Baltic (Lampe, 1999; Meyer and Lampe, 1999). Nutrient concentration in water of this specific area with emphasis on the flood event has been reported by Pastuszak et al. (1998, 2000).

Comparison of the pollution loads entering the Baltic Sea in 1990 and 1995 shows an increase in N and organic matter by 15 and 10%, respectively, but a decrease in P by 18% (HELCOM, 1993, 1998a). However, because of considerable inter-annual variations in the riverine run-off, these estimates are hardly comparable. More complete and representative data were collected during the 1995 trial.

Pollution discharged by the Polish and Lithuanian rivers also includes loads originating from other countries situated upstream or on the other side of the river, e.g. Poland and Germany along the Oder River.

In the case of organic matter, rivers contribute 88% to the total load (of the Baltic Proper), while municipal and industrial direct sources contribute 8 and 4%, respectively. Similar relations are observed for P. 94% of the total N load is contributed by rivers, about 5% by municipalities and only 1% by industry. A major part of the municipal and industrial wastes undergo some form of treatment. Industrial factories consist mainly of chemical, pulp and paper, petrochemical and food processing plants.

(v) General Remarks and Recommendations

According to HELCOM (1998a), 80–95% of N in the rivers discharging into the Baltic Proper is of the anthropogenic origin and this N is mainly derived from diffuse sources. These calculations, as well as those for unit discharges, reveal serious limitations in the implementation of the 1988 Ministerial Declaration which recommended a 50% overall reduction of pollution input into the Baltic Sea. In fact, there was considerable decrease in organic matter and P loads in Poland in 1995 as compared with the 1988–1989 period (Rybiński et al., 1992) but the decrease in the riverine N load was negligible. Nevertheless, the environmental investments in Poland, which increased from 0.6% of the gross domestic product in 1989 to 1.3% in 1993, did result in a substantial reduction of N and P from point sources. Input of the airborne N to the Baltic Sea has increased gradually during the Twentieth century, reaching its highest level in the 1980s. The atmospheric load of P is assumed to be negligible. The mean deposition of total N in the reduced and oxidised forms in the entire Baltic was estimated as 324 000 tonnes annually in the 1986–1990 (HELCOM, 1991). The relevant value for the Baltic Proper was 100 000 t a^{-1}, thereby contributing almost ¼ of the total annual load of the riverine, direct point sources and airborne N. It follows from the combined emission/input models that only 65% of the airborne N originated from countries bordering the Baltic Sea. Recent estimates show that there was a decrease in the N deposition into the Baltic Proper by about 20–30% and into its drainage area by about 10–25% from the mid-1980s to 1995 (HELCOM, 1997). The main reason was a decline of approximately 20% in the emission of N oxides and ammonia from the five largest contributors, *viz.* Denmark, Germany, Poland, Sweden and United Kingdom. Based on measurements carried out in the Polish coastal zone, the flux of the airborne assimilable N compounds decreased by 35% for the oxidised forms and by 50% for ammonium during 1987–1994 (IMGW, 1997–1998). In 1993–1994, three successive North Sea inflows renewed the deep waters in the Baltic Proper. These salt water inflows have suppressed the mixing across the halocline and limited the downward transport of oxygen. In 1994–1996, the deep basins of the Baltic Proper become anoxic again, first of all in the southern regions. These frequently alternating oxygen conditions are attributed to the growing pool of organic matter, which has accumulated at the bottom as a result of the intensive phytoplankton blooms and the input of suspended matter from rivers. The occurrence of large areas of the seafloor which are anoxic has a great impact on the cycling of metals and nutrients and their budget in the Baltic Proper. Below the redoxcline, nitrate is rapidly removed by denitrification, whereas phosphate is released from the bottom sediments into the water column.

C. SEAWATER

(i) Introduction

Overview of Worldwide Literature

It has been suggested that the short and long-range transport of particulate matter plays an important role in the cycle of selected trace elements in the

North Sea (Kersten et al., 1991b). Particle transport mostly depends on the water circulation and mixing, although suspended particles can exhibit a non-conservative behaviour resulting in an increase in concentration from north to south (Eisma and Kalf, 1987; Sündermann, 1994). Enhanced wind and wave conditions may occasionally be responsible for occurrence of short-term and more localised elevated levels caused by resuspension of bottom sediments. An extensive studies of dissolved and particulate trace elements in the North Sea including estuarine and coastal waters of England were performed by several authors (e.g. Nolting and Eisma, 1988; Tappin et al., 1995; McManus and Prandle, 1996; Millward et al., 1996, 1999; Muller, 1996; Jambers et al., 1999; Nolting et al., 1999). Variations in salinity are mostly accounted for the variance of the levels of the less particle-reactive elements such as Ni, Cu, Zn and Cd. For the typically particle-reactive elements, i.e. Co, Fe, Mn and Pb, transport in the suspended particulate phase is a significant, in some cases predominant factor (Tappin et al., 1995). It was shown that estuarine, porewater and atmospheric inputs of As species were relatively small and that almost all of the methylated As compounds in the North Sea are formed as a result of decaying algal tissue (Millward et al., 1996).

The worldwide reports indicate relatively great variations in metal concentrations and REE in seawater and particulate matter (Chester et al., 1978; Buat Menard and Chesselet, 1979; Blomqvist and Larsson, 1994; Tappin et al., 1995; Kuss and Kremling, 1999; Kuss et al., 2001). In the Northern Hemisphere, a great attention has been paid to the potential effects on both the North Sea and Baltic Sea pollutants and contaminants introduced from the industrialised countries along their coastline (Tappin et al., 1995; HELCOM, 1993, 1996, 1997, 1998a, 1998b). Among these chemical substances are several trace elements and radionuclides which can exert harmful influence on flora and fauna even at low levels. According to Tappin et al. (1995) the factors governing over the distribution of trace metals in the North Sea environment are boundary inputs, particle-water exchanges and advection and mixing within the basin. Physical mixing of fluvial and marine particulates leads to a continuous decrease in the trace element concentrations of the particulate matter with increasing salinity (Zwolsman and van Eck, 1999). A strong correlation between organic- and trace element-rich particles may indicate that organic matter effectively forms complexes with heavy metals (Jambers et al., 1999).

Main sources of minor and major elements in waters the Baltic Sea are direct atmospheric fallout and the riverine material with other point sources or diffuse leakage along the coasts. Additional indirect source is a transport of water-borne elements into the North Sea, via the Danish Straits – Skagerrak and Kattegat and further to the Baltic Sea (HELCOM, 1993).

The continuos monitoring of radionuclide concentrations in seawater is very important task; knowledge of their chemical properties and decay gives useful information on the movement of water masses, interaction between the sea and the atmosphere as well as the chronology of different processes taking place in the marine environments. Furthermore, knowledge of the actual concentrations and isotopic composition of radionuclides in seawater (Leonard et al., 1997; Assinder, 1999; Egorov et al., 1999; Kanivets et al., 1999; Leonard et al., 1999; Swarzenski et al., 1999a, 1999b) provides the base for radioecological recognition of the marine ecosystems and the undertaking of preventive measures against potential hazards resulting from the use of nuclear techniques in man's activity (Szefer, 1981a, 1981b). The first measurements of Ra, Th and U in the ocean have been reported by Koczy (1950, 1956), Koczy et al. (1956, 1957), Sackett et al. (1958) and Moore (1969). Extensive radiochemical studies of seawater have been performed since 1969/70's (Bhat et al., 1969; Kaufman, 1969; Miyake et al., 1970, 1973, 1977; Krishnaswami et al., 1972; Broecker et al., 1973; Imai and Sakanoue, 1973; Kaufman et al., 1973; Amin et al., 1974; Matsumoto, 1975; Santschi et al., 1979; Li et al., 1980; Moore et al., 1980; Moore, 1981; Anderson, 1982; Bacon and Anderson, 1982 and others). Among numerous papers on radionuclides in sea water are also those reported recently the occurrence and behaviour of ^{237}Np and ^{99}Tc in UK coastal waters, and ^{137}Cs and ^{90}Sr in the Black Sea after the Chernobyl NPP accident (Leonard et al., 1997; Assinder, 1999; Egorov et al, 1999; Kanivets et al., 1999). Szefer (1981a, 1981b) overviewed worldwide data on the distribution and behaviour of U and Th in seawater.

Estuaries, shallow coastal and continental shelf waters account for third to half of the global oceanic primary production (Mackenzie et al., 1991; Mantoura et al., 1991; Wollast, 1991). Accelerating primary production or eutrophication appears to be attributed to increasing anthropogenic nutrient loading caused by urban, industrial, agricultural and shrimp farming growth in coastal waters (Ryther and Dunstan, 1971; D'Elia, 1987; Brockmann et al., 1988; Smetacek et al., 1991; Paerl, 1995; Guerrero-Galván et al., 1999; Páez-Osuna et al., 1999; Alonso-Rodríguez et al., 2000). Estuarine and coastal ecosystems have recently received a great attention which exhibit the most obvious symptoms of eutrophication, i.e. unprecedented harmful algal blooms (Grönlund and Leppänen, 1990; Conley et al., 1993; Conley and Johnstone, 1995; Paerl, 1995, 1997; Grönlund et al., 1996; Rahm et al., 1996; Paerl and Whitall, 1999). According to Hydes et al. (1999) high productivity of the North Sea is maintained by both the total amount of nitrate that is supplied to the ecosystem and recycling in its shallow waters. Studies of total carbohydrates and total organic C are useful in a qualitative or semi-quantitative indication of the amounts of sewage-derived organic matter as well as organic effluents from agriculture and sugar-cane industry (Páez-Osuna et al., 1999).

(ii) Chemical Elements in Seawater

There are relatively numerous available data for trace element concentration in Baltic Sea waters. However a significant pool of the concentration data is obtained by unsatisfactory analytical methods or is limited to very narrow coastal areas. Therefore only selected number of measures can be accepted as representative for the "open" Baltic Sea (Brügmann, 1981). Extensive investigations of both the dissolved and/or particulate species of trace elements including their balance have been performed since 1970's (Bojanowski and Samuła-Koszałka, 1974; Brügmann, 1979, 1981, 1984, 1986a, 1986b, 1988; Olausson et al., 1977; Kremling and Petersen, 1978, 1984; Magnusson and Westerlund, 1980; Schmidt, 1980, 1992; Kremling et al., 1981, 1986, 1997; Magnusson and Rasmussen, 1982; Boström et al., 1983; Kremling, 1983; Wrembel, 1983; Andreae and Froelich, 1984; Dyrssen, 1985; Prange and Kremling, 1985; Stoeppler et al., 1986; Cheng, 1987; Osterroht et al., 1988; Björklund, 1989; Skwarzec et al., 1988; Kremling and Pohl, 1989; Dyrssen and Kremling, 1990; Brügmann, et al., 1991/1992, 1997, 1998; Schultz Tokos et al., 1993; Oehlmann et al., 1994; Kravtsov and Emelyanov, 1995; Schneider, 1995, 1996; Schneider and Pohl, 1996; Bauer et al., 1997; Brügmann, and Matschullat, 1997; Lepland and Stevens, 1998; Pohl and Hennings, 1999; Kremling and Streu, 2000; Hou et al., 2001; Truesdale et al., 2001; Sokolowski et al., 2001). The concentrations of trace elements in the German Bight, i.e. adjacent area to the coastal waters of the Baltic Sea, have been also reported by several authors (e.g. Mart and Nürnberg, 1986; Puls et al., 1997).

The abundance, geochemical and mineralogical composition as well as distribution pattern of particulate matter in the Baltic Sea depend on the biogeography, hydrography and topography of brackish waters. It should be emphasised that this material, is particularly specific in its nature and that the different methods and materials used for the sampling, pre-treatment and final analysis. Therefore, results reported by several authors can not be always compared straightforwardly (Bernard et al., 1989).

Investigations of particulate trace element concentrations have been carried out mostly together with analysis of dissolved their species. Most available information concerns the geochemistry of the bulk of particulate matter (Emelyanov, 1974, 1976; Emelyanov and Pustelnikov, 1975a, 1975b, 1977; Weigel, 1976, 1977; Pustelnikov, 1977; Boström et al., 1981, 1988; Gustavsson, 1981; Emelyanov and Pustelnikov, 1982; Brzezińska et al., 1984; Gordeev et al., 1984; Brügmann, 1986a, 1986b, 1988; Ingri et al., 1991; Brügmann et al., 1992; Kravtsov and Emelyanov, 1995; Lithner et al., 1996; Kremling et al., 1997; Leivouri and Vallius, 1998; Pohl et al., 1998; Siegel et al., 1998; Pohl and Hennings, 1999; Laima et al., 2001). Studies on the composition of individual particle using automated electron microprobe have been performed by Bernard et al. (1989). The concentration of Cd in coastal water from the German Bight have been reported by Sperling (1982) and Mart and Nürnberg (1986). Pohl and Hennings (1999) studied the effect of redox processes on the partitioning of Cd, Pb, Cu and Mn between dissolved and particulate phases in the Baltic Sea.

Trace metal speciation studies of the Baltic Sea have been performed by Andreae and Froelich (1984), Brügmann (1984), Bordin et al. (1988), Brügmann et al. (1997, 1998). Chemical speciation analysis of I in the Baltic Sea has been reported recently by Hou et al. (2001) and Truesdale et al. (2001). Abdullah et al. (1995) studied As and Se species in the oxic and anoxic waters of the Oslofjord, Norway.

Horizontal trends

As can be seen in Table 2.7 trace elements in Baltic waters show great variations of their levels depending upon metal, its chemical species and sampling site. For instance, spatial trends are observed for Cd and Pb which reach higher levels in waters of the Gulf of Finland than in other subareas, especially the Baltic Proper. However in the case of other trace elements, spatial differences of their levels are generally not statistically significant and they may be related to relatively large sample-to-sample variations.

Although coastal and estuarine areas, generally much affected by anthropogenic input of trace elements, seem to be most polluted however their seawater levels detected not always must reflect the expected elevated values. It has been reported (Brügmann, 1979, 1981) that relatively great concentration of dissolved organic matter in the Baltic Sea reduces the fraction of 'free' or 'labile bound' species of heavy metals, potentially harmful to organisms, to ca. 20% (Cu, Hg) or 50% (Pb, Zn) of their total concentration. In coastal areas adjacent to potential trace element sources, the concentrations of suspended and colloid-dispersed matter or dissolved organic compounds reach also maximum values. The major fraction of any potentially dangerous inorganic species of trace elements would be converted there into their relatively harmless organically complexed forms; next they can coagulate with colloidal particles or may be adsorbed by other dispersed matter and finally deposited.

According to Oehlmann et al. (1994) and Bauer et al. (1997), the concentrations of tributyltin (TBT) in German coastal waters ranged highly from 10 to 300 ng dm^{-3}. Similar range of values (< 50–430 ng Sn dm^{-3}) has been detected in waters of Swedish west coast during 1986–88. Waters of Belt Sea and Kattegat contained in 1987 lower levels of organotin amounting to < 40 and < 40–80 ng dm^{-3}, respectively (Cheng, 1987; Björklund, 1989). Such great variability may be explained either by natural factors such as inter-tidal differences in TBT concentrations in water or by anthropogenic influences as seasonally inputs of this compound.

The extensive flooding of the Oder catchment in 1997 resulted in transport of large quantities of industrial and municipal wastes to the Szczecin Lagoon and the Pomeranian Bay (Fenske et al., 1998; Müller, 1998; Glasby et al., 2001; Szefer et al., 2000). Potential consequences of this event have inspired several authors with undertaking detailed studies in this respect. The concentration data obtained for this area before and after flood were compared and discussed (Siegel et al., 1998). Concentrations of dissolved and particle-bound trace metals at different stations during the Oder flood are presented in Fig. 2.8. It is pointed out that just

TABLE 2.7.
Concentrations of chemical elements ($\mu g\ dm^{-3}$) in water of the Baltic Sea and other northern areas

Region	Sampling date	Sample depth (m)	Fraction (μm)	N	Salinity (PSU)	Al	As	Ba	Ca	Cd	Co	References
Baltic Sea (whole)	1972–75			644						0.21		Brügmann, 1979
										0.06–1.99		
Baltic Proper												
Arkona	1978	10	< 0.45	1						0.045		Magnusson and Westerlund, 1980
		20		1						0.044		
		30		1						0.043		
Bornholm	1978	10	< 0.45	5						0.05		Magnusson and Westerlund, 1980
										0.039–0.063		
		25–60		13						0.056		
										0.033–0.086		
		80		3						0.067		
										0.059–0.072		
	1981	10		1	7.91		8.62^					Andreae and Froelich, 1984
		15–85		8	8.66		9.63^					
					7.91–13.6		8.11–12.6					
		95		1	15.08		15.1^					
S. Gotland	1978	10	< 0.45	3						0.042		Magnusson and Westerlund, 1980
										0.040–0.045		
		30–100		9						0.043		
										0.034–0.079		
		120		1						0.037		
E. Gotland	1978	10	< 0.45	3						0.04		Magnusson and Westerlund, 1980
										0.037–0.042		
		30–160		14						0.04		
										0.030–0.062		
		200		1						0.027		
	1981	10		2	7.71		7.84^					Andreae and Froelich, 1984
					7.68–7.74		7.78–7.89					
		20–230		42	10.53		12.8^					
					7.78–12.46		8.05–17.9					
		235		1			13.6^					
	1991	5	< 0.10	1		ND**			107**			Andersson et al., 1994
		125		1		ND**			133**			
		225		1		ND**			152**			
	1992	10	< 0.4	1						0.152*		Pohl and Hennings, 1999
		50–225		5						0.15*		
										0.045-0.224		
		240		1						0.066*		
	1993	10	< 0.4	1						0.118*		
		50–225		4						0.09*		
										0.037–0.143		

Region	Sampling date	Sample depth (m)	Fraction (μm)	N	Salinity (PSU)	Al	As	Ba	Ca	Cd	Co	References
		240		1						0.058*		
	1994	10	< 0.4	1						0.092*		
		50–225		4						0.1*		
										0.079–0.11		
		240		1						0.132*		
	1995	10	< 0.4	1						0.125*		
		50–225		7						0.135*		
										0.117–0.153		
		240		1						0.167*		
	1996	10	< 0.4	1						0.11*		
		50–225		7						0.104*		
										0.090–0.117		
		230–240		1						0.109*		
	1995	30	< 10 kD ultrafiltrate[a]	1					100**			Porcelli et al., 1997
			> 10 kD colloid conc.[b]	1					0**			
			10 kD filter rinse[c]	1					0.9**			
			< 3 kD ultrafiltrate[a]	1					98**			
			> 3 kD colloid conc.[b]	1					0.1**			
			3 kD filter rinse[c]	1					1.7**			
	1995	30	< 0.45		7.231				2.52^			Andersson et al., 1998a
		175			11.87				3.57^			
W. Gotland	1978	10	< 0.45	4						0.044		Magnusson and Westerlund, 1980
										0.043–0.046		
		30–300		15						0.039		
										0.025–0.066		
		400		1						0.032		
N. Gotland	1978	10	< 0.45	3						0.043		Magnusson and Westerlund, 1980
										0.036–0.055		
		30–110		9						0.033		
										0.024–0.045		
		160		1						0.029		
	1981	10		1	6.71		8.45^					Andreae and Froelich, 1984
		20–80		6	7.02		9.16^					
					6.71–7.96		7.05–11.5					
		90		1	8.46		12^					
Gotland Deep	1980–84	10	< 0.4	1	7.39					0.018	1.0***	Brügmann, 1988
		50–200		4	10.23					0.02	19.9***	
					7.39–12.36					0.001–0.049	1.0–71.0	
		235		1	12.36					0.001	49.2***	
	1985	200–233	< 0.4	3						0.027*	0.97*	Dyrssen and Kremling, 1990
Central Baltic	1979–81	10–235	< 0.4	34	7.68–12.58					0.051–0.388*	< 0.08–1.78*	Kremling, 1983
Northern Baltic	1985–86	10	< 0.45	5–6 (total)	6.92±0.09					0.039±0.013		Bordin et al., 1988
Proper Baltic				5–6 (free)						0.035±0.013		
		140	< 0.45	5–6 (total)	8.67±1.02					0.024±0.007		

				5–6 (free)						0.017±0.018	
Northern Baltic											
Bothnian Bay	1985–86	10	< 0.45	6 (total)	3.34±0.10					0.039±0.006	Bordin et al., 1988
				6 (free)						0.028±0.003	
	1995	80	< 0.45		3.276				49.7**		Porcelli et al., 1997
Bothnian Sea	1985–86	10	< 0.45	6 (total)	5.59±0.25					0.036±0.008	Bordin et al., 1988
				6 (free)						0.037±0.010	
		140	< 0.45	6 (total)	6.56±0.18					0.023±0.004	
				6 (free)						0.026±0.004	
Gulf of Finland	1981	10		1	6.17		7.90 ^				Andreae and Froelich, 1984
		20–50		4	7.09		7.80 ^				
					6.31–7.80		7.16–8.93				
		65		1	8.22		7.61 ^				
	1985–86	10	< 0.45	5 6 (total)	6.40±0.24					0.036±0.005	Bordin et al., 1988
				5–6 (free)						0.026±0.005	
		70	< 0.45	5–6 (total)	6.86±0.39					0.035±0.003	
				5–6 (free)						0.022±0.007	
	1991	0–1	< 0.45	53						0.05–0.90	Kravtsov and Emelyanov, 1997
Baltic Sea	1988	Micro-- layer	< 0.4	9						0.067±0.027	Brügmann, 1991/92
Western Baltic	1982		< 0.45	5			0.7				
							0.65–0.84				Stoeppler et al., 1986
	1983		< 0.45	17			0.76				
							0.45–1.10				
Coast of Warnemünde	1991–94		< 0.4	153						204±78[d]	Schultz Tokos et al., 1993
Kiel/Mecklenburg Bights			< 0.4	149						131±52[d]	Schneider and Pohl, 1996
Kattegat	1980	10	< 0.45	1						0.062	Magnusson and Westerlund, 1980
		20		1						0.1	
		30		1						0.062	
Kattegat-Bothnian Bay	1988	0.2	< 0.4	11	6.95±3.27					0.020±0.005	Brügmann et al., 1991/92
		6–200		21	12.1±10.6					0.024±0.009	
		80–400		15						0.018±0.008	
Oresund	1980	10	< 0.45	1						0.075	
		20–30		2						0.083	Magnusson and Westerlund, 1980
										0.081–0.084	
		40		1						0.075	
Baltic Proper	1984–85	Surface	< 0.45	13–15		2.3±1.0		22.1±4.5			Ingri et al., 1991
		Bottom		14		2.2±1.0		22.1±4.8			
Aland Sea		Surface		2		2.5±0.2		21.5±0.7			
		Bottom		2		2.2±0.2		25.5±0.7			
Belt Sea		Surface		11–13		3.8±1.5		18.1±5.3			
		Bottom		13		2.8±1.1		16.6±5.2			

Region	Sampling date	Sample depth (m)	Fraction (μm)	N	Salinity (PSU)	Al	As	Ba	Ca	Cd	Co	References
Bothnian Bay-Skagerrak	1984		< 0.4	36						0.024	0.009	Brügmann et al., 1992
										0.001–0.096	<0.001–0.165	
Southern Baltic												
Gulf of Gdansk	1980	Surface	< 0.45	11–20						0.030±0.010		
		Bottom								0.030±0.010		
Gdansk Deep		Surface		18–20						0.030±0.010		
		Bottom								0.030±0.010		
Slupsk Furrow		Surface		17–19						0.020±0.010		
		Bottom								0.030±0.010		
Bornholm Basin		Surface		13–14						0.020±0.010		
		Bottom										
Baltic Sea (whole)						40		10				Boström et al., 1983
German coast												
Helgoland	1977		< 0.45	5						0.028		Mart and Nürnberg, 1986
										0.025–0.034		
	1983		< 0.45	3						0.028	0.015	
										0.018–0.038	0.007–0.026	
German Bight	1983		< 0.45	13						0.028	0.011	

* – nmol kg^{-1}
** – mg dm^{-3}
*** – ng dm^{-3}
^ – nM
[a] – Measured concentrations are ±5%, except for K, which are ±10%, and where noted otherwise.
[b] – The measured concentrations in the colloid concentrates have been corrected for the concentrations of < 3 kD / < 10 kD solutes and normalised to the total sample weight.
[c] – The measured concentrations in the acid rinse have been normalized to the total sample weights. Errors are ca. 7% of the reported concentration.
[d] – pmol dm^{-3}

TABLE 2.7.– continued

Region	Sampling date	Sample depth (m)	Fraction (μm)	N	Salinity (PSU)	Cu	Fe	Ge	Hg	Ir	References
Baltic Sea (whole)	1972–75			644					0.0093		Brügmann, 1979
									0.0005–0.065		
Baltic Proper											
Arkona	1978	10	< 0.45			0.84	0.6				Magnusson and Westerlund, 1980
		20				0.92	0.7				
		30				0.86	0.8				
	1979	3	< 0.4	1	8.64						Prange and Kremling, 1985
		8–13.0		2	9.72						
					9.04–10.4						
		21		1	14.03						
Bornholm	1978	10	< 0.45	5		0.91	0.52				Magnusson and Westerlund, 1980
						0.71–1.18	0.2–0.9				
		25–60		13		0.9	0.45				
						0.74–1.07	0.2–1.1				
		80		3		0.74	1.4				
						0.65–0.93	0.8–2.5				
	1979	5	< 0.4	1	8						Prange and Kremling, 1985
		10.0–70		7	9.46						
					8.00–13.5						
		80		1	14.6						
	1981	10		1	7.91			16.1**			Andreae and Froelich, 1984
		15–85		8	8.66			96.5**			
					7.91–13.6			10.2–260			
		95		1	15.08			373**			
S. Gotland	1978	10	< 0.45	3		0.74	0.27				Magnusson and Westerlund, 1980
						0.68–0.83	0.20–0.30				
		30–100		9		0.77	0.87				
						0.55–1.14	0.30–2.9				
		120		1		0.63	1.6				
E. Gotland	1978	10	< 0.45	3		0.88					Magnusson and Westerlund, 1980
						0.76–0.99					
		30–160		14		0.73					
						0.45–1.04					
		200		1		0.38					
	1981	10		2	7.71			40**			Andreae and Froelich, 1984
					7.68–7.74			33.5–45.6			
		20–230		42	10.53			301**			
					7.78–12.46			34.5–475			
		235		1				422**			
	1991	5	< 0.1	1			ND^ ^				Andersson et al., 1994
		125		1			ND^ ^				

Region	Sampling date	Sample depth (m)	Fraction (μm)	N	Salinity (PSU)	Cu	Fe	Ge	Hg	Ir	References
		225		1			0.1^ ^				
	1992	5	< 0.1	1						10.0±1.0[f]	Anbar et al., 1996
		30–150		2						24.8[f]	
										10.9–38.6	
		225		1						38.9±2.2[f]	
	1992	10	< 0.4	1		9.7*					Pohl and Hennings, 1999
		50–225		5		5.76*					
						0.6–10.6					
		240		1		1.1*					
	1993	10	< 0.4	1		6.7*					
		50–225		4		3.95*					
						1.5–6.9					
		240		1		2.1*					
	1994	10	< 0.4	1		8.4*					
		50–225		4		5.9*					
						3.8–9.0					
		240		1		5.6*					
	1995	10	< 0.4	1		8.4*					
		50–225		7		7.13*					
						6.1–8.7					
		240		1		6.8*					
	1996	10	< 0.4	1		9.5*					
		50–225		7		6.46*					
						5.4–9.5*					
		230–240		1		3.6*					
	1995	30	< 0.45	1	7.232						Porcelli et al., 1997
		175	< 0.45	1	11.87						
	1995	30	< 10 kD ultrafiltrate[a]	1							
			> 10 kD colloid conc.[b]	1							
			10 kD filter rinse[c]	1			2				
			< 3 kD ultrafiltrate[a]	1							
			> 3 kD colloid conc.[b]	1							
			3 kD filter rinse[c]	1			0.3				
	1995	30	< 0.45		7.231						Andersson et al., 1998a
		175			11.87						
W. Gotland	1978	10	< 0.45	4		1					Magnusson and Westerlund, 1980
						0.88–1.19					
		30–300		15		0.75					
						0.48–1.06					
		400		1		0.35					
N. Gotland	1978	10	< 0.45	3		0.77					Magnusson and Westerlund, 1980
						0.73–0.80					
		30–110		9		0.57					
						0.39–0.86					

		160		1		0.35					
	1981	10		1	6.71			34.8**			Andreae and Froelich, 1984
		20–80		6	7.02			105**			
					6.71–7.96			37.7–249			
		90		1	8.46			258**			
	1990–91	75	< 0.45^	1	7.93					10.9±1.1[f]	Anbar et al., 1996
		50	< 0.45^ ^	1	7.31					14.2±2.0[f]	
Gotland Deep	1980–84	10	< 0.4	1	7.39	0.68	1.2		2.4^		Brügmann, 1988
		50–200		4	10.23	0.36	35		2.28^		
					7.39–12.36	0.05–0.72	1.0–131		2.1–2.4		
		235		1	12.36	0.2	114		2.4^		
	1985	200–233	< 0.4	3		0.6*	1900*				Dyrssen and Kremling, 1990
Central Baltic	1979–81	10–235	< 0.4	34	7.68–12.58	0.1–14[d]	< 0.05–2.34[e]				Kremling, 1983
Northern Baltic Proper	1985–86	10	< 0.45	5–6 (total)	6.92±0.09^	0.478±0.057					Bordin et al., 1988
				5–6 (free)		0.361±0.078					
		140	< 0.45	5–6 (total)	8.67±1.02^	0.379±0.054					
				5–6 (free)		0.189±0.064					
Northern Baltic											
Bothnian Bay Proper	1985–86	10	< 0.45	6 (total)	3.34±0.10^	0.501±0.010					Bordin et al., 1988
				6 (free)		0.282±0.041					
	1995	80	< 0.45		3.276		7±3				Porcelli et al., 1997 Andersson et al., 1998a
Bothnian Sea	1985–86	10	< 0.45	6 (total)	5.59±0.25	0.521±0.057					Bordin et al., 1988
				6 (free)		0.351±0.042					
		140	< 0.45	6 (total)	6.56±0.18	0.537±0.032					
				6 (free)		0.330±0.035					
Gulf of Finland	1981	10		1	6.17			23.3**			Andreae and Froelich, 1984
		20–50		4	7.09			62.7**			
					6.31–7.80			34.2–90.1			
		65		1	8.22			121**			
	1985–86	10	< 0.45	5 6 (total)	6.40±0.24	0.535±0.052					Bordin et al., 1988
				5–6 (free)		0.333±0.039					
		70	< 0.45	5–6 (total)	6.86±0.39	0.458±0.057					
				5–6 (free)		0.216±0.037					
	1991	0–1	< 0.45	53		0.8–5.8					Kravtsov and Emelyanov, 1997
Baltic Sea	1988	Micro-layer	< 0.4	9		4.00±1.38	4.28±2.23				Brügmann et al., 1991/92
Kattegat	1980	10	< 0.45	1		0.6	0.4				Magnusson and Westerlund, 1980
		20		1		0.52	0.7				
		30		1		0.38	0.8				
Kattegat-Bothnian Bay		0.2		11	6.95±3.27	0.69±0.16	2.00±1.26		0.020±0.015		Brügmann et al., 1991/92
		6–200		21	12.1±10.6	0.59±0.20	1.62±1.61		0.018±0.016		
		80–400		15		0.49±0.21	6.33±9.79		0.018±0.013		

Region	Sampling date	Sample depth (m)	Fraction (μm)	N	Salinity (PSU)	Cu	Fe	Ge	Hg	Ir	References
Oresund	1980	10	< 0.45	1		0.88	0.6				Magnusson and Westerlund, 1980
		20–30		2		0.97	0.55				
						0.86–1.08	0.4–0.7				
				1		0.88	0.7				
		40									
Baltic Sea											
Baltic Proper	1984–85	Surface	< 0.45	13–15			1.4±1.2				Ingri et al., 1991
		Bottom		14			1.5±1.4				
Aland Sea		Surface		2			1.4±0.4				
		Bottom		2			6.5±6.4				
Belt Sea		Surface		11–13			1.0±0.7				
		Bottom		13			1.0±0.8				
Baltic Sea											
Bothnian Bay-	1984		< 0.4	36		0.61	10				Brügmann et al., 1992
Skagerrak						< 0.01–2.18	0.3–131				
Southern Baltic											
Gulf of Gdansk	1980	Surface	< 0.45	11–20		0.36±0.03					
		Bottom				0.42±0.04					
Gdansk Deep		Surface		18–20		0.34±0.02					
		Bottom				0.33±0.05					
Slupsk Furrow		Surface		17–19		0.63±0.15					
		Bottom				0.61±0.12					
Bornholm Basin		Surface		13–14		0.39±0.14					
		Bottom				0.42±0.09					
Baltic Sea (whole)							60				Boström et al., 1983
Kattegat-Transition Region	1995	15	< 0.45	1	19.92					25.5±1.5[f]	Anbar et al., 1996
German coast											
Heligoland	1977		< 0.45	5		0.772					Mart and Nürnberg, 1986
						0.48–1.48					
	1983		< 0.45	3		0.224					
						0.12–0.327					
German Bight	1983		< 0.45	13		0.235					
						0.077–0.444					

* – nmol kg^{-1}
** – pM
^ – ng dm^{-3}
^ ^ – mg dm^{-3}
[a] – Measured concentrations are ±5%, except for K, which are ±10%, and where noted otherwise.
[b] – The measured concentrations in the colloid concentrates have been corrected for the concentrations of < 3 kD / < 10 kD solutes and normalised to the total sample weight.
[c] – The measured concentrations in the acid rinse have been normalized to the total sample weights. Errors are ca. 7% of the reported concentration.
[d] – nmol kg^{-1}
[e] – μmol kg^{-1}
[f] – 10^8 atoms kg^{-1}

TABLE 2.7. – continued

Region	Sampling date	Sample depth (m)	Fraction (μm)	N	Salinity (PSU)	K	Mg	Mn	Mo (nmol kg^{-1})	N (mM)*	References
Baltic Proper											
Arkona	1979	3	< 0.4	1	8.64				28.8		Prange and Kremling, 1985
		8–13.0		2	9.72				30.9		
					9.04–10.4				26.6–35.2		
		21		1	14.03				66.7		
Bornholm	1979	5	< 0.4	1	8				23.2		Prange and Kremling, 1985
		10.0–70		7	9.46				30.1		
					8.00–13.5				25.1–40.2		
		80		1	14.6				42.1		
	1981	10		1	7.91					< 0.05	Andreae and Froelich, 1984
		15–85		8	8.66					< 0.05	
					7.91–13.6						
		95		1	15.08					< 0.05	
E. Gotland	1979	5	< 0.4	2	7.1				19.2		Prange and Kremling, 1985
					6.84–7.36				17.1–21.3		
		10–218		19	10.6				30		
					6.88–12.7				18.9–41.7		
		218		1	12.8				45.5		
	1981	10		2	7.71					0.09	Andreae and Froelich, 1984
					7.68–7.74					0.05–0.12	
		20–230		42	10.53					1.88	
					7.78–12.46					< 0.05–8.17	
	1991	5	< 0.1	1		84***	273***	< 0.005***			Andersson et al., 1994
		125		1		112***	370***	0.15***			
		225		1		132***	427***	0.53***			
	1995	10	< 0.4	1				10**			Pohl and Hennings, 1999
		50–225		7				35.7**			
								9.0–102			
		240		1				107**			
	1996	10	< 0.4	1				12**			
		50–225		7				122**			

Region	Sampling date	Sample depth (m)	Fraction (μm)	N	Salinity (PSU)	K	Mg	Mn	Mo (nmol kg⁻¹)	N (mM)*	References
								15–481			
		230		1				20929**			
	1997	10	< 0.4	1				15**			
		50–200		7				2780**			
								7.4–9337			
		230		1				6987**			
	1995	30	< 0.45	1	7.232	84.5***	273***	2			Porcelli et al., 1997
		175	< 0.45	1	11.87	137***	436***	2			
	1995	30	< 10 kD ultrafiltrate[a]	1		80***	273***	< 1			
			> 10 kD colloid conc.[b]	1		0***	0***	< 1			
			10 kD filter rinse[c]	1		0.7***	2.3***	< 1			
			< 3 kD ultrafiltrate[a]	1		79***	272***	< 1			
			> 3 kD colloid conc.[b]	1		0***	0.2***	< 1			
			3 kD filter rinse[c]	1		1.4***	4.4***	< 1			
	1995	30	< 0.45		7.231	2.16^ ^	11.2^ ^	0.04^ ^			Andersson et al., 1998a
		175			11.87	3.50^ ^	17.9^ ^	0.04^ ^			
N. Gotland	1979	5	< 0.45	1	6.75				18.5		Prange and Kremling, 1985
		10–130		8	8.28				21.9		
					6.75–9.96				17.5–29.6		
		140		1	10.57				38.3		
	1981	10		1	6.71					< 0.05	Andreae and Froelich, 1984
		20–80		6	7.02					2.18	
					6.71–7.96					< 0.05–4.72	
		90		1	8.46						
Gotland Deep	1980–84	10	< 0.4	1	7.39					2	Brügmann, 1988
		50–200		4	10.23					3.36	
					7.39–12.36					0.04–7.66	
		235		1	12.36					0.07	
	1985	200–233	< 0.4	3				9400**			Dyrssen and Kremling, 1990
Central Baltic	1979–81	10–235	< 0.4	34	7.68–12.58			< 0.1–14.5^ ^			Kremling, 1983
Bothnian Bay	1979	5	< 0.4	1	2.89				9.3		Prange and Kremling, 1985
		10.0–30		3	3.3				9.6		
		50			2.87–3.58				8.9–10.3		
				1	3.86				11.6		
	1995	80	< 0.45		3.276	38.1***	126***	3			Porcelli et al., 1997

						0.97^ ^	5.18^ ^	0.05^ ^			Andersson et al., 1998a
Bothnian Sea	1979	5	< 0.4	1	5.07				14		Prange and Kremling, 1985
		10.0–30		3	6.02				16.2		
		50			5.08–6.82				14.0–17.9		
				1	6.95				19.2		
Gulf of Finland	1979	3	< 0.4	1	5.89				17.5		Prange and Kremling, 1985
		13–48		4	6.62				17.9		
					6.34–7.21				16.2–18.9		
		58		1	7.69				20.2		
	1981	10		1	6.17					< 0.05	Andreae and Froelich, 1984
		20–50		4	7.09					1.18	
					6.31–7.80					< 0.05–4.07	
		65		1	8.22					5.22	
Baltic Proper	1984-85	Surface	< 0.45	13-15				0.8±0.3			Ingri et al., 1991
		Bottom		14				34.9±79.7			
Aland Sea		Surface		2				0.6±0.1			
		Bottom		2				6.4±5.16			
Belt Sea		Surface		11-13				1.6±1.11			
		Bottom		13				10.4±15.41			
Baltic Sea (whole)								10			Boström et al., 1983

* – mM (calculated as NO_3)
** – nmol kg^{-1}
*** – mg dm^{-3}
^ ^ – Concentration of Mn (μmol kg^{-1}) and K, Mg and Na (mmol kg^{-1})
[a] – Measured concentrations are ±5%, except for K, which are ±10%, and where noted otherwise.
[b] – The measured concentrations in the colloid concentrates have been corrected for the concentrations of < 3 kD / < 10 kD solutes and normalised to the total sample weight.
[c] – The measured concentrations in the acid rinse have been normalized to the total sample weights. Errors are ca. 7% of the reported concentration.

TABLE 2.7. – continued

Region	Sampling date	Sample depth (m)	Fraction (μm)	N	Salinity (PSU)	Na	Ni	P (μmol dm^{-3})	Pb	References
Baltic Sea (whole)	1972–75			644					0.64	Brügmann, 1979
									0.4–2.1	
Baltic Proper										
Arkona	1978	10	< 0.45	1			0.77		0.14	Magnusson and Westerlund, 1980
		20		1			0.77		0.16	
		30		1			0.72		0.14	
	1979	3	< 0.4	1	8.64			0.14		Prange and Kremling, 1985
		8–13.0		2	9.72			0.31		
					9.04–10.4			0.15–0.46		
		21		1	14.03			1.37		
Bornholm	1978	10	< 0.45	5			0.74		0.21	Magnusson and Westerlund, 1980
							0.64–0.85		0.13–0.31	
		25–60		13			0.75		0.19	
							0.66–1.08		0.13–0.29	
		80		3			0.71		0.15	
							0.70–0.73		0.12–0.18	
	1979	5	< 0.4	1	8			0.17		Prange and Kremling, 1985
		10.0–70		7	9.46			0.54		
					8.00–13.5			0.19–0.93		
		80		1	14.6			1.8		
	1981	10		1	7.91			0.14		Andreae and Froelich, 1984
		15-85		8	8.66			0.53		
					7.91–13.6			0.14–0.90		
		95		1	15.08			0.78		
S. Gotland	1978	10	< 0.45	2			0.7		0.12	Magnusson and Westerlund, 1980
							0.60–0.79		0.12–0.12	
		30–100		9			0.71		0.14	
							0.56–0.85		0.08–0.25	
		120		1			0.78		0.07	
E. Gotland	1978	10	< 0.45	3			0.77		0.21	Magnusson and Westerlund, 1980
							0.75–0.78		0.12–0.28	
		30–160		14			0.76		0.15	
							0.66–0.92		0.09–0.29	
		200		1			0.54		0.15	
	1979	5	< 0.4	2	7.1			0.08		Prange and Kremling, 1985
					6.84–7.36			0.04–0.11		
		10–218		20	10.6			2.72		
					6.88–12.7			0.11–5.51		
		233		1	12.8			6.2		
	1981	10		2	7.71			0.12		Andreae and Froelich, 1984
					7.68–7.74			0.06–0.17		
		20–230		42	10.53			3.5		

					7.78–12.5			0.06–11.3		
		235		1						
	1991	5	< 0.10	1		2270***				Andersson et al., 1994
		125		1		3070***				
		225		1		3520***				
	1992	10	< 0.4	1					0.224**	Pohl and Hennings, 1999
		50–225		5					0.173**	
									0.098–0.353	
		240		1					0.124**	
	1993	10	< 0.4	1					0.149**	
		50–225		4					0.063**	
									0.022–0.109	
		240		1					0.053**	
	1994	10	< 0.4	1					0.051**	
		50–225		4					0.06**	
									0.045–0.08	
		240		1					0.052**	
	1995	10	< 0.4	1					0.075**	
		50–225		7					0.1**	
									0.055–0.156	
		240		1					0.096**	
	1996	10	< 0.4	1					0.083**	
		50–225		7					0.073**	
									0.031–0.228	
		230–240		1					0.083**	
	1995	30	< 0.45			2230***				Porcelli et al., 1997
		175	< 0.45			3730***				
	1995	30	< 10 kD ultrafiltrate[a]	1		2200***				Porcelli et al., 1997
			> 10 kD colloid conc.[b]	1		0***				
			10 kD filter rinse[c]	1		20***				
			< 3 kD ultrafiltrate[a]	1		2200***				
			> 3 kD colloid conc.[b]	1		0.7***				
			3 kD filter rinse[c]	1		40***				
	1995	30	< 0.45	1	7.23	97**				Andersson et al., 1998a
		175		1	11.87	162**				
W. Gotland	1978	10	< 0.45	4			0.78		0.14	Magnusson and Westerlund, 1980
							0.69–0.88		0.08–0.25	
		30–300		15			0.65		0.17	
							0.49–0.82		0.04–0.31	
		400		1			0.66		0.09	
N. Gotland	1978	10	< 0.45	3			0.8		0.09	Magnusson and Westerlund, 1980
							0.75–0.88		0.08–0.11	
		30–110		9			0.73		0.08	
							0.68–0.88		0.03–0.15	
		160		1			0.69		0.03	
	1979	5	< 0.4	1	6.75			0.08		Prange and Kremling, 1985

Region	Sampling date	Sample depth (m)	Fraction (μm)	N	Salinity (PSU)	Na	Ni	P (μmol dm⁻³)	Pb	References
		10–130		8	8.28			1.04		
					6.75–9.96			0.04–3.19		
		140		1	10.57			8.15		
	1981	10		1	6.71			0.06		Andreae and Froelich, 1984
		20–80		6	7.02			0.77		
					6.71–7.96			< 0.02–1.05		
		90		1	8.46					
Gotland Deep	1980–84	10	< 0.4	1	7.39		0.77*	0.34	0.042	Brügmann, 1988
		50–200		4	10.23		0.58*	1.83	0.021	
					7.39–12.36		0.48–0.64	0.37–2.88	0.002–0.037	
		235		1	12.36		0.49*	1.49	0.027	
Central Baltic	1979–81	10–235	< 0.4	34	7.68–12.6		9.5**			Kremling, 1983
							7.9–17.7			
Northern Baltic Proper	1985–86	10	< 0.45	5–6 (total)	6.92±0.09				0.189±0.024	Bordin et al., 1988
				5–6 (free)					0.166±0.040	
		140	< 0.45	5–6 (total)	8.67±1.02				0.168±0.070	
				5–6 (free)					0.132±0.035	
Northern Baltic										
Bothnian Bay	1979	5	< 0.4	1	2.89			< 0.04		Prange and Kremling, 1985
		10–100		6	3.3			< 0.04		
					2.87–3.58					
		125		1	3.86			< 0.04		
	1985–86	10	< 0.45	6 (total)	3.34±0.10				0.246±0.066	Bordin et al., 1988
				6 (free)					0.224±0.027	
	1995	80	< 0.45		3.276	1040***				Porcelli et al., 1997
						45.2[d]				Andersson et al., 1998a
Bothnian Sea	1979	5	< 0.4	1	5.07			0.04		Prange and Kremling, 1985
		10.0–30		3	6.02			0.1		
		50		1	5.08–6.82			0.04–0.22		
					6.95			0.16		
	1985–86	10	< 0.45	6 (total)	5.59±0.25				0.120±0.029	Bordin et al., 1988
				6 (free)					0.105±0.025	
		140	< 0.45	6 (tatal)	6.56±0.18				0.180±0.032	
				6 (free)					0.208±0.050	
Gulf of Finland	1979	3	< 0.4	1	5.89			0.04		Prange and Kremling, 1985
		13–48		4	6.62			0.23		
					6.34–7.21			0.16–0.33		
		58		1	7.69			0.93		
	1981	10		1	6.17			0.09		Andreae and Froelich, 1984
		20–50		4	7.09			0.4		
					6.31–7.80			0.06–0.84		
		65		1	8.22			1.33		
	1985–86	10	< 0.45	5 6 (total)	6.40±0.24				0.324±0.024	Bordin et al., 1988
				5–6 (free)					0.184±0.037	

		70	< 0.45	5–6 (total)	6.86±0.39		0.497±0.054	
				5–6 (free)			0.189±0.039	
	1991	0–1	< 0.45	53			0.1–3.6	Kravtsov and Emelynov, 1997
Baltic Sea	1988	Micro-layer	< 0.4	9		0.57±0.05	0.075±0.032	Brügmann et al., 1991/92
Kattegat	1980	10	< 0.45	1		0.56	0.15	Magnusson and Westerlund, 1980
		20		1		0.4	0.16	
		30		1		0.43	0.08	
Kattegat-Bothnian		0.2		11	6.95±3.27	0.52±0.06	0.007±0.002	Brügmann et al., 1991/92
Bay		6–200		21	12.1±10.6	0.64±0.31	0.013±0.012	
		80–400		15		0.52±0.06	0.017±0.031	
Oresund	1980	10	< 0.45	1		0.86	0.14	Magnusson and Westerlund, 1980
		20–30		2		0.87	0.11	
						0.85–0.89	0.11–0.11	
		40		1		0.73	0.11	
Bothnian Bay-	1984		< 0.4	36		0.57		Brügmann et al., 1992
Skagerrak						0.09–0.96	0.06	
Southern Baltic							< 0.001–0.39	
Gulf of Gdansk	1980	Surface	< 0.45	11–20			0.100±0.010	
		Bottom					0.160±0.040	
Gdansk Deep		Surface		18–20			0.130±0.020	
		Bottom					0.130±0.020	
Slupsk Furrow		Surface		17–19			0.140±0.050	
		Bottom					0.170±0.020	
Bornholm Basin		Surface		13–14			0.130±0.010	
		Bottom					0.230±0.050	
German coast								
Helgoland	1977		< 0.45	5			0.053	Mart and Nürnberg, 1986
							0.013–0.166	
	1983		< 0.45	3		0.35	0.033	
						0.263–0.417	0.015–0.043	
German Bight	1983		< 0.45	13		0.471	0.011	
						0.126–0.921	0.003–0.041	

* – ng dm^{-3}
** – nmol kg^{-1}
*** – mg dm^{-3}
[a] – Measured concentrations are ±5%, except for K, which are ±10%, and where noted otherwise.
[b] – The measured concentrations in the colloid concentrates have been corrected for the concentrations of < 3 kD / < 10 kD solutes and normalised to the total sample weight.
[c] – The measured concentrations in the acid rinse have been normalized to the total sample weights. Errors are ca. 7% of the reported concentration.
[d] – mmol kg^{-1}

TABLE 2.7. – continued

Region	Sampling date	Sample depth (m)	Fraction (μm)	N	Salinity (PSU)	S (μmol dm⁻³)	Sb	Se	Si (μmol kg⁻¹)	References
	1979	3	< 0.4	1	8.64				5.4	Prange and Kremling, 1985
		8–13.0		2	9.72				10.9	
					9.04–10.4				10.8–11.0	
		21		1	14.03				27.7	
	1979	5	< 0.4	1	8				5.3	Prange and Kremling, 1985
		10.0–70		7	9.46				12	
					8.00–13.5				4.7–24.7	
		80		1	14.6				51.9	
	1981	10		1	7.91		0.87^		5.6^ ^	Andreae and Froelich, 1984
		15–85		8	8.66		0.56^		21.5^ ^	
					7.91–13.6		0.31–0.83		4.5–64.9	
		95		1	15.08		0.59^		83.8^ ^	
	1979	5	< 0.4	2	7.1				6.2	Prange and Kremling, 1985
					6.84–7.36				5.7–6.7	
		10–218		20	10.6				57	
					6.88–12.7				6.5–85.7	
		233		1	12.8				93	
	1981	10		2	7.71		0.64^		4.55^ ^	Andreae and Froelich, 1984
					7.68–7.74		0.60–0.68		3.2–5.9	
		20–230		42	10.53	11	0.38^		56.7^ ^	
					7.78–12.5	< 0.1-52.0	0.22-0.68		8.3-98.7	
		235		1			0.15^			
	1991	5	< 0.10	1					0.39**	Andersson et al., 1994
		125		1					1.6**	
		225		1					3.6**	
	1995	30	< 10 kD ultrafiltrate[a]	1					0.5**	Porcelli et al., 1997
			> 10 kD colloid conc.[b]	1					0.01**	
			10 kD filter rinse[c]	1					0**	
			< 3 kD ultrafiltrate[a]	1					0.4**	
			> 3 kD colloid conc.[b]	1					0**	
			3 kD filter rinse[c]	1					0.3**	
	1995	30	< 0.45	1	7.23				14.7*	Andersson et al., 1998a
		175		1	11.87				42.0*	
	1979	5	< 0.4	1	6.75				6.8	Prange and Kremling, 1985
		10–130		8	8.28				26.8	
					6.75–9.96				6.8–61.4	

		140		1	10.57		93.1	
	1981	10		1	6.71	0.67^	5.9^ ^	Andreae and Froelich, 1984
		20–80		6	7.02	0.56^	18.5^ ^	
					6.71–7.96	0.36–0.78	5.9–47.5	
		90		1	8.46	0.35^		
Northern Baltic								
Bothnian Bay	1979	5	< 0.4	1	2.89		30	Prange and Kremling, 1985
		10–100		6	3.3		31.1	
					2.87–3.58		30.0–33.8	
		125		1	3.86		33.1	
	1995	80	< 0.45		3.276		38.8*	Andersson et al., 1998a
Bothnian Sea	1979	5	< 0.4	1	5.07		12	Prange and Kremling, 1985
		10.0–30		3	6.02		18.8	
		50		1	5.08–6.82		12.3–32.4	
					6.95		33.3	
Gulf of Finland	1979	3	< 0.4	1	5.89		1.5	Prange and Kremling, 1985
		13–48		4	6.62		4.9	
					6.34–7.21		2.0–9.8	
		58		1	7.69		24	
	1981	10		1	6.17	0.53^	5^ ^	Andreae and Froelich, 1984
		20–50		4	7.09	0.51^	11.5^ ^	
					6.31–7.80	0.47–0.55	5.0–18.7	
		65		1	8.22	0.38	26.7^ ^	

* – μmol kg^{-1}
** – mg dm^{-3}
^ – nM
^ ^ – μM
[a] – Measured concentrations are ±5%, except for K, which are ±10%, and where noted otherwise.
[b] – The measured concentrations in the colloid concentrates have been corrected for the concentrations of < 3 kD / < 10 kD solutes and normalised to the total sample weight.
[c] – The measured concentrations in the acid rinse have been normalized to the total sample weights. Errors are ca. 7% of the reported concentration.

TABLE 2.7. – continued

Region	Sampling date	Sample depth (m)	Fraction (μm)	N	Salinity (PSU)	Sr	Ti	V	Zn	References
								(nmol kg^{-1})		
Baltic Proper										
Arkona	1978	10	< 0.45						3.3	Magnusson and Westerlund, 1980
		20							4.2	
		30							3.8	
	1979	3	< 0.4	1	8.64			2.4		Prange and Kremling, 1985
		8–13.0		2	9.72			4.3		
					9.04–10.4			3.6–5.0		
		21		1	14.03			4.2		
Bornholm	1978	10	< 0.45	5					2.98	Magnusson and Westerlund, 1980
									1.5–7.0	
		25–60		13					2.44	
									1.5–3.7	
		80		3					3.23	
	1979	5	< 0.4	1	8			2.9	2.6–4.0	Prange and Kremling, 1985
		10.0–70		7	9.46			4.6		
					8.00–13.5			2.8–6.7		
		80		1	14.6			2.3		
S. Gotland	1978	10	< 0.45	3					1.8	Magnusson and Westerlund, 1980
									1.7–1.9	
		30–100		9					2.86	
									1.8–5.4	
		120		1					1.9	
E. Gotland	1978	5	< 0.45	3					2.73	Magnusson and Westerlund, 1980
									2.5–3.2	
		30–160		14					3.49	
									2.1–5.3	
		200		1					2.4	
	1979	5	< 0.4	2	7.1			2.9		Prange and Kremling, 1985
					6.84–7.36			2.7–3.1		
		10–218		20	10.6			2.03		
					6.88–12.7			ND–3.0		
		218		1	12.8			ND		
	1990	5	< 0.45	1	7.32	1.67*				Andersson et al., 1992
		75–150		2	9.13	2.10*				
					8.14–10.11	1.756–2.442				
		225		1	11.18	2.56*				
	1991	5	< 0.10	1		1.68**				Andersson et al., 1994
		125		1		2.20**				
		225		1		2.59**				
W. Gotland	1978	10	< 0.45	4					3	Magnusson and Westerlund. 1980
									2.6–3.4	
		30–300		15					2.75	

									1.7–5.7	
		400		1					1.6	
N. Gotland	1978	10	< 0.45	3					2.27	Magnusson and Westerlund, 1980
									1.8–2.5	
		30–110		9					1.91	
									1.4–2.8	
		160		1					1.6	
	1979	5	< 0.4	1	6.75			2.8		Prange and Kremling, 1985
		10–130		8	8.28			2.3		
					6.75–9.96			ND–3.4		
		140		1	10.57			ND		
Gotland Deep	1980–84	10	< 0.4	1	7.39				1.4	Brügmann, 1988
		50–200		4	10.23				1.75	
					7.39–12.36				1.2–2.0	
		235		1	12.36				2.7	
Northern Baltic										
Bothnian Bay	1979	5	< 0.4	1	2.89			2.7		Prange and Kremling, 1985
		10–100		6	3.3			1.67		
					2.87–3.58			0.8–2.6		
		125		1	3.86			1.8		
Bothnian Bay	1982	13	< 0.4		3.19–3.87				19.8±5.0°	Kremling and Petersen, 1984
Bothnian Bay – northern part	1990	5	< 0.45	1	2.46	0.566*				Andersson et al., 1992
		25–50		1	3.39	0.771*				
				1	3.32–3.46	0.750–0.792				
		80		1	3.48	0.794*				
	1991	80		1	3.63	0.819*				
central part	1990	5	< 0.45	1	5.12	1.146*				
		175		1	6.36	1.422*				
Bothnian Sea	1979	5	< 0.4	1	5.07			0.8		Prange and Kremling, 1985
		10.0–30		3	6.02			ND–1.3		
		50		1	5.08–6.82			0.4		
					6.95					
Gulf of Finland	1979	3	< 0.4	1	5.89			2.7		Prange and Kremling, 1985
		13–48		4	6.62			1.5		
					6.34–7.21			0.8–2.1		
		58		1	7.69			0.9		
Kattegat	1980	10	< 0.45	1			0.15		2.1	Magnusson and Westerlund, 1980
		20		1			0.16		2.9	
		30		1			0.08		1.5	
Oresund	1980	10	< 0.45	1			0.14		3.9	Magnusson and Westerlund, 1980
		20–30		2			0.11		3.7	
							0.11–0.11		2.9–4.5	
		40		1			0.11		3.7	

* – mg kg^{-1}
** – mg dm^{-3}
° – nmol kg^{-1}
ND – not detected

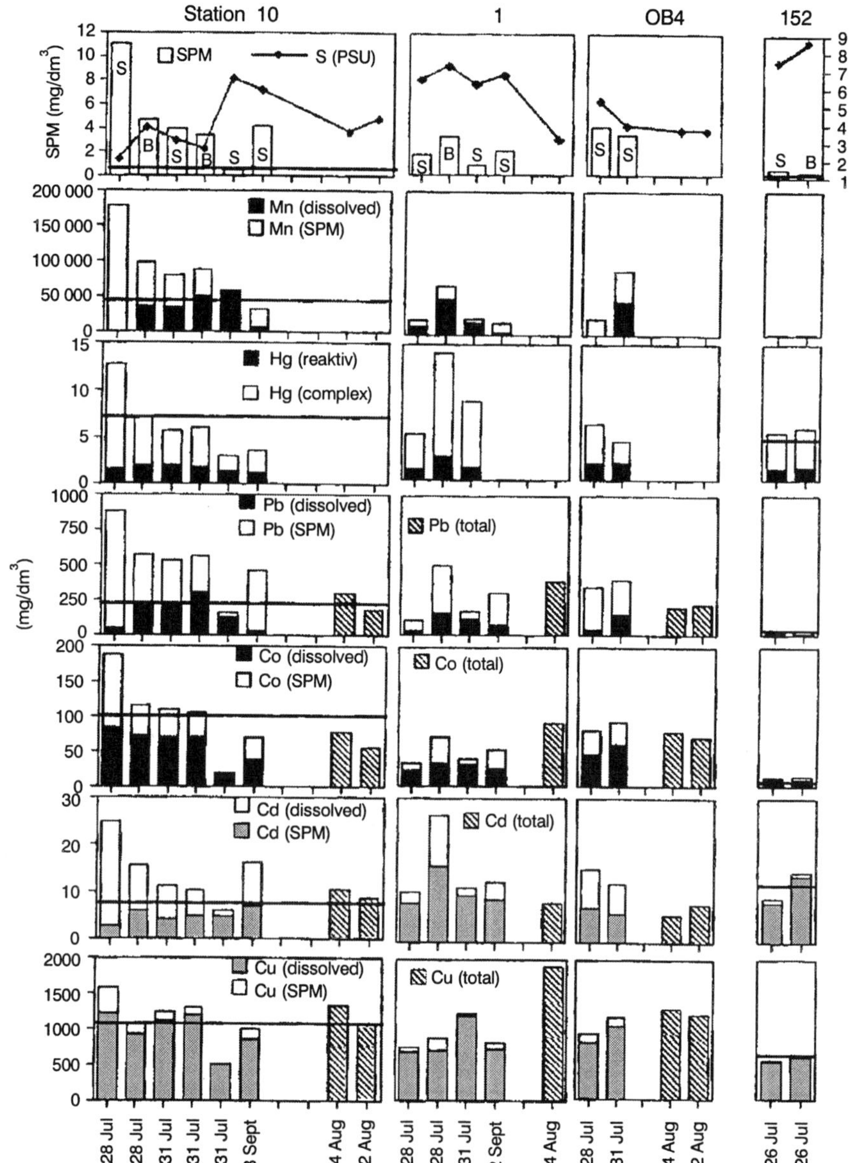

Fig. 2.8. Concentration of dissolved and particle-bound trace metals and SPM at different stations during the Oder flood. The straight lines indicate the mean values of the plume during a TRUMP-experiment in June 1995. S-Surface water; B-Bottom water. After Siegel et al. (1998); modified.

after Oder flood concentrations of Hg, Pb, Cd and Mn increased respectively ca 2-, 4-, 3- and 3.5-times greater than before the flooding (Pohl et al., 1998; Siegel et al., 1998). It is reported that fluffy material from the Oder estuary appears to be the main source of heavy metals in the muddy sediments of the Arkona Basin and Bornholm Deep (Laima et al., 1999; Löffler et al., 2000; Witt et al., 2001; Christiansen et al., 2001; Emeis et al., 2001). Several authors (Miltner and Emeis, 1999, 2000, 2001; Leipe et al., 2000) studied the distribution, composition, origin and transport of terrestrial organic matter from the Oder River to sediments in the Pomeranian Bay, Baltic Sea. It is concluded that most terrestrial organic material is transported near the sediment-water interface and that transport of terrestrial organic matter between the individual basins is less important than the direct input from the rivers (Miltner and Emeis, 2001). Model simulation of the transport of Odra flood water through the Szczecin Lagoon into the Pomeranian Bight in July/August 1997 has been presented (Müller-Navara et al., 1999).

Vertical trends in respect to redox conditions and metal speciation

Vertical distribution of trace metals in Baltic water column has been studied by several authors (Kremling, 1983; Brügmann et al., 1997, 1998). According to Brügmann et al. (1997, 1998) a few factors have a great influence on the concentration, speciation and fate of trace elements in the Baltic Sea. The mean residence time of the brackish Baltic waters is estimated to be between 20–40 years but the mean residence time of trace metals in the water column is an order of magnitude lower. In consequence they are enriched in different compartments of the marine ecosystem reaching sometimes toxic levels there. Taking into account the specific biogeographical characteristics of the Baltic Sea, its pollutant dilution capacity is relatively low. In contrast to the transition area to the North Sea, the brackish waters of the Baltic Sea are favourable to extend the lifetime of trace metals associated with organic compounds before their ultimate flocculation and deposition. Mainly stable vertical stratification leads to stagnant and anoxic waters in the central deep basins, e.g. the Gotland Basin. It is resulted in immobilisation metallic toxicants such as Cd, Hg, Pb and Cu. To other factors having pronounced effect on metal speciation are a high suspension load responsible for their rapid sedimentation, high precipitation accelerated the wash-out of trace elements from the atmosphere and the eutrophication of the Baltic Sea, strongly linked to the fate of trace metals. The concentration and speciation of trace elements such as Cd, Co, Cu, Fe, Mn, Ni, Pb and Zn have been extensively studied in the Gotland Deep water column in 1991 and after the salt water inflow in 1994 (Brügmann et al., 1997, 1998). Below the depth of 125 m dramatic variations in the total 'dissolved' metal concentrations as well as in their speciation composition were noted (Brügmann et al., 1997, 1998).

Iridium being one of platinum group elements is used as a tracer of extraterrestrial material since this element is enriched in meteorites relative to Earth's crust material. Study of Ir transport in seawater showed that it is less abundant (mean concentration is 4×10^8 atoms kg^{-1}; 10^8 atoms kg^{-1} = 1.66×10^{-16} mol kg^{-1}) in this medium than Os, Pd, Pt, Rh, Ru and Au suggesting that Ir is presumably the rarest stable element in the oceans (Anbar et al., 1996). Concentration of Ir was determined in oxic and anoxic waters of the Baltic Sea, Kattegat-Transition region and rivers entering the Baltic Sea, i.e. the Kalixälven, Neva and Vistula rivers. The concentration of Ir fell well below the North Sea and the average Baltic river input (see Chapter 2B) indicating that most of the dissolved riverine Ir is effectivelly removed from solution. It has been found (Anbar et al., 1996) that Fe–Mn oxyhydroxides scavenge Ir under oxidising conditions while anoxic environment is not a major sink for Ir in the Baltic Sea; ca. 30% labile Ir is associated with particles > 0.45 μm with Mn-oxyhydroxides as their substantial component (Andersson et al., 1992, 1995; Anbar et al., 1996). The distribution pattern of trace elements, i.e. Cd, Co, Cu, Fe, Mn, Ni, Pb and Zn at the oxic-anoxic interface and in sulfidic water of the Drammensfjord, Norway was presented by Öztürk (1995).

Temporal trends

In order to study of trace elements distribution in Baltic water as a function of time the concentrations of dissolved and particulate forms of Cd, Cu and Zn were processed statistically (Schneider, 1996). The estimated temporal trend curves for the Mackleburg Bight/Arkona Sea and for the surficial and deep waters of the Bornholm Sea/Gotland Sea are presented in Figures 2.9–2.11. A comparable negative trend of ca. 7% yr^{-1} was observed for dissolved Cd in the Mackleburg Bight/Arkona and in the surficial waters of the Bornholm Sea/Gotland Sea during 1980–1992. Also a negative trend of ca. 11% yr^{-1} for particulate Cd was found. It is important to note that no temporal changes were detected for dissolved Cd in the deep waters of Bornholm Sea/Gotland Sea; its mean concentrations were from two to four times smaller than those in surficial waters. Such difference can be explained by the occurrence of H_2S in the bottom waters which forms CdS precipitate. The deposition of this sulphide particles to sea bottom has place and therefore the concentration of dissolved Cd is stabilised at a low its level (Schneider, 1996). The trend analyses for dissolved and particulate Cu (Fig. 2.10) are very similar to that for Cd (Fig. 2.9). The bottom waters of the Bornholm Sea/Gotland Sea are also kept at a low and steady level of dissolved Cu owing to its chemical affinity to H_2S.

Dissolved Zn, in contrast to Cu and Cd, did not show any temporal trend for the Mackleburg Bight/Arkona and the surficial waters of the Bornholm Sea/Gotland Sea. However a positive trends of 3.6 and 8.5% yr^{-1} are detected for particulate Zn in these both areas (Fig. 2.11) but due to the minor contribution of the particulate fraction these temporal variations during 1980–1992 are not significant in respect to the total Zn inventory. There is insignificant difference between concentration of Zn in waters above and below halocline which is a result of less effective, in contrast to Cd and Cu, formation of ZnS precipitate.

Decreasing levels of Cd and Cu in the surficial Baltic waters (Figs. 2.9 and 2.10) may correspond to a reduced input of these elements into surface waters. Other explanation for this deficiency is that these two elements may be removed from surficial layers by settling particles, e.g. phytoplankton which is known as

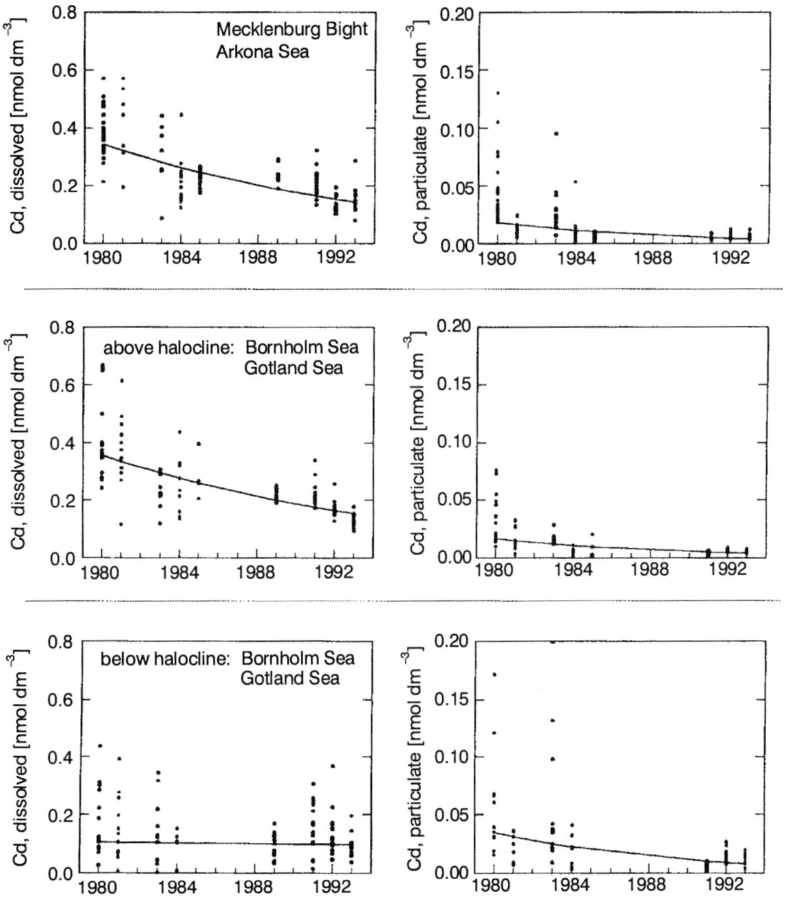

Fig. 2.9. Concentrations of dissolved and particulate Cd – in nmol dm^{-3} – and calculated trend curves. After Schneider (1996); modified.

a carrier for trace elements into deeper water layers. According to Kremling and Streu (2000) the negative temporal trend pattern is probably a result of reduced riverine and atmospheric inputs of these metals to the Baltic waters, especially in shallower and seasonally mixed areas of the Arkona and Belt Seas. For Mn, however, in contrast to the fate of the more "nutrient-like' trace elements (Cd, Cu, Ni, Zn) more significant short-term (intra-annual) variabilities are observed in surficial waters of open Baltic Sea. These fluctuations seem to be associated with geochemical redox processes combined with the hydrographic and morphological conditions dominating in the Baltic Sea (Kremling and Streu, 2000).

According to Kremling and Wilhelm (1997) a mean Ca concentrations increase of ca. 4% corresponding to an increase of the overall average Ca flux via the freshwater from 3.1 to 4.5 g m^{-2} yr^{-1} within the past ca. 25 years. It is suggested that this significant positive temporal trend of Ca flux in the run-off is mainly

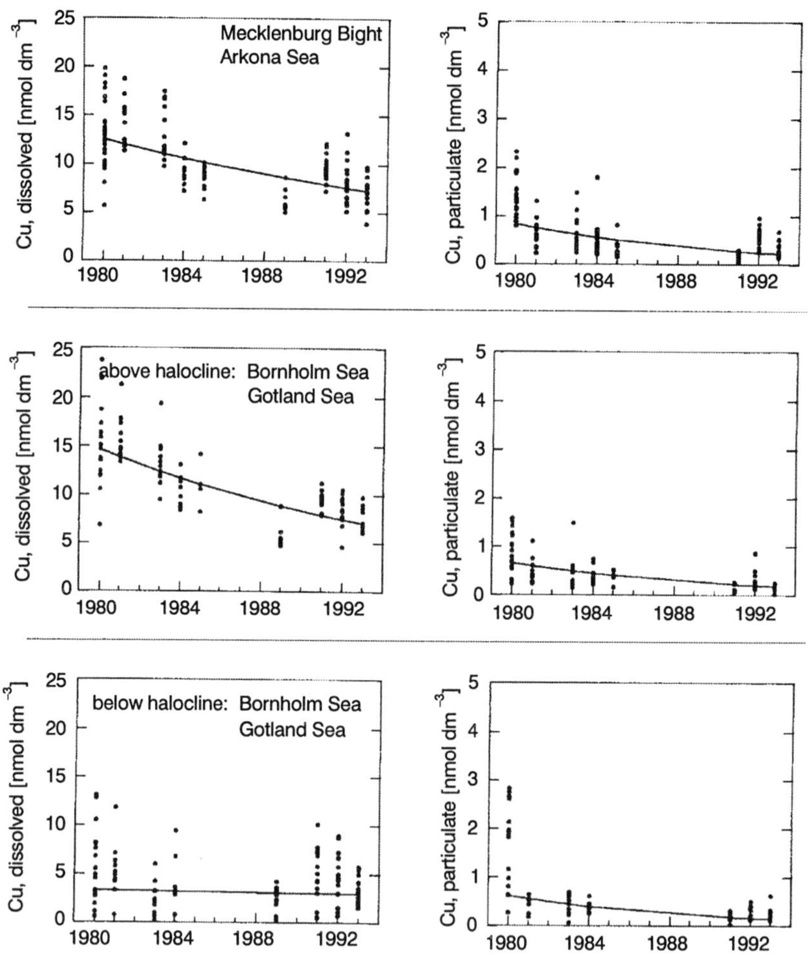

Fig. 2.10. Concentrations of dissolved and particulate Cu – in nmol dm^{-3} – and calculated trend curves. After Schneider (1996); modified.

caused by the combined impact of the deposition of acidifying air pollutants and the agricultural use of nitrogen fertilisers as well as by the elevated draining process (Kremling and Wilhelm, 1997). However this latter process, is not yet in a steady state, and that other constituents could be contributed to long-term changes of the major chemical composition of Baltic waters (Kremling and Wilhelm, 1997).

Trace element speciation

As most trace elements, the metalloids such as As, Sb and Se can be toxic and essential to marine organisms (Andreae and Klumpp, 1979; Maher and Butler, 1988; Vandermeulen and Foda, 1988; Cutter and Cutter, 1995). According to

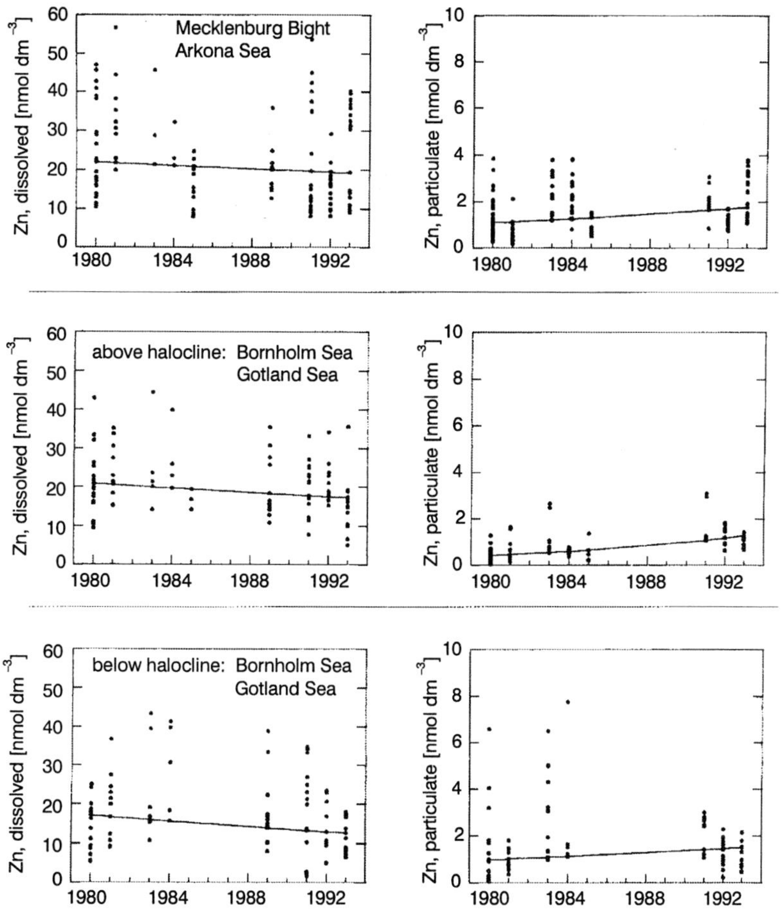

Fig. 2.11. Concentrations of dissolved and particulate Zn – in nmol dm^{-3} – and calculated trend curves. After Schneider (1996); modified.

Nriagu (1989) the anthropogenic emissions of Sb and As to the atmosphere exceed their natural inputs. Andreae and Froelich (1984) have reported As, Ge and Sb species concentrations determined from five hydrographic stations along the central axis of the Baltic Sea, i.e. from the Bornholm Basin to the Gulf of Finland. In the oxic waters the As(V) and Sb(V) predominated, while in the anoxic basins mainly trivalent species occurred and possibly some sulpho-complexes in the sulphide zone. Methylated As species constituted a large fraction of dissolved As in the surface waters and methylated species of As, Sb and Ge were detected throughout the water column. The methylated species showed essentially conservative mixing behaviour with no evidence of inputs by methylation processes in the anoxic waters. Germanium is present as dissolved germanic acid $Ge(OH)_4^0$ and as mono- and dimethyl-Ge species. Germanic acid levels in the Baltic water were an order of magnitude higher than in the ocean and much higher than in fluvial input.

Inorganic Ge species were closely correlated with dissolved Si (Andreae and Froelich, 1984). The speciation of As and Sb in the Baltic Sea is controlled by the biogeochemical cycling of these elements. As is removed by biological uptake while Sb by particulate scavenging along the water column. Both these elements are only partly regenerated in the anoxic zone. Methylated and reduced forms occur in the biologically active surface layer of water. It has been suggested (Andreae and Froelich, 1984) that anthropogenic factor was mainly responsible for significant contribution to the total elemental load of these three elements, with atmospheric component dominating the input of Ge to the Baltic Sea.

Brügmann et al. (1997, 1998) discussed results from metal speciation studies in the Gotland Deep. The studies were performed before and after a redox turnover in 1991 and 1994, respectively. After a stagnation period of almost 15 years, in 1993/94 very large volume of saline North Sea-derived water intruded into the Baltic Sea. This event inspired several investigators to make environmental studies of chemical elements in the Gotland Deep as a model area. The concentrations and speciation (particulate, colloidal, anionic and cationic forms) of Cd, Co, Cu, Fe, Mn, Ni, Pb and Zn have been investigated in the Gotland water column (Fig. 2.12). Below the depth of 125 m, extreme changes in the total 'dissolved' metal concentrations as well as in the ratios between different chemical species were detected. This mainly concerns those elements for which solubility differs significantly with the redox state, i.e. Co, Fe and Mn, but also for those elements which form rather insoluble sulphides, i.e. Cd, Cu, Pb and Zn, and/or stable complexes with organic ligands, i.e. Cu (Brügmann et al., 1997, 1998). As can be seen in Fig. 2.13 below 200 m depth, almost the total Mn concentration existed in particulate forms. Colloidal and anionic species do not seem to have a key contribution to the Mn speciation under the present post-inflow environmental state. As regards Fe, except the 10 m depth, its particulate species predominated. The behaviour of Fe, in contrast to Mn, indicated faster re-oxidation and settling kinetics of the different Fe forms. In the euphotic layer, the colloidal Fe represented the major contribution, possibly resulting from organic associates arising from the primary production (Brügmann et al., 1997). The cationic fraction of Zn dominated in the Gotland Deep water, except a depth water of 50 m, significant quantities of Zn occurred also in particulate forms, reaching maximum value at 236 m depth. Colloidal Zn exhibited higher levels in the halocline and at 220 m while anionic species showed their maximum levels in the euphotic layer. As for Cd, the cationic forms of Cd clearly dominated and below 200 m its particulate fraction became significant increasing toward the bottom (Fig. 2.13). The colloidal fraction of Cu was mostly much lower than the particulate one although it still occurred in noticeable amounts, except 50 m. Below 200 m depth cationic forms dominated over the dissolved Cu fraction with the exception of the extremely enriched in Cu water layer around 220 m with the oxygen deficit. This could be explained by the presence of remains from anionic Cu complexes formed under the conditions of high levels of H_2S. Supposedly these complex compounds survived in the relict waters, still contained minimum concentrations of oxygen (Brügmann

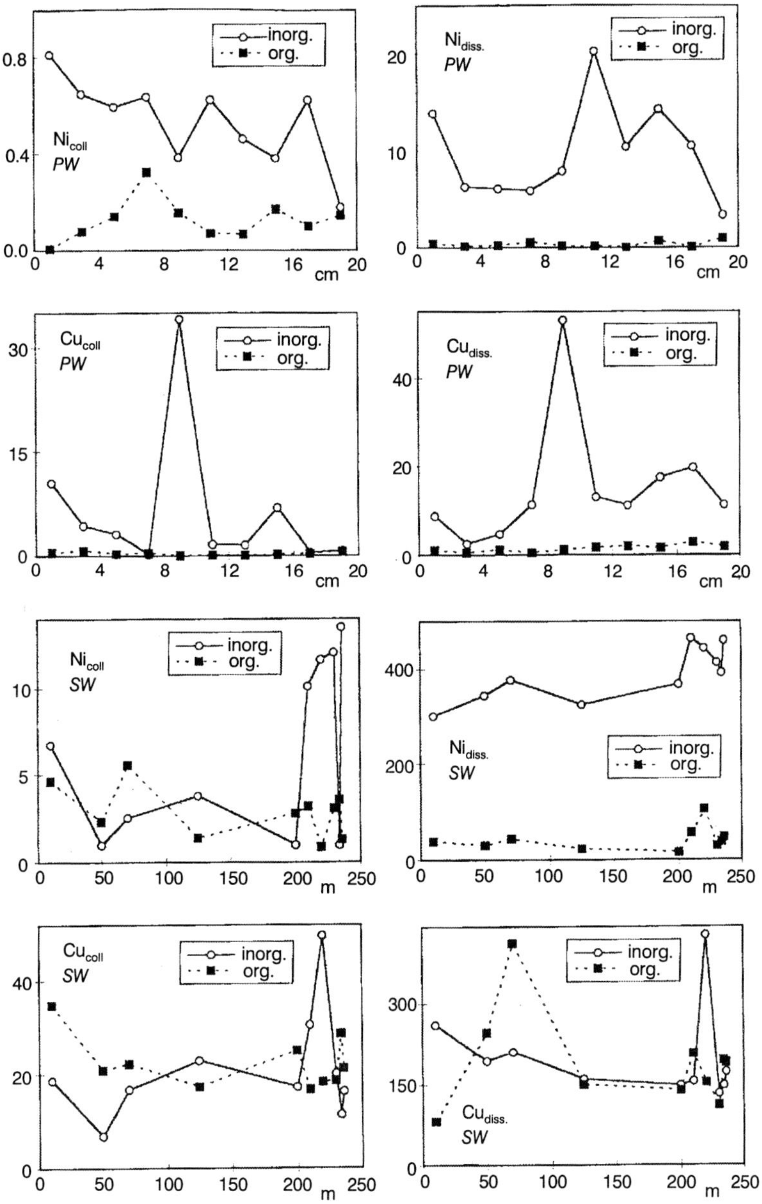

Fig. 2.12. Comparison of the "inorganic" vs "organic" speciation of Cu and Ni in sea water (*SW*) and pore water samples (*PW*) collected at station *F-81*/Gotland Deep, June 1994. $Me_{coll.}$/inorg. – *TSK* separation and 2 M HCl/1 M HNO_3 elution. $Me_{coll.}$/org. – *TSK* separation and 30% ACN elution. $Me_{diss.}$/inorg. – Σ *DEAE* and *8-HQ* separation and 2 M HCl/1 M HNO_3 elution. $Me_{diss.}$/org. – C_8-*SepPak* separation and 30% ACN elution. After Brügmann et al. (1998); modified.

et al., 1997). The speciation of Pb and Ni was mostly dominated by the cationic forms; the Pb cationic species closely correlated with its distribution pattern of the particulate matter (Fig. 2.13). The geochemical distribution pattern of Mo appeared to be similar to that of Mn showing increasing of its concentrations toward the bottom. The particulate forms may be due to the adsorption of dissolved Mo by freshly precipitated Mn oxyhydroxides (Brügmann et al., 1997).

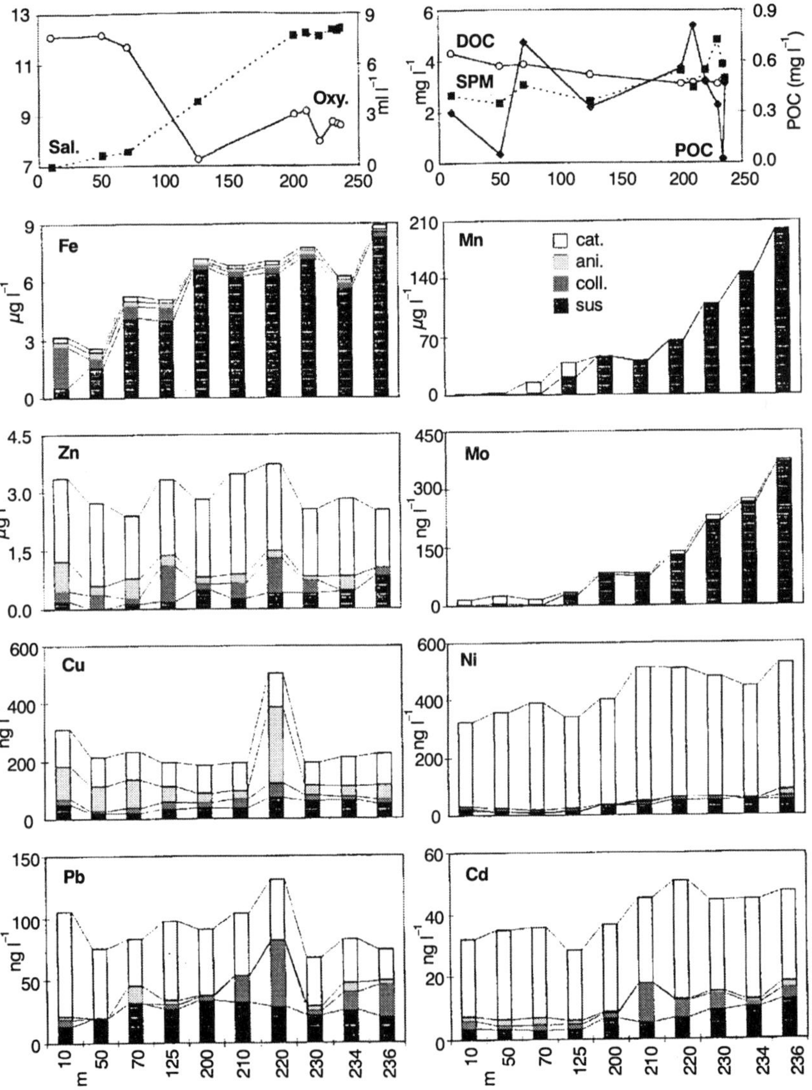

Fig. 2.13. "Inorganic" trace metal speciation in the water column, station F-81/Gotland Deep, June 1994. DOC – dissolved organic carbon, POC – particulate organic carbon, SPM – suspended particulate matter, cat. – cationic forms, ani. – anionic forms, coll. – colloidal forms, sus – particulate forms. After Brügmann et al. (1998); modified.

The Bornholm Deep sediments exhibit anomalously high enrichments of Mo and, to a lesser extent, Sb and As compared to the other sediments reported here. Prange and Kremling (1985) have suggested that Mo is most probably removed from Baltic Sea waters by adsorption on particulate organic matter after the reduction of MoO_4^{2-} to MoO^{2+}. However, Erickson and Helz (2000) have shown that MoO_4^{2-} can be reduced to MoS_4^{2-} in anoxic waters where the $H_2S(aq)$ concentration exceeds 11 μM. ΣH_2S concentrations in the anoxic waters of the Gotland Deep are in the range 7.9-52 μM demonstrating that this condition is frequently met, especially in the deepest waters of this deep (Kremling 1983). MoS_4^{2-} can be scavenged by Fe-bearing minerals (Erickson and Helz, 2000). Pyrite has been reported at depths below 4 mm in the sediments of the Arkona Basin (Neumann et al. 1998) and presumably also occurs in the sediments of the Bornholm Deep. Adsorption of MoS_4^{2-} on pyrite could therefore explain the very high EF for Mo (32) in the Bornholm Deep sediments (Szefer et al., 2001). Stability field data also indicate that Sb and As can occur as sulphides ($Sb_2S_4^{2-}$ and As_2S_3) in anoxic marine environments (Brookins, 1998; Glasby and Schulz, 1999). This may also explain their high EFs in Bornholm Deep sediments.

According to Hou et al. (2001) there is no significant difference for ^{129}I and ^{127}I in iodide (I^-) and iodate (IO_3^-) forms between the bottom and surface Baltic water. IO_3^- is the predominant species of iodine in the bottom water in the Kattegat (Hou et al., 2001). The ratio of $^{129}I/^{127}I$ for IO_3^- in Baltic water is much higher than that for I^- and close to the level in the Kattegat. This means that both the ^{129}I and ^{127}I in IO_3^- form in Baltic water origin from the Kattegat. The lower $^{129}I/^{127}I$ ratio for I^- in Baltic water may be attributed to the extensive dilution of $^{129}I^-$ (Hou et al., 2001). IO_3^- level is high in the saline bottom, water, especially in the Kattegat, but low in surface waters of the Baltic Sea (Hou et al., 2001; Truesdale et al., 2001). In the Kattegat IO_3^- level was higher in the bottom water than that in surficial water (Hou et al., 2001; Truesdale et al., 2001) while $^{127}I^-$ levels were similar for the bottom and surfical waters. This distribution pattern can be explained by fact that highly saline water from the North Sea mixes with less saline outflow water from the Baltic Sea in the Kattegat; the North Sea water goes down to the bottom and in consequence most of Baltic water remains on the surface (Hou et al., 2001). After Truesdale et al. (2001) IO_3^- ions are more concentrated in surficial than in bottom waters, suggesting their reduction in the deeper waters. The concentration of organic-I in Baltic water was very low and ranged from ND to 0.040 μM (Truesdale et al., 2001). Such low levels are probably caused by promoting greater decomposition in the Baltic Sea characterised by much greater depth as compared with other estuaries (Truesdale et al., 2001).

(iii) Radionuclides in Seawater

Radiological data of investigations of the Baltic Sea including the Danish Straits have been reported by several authors (Salo and Voipio, 1966, 1978;

Voipio and Salo, 1971; Ivanova, 1978; Aarkrog et al., 1980, 1986; Kautsky, 1981; Kautsky and Eicke, 1982; Miettinen et al., 1982; Lazarev et al., 1983a, 1983b, 1986; Holm et al, 1986; Ilus et al, 1986, 1987, 1988, 1992, 1993; Jaworowski et al, 1986; Kautsky et al, 1986; Kowalewska, 1986; Salo et al., 1986; Tuomainen et al, 1986; Weiss and Moldenhawer, 1986; Leskinen et al., 1987; Nies, 1988, 1989, 1994; Nies and Wedekind, 1988; Ribbe et al., 1991; Bojanowski et al, 1995; HELCOM, 1995; Herrmann et al., 1995; Skwarzec, 1995; Nies and Nielsen, 1996; Herrmann, 2000; Isajenko et al., 2000; Tishkov et al., 2000; Styro et al., 2001). The concentration data for radionuclides in water of the Baltic Sea are listed in Table 2.8. Before 1996, the Baltic Sea as a semi-closed and shallow brackish water basin has been mainly contaminated by tritium (^{3}H), strontium (^{90}Sr), caesium (^{137}Cs) and plutonium ($^{239+240}Pu$) isotopes originating from global fallout. Since the 1970s had place additional radioactive contamination of Baltic water caused by entering North Sea waters transporting radionuclides from the Sellafield nuclear reprocessing plant (Panteleev et al., 1995). As can be seen in Figure 2.14, the Chernobyl fallout changed dramatically pre-1986 distribution of radiocaesium in water of the Baltic Sea. The concentrations of this radionuclide were generally smallest in the southern Baltic and the Bothnian Bay and they were the greatest in waters of the Gulf of Finland, the Bothnian Sea and the Mecklenburg Bight (Fig. 2.15). More contaminated were rather coastal than open sea waters. According to Report (IAEA, 1986) post-Chernobyl $^{134}Cs/^{137}Cs$ activity ratio amounting on 0.5 corresponded to the theoretical value for the fuel of the Chernobyl nuclear power plant (Panteleev et al., 1995). The distribution of artificial radionuclides in the English Channel, southern North Sea, Skagerrak and Kattegat during 1990–1993 has been reported by Herrmann et al. (1995). The compilation of environmental measurements of ^{137}Cs and ^{90}Sr in Baltic seawater from 1961 to 1995 has been made by Herrmann (2000).

Besides horizontal also vertical distribution of several radionuclides has been studied. For instance, during 1984–1991, the concentrations of radiocaesium in surficial Baltic waters differed, in most cases significantly, from those in near-bottom waters. In 1991, the radiocaesium concentrations were smaller in deep than surficial water layers of the Baltic Proper. However, the initial contamination of the surficial water body penetrated rapidly to deeper layers in areas lacking a stable halocline (Panteleev et al., 1995). For radioactive contamination of Baltic waters were also responsible more than 20 other the Chernobyl-derived radionuclides. A numerous group of radionuclides, e.g. ^{89}Sr, ^{132}Te and ^{140}Ba was also observed in seawater after the Chernobyl accident, but due to their short half-lives, decayed within several days or months (Panteleev et al., 1995). Radioactive contamination of the Baltic Sea in vicinity of the Leningrad Nuclear Power Plant in 1971–1996 has been evaluated by Gedeonov et al. (1998). Variations of radiocaesium concentrations were also investigated in the southeastern part of the Baltic, following the Chernobyl power plant accident (Styro et al., 2001). The rate of 'self-cleaning' was estimated as very slow, the mean concentration of ^{137}Cs

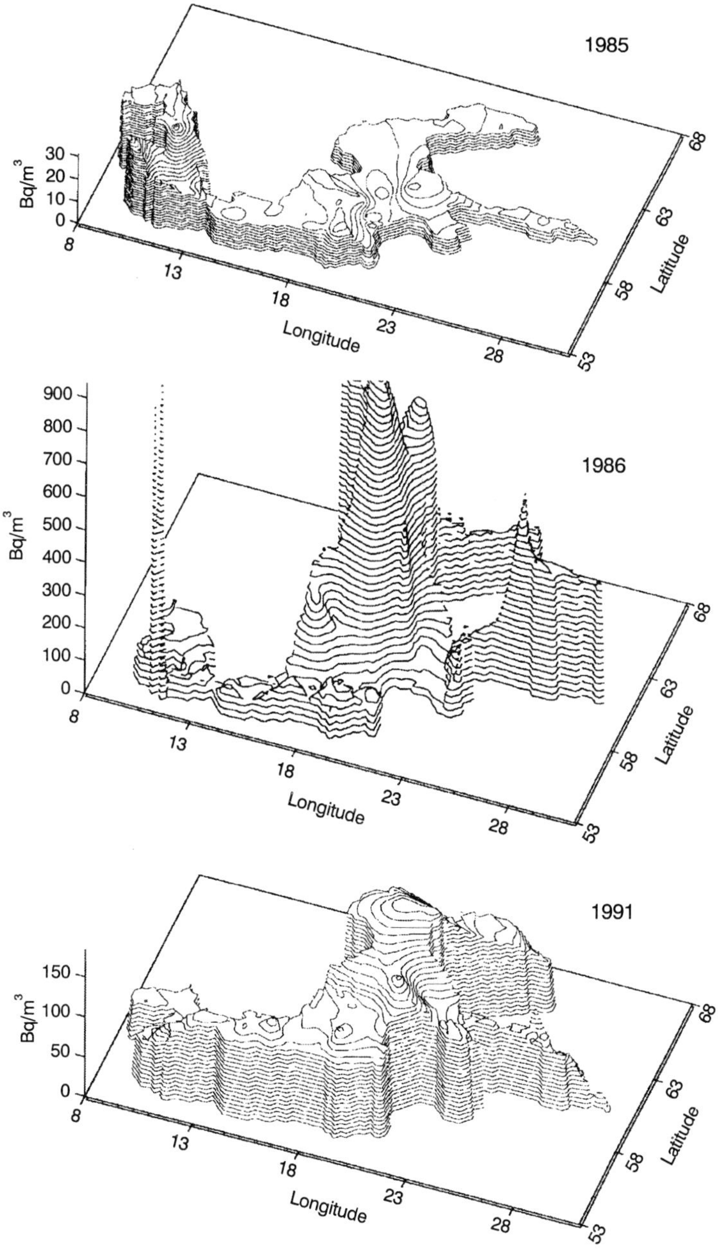

Fig. 2.14. Distributions of Cs-137 activity concentration in the Baltic Sea. After Panteleev et al. (1995); modified.

in 1996 was almost the same as that detected directly after the accident in 1986. The data display a continuing significant pollution of the waters of the Baltic Sea as a result of the Chernobyl accident (Styro et al., 2001).

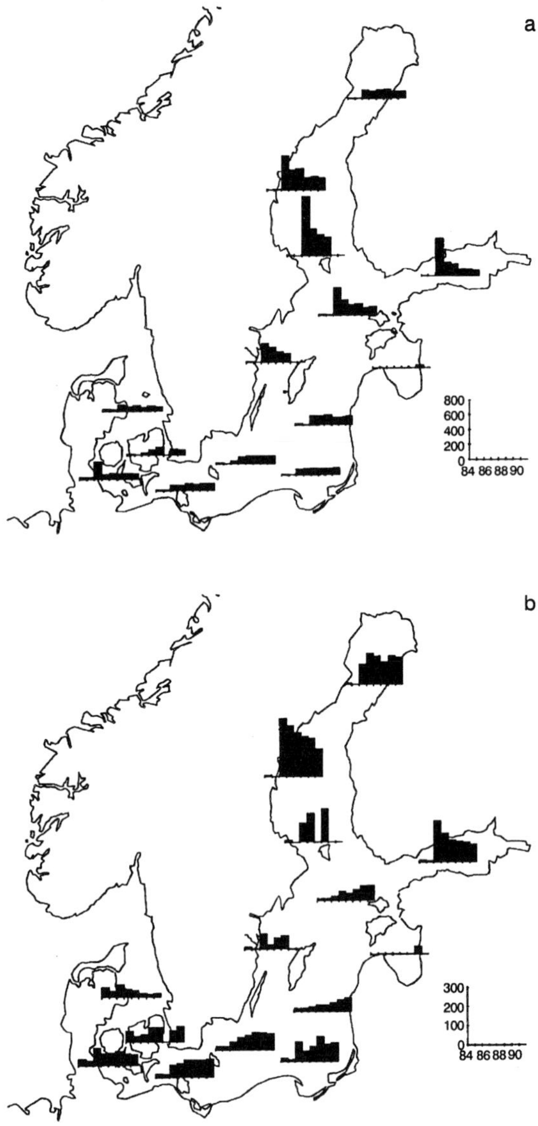

Fig. 2.15. Temporal evolution of Cs-137 concentration in (a) surface water, (b) near bottom water 1984-1991 in Baltic Sea. (Annual mean 10 m above bottom, Bq/m^3). After Panteleev et al. (1995); modified.

Concentrations of Pu in surface Baltic water were low (Holm, 1995) and did not exceed 10 mBq m^{-3}. Plutonium has a great affinity to particulate matter, especially to its organic matter component (Östlund, 1991), therefore the concentration ratio of $^{239+240}$Pu in suspended matter to that in a seawater phase is significantly higher than unity (4×10^5) (Skwarzec and Bojanowski, 1992; Skwarzec, 1997). According to Östlund (1991) possible two step mechanism governs over

TABLE 2.8.

Concentrations of total (t), dissolved (d) and suspended (s) radionuclides (Bq m^{-3}) in water of the Baltic Sea and other northern areas

Region	Sampling date	Depth (m)	N	Salinity (PSU)	110m-Ag	241-Am	140-Ba	141-Ce	144-Ce	60-Co	References
Nothern Baltic											
Bothnian Sea	1986	0	1	5.96	3.2						Ilus et al., 1987
		130	1	6.05	3.1						
			2	5.37–5.67	4.6		220–260	ND-110	ND-60	ND-5.5	
			1	3.34	30		860	9.3	ND	ND	
Gulf of Finland	1986		1	3.34	30		860	9.3	ND	ND	Ilus et al., 1992, 1993
	1989	0	1	3.06		0.0046					
Finnish coast	1982–83		11	5.88		4.4*					Tuomainen et al., 1986
				3.8–6.8		0.4–10					
North Gotland	1986	0–150		8.76		ND-0.006					Ilus et al., 1987
				6.83–10.24							
		0	2	6.03–6.90		0.003–0.006	200				
Baltic Proper	1989	0		6.41		0.0027					Ilus et al., 1993
		135		9.34		0.0053					
Danish Straits	1983	0	7	13.2		0.38*					Aarkrog et al., 1986
				8.5–16		0.21-0.67					
		4	1	8		0.58*					
		5	1	32.6		1.75*					

* – mBq m^{-3}

TABLE 2.8. – continued

Region	Sampling date	Depth (m)	N	Salinity (PSU)	134-Cs	136-Cs	137-Cs	3-H	131-I	40-K	140-La	95-Nb	References
Open Baltic	1982–83		70	5.01–20				6.34**					Weiss and Moldenhawer, 1986
								4.0–9.7					
Arkona Basin													
Pomeranian Bay	1993						99±6*						Bojanowski et al., 1995
Baltic Proper													
	1990	0–230	8	9.27	11.1		71.4 (7)			3050			Ilus et al., 1993
				5.89–12.09	1.4–16.0		26–110			2100–4200			
Nothern Baltic													
Bothnian Sea	1986	0		5.96	270		570			2200		ND	Ilus et al., 1987
		130		6.05	240		490			2200		ND	
			2	5.37–5.67	20–590	ND-35	48–1100		220–1900	1900–2100	180–210	3.9–230	
	1990–95	Surface	6				123–216						Herrmann, 2000
Gulf of Finland													
	1986		1	3.34	1700	370	3100		7400	1200	820	19	Ilus et al., 1987
	1990	0–73	19	5.0 (17)	13.8		98.2			1930			Ilus et al., 1993
				1.22–7.46	5.3–20		37–130			680–2700			
	1989		16	3.29–4.72	25–35		120–160	ND-22000		1300–1900			Ilus et al., 1992
			16	3.62–4.27	15–20		100–140	ND-10000		670–1600			
	1990		13	5.53–6.00	40–51		200–240	ND-7600		2000–2400			
			13	5.36–5.95	21–37		130–260	ND-16000		1900–2500			
	1990–95	Surface	6		40–51		56–88						Herrmann, 2000
		Deep	6		21–37		53–42						
Finnish coast	Pre-1988		13		61–170		160–470			1300–2800			Ilus et al., 1988
North Gotland	1986	0–150		8.76	21.0 (5)		51.7			3280		ND-150	Ilus et al., 1987
				6.83–10.24	3.4–47		21–110			2500–3900			
Eastern Baltic	1990–95	Surface	6				85–108						Herrmann, 2000
		Deep	6				81–95						
Western Baltic	1990–95	Surface	6				78–117						Herrmann, 2000
		Deep	6				70–100						
Danish Straits	1983	0						2.1±0.7**					Aarkrog et al., 1986
		5						1.2±0.4**					
		10						1.8±0.4**					
	1983	0	88		ND-6.9		3.5–189						Aarkrog et al., 1986
		2.0–30	8		ND-2.8		16.2–65						
		34–60	3		ND		14.4–56						
Kattegat	1990–95	Surface	6				50–70						Herrmann, 2000
		Deep	6				17–29						
Belt	1990–96	Surface	6				60–84						Herrmann, 2000
		Deep	6				55–67						

* – Not filtered
** – kBq m^{-3}
ND – Not detected

TABLE 2.8. – continued

Region	Sampling date	Depth (m)	N	Salinity (PSU)	210-Po (d)	210-Po (s)	239+240-Pu (d) (mBq m^{-3})	239+240-Pu (s) (mBq m^{-3})	239+240-Pu (t) (mBq m^{-3})	125-Sb	90-Sr	References
Southern Baltic												
Gulf of Gdansk					0.26±0.03	0.12±0.01	2.4–4.5	1.3–1.5				Skwarzec, 1995
Gdansk Deep					0.33±0.11		3.6±0.6	1.1±0.5				Skwarzec and Bojanowski,
Slupsk Furrow					0.57±0.17	0.07±0.02						1988, 1992
Bornholm Deep					0.17±0.07							
Arkona Basin					0.34±0.09							
Pomeranian Bay	1993				0.33±0.09*							Bojanowski et al., 1995
Lübeck Bight					0.52±0.03	0.11±0.03						Skwarzec, 1995
Baltic Proper	1980–88				0.36±0.18	0.08±0.03	3.5±1.4	1.3±0.8				Skwarzec and Bojanowski, 1988, 1992
Baltic Proper							2.0–7.0					
	1989	0–135	4	6.41–13.71					2.4–3.2		22	Ilus et al., 1992, 1993
	1990	0–230	8	9.27					3		17.7 (6)	
				5.89–12.09					2.0–4.0		17–18	
Northern Baltic												
Bothnian Sea	1986	0		5.96								
		130		6.05								
	1990–95		6								15–21	Herrmann, 2000
Gulf of Finland							8.0					
	1986									43	300	Ilus et al., 1987
			1	3.34								Ilus et al., 1993
	1990	0–73	19	5.0 (17)					3.0 (1)			
				1.22–7.46								
	1990–95	Surface	6		40–51						18–22	Herrmann, 2000
		Deep	6		21–37						16–20	
North Gotland	1986	0–150		8.76					2.5			
				6.83–10.24					1.3–4.4			
Eastern Baltic	1990–95	Surface	6								17–18	Herrmann, 2000
		Deep	6								16–19	
Western Baltic	1990–95	Surface	6								13–19	Herrmann, 2000
		Deep	6								16–18	
Danish Straits	1983	0	12	23.6				3.27			22.1	Aarkrog et al., 1986
				7.8–34.7				2.8–4.7 (N=7)			8.5–32	
		5	1					3.1			18.8	
		10.0–34	3								28.6	
Kattegat	1990–95	Surface	6								10.0–15	Herrmann, 2000
		Deep	6								3.0–9.0	
Belt	1990–96	Surface	6								10.0–14	Herrmann, 2000
		Deep	6								4.0–20	

* Not filtered

TABLE 2.8. – continued

Region	Sampling date	Depth (m)	Fraction (μm)	N	Salinity (PSU)	230-Th (ng kg^{-1} (x10^5))	232-Th (ng dm^{-3})	234-Th (d) (mBq dm^3)	234-Th (s) (mBq dm^3)	U (μg dm^3)	234-U (d)*	235-U (d) (mBq dm^{-3})	238-U (d) (mBq dm^{-3})	References
Southern Baltic				13			0.034							Szefer and Bojanowski, 1981
							0.013–0.055							
Baltic Proper	1977	surface		15						0.87				Bojanowski and Szefer, 1979
										0.65–1.06				
North Gotland	1953	5–100		2	6.75					0.8				Koczy et al., 1957
	1979	5	< 0.4	1	6.75								1.92*	Prange and Kremling, 1985
		10–130		8	8.28								2.71*	
					6.75–9.96								1.98–3.48	
		140		1	10.57								4.88*	
Gotland	1953	5–200		3	9.74					3.27				Koczy et al., 1957
					7.6–13.7					1.9–5.9				
East Gotland	1979	5	< 0.4	2	7.1								2.65*	Prange and Kremling, 1985
					6.84–7.36								2.36–2.93	
		10–218		20	10.6								3.56*	
					6.88–12.7								2.57–5.12	
		233		1	12.8								6.24*	
Landsort	1953	5–455		6	9.32					1.58				Koczy et al., 1957
					7.16–11.7					0.8–4.0				
	1980	13	< 0.45	1	7.69					0.89±0.03				Gellermann et al., 1983
		90	< 0.45	1	10.53					1.01±0.04				
Bornholm	1953	5.0–75		2	12.1					1.65				Koczy et al., 1957
					7.5–16.7					1.1–2.1				
		5	< 0.45	1	7.99					0.83±0.03				
	1979	3	< 0.4	1	8.64								3.17*	Prange and Kremling, 1985
		8–13.0		2	9.72								4.06*	
					9.04–10.4								3.45–5.69	
		21		1	14.03								6.1*	

Arkona		5	< 0.45	1	8.1			0.82±0.03			Gellermann et al., 1983
		41	< 0.45	1	16.56			1.51±0.07			
Arkona	1979	3	< 0.4	1	8.64					3.39*	Prange and Kremling, 1985
		8–13.0		2	9.72					3.9*	
		21		1	14.03					9.43*	
North Arkona		5	< 0.45	1	7.99			0.82±0.03			Koczy et al., 1957
		35		1	12.04			1.16±0.05			
Gotland Deep	1991	5	< 0.1			10±2	0.136±0.001		175±6	0.780±0.001**	Andersson et al., 1995
		125				0.4±0.4	0.088±0.001		175±6	0.847±0.001**	
		225				1.7±0.2	0.394±0.001		170±15	0.607±0.002**	
	1995	30	< 0.45						173±5	3280±8^	Andersson et al., 1998a
		175							163±7	4751±21^	
	1995	30	Unfiltered						181±5	776±2^^	Porcelli et al., 1997
		175	Unfiltered						164±5	1133±3^^	
		30	< 10 kD ultrafiltrate[a]						180±6	695±2^^	
			> 10 kD colloid conc.[b]						179±6	4^^	
			10 kD filter rinse[c]						177±7	56^^	
			< 3 kD ultrafiltrate[a]						168±3	441±1^^	
			> 3 kD colloid conc.[b]						174±6	22^^	
			3 kD filter rinse[c]						173±5	209^^	
		175	< 10 kD ultrafiltrate						160±5	1057±3^^	Porcelli et al., 1997
			< 3 kD ultrafiltrate						159±6	486±1^^	
Northern Baltic	1983–85	0–430		41				0.64			Löfvendahl, 1987
								0.13–0.83			
Southern Baltic	1984–85	3.0–70		45				1.36			Löfvendahl, 1987
								0.67–2.60			
	1991	5	< 0.45			1.7±0.2	0.160±0.001		170±7	0.780±0.002**	Andersson et al., 1995

Region	Sampling date	Depth (m)	Fraction (μm)	N	Salinity (PSU)	230-Th (ng kg^{-1} (x10^5))	232-Th (ng dm^{-3})	234-Th (d) (mBq dm^{3})	234-Th (s) (mBq dm^{3})	U (μg dm^{3})	234-U (d)[a]	235-U (d) (mBq dm^{-3})	238-U (d) (mBq dm^{-3})	References
Gulf of Gdansk										0.68±0.01	10.0±0.16[d]	0.38±0.03	8.37±0.16	Skwarzec, 1995
Gdansk Deep										0.83±0.03	12.2±0.35[d]	0.32±0.06	10.2±0.32	
Bornholm Deep										0.85±0.01	11.8±0.17[d]	0.36±0.03	10.4±0.14	
Lübeck Bight														
Mecklenburg Bay	1994	3.0–24.0		25				1.38–6.87	0.87–9.32				15.5–29.1	Kersten et al., 1998
Nothern Baltic														
Bothnian Bay														
	1979	5	< 0.4	1	2.89								1.15*	Prange and Kremling, 1985
		10–100		6	3.3								1.20*	
					2.87–3.58								1.06–1.27	
		125		1	3.86								1.35*	
Gulf of Finland	1953	3								0.7				Koczy et al., 1957
	1979	3	< 0.4	1	5.89								2.1*	Prange and Kremling, 1985
		13–48		4	6.62								2.55*	
					6.34–7.21								1.94–2.80	
		58		1	7.69								2.82*	
Gulf of Bothnia														
Northmost part	1991	5	< 0.45			44±5	7.7±0.2*				331±7		0.27±0.001**	Andersson et al., 1995
		80	< 0.45			33±6	0.50±0.003*				232±8		0.37±0.001**	
	1995	80	< 0.45								247±6		1495±4^	Porcelli et al., 1997
			< 0.45								247±6		356±1^^	Porcelli et al., 1997
			< 10 kD								245±7		170±0.6^^	
	1995	80		1	3.28					0.36				
Central part	1991	5	< 0.45			39±2	5.26±0.04				188±6		0.620±0.002**	Andersson et al., 1995
Bothnian Sea	1979	5	< 0.4	1	5.07			0.04					1.85*	Prange and Kremling, 1985
		10.0–30		3	6.02			0.1					2.21*	
		50		1	5.08–6.82			0.04–0.22					1.86–2.73	
					6.95			0.16					2.60*	

East Baltic Proper	1995	30–175		3				0.96			Porcelli et al., 1997;
								0.78–1.13			Andersson et al., 1998a
Öresund	1953	3.0–27		4	19.6 9.48–29.7			1.35 0.7–1.8			Koczy et al., 1957
Skagerrak	1953	15–120		2	32.2–35.0			1.45			Koczy et al., 1957
		15–120		3				1.2–1.7 1.45 1.2–1.7			
		0–600		4				3.23 2.85–3.59			Bojanowski and Szefer, 1979
Northern Skagerrak	1991	100	< 0.40				0.239±0.004		155±7	3.245±0.008**	Andersson et al., 1995
Kattegat		0–50		7				2.95 1.96–3.75			Bojanowski and Szefer, 1979
Kattegat - Fladen	1991	5	< 0.45			2.0±0.1	0.275±0.001		158±8	2.34±0.009**	Andersson et al., 1995

* – nmol kg^{-1}
** – μg kg^{-1}
^ – pmol kg^{-1}
^ ^ – pg g^{-1}
[a] – $\delta^{234}U = [(^{234}U/^{238}U)_r/(^{234}U/^{238}U)_{eq} - 1] \times 10^3$, where $(^{234}U/^{238}U)_{eq}$ is the secular equilibrium ratio of 5.472×10^{-5}
[a] – Measured concentrations are ±5%, except for K, which are ±10%, and where noted otherwise.
[b] – The measured concentrations in the colloid concentrates have been corrected for the concentrations of < 3 kD / < 10 kD solutes and normalised to the total sample weight.
[c] – The measured concentrations in the acid rinse have been normalized to the total sample weights. Errors are ca. 7% of the reported concentration.
[d] – mBq dm^{-3}

the water/sediment $^{239+240}Pu$ partition in the Baltic Sea, namely the adsorption of Pu(V) on dissolved organic carbon and goethite (α-FeOOH). The adsorbed plutonium is subsequently reduced, forming stable particle Pu (IV), hence its complexes with humic substances are considered as stable (Östlund, 1991).

Inputs of tritium as well as manganese (^{54}Mn) and cobalt (^{60}Co) isotopes in Baltic waters are mainly detected near the nuclear power stations although additional source of tritium can be also direct atmospheric fallout.

Studies on organically-bound ^{210}Po in the southern Baltic have been performed by Bojanowski et al. (1981). The mean concentration of Po in Baltic water was 0.6 mBq m^{-3}, 80% of which is present in dissolved forms (Skwarzec and Bojanowski, 1988; Skwarzec, 1997). There are significant spatial variations in concentrations of dissolved polonium, for example its levels in waters of the Słupsk Furrow (0.57 mBq m^{-3}) and the Lübeck Bay (0.52 mBq m^{-3}) were more than three times higher than those reported for waters of the Bornholm Deep (0.17 mBq m^{-3}). The concentration ratio of ^{210}Po in suspended matter to that in a seawater phase is similar to the ratio reported for plutonium amounting on 2×10^5 (Skwarzec, 1997). The distribution of $^{239+240}Pu$, ^{137}Cs and ^{210}Po in water of the Pomeranian Bay is reported by Bojanowski et al. (1995).

The concentrations of U or ^{238}U and ^{234}U in Baltic waters were reported by several authors (Koczy, 1950; Koczy et al., 1957; Bojanowski and Szefer, 1979; Gellermann et al., 1983; Duniec et al., 1984; Prange and Kremling, 1985; Löfvendahl, 1987; Skwarzec, 1995; Porcelli et al., 1997; Andersson et al., 1998a, 1998b; Andersson et al., 2001b). Data of Th (^{232}Th) and ^{234}Th have been reported sporadically (Szefer and Bojanowski, 1981; Andersson et al., 1995; Kersten et al., 1998). The U concentration in water of the Baltic Sea shows a strong correlation to salinity (Bojanowski and Szefer, 1979; Gellermann et al., 1983; Duniec et al., 1984; Löfvendahl, 1987; Skwarzec, 1995; Porcelli et al., 1997); consequently, the concentration of U increases from 0.15 $\mu g\ kg^{-1}$ in the northern part, dominated by fresh water, to above 1.0 $\mu g\ kg^{-1}$ in the Belt Sea (Löfvendahl, 1987). However, it is also reported that dissolved U is not strictly conservative in the Baltic Sea and in specific conditions may be removed from the water phase and incorporated into the sediments. This mechanism is proposed for precipitation of reduced form of U followed by its adsorption onto organic material in anoxic waters of the Gotland Deep (Prange and Kremling, 1985). However, in oxic and low-saline waters other parameters are responsible for not conservative behaviour of U in the Baltic Sea. For instance, approximately half of U is removed at low salinities within the Baltic Sea attributed to rapid flocculation of colloid-bound U during estuarine mixing (Porcelli et al., 1997). Figure 2.16 clearly illustrates that U from the Kalix River estuarine waters (F-2) in contrast to Gotland Deep water (BY-15 – see location in Fig. 2.3), falls below the "conservative mixing" line indicating the lost of U at salinity 3.3‰. Other route of the removing U in oxic Baltic waters is adsorption of U onto secondary iron-oxyhydroxides [Fe(OOH)] particles supported by strong correlation between U and Fe concentrations (Anders-

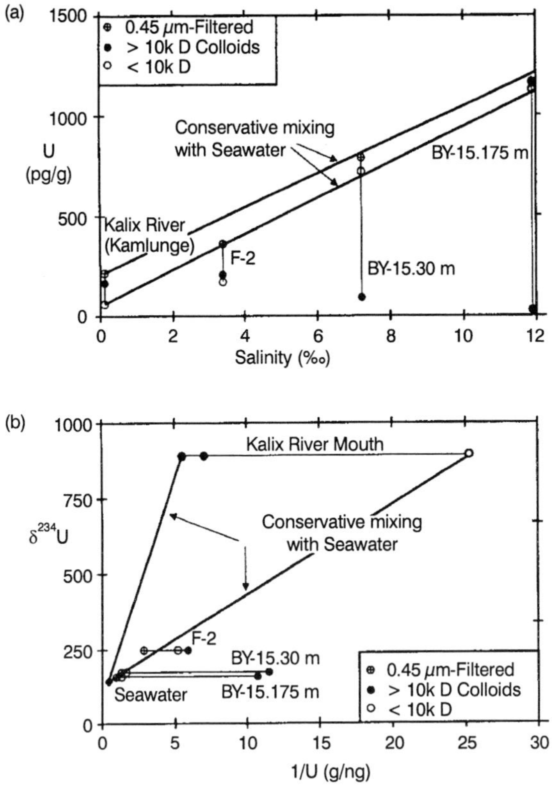

Fig. 2.16. (a) Kalix River and Baltic Sea uranium concentrations are plotted against salinity. Uranium in < 0.45 μm-filtered waters from the Kalix and BY-15 fall on a conservative mixing line with seawater, while that from F-2 falls below the line, indicating that uranium is lost at salinities < 3.3‰. The uranium in 0.45 μm-filtered water from F-2 can be interpreted as due to conservative mixing between seawater and "solute" riverine uranium, consistent with the lower mixing line, while riverine colloidal uranium is removed. (b) The δ^{234}U values are plotted against the 1/U concentration. The < 0.45 μm filtered waters from the Baltic Sea are consistent with mixing between 10 kD-filtered Kalix River water and seawater. After Porcelli et al. (1997); modified.

son et al., 1998a, 1998b) (Fig. 2.17). The $^{234}U/^{238}U$ activity ratio in the Bothnian Bay was high and ranged from 1.21 to 1.60 while in the remaining subareas of the Baltic Sea showed a rather homogeneous distribution pattern with values of 1.15±0.10 (Löfvendahl, 1987). It demonstrates the prevailing influence of seawater having a mean ratio of 1.14–1.15 (Koide and Goldberg, 1965; Veeh, 1968; Ku et al., 1977; Sugimura and Mayeda, 1980). It is concluded that the larger rivers entering the Bothnian Bay have high $^{234}U/^{238}U$ activity ratio; surface waters with lower salinity are characterised also by the high values as a consequence of higher proportion of river water (Löfvendahl, 1987). Particles have high the activity ratio closed to that of dissolved U in the adjacent water indicating that U on particles is predominantly nondetrital and isotopically exchanges rapidly with the ambient dissolved U (Andersson et al., 1998a).

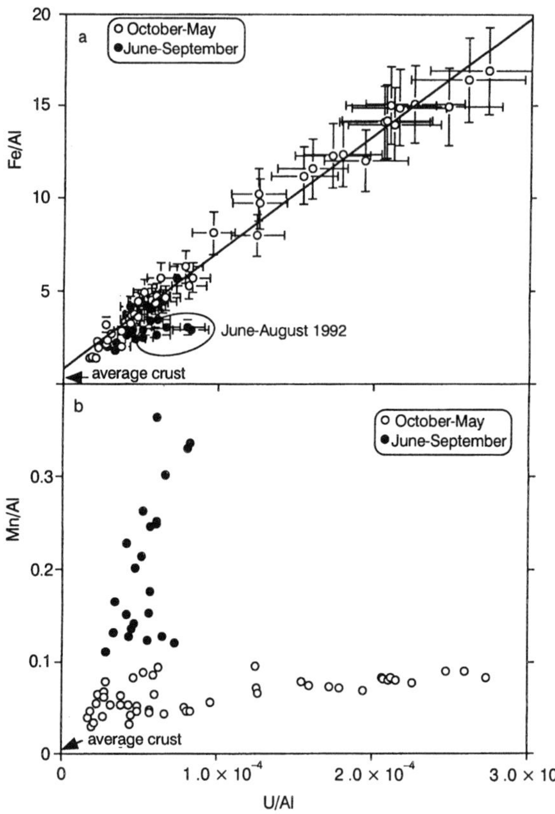

Fig. 2.17. (a) Particulate U/Al and Fe/Al ratios at Kamlunge during 1991-1992. The U/Al and Fe/Al ratios are always much greater than those of crustal material and are strongly correlated, indicating that authigenic Fe is the major carrier of nondetrital U. The data for June-August 1992 have unusually high U/Fe ratios and lie off the correlation line. (b) The U/Al and Mn/Al ratios in the Kamlunge particles. All samples are enriched in Mn and U relative to crustal material. There is no correlation evident between Mn/Al ratios, which varies considerably only during the summer, and U/Al ratios, which varies substantially from October to June. This indicates that authigenic Mn phases are not major hosts for nondetrital U in the Kalix. After Andersson et al. (1998a); modified.

The behaviour of Th in the Baltic Sea has been investigated sporadically (Andersson et al., 1995; Kersten et al., 1998); the latter reported activities of 'dissolved' and particulate ^{234}Th in the ranges of 1.4–6.9 and 0.9–9.3 mBq dm^{-3}, respectively. Based on the "colloidal pumping" model, Kersten et al. (1998) predicted that 98% of the 'dissolved' ^{234}Th in the Mecklenburg Bay is associated with colloids rather than is truly dissolved. According to Szefer (1977) the concentrations of total Th (^{232}Th) in coastal water of the Baltic Sea (Gulf of Gdańsk) ranged from 0.062 to 0.073 μg dm^{-3}. For comparison, concentration of total ^{232}Th in surficial 'open' waters of the North Sea amounted on the average 0.001 μg dm^{-3}. Bearing in mind that Th, in contrast to U, is highly adsorbed onto particles, the observed difference is a result of greater concentration of suspended matter (enriched in Th) in coastal Baltic waters than in open waters of the North Sea (Szefer et al., 1981a). Kowalewska (1986) discussed the distribution of ^{226}Ra in water of the southern Baltic.

Measurements of Sr isotopes, i.e. $^{87}Sr/^{86}Sr$ ratio in water of the Baltic Sea (Andersson et al., 1992, 1994) indicated significant correlation with salinity, however distinct deviations from a single mixing line were detected corresponding to the many rivers draining to the Baltic Sea. Studies were conducted on a profile across an oxic-anoxic boundary in the Baltic Sea and in the river drainage basin.

The concentrations of ^{143}Nd and ^{144}Nd in the Baltic Sea were also determined by Andersson et al. (1992). There is no correlation between Nd and Sm concentrations and salinity in water of the Baltic Sea; it means that these isotopes are nonconservative in their behaviour. The highest levels of Nd and Sm were found in the bottom waters indicating either resuspension of bottom sediments or scavenging by sinking particles in the water column. The increase in concentration with depth is similar to that detected in the oceans (Piepgras and Wasserburg, 1987) but this change in the Baltic Sea is more extreme.

(iv) Nutrients in Seawater

Nutrients in the Baltic Sea have been studied by several authors (Voipio, 1961; Nehring, 1984; Rosenberg, 1985; STUK, 1987; Elmgren, 1989; Wulff and Rahm, 1989; Wulff and Stigebrandt, 1989; Trzosińska, 1990, 1992; Wulff et al., 1990; Conley et al., 1993; Falkowska et al., 1993; Sandén and Rahm, 1993; Jonsson and Carman, 1994; Majewski and Lauer, 1994; Toompuu and Wulff, 1995; Rahm et al., 1996; Wulff et al., 1996; Conley et al., 1997; Laanemets et al., 1997; Stockenberg and Johnstone, 1997; HELCOM, 1990, 1993, 1996, 1998a; Pitkänen, 1991; Humborg et al., 1998; Siegel et al., 1998; Pihl et al., 1999; Savchuk and Wulff, 1999; Laima et al., 2001). Primary productivity in the southern Baltic has been estimated by some authors (Renk, 1990; IMGW, 1997–1998; Falandysz et al., 2000). Nutrient budget has been reported by Yurkovskis et al. (1993) and Wulff et al. (1996).

The eutrophication in the Baltic Sea has grown during the last 25 years (Wulff and Rahm, 1988; Grönlund and Leppänen, 1990; Kahma and Voipio, 1990; Nehring and Matthäus, 1990) which favours O_2 depletion and H_2S formation in stagnant deep waters (Nehring, 1996). The nutrient levels and phytoplankton growth have indicated increasing trends in the northern Baltic Sea (Wulff et al., 1990; Perttilä et al., 1995; Rahm et al., 1996) reflecting a large input of nutrients of anthropogenic in origin (Tuominen et al., 1998). According to Larsson et al. (1985) during the last century the loads of P and N to the Baltic Sea increased ca. 9- and 4-times, respectively. The study of nutrient budget in the Sea and its subareas has been started by Stigebrandt and Wulff (1987), Wulff and Rahm (1988) and Jonsson et al. (1990). The spatial trends in the concentrations of nutrients covering the whole Baltic Sea have been assayed sporadically (Wulff and Rahm, 1988; Wulff et al., 1993; Sandén and Danielsson, 1995). Toompuu and Wulff (1996) performed spatial analysis of monitoring data for evaluating of nutrient distribution in the Baltic Proper. From data obtained by Sandén and Danielsson (1995) clearly results that both the Gulf of Bothnia and the Gulf of Finland differ from other the Baltic subareas and that the spatial distribution of the nutrients depends on processes such as nature of phytoplankton growth, the upwelling of nutrient-rich water from deeper layers of water as well as on the large-scale currents in the sea

Spatial and seasonal trends

Significant differences in nutrient concentrations are observed between some subareas of the Baltic Sea. For instance, higher NO_3^- levels have been found in the northern part of the Baltic Sea, i.e. the Bothnian Bay and the Gulf of Finland; lower levels of PO_4^{3-} and SiO_3^{2-} have been detected in the Bothnian Bay and Bothnian Sea (Sandén and Danielsson, 1995). The concentrations of PO_4^{3-} during winter and autumn were significantly smaller in the Gulf of Bothnia and in spring they were also smaller in the North Baltic Proper. During this season the concentrations of PO_4^{3-} were generally remarkably greater in the Gulf of Finland than in the remaining subareas. The small winter and autumn PO_4^{3-} concentrations in the Gulf of Bothnia are partly caused by smaller load of P into this Baltic subarea (HELCOM, 1996; Larsson et al., 1995; Sandén and Danielsson, 1995). The precipitation of PO_4^{3-} mainly in the form of Fe and Mn phosphates is also responsible for supporting the concentration at low level in this region. Bearing in mind that P is the limiting nutrient for production in the Bothnian Bay with low overall production it is concluded that this leads to a low organic load on the sediment and in consequence to relatively high oxygen concentrations in the deep waters keeping the P recirculation to the water mass at a low level (Sandén and Danielsson, 1995). The greater concentration of PO_4^{3-} in the Gulf of Finland could be a result of a great load of P into the subarea causing higher primary production; more efficient recirculation of P from the sediments is then observed. Small concentration of P in spring and summer is attributed to the biological productivity in the entire Baltic Sea (Sandén et Danielsson, 1995).

According to Sandén and Danielsson (1995) concentrations of NO_3^- were mostly significantly greater in the Gulf of Finland and Bothnian Sea during winter and autumn. During spring and summer, however the Bothnian Bay was characterised by the greater concentrations of NO_3^- as compared to other the Baltic subareas; the Baltic Proper showed during summer the lowest values, predominantly being below the detection limit. It has been reported (Alasaarela et al., 1986; Granéli et al., 1990; Kivi et al. 1993) that either N in the Baltic Proper and the Gulf of Finland or P in the Bothnian Bay are the most limiting nutrients. High levels of NO_3^- in the Bothnian Bay are attributed to a low primary production and relatively great load of N to this subarea. Other parameters such as both oxic sediments and underlaying waters cause low denitrification of NO_3^- keeping its concentration at high level in the subarea. The exchange of water with the

Bothnian Sea subarea acts in the opposite direction because the concentrations of NO_3^- during winter and autumn are remarkably smaller in this basin (Sandén and Danielsson, 1995). The above mentioned term 'denitrification', i.e. bacterial reduction of NO_3^- to nitrogenous gases primarily N_2, is a very important process which may counteract the eutrophication process by removing dissolved inorganic N from the ecosystem (Schaffer and Rönner, 1984; Rönner, 1985; Rönner and Sörensson, 1985; Seitzinger and Nixon, 1985). Phosphorus vs. N abatement in the Gulf of Riga has been studied by Dahlberg et al. (1995).

Recently, the conditions in the Gulf of Finland are considered to be largely oxic; however, variations in concentrations of O_2 in deep water, predominantly caused by changes in advectional inflows of more saline and oxygen-poor waters from the Baltic Proper, are probably favourable for nutrient recycling processes in the Gulf of Finland (Andersin and Sandler, 1991; Conley et al., 1997). A result of the processes would be the loss of denitrification with deficiency of O_2 (Smith and Hollibaugh, 1989; Conley et al., 1997). Provided a significant amounts of inorganic P enrich Gulf of Finland sediments and reduced concentrations of O_2 would greatly increase sediment-water P fluxes and may deteriorate summer blue-green algal blooms in the Gulf of Finland (Pitkänen and Tamminen, 1995).

(v) General Remarks and Recommendations

According to Schneider (1996) the temporal trends for several trace elements in Baltic waters could be studied successfully after taking into account exclusively concentration data matrix for Cd, Cu and Zn. The concentrations of Hg and other elements are not included in the assessment since no data quality assurance could be made because of a lack of seawater reference material with their certified values. Difficulties with utilisation of the Pb data are caused by unsatisfied results from the analysis of seawater reference material characterised by insufficient precision and accuracy of Pb measurements.

The data obtained for Cd, Cu and Zn imply that analysis of trace elements in water is appropriate tool for monitoring metallic pollutants in the Baltic Sea. The temporal trends of less than even 10% yr^{-1} may be detected within a decade at a 95% probability level. It should be stressed however that these trends are generally not consistent with those resulted from analysis of trace elements in marine organisms. Therefore, biota levels of metals do not necessarily reflect their concentrations in water and can not by used as a surrogate to monitor the metallic pollutants in the Baltic Sea. Schneider (1996) strongly recommends including the complete priority list of trace metals (Cd, Cu, Hg, Pb, Zn) in water in the mandatory part of the BMP contributions of all Contracting Parties. Developing of modern analytical techniques leads to improve quality of the measurements and therefore gives hope that this problem will be solved in the near future. Since trace elements in surficial water, especially those named as nutrient-like, are involved into a seasonal cycle hence sampling strategy should be carefully designed and standardised. It is suggested (Schneider, 1996; Kremling et al., 1997) that water sampling should be restricted to selected offshore stations only and to the winter season (end of November–end of February) when concentration of trace elements is great and when deep convective mixing results in their even vertical distribution from the surficial layers to halocline. Therefore, samples should be collected from mid-water depths above the halocline. Due to potential risk of contamination during sample processing and taking into account low particulate metal levels (excluding Pb) unfiltered and acidified samples should be analysed exclusively. Monitoring programmes should be useful to detect changes in pollutant levels within the intra-annual variances (Kremling et al., 1997). Some secondary effects of the eutrophication are recently observed, e.g. drastic changes in water transparency in the Baltic Proper. For instance. in the southern Baltic during some 30 years, water transparency has been gradually decreased, from about 8–10 m to about 5–7 m. Similar changes in water transparency were found in the northern regions as well as in the Swedish coastal area of the Baltic Proper (HELCOM, 1993). The eutrophication of the Baltic is strongly associated with the distribution and fate of chemical elements. The concentrations of dissolved elements in the euphotic layer are diminished as a result of high organic production in the Baltic Sea. In addition to this dilute effect, biota may exude compounds being able to chelate trace elements or may enhance their deposition. Owing to organic associations, trace elements might become more accessible for uptake by phytoplankton homeostatically controlled or transformed to less toxic states (Brügmann et al., 1997). Low unit discharge of N and P reaching the Baltic Proper from the southern and eastern coasts could be attributed to the drastic reduction of the fertiliser consumption in the former countries in transition since 1989/1990, although according to Rosenberg et al. (1990) Poland and the Baltic states may provide a substantial contribution to the east Baltic Proper. Other explanation for low concentration of the nutrient in the Baltic Proper is given by Sandén and Danielsson (1995) who postulated that this subarea has a shortcoming in the location of monitoring stations as compared to the remaining Baltic subareas. Possible improvement of the economic situation of these countries in the future does not create a good perspective for the Baltic Sea, as far as the eutrophication processes are concerned (Falandysz et al., 2000).

D. PARTICULATE MATTER

(i) Introduction

Estuarine and coastal waters are influenced by suspended material, partly allochtonous in origin, which is transported through large rivers to the Baltic basin from the drainage area. Particles of Fe- and Mn-oxides and organic matter are effective sorbents for trace elements as well as are vehicles for their transport to the bottom sediments (Krauskopf, 1956; Lithner et al., 1996). The concentrations of metals and metalloids in the particulate matter depend on concentrations of the chemical element in the ambient water and major sorbents in the solids. This is partly reflected by the positive relationship often found between concentrations of trace elements and organic matter in Baltic sediments. Increasing salinity from the pelagial with settling particulate matter to the bottom with deposited sediments may be resulted in desorption of trace elements by ionic exchange (Lithner et al., 1996).

(ii) Chemical Elements in Suspended Matter

Geochemical composition of suspended matter

Concentrations of chemical elements in suspended matter of the Baltic Sea are presented in Table 2.9. Relative to global background levels, the particulate matter contained metal 'excesses' amounting to more than 90% of the total contents (Cd, Mn, Pb and Zn). Automated electron probe X-ray microanalysis (EPXMA) revealed that the elemental composition of Baltic sediments is mainly governed by post-depositional processes of early diagenesis and is only weakly related to the composition of suspended matter in the overlying water body (Bernard et al., 1989). For instance, in relation to surface mud sediments of the central Baltic net-sedimentation basins, Zn, Cd, Cu and Mn had 30–100% higher levels in the suspended materials. The general pattern of metal contents of particulate matter taken from the depth of 10 m on a transect between the Bothnian Bay and the North Sea was – possibly as a result of anthropogenic inputs – rather similar for Cu, Pb, and Zn. The distribution of Fe and Mn along the transect was probably governed by the natural loading pattern and by the biogeochemistry of those elements (Bernard et al. 1989). According to Boström et al. (1981) particulate transport is probably a major pathway for many elements which can end in the pelagic sediments. As can be seen in Fig. 2.18 there are the striking similarities between the Baltic suspended matter, Atlantic pelagic sediments and Pacific sediments settling in vicinity of the continents.

Dissolved and suspended concentrations of Al, Ba, Fe, Mn, and Si and suspended P and Ti have been studied in the Baltic Proper, the Belt Sea-Kattegat and the Åland Sea (Boström et al., 1981; Ingri et al. 1991). Three major components were distinguished: a detrital, a Mn rich and an organic phase, i.e. suspended Al, Ti, most Fe and partly Si (50%) were present in detrital phase while the amount of P in the detrital component was negligible. Suspended P showed a positive correlation to the non-detrital Fe concentration. Non-detrital Mn was strongly enriched in the suspended phase.

Detrital phase

Detrital particles originate from resuspended sediments within the Baltic Sea and from suspended material added by river discharge. The detrital fraction is

TABLE 2.9.

Concentrations of chemical elements ($\mu g\ g^{-1}$) in suspended matter of the Baltic Sea and other northern areas

Region	Sampling date	Sample depth (m)	Fraction (μm)	N	Salinity (PSU)	Al (% d.w.)	As	Ba	Be	Ca	References
Baltic	1965–72		> 0.5	41-193				210–710			Emelyanov, 1974
			> 0.5			1.1					Emelyanov and
						8.0–88					Pustelnikov, 1975a
						1.6		220			Boström et al., 1983
	1973					1.33–15.98^ ^		200–1170	< 1.0		Emelyanov, 1976
	1978		> 0.5	25–30				430			Gordeev et al., 1984
	1980–81		> 0.4	230		2.5					Brügmann, 1986
						14*					
	1984		> 0.4			4.19		1070			Brügmann et al, 1992
						0.39–12.9		33–3760			
	1984		> 0.4	15		2.0–50***		0.7–39***			Bernard et al., 1989
Baltic Proper	1984–85	Surface	> 0.45	/14/		7.8±5.5*		0.2±0.1* (N=10)			Ingri et al., 1991
		Bottom				8.5±7.0*		0.3±0.1* (N=14)			
Aland Sea	1984–85	Surface	> 0.45	2		10.1±4.9*		0.2±0.1*			
		Bottom		2		49.7±30.8*		0.4±0.3*			
Belt Sea	1984–85	Surface	> 0.45	13		8.0±4.6*		0.4±0.4*			
		Bottom		13		36.6±33.8*		0.4±0.3*			
Landsort Deep	1982	5–400	> 0.45	8		2.17–5.19		320–800		2.05–2.94^	Boström et al., 1988
E. Gotland	1991	5	> 0.45	1		0.5*				1.4*	Andersson et al., 1994
		125		1		0.6*				1.0*	
		225		1		3.3*				1.8*	
	1995	30	> 0.45	1	7.231	0.8^				3.8^	Andersson et al., 1998a
Swedish coast	1988–89			20			6.0–417				Lithner et al., 1996
Bothnian Bay and	1979			39		3.1±1.0**		470±220**			Boström et al., 1981
Northern Baltic											
	1995	80	> 0.45		3.276	2.9^				1.7^	Andersson et al., 1998a
	1996	64–73		12		1.98–6.27	4.5–25.8				Leivuori and Vallius, 1998.
Western Baltic											

* – $\mu g\ dm^{-3}$

** – Concentration in ashed material.

*** – Normalized concentration to 100 wt. %.

^ – %

^ ^ – Expressed as oxide (%).

TABLE 2.9. – continued

Region	Sampling date	Sample depth (m)	Fraction (μm)	N	Salinity (PSU)	Cd	Co	Cr	Cu	References
Baltic	1965–72		> 0.5	41–193			6	92	143	Emelyanov, 1974
							0.014*			
			> 0.5			5.9				Emelyanov and
						0.012*				Pustelnikov, 1975a
			> 0.4	100		5.8			61	Weigel, 1976, 1977
						0.0064*			0.078*	
								190	53	Boström et al., 1983
	1973					3.0–4.5	10.0–23.0		58–700	Emelyanov, 1976
	1980–81		> 0.4	72		1.7			91	Gustavsson, 1981
						0.026*			0.69*	
	1978		> 0.5	25–30			12	230	200	Gordeev et al., 1984
									0.32*	
	1980–81		> 0.4	230		2.9	7.9		7.3	Brügmann, 1986
						0.0051*	0.0037	0.13*	0.079*	
						4.6			99	Brügmann et al, 1992
						0.03–74.2*			17–1100*	
Southern Baltic	1977–80		> 0.4	13–68		15	6.8	130	760	Brzezińska et al., 1984
						0.026*	0.012*	0.29*	1.3*	
Gulf of Gdansk	1980	0–78		12		74			105	Skwarzec et al., 1988
						0.06*			0.12*	
Gdansk Deep	1980	0–105		20		66			259	
						0.06*			0.25*	
Slupsk Furrow	1980	0–89		18		68			103	
						0.06*			0.08*	
Bornholm Basin	1980	0–88		14		44			80	
						0.06*			0.11*	
	1992	10	> 0.4	1		2.8			64	Pohl and Hennings, 1999
		50–225		5		7.1			80	
						2.6–12.8			32–141	
		240		1		10.7			104	
	1993	10	> 0.4	1		1.6			4	
		50–225		4		9			142	
						2.3–17.9			20–241	
		240		1		2.2			33	
	1994	10	> 0.4	1		1.8			31	

Region	Sampling date	Sample depth (m)	Fraction (μm)	N	Salinity (PSU)	Cd	Co	Cr	Cu	References
		50–225		4		4.68			82.5	
						1.8–8.3			39–125	
		240		1		5.2			104	
	1995	10	> 0.4	1		0.6			20	
		50–225		7		4.47			126	
						0.6–10.5			38–287	
		240		1		8.7			88	
	1996	10	> 0.4	1		3.2			46	
		50–225		7		8.87			67	
						3.1–20.1			52–120	
		230–240		1		19			352	
Gotland Deep	1984	10	> 0.4	1	7.39	0.0007* (5.3)			0.04* (323)	Brügmann, 1988
		50–200		4	10.23	0.0017* (16.1)			0.02* (179)	
					7.39–12.4	0.0004–0.0046 (3.1–50.7)			0.02–0.03 (90–366)	
		235		1	12.36	0.0036* (13.3)			0.41* (1450)	
	1985	200–233	> 0.4	3		0.031**	0.05**		0.3**	Dyrssen and Kremling, 1990
Swedish coast	1988–89			20		0.6–2.5	17.0–29.0	28–46	39–374	Lithner et al., 1996
	1996	64–73		12		0.27–4.6		22.0–185	12.0–53	Leivuori and Vallius, 1998.
Western Baltic										
Coast of Warnemünde	1991–94		> 0.4	153		1.8±1.4				Schultz Tokos et al., 1993
Kiel/Mecklenburg Bights			> 0.4	149		2.1±2.0				Schneider and Pohl, 1996

* – $\mu g\ dm^{-3}$
** – $nmol\ kg^{-1}$

TABLE 2.9. – continued

Region	Sampling date	Sample depth (m)	Fraction (μm)	N	Salinity (PSU)	Fe (% d.w.)	Ge	K	Mg	Mn (% d.w.)	References
Baltic	1965–72		> 0.5	41–193		1.8				0.11	Emelyanov, 1974
						69*				3.0*	
			> 0.5			1.8				0.1	Emelyanov and
						49*				2.5*	Pustelnikov, 1975a
			> 0.4	100		0.4					Weigel, 1976, 1977
						5.2*					
						1.1**				0.16	Boström et al., 1983
	1973					0.41–7.66	< 5.0			0.02–0.44	Emelyanov, 1976
	1980–81		> 0.4	72		1.6				0.15	Gustavsson, 1981
						220*				18*	
	1978		> 0.5	25–30		1.7				0.45	Gordeev et al., 1984
										7.9*	
	1980–81		> 0.4	230		1.5				0.17	Brügmann, 1986
						8.3*				2.9*	
						2.49				0.63	Brügmann et al, 1992
						0.36–9.88				0.03–19.6	
	1984		> 0.4			1.2–49***				0.8–51***	Bernard et al., 1989
Southern Baltic	1977–80		> 0.4	13–68		0.4					Brzezińska et al., 1984
						8.3*					
Baltic Proper	1984–85	Bottom	> 0.45	12		7.1±5.4*				2.5±1.7*	Ingri et al., 1991
		Surface		14		8.2±6.7*				73.5±104.5*	
Aland Sea	1984–85	Bottom	> 0.45	2		8.0±3.9*					
		Surface		2		33.6±18.6*				8.5±6.0*	
Belt Sea	1984–85	Bottom	> 0.45	9		6.9±4.4*				1.3±1.1*	
				13		24.8±23.6*				3.9±2.3*	
Landsort Deep	1982	5–400	> 0.45	8		2.86±6.72				0.19±6.43	Boström et al., 1988
E. Gotland	1991	5	> 0.45			0.2		1.7*	1.6*	0.04*	Andersson et al., 1994
		125				1.3		0.9*	2*	7.43*	
		225				2.4		2.3*	3.2*	0.03*	
	1995	30	> 0.45	1	7.231	0.56		4.3'	5.5'	0.64'	Andersson et al., 1998a

Region	Sampling date	Sample depth (m)	Fraction (μm)	N	Salinity (PSU)	Fe (% d.w.)	Ge	K	Mg	Mn (% d.w.)	References
	1992	10	> 0.4	1						383^	Pohl and Hennings, 1999
		50–225		5						6990^	
										81–31155	
		240		1						147^	
	1993	10	> 0.4	1						420^	
		50–225		4						20560^	
										580–44268	
		240		1						775^	
Gotland Deep	1984	10	> 0.4	1	7.39	1.3* (1.0)				0.31* (0.23)[a]	Brügmann, 1988
		50–200		4	7.39–12.22	1.2–7.9 (0.8–5.0)				0.14–27.3 (0.06–17.7)[a]	
		235		1	12.36	1.3* (0.5)				0.09* (0.03)[a]	
Swedish coast	1988–89			20		0.43–0.52				0.09–0.34	Lithner et al., 1996
Bothnian Bay and Northern Baltic	1979			39		23.0±12**				1.6±0.2**	Boström et al., 1981

* – $\mu g\ dm^{-3}$
** – Expressed as oxide.
*** – Normalized concentration to 100 wt. %.
^ – $\mu g\ g^{-1}$
[a] – (%)

TABLE 2.9. – continued

Region	Sampling date	Sample depth (m)	Fraction (μm)	N	Salinity (PSU)	Mo	Na	Ni	P	Pb	References
Baltic	1965–72		> 0.5	41–193				100			Emelyanov, 1974
								0.32*			
			> 0.5					110			Emelyanov and
								0.28*			Pustelnikov, 1975a
			> 0.4	100						120	Weigel, 1976, 1977
										0.140*	
								16			Boström et al., 1983
	1973					< 5.0–11		25–200	0.03–0.25**		Emelyanov, 1976
	1980–81		> 0.4	72						120	Gustavsson, 1981
										0.81*	
	1980–81		> 0.4	230				27		140	Brügmann, 1986
								0.016*		0.063*	
										92	Brügmann et al, 1992
										20–292	
	1984		> 0.4						0.6–22.0***		Bernard et al., 1989
Southern Baltic	1977–80		> 0.4	13–68						120	Brzezińska et al., 1984
										0.18*	
Gulf of Gdansk	1980	0–78		12						240	Skwarzec et al., 1988
										0.18*	
Gdansk Deep	1980	0–105		20						952	
										0.80*	
Slupsk Furrow	1980	0–89		18						245	
										0.35*	
Bornholm Basin	1980	0–88		14						251	
										0.34	
Baltic Proper	1984–85	Bottom	> 0.45	15					13.9±8.86.4		Ingri et al., 1991
		Surface		14					4.6±3.1		
Aland Sea	1984–85	Bottom	> 0.45	2					12.6±4.5*		
		Surface		2					2.7±0.3*		

Region	Sampling date	Sample depth (m)	Fraction (μm)	N	Salinity (PSU)	Mo	Na	Ni	P	Pb	References
Belt Sea	1984–85	Bottom	> 0.45	13					10.4±8.8*		
				13					5.3±1.7*		
E. Gotland	1991	5	> 0.45				7*				Andersson et al., 1994
		125					8*				
		225					16*				
	1995	30	> 0.45	1	7.231		22[b]		7.48[b]		Andersson et al., 1998a
	1992	10	> 0.4	1						32	Pohl and Hennings, 1999
		50–225		5						32.4	
										24–38	
		240		1						26	
	1993	10	> 0.4	1						5	
		50–225		4						56.5	
										17–116	
		240		1						21	
Gotland Deep	1984	10–235	> 0.45	6	7.39–12.36			30–218	0.34–2.88[a]	28–138	Boström, 1988
								0.007–0.026*		0.005–0.022*	
Swedish coast	1988–89			20				5.0–27		31–350	Lithner et al., 1996

* – μg dm^{-3}
** – Expressed as oxide (%).
*** – Normalized oxide concentration to 100 wt. %.
[a] – μmol dm^{-3}
[b] – %

TABLE 2.9. – continued

Region	Sampling date	Sample depth (m)	Fraction (μm)	N	Salinity (PSU)	S	Si	Sn	Sr	Ti	V	Zn	Zr	References
			> 0.4	100								730		Weigel, 1976, 1977
												0.91*		
												300		Boström et al., 1983
	1973						24.5–93.7**	< 6.0–12		0.06–93.7**	18–150	225–1080	< 40–180	Emelyanov, 1976
	1980–81		> 0.4	72								950		Gustavsson, 1981
												1.4*		
	1978		> 0.5	25–30								750		Gordeev et al., 1984
												1.4*		
	1980–81		> 0.4	230								270		Brügmann, 1986
												0.10*		
												424		Brügmann et al, 1992
												110–6410		
	1984		> 0.4			1.6–38**	10.0–86.0***			0.8–58***				Bernard et al., 1989
Southern Baltic	1977–80		> 0.4	13–68								1200		Brzezińska et al., 1984
												1.6*		
Gulf of Gdansk	1980	0–78		12								1122		Skwarzec et al., 1988
												1.3*		
Gdansk Deep	1980	0–105		20								1935		
												1.7*		
Slupsk Furrow	1980	0–89		18								3200		
												2.8*		
Bornholm Basin	1980	0–88		14								1770		
												2.0*		
Baltic Proper	1984–85	Bottom	> 0.45	15			66.0±0.50*			0.5±0.4*				Ingri et al., 1991
		Surface		14			46.0±37.0*			0.5±0.4*				
Aland Sea	1984–85	Bottom	> 0.45	2			115±1.0*			0.8±0.2*				
		Surface		2			155±93.0*			2.7±1.5*				

Region	Sampling date	Sample depth (m)	Fraction (μm)	N	Salinity (PSU)	S	Si	Sn	Sr	Ti	V	Zn	Zr	References
Belt Sea	1984–85	Bottom	> 0.45	13			77.0±55.0*			0.5±0.3*				
				13			150±90.0*			2.0±1.8*				
E. Gotland	1991	5	> 0.45				7.2*		0.02*					Andersson et al., 1994
		125					7.3*		0.12*					
		225					16.4*		0.03*					
	1995	30	> 0.45	1	7.231		6.1[a]			0.064[a]				Andersson et al., 1998a
Gotland Deep	1984	10	> 0.4	1	7.39							0.05* (0.04)[a]		Brügmann, 1988
		50–200		4	10.23							0.38* (0.28)[a]		
					7.39–12.36							0.10–0.51 (0.06–0.064)		
		235		1	12.36							0.78* (0.28)[a]		
	1988–89			20								161–529		Lithner et al., 1996
Gulf of Finland	1996	64–73		12						743–2760	29–98	78–317		Leivuori and Vallius, 1998
Bothnian Bay and Northern Baltic	1979			39					370±33[b]	1600±400[b]	76±2[b]			Böstrom et al., 1981
Western and southern Baltic	1995	80	> 0.45		3.276		12.6[a]			0.16[a]				Andersson et al., 1998a

* – mg g^{-1}
** – Expressed as oxide.
*** – Normalized concentration to 100 wt. %.
[a] – (%)

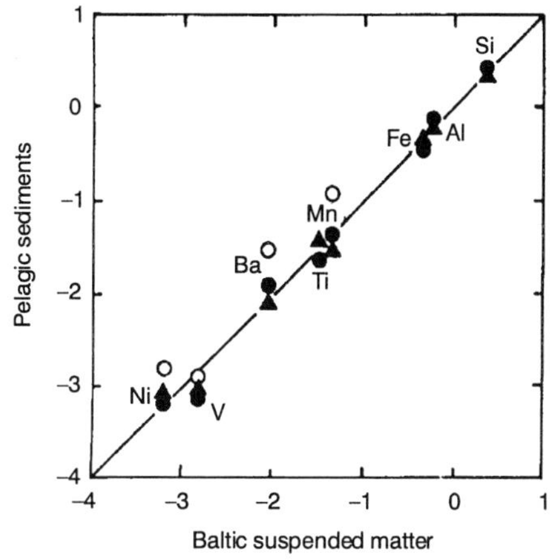

Fig. 2.18. Comparison of Baltic suspended matter with pelagic sediments: (▲) = mean Atlantic Ocean pelagic sediment; (●) = mean Pacific Ocean pelagic sediments, formed close to continents; (○) = mean data for total Pacific pelagic sediments (only shown for Mn, Ba, V and Ni since Al, Ti, Fe and Si values are similar). Before plotting all data have been normalized to a constant Σ(Al + Fe); all values in logarithmic abundances. After Boström et al. (1981); modified.

consisted mainly from quartz, K-rich and Fe-rich aluminosilicates (Bernard et al. 1989). According to Boström et al. (1981) there is a significant correlation between concentrations of Fe, Si, Ti, and Al in Baltic suspended matter suggesting that these elements are mostly present in a detrital component. Similar distribution pattern has been reported by Ingri et al. (1991) who also found suspended Al, Ti, most Fe and 50% of suspended Si in detrital component.

Normalisation to Al therefore indicates to what extent other elements are enriched in suspended matter because of organic and other authigenic material (Sholkovitz and Price, 1980; Guo et al., 2000). Some particulate Fe/Al ratios for the Baltic Proper were higher than range of 0.5–0.7 suggesting predominant contribution of Fe in the detrital fraction, although a sample from the Landsort Deep, strongly enriched in Fe, was described by very high value of an Fe/Al ratio amounting to 16 (Ingri et al., 1991). Therefore, distribution of chemical elements in geochemical components other than detrital, is described below using concentration ratio of given element to Al (Ingri et al., 1991).

Detrital-authigenic phase

According to several authors (Boström et al., 1988; Bernard et al., 1989) the abundance of the Fe-rich suspended phase is highly variable (< 7%), however in the Skagerrak deep water and especially under nearly anoxic conditions of the Bornholm Basin, much higher relative values were obtained. The bottom of the latter area is favourable for authigenic formation of the Fe phosphates and Fe oxides/hydroxides at or near the oxic/anoxic boundary (Davison et al., 1980, Bernard et al., 1989). The concentrations of suspended non-detrital Fe significantly correlated with those of P both in subsurface and bottom water (Ingri et al., 1991). Besides detrital fraction, non-detrital Fe in estuarine suspended matter has been suggested to be present as oxyhydroxide, ferriphosphate and in organic matter (Price and Calvert, 1973; Ingri et al., 1991). Ingri et al. (1991) concluded that it is not possible to distinguish whether the Fe-P relation was a result of scavenging of P by Fe-oxyhydroxide or/and the presence of P together with Fe in the organic fraction. According to several authors (Emelyanov and Pustelnikov, 1975a. 1975b; Görlich et al., 1989; Szefer et al., 1995) non-detrital Fe was present as an oxyhydroxide in the Baltic Sea. It is postulated (Szefer et al., 1995; Szefer, 1998) that Fe-Mn phase is responsible mainly for the deposition of labile, easily extractable forms of Ag, Cd, Cu, Pb, Zn, and P in the Vistula estuary. These elements are most probably scavenged by Fe- and Mn-oxyhydroxides at the hydrological front where mixing of the Vistula river water with the brackish Baltic Sea water takes place.

The suspended Si/Al ratio suggests two different trends (Ingri et al., 1991). Except for subsurface samples, concentrations of suspended Si increased with those of suspended Al. This is postulated that Si, to a large extent, was present in detrital particles. Many subsurface samples were high in suspended Si without a corresponding enhancement in suspended Al. The Si/Al ratios in the Baltic Proper and Belt Sea-Kattegat were twice the ratio in average Earth's crust, indicating that a large authigenic phase was present in subsurface samples. Microscopic investigation of the particulate fraction in the Baltic Sea has shown that diatoms are abundant (Emelyanov and Pustelnikov, 1975a). It thus seems reasonable to suggest that most of the non-detrital Si in subsurface samples was present as diatoms.

Non detrital phase

According to Ingri et al. (1991) most Mn in oxygenated water in the Baltic Proper was in the suspended matter, whereas in the Belt Sea the major portion was in the dissolved phase. In contrast to suspended Fe, the major fraction of suspended Mn was in a non-detrital form. The average particulate Mn/Al ratio in the Baltic Proper including most samples from deeper basins was 27.5, i.e. almost three orders of magnitude higher than the ratio for average Earth's crust. This is in an agreement with data obtained by Boström et al. (1981) resulting in insignificant correlation of Mn with Al or Fe. It means that most Mn is probably admixed in suspended matter as a non-terrigenous phase. Likewise Ni is not significantly correlated with Al suggesting that it is partly present in biogenous phase. The identified Mn-rich particles are suspected to have Mn-oxides/hydroxides and/or carbonates (Fig. 2.19), some of them contain significant quantities of Si and Fe (Bernard et al., 1989). The relative concentration of Mn in Fe-rich particles is controlled by the redox conditions. The Mn^{+2} migrates out of the reducing sediment and anoxic adjacent water layer and next is oxidised under oxic conditions to particulate Mn component (Bernard et al., 1989). During anoxic conditions in

Fig. 2.19. Electron micrograph of (A) $BaSO_4$ particle, (B) Fe-rich particle, (C) Mn-rich particle and (D) Zn-rich particle. The bar on each photograph represents 1 μm. After Bernard et al. (1989).

the deep water layers, e.g. in the Gotland Basin the presence of Mn and S dominates trace element distribution. During anoxic conditions in deep water layer of the Gotland Basin the formation of metal sulphides on the surfaces of clay minerals took place, however this process is reversible resulting in metals release from surface sediments to the water column under oxic conditions. However, the dissolved species of Cd and Pb are scavenged out of the water column again with Mn precipitates (Pohl and Hennings, 1999). According to Boström et al. (1988) any sinking Mn-rich particles would dissolve in the anoxic zone leading to new upward migration of dissolved Mn and renewed precipitation at the redoxcline. It is much probable that such formed Fe-Mn-rich particles are deposited as sediments and Fe-Mn-concretions where the redoxcline layer reaches hilly bottom. In estuarine Baltic areas such as the northern Bothnian Bay or the Gulf of Gdańsk the hydrogeochemical behaviour of Mn is also similar to that for Fe. The Swedish rivers and the Polish Vistula River transport the non-detrital suspended Mn phase to the Baltic Sea (Pontér et al., 1990; Szefer et al., 1995). According to Pontér et al. (1990)

a combination of increased pH, temperature and particulate Mn triggered the precipitation of dissolved Mn.

According to Ingri et al. (1991) an additional non-detrital phase was present in the Baltic suspended matter to account for the enhanced Ba concentration. Hence, particulate Ba seems to be distributed between a detrital, a Mn-rich and at least one more authigenic phase. Boström et al. (1981) also demonstrated that much Ba in Baltic suspended matter must have a non-terrigenous origin. It has been shown by Bernard et al. (1989) that Ba-S rich particles identified in the Baltic Sea are also barite mineral grains (Fig. 2.19). Most of samples had higher relative concentration of Ba+S than 2% whereas this component was generally low in the Gulf of Bothnia and the Gulf of Finland. However, anthropogenic input of barite as a constituent of oil-drilling mud is also possible (Holmes, 1982) because drilling activities have increased during last 20 years in the Baltic Sea (Bernard et al., 1989).

Most suspended P was present in organic matter, although scavenging by non-detrital Mn and Fe also takes place. According to Bernard et al. (1989) P is present in significant relative quantities in suspended particles classified partly as organic. There is a strong linear correlation between concentrations of P and Fe oxyhydroxides in Baltic ferromanganese nodules and surface sediments (Winterhalter and Siivola, 1967).

Spatial and temporal (depth) trends

With the exception of most industrialised areas, i.e. Oxelösund and Rönnskär smelters, the Pb levels in particulate matter increase towards south, similarly to the atmospheric deposition pattern for Pb. Concentrations of particulate As increase towards the north in the vicinity of the Rönnskär smelter. Lithner et al. (1996) reported significant correlation between concentrations of Pb and As in particulate matter and known temporal trends from the Baltic Sea. The Pb concentration in particulate matrixes was 37% (after normalisation 50%) lower than in surficial sediments of the Bothnian Sea. Bearing in mind that a net deposition rate is estimated to be 1 mm yr^{-1}, this means that it corresponds to the years 1979–1988. Data on land mosses have indicated that the atmospheric fallout of Pb was reduced in Sweden by ca. 40–50% from late 1970's to late 1980's (Rühling et al., 1992). This great agreement between atmospheric and suspended matter data supports potential abilities of the latter material to monitor temporal variations in pollutant levels, e.g. Pb in the Baltic environment. The concentration of As in the water was 30% lower in 1987 than in 1981; similar temporal trend was also detected for As in suspended matter in the Bothnian Sea (Lithner et al., 1991, 1996). Concentration of Cd in the Bothnian Sea, except locally polluted sites such as Oxelösund and Rönnskär, was mostly higher in suspended matter than in surficial sediments. There was no significant difference between distribution patterns in remote areas and other areas, possibly caused by a recent increasing of Cd in the whole area. This finding is agreeable to the temporal trends for Cd distribution in herring liver showing a gradual increase during the 1980's in the Baltic Proper and also in the Gulf of Bothnia (Lithner et al., 1996).

Broman et al. (1994) observed the spatial and the seasonal variations of flux and concentration of elements such as Al, Cd, Co, Cr, Cu, Fe, Hg, Mn, Ni, Pb and Zn, organic matter and C and N in settling particulate matter collected with sediment traps during seven inter-connective, continuous periods totalling 15 months. Among the elements studied, Cu, Hg, Pb and Cd exhibited the most elevated levels in the interior of the area explored which decreased markedly further out in the Stockholm Archipelago, indicating local anthropogenic input. Zn, Cr and Fe also displayed signs of supply from the urbanised area. The flux of most the elements studied revealed both spatial and seasonal relationship with the weight of the particulate matter (Broman et al., 1994).

Elemental partitioning between dissolved and particulate fraction

According to Santschi et al. (1997) the solution/particle partitioning of element ion is controlled by solution/particle partitioning of the organic ligand. The partition coefficient (K_d) is the ratio between the metal concentration of the suspended matter and the dissolved metal concentration (Li et al., 1984b; Balls, 1989; Brügmann et al., 1992; Turner, 1996):

$$K_d = Me_{part.} / Me_{diss.}$$

Ideally, the coefficient should reflect the distribution of metal in equilibrium state between the two phases, and the exchange reactions, e.g. adsorption/desorption, oxidation/reduction, precipitation/dissolution and ingestion/excretion, should be reversible within some reasonable time scale (Kremling et al., 1997). This coefficient is dependent on pH, salinity, oxygen concentration and particle concentration as well as the particle size and nature (Bourg, 1987; Pohl and Hennings, 1999). The oxygen influence is well reflected by changes of the Fe/Mn ratio in the bottom layer water. When Fe and Mn are released from sediments under anaerobic conditions, increase of oxygen concentration causes re-oxidation and precipitation of Fe and Mn, however with more higher oxidation rate for Fe (Fig. 2.20) (Kremling et al., 1997). Pohl and Hennings (1999) discussed partition dynamics of trace elements in the eastern Gotland showing significant temporal trends for Cd, Pb and Cu. In the interpretation of observed seasonal changes of the metal concentrations, hydrographic changes between 1992 and 1996 in this area should be taken into account. The influx of dense, oxygen-rich waters of the North Sea via the Danish Straits in January 1993 caused increase both the salinity and the oxygen content in the deep water of the Bornholm and Gdańsk Basin (Nehring et al., 1994; Brügmann et al., 1997, 1998). This event had less impact to the eastern area of the Gotland Basin; only in the beginning of July 1993 the water column was practically free of H_2S, however in November in the same year the bottom water had again becomes anoxic. The next inflows of oxygen-rich water in December 1993 and throughout 1994 lead to a significant improvement in the oxygen conditions in the deep waters of the eastern Gotland Basin (Matthäus, 1994;

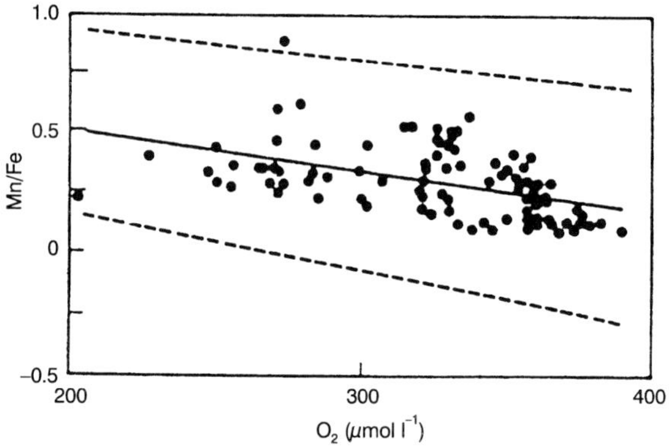

Fig. 2. 20. Distribution of Mn/Fe mass ratio in SPM vs oxygen concentration in winter samples, with indicated 95% confidence limits; Mn/Fe = 0.85 – 0.0017 x oxygen concentration; $p < 0.01$; n = 94. After Kremling et al. (1997); modified.

Nehring et al., 1994, 1995; Pohl and Hennings, 1999). These hydrographic changes have influenced significantly partition dynamics of trace elements. For instance, the Cd concentration increased with depth water in 1993, however because of oxygen-rich saltwater inflow in the deep layers Cd quantities have been possibly adsorbed onto Mn-oxyhydroxides precipitated in the water column upon oxic conditions. This process is well reflected by the vertical distribution pattern of particulate Cd showing a 3-fold increase with depth in 1994 and 5-fold increase in 1996. The vertical increasing the K_d values demonstrates a stronger affinity to the particulate matter (Pohl and Hennings, 1999). As regards Pb, in the anoxic bottom waters ca. 2-fold decrease its concentration was observed during 1992 and 1993. This difference can be explained by the episodic influx of oxygen-rich water from the North Sea resulting in changes of the partitioning between both dissolved and particulate Pb in favour of this latter fraction. It is assumed that Pb is adsorbed by Mn precipitates (Pohl and Hennings, 1999). Partitioning of Cu has been also resulted from the changes of hydrographic conditions in the bottom of the central deep basins. As a result, in the years 1993–1996 the increase in dissolved Cu in the deep waters of the Gotland Basin took place. The low concentration of dissolved Cu in 1993 in the Basin reflected primordial concentration of this element in the inflow of North Sea water. The higher levels detected in the subsequent years may be caused by the dissolution of Cu compounds from the bottom sediments under oxic conditions resulting in the enrichment of dissolved Cu in the deep waters. Approximately 40% of total Cu concentration in the deep layer was bound to suspended matter in 1992 probably in the form of insoluble Cu-S precipitates; the proportion of particulate Cu decreased gradually in subsequent years reaching minimum value of < 5% in 1995 and 1996. These changes are reflected by the K_d values dependent on the formation of dissolved Cu species, i.e. hydroxy-, chloro- and organic-complexes (Pohl and Hennings, 1999). For Cd, Fe and Pb, the K_d values decrease with the water depth (Brügmann et al.,

1992). This may be a result of the major pathway of entry of these elements into the Baltic Sea via the atmosphere (Cd, Pb), the effective binding by phytoplankton in the euphotic layer (Cd) and the release of trace elements from the particulate matter at depth caused by altered redox conditions (Fe). Suspended matter from the microlayer and from 0.2 m depth exhibited high K_d values for Cd, Cu, Fe and Pb, similarly to particulates having very high Mn levels. This can be explained by adsorption, absorption and/or co-precipitation of these elements with Mn oxide (Brügmann et al., 1992).

According to Sokolowski et al. (2001) in the deep zone of the Vistula estuary, Gulf of Gdańsk, desorption from detrital and/or resuspended particles by aerobic decomposition of organic material may be the main mechanism responsible for enrichment of particle-reactive metals (Cu, Pb, Zn) in the overlying bottom waters. The increased concentrations of dissolved Fe may have been due to reductive dissolution of Fe oxyhydroxides within the deep sediments by which dissolved Ni was released to the water. The distribution of Mn was related to dissolved oxygen concentrations, indicating that Mn is released to the water column under oxygen reduced conditions. However, Mn transfer to the dissolved phase from anoxic sediments in deeper part of the Vistula plume was hardly evidenced suggesting that benthic flux of Mn occurs under more severe reductive regime than is consistent with mobilisation of Fe. Behaviour of Mn in a shallower part has been presumably affected by release from pore waters and by oxidisation into less soluble species resulting in seasonal removal of this metal (e.g. in April) from the dissolved phase. The particulate fractions, varied from ca. 6% (Ni) and 33% (Mn, Zn, Cu) to 80% (Fe) and 89% (Pb) of the total (labile particulate plus dissolved) concentrations. The affinity of the metals for particulate matter decreased in the following order: Pb > Fe > Zn $\geq$ Cu > Mn > Ni. Significant relationships between particulate Pb-Zn-Cu reflected the affinity of these metals for organic matter, and the significant relationship between Ni-Fe reflected the adsorption of Ni onto Fe-Mn oxyhydroxides in estuarine waters of the Gulf of Gdansk (Sokolowski et al., 2001).

(iii) Radionuclides in Suspended Matter

The activity concentrations of several radionuclides in Baltic suspended particulate matter are mostly greater than those in the top segments of sediments at the same sampling site. Therefore particulate matter may be more sensitive monitor of some radionuclides occurred at low levels in the marine environments. Analysis of this material collected from two coastal areas close to the Finnish nuclear power plants (NPPs) indicated that fraction of the Chernobyl-derived radiocaesium in samples highly exceeded its local NPPs contribution (Ilus and Ilus, 2000). Particulate matter from the Baltic coastal zone has been studied in respect to sorption and release of radiocaesium (Knapińska-Skiba et al., 1997).

Recent work in the Baltic Sea has shown evidently that U is removed from both anoxic and oxic waters (Anderson et al., 1995) suggesting potential important role of Fe-Mn-oxyhydroxide phase in redistributing U in Baltic water column (Andersson et al., 1998a). Apparent partition coefficient (K_d) was calculated for U between the authigenic Fe on particles and the solution. This value appeared to be relatively constant throughout the year indicating possible equilibrium between Fe in solution and authigenic Fe-oxyhydroxides on detrital particles. High values of K_d computed for one summer with simultaneous high concentration of U in brackish waters can be explained by U scavenging by biogenic phases with low authigenic Fe concentration (Andersson et al., 1998a). A complementary study of U transport in river watershed and the Baltic Sea has been performed by Porcelli et al. (1997). Within the Baltic Sea ca. 50% of U is removed at low salinity. The proportion that is lost corresponds to that of river-derived colloid-bound U; it means that while the dissolved form of this radionuclide behaves conservatively during estuarine mixing process, colloidal form of U is lost due to rapid flocculation of colloidal material. Hence, the association of U with colloids may be an important parameter in tracing U behaviour in estuarine systems (Porcelli et al., 1997; Andersson et al., 1998b).

Andersson et al. (1992, 1994) conducted studies on a profile across an oxic-anoxic boundary in the Baltic Sea and on inflowing rivers in respect to behaviour

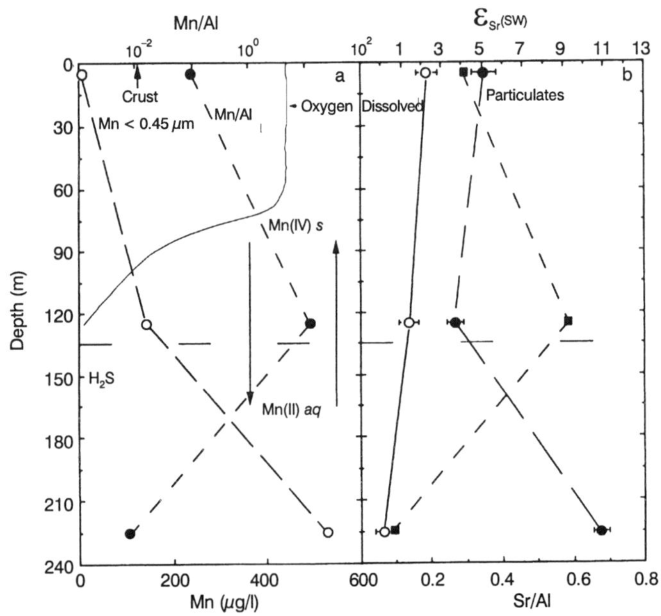

Fig. 2.21. Vertical profile at station BY-15 in the Baltic Sea (see Fig. 2.3). (a) Dissolved Mn load in μg/l (circles) and Mn/Al ratio in the particulate load (dots). Thin solid line shows the dissolved oxygen, varying from 100% saturation in surface waters to ~ 3% at 125 m. The redoxboundary (horizontal dashed line) is drawn between 125 m, where O_2 ~ 0.2 ml/l, and 150 m, where H_2S is present. The arrows show the migration of dissolved (*aq*) Mn(II) from the anoxic water into the oxic, where it oxidizes to insoluble (*s*) Mn(IV). These oyhydroxides fall in the water and redissolve in the anoxic water. (b) ε_{Sr}(SW) in dissolved (circles) and particulate load (dots) and particulate Sr/Al (squares). After Andersson et al. (1994); modified.

of Sr isotopes over an annual cycle. The $^{87}Sr/^{86}Sr$ ratio generally differed between particulate and dissolved fractions, with greater contribution of radiogenic Sr to the particulate loads, attributing to differential weathering of minerals. It is found that minerals with high Rb/Sr ratio predominantly occurred in the particulate load in contrast to dissolved load characterised by its low value (Andersson et al., 1994). A strong correlation was reported for the pairs Sr-Al, Fe-Al and Mn-Al in the particulate matter in brackish Baltic waters and fresh waters. Sr is removed from water phase both in rivers and the Baltic Sea in the presence of Fe- and Mn-oxyhydroxide particulates. The settling particles are dissolved in anoxic waters resulting in Sr release (Fig. 2.21); hence is considered as only quasi-conservative whenever there is formation or dissolution of Fe- and Mn(OOH) (Andersson et al., 1992, 1994).

References

Aarkrog, A., L. Bøtter-Jensen, H. Dahlgaard, H. Hansen, J. Lippert, S.P. Nielsen and K. Nilsson, 1980. Environmental Radioactivity in Denmark in 1979. Risø-R-403 (Risø National Laboratory, Denmark).

Aarkrog, A., H. Dahlgaard and S. Boelskifte, 1986. Transfer of radiocesium and ^{90}Sr from Sellafield to the Danish Straits, in: Study of Radioactive materials in the Baltic Sea. (International Atomic Energy Agency, Vienna). Report (IAEA-TECDOC-362) of the Final Research Co-ordination Meeting on the Study of Radioactive Materials in the Baltic Sea organized by the IAEA and held in Helsinki, Finland 24–28 September, 1984, pp. 32–51.

Abaychi, J.K., and A.A.Z. DouAbal, 1985. Trace metals in Shatt Al-Arab River, Iraq. Water Res. 19, 457–462.

Abdullah, M.I., Z. Shiyu and K. Mosgren, 1995. Arsenic and selenum species in the oxic and anoxic waters of the Oslofjord, Norway. Mar. Pollut. Bull. **31**, 116–126.

Adams, F., M. Van Craen, P. Van Espen and D. Andreuzzi, 1980. The elemental composition of atmospheric aerosol particles at Chacaltaya, Bolivia. Atmos. Environ. **14**, 879–893.

Ahl, T., 1977. River discharges of Fe, Mn, Cu, Zn and Pb into the Baltic Sea from Sweden. Ambio Spec. Rep. **5**, 219–228.

Alasaarela, E., E. Tolonen and V. Eloranta, 1986. Nutrients regulating algal growth in the Bothnian Bay. Ophelia Suppl. **4**, 323–328.

Alonso-Rodríguez, R., F. Páez-Osuna, F. and R. Cortés-Altamirano, 2000. Trophic conditions and stoichiometric nutrient balance in subtropical waters influenced by municipal sewage effluents in Mazatlán Bay (SE Gulf of California). Mar. Pollut. Bull. **40**, 331–339.

Amin, B.S., S. Krishnaswami and B.L.K. Somayajulu, 1974. $^{234}Th/^{238}U$ activity ratios in Pacific Ocean bottom waters. Earth Planet. Sci. Letters **21**, 342–344.

Anbar, A.D., G.J. Wasserburg, D.A. Papanastassiou and P.S. Anderson, 1996. Iridium in natural waters. Science **273**, 1524–1528.

Andersin, A.-B., and H. Sandler, 1991. Macrobenthic fauna and oxygen deficiency in the Gulf of Finland. Memb. Soc. Fauna Flora Fennica **67**, 3–10.

Anderson, R.F., 1982. Concentration, vertical flux, and remineralization of particulate uranium in seawater. Geochim. Cosmochim. Acta **46**, 1293–1299.

Andersson, P.S., G.J. Wasserburg and J. Ingri, 1992. The sources and transport of Sr and Nd isotopes in the Baltic Sea. Earth Planet. Sci. Letters **113**, 459–472.

Andersson, P.S., G.J. Wasserburg, J. Ingri and M.C. Stordal, 1994. Strontium, dissolved and particulate loads in fresh and brackish waters: the Baltic Sea and Mississippi Delta. Earth Planet. Sci. Letters **124**, 195–210.

Andersson, P.S., G.J. Wasserburg, J.H. Chen, D.A. Papanastassiou and J. Ingri, 1995. ^{238}U-^{234}U and ^{232}Th-^{230}Th in the Baltic Sea and in river water. Earth Planet. Sci. Letters **130**, 217–234.

Andersson, P.S., D. Porcelli, G.J. Wasserburg and J. Ingri, 1998a. Particle transport of ^{238}U-^{234}U in the Kalix River and in the Baltic Sea. Geochem. Cosmochim. Acta **62**, 385–392.
Andersson, P.S., D. Porcelli, O. Gustafsson, J. Ingri, G.J. Wasserburg, 1998b. Colloid and particle transport of U isotopes in the low salinity zone of a stable estuary. Mineral. Mag. **62A** (Pt. 1), 57–58.
Andersson, P.S., R. Dahlqvist, J. Ingri, and Ö. Gustafsson, 2001a. The isotopic composition of Nd in a boreal river: a reflection of selective weathering and colloidal transport. Geochim. Cosmochim. Acta **65**, 521-527.
Andersson, P.S., D. Porcelli, Ö. Gustafsson, J. Ingri and G.J. Wasserburg, 2001b. The importance of colloids for the behavior of uranium isotopes in the low-salinity zone of a stable estuary. Geochim. Cosmochim. Acta **65**, 13-25.
Andreae, M.O., 1982. Marine aerosol chemistry at Cape Grim, Tasmania and Townsville, Queensland. J. Geophys. Res. **87**, 8875–8885.
Andreae, M.O., and D.W. Klumpp, 1979. Biosynthesis and release of organoarsenic compounds by marine algae. Environ. Sci. Technol. **13**, 738–741.
Andreae, M.O., and P.N. Froelich, Jr, 1984. Arsenic, antimony, and germanium biogeochemistry in the Baltic Sea. Tellus **36B**, 101–117.
Annegarn, H.J., R.E. van Grieken, P. van Espen, F. von Blottnitz, J.P.F. Sellschop, J.W. Winchester and W. Maenhaut, 1978. Background aerosol composition at Gobabeb, south west Africa. Madoqua **II**, 107–118.
Arimoto, R., R.A. Duce, B.J. Ray and C.K. Unni, 1985a. Atmospheric trace elements at Enewetak Atoll; 2. Transport to the ocean by wet and dry deposition. J. Geophys. Res. **90**, 2391–2408.
Arimoto, R., R.A. Duce, B.J. Ray, A.D. Hewitt and J. Williams, 1985b. Trace elements in the atmosphere of American Samoa: concentrations and deposition to the Tropical South Pacific. J. Geophys. Res. **92**, 8465–8479.
Ashawa, S.C., D.N. Kelkar and U.C. Mishra, 1985. Wet deposition of cadmium, lead and copper at Deonar (Bombay) and its relationship with rainfall. Water Air Soil Pollut. **25**, 63–71.
Assinder, D.J., 1999. A review of the occurrence and behaviour of neptunium in the Irish Sea. J. Environ. Radioactivity **44**, 335–347.
Ayling, G.M. and H. Bloom, 1976. Heavy metals analyses to characterize and estimate distribution of heavy metals in dust fallout. Atmos. Environ. **10**, 61–64.
Åberg, G., 1995. The use of natural strontium isotopes as tracers in environmental studies. Water Air Soil Pollut. **79**, 309-322.
Åberg, G., and F.E. Wickman, 1987. Variations of $^{87}Sr/^{86}Sr$ in water from streams discharging into Bothnian Bay, Baltic Sea. Nord. Hydrol. **18**, 33–42.
Åström, M., 1998. Partitioning of transition metals in oxidised and reduced zones of sulphide-bearing fine-grained sediments. Appl. Geochem. **13**, 607–617.
Åström, M., 2001. The effect of acid soil leaching on trace element abundance in a medium-sized stream, W. Finland. Appl. Geochem. **16**, 387–396.
Åström, M., and J. Åström, 1997. Geochemistry of stream water in a catchment in Finland affected by sulphide fine sediments. Appl. Geochem. **12**, 593–605.
Åström, M., and K. Nylund, 2000. Impact of historical metalworks on the concentrations of major and trace elements in sediments: a case study in Finland. Appl. Geochem. **15**, 807–817.
Äyräs, M., V. Pavlov, and C. Reimann, 1997. Comparison of sulphur and heavy metal content and their regional distribution in humus and moss samples from the vicinity of Nikel and Zapoljarnij, Kola Peninsula, Russia. Water Air Soil Pollut. **98**, 361–380.
Bacon, M.P., and R.F. Anderson, 1982. Distribution of thorium isotopes between dissolved and particulate forms in the deep sea. J. Geophys. Res. **87**, 2045–2056.
Balls, P.W., 1989. The partition of trace metals between dissolved and particulate phases in European coastal waters: a compilation of field data and comparison with laboratory studies. Neth. J. Sea Res. **23**, 7–14.
Baturin, G.N., and A.V. Kočenov, 1969. Migration of uranium in rivers and its residence time in World Ocean, seas and lakes. Geokhimia **6**, 715–723 (in Russian).

Bauer, B., P. Fioroni, U. Schulte-Oehlmann, J. Oehlmann and W. Kalbfus, 1997. The use of *Littorina littorea* for tributyltin (TBT) effect monitoring – results from the German TBT survey 1994/1995 and laboratory experiments. Environ. Pollut. **96**, 299–309.

Berg, T., and E. Steinnes, 1997. Use of mosses (*Hylocomium splendens* and *Pleurozium schreberi*) as biomonitors of heavy metal deposition: from relative to absolute deposition values. Environ. Pollut. **98**, 61–71.

Bergametti, G., A.-L. Dutut, P. Buat-Menard, R. Losno and E. Remoudaki, 1989. Seasonal variability of the elemental composition of atmospheric aerosol particles over the nortwestern Mediterranean. Tellus **41B**, 353–361.

Bernard, P. C., van R.E. Grieken and L. Brügmann, 1989. Geochemistry of suspended matter from the Baltic Sea. 1. Results of individual particle characterization by automated electron microprobe. Mar. Chem. **26**, 155–177.

Bhat, S.G., S. Krishnaswamy, D. Lal, Rama and W.S. Moore, 1969. $^{234}Th/^{238}U$ ratios in the ocean. Earth Planet. Sci. Letters **5**, 483–491.

Björklund, 1989. Organotin in the Swedish aquatic environment KEMI Rep. 8/89, Stockholm, 50 pp.

Blank, M., M. Leinen and J. Prospero, 1985. Major Asian eolian inputs indicated by the mineralogy of aerosols and sediments in the western North Pacific. Nature **314**, 84–86.

Blomqvist, S. and U. Larsson, 1994. Detrital bedrock elements as tracers of settling resuspended particulate matter in a coastal area of the Baltic Sea. Limnol. Oceanogr. **39**, 880-896.

Bogen, J., 1973. Trace elements in atmospheric aerosol in the Heidelberg area, measured by instrumental neutron activation analysis. Atmos. Environ. **7**, 1117–1125.

Bogen, J., 1974. Trace elements in precipitation and cloud water in the area of Heidelberg, measured by instrumental neutron activation analysis. Atmos. Environ. **8**, 835–844.

Bojanowski, R., and T. Samuła-Koszałka, 1974. The simultaneous spectrophotometric determination of manganese, nickel, lead, copper, cobalt and zinc in sea water, in dissolved form and suspended matter. Stud. Mater. Oceanol. **No. 8**, 139–176.

Bojanowski, R., and T. Koszałka, 1975. Trace elements discharged to the Baltic by Vistula waters. Stud. Mater. Oceanol. **14**, 199–216 (in Polish, with English abstract).

Bojanowski, R., and P. Szefer, 1979. Uranium in water of the Baltic Sea. Stud. Mater. Oceanol. **26**, 253–274 (in Polish with English abstract).

Bojanowski, R., J. Pempkowiak and P. Szefer, 1981. Organic-bound polonium-210 in waters of the Baltic Sea and the Vistula River Stud. Mater. Oceanol. **34**, 93–112 (in Polish with English abstract).

Bojanowski, R., Radecki, Z. and Knapińska-Skiba, D., 1995. The distribution of ^{137}Cs, $^{239+240}Pu$ and ^{210}Po in the Pomeranian Bay (southern Baltic) ecosystem. Biul. Mors. Inst. Ryb. **3**, 15–24.

Bolałek, J., 1985. The input of metals with rain into the southern Baltic. Stud. Mater. Oceanol. **48**, 29–42 (in Polish, with an English abstract).

Bordin, G., M. Perttilä and H. Scheinen, 1988. Distribution of total and ASV-labile cadmium, lead, and copper in sea water in the northern parts of the Baltic Sea in 1986–86. Mar. Pollut. Bull. **19**, 325–327.

Borg, H., and P. Jonsson, 1996. Large-scale metal distribution in Baltic Sea sediments. Mar. Pollut. Bull. **32**, 8–21.

Borole, D.V., S. Krishnaswami and B.L.K. Somayajulu, 1982. Uranium isotopes in rivers, estuaries and adjacent coastal sediments of western India: theit weathering, transport and oceanic budget. Geochim. Cosmochim. Acta **46**, 125–137.

Boström, K., J.-O. Burman, C. Pontér and J. Ingri, 1981. Selective removal of trace elements from the Baltic by suspended matter. Mar. Chem. **10**, 335–354.

Boström, K., J.-O. Burman and J. Ingri, 1983. A geochemical massbalance for the Baltic. Environ. Biogeochem. Ecol. Bull. (Stockholm) **35**, 39–58.

Boström, K., J. Ingri, C. Pontér, 1988. Origin of iron-manganese rich suspended matter in the Landsort Deep, N.W. Baltic Sea. Mar. Chem. **24**, 93–98.

Boström, K., B. Boström and P. Andersson, 1989. Natural and anthropogenic components in bulk precipitation at Blidö (Archipelago of Stockholm). Water Res. **25**, 1291–1301.

Bourg, A.C.M., 1987. Trace metal adsorption modelling and particle – water interactions in estuarine environments. Cont. Shelf Res. **7**, 1319–1332.

Boutron, C., 1979a. Past and present day tropospheric fallout fluxes of Pb, Cd, Cu, Zn and Ag in Antarctica and Greenland. Geophys. Res. Letters **6**, 159–162.
Boutron, C., 1979b. Alkali and alkaline earth enrichments in aerosols deposited in Antarcic snows. Atmos. Environ. **13**, 919–924.
Boutron, C., 1980. Trace metals in remote Arctic snow: natural or anthropogenic? Nature **284**, 574–576.
Boutron, C., and C. Lorius, 1975. Trace element in east Antarctica snows samples, in: Isotopes and Impurities in Snow and Ice, Proc. Grenoble Symp., Grenoble, August; September 1975 (IAHS Publ.) No. **118**, 164–171.
Boutron, C., and C. Lorius, 1979. Trace metals in Antarctic snows since 1914. Nature **277**, 551–554.
Bricker, S.B., 1993. The history Cu, Pb, and Zn inputs to Narragansett Bay, Rhode Island, as recorded by salt-marsh sediments. Estuaries **16**, 589–607.
Brockmann, V., G. Billen and W.W.C. Gieskes, 1988. North Sea nutrients and eutrophication, in: Pollution of the North Sea: an Assessment, eds. W. Salomons, B.L. Baye, E.K. Duursma and U. Förstner (Springer, Berlin) pp. 348–389.
Broecker, W.S., A. Kaufman and R.M. Trier, 1973. The residence time of thorium in surface water and its implications regarding the rate of reactive pollutants. Earth Planet. Sci. Letters **20**, 35–44.
Broman, D., I. Lundbergh and C. Naf, 1994. Spatial and seasonal variation of major and trace elements in settling particulate matter in an estuarine-like archipelago area in the Northern Baltic Proper. Environ. Pollut. **85**, 243–257.
Brookins, D.G., 1998. Eh, pH Diagrams for Geochemistry (Springer-Verlag, Berlin), 176 pp.
Bruneau, L., 1980. Pollution from industries in the drainage area of the Baltic. Ambio **9**, 145–152.
Brügmann, L., 1979. Zum Spurenmetallgehalt des Ostseewassers. Chem. Erde **38**, 292–297.
Brügmann L., 1981. Heavy metals in the Baltic Sea. Mar. Pollut. Bull. **12**, 214–218.
Brügmann, L., 1984. Electrochemical speciation of trace metals in seawater. Sci. Total Environ. **37**, 41–60.
Brügmann, L., 1986a. Particulate trace metals in waters of the Baltic Sea and parts of the adjacent NE Atlantic. Beiträge zur Meereskunde **55**, 3–18.
Brügmann, L., 1986b. The influence of coastal zone processes on mass balances for trace metals in the Baltic Sea. Rapp. P.-v. Reun. Cons. int. Explor. Mer **186**, 329–342.
Brügmann, L., 1988. Some peculiarities of the trace-metal distribution in Baltic waters and sediments. Mar. Chem. **23**, 425–440.
Brügmann, L., 1990. Bestimmung gelöster und suspendierter Schwermetalle in der Elbe. Dtsch. Hydrogr. Z. **43**, 127–135.
Brügmann, L., 1995. Metals in sediments and suspended matter of the river Elbe. Sci. Total Environ. **159**, 53–65.
Brügmann, L, and J. Matschullat, 1997. Zur Biogeochemie und Bilanzierung von Schwermetallen in der Ostsee, in: Geochemie und Umwelt-Umweltrelevante Prozesse in Atmo-, Pedo- and Hydrosphäre, eds. J. Matschullat, H.J. Tobschall, H.J. Vogt (Springer Verlag, Berlin) pp. 267–289 (in German).
Brügmann, L, and U. Hennings, 2000. Trace metals in the Baltic Sea atmosphere, 1980–91. (in preparation).
Brügmann, L., H. Gaul, K.-H. Rohde and U. Ziebarth, 1991/92. Regional distribution and temporal trends of some contaminants in the water of the Baltic Sea. Dtsch. Hydrogr. Z. **44**, 161–184.
Brügmann, L., P.C. Bernard and R.E. van Grieken, 1992. Geochemistry of suspended matter from the Baltic Sea. 1. Results of bulk trace metal analysis by AAS. Mar. Chem. **38**, 303–323.
Brügmann, L., R. Hallberg, C. Larsson and A. Löffler, 1997. Changing redox conditions in the Baltic deep basins: Impacts on the concentration and speciation of trace metals. Ambio **26**, 107–112.
Brügmann, L., R. Hallberg, C. Larsson and A. Löffler, 1998. Trace metal speciation in sea and pore water of the Gotland Deep, Baltic Sea, 1994. Appl. Geochem. **13**, 359–368.
Brzezińska, A., and C. Garbalewski, 1980. Atmosphere as a source of trace metals input to the Gdańsk Basin. Oceanologia **12**, 43–58.
Brzezińska, A., A. Trzosińska, W. Żmijewska and L. Wódkiewicz, 1984. Trace metals in suspended matter and surficial bottom sediments from the Southern Baltic. Oceanologia **18**, 59–77.

Buat-Menard, P., and R. Chesselet, 1979. Variable influence of the atmospheric flux on the trace metal chemistry of oceanic suspended matter. Earth Planet. Sci. Letters **42**, 399–411.

Burman, J.-O., 1983. Element transport in suspended and dissolved phases in the Kalix River. Environ. Biogeochem. Ecol. Bull. (Stockholm) **35**, 99–113.

Callaway, J.C., R.D. Delaune and Jr W.H. Patrick, 1998. Heavy metal chronologies in selected coastal wetland from northern Europe. Mar. Pollut. Bull. **36**, 82–96.

Cambray, R.S., D.F. Jefferies and G. Topping, 1979. The atmospheric input of trace elements to the North Sea. Mar. Sci. Commun. **5**, 175–194.

Carpenter, R., M.L. Peterson, J.T. Bennet and B.L.K. Somayajulu, 1984. Mixing and cycling of uranium, thorium and ^{210}Pb in Puget Sound sediments. Geochim. Cosmochim. Acta **48**, 1949–1963.

Carpi, A., L.H. Weinstein and D.W. Ditz, 1994. Bioaccumulation of mercury by *Sphagnum* moss near a municipal solid waste incinerator. Air Waste **44**, 669–672.

Cawse, P.A., 1978. A survey of atmospheric trace elements in the U.K.: results for 1977. AERE Harwell Report R 9164. HMSO London.

Cheng, Z., 1987. Monitoring and assessment of trace metals in sewater and marine organisms. Miljöstyrelsens havforureninglab, Charlottenlund.

Cheng M.D., P.K. Hopke, S. Landsberger and L.A. Barrie, 1991. Distribution characteristics of trace elements and ionic species of aerosol collected at Canadian high Arctic. Atmos. Environ. **25A**, 2903–2909.

Chester, R., and G.F. Bradshow, 1991. Source control on the distribution of particulate trace metals in the North Sea atmosphere. Mar. Pollut. Bull. **22**, 30–36.

Chester, R., A. Griffiths and J.H.Stoner, 1978. Minor metal content of surface seawater particulates and organic-rich shelf sediments. Nature (London) **275**, 308–309.

Chester, R., A.G. Griffin and J.M. Hirst, 1979. The influence of soil-sized particulates on the elemental chemistry of the deep-sea sediments of the northeastern Atlantic. Mar. Geol. **32**, 142–154.

Chester, R., E.J. Sharples and G.S. Sanders, 1984. Saharan dust incursion over the Tyrrhenian Sea. Atmos. Environ. **18**, 929–935.

Chester, R., A.S. Berry and K.J.T. Murphy, 1991. The distribution of particulate atmospheric trace metals and mineral aerosols over the Indian Ocean. Mar. Chem. **34**, 261–290.

Chester, R., M. Nimmo and M.R. Preston, 1999. Trace metal chemistry of atmospheric dry deposition samples collected at Cap Ferrat: a coastal site in the Western Mediterranean. Mar. Chem. 68, 15–30.

Chester, R., M. Nimmo, G.R. Fones, S. Keyse and Z. Zhang, 2000. Trace metal chemistry of particulate aerosols from the UK mainland coastal rim of the NE Irish sea. Atmos. Environ. **34**, 949–958.

Christiansen, C., Edelvang, K., Emeis, K., Graf, G., Jähmlich, S., Kozuch, J., Laima, M., Leipe, T., Löffler, A., Lund-Hansen, L.C., Miltner, A., Pazdro, K., Pempkowiak, J., Shimmield, G., Shimmield, T., Smith, J., Voss, M. and Witt, G., 2001. Material transport from the nearshore to the basinal environment in the Southern Baltic Sea, I: Processes and mass estimates. J. Mar. Systems (in press).

Conley, D.J., and R.W. Johnstone, 1995. Biogeochemistry of N, P and Si in Baltic Sea sediments: response to a simulated deposition of a spring diatom bloom. Mar. Ecol. Prog. Ser. **122**, 265–276.

Conley, D.J., C.L. Schelske and E.F. Stoermer, 1993. Modification of the biochemical cycle of silica with eutrophication. Mar. Ecol. Prog. Ser. **101**, 179–192.

Conley, D.J., A. Stockenberg, R. Carman, R.W. Johnstone, L. Rahm and F. Wulff, 1997. Sediment-water nutrient fluxes in the Gulf of Finland, Baltic Sea. Ester. Coast. Shelf Sci. **45**, 591–598.

Cutter, G.A., and L.S. Cutter, 1995. Behaviour of dissolved antimony, arsenic, and selenium in the Atlantic Ocean. Mar. Chem. **49**, 295–306.

D'Elia, C.F., 1987. Nutrient enrichment of the Chesapeake Bay: too much of a good thing. Environment **29**, 2.

Dahlberg, K., A.-K. Hallin and F. Wulff, 1995. Phosphorus versus nitrogen abatement in the Gulf of Riga – a nutrient budget model approach, in: International Congress on Modelling and Simulation Proceedings, eds. P. Bining, H. Bridgman and B. Williams. Water Resour. Ecol. **vol. 3**, 113–118.

Davies-Colley, R.J., P.O. Nelson and K.J. Williamson, 1985. Sulphide control of cadmium and copper concentrations in recent Baltic Sea sediments. Geochim. Cosmochim. Acta **38**, 823–842.

Davison, W., C. Woof and E. Rigg, 1980. The dynamics of iron and manganese in a seasonally anoxic freshwater lake; direct measurements of fluxes using sediment traps. Limnol. Oceanogr. **27**, 987–1003.

Dick, A.L., 1991. Concentrations and sources of metals in the Antarctic Peninsula aerosol. Geochim. Cosmochim. Acta **55**, 1827–1836.

Douglas, G.B., and J.A. Adeney, 2000. Diagenetic cycling of trace elements in the bottom sediments of the Swan River Estuary, Western Australia. Appl. Geochem. **15**, 551–566.

Dörr, H., A. Mangini, W. Schmitz, F. Weber and K.O. Münnich, 1991. Stable lead and ^{210}Pb in German lake sediments and soils: a tracer for anthropogenic lead emission, in: Heavy Metals in the Environment, ed. J.P. Vernet (Elsevier, Amsterdam), pp. 247–260.

Duce, R.A., G.L. Hoffman and W.H. Zoller, 1975. Atmospheric trace metals at remote northern and southern hemisphere sites: pollution or natural? Science **187**, 59–61.

Duce, R.A., B.J. Ray, G.L. Hoffman and P.R. Walsh, 1976. Trace metal concentration as a function of particle size in marine aerosols from Bermuda. J. Geophys. Res. **3**, 339–342.

Duce, R.A., C.K. Unni, B.J. Ray, J.M. Prospero and J.T. Merrill, 1980. Long-range atmospheric transport of soil dust from Asia to the tropical North Pacific. Science **209**, 1522–1524.

Duce, R.A., R. Arimoto, B.J. Ray, C.K. Unni and P.J. Harder, 1983. Atmospheric trace elements at Enewetak Atoll: 1. Concentrations, sources, and temporal variability. J. Geophys. Res. **88**, 5321–5342.

Dulac, F., P. Buat-Mnard, M. Amold, U. Ezat and D. Martin, 1987. Atmospheric input of trace metals to the western Mediterranean Sea: 1. Factors controlling the variability of atmospheric concentrations. J. Geophys. Res. **92**, 8437–8453.

Dulac, F., P. Buat-Mnard, U. Ezat, S. Melki and G. Bergametti, 1989. Atmospheric input of trace metals to the western Mediterranean: uncertainties in modelling dry deposition from cascade impactor data. Tellus **41B**, 362–378.

Duniec, S., L. Hallstadius, J. Rioseco, E. Holm, G. Guerrin and B. Skwarzec, 1984. A radiological investigation of the Baltic Sea, in: LUNFD6/(NFRA-3050) (Radiation Physics Department, University of Lund, Sweden), 1-12.

Dyrssen, D., 1985. Metal complex formation in sulphide seawater. Mar. Chem. **15**, 285–293.

Dyrssen, D., and K. Kremling, 1990. Increasing hydrogen sulfide concentration and trace metal behavior in the anoxic Baltic water. Mar. Chem. **30**, 193–204.

Ecker, F.-J., E. Hirai and T. Chohji, 1990. Airborne trace metals in snow of the Japan Sea side of Japan. Atmos. Environ. **24A**, 2593–2600.

Edén, P., and A. Björklund, 1996. Applicability of overbank sediment for environmental assessment according to wide-spaced sampling in Fennoscandia. Appl. Geochem. **11**, 271–276.

Egorov, V.N., P.P. Povinec, G.G. Polikarpov, N.A. Stokozov, S.B. Gulin, L.G. Kulebakina and I. Osvath, 1999. ^{90}Sr and ^{137}Cs in the Black Sea after the Chernobyl NPP accident: inventories, balance and tracer applications. J. Environ. Radioactivity **43**, 137–155.

Eisma, D., and J. Kalf, 1987. Dispersal, concentration and deposition of suspended matter in the North Sea. J. Geol. Soc. **144**, 161–178.

Eisma, D., S.J. van der Gaast, J.M. Martin and A.J. Thomas, 1978. Suspended matter and bottom deposits of the Orinoko Delta: Turbidity, mineralogy and elementary composition. Neth. J. Sea Res. **12**, 224–251.

Elmgren, R., 1989. Man's impact on the ecosystem of the Baltic Sea: energy flows today and at the turn of the century. Ambio **18**, 326–332.

Emeis, K., C. Christiansen, K. Edelvang, S. Jähmlich, J. Kozuch, M. Laima, T. Leipe, A. Löffler, L.C. Lund-Hansen, A. Miltner, K. Pazdro, J. Pempkowiak, G. Shimmield, T. Shimmield, J. Smith, M. Voss and G. Witt, 2001. Material transport from the nearshore to the basinal environment in the Southern Baltic Sea, II: Origin and properties of material. J. Mar. Systems (in press).

Emelyanov, E.M., 1974. Raspredelenie Fe, Al, Ti, Mn, Cu, Ni, Co wo wzwiesi Baltijskogo Morja. Woprosy Czetwerticznoj Geologii **7**, 81–98.

Emelyanov, E.M., 1976. Die Entstehung des chemischen Gehalts von Beimengungen und Ablagerungen in der Ostsee. Betrge Meersk. **38**, 61–79.

Emelyanov, E.M., and O.S. Pustelnikov, 1975a. Amount of suspended forms of elements (Org. C, SiO_2 amorph., Fe, Al, Ti, Mn, Ni, Cu, Co) in the waters of the Baltic Sea. Geokhimiya **7**, 1049–1063.

Emelyanov, E.M., and O.S. Pustelnikov, 1975b. Compositions of river and marine forms of suspended matter in the Baltic. Geochem. Int. **12**, 195–208.

Emelyanov, E.M., and O.S. Pustelnikov, 1977. Żelezo wo wzwesi i donnych osadkach Baltijskogo Morja. Baltica **6**, 127–136.

Emelyanov, E.M., and O.S. Pustenikov, 1982. Suspended matter, its composition and the balance of sediments in the Baltic, in: Geology of the Baltic Sea, eds. V.K. Gudelis and E.M. Emelyanov (Wydawnictwo Geologiczne, Warsaw), 161–187 (in Polish).

Erickson, B.E., and G.R. Helz, 2000. Molybdenum (VI) stability in sulfidic waters: Stability and lability of thiomolybdates. Geochim. Cosmochim. Acta **64**,1149-1158.

Essien, I.O., D.N. Sandoval and P.K. Kuroda, 1985. Deposition of excess amount of natural U from the atmosphere. Health Phys. **48**, 325–331.

Falandysz, J., A. Trzosińska, P. Szefer, J. Warzocha and B. Draganik, 2000. The Baltic Sea, in: Seas at the Millennium: An Environmental Evaluation, Vol. I: Europe, The Americas and West Africa, ed. C.R.C. Sheppard (Pergamon, Elsevier), **Chapt. 7**, pp. 99–120.

Falkowska, L., J. Bolalek and J. Nowacki, 1993. Nutrients and oxygen in the Gulf of Gdansk. Mar. Pollut. (National Scientific Committee on Oceanic Research, Polish Academy of Sciences, Gdansk) **64** (3), 131–162.

Fenske, Ch, H. Westphal, A. Bachor, E. Breitenbach, H. Meyer, W. Bucholz and W.-D. Jülich, 1998. Investigations in the Szczecin Lagoon during the Oder flood (July – August 1997) from a biogeochemical perspective. Deutsche Hydrogr. Z. (German J. Hydrogr.) **50**, 231–243.

Fernández, J.A., A. Rey and A. Carballeira, 2000. An extended study of heavy metal deposition in Galicia (NW Spain) based on moss analysis. Sci. Total Environ. **254**, 31–44.

Ferry, B.W., M.S. Baddeley and D.L. Hawksworth, 1973. Air Pollution and Lichens. Toronto, University of Toronto Press.

Fisher, R.S., and A.M. Stueber, 1976. Strontium isotopes in selected streams within the Susquehanna River basin. Water Resour. Res. **12**, 1061–1068.

Förstner, U, W. Ahlf, W. Calmano, M. Kersten and W. Salomons, 1986. Mobility of heavy metals in dredged harbor sediments, in: Sediments and Water Interactions, ed. P.G. Sly. Proc. of the 3rd International Symp. (Springer Verlag), Geneva, pp. 371–380.

Garty, J., 1993. Lichens as biomonitors for heavy metal pollution, in: Plants as Biomonitors, ed. B. Markert (VCH Publishers) pp. 194–263.

Gedeonov, L.I., M.I. Belen'kii, V.M. Gavrilov, Z.G. Gritchenko, L.M. Ivanova, L.N. Lazarev, T.E. Orlova, Yu.A. Panteleev and V.P. Tishkov, 1998. Radioactive contamination of the Baltic Sea in the region of the Leningrad Nuclear Power Plant in 1971–1996. Radiochemistry (Moscow) **40**, 385–390.

Gellermann, R., and K. Fröhlich, 1984. Nutzung der Uraniuisotope ^{234}U und ^{238}U in der Hydrologie und Hydrogeologie. Freiberg Forschungsheft C 397 (Leipzig: VEB Deutscher Verlag für Grundstoffindustrie), 101 pp.

Gellermann, R., and W. Stolz, 1997. Uran in Wässern. UWSF – Z. Umweltchem. Ökotox. **9**, 87–92.

Gellermann, R., K. Fröhlich and D. Lange, 1983. Untersuchungen zur Geochemie und Isotopengeochemie des Uraniums in der Ostsee. Beitr. Meereskd. **49**, 65–75.

Glasby G.P., and P. Szefer, 1998. Marine pollution in Gdańsk Bay, Puck Bay, and the Vistula Lagoon, Poland – An overview. Sci. Total Environ. **212**, 49–57.

Glasby, G.P., and H.D. Schulz, 1999. E_H, pH diagrams for Mn, Fe, Co, Ni, Cu and As under seawater conditions: Application of two new types of E_H, pH diagrams to the study of specific problems in marine geochemistry. Aquat. Geochem. **5**, 227-248.

Glasby, G.P., P. Szefer, J. Gełdon and J. Warzocha, 2001. Heavy-metal pollution of sediments from Szczecin Lagoon and the Gdansk Basin, Poland. Mar. Pollut. Bull. (submitted).

Goldstein, S.J., and S.B. Jacobsen, 1987. The Nd and Sr isotopic systematics of river-water dissolved material: Implications for sources of Nd and Sr in seawater. Chem. Geol. (Isot. Geosci) **66**, 245–272.

Goodman, H.S., B.N. Noller, G.I. Pearman and H. Bloom, 1976. The heavy metal composition of atmospheric particulates in Hobart, Tasmana. Clean Air **10**, 38–41.

Gordeev, V.V., A.S. Miklishansky and S.B. Tambiev, 1984. Geochemistry of particulates and waters of the Gulf of Riga, in: Geological History and Geochemistry of the Baltic Sea (Nauka, Moscow), 18–32 (in Russian).

Görlich, K., E.A. Görlich, K. Tomala, A.Z. Hrynkiewicz and P.Q. Hung, 1989. ^{57}Fe Mössbauer study of a sediment column in the Gdansk Basin, Baltic Sea: Palaeoenvironmental application. Mar. Geol. **88**, 49–69.

Granéli, E., K. Wallström, U. Larsson, W. Granéli and R. Elmgren, 1990. Nutrient limitation of primary production in the Baltic Sea. Ambio **19**, 142–151.

Granskog, M.A., 1999. Baltic Sea ice as a medium for sediment, nutrient and pollutant storage and transport – some preliminary results. NHP Report No. 44, in: Proc. of XX Nordic Hydrological Conference, Helsinki, Finland, 10–13.08.1999, pp. 21–31.

Grelowski, A., M. Pastuszak, S. Sitek and Z. Witek, 2000. Budget calculations of nitrogen, phosphorus and BOD_5 passing through the Oder Estuary. J. Mar. Systems, **25**, 221–237.

Grimvall, A., P. Stålnacke, K. Sundblad, E. Niemirycz, H. Pitkånen and A. Bruhn, 1991. Trend analysis of nutrient concentrations in Baltic Sea rivers. Proceedings of 3rd International Conference on System Analysis Approach to Environment, Energy and Natural Resource Management in the Baltic Region (Copenhagen, Denmark 1991).

Grodzinska, K., and B. Godzik, 1991. Heavy metals and sulphur in mosses from southern Spitsbergen. Polar Res. **9**, 133–140.

Grönlund, L., and J.-M. Leppänen, 1990. Long-term changes in the nutrient reserves and pelagic production in the western Gulf of Finland. Finnish Mar. Res. **257**, 15–27.

Grönlund L., K. Kononen, E. Lahdes and K. Mäkelä, 1996. Community development and modes of phosphorus utilization in a late summer ecosystem in the central Gulf of Finland, the Baltic Sea. Hydrobiologia **331**, 97–108.

Guerrero-Galván, S.R., F. Páez-Osuna, A.C. Ruiz-Fernández and R. Espinoza-Angulo, 1999. Seasonal variation in the water quality and chlorophyll *a* of semi-intensive shrimp ponds in a subtropical environment. Hydrobilogia **391**, 33–45.

Guo, L., P.H. Santschi and K.W. Warnken, 2000. Trace metal composition of colloidal organic material in marine environments. Mar. Chem. **70**, 257–275.

Gustavsson, I., 1981. Koncentrationen av nagra tungmetaller i kustvattnen untanfr Oxelsund och Vallvik 1980/81. Naturvardsverket Rapport SNV PM 1465, Solna.

Gustafsson, J.P., and G. Jacks, 1995. Arsenic geochemistry in forested soil profiles as revealed by solid-phase studies. Appl. Geochem. **10**, 307–315.

Gustafsson, M.E.R., and L.G. Franzén, 2000. Inland transport of marine aerosols in southern Sweden. Atmos. Environ. **34**, 313–325.

Gustafsson, Ö., A. Widerlund, P.S. Andersson, J. Ingri, P. Roos and A. Ledin, 2000. Colloid dynamics and transport of major elements through a boreal river – brackish bay mixing zone. Mar. Chem. **71**, 1–21.

Hallberg, R.O., 1991. Environmental implications of metal distribution in Baltic Sea sediments. Ambio **20**, 309–316.

Hasanen, E., M. Lipponen, R. Kattainen, K. Markkanen, P. Minkkinen and P. Brjukkhanov, 1990. Elemental concentrations of aerosol samples from the Baltic Sea area. Chemosphere **21**, 339–347.

Hawksworth, D.L., 1971. Lichens as litmus for air pollution: a historical review. Int. J. Environ. Stud. **1**, 281–296.

Heidam, N.Z., 1981. On the origin of the Arctic aerosol: a statistical approach. Atmos. Environ. **15**, 1421–1427.

Heidam, N.Z., 1984. The components of the Arctic aerosol Atmos. Environ. **18**, 329–343.

HELCOM, 1990. Second Periodic Assessment of the State of the Marine Environment of the Baltic Sea, 1984–1988; Background Document (Baltic Marine Environment Protection Commission, Helsinki), Baltic Sea Environment Proceedings **No. 35B**.

HELCOM, 1991. Airborne pollution load to the Baltic Sea 1986–1990. Baltic Sea Environment Proceedings No. **39**.

HELCOM, 1993. First assessment of the state of the coastal waters of the Baltic Sea (Baltic Marine Environment Protection Commission, Helsinki), Baltic Sea Environment Proceedings No. **54**.

HELCOM, 1995. Radioactivity in the Baltic Sea 1984–1991. Baltic Sea Environment Proceedings (Baltic Marine Environment Protection Commission, HELCOM) **No. 61**.

HELCOM, 1996. Third Periodic Assessment of the State of the Marine Environment of the Baltic Sea, 1989–93; Background Document. Baltic Sea Environment Proc. (Baltic Marine Environment Protection Commission, Helsinki) **No. 64B**.

HELCOM, 1997. Airborne pollution load to the Baltic Sea 1991–1995. Baltic Sea Environment Proceedings No. **69**.

HELCOM, 1998a. The third Baltic Sea Pollution Load Compilation (PLC-3) (Baltic Marine Environment Protection Commission, Helsinki), Baltic Sea Environment Proceedings **No. 70**.

HELCOM, 1998b. Final Report on the Implementation of the 1988 Ministerial Declaration. Baltic Sea Environment Proceedings **No. 71**.

Helios Rybicka, E., 1983. The content and chemical forms of heavy metals in the river sediments of the Cracow area: the role of clay minerals. Environ. Techn. Lett. **4**, 515–520.

Helios Rybicka, E., 1991. Application of clay minerals in the synthesis of low-temperature sulfide minerals in relation to Kupferschiefer, ed. G.H. Moh. Neues Jahrbuch Miner. Abh. **163**, 231–233.

Helios Rybicka, E., 1992. Heavy metal partitioning in polluted river and sea sediments: clay minerals effects. Miner. Petrogr. Acta **XXXV-A**, 297–305.

Helios Rybicka, E., 1993. Phase-specific bonding of heavy metals in sediments of the Vistula River, Poland. Appl. Geochem. Suppl. Issue **No. 2**, 45–48.

Helios Rybicka, E., 1996a. Impact of mining and metallurgical industries on the environment in Poland. Appl. Geochem. **11**, 3–9.

Helios Rybicka, E., 1996b. Geochemical control of mining operations, in: Geochemical Approaches to Environmental Engineering of Metals, ed. R. Reuther (Springer – Verlag, Berlin) 221 pp.

Helios Rybicka, E., 1996c. Environmental impact of mining and smelting industries in Poland, in: Environmental Geochemistry and Health, eds. J.D. Appleton, R. Fuge, and G.J.H. McCall, Geol. Soc. Spec. Publ. **No. 113**, 183–193.

Helios Rybicka, E., and M. Strzebońska, 1999. Distribution and chemical forms of heavy metals in the flood 1997 sediments in the Upper and Middle Odra River and its tributaries, Poland. Acta Hydrochim. Hydrobiol. **27**, 331-337.

Helios Rybicka, E., M.J. Wilson and W.J. McHardy, 1994. Chemical and mineralogical forms and mobilization of copper and lead in soils from a Cu-smelting area in Poland. J. Environ. Sci. Health **A29**, 531–546.

Herpin, U., J. Berlekamp, B. Markert, B. Wolterbeek, K. Grodzinska, U. Siewers, H. Lieth and V. Weckert, 1996. The distribution of heavy metals in a transect of three states the Netherlands, Germany and Poland, determined with the aid of moss monitoring. Sci. Total Environ. **187**, 185–198.

Herrmann, J., P.J. Kershaw, P. Bailly du Bois and P. Gueguenìat, 1995. The distribution of artificial radionuclides in the English Channel, southern North Sea, Skagerrak and Kattegat, 1990–1993. J. Mar. Systems **6**, 427–456.

Herrmann, J., 2000. Levels of radioactivity, in: The Radiological Exposure of the Population of the European Community to Radioactivity in the Baltic Sea. Marina-Balt Project, ed. S.P. Nielsen. Proceedings of a Seminar held at Hasseludden Conference Centre, Stockholm, 9–11 June 1998, European Commission, Directorate-General Environment, EUR 19200 EN (European Communities, 2000, Belgium), pp. 77–129.

Heybowicz, E., and T. Borkowski, 1997. Water quality. Riverine input of pollutants, in: Environmental Conditions in the Polish Zone of the Southern Baltic Sea during 1996. Maritime Branch Materials (Institute of Meteorology and Water Management, Gdynia), pp. 165–177.

Holm, E., 1995. Plutonium in the Baltic Sea. Appl. Radiat. Isot. **46**, 1225–1229.

Holm, E., 2001. The Swedish nuclear dilemma. J. Environ. Radioactivity. 52, 113–115.

Holm, E., S. Duniec, L. Hallstadius, S. Mattsson, J. Rioseco and C. Samuelsson, 1986. Releases distribution and pathways for radionuclides in the Baltic Sea, in: Study of Radioactive materials in the Baltic Sea. (International Atomic Energy Agency, Vienna). Report (IAEA-TECDOC-362) of the Final Research Co-ordination Meeting on the Study of Radioactive Materials in the Baltic Sea organized by the IAEA and held in Helsinki, Finland 24–28 September, 1984, pp. 152–161.

Holmes, C.W., 1982. Geochemical indices of fine sediment transport in the North West Gulf of Mexico. J. Sediment. Petrol. **52**, 307–321.

Hopke, P.K., E.S. Gladney, G.E. Gordon, W.H. Zoller and A.G. Jones, 1976. The use of multivariate analysis to identify sources of selected elements in the Boston urban aerosol. Atmos. Environ. **10**, 1015–1025.

Hou, X., H. Dahlgaard and S.P. Nielsen, 2001. Chemical speciation analysis of ^{129}I in seawater and a preliminary investigation to use it as a tracer for geochemical cycle study of stable iodine. Mar. Chem. **74**,145-155.

Humborg, C., G. Nausch, T. Neumann, F. Pollehne and N. Wasmund, 1998. The exceptional Oder flood in summer 1997 – The fate of nutrients and particulate organic matter in the Baltic Sea. Deutsche Hydrogr. Z. (German J. Hydrogr.) **50**, 169–181.

Humborg, C., K. Fennel, M. Pastuszak and W. Fennel, 2000. A box model approach for a long-term assessment of estuarine eutrophication, Szczecin Lagoon, southern Baltic. J. Mar. Systems, **25**, 387–403.

Hydes, D.J., B.A. Kelly-Gerreyn, A.C. Le Gall and R. Proctor, 1999. The balance of supply of nutrients and demands of biological production and denitrification in a temperate latitude shelf sea – a treatment of the southern North Sea as an extended estuary. Mar. Chem. **68**, 117–131.

IAEA, 1986. Summary Report on the Post-Accident Review Meeting on the Chernobyl Accident. Report by the International Nuclear Safety Advisory Group. Safety Series No. 75-INSAG-1.

Ilus, E., and T. Ilus, 2000. Sources of radioactivity, in: The Radiological Exposure of the Population of the European Community to Radioactivity in the Baltic Sea. Marina-Balt Project, ed. S.P. Nielsen. Proceedings of a Seminar held at Hasseludden Conference Centre, Stockholm, 9–11 June 1998, European Commission, Directorate-General Environment, EUR 19200 EN (European Communities, 2000, Belgium), pp. 9–76.

Ilus, E., R. Saxén and T.K. Taipale, 1986. Accumulation of certain long-lived radionuclides by fish, in: Study of Radioactive materials in the Baltic Sea. (International Atomic Energy Agency, Vienna). Report (IAEA-TECDOC-362) of the Final Research Co-ordination Meeting on the Study of Radioactive Materials in the Baltic Sea organized by the IAEA and held in Helsinki, Finland 24–28 September, 1984, pp. 63–78.

Ilus, E., K.-L. Sjöblom, R. Saxén, H. Aaltonen and K. Taipake, 1987. Finnish studies in the Baltic Sea after the Chernobyl accident in 1986. Suppl. 11 to Annual Rep. STUK-A55. STUK-A66, April 1987 (Finnish Centre for Radiation and Nuclear Safety, Helsinki, Finland) 40 pp.

Ilus, E., S. Klemola, K.-L. Sjöblom and T. Ikäheimonen, 1988. Radioactivity of *Fucus vesiculosus* along the Finnish coast in 1987. Suppl. 9 to Annual Rep. 1987 (STUK-A74). STUK-A83, May 1988 (Finnish Centre for Radiation and Nuclear Safety, Helsinki, Finland) 36 pp.

Ilus, E., K.-L. Sjöblom, S. Klemola and H. Arvela, 1992. Monitoring of radionuclides in the environs of Finnish nuclear power plants in 1989–1990. Suppl. 9 to Annual Rep. STUK-A89. STUK-A102, January 1992 (Finnish Centre for Radiation and Nuclear Safety, Helsinki, Finland) 91 pp.

Ilus, E., K.-L. Sjöblom, T.K. Ikäheimonen, R. Saxén and S. Klemola, 1993. Monitoring of radionuclides in the Baltic Sea in 1989–1990. Suppl. 10 to Annual Rep. STUK-A89. STUK-A103, February 1993 (Finnish Centre for Radiation and Nuclear Safety, Helsinki, Finland) 35 pp.

Imai, T., and M. Sakanoue, 1973. Content of plutonium, thorium and protactinium in sea water and recent coral in the North Pacific. J. Oceanogr. Soc. Japan **29**, 30–36.

IMGW, 1997–1998. Environmental conditions in the Polish zone of the southern Baltic Sea Institute of Meteorology and Water Management, Maritime Branch Materials, Gdynia (in Polish).

Infante, R., and I.L. Acosta, 1991. Size distribution of trace metals in Ponce, Puerto Rico air particulate matter. Atmos. Environ. **26B**, 121–131.

Injuk, J., and R. Van Grieken, 1995. Atmospheric concentrations and deposition of heavy metals over the North Sea: a literature review. J. Atmos. Chem. **20**, 179–212.

Ingri, J., and A. Widerlund, 1994. Uptake of alkali and alkaline-earth elements on suspended iron and manganese in the Kalix River, northern Sweden. Geochim. Cosmochim. Acta **58**, 5433–5442.

Ingri, J., R. Löfvendahl and K. Boström, 1991. Chemistry of suspended particles in the southern Baltic Sea. Mar. Chem. **32**, 73–87.

Ingri, J., P. Torssander, P.S. Andersson, C.-M. Mörth and C.-M. Kusakabe, 1997. Hydrogeochemistry of sulfur isotopes in the Kalix River catchment, northern Sweden. Appl. Geochem. **12**, 483–496.

Ingri, J., A. Widerlund, M. Land and B. Öhlander, 1998. Seasonal variations of the rare earth elements in a boreal river. Abstract Vol. Eight Annual V.M. Goldschmidt Conf., 30 August-3 September, 1998, Toulouse, France. Mineral. Mag. **62A**, 681–682.

Ingri, J., A. Widerlund, M. Land, Ö. Gustaffson, P. Andersson and B. Öhlander, 2000. Temporal variations in the fractionation of the rare earth elements in a boreal river; the role of colloidal particles. Chem. Geol. **166**, 23–45.

Injuk, J., and R. Van Grieken, 1995. Atmospheric concentrations and deposition of heavy metals over the North Sea. J. Atmos. Chem. **20**, 179–212.

Isajenko, K., M. Biernacka, M. Bysiek, J. Henschke and P. Lipiński, 2000. Radiological monitoring of the coastal area of the Baltic Sea in Poland, in: The Radiological Exposure of the Population of the European Community to Radioactivity in the Baltic Sea. Marina-Balt Project, ed. S.P. Nielsen. Proceedings of a Seminar held at Hasseludden Conference Centre, Stockholm, 9–11 June 1998, European Commission, Directorate-General Environment, EUR 19200 EN (European Communities, 2000, Belgium), pp. 349–358.

Ivanova, L.M., 1978. Investigations on the radioactive pollution in the Baltic Sea (V.G. Khlopin Radium Institute, Leningrad).

Jalkanen, L., A. Mäkinen, E. Häsänen and J. Juhanoja, 2000. The effect of large anthropogenic particulate emissions on atmospheric aerosols, deposition and bioindicators in the eastern Gulf of Finland region. Sci. Total Environ. **262**, 123–136.

Jambers, W., V. Dekov and R. Van Grieken, 1999. Single particle characterisation of inorganic and organic North Sea suspension. Mar. Chem. **67**, 17–32.

Jambers, W., V. Dekov and R. Van Grieken, 2000. Single particle and inorganic characterization of rainwater collected above the North Sea. Sci. Total Environ. **256**, 133–150.

Jaworowski, Z., A. Barański, D. Grzybowska, A. Pawlak, J. Tomczak, A. Pietruszewski, M. Suplińska, D. Skiba and T. Wardaszko, 1986. Monitoring of the radioactive materials in the Baltic Sea, in: Study of Radioactive materials in the Baltic Sea. (International Atomic Energy Agency, Vienna). Report (IAEA-TECDOC-362) of the Final Research Co-ordination Meeting on the Study of Radioactive Materials in the Baltic Sea organized by the IAEA and held in Helsinki, Finland 24–28 September, 1984, pp. 141–151.

Jickells, T., 1995. Atmospheric inputs of metals and nutrients to the oceans: their magnitude and effects. Mar. Chem. **48**, 199–214.

Johnson-Pyrtle, A., M.R. Scott, T.E. Laing and J.P. Smol, 2000. ^{137}Cs distribution and geochemistry of Lena River (Siberia) drainage basin lake sediments. Sci. Total Environ. **255**, 145–159.

Jonsson, P., and R. Carman, 1994. Changes in deposition of organic matter and nutrients in the Baltic Sea during the twentieth century. Mar. Pollut. Bull. **28**, 417–426.

Jonsson, P., R. Carman and F. Wulff, 1990. Laminated sediments in the Baltic – A tool for evaluating nutrient mass balances. Ambio **19**, 152–158.

Kahma, K., and A. Voipio, 1990. Elimination and seasonal variation from long-term changes of some nutrients in the Baltic Sea. Finnish Mar. Res. **257**, 3–14.

Kanivets, V.V., O.V. Voitsekhovitch, V.G. Simov and Z.A. Golubeva, 1999. The post-Chernobyl budget of ^{137}Cs and ^{90}Sr in the Black Sea. J. Environ. Radioactivity **43**, 121–135

Kaufman, A., 1969. The Th^{232} concentration of surface ocean water. Geochim. Cosmochim. Acta **33**, 717–724.

Kaufman, A., R.M. Trier and W.S. Broecker, 1973. Distribution of ^{228}Ra in the World Ocean. J. Geophys. Res. **36**, 8827–8848.

Kautsky, H., 1981. Radiological investigations in the Western Baltic Sea including Kattegat during the years 1975 to 1980. Dtsch. Hydrogr. Z. **34**, 125–149.

Kautsky, H., and H.-F. Eicke, 1982. Radiological investigations in the Western Baltic Sea and the Baltic Proper, including the Danish Straits, during the years 1981. Dtsch. Hydrogr. Z. **35**, 211–221.

Kautsky, H., C. Wedekind and H.-F. Eicke, 1986. Radiolological investigations in the Baltic Sea, iclud-ing the Danish Straits and the Kattegat during 1982 and 1983, in: Study of Radioactive materials in the Baltic Sea. (International Atomic Energy Agency, Vienna). Report (IAEA-TECDOC-362) of the Final Research Co-ordination Meeting on the Study of Radioactive Materials in the Baltic Sea organized by the IAEA and held in Helsinki, Finland 24–28 September, 1984, pp. 110–132.

Kerminen, V.-M., K. Teinilä and R. Hillamo, 2000. Chemistry of sea-salt particles in the summer Antarctic atmosphere. Atmos. Environ. **34**, 2817–2825.

Kersten, M., M. Kriews and U. Förstner, 1991a. Partitioning of trace metals released from polluted marine aerosols in coastal seawater. Mar. Chem. **36**, 165–182.

Kersten, M., G. Irion and U. Förstner, 1991b. Particulate trace metals in surface waters of the North Sea, in: Trace Metals in the Environment, ed. J.-P. Vernet (Elsevier, London), pp. 137–159.

Kersten, M., S. Thomsen, W. Priebsch and C.-D. Garbe-Schönberg, 1998. Scavenging and particle residence times determined from ^{234}Th/^{238}U disequilibria in the coastal waters of Mecklenburg Bay, in: Geochemical Investigations of the Baltic Sea and Surrounding Areas, eds. P. Szefer and G.P. Glasby (Elsevier Sci. Ltd). Spec. Issue, Appl. Geochem. **13**, 339–347

Khemani, L.T., G.A. Momin, M.S. Naik, P.S. Prakasa Rao, R. Kumar and Bh. V. Ramana Murty, 1985. Trace elements and sea salt aerosols over the sea areas around the Indian sub-continent. Atmos. Environ. **19**, 277–284.

Kivi, K., S. Kaitala, H. Kuosa, J. Kuparinen, E. Leskinen, R. Lignell, B. Marcussen and T. Tamminen, 1993. Nutrient limitation and grazing control of Baltic plankton community during annual succession. Limnol. Oceanogr. **38**, 893–905.

Kļaviņš, M., A. Briede, V. Rodinov, I. Kokorīte, E. Parele and I. Kļaviņa, 2000. Heavy metals in rivers of Latvia. Sci. Total Environ. **262**, 175–183.

Knapińska-Skiba, D., R. Bojanowski and Z. Radecki, 1997. Sorption and release of radiocesium from particulate matter of the Baltic coastal zone. Neth. J. Aquat. Ecol. **28**, 413–419.

Koczy, G., 1950. Weitere Uranbestimmungen an Meerwasserproben. Sitz-ber. Mathem.-Naturw. Kl. Abt. IIa. **158 Bd.** (Mitteilungen des Institutes für Radiumforschung Nr. 463) 113–121.

Koczy, F.F., 1956. The content of uranium and radium in ocean water. Rep. Swed. Deep-Sea Expedition 10.

Koczy, F., E. Picciotto, G. Poulaert and S. Wilgain, 1956. Mesure les isotopes du thorium dans l'eau de mer. Geochim. Cosmochim. Acta **11**, 103–129.

Koczy, F.F., E. Tomic and F. Hecht, 1957. Zur Geochemie des Urans im Ostseebecken. Geochim. Cosmochim. Acta **11**, 86–102.

Koide, M., and E.D. Goldberg, 1965. Uranium-234/uranium-238 ratios in sea water. Progress in Oceanography **3**, 173–177.

Kowalczyk, G.S., C.E. Choquette and G.E. Gordon, 1978. Chemical element balances and identification of air pollution sources in Washington, D.C. Atmos. Environ. **12**, 1143–1153.

Kowalewska, G., 1986. Radium-226 in water and sediments of the southern Baltic Sea. Oceanologia **23**, 65–76.

Krauskopf, K.B., 1956. Factors controlling the concentration of thirteen rare metals in sea-water. Geochim. Cosmochim. Acta **9**, 1–32.

Kravtsov, V.A., and E.M. Emelyanov, 1995. Trace metal behaviour in the waters of the Gulf of Finland, eds. I. Cato and F. Klingberg, Proc. fourth Mar. Geol. Conference – the Baltic, Upssala 1995 (Sveriges Geologiska Underskning) Ser. **Ca 86**, pp. 99–109.

Kremling, K., 1983. The behaviour of Zn, Cd, Cu, Ni, Co, Fe and Mn in anoxic Baltic waters. Mar. Chem. **13**, 87–108.

Kremling, K., and H. Petersen, 1978. The distribution of Mn, Fe, Zn, Cd and Cu in Baltic seawater, a study on the basis of anchor station. Mar. Chem. **6**, 155–170.

Kremling, K., and H. Petersen, 1984. Synoptic survey on dissolved trace metal levels in Baltic surface waters. Mar. Pollut. Bull. **15**, 329–334.

Kremling, K., and C. Pohl, 1989. Studies on the spatial and seasonal variability of dissolved cadmium, copper and nickel in northeast Atlantic surface waters. Mar. Chem. **27**, 43–60.

Kremling, K., and G. Wilhelm, 1997. Recent increase of the calcium concentrations in Baltic Sea waters. Mar. Pollut. Bull. **34**, 763–767.

Kremling, K., and P. Streu, 2000. Further evidence for a drastic decline of potentially hazardous trace metals in Baltic Sea surface waters. Mar. Pollut. Bull. **40**, 674–679.

Kremling, K., A. Wenck and C. Osterroht, 1981. Investigations on dissolved copper-organic substances in Baltic waters. Mar. Chem. **10**, 209–219.

Kremling, K., L. Brgmann and A. Jenssen, 1986. Trace metals (TM) in the Baltic, in: Baltic Marine Environment Protection Commission – Helsinki Commission, First Periodic Assessment of the State of the Marine Environment of the Baltic Sea Area, 1980–1985. Baltic Sea Environment Proc. (Helsinki) **No. 17B**, 82–130.

Kremling, K., J.J.S. Tokos, L. Brügmann and H.-P. Hansen, 1997. Variability of dissolved and particulate trace metals in the Kiel and Mecklenburg Bights of the Baltic Sea, !990–1992. Mar. Pollut. Bull. **34**, 112–122.

Krishnaswami, S., D. Lal, B.L.K. Somayajulu, F.S. Dixon, S.A. Stonecipher and H. Craig, 1972. Silicon, radium, thorium, and lead in seawater: in-situ extraction by synthetic fibre. Earth Planet. Sci. Letters **16**, 84–90.

Ku, T.L., K.G. Knauss and G.G. Mathieu, 1977. Uranium in open ocean: concentration and isotopic composition. Deep-Sea Res. **24**, 1005–1017.

Kubin, E., and H. Lippo, 1996. Atmospheric heavy metal deposition in Finland from 1985–1990. Appl. Geochem. **11**, 155–161.

Kuss, J., and K. Kremling, 1999. Spatial variability of particle associated trace elements in near-surface waters of the North Atlantic (30°N/60°W to 60°N/2°W), derived by large volume sampling. Mar. Chem. **68**, 71–86.

Kuss, J., C.-D. Garbe-Schönberg and K. Kremling, 2001. Rare earth elements in suspended particulate material of North Atlantic surface waters. Geochim. Cosmochim. Acta **65**, 187-199.

Laanemets, J., K. Kononen and J. Pavelson, 1997. Nutrient intrusions at the entrance to the Gulf of Finland. Boreal Environ. Res. **2**, 337–344.

Laaksoharju, M., E.-L. Tullborg, P. Wikberg, B. Wallin and J. Smellie, 1999. Hydrogeochemical conditions and evolution at the Äspö HRL, Sweden. Appl. Geochem. **14**, 835–859.

Lahermo, P., J. Mannio and T. Tarvainen, 1995. The hydrogeochemical comparison of streams and lakes in Finland. Appl. Geochem. **10**, 45–64.

Laima, M.J.C., L.C. Lund-Hansen, K. Pazdro, C. Christiansen and K.-C. Emeis, 1999. Near-bottom fluxes and composition of suspended matter in the Pomeranian Bight. Oceanologia **41**, 335–353.

Laima, M.J., H. Matthiesen, L.C. Lund-Hansen, C. Christiansen, S. Jähmlich, and K.-C. Emeis, 2001. Transport of P, Fe and Mn along a depth gradient in the SW Baltic. Boreal Environ. Res. (in press).

Lampe, R., 1999. The Odra Estuary as a filter and transformation area. Acta Hydrochim. Hydrobiol. **27**, 292–297.

Land, M., and B. Öhlander, 1997. Seasonal variations in the geochemistry of shallow groundwater hosted in granitic till. Chem. Geol. **143**, 205–216.

Land, M., J. Ingri and B. Öhlander, 1999a. Past and present weathering rates in northern Sweden. Appl. Geochem. **14**, 761–774.

Land, M., B. Öhlander, J. Ingri and J. Thunberg, 1999b. Solid speciation and fractionation of rare earth elements in a spodosol profile from northern Sweden as revealed by sequential extraction. Chem. Geol. **160**, 121–138.

Land, M., J. Ingri, P.S. Andersson and B. Öhlander, 2000. Ba/Sr, Ca/Sr and $^{87}Sr/^{86}Sr$ ratios in soil water and groundwater: implications for relative contributions to stream water discharge. Appl. Geochem. **15**, 311–325.

Larsson, U., R. Elmgren and F. Wulff, 1985. Eutrophication and the Baltic Sea: causes and consequences. Ambio **14**, 9–14.

Lawson, D.R., and J.W. Winchester, 1978. Sulfur and trace element concentration relationships in aerosols from the South America continent. Geophys. Res. Letters **5**, 195–198.

Lazarev, L.N., L.T. Gedeonov, L.M. Ivanova, N.A. Oleneva, T.E. Orlova, S.M. Vakulovsky, A.I. Nikitin and V.B. Chumishev, 1983a. The content of strontium-90 and caesium-137 in the water of the Baltic Basin 1979–81. Radiohimija **No. 5**, 669–675.

Lazarev, L.N., L.T. Gedeonov, L.M. Ivanova, T.E. Orlova, N.A. Tishkova, S.M. Vakulovsky, A. Salo, M. Puhakainen, R. Saxén and K. Tuomainen, 1983b. Study of radioactive contamination of the Baltic Sea in 1979–80. STL-B-TUTO-23 (Institute of Radiation Protection, Helsinki).

Lazarev, L.N., Yu.V. Kuznetsov, L.I. Gedeonov, D.I. Gusev, V.V. Anisimov, V.A. Blinov, L.M. Ivanova, T.E. Orlova, A.V. Stepanov, N.A. Tishkova, V.P. Tishkov, V.M. Flegontov, S.M. Vakulovskiy, A.I. Nikitin, I.Yu. Katrich, V.B. Chumichev, O.A. Pavlovskiy and V.D. Stepanova, 1986. Investigation of the radioactive contamination of the Baltic Sea in 1978–83, in: Study of Radioactive materials in the Baltic Sea. (International Atomic Energy Agency, Vienna). Report (IAEA-TECDOC-362) of the Final Research Co-ordination Meeting on the Study of Radioactive Materials in the Baltic Sea organized by the IAEA and held in Helsinki, Finland 24–28 September, 1984, pp.162–175.

Lehmann, J., Th. Puff, H. Damke, J. Eidam, K.-H. Henning, W.-D. Jülich and H. Rossberg, 1999. The Odra River load of heavy metals at Hohenwutzen during the flood of 1997. Acta Hydrochim. Hydrobiol. **27**, 321-324.

Leipe, Th., A. Loeffler, K.-C. Emeis, S. Jaehmlich, R. Bahlo and K. Ziervogel, 2000. Vertical patterns of suspended matter characteristics along a coastal-basin transect in the western Baltic Sea. Estuar. Coast. Shelf Sci. **51**, 789-804.

Leivuori, M., and H. Vallius, 1998. A case study of seasonal variation in the chemical composition of accumulating suspended sediments in the central Gulf of Finland. Chemosphere **36**, 2417–2435.

Leonard, K.S., D. McCubbin, J. Brown, R. Bonfield and T. Brooks, 1997. Distribution of technetium-99 in UK coastal waters. Mar. Pollut. Bull. **34**, 628–636.

Leonard, K.S., D. McCubbin, P. Blowers and B.R. Taylor, 1999. Dissolved plutonium and americium in surface waters of the Irish Sea, 1973–1996. J. Environ. Radioactivity **44**, 129–158.

Lepland, A., and R.L. Stevens, 1998. Manganese authigenesis in the Landsort Deep, Baltic Sea. Mar. Geol. **151**, 1–25.

Leppäranta, M., M. Tikkanen and P. Shemeikka, 1998. Observations of ice and its sediments on the Baltic Sea coast. Nordic Hydrol. **29**, 199–220.

Leskinen, S., J.K. Miettinen and T. Jaakkola, 1987. Behaviour of $^{239}Pu+^{240}Pu$ and ^{241}Am in the Baltic Sea; measurements and interpretation in 1980–1984. J. Radioanal. Nucl. Chem. **115**, 289–298.

Li, Y.-H., 1981. Geochemical cycles of elements and human perturbation. Geochim. Cosmochim. Acta **46**, 2073–2084.

Li, Y.-H., H. Teraoka, T.-S. Yang and J.-S. Chen, 1984a. The elemental composition of suspended particles from Yellow and Yangtze Rivers. Geochim. Cosmochim. Acta **48**, 1561–1564.

Li, Y.-H., H.W. Feely and J.R. Toggweiler, 1980. ^{228}Ra and ^{228}Th concentrations in GEOSECS Atlantic surface waters. Deep-Sea Res. **27A**, 545–555.

Li, Y.-H., L. Burkhardt, M. Buchholtz, P. O'Hara and P.H. Santschi, 1984b. Partition of radiotracers between suspended particles and seawater. Geochim. Cosmochim. Acta **48**, 2011–2019.

Lindberg, S.E., and R.C. Harris, 1983. Water and acid soluble trace metals in atmospheric particles. J. Geophys. Res. **88**, 5091–5100.

Lithner, G., K. Holm, M.-B.Larsson and K. Holmgren, 1991. Metaller i djurplankton och arsenik i havsvatten i Östersjöomrädet 1975–1987 (Metal content of zooplankton and aresnic content of sea water in the Baltic Sea area 1975–1987). (The Swedish Environmental Protection Agency), Report **No. 3895** (in Swedish, with English summary).

Lithner, G., D. Broman, C. Naef, H. Borg, A.-M. Johansson, P. Kaerrhage and M.-B. Larsson, 1996. Metals in settling particles and surficial sediments of the Swedish Baltic coast 1988–1989, in: Dev. Prog. Sediment Qual. Assess, eds. M. Munawar and G. Dave. Int. Symp. Sediment Qual. Assess. (SPB Academic Publishing: Amsterdam, Netherlands), 27–48.

Löffler, A., T. Leipe and K.-C. Emeis, 2000. The "fluffy layer" in the Pomeranian Bight (western Baltic Sea): geochemistry, mineralogy and environmental aspects. Meyniana **52**, 85-100.

Löfvendahl, R., 1987. Dissolved uranium in the Baltic. Mar. Chem. **21**, 213–227.

Löfvendahl, R., 1990. Changes in the flux of some major dissolved components in Swedish rivers during the present century. Ambio **19**, 210–219.

Löfvendahl, R., G. Åberg and P.J. Hamilton, 1990. Strontium in rivers of the Baltic Basin. Aquat. Sci. **52**, 315–329.

Macdonald, R.W., D.M. Macdonald, M.C. O'Brien and C. Gobeil, 1991. Accumulation of heavy metals (Pb, Zn, Cu, Cd), carbon and nitrogen in sediments from Strait of Georgia B.C., Canada. Mar. Chem. **34**, 109–135.

Macdonald, R.W., L.A. Barrie, T.F. Bidleman, M.L. Diamond, D.J. Gregor, R.G. Semkin, W.M.J. Strachan, Y.F. Li, F. Wania, M. Alaee, L.B. Alexeeva, S.M. Backus, R. Bailey, J.M. Bewers, C. Gobeil, C.J. Halsall, T. Harner, J.T. Hoff, L.M.M. Jantunen, W.L. Lockhart, D. Mackay, D.C.G. Muir, J. Pudykiewicz, K.J. Reimer, J.N. Smith, G.A. Stern, W.H. Schroeder, R. Wagemann and M.B. Yunker, 2000. Contaminants in the Canadian Arctic: 5 years of progress in understanding sources, occurrence and pathways. Sci. Total Environ. **254**, 93–234.

Mackenzie, F.T., J.M. Brewers, R.J. Charlson, E.E. Hoffman, G.A. Knauer, J.C. Kraft, E.-M. Nöthig, B. Quack, J.J. Walsh, M. Whitfield and R. Wollast, 1991. What is the importance of ocean margin processes in global change? in: Ocean Margin Processes in Global Change, eds. R.F.C. Mantoura, J.-M. Martin and R. Wollast (John Wiley & Sons, Chichester), pp. 433–454.

Macklin, M.G., and K. Klimek, 1992. Dispersal, storage and transformation of metal-contaminated alluvium in the upper Vistula basin, southwest Poland. Appl. Geography **12**, 7–30.

Maenhaut, W., W.H. Zoller, R.A. Duce and G.L. Hoffman, 1979. Concentrations and size distribution of particulate trace elements in the South Polar atmosphere. J. Geophys. Res. **84**, 2421–2431.

Maenhaut, W., M. Darzi and J.W. Winchester, 1981a. Seawater and nonseawater aerosol components in the marine atmosphere of Samoa. J. Geophys. Res. **86**, 3187–3193.

Maenhaut, W., A. Selen, P. van Espen and R. Van Grieken, 1981b. PIXE analysis of aerosol samples collected over the Atlantic Ocean from a sailboat. Nucl. Instr. and Meth. **181**, 399–405.

Maenhaut, W., H. Raemdonck, A. Selen, R. Van Grieken and J.W. Winchester, 1983. Characterization of the atmospheric aerosol over the eastern Equatorial Pacific. J. Geophys. Res. **88**, 5353–5364.

Magnusson, B., and S. Westerlund, 1980. The determination of Cd, Cu, Fe, Ni, Pb and Zn in Baltic sea water. Mar. Chem. **8**, 231–244.

Magnusson, B., and L. Rasmussen, 1982. Trace metal levels in coastal sea water. Investigation of Danish waters. Mar. Pollut. Bull. **13**, 81–84.

Maher, W., and E. Butler, 1988. Arsenic in the marine environment. Appl. Organomet. Chem. **2**, 191–214.

Majewski, A., and Z. Lauer (eds.), 1994. Atlas of the Baltic Sea (Institute of Meteorology and Water Management, Warszawa) (in Polish).

Mantoura, R.F.C., J.-M. Martin and R. Wollast (eds.), 1991. Ocean Margin Processes in Global Change (John Wiley & Sons, Chichester).

Maring, H.B., and R.A. Duce, 1989. The impact of atmospheric aerosols on trace metal chemistry in open ocean surface seawater: 2. Copper. J. Geophys. Res. **94**, 1039–1045.

Markert, B., U. Herpin, J. Berlekamp, J. Oehlmann, K. Grodzinska, B. Mankovska, I. Suchara, U. Siewers, V. Weckert and H. Lieth, 1996. A comparison of heavy metal deposition in selected Eastern European countries using the moss monitoring method, with special emphasis on the 'Black Triangle'. Sci. Total Environ. **193**, 85–100.

Mart, L., and H.W. Nürnberg, 1986. Cd, Pb, Cu, Ni and Co distribution in the German Bight. Mar. Chem. **18**, 197–213.

Mart, L., H.W. Nürnberg and H. Rützel, 1985. Levels of heavy metals in the tidal Elbe and its estuary and the heavy metal input into the sea. Sci. Total Environ. **44**, 35–49.

Martin, J.-M., and M. Meybeck, 1978. Trace element composition of Zaire suspended sediments. Neth. J. Sea Res. **12**, 414–420.

Martin, J.-M., and M. Meybeck, 1979. Elemental mass balance of material carried by major world rivers. Mar. Chem. **7**, 173–206.

Martin, J.-M., and M. Whitfield, 1983. The significance of the river input of chemical elements to the ocean, in: Trace Metals in Sea Water, eds. C.S. Wong, E. Boyle, K.W. Bruland, J.D.Burton and E.D. Goldberg (Plenum Publishing Corporation) 265–296.

Matschullat, J., 1997. Trace element fluxes to the Baltic Sea: Problems of input budgets. Ambio **26**, 363–368.

Matschullat, J., and E. Bozau, 1996. Atmospheric element input in the Eastern Ore Mountains. Appl. Geochem. **11**, 149–154.

Matschullat, J., W. Maenhaut, F. Zimmermann and J. Fiebig, 2000. Aerosol and bulk deposition trends in the 1990's, Eastern Erzgebirge, Central Europe. Atmos. Environ. **34**, 3213–3221.

Matsumoto, E., 1975. ^{224}Th-^{238}U radioactive disequilibrium in the surface layer of the ocean. Geochim. Cosmochim. Acta **39**, 205–212.

Matthäus, W., 1994. Auswirkungen der Salzeinströme 1993/1994 aud die ozeanographischen Bedingungen in der zentralen Ostsee. Inf. Fischwirtsch. Hamburg **41**, 142–147.

McCarthy, L.H., T.G. Williams, G.R. Stephens, J. Peddle, K. Robertson and D. Gregor, 1997. Baseline studies in the Slave River, NWT 1990–1994: evaluation of the chemical quality of water and suspended sediment from the Slave River, NWT. Sci. Total Environ. **197**, 21–53.

McManus, J.P., and D. Prandle, 1996. Determination of source concentrations of dissolved and particulate trace metals in the southern North Sea. Mar. Pollut. Bull. **32**, 504–512.

Meyer, H., and R. Lampe, 1999. The restricted buffer capacity of a south Baltic estuary – the Oder estuary. Limnologica **29**, 242–248.

Mierzwiński, A., and E. Niemirycz, 1997a. State Environmental Monitoring Program in Poland, in: River Quality, Dynamics and Restoration, eds. A. Laenen and D.A. Dunnette (Lewis Publishers, Boca Raton), pp. 241–251.

Mierzwiński, A., and E. Niemirycz, 1997b. National Environmental Monitoring Programme, in: International River Water Quality, Dynamics and Restoration, eds. G. Best. T. Bogacka, E. Niemirycz (E & FN Spon, London), pp. 280–290.

Miettinen, J.K., S. Leskinen and T. Jaakkola, 1982. Studies on distribution of actinides between seawater and particulate fractions in the Baltic Sea and its gulfs, in: Transuranic Cycling Behaviour in the Marine Environment. IAEA-TECDOC-265 (IAEA, Vienna).

Millward, G.E., H.J. Kitts, S.D.W. Comber, L. Ebdon and A.G. Howard, 1996. Methyled arsenic in the southern North Sea. Estuar. Coast. Shelf Sci. **43**, 1–18.

Millward, G.E., T.K. Sands and C.F. Jago, 1999. Particulate metals and their settling velocities in the Humber Estuary, UK. Mar. Chem. **68**, 145–168.

Miltner, A., and K.-C. Emeis, 1999. Input of terrestrial organic matter to the Baltic Sea – The Pomeranian Bight as a case study, in: Geochemistry of the Earth's Surface, ed. H. Ármannsson (Balkema, Rotterdam), pp. 327-330.

Miltner, A., and K.-C. Emeis, 2000. Origin and transport of terrestrial organic matter from the Oder Lagoon to the Arkona Basin, Southern Baltic Sea. Organic Geochem. **31**, 57-66.

Miltner, A., and K.-C. Emeis, 2001. Terrestrial organic matter in surface sediments of the Baltic Sea, Northwest Europe, as determined by CuO oxidation. Geochim. Cosmochim. Acta **65**, 1285-1299.

Miyake, Y., K. Saruhashi and Y. Sugimura, 1973. The excess of ^{228}Th over ^{232}Th in ocean water. Records of Oceanographic Works in Japan **12**, 27–30.

Miyake, Y., Y. Sugimura and T. Yasujima, 1970. Thorium concentration and the activity ratios ^{230}Th/^{232}Th and ^{228}Th/ ^{232}Th in the western North Pacific. J. Ocenograph. Soc. Japan **26**, 130–136.

Miyake, Y., Y. Sugimura and T. Yasujima, 1970. Concentration of thorium isotopes and the activity ratios ^{230}Th/^{232}Th and ^{228}Th/ ^{232}Th in seawater in the Pacific. Papers in Meteorology and Geophysics **29**, 75–81.

Mohrholz, V., M. Pastuszak, S. Sitek, K. Nagel and H.U. Lass, 1998. The exceptional Oder flood in summer 1997 – Riverine mass and nutrient transport into the Pomeranian Bight. Deutsche Hydrogr. Z. (German J. Hydrogr.) **50**, 129–144.

Moore, W.S., 1967. Amazon and Mississippi River concentrations of uranium, thorium, and radium isotopes. Earth Planet. Sci. Letters **2**, 231–234.

Moore, W.S., 1969. Measurements of Ra^{228} and Th^{228} in sea water. J. Geophys. Res. **74**, 694–704.

Moore, W.S., 1981. The thorium isotope content of ocean water. Earth Planet. Sci. Letters **53**, 419–426.

Moore, W.S., H.W. Feely and Y.-H. Li, 1980. Radium isotopes in sub-Arctic waters. Earth Planet. Sci. Letters **49**, 329–340.

Muller, F.L.L., 1996. Interactions of copper, lead and cadmium with the dissolved, colloidal and particulate components of estuarine and coastal waters. Mar. Chem. **52**, 245–268.

Müller, G., and U. Förstner, 1975. Heavy metals in sediments of the Rhine and Elbe estuaries: mobilization or mixing effects? Environ. Geol. **1**, 33–39.

Müller, A., 1998. The flood in the Oder River 1997 – Impact on water quality. Deutsche Hydrogr. Z. (German J. Hydrogr.) **50**, 245–251.

Müller, A., and M. Wessels, 1999. The flood in the Odra River 1997 – Impact of suspended solids on water quality. Acta Hydrochim. Hydrobiol. **27**, 316–320.

Müller, A., and P. Heininger, 1999. On sediment pollution in selected German coastal waters of the Baltic Sea. Limnologica **29**, 255–261.

Müller-Navarra, S.H., K. Hüber and H. Komo, 1999. Model simulations of the transport of Odra flood water through the Szczecin Lagoon into the Pomeranian Bight in July/August 1997. Acta Hydrochim. Hydrobiol. **27**, 364-373.

Nagaitsev, Yu.V., 1996. Mineralogy and geochemistry of finely dispersed systems in present day river sediments (southern shore of Gulf of Finland). Vestn. S.-Petersb. Univ. (Izdatel'stvo Sankt-Petersburgskogo Universiteta) Ser. 7: Geol. Geogr. **1**, 30–35.

Neal, C., W.A. Hause, G.J.L. Leeks, B.A. Whitton and R.J. Williams (Guest Editors), 2000. Water Quality of UK Rivers Entering the North Sea (LOIS). Sci. Total Environ. **251/252**, 703 pp.

Nehring, D., 1984. Variations in the nutrient situation of the Baltic proper. Limnologica (Berlin) **15**, 277–287.

Nehring, D., 1996. Gefährdung und Schutz der Ostsee vor Umweltbelastungen. Chemie in unserer Zeit **30**, 300–307.

Nehring, D., and W. Matthäus, 1990. Aktuelle Trends hydrographischer und chemischer Parameter in der Ostsee. 1958–1989 (Actual trends of hydrographic and chemical parameters in the Baltic Sea, 1958–1989). Institut für Meereskunde, Warnemünde Meereswissenschaftliche Berichte 2 (in German).

Nehring, D., W. Matthäus, H.U. Lass, and G. Nausch, 1994. Die hydrographisch-chemischen bedingungen in der westlichen und zentralen Ostsee im Jahre 1993. Deutsche Hydrogr. Z. **46**, 151–162.

Nehring, D., W. Matthäus, H.U. Lass, G. Nausch and K. Nagel, 1995. The Baltic Sea 1994 – consequences of the hot summer and inflow events. Deutsche Hydrogr. Z. **47**, 131–144.

Neumann, T., T. Leipe, T. Brand and G. Shimmield, 1996. Accumulation of heavy metals in the Oder Estuary and its off-shore basins. Chem. Erde **56**, 207–222.

Neumann, T., T. Leipe and G. Shimmield, 1998. Heavy-metal enrichment in surficial sediments in the Oder River discharge area: source or sink for heavy metals? Appl. Geochem. **13**, 329–337.

Niemirycz, E., 1999. The pollution load from the River Odra in comparison to that in other Polish rivers in 1988–1997. Acta Hydrochim. Hydrobiol. **27**, 286–291.

Niemirycz, E., and T. Bogacka, 1997. Pollution outflow into the Baltic Sea. Riverine input of pollutants, in: Environmental Conditions in the Polish Zone of the Southern Baltic Sea during 1996. Maritime Branch Materials (Institute of Meteorology and Water Management, Gdynia), pp. 177–200.

Nies, H., 1988. The radioactive contamination of the Baltic Sea during the years 1983 to 1987 and its radiological consequences. Deutsche Hydrogr. Z. **41**, 39–44.

Nies, H., 1989. The distribution of the Chernobyl fallout over the Baltic Sea and its change during 1987 and 1988 in seawater. Baltic Sea Environment Proc., Three Years Observation of the Levels of some Radionuclides in the Baltic Sea after the Chernobyl Accident (HELCOM, Helsinki) **No. 31**, 31–51.

Nies, H., 1994. Monitoring of radioactive substances in the Baltic Sea, in: HELCOM Seminar for Experts from Estonia, Latvia, Lithuania and Russia on the Implementation of HELCOM Arrangements, other International Instruments and Related Matters. Baltic Sea Environment Proc. (HELCOM, Helsinki) Riga, September 1993, **No. 59**, 61–70.

Nies, H., and C. Wedekind, 1988. The contamination of the North Sea and Baltic Sea by the Chernobyl fallout, in: Radionuclides: A Tool for Oceanography, eds. J.C. Guary, P. Guéguéniat and R.J. Pentreath (Elsevier Applied Science, Amstrdam) pp. 227–239.

Nies, H., and S.P. Nielsen, 1996. Radioactivity in the Baltic Sea, in: Radionuclides in the Oceans, coordinators P. Guéguéniat, P. Germain and H. Métivier (Les Editions de Physique, France) pp. 219–231.

Nimis, P.L., M. Castello, and M. Perotti, 1993. Lichens as bioindicators of heavy metal pollution: a case study at La Spezia (N Italy), in: Plants as Biomonitors, ed. B. Markert (VCH Publishers) pp. 265–284.

Nimis, P.L., G. Lazzarin, A. Lazzarin and N. Skert, 2000. Biomonitoring of trace elements with lichens in Veneto (NE Italy). Sci. Total Environ. **255**, 97–111.

Nolting, R.F., and D. Eisma, 1988. Elementary composition of suspended particulate matter in the North Sea. Neth. J. Sea Res. **22**, 219–236.

Nolting, R.F., H.J.W. Hein de Baar, K.R. Timmermans and K. Bakker, 1999. Chemical fractionation of zinc versus cadmium among other metals nickel, copper and lead in the northern North Sea. Mar. Chem. **67**, 267–287.

Nriagu, J.O., 1979. Global inventory of natural and anthropogenic emissions of trace metals to the atmosphere. Nature **279**, 409–411.

Nriagu, J.O., 1989. A global assessment of natural sources of atmospheric trace metals. Nature **338**, 47–51.

Nriagu, J.O., and J.M. Pacyna, 1988. Quantitative assessment of worldwide contamination of air, water and soils by trace metals. Nature **333**, 134–139.

Oehlmann, J., S. Liebe, B. Watermann, E. Stroben, P. Fioroni and U. Deutsch, 1994. New perspectives of sensitivity of littorinids to TBT pollution. Cahiers Biologir Marine **35**, 254–255.

Olausson, E., O. Gustafsson, T. Mellin and R. Svensson, 1977. Current level of heavy metal pollution and eutrophication in the Baltic Proper. Meddeland fran Maringeologiska laboratoriet, Gothenburg, No. 8.

Osika, R., 1986. Poland, in: Mineral Deposits of Europe, Central Europe, eds. F.W. Dunning and A.M. Evans (The Institution of Mining and Metallurgy, The Mineralogical Society, London) **Vol. 3**, pp. 55–97.

Osterroht, C., K. Kremling and A. Wenck, 1988. Small-scale variations of dissolved organic copper in Baltic waters. Mar. Chem. **23**, 153–165.

Ostrowski, S., 1963. Estimation of the quantity of dissolved solids in the Vistula runoff. Acta Geophys. Pol. **11**, 235–238.

Öhlander, B., J. Ingri and C. Pontér, 1991. Geochemistry of till weathering in the Kalix River watershed, northern Sweden, in: Chemical Weathering Under Field Conditions, ed. K. Rosén. Reports in Forest Ecology and Forest Soils (Swedish University of Agricultural Sciences), **63**, pp. 1–18.

Öhlander, B., M. Land, J. Ingri, and A. Widerlund, 1996. Mobility of rare earth elements during weathering of till in northern Sweden. Appl. Geochem. **11**, 93–99.

Öhlander, B., J. Ingri, M. Land and H. Schöberg, 2000. Change of Sm-Nd isotope composition during weathering of till. Geochim. Cosmochim. Acta **64**, 813–820.

Östlund, P., 1991. Interrelations between plutonium and some transition elements in Gulf of Bothnian sediments. Ambio **20**, 38–42.

Öztürk, M., 1995. Trends of trace metal (Mn, Fe, Co, Ni, Cu, Zn, Cd and Pb) distributions at the oxic-anoxic interface and in sulfidic water of the Drammensfjord. Mar. Chem. **48**, 329–342.

Pacyna, J.M., 1983. Trace element emission from anthropogenic sources in Europe. NILU Techn. Rep. Lillestrom, **No. 10/82**, 107 pp.

Pacyna, J.M., 1984. Estimation of the atmospheric emissions of trace elements from anthropogenic sources in Europe. Atmos. Environ. **18**, 41–50.

Pacyna, J.M., 1992. The Baltic Sea environmental programme. The topical area study for atmospheric deposition of pollutants. Final Technical Report and Final Synthesis Report. NILU Rep. **No. 46**, 141 pp.

Pacyna, J.M., 1993. Atmospheric deposition of heavy metals to the Baltic Sea, in: Intern. Conf. Heavy Metals in the Environment, eds R.J. Allen and J.O. Nriagu (CEP Consultants, Toronto), **1**, pp. 93–96.

Pacyna, J.M., and K. Tørseth, 1997. Central European hot spots of air pollution, in: Restoration of Forests, eds. R.M. Gutkowski and T. Winnicki (Kluwer Academic Publishers), pp. 15–35.

Pacyna, J.M., A. Semb and J.E. Hanssen, 1984. Emission and long-range transport of trace elements in Europe. Tellus, **36B**, 163–178.

Paerl, H.W., 1995. Coastal eutrophication in relation to atmospheric nitrogen deposition: Current perspectives. Ophelia **41**, 237–259.

Paerl, H.W., 1997. Coastal eutrophication and harmful algal blooms: Importance of atmospheric deposition and groundwater as "new" nitrogen and other nutrient sources. Limnol. Oceanogr. **42**, 1154–1165.

Paerl, H.W., and D.R. Whitall, 1999. Anthropogenically-derived atmospheric nitrogen deposition, marine eutrophication and harmful algal bloom expansion: is there a link. Ambio **28**, 307–311.

Páez-Osuna, F., S.R. Guerrero-Galván and A.C. Ruiz-Fernández, 1999. Discharge of nutrients from shrimp farming to coastal waters of the Gulf of California. Mar. Pollut. Bull. **38**, 585–592.

Panteleev, Y., H. Nies, T.K. Ikäheimonen and V. Tishkov, 1995. Radionuclides in seawater, in: HELCOM, 1995. Radioactivity in the Baltic Sea 1984–1991. Baltic Sea Environment Proceedings (Helsinki Commission, Baltic Marine Environment Protection Commission, Helsinki) **No. 61**, 59–68.

Pastuszak, M., 1995a. The hydrochemical and biological impact of the river Vistula on the pelagic system of the Gulf of Gdańsk in 1994. Part 1. Variability in nutrient concentrations. Oceanologia **37**, 181–205.

Pastuszak, M., 1995b. The hydrochemical and biological impact of the river Vistula on the pelagic system of the Gulf of Gdańsk in 1994. Part 1. Variability in nutrient concentrations. Oceanologia **37**, 181–205.

Pastuszak, M., and K. Nagel, 1996. Variability in nutrient distribution in the Pomeranian Bay in September 1993. Oceanologia **38**, 195–225.

Pastuszak, M., S. Sitek and A. Grelowski, 1998. The exceptional Oder flood in summer 1997. Nutrient concentrations in the Świna Strait during the years 1996 and 1997 – with emphasis on the flood event. Deutsche Hydrogr. Z. (German J. Hydrogr.) **50**, 183–202.

Pastuszak, M., H. Siegel, S. Sitek, M. Gerth, G. Tscherich and A. Grelowski, 2000. Impact of water temperature on nutrient concentrations in the Oder estuary in 1996–1998. Deutsche Hydrogr. Z. (German J. Hydrogr.) (in press).

Peirson, D.H., P.A. Cawse and R.S. Cambray, 1974. Chemical uniformity of airborne particulate material, and maritime effect. Nature **251**, 675–679.

Perttilä, M., L. Niemistö and K. Mäkelä, 1995. Distribution, development and total amounts of nutrients in the Gulf of Finland. Estuar. Coast. Shelf Sci. **41**, 345–360.

Petersen, G., 1996. The current state and future direction of numerical models in simulating atmospheric long-range transport of heavy metals over Europe: A review. GKKS-Forschungszentrum Geesthacht GmbH - Geesthacht, pp. 123–132.

Petersen, G., 1999. Airborne heavy metals over Europe: emissions, long-range transport and deposition fluxes to natural ecosystems, in: Contaminated Forests – Recent Developments in Risk Identification and Future Perspectives, eds. I. Linkov and W.R. Schell (Kluwer Academic Publishers, Dordrecht), 20 pp.

Petersen, G., H. Weber and H. Graßl, 1989. Modelling the atmospheric transport of trace metals from Europe to the North Sea and Baltic Sea, in: Control and Fate of Atmospheric Trace Metals, eds. J.M. Pacyna and B. Ottar (Kluwer Academic Publishers, Dordrecht), NATO-ASI Series, Series C: Mathematical and Physical Sciences **Vol. 68**, pp. 57–83.

Petersen, G., Å. Iverfeldt and J. Munthe, 1995. Atmospheric mercury species over central and northern Europe. Model calculations and comparison with observations from the Nordic air and precipitation network for 1987 and 1988. Atmos. Environ. **29**, 47–76.

Petersen, G., J. Munthe, K. Pleijel, R. Bloxam and A. Vinod Kumar, 1998. A comprehensive Eulerian modeling framework for airborne mercury species: development and testing of the tropospheric chemistry module (TCM). Atmos. Environ. **32**, 829–843.

Pettersson, C., B. Allard and H. Boren, 1997. River discharge of humic substances and humic-bound metals to the Gulf of Bothnia. Estuar. Coast. Shelf Sci. **44**, 533–541.

Pettersson, U.T. and J. Ingri, 2001. The geochemistry of Co and Cu in the Kafue River as it drains the Copperbelt mining area, Zambia Chem. Geol. **177**, 399-414.

Piepgras, D.J., and G.J. Wasserburg, 1987. Rare earth element transport in the western Atlantic inferred from Nd isotopic observations. Geochim. Cosmochim. Acta **51**, 1257–1271.

Pihl, L., A. Svenson, P.-O. Moksnes and H. Wennhage, 1999. Distribution of green algal mats throughout shallow soft bottoms of the Swedish Skagerrak archipelago in relation to nutrient sources and wave exposure. J. Sea Res. **41**, 281–294.

Piña, A.A., G.T. Villaseñor, M.M. Fernández, A. Luszczewski Kudra and R.L. Ramos, 2000. Scanning electron microscope and statistical analysis of suspended heavy metal particles in San Luis Potosi, Mexico. Atmos. Environ. **34**, 4103–4112.

Pitkänen, H., 1991. Nutrient dynamics and trophic conditions in the eastern Gulf of Finland: The regulatory role of the Neva Estuary. Aqua Fennica **21**, 105–115.

Pitkänen, H., and T. Tamminen, 1995. Nitrogen and phosphorus as production limiting factors in the estuarine waters of the eastern Gulf of Finland. Mar. Ecol. Progr. Ser. **129**, 283–294.

Pohl, C., and U. Hennings, 1999. The effect of redox processes on the partitioning of Cd, Pb, Cu, and Mn between dissolved and particulate phases in the Baltic Sea. Mar. Chem. **65**, 41–53.

Pohl, C., U. Hennings, I. Petersohn and H. Siegel, 1998. Trace metal budget, transport, modification and sink in the transition area between the Oder and Peene Rivers and the southern Pomeranian Bight. Mar. Pollut. Bull. **36**, 598–616.

Polkowska, Ż., M. Grynkiewicz, T. Górecki and J. Namieśnik, 2001. Levels of lead in atmospheric deposition in a large urban agglomeration in Poland. J. Environ. Monit. **3**, 146-149.

Pontér, C., J. Ingri, J.-O. Burman and K. Boström, 1990. Temporal variations in dissolved and suspended iron and manganese in the Kalix River, northern Sweden. Chem. Geol. **81**, 121–131.

Pontér, C., J. Ingri and J. Boström, 1992. Geochemistry of manganese in the Kalix River, northern Sweden. Geochim. Cosmochim. Acta **56**, 1485–1494.

Porcelli, D., P.S. Andersson, G.J. Wasserburg, J. Ingri and M. Baskaran, 1997. The importance of colloids and mires for the transport of uranium isotopes through the Kalix River watershed and Baltic Sea. Geochim. Cosmochim. Acta **61**, 4095–4113.

Pöllänen, R., and H. Toivonen, 1994. Transport of large uranium fuel particles released from a nuclear power plant in a severe accident. J. Radiolog. Protect. **14**, 55–65.

Pöllänen, R., I. Valkama and H. Toivonen, 1997. Transport of radioactive particles from the Chernobyl accident. Atmos. Environ. **31**, 3575–3590.

Prange, A., and K. Kremling, 1985. Distribution of dissolved molybdenum, uranium and vanadium in Baltic Sea waters. Mar. Chem. **16**, 259–274.

Price, N.B., and S.E. Calvert, 1973. A study of the geochemistry of suspended particulate matter in coastal waters. Mar. Chem. **1**, 169–189.

Protasowicki, M., E. Niedźwiecki, W. Ciereszko, A. Perkowska and E. Meller, 1999. The comparison of sediment contamination in the area of estuary and the lower course of the Odra before and after the flood of summer 1997. Acta Hydrochim. Hydrobiol. **27**, 338–342.

Puckett, K.J., and E.J. Finnegan, 1980. Analysis of the element content of lichens from the Northwest Territories, Canada. Can. J. Bot. **58**, 2073–2089.

Puls, W., W. Gerwinski, M. Haarich, M. Schirmacher and D. Schmidt, 1997. Lead budget for the German Bight. Mar. Pollut. Bull. **34**, 410–418.

Pustelnikov, O.S., 1977. Geochemical features of suspended matter in connection with recent sedimentation process in the Baltic Sea. Ambio Spec. Rep. **5**, 157–162.

Queirolo, F., S. Stegen, J. Mondaca, R. Cortés, R. Rojas, C. Contreras, L. Munoz, M.J. Schwuger and P. Ostapczuk, 2000. Total arsenic, lead, cadmium, copper, and zinc in some salt rivers in the northern Andes of Antofagasta, Chile. Sci. Total Environ. **255**, 85–95.

Rahm, L, D. Conley, P. Sandén, F. Wulff and P. Stålnacke, 1996. Time series analysis of nutrient inputs to the Baltic Sea and changing DSi:DIN ratios. Mar.Ecol. Prog. Ser. **130**, 221–228.

Reimann, C., P. de Caritat, J.H. Halleraker, T. Volden, M. Åyräs, H. Niskavaara, V.A. Chekushin and V.A. Pavlov, 1997a. Rainwater composition in eight Arctic catchments in northern Europe (Finland, Norway and Russia). Atmos. Environ. **31**, 159–170.

Reimann, C., P. de Caritat, J.H. Halleraker, T.E. Finne, R. Boyd, O. Jćger, T. Volden, G. Kasulina, I. Bogatyrev, V.A. Chekushin, V.A. Pavlov, M. Åyräs, M.L. Räisänen and H. Niskavaara, 1997b. Regional atmospheric deposition patterns of Ag, As, Bi, Cd, Hg, Mo, Sb and Tl in a 188,000 km^2 area in the European Arctic as displayed in terrestrial moss samples – long range atmospheric transport vs. local impact. Atmos. Environ. **31**, 3887-3901.

Reimann, C., U. Siewers, T. Tarvainen, L. Bityukova, J. Eriksson, A. Gilucis, V. Gregorauskiene, V. Lukashev, N.N. Matinian and A. Pasieczna, 2000. Baltic soil survey: total concentrations of major and selected trace elements in arable soils from 10 countries around the Baltic Sea. Sci. Total Environ. **257**, 155–170.

Remoudaki, E., G. Bergametti and P. Buat-Menard, 1991. Temporal variability of atmospheric lead concentrations and fluxes over the nortwestern Mediterranean Sea. J. Geophys. Res. **96**, 1043–1055.

Renk, H., 1990. Primary production of the southern Baltic in 1979–1983, Oceanologia **29**, 51–75.

Ribbe, J., S.H. Müller-Navarra and H. Nies, 1991. A one-dimensional dispersion model for radionuclides in the marine environment applied to the Chernobyl fallout over the Northern Baltic Sea. J. Environ. Radioactivity **14**, 55–74.

Riget, F., G. Asmund and P. Aastrup, 2000. The use of lichen (*Cetraria nivalis*) and moss (*Rhacomitrium lanuginosum*) as monitors for atmospheric deposition in Greenland. Sci. Total Environ. **245**, 137–148.

Rodhe, H., R. Söderlund and J. Ekstedt, 1980. Deposition of air-borne pollutants on the Baltic Sea. Ambio **9**, 168–173.

Rosenberg, R., 1985. Eutrophication – The future marine coastal nuisance. Mar. Pollut. Bull. **16**, 227–231.

Rosenberg, R., R. Elmgren, S. Fleischer, P. Jonsson, G. Persson and H. Dahlin, 1990. Marine eutrophication case studies in Sweden. Ambio **19**, 102–108.

Rosner, G., and R. Winkler, 2001. Long-term variation (1986-1998) of post-Chernobyl ^{90}Sr, ^{137}Cs, ^{238}Pu, and ^{239,240}Pu concentrations in air, depositions to ground, resuspension factors and resuspension rates in south Germany. Sci. Total Environ. **273**, 11-25.

Rönner, U., 1985. Nitrogen transformation in the Baltic Proper: denitrification counteracts eutrophication. Ambio **14**, 134–138.

Rönner, U., and F. Sörensson, 1985. Denitrification rates in the low-oxygen waters of the stratified Baltic Proper. Appl. Environ. Microbiol. **50**, 801–806.

Rühling, Å., G. Brumelis, N. Goltsova K. Kvietkus, E. Kubin, S. Liivs, S. Magnusson, A. Mäkinen, K. Pilegaard, I. Rasmusson, E. Sander and E. Steinnes, 1992. Atmospheric heavy metal deposition in northern Europe 1990. Nordic Council of Ministers, NORD 12.

Rybinski, J., E. Niemirycz and Z. Makowski, 1992. Pollution load. (National Scientific Committee on Oceanic Research, Polish Academy of Sciences, Gdansk). Mar. Pollut. **2**, 21–52.

Ryther, J.H., and W.M. Dunstan, 1971. Nitrogen, phosphorus and eutrophication in the coastal marine environment. Science **171**, 1008–1112.

Sackett, W.M., H.A. Potratz and E.D. Goldberg, 1958. Thorium content of ocean water. Science **3317**, 204–205.

Sackett, W.M., T. Mo, R.F. Spalding and M.E. Exner, 1973. A revaluation of the marine geochemistry of uranium, in: Radioactive Contamination of the Marine Environment (IAEA, Vienna), pp. 757–769.

Sadasivan, S., 1978. Trace elements in size separated aerosols over sea. Atmos. Environ. **12**, 1677–1683.

Sadasivan, S., 1980. Trace constituents in cloud water, rainwater and aerosol samples collected near the west coast of India during the southwest monsoon. Atmos. Environ. **14**, 33–38.

Sahu, K.C., 1990. Toxic trace metals in settled dust of north-east Bombay. Ind. J. Earth Sci. **17**, 91–107.

Sakuragi, Y., J.L. Meason and P.K. Kuroda, 1983. Uranium and plutonium isotopes in the atmosphere. J. Geophys. Res. **88**, 3718–3724.

Salo, A., and A. Voipio, 1966. On the ^{137}Cs and ^{90}Sr ratio in Baltic waters, in: Radioecological Concentration Processes, eds. B. Åberg and F. Hungate (Pergamon Press, Oxford and New York) pp. 827–833.

Salo, A., and A. Voipio, 1978. Balance of Sr-90 and Cs-137 in the Baltic Sea., revised 1978. Mar. Pollut. Bull. **12**, 218–224.

Salo, A., R. Saxén and M. Puhakainen, 1984. Transport of airborne ^{90}Sr and ^{137}Cs deposited in the basins of the five largest rivers in Finland. Aqua Fennica **14**, 21–31.

Salo, A., K. Tuomainen and A. Voipio, 1986. Inventories of some long-lived radionuclides in the Baltic Sea. Sci. Total Environ. **54**, 247–260.

Salomons, W., and W.D. Eysink, 1979. Pathways of mud and particulate trace metals from rivers to the southern North Sea. International Meeting on Holocene Marine Sedimentation in the North Sea Basin (Texel, The Netherlands), 1–36 p. + Figures.

Sandén, P., and Å. Danielsson, 1995. Spatial properties of nutrient concentrations in the Baltic Sea. Environ. Monit. Asses. **34**, 289–307.

Sandén, P., and L. Rahm, 1993. Nutrient trends in the Baltic Sea. Environmetrics **4**, 75–103.

Santschi, P.H., Y.-H. Li and J. Bell, 1979. Natural radionuclides in the water of Narragansett Bay. Earth Planet. Sci. Letters **45**, 201–213.

Santschi, P.H., J.J. Lenhart and B.D. Honeyman, 1997. Heterogeneous processes affecting trace contaminant distribution in estuaries: The role of natural organic matter. Mar. Chem. **58**, 99–125.

Savchuk, O., and F. Wulff, 1999. Modelling regional and large-scale response of Baltic Sea ecosystem to nutrient load reductions. Hydrobiologia **393**, 35–43.

Saxén R., and E. Illus, 2000.Discharge of ^{137}Cs by Finnish rivers to the Baltic Sea in 1986–1996, in: The Radiological Exposure of the Population of the European Community to Radioactivity in the Baltic Sea. Marina-Balt Project, ed. S.P. Nielsen. Proceedings of a Seminar held at Hasseludden Conference Centre, Stockholm, 9–11 June 1998, European Commission, Directorate-General Environment, EUR 19200 EN (European Communities, 2000, Belgium), pp. 333–347.

Sayles, F., H. Livingston and G. Panteleyev, 1997. The history and source of particulate ^{137}Cs and ^{239,240}Pu deposition in sediments of the Ob River Delta, Siberia. Sci. Total Environ. **202**, 25–41.

Schaffer, G., and U. Rönner, 1984. Denitrification in the Baltic proper deep-water. Deep-Sea Res. **31**, 197–220.

Schaug, J., J.P. Rambæk, E. Steinnes and R.C. Henry, 1990. Multivariate analysis of trace element data from moss samples used to monitor atmospheric deposition. Atmos. Environ. **24A**, 2625–2631.

Schimmack, W., K. Auerswald and K. Bunzl, 2001. Can $^{239+240}$Pu replace ^{137}Cs as an erosion tracer in agricultural landscapes contaminated with Chernobyl fallout. J. Environ. Radiactivity **53**, 41–57.

Schmidt, D., 1980. Comparison of trace heavy-metal levels from monitoring in the German Bight and the southwestern Baltic Sea. Helgoländer Meeresunters. **33**, 576–586

Schmidt, D., 1992. Mercury in Baltic and North Sea waters. Water Air Soil Pollut. **62**, 43–55.

Schneider, B., 1984. Investigations on atmospheric trace metals over Kiel Bight area. Proc. of the 14th Conf. of the Baltic Oceanogr., Gdynia, Poland.

Schneider, B., 1987. Source characterization for atmospheric trace metals over Kiel Bight. Atmos. Environ. **21**, 1275–1283.

Schneider, B., 1993. Untersuchung und Bewertung des Schadstoffeintrags über die Atmosphäre im Rahmen von PARCOM (Nordsee) und HELCOM (Ostsee) – Teilvorhaben: Messungen von Spurenmetallen. GKSS Research Center, Gesthacht, Germany, GKSS 93/E/53.

Schneider, B., 1995. Bilanzen und Kreisläufe von Spurenmetallen in der Ostsse. Geowissenschaften **13**, 464–469.

Schneider, B., 1996. Heavy metals – Water, in: HELCOM, 1996. Third Periodic Assessment of the State of the Marine Environment of the Baltic Sea, 1989–93; Background Document. Baltic Sea Environment Proc. (Baltic Marine Environment Protection Commission, Helsinki) **No. 64B**, pp. 144–149.

Schneider, B., and C. Pohl, 1996. Time series for dissolved cadmium at a coastal station in the Western Baltic Sea. J. Mar. Systems **9**, 159–170.

Schneider, B., D. Ceburnis, R. Marks, J. Munthe, G. Petersen and M. Sofiev, 2000. Atmospheric Pb and Cd input into the Baltic Sea: a new estimate based on measurements. Mar. Chem. **71**, 297–307.

Seaward, M.R.D., 1995. Use and abuse of heavy metal bioassays in environmental monitoring. Sci. Total Environ. **176**, 129–134.

Seitzinger, S.P., and S.W. Nixon, 1985. Eutrophication and the rate of denitrification and N_2O production in coastal marine sediments. Limnol. Oceonogr. **30**, 1332–1339.

Sholkovitz, E., 1993. The geochemistry of rare earth elements in the Amazon River estuary. Geochim. Cosmochim. Acta **57**, 2181–2190.

Sholkovitz, E., 1995. The aquatic chemistry of rare earth elements in rivers and estuaries. Aquat. Geochem. **1**, 1–34.

Sholkovitz, E.R., and N.B. Price, 1980. The major element chemistry of suspended matter in the Amazon estuary. Geochim. Cosmochim. Acta **44**, 163–171.

Schultz Tokos, J.J., K. Kremling, P. Streu, L. Brügmann and U. Hennings, 1993. Spatial and temporal variability of selected trace metals in Kiel Bight and Mecklenburg Bight. Ber. Inst. Meereskd. Christian-Albrechts-Univ. Kiel 238.

Siegel, H., W. Matthäus, R. Bruhn, M. Gerth, G. Nausch, T. Neumann and C. Pohl, 1998. The exceptional Oder flood in summer 1997 – Distribution patterns of the Oder discharge in the Pomeranian Bight. Deutsche Hydrogr. Z. (German J. Hydrogr.) **50**, 145–167.

Singer, D.A., 1995. World class base and precious metal deposits – a quantitative analysis. Econ. Geol. **90**, 88–104.

Skwarzec, B., 1995. Polonium, uranium and plutonium in ecosystem of the southern Baltic Sea. Rozprawy i monografie (Polska Akademia Nauk, Instytut Oceanologii w Sopocie), 183 pp. (in Polish with English summary).

Skwarzec, B., 1997. Polonium, uranium and plutonium in the southern Baltic Sea. Ambio **26**, 113–117.

Skwarzec, B., and R. Bojanowski, 1988. ^{210}Po content in sea water and its accumulation in southern Baltic plankton. Mar. Biol. **97**, 301–307.

Skwarzec, B., and R. Bojanowski, 1992. Distribution of plutonium in selected components of the Baltic ecosystem within the Polish economic zone. J. Environ. Radioactivity **15**, 249–263.

Skwarzec, B., R. Bojanowski and J. Bolałek, 1988. The determination of Cu, Pb, Cd and Zn in the southern Baltic water, suspension and sediments. Oceanologia **25**, 75–85.

Smetacek, V., U. Bathmann, E.-M. Nöthig and R. Scharek, 1991. Coastal eutrophication: Causes and consequences, in: Ocean Margin Processes in Global Change, eds. R.F.C. Mantoura, J.-M. Martin and R. Wollast (John Wiley & Sons, Chichester), pp. 251–279.

Smith, J.T., R.T. Clarke and R. Saxén, 2000. Time-dependent behaviour of radiocaesium: a new method to compare the mobility of weapons test and Chernobyl derived fallout. J. Environ. Radioactivity **49**, 65–83.

Sofiev, M., G. Petersen, Krüger, B. Schneider, M. Hongisto and K. Jylha, 2000. Model simulation of atmospheric trace metal concentrations and deposition over the Baltic Sea. Atmos. Environ. (in press).

Sokolowski, A., M. Wolowicz and H. Hummel, 2001. Distribution of dissolved and labile particulate trace metals in the overlying bottom water in the Vistula River plume (southern Baltic Sea). Mar. Pollut. Bull. (in press).

Spalding, R.F., and W.M. Sackett, 1972. Uranium in runoff from the Gulf of Mexico distributive province: Anomalous concentrations. Science **175**, 629–631.

Spalding, R.F., and M.E. Exner, 1976. Temporal uranium variations in the Brazos River. Proc. Fourth Nat. Symp. Radioecol (Dowder, Hutcherson-Ross) 227–230.

Sperling, K.R., 1982. Cadmium determination in coastal water samples from the German Bight. Vom Wasser **58**, 113–142.

Stahlschmidt, T., M. Schulz and W. Dannecker, 1997. Application of total-reflection x-ray fluorescence for the determination of lead, calcium and zinc in size-fractionated marine aerosols. Spectrochim. Acta **52B**, 995–1001.

Steinnes, E., 1995. A critical evaluation of the use of naturally growing moss to monitor the deposition of atmospheric metals. Sci. Total Environ. **160/161**, 143–249.

Stepanets, O., A. Borisov and G. Solv'eva, 1999. Distribution of anthropogenic radionuclides in the estuaries of the Ob and Yenisey rivers and adjacent Kara Sea. Rep. Polar Res. **300**, 132–139.

Stigebrandt, A., and F. Wulff, 1987. A model for the dynamics of nutrients and oxygen in the Baltic proper. J. Mar. Res. **45**, 729–759.

Stockenberg, A., and R.W. Johnstone, 1997. Benthic denitrification in the Gulf of Bothnia. Estuar. Coast. Shelf Sci. **45**, 835–843.

Stoeppler, M., M. Burow, F. Backhaus, W. Schramm and H.W. Nürnberg, 1986. Arsenic in seawater and brown algae of the Baltic and the North Sea. Mar. Chem. **18**, 321–334.

Struck, B.D., M. Froning, R. Pelzer, I. Sistemich and P. Ostapczuk, 1996. Atmospheric dispersion and physicochemical behaviour of Cd and Pb in rainwater after emission by a lead works. Sci. Total Environ. **182**, 85–91.

Struyf, H., and R. Van Grieken, 1993. An overview of wet deposition of micropollutants to the North Sea. Atmos. Environ. **27A**, 2669–2687.

STUK, 1987. Finnish studies on radioactivity in the Baltic Sea after the Chernobyl accident in 1986. Supplement 11 to Annual Report STUK-A55. Finnish Centre for Radiation and Nuclear Safety, Helsinki.

Styro, D., Zh. Bumeliene, M. Lukinskiene and R. Morkuniene, 2001. ^{137}Cs and ^{90}Sr behavioural regularities in the souteastern part of the Baltic Sea. J. Environ. Radiactivity **53**, 27–39.

Sugimura, Y., and M. Mayeda, 1980. The uranium content and the activity ratio ^{234}U/^{238}U in sea water in the Pacific Ocean. Isot. Mar. Chem. (Uchida Rokakuho Pub. Co. Ltd., Tokyo) **Chapt. 12**, 211–246.

Suszkin, A., 1990. Survey of trace elements in the precipitation collected in Geesthacht (Schleswig-Holstein) in the years of 1976, 1983, and 1987. Wissenschaft und Umwelt **2**, 85–87.

Sündermann, J., 1994. Circulation and Contaminant Fluxes in the North Sea (Springer-Verlag, Berlin), 654 pp.

Swarzenski, P.W., B.A. McKee, J.M. Skei and J.F. Todd, 1999a. Uranium biogeochemistry across the redox transition zone of a permanently stratified fjord: Framvaren, Norway. Mar. Chem. **67**, 181–198.

Swarzenski, P.W., B.A. McKee, K. Sørensen and J.F. Todd, 1999b. ^{210}Pb and ^{210}Po, manganese and iron cycling across the O_2/H_2S interface of a permanently anoxic fjord: Framvaren, Norway. Mar. Chem. **67**, 199–217.

Szefer, P., 1977. Investigation on the uranium and thorium contents of Vistula estuary water, also inshore water of the Gulf of Gdańsk. Stud. Mater. Oceanol. **17**, 379–392 (in Polish, with an English abstract).

Szefer, P., 1981a. Occurrence and distribution of thorium isotopes in seas and oceans. Stud. Mater. Oceanol. **34**, 153–182 (in Polish with English abstract).

Szefer, P., 1981b. Distribution and migration of ^{234}U i ^{238}U isotopes in seawater. Stud. Mater. Oceanol. **34**, 183–228 (in Polish with English abstract).

Szefer, P., 1987. Distribution and migration of uranium isotopes in continental and estuarine waters. Stud. Mater. Oceanol. **51**, 133–193 (in Polish with English abstract).

Szefer, P., 1989. Distribution and possible sources of selected chemical elements in the Baltic Sea ecosystem. (Medical University of Gdańsk), 170 pp. (in Polish).

Szefer, P., 1990. Mass-balance of metals and identification of their sources in both river and fallout fluxes near Gdańsk Bay, Baltic Bay. Sci. Total Environ. **95**, 131–139.

Szefer, P., 1998. Distribution and behaviour of selected heavy metals in various components of the southern Baltic ecosystem, in: Geochemical Investigations of the Baltic Sea and Surrounding Areas, eds. P. Szefer, P. and G.P. Glasby (Elsevier Science Ltd, Great Britain). Appl. Geochem. (Spec. Issue) **13**, 287–292.

Szefer, P., and R. Bojanowski, 1981. Mikrometoda oznaczania toru w wodzie morskiej i oceanicznej. (Micromethod of thorium determination in sea and oceanic waters). Materiały Dorocznego Zjazdu Naukowego Polskiego Towarzystwa Chemicznego, Stowarzyszenia Inżynierów i Techników Przemysłu Chemicznego oraz Sympozjum Materiałów Nieorganicznych o Specjalnych Własnościach. Gdańsk, 8–12.09.1981, p. 163 (in Polish).

Szefer, P., and K. Szefer, 1986. Some metals and their possible sources in rain water of the southern Baltic coast, 1976 and 1978–80. Sci. Total Environ. **57**, 79–89.

Szefer, P., G.P. Glasby, J. Pempkowiak and R. Kaliszan, 1995. Extraction studies of heavy-metal pollutants in surficial sediments from the southern Baltic Sea off Poland. Chem. Geol. **120**, 111–126.

Szefer, P., G.P. Glasby, J. Geldon, R.M. Renner, E. Björn, J. Snell and W. Frech, 2000. Distribution and relationships between heavy metals and rare earth elements in surficial sediments from the southern Baltic (in preparation).

Tanaka, S., M. Darzi and J.W. Winchester, 1980. Short term effect of rainfall on elemental composition and size distribution of aerosols in north Florida. Atmos. Environ. **14**, 1421–1426.

Tanizaki, Y., and S. Nagatsuka, 1983. Distribution of chemical behavior of 21 elements in river water in the Yakushima Island. Bull. Chem. Soc. Jpn. **56**, 619–624.

Tanizaki, Y., M. Yamazaki and S. Nagatsuka, 1984. Physicochemical speciation of trace elements in river water by size fractionation. Bull. Chem. Soc. Jpn. **57**, 1545–1550.

Tanizaki, Y., M. Yamazaki and S. Nagatsuka, 1985. Physicochemical speciation of trace elements in river water by means of ultrafiltration. Bull. Chem. Soc. Jpn. **58**, 2995–3002.

Tappin, A.D., G.E. Millward, P.J. Statham, J.D. Burton and A.W. Morris, 1995. Trace metals in the central and southern North Sea. Estuar. Coast Shelf Sci. **41**, 275–323.

Taylor, R., 1984. The runoff of nitrogen and phosphorous compounds from selected agricultural regions in the Vistula and Odra drainage basins. Oceanologia 18, 135–148.

Tishkov, V.P., L.M. Ivanova, V.M. Gavrilov, A.V. Stepanov and Yu.A. Panteleev, 2000. Radioactive contamination of the Baltic Sea water in the region of the Leningrad Nuclear Power Plant (LNNP), 1971–1997, in: The Radiological Exposure of the Population of the European Community to Radioactivity in the Baltic Sea. Marina-Balt Project, ed. S.P. Nielsen. Proceedings of a Seminar held at Hasseludden Conference Centre, Stockholm, 9–11 June 1998, European Commission, Directorate-General Environment, EUR 19200 EN (European Communities, 2000, Belgium), pp. 373–410.

Tomza, U., W. Maenhaut and J. Cafmeyer, 1982. Trace elements in atmospheric aerosols at Katowice, Poland, in: Trace Substances in Environmental Health-XVI, Symposium, ed. D.D. Hemphill, (University of Missouri, Columbia) 105–115.

Toompuu, A., and F. Wulff, 1995. Spatial large-scale correlation of temperature, salinity and nutrient concentrations in the Gulf of Finland. Environmetrics **6**, 55–72.

Toompuu, A., and F. Wulff, 1996. Optimum spatial analysis of monitoring data on temperature, salinity and nutrient concentrations in the Baltic Proper. Environ. Monit. Assess. **43**, 283–308.

Trefry, J.H., and B.J. Presley, 1976. Heavy metal transport from the Mississippi River to the Gulf of Mexico, in: Marine Pollutant Transfer, eds. H.L. Windom and R.A. Duce (D.C. Heath) 39–76.

Trefry, J.H., S. Metz, R.P. Trocine and T.A. Nelsen, 1985. A decline in lead transport by the Mississippi River. Science **230**, 439–441.

Truesdale, V.W., G. Nausch and A. Baker, 2001. The distribution of iodine in the Baltic Sea during summer. Mar. Chem. **74**, 87-98.

Trzosińska, A., 1990. Seasonal fluctuations and long-term trends of nutrient concentrations in the Polish zone of the Baltic Sea. Oceanologia **29**, 27–50

Trzosińska, A., 1992. Nutrients, in: An Assessment of the Effects of Pollution in the Polish Coastal Area of the Baltic Sea, 1984–1989. Studia i Materiały Oceanologiczne **61**, Marine Pollution (2), 107–130.

Tuomainen, K., E. Ilus and T.K. Taipale, 1986. Accumulation of certain long-lived radionuclides by litoral algae and bottom animals, in: Study of Radioactive materials in the Baltic Sea. (International Atomic Energy Agency, Vienna). Report (IAEA-TECDOC-362) of the Final Research Coordination Meeting on the Study of Radioactive Materials in the Baltic Sea organized by the IAEA and held in Helsinki, Finland 24–28 September, 1984, pp. 69–78.

Tuominen, L., A. Heinänen, J. Kuparinen and L.P. Nielsen, 1998. Spatial and temporal variability of denitrification of the northern Baltic Proper. Mar. Ecol. Prog. Ser. **172**, 13–24.

Turner, A., 1996. Trace-metal partitioning in estuaries: importance of salinity and particle concentration. Mar. Chem. **54**, 27–39.

Urba, A., K. Kvietkus and R. Marks, 2000. Gas-phase mercury in the atmosphere over the southern Baltic Sea coast. Sci. Total Environ. **259**, 203–210.

Vandermeulen, J.H., and A. Foda, 1988. Cycling of selenite and selenate in marine phytoplankton. Mar. Biol. **98**, 115–123.

Valette-Silver, N.J., S.B. Bricker and W. Salomons, 1993. Historical trends in contamination of estuarine and coastal sediments: an introduction to the dedicated issue. Estuaries **16**, 575–576.

Veeh, H.H., 1968. $^{234}U/^{238}U$ in the East Pacific sector of the Antarctic Ocean and in the Red Sea. Geochim. Cosmochim. Acta **32**, 117–119.

Vital, H., and K. Stattegger, 2000. Major and trace elements of stream sediments from the lowermost Amazon River. Chem. Geol. **168**, 151–168.

Voß, M., and U. Struck, 1997. Stable nitrogen and carbon isotopes as indicator of eutrophication of the Oder river (Baltic Sea). Mar. Chem. **59**, 35–49.

Voipio, A., 1961. The silicate in the Baltic Sea. Ann. Acad. Sci. Fenn. Ser., **A.II.106**, 1–15.

Voipio, A., and A. Salo, 1971. On the balances of ^{90}Sr and ^{137}Cs in the Baltic Sea. Nord. Hydrol. **2**, 57–63.

Weigel, H.-P., 1976. Atomabsorptionsmessungen von Blei, Cadmium, Kupfer, Eisen und Zink in Seston der Ostsee. Helgolnder wiss. Meeresunters. **28**, 206–216.

Weigel, H.-P., 1977. On the distribution of particulate metals, chlorophyll and seston in the Baltic Sea. Mar. Biol. **44**, 217–222.

Weisel, C.P., R.A. Duce, J.L. Fasching and R.W. Heaton, 1984. Estimates of the transport of trace metals from the ocean to the atmosphere. J. Geophys. Res. **89**, 11,607–11,618.

Weiss, D., and F. Moldenhawer, 1986. Results of the radiological Baltic-Monitoring Programme of the GDR during 1975–1983, in: Study of Radioactive materials in the Baltic Sea. (International Atomic Energy Agency, Vienna). Report (IAEA-TECDOC-362) of the Final Research Co-ordination Meeting on the Study of Radioactive Materials in the Baltic Sea organized by the IAEA and held in Helsinki, Finland 24–28 September, 1984, pp. 89–109.

Whitfield, M., and D.R. Turner, 1979. Water-rock partition coefficients and the composition of seawater and river water. Nature **278**, 132–137.

Widerlund, A., and J. Ingri, 1995. Early diagenesis of arsenic in sediments of the Kalix River estuary, Northern Sweden. Chem. Geol. **125**, 185–196.

Widerlund, A., and J. Ingri, 1996. Redox cycling of iron and manganese in sediments of the Kalix River estuary, Northern Sweden. Aquat. Geochem. **2**, 185–201.

Williams, T.P., J.M. Bubb and J.N. Lester, 1994. Metal accumulation within marsh environments: a review. Mar. Pollut. Bull. **38**, 277–290.

Windom, H.L., 1990. Flux of particulate metals between east coast North American revers and the North Atlantic Ocean. Sci. Total Environ. **97/98** 115–124.

Windom, H., R. Smith, F. Niencheski and C. Alexander, 2000. Uranium in rivers and estuaries of globally diverse, smaller watersheds. Mar. Chem. **68**, 307–321.

Winkels, H.J., S.B. Kroonenberg, M.Y. Lychagin, G. Marin, G.V. Rusakov and N.S. Kasimov, 1998. Geochronology of priority pollutants in sedimentation zones of the Volga and Danube delta in comparison with the Rhine delta. Appl. Geochem. **13**, 581–591.

Winterhalter, B., and J. Siivola, 1967. An electron microprobe study of the distribution of iron, manganese and phosphorus in concretions from the Gulf of Bothnia, northern Baltic Sea. Comptes Rendus de la Societéte Geologique de la Finlande **39**, 161–172.

Witt, G., T. Leipe and K.-C. Emeis, 2001. Using fluffy layer material to study the fate of particle bound organic pollutants in the southern Baltic Sea. Environ. Sci. Technol. **35**, 1567-1573.

Wollast, R., 1991. The coastal organic carbon cycle: fluxes, sources, and sinks, in: Ocean Margin Processes in Global Change, eds. R.F.C. Mantoura, J.-M. Martin and R. Wollast (John Wiley & Sons, Chichester), pp. 365–382.

Wrembel, H.Z., 1983. An estimation of the mercury content in the waters of the Pomeranian Baltic shore area. Acta Hydrochim. Hydrobiol. **11**, 523–538.

Wrembel, H.Z., 1993. Atmosphere and rivers as the major sources of mercury to the Baltic Sea. Acta Geophys. Polon. **41**, 1–48.

Wrembel, H.Z., 1994. Atmosphere as a mercury source for the Polish Baltic zone. Stud. Mater. Oceanol. **No. 66**, 95–135.

Wulff, F., and L. Rahm, 1988. Long-term, seasonal and spatial variations of nitrogen, phosphorus and silicate in the Baltic: an overview. Mar. Environ. Res **26**, 19–37.

Wulff, F., and L. Rahm, 1989. Optimizing the Baltic sampling programme: the effects of using different stations in calculations of total amounts of nutrients. Beitr. Meereskd. **60**, 61–66.

Wulff, F., and A. Stigebrandt, 1989. A time-dependent budget model for nutrients in the Baltic Sea. Global Biogeochem. Cycles **3**, 63–78.

Wulff, F., A. Stigebrandt and L. Rahm L., 1990. Nutrient dynamics of the Baltic Sea. Ambio **29**, 126–133.

Wulff, F., L. Rahm, P. Jonsson, L. Brydsten, T. Ahl and Å. Granmo, 1993. A mass-balance model of chlorinated organic matter for the Baltic Sea – a challenge for ecotoxicology. Ambio **22**, 27–31.

Wulff, F., M. Perttilä and L. Rahm, 1996. Monitoring, mass balance calculation of nutrients and the future of the Gulf of Bothnia. Ambio Spec. Rep. **8**, 28–35.

Yurkovskis, A., F. Wulff, L. Rahm, A. Andruzaitis and M. Rodriguez-Medina, 1993. A nutrient budget of the Gulf of Riga, Baltic Sea. Estuar. Coast. Shelf Sci. **37**, 113–127.

Zwolsman, J.J.G., G.W. Berger and G.T.M. VanEck, 1993. Sediment accumulation rates, historical input, post-depositional mobility and retention of major elements and trace metals in salt marsh sediments of the Scheldt estuary SW Netherlands. Mar. Chem. **44**, 73–94.

Zwolsman, J.J.G., and G.T.M. van Eck, 1999. Geochemistry of major elements and trace metals in suspended matter of the Scheldt estuary, southwest Netherlands. Mar. Chem. **66**, 91–111.

Chapter 3

Biota as a Medium for Chemical Elements

A. PHYTOBENTHOS

(i) Introduction

General Characteristics and Taxonomy

The Baltic Proper, especially southern part with its predominantly sandy bottoms, does not favour development of much phytobenthos, represented by macrophytes such as: brown algae Phaeophyta, red algae Rhodophyta, green algae Chlorophyta and Charophyta. In the littoral zone some vascular plants are found, e.g. sea grass *Zostera marina*, *Chara*, *Potamogeton* and *Phragmites communis*. Along the shore, green algae from *Enteromorpha* and *Cladophora* genus grow on stones moistened by water. Species of fresh water origin, i.e. Charophyta (*Chara, Tolypella, Nitella*) grow on muddy bottoms where wave action is limited. The most typical Phaeophyta species is *Fucus vesiculosus* although in some areas (Gulf of Gdańsk) filamentous brown algae *Ectocarpus* sp. and *Pilayella littoralis* are very abundant. *Ceramium* sp. is a very common red algae (Rhodophyta) growing on underwater piles, stones and other plants (Falandysz et al., 2000). *Furcellaria fastigiata, Phyllophora brodiaei* and *Ahnfeltia plicata* are less common. *Zostera marina* is the most common vascular plant on sandy bottoms, forming underwater meadows in the littoral zone. In low salinity waters of coastal bays, typically fresh water species are also noted, namely: *Potamogeton* sp., *Zannichella palustris*, *Ceratophyllum demersum* and *Myriophyllum spicatum*. Seaweeds, and their environment, phycology, biogeography, and ecophysiology have been described by several authors (Podbielkowski and Tomaszewicz, 1979; Lüning, 1990; Hoek van den et al., 1995;

Lee, 1999). Phytobenthic zone biodiversity in the Baltic Sea has been monitored by Bäck et al. (1998).

Phylum: Green algae Chlorophyta
Family: Chlorophyceae

Along the shore, green algae from *Enteromorpha* and *Cladophora* genus grow on stones moistened by water.

Enteromorpha

This is known ca. 25 species which occur in sea and brackish waters of European coasts; a few species inhabit freshwaters. They can grow in great abundance on rocky coasts but can also form free floating masses in lagoons and brackish pools.

Cladophora

Distributed in see and fresh waters (in Europe is identified 9 freshwater species and 25 marine species) *Cladophora* is widespread in temperate and tropical seas but it is virtually absent in polar waters.

Chara

Stoneworts, Charophyceae, are mainly freshwater plants, some of them inhabit brackish waters (*Ch. baltica, Ch. aspera*, *Ch. crinita*). Distributed in big amounts in waters enriched in calcium, almost eutrophic waters. Common species in all over the world.

Tolypella

Algae have relatively small light requirements and therefore live in deeper freshwater. Some species, e.g. *T. nidifica* are present in brackish waters.

Nitella

Mainly freshwater species inhabit more acid waters (pH 6–8) than *Chara* (pH 7–8); prefer oligotrophic waters.

Phylum: Brown algae Phaeophyta
Family: Phaeophyceae

The most typical Phaeophyta species is *Fucus vesiculosus*. In some areas (Gulf of Gdańsk) filamentous brown algae *Ectocarpus* sp. and *Pilayella littoralis* are very abundant.

Bladder wrack, *Fucus vesiculosus*

Bladder wrack is a common intertidal species distributed along the temperate rocky coasts of the North Atlantic; is eurythermal and also euryhaline species, since it penetrates into the Baltic.

Ectocarpus siliculosus

It is very common cosmopolitan species in the European coasts of the Atlantic and the Mediterranean.

Pilayella littoralis

Species presents in the North Atlantic (Spitsbergen, Greenland, Novaya Zemlya, the Baltic Sea).

Ascophyllum nodosum

This species is restricted to the North Atlantic where is very common fucoid; the southern limits along the European coasts are situated on the coast of northern Portugal.

Phylum: Red algae Rhodophyta
Family: Rhodophyceae

Ceramium sp. is a very common red algae growing on underwater piles, stones and other plants. *Furcellaria fastigiata, Phyllophora brodiaei* and *Polysiphonia* are less common.

Ceramium

Ceramium species are common everywhere on sea coasts in littoral and sublittoral zones; the Arctic, the North Atlantic and Pacific to 30^0 N.

Furcellaria lumbricalis = F. fastigiata

Cool water seaweed genera is endemic in the Arctic – North Atlantic and lives in sublittoral zone.

Phyllophora truncata = P. brodiaei

It is cold temperate North Atlantic species which reaches southern Alaska via the Arctic region.

Ahnfeltia plicata

Arctic-cold temperate alga; appears in the North Atlantic and Pacific Oceans. Northern limits: south Arctic, Southern limits: in the Atlantic – south Portugal, Connecticut; in the Pacific – the North America – Mexico, the North Asia – Korea.

Polysiphonia

On European coasts there are ca. 25 species and the genus is found on sea coasts world-wide.

Rhodomela subfusca = R. confervoides

Arctic-cold temperate alga; Northern limits: north Arctic; Southern limits: in the Atlantic – Europe/Africa: in the North America: Connecticut; in the Pacific – the North America: North Washington.

Phylum: Spermatophyta
Family: Spermatophyceae

Eel grass, *Zostera marina*

Zostera marina is the most common vascular plant; it has a middle distribution in all temperate regions of the Northern Hemisphere and may be found on sandy and muddy substrata in the upper sublittoral zone. It is distributed in the Baltic Sea except the Gulf of Bothnia because of too small its water salinity.

Pondweed, Genus *Potagometon*
This cosmopolitan perennial plant inhabits fresh euthrophic waters in both the Hemispheres; occurs also in estuary or bay of brackish and seawater. Some species are distributed in brackish coastal seawater.

Zanichella palustris
It is annual plant distributed in strongly eutrophised saline or brackish waters, almost cosmopolitan species, not observed in Australia.

Ruppia maritima
Aquatic perennial plant almost cosmopolitan, inhabits shallow sea and brackish waters, observed in estuaries, gulfs and lagoons.

Hornwort, *Ceratophyllum demersum*
This almost cosmopolitan perennial plant inhabits euthrophic waters.

Water milfoil, *Myriophyllum spicatum*
This underwater perennial plant inhabits brackish waters up to salinity 9 PSU, mainly in fresh waters on all continents

Water thyme, *Elodea canadensis*
Perennial plant occurs in almost all types of waters except extremely dystrophic and oligotrophic and saline waters. It is common plant distributed in Europe, the North and South Americas, Asia and Australia.

Sweet flag, *Acorus calamus*
This perennial plant is halofite, grows in eutrophic waters on muddy bottom. Inhabits waters of ponds, lakes and rivers. Common species distributed in Europe, Asia, the North America

Overview of Worldwide Literature

Macroalgae have been studied extensively for trace metal concentrations in respect to their potential use as biomonitor of metallic pollutants in the marine environments. The most commonly used seaweed groups were: Phaeophyta, Chlorophyta, Rhodophyta and Spermatophyta (Black and Mitchell, 1952; Lunde, 1970; Butterworth et al., 1972; Preston et al., 1972; Bryan and Hummerstone, 1973c, 1977; Fuge and James, 1973, 1974; Hägerhäll, 1973; Haug et al., 1974; Stenner and Nickless, 1974; Bok and Keong, 1976; Foster, 1976; Saenko et al., 1976; Zingde et al., 1976; Lande, 1977; Romeril, 1977; Agadi et al., 1978; Melhuus et al., 1978; Munda, 1978; Myklestad and Eide, 1978; Shiber and Washburn, 1978; Sivalingam, 1978; Bohn, 1979; Drifmeyer et al., 1980; Eide et al., 1980; Khristoforova and Bogdanova, 1980; Hornung et al., 1981; Jahnke et al., 1981; Julshamn, 1981a, 1981b; Burdon-Jones et al., 1982; Luoma et al., 1982; Bryan, 1983, 1984,

1985; Munda, 1984; Wahbeh, 1984; Barnett and Ashcroft, 1985; Bryan et al., 1985; Di Giulio and Scanlon, 1985; Wahbeh et al., 1985; Sears et al., 1985; Langston, 1986; Sawidis and Voulgaropoulos, 1986; Sharp et al., 1988; Ramirez et al., 1990; Bryan and Langston, 1992; Gnassia-Barelli et al., 1995; Riget et al., 1995; Sfriso et al., 1995; Jayasekera and Rossbach, 1996; Warnau et al., 1996; Nicolaidou and Nott, 1998; Pergent-Martini, 1998; Brown et al., 1999; Filho et al., 1999; Muse et al., 1999; Caçador et al., 2000; Páez-Osuna et al., 2000; Campanella et al., 2001). It results from these reports that some phytobenthos can adsorb selected metals from water especially actively, thus reflecting their levels in the surrounding environment. A concentration of trace elements depends not only on particular systematic groups the plants belong to, and current physiological condition; great influence have also environmental conditions. The most papers pertaining to this topic dealt with the occurrence of chemical elements in brown algae, especially predisposed to bioaccumulate of trace metals from the aquatic environment (Phillips, 1980; Bryan et al., 1985) because of linear relationships between metal concentrations, e.g. Cu, Zn and Mn in algae tissue and the ambient seawater (Fuge and James, 1974; Morris and Bale, 1975; Seeliger and Edwards, 1977; Bryan and Gibbs, 1983; Bryan et al., 1985). Therefore, seaweeds are useful and effective organisms for biomonitoring of dissolved species of metals since, in contrast to animals, the dietary route for some metals uptake is not involved (Phillips, 1977c, 1980; Bryan et al., 1985). It indicates a little regulation of metal bioaccumulation (Bryan, 1969) suggesting a constant concentration factor (CF) (Bryan, 1983).

Edible seaweed products have been used in various countries as a food item. However. ineffective control exists over the chemical composition of these products which could contain elevated levels of heavy metals and radionuclides. According to van Netten et al. (2000) most of imported seaweed products had Hg levels orders of magnitude higher than unpolluted local products. It has been found that the content of I in imported seaweed product varied widely reaching the highest values in Japanase *Laminaria japonica* (van Netten et al., 2000).

It is well known that macroalgae are good bioindicators for dissolved species of radionuclides such as ^{106}Ru, $^{239+240}$Pu, ^{238}Pu, ^{241}Am, ^{99}Tc and ^{137}Cs in the marine environments (Hamilton and Clifton, 1980; Woodhead, 1984; Dahlgaard et al., 1986; Aarkrog et al., 1987).

The radionuclides, like heavy metals, may be introduced to the marine ecosystems as dangerous contaminants for public health. According to several authors (Hamilton, 1980; Hamilton and Clifton, 1980; Woodhead, 1984) radionuclides such as ^{106}Ru, $^{239+240}$Pu, ^{241}Am and ^{238}U are accumulated in seaweeds such as *Porphyra umbilicalis* and *Fucus vesiculosus* in the Sellafield (Windscale), north-east England where nuclear reprocessing plants are located. It is pointed out that a reduction in ^{106}Ru uptake to *Porphyra umbilicalis* over distance indicated the removal of the radionuclide to bottom sediment and radioactive decay (Woodhead, 1984). Specimens of *F. vesiculosus* from the same region exhibited comparable ratio of Am/Pu to that found in *Mytilus* byssus, suggesting similar origin of both the nuclides from the ambient seawater (Hamilton and Clifton, 1980). Druehl et al. (1988) registered trace quantities of ^{131}I in brown seaweed reflected the Chernobyl-derived radioactivity. According to van Netten et al. (2000) some of imported seaweed products showed traces of ^{137}Cs likely related to the Chernobyl accident. The detected traces of ^{226}Ra in these products may correspond to naturally occurring uranium decay. The data reported by Yamada et al. (1999) suggested that the enhanced accumulation of $^{239+240}Pu$ and ^{137}Cs in algae due to radioactive dumping into the Japan Sea by the former USSR and Russia is not significant.

(ii) Occurrence of Chemical Elements in Seaweeds

The concentration data in respect to heavy metals have been reported also for different species of seaweeds from the Baltic Sea and adjacent regions (Bojanowski, 1972; Hägerhäll, 1973; Phillips, 1979; Brix and Lyngby, 1982, 1983; Brix et al., 1983; Lyngby and Brix, 1982, 1984, 1987; Lyngby et al., 1982; Kangas and Autio, 1986; Stoeppler et al., 1986; Forsberg et al., 1988; Jankovski et al., 1988; Söderlund et al., 1988; Szefer and Skwarzec, 1988; Szefer and Szefer, 1991; Szefer et al., 1994a; Falandysz, 1994; Ostapczuk et al., 1997b; Struck et al., 1997a). The concentrations of selected chemical elements in particular species of seaweeds from coastal waters of the Baltic Sea and other northern regions are presented in Table 3.1.

Inter-species trends

Several authors (Bojanowski, 1972; Kangas and Autio, 1986, Szefer and Skwarzec, 1988; Szefer and Szefer, 1991) detected species-dependent differences in trace metal concentrations in seaweeds inhabited the Baltic Sea and adjacent areas. The concentration of Zn was smaller in *Cladophora glomerata* than in *Ceramium tenuicorne* and *Pilayella littoralis* and smaller in these three species of annual filamentous algae than in perennial *Fucus vesiculosus* from the Tvärminne area, Northern Baltic Sea (Kangas and Autio, 1986). The concentrations of Cu were characterised by similar values in all the annual species but were smaller than in *F. vesiculosus* from the same region. As for Fe, its levels were ca. three times higher in *C. glomerata* and *P. littoralis* than in *C. tenuicorne*.

Species dependent changes were also assayed for trace elements Cd, Co, Cu, Mn, Ni, Pb, Ti, Zn and macroelements Al, Ca, Fe, K, Mg and Na in samples of seaweeds from the coastal region of the southern Baltic and from Żarnowiec Lake (Szefer and Skwarzec, 1988). It is pointed out that *Cladophora rupestris* collected in southern Baltic shore contained more Fe, Ni and Zn and less Mn and Pb as compared to *Potamogetom pectinatus* from the same region.

Intra-tissue/age dependent trends

According to Bojanowski (1972) the distribution of elements in the various parts of species *F. vesiculosus* from the southern Baltic differed considerably depending upon age. The young parts of the plant had a higher ash content and a greater content of the main ions (on average 10–30%), whereas the older parts of plants had almost twice the content of trace elements. This finding is in an agreement with the distribution of selected trace elements in *F. vesiculosus* from the Tvärminne area, Northern Baltic Sea (Kangas and Autio, 1986). The concentrations of Zn consistently increased from the tops to the stipes of *Fucus*; the distribution pattern of Cu showed the same intraspecies trend while the concentration of Fe changed irregularly (Fig. 3.1). Having in mind the reported data, apparently during the reproduction period, the apical parts of species *F. vesiculosus* with the receptacles, accumulate considerable amounts of trace elements and almost double the amount of the main mineral components. Especially large disproportion has been observed in the distribution of Ni, which may suggest that this element participates in the reproduction cycle. The young parts of the plants show a greater fluctuation of trace elements suggesting their significant mobility (Bojanowski, 1972). The high levels of trace elements in *Fucus* stipes are explained by the relatively slow and irreversible their accumulation and the synthesis of more binding sites with age (Bryan and Hummerstone, 1973c; Bohn, 1979; Bryan et al., 1985). It seems that trace metals are not transferred along the thallus from older parts to younger (Strömgren, 1979). According to Kangas and Autio (1986) the irregular trend for Fe in Baltic *Fucus* may be a result of pollution of the thallus surface by various elements (Bryan and Hummerstone, 1973c). It is postulated that the mucus covering the thallus may have additional Fe amounts of outside origin; hence its total concentration measured is not exclusively corresponded to bound to the algal tissues (Romeril 1977). According to Forsberg et al. (1988) the content of Al, Fe, Mn, Ni, Zn and Co in the older parts of tallus of *F. vesiculosus* from the Archipelago of Stockholm, Baltic Sea, significantly exceeded those of the growing tips. Chromium showed a similar trend but such tendency

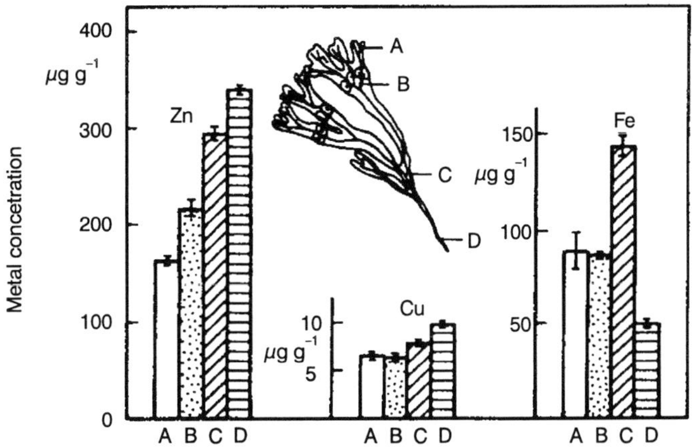

Fig. 3.1. Concentrations of zinc, copper and iron in different parts of the *Fucus* thalus; individuals sampled at Lingskir in July 1979. After Kangas and Autio (1986); modified.

TABLE 3.1.

Concentrations of chemical elements ($\mu g\ g^{-1}$ dry wt.) in seaweeds of the Baltic Sea and other northern areas

Region	Sampling date	N	Ag	Al	As	Ba	Ca	Cd	Co	Cr	Cu	Fe	References
PHAEOPHYCEAE													
Bladder wrack (*Fucus vesiculosus*)													
Southern Baltic													
Gulf of Gdansk and open waters	1965–68	14					17.5*		1.44		5.6	380	Bojanowski, 1972
							14.2–19.6		0.92–2.49		3.6–9.4	170–730	
Western Baltic		22			22		16.4*	0.79	2.24		2.03	455	Struck et al., 1997
	1982	11			17.5								Stoeppler et al., 1986
					10.2–22.6								
	1983	11			16.8								
					12.1–19.4								
Northern Baltic, Tvärminne	1979–81	54						1.3–5.6			11.8	262	Kangas and Autio, 1986
											6.0–33.0	25.0–1400	
Öresund													
Area I[a]	Pre–1973							1.02		5.16	9.33		Hägerhäll, 1973
								ND–2.8		0.11–8.81	2.22–17.0		
Area II[b]								4.4		9.11	19.1		
								0.98–7.0		1.22–16.6	3.21–66.4		
Swedish coast	1977	6^						2.6				105	Phillips, 1979
								1.9–3.9				53–151	
Danish coast	1977	3^						2.33				91.3	
								1.9–2.7				52–148	
Archipelago of Stockholm													
Skotkobb	1933	3		114–310				13.1–11.9	0.47–0.58	0.49–0.68	8.0–8.1	127–230	Forsberg et al., 1988
	1984	14		24–192				10.8–12.3	0.69–0.93	0.44–0.59	6.3–6.5	68–208	
Hogkobben	1933	4		111–104				13.3–14.7	0.48–0.65	0.32–0.52	9.0–7.7	93–182	
	1984	14		117–238				8.3–10.4	0.88–1.19	0.50–0.63	5.4–5.1	123–261	
Angskar	1984	19		59–154				6.6–6.0	0.32–0.53	0.34–0.43	4.3–4.4	86–178	
Soderarm	1984	16		16–100				10.8–10.0	0.45–0.69	0.29–0.49	5.3–4.6	67–157	
Baltic Proper													
Swedish coast	1986	12^		51				6.97	0.61	0.29	4.21	81.8	Söderlund et al., 1988
				17.0–130				4.4–8.1	0.22–2.51	0.16–0.71	2.6–7.0	48.0–169	
		12^^		116				7	0.94	0.47	3.82	176	
				32.0–382				4.1–10.5	0.42–3.17	0.11–1.44	2.1–8.0	65–522	
Southern Bothnian Sea	1986	5^		142				8.68	0.58	0.4	5.44	266	Söderlund et al., 1988
				51.0–291				4.6–12.9	0.40–0.84	0.22–0.61	4.3–6.7	105–495	
		5^^		102				13.3	0.97	0.36	4.63	203	
				92.0–110				10.8–17.2	0.64–1.39	0.24–0.47	3.0–6.1	186–214	

North Sea, German coast					47		12.0*	0.56	3.12		3.37	604	Struck et al., 1997
	1986–94				45–450	28–32							Ostapczuk et al., 1997a
Norwegian coast,	1972	1	2					1		4	35	140	Lande, 1977
Tronheimsfjord	1973	1	20							1	85	1170	
UK estuaries and coasts	1976–80	5	0.5–2.2					1.0–28.0	0.9–7.8	1.0–5.7	4.0–293	90–967	Bryan, 1983
UK coastal waters	1980–84		0.3±0.1		29.2±9.2			1.4±0.6		3.7±2.5	28.0±10.0	770±400	Langston, 1986
			0.12–0.46		12.1–48.5			0.5–2.5		0.7–66.0	10.0–42.0	230–1530	Langston, 1986
UK estuaries			< 0.1–4.1		11–382			0.73–75.0		0.15–5.3	7.3–302	104–2080	
Fucus serratus													
Öresund													
Area I	Pre-1973							1.76		4.65	9.9		Hägerhäll, 1973
								ND–3.40		1.56–7.85	2.61–11.6		
Area II								2.91		6.91	39.7		
								0.16–6.85		1.96–10.5	2.90–85.1		
Fucus inflatus													
Öresund													
Area I	Pre-1973							1.42		7.77	4.53		Hägerhäll, 1973
								ND–4.11		0.66–9.61	1.70–7.90		
Area II								0.78		6.61	46.3		
								ND–2.90		1.68–8.61	18.5–133		
Ectocarpus siliculosus													
Gulf of Gdansk	1978			1400			18.8*	0.61	0.5		6.4	230	Szefer and Skwarzec, 1988
Pilayella littoralis													
Gulf of Gdansk	1987	12					23.6*	2	3		6.8	3400	Szefer et al., 1994a
							6.7–46.3	1.0–3.1	0.5–4.6		3.5–11.0	1200–6700	
Laminaria saccharina													
Öresund													
Area I	Pre-1973							0.18		0.41	4.13		Hägerhäll, 1973
								ND–0.96		ND–0.86	1.50–11.8		
Area II								4.08		15.1	11		
								2.02–10.6		0.96–50.5	3.63–20.2		
Western Baltic	Pre-1986				54								Stoeppler et al., 1986
Laminaria digitata													
Öresund													
Area I	Pre-1973							0.54		0.31	6.92		Hägerhäll, 1973
								0.18–1.31		ND–0.90	2.15–12.2		
Area II								0.33		11.2	30.1		
								0.30–1.78		0.90–20.5	8.88–100		
Chorda filum													
Öresund													
Area I	Pre-1973							0.48		ND	3.56		Hägerhäll, 1973

Region	Sampling date	N	Ag	Al	As	Ba	Ca	Cd	Co	Cr	Cu	Fe	References
								ND–1.51			0.98–5.90		
Area II								1.63		1.1	90.6		
								0.55–2.02		0.65–2.0	50.1–118		
Knotted wrack (*Ascophyllum nodosum*)													
Öresund													
Area I	Pre-1973							0.56		0.91	10.4		Hägerhäll, 1973
								ND–1.33		ND–2.6	2.25–45.7		
Area II								0.81		6.83	18		
								ND–1.50		1.18–9.61	3.80–52.5		
Norwegian coast,	1972	14	1.43								31.6	157	Lande, 1977
Tronheimsfjord			< 1.0–2.0								6.0–123	51–467	
	1973	1	1								38	302	
Fal Estuary, UK		12						0.19–0.99		ND–0.13	3.9–381	36–132	Bryan et al., 1985
BRYOPHYCEAE													
Southern Bothnian Sea	1986	3		1720				2.4	1.27	1.37	15.4	2153	Söderlund et al., 1988
				1520–2070				2.1–2.6	1.0–1.5	0.8–1.8	14.4–16.3	1460–2830	

* – mg g^{-1}
^ – Growing tips.
^ ^ – Old thallus.
[a] – Waters with trace elements concentrations that were normal for coastal areas.
[b] – Waters with high trace element concentrations.

TABLE 3.1. – continued

Region	Sampling date	N	Hg	K	Mg	Mn	Na	Ni	P	Pb	References
						PHAEOPHYCEAE					
Bladder wrack (*Fucus vesiculosus*)											
Southern Baltic											
Gulf of Gdansk and open	1965–68	14		24.6*	8.9*	1060	16.4*	15.1	1.3*		Bojanowski, 1972
waters				8.3–32.5	6.7–11.1	280–1620	8.3–22.1	9.5–22.4	0.8–1.7		
Western Baltic		22	0.0018	33.2*	10.5*	747	26.7*	8.22	26.04*	1.04	Struck et al., 1997
Northern Baltic,	1979–81	54								0.05–19.0	Kangas and Autio, 1986
Tvärminne											
Öresund											
Area I[a]	Pre-1973							18.7		0.12	Hägerhäll, 1973
								7.0–46.3		ND–2.90	
Area II[b]								22.2		3.66	
								7.61–40.7		0.93–15.1	
		6^								17.5	Phillips, 1979
										14.0–21.3	
Danish coast	1977	3^								22.5	
										15.5–27.4	
Archipelago of Stockholm											
Skotkobb	1933	3				84–123		3.4–7.6		6.3–6.9	Forsberg et al., 1988
	1984	14				132–264		8.3–25.5		2.3–1.8	
Hogkobben	1933	4				93–182		3.4–12.0		6.6–5.5	
	1984	14				123–261		9.3–30.3		2.0–1.7	
Angskar	1984	19				130–204		5.4–12.4		2.5–4.0	
Soderarm	1984	16				106–211		6.6–19.2		2.2–2.8	
Swedish coast	1983–84	6^		33.47*			17.43*		0.76*		Forsberg et al., 1988
				29.5–39.86			14.65–21.18		0.17–1.81		
		6^ ^		23.63*			13.93*		0.14*		
				15.71–26.39			10.98–18.45		0.05–0.31		
Baltic Proper											
Swedish coast	1986	12^				96.3		8.51		3.03	Söderlund et al., 1988
						79–139		4.2–29.1		2.0–6.4	
		12^ ^				235		18.3		2.95	
						168–306		7.5–46.4		2.1–3.7	
Southern Bothnian Sea	1986	5^				129		6.36		5.32	Söderlund et al., 1988
						108–152		5.1–8.1		3.0–11.7	
		5^ ^				252		22.9		4.33	

Region	Sampling date	N	Hg	K	Mg	Mn	Na	Ni	P	Pb	References
						187–290		18.1–31.8		2.7–7.2	
North Sea, German coast			0.01	41.4*	8.02*	356	32.0*	9.39	3.1*	1.87	Struck et al., 1997
Norwegian coast,	1973	1						7			Lande, 1977
Tronheimsfjord	1973	1	0.5					2			
UK estuaries and coasts	1976–80	5				108–230		4.5–36		1.6–29.0	Bryan, 1983
UK coastal waters	1980–84		0.21±0.10			168±67		12.9±7.2		10.9±7.9	Langston, 1986
			0.07–0.42			69–264		4.1–22.5		1.1–15.6	
UK estuaries			0.03–0.24			51–573		2.6–53.0		1.3–21.6	Langston, 1986
Fucus serratus											
Öresund											
Area I	Pre-1973							16.3		0.99	Hägerhäll, 1973
								4.60–23.2		ND–2.41	
Area II								24.1		2.32	
								6.66–72.1		0.15–25.9	
Fucus inflatus											
Öresund											
Area I	Pre-1973							9.8		2.95	Hägerhäll, 1973
								2.20–17.5		ND–7.31	
Area II								10.4		2.63	
								3.36–42.8		0.51–5.55	
Ectocarpus siliculosus											
Gulf of Gdansk	1978			2.4*	15.4*	70	1.5*	5.2		15	Szefer and Skwarzec, 1988
Pilayella littoralis											
Gulf of Gdansk	1987	12		8.2*	4.8*	1000	8.3*	7.1		13.1	Szefer et al., 1994a
				3.1–23.6	2.2–7.2	100–2500	2.2–26.7	3.6–9.2		7.3–29.0	
Laminaria saccharina											
Öresund											
Area I	Pre-1973							4.54		0.71	Hägerhäll, 1973
								1.95–8.02		ND–4.30	
Area II								11.2		25.5	
								8.01–20.2		2.96–40.2	
Laminaria digitata											
Öresund											
Area I	Pre-1973							13.6		0.1	Hägerhäll, 1973
								5.6–19.6		ND–0.6	
Area II								15.5		2.96	

					7.00–44.4	0.18–7.20	
Chorda filum							
Öresund							
Area I	Pre-1973				11.4	0.9	Hägerhäll, 1973
					1.96–18.5	ND–1.63	
Area II					13.4	2	
					12.2–25.6	0.98–3.0	
Knotted wrack (*Ascophyllum nodosum*)							
Öresund							
Area I	Pre-1973				10.1	0.86	Hägerhäll, 1973
					2.40–29.2	ND–7.06	
Area II					16.1	5.23	
					1.11–38.4	ND–18.4	
Norwegian coast,	1972	14			6.79		Lande, 1977
Tronheimsfjord					1.0–22.0		
	1973	1	0.1		3		
Fal Estuary, UK		12		8.8–91.6	0.37–1.82	0.61–1.86	Bryan et al., 1985
				BRYOPHYCEAE			
Fontinalis dalecarlica							
Southern Bothnian Sea	1986	3		189	5.27	13.4	Söderlund et al., 1988
				127–264	4.6–5.7	9.0–20.3	

* – mg g^{-1} dry wt.
^ – Growing tips.
^ ^ – Old thallus.
[a] – Waters with trace elements concentrations that were normal for coastal areas.
[b] – Waters with high trace element concentrations.

TABLE 3.1. – continued

Region	Sampling date	N	S	Se	Sn	Sr	Ti	V	Zn	References
					PHAEOPHYCEAE					
Bladder wrack (*Fucus vesiculosus*)										
Southern Baltic										
Gulf of Gdansk and	1965–68	14	74.5**			0.98*			310	Bojanowski, 1972
open waters			64.3–90.2			0.69–1.55			150–500	
Western Baltic		22	27.0*			0.968*			61.5	Struck et al., 1997
Northern Baltic	1979–81	54							379	Kangas and Autio, 1986
Tvärminne									57–1190	
Öresund										
Area I[a]	Pre-1973								86.1	Hägerhäll, 1973
									43.6–122.7	
Area II[b]									180	
									46.7–200	
Swedish coast	1977	6^							118	Phillips, 1979
									73–270	
Danish coast	1977	3^							143	
									100–203	
Archipelago of Stockholm										Forsberg et al., 1988
Skotkobb	1933	3						0.46–0.80	431–547	
	1984	14						0.04–0.36	438–815	
Hogkobben	1933	4						0.54–0.69	503–631	
	1984	14						0.17–0.68	450–779	
Angskar	1984	19						0.11–0.29	255–383	
Soderarm	1984	16						0.09–0.33	403–625	
Baltic Proper										
Swedish coast	1986	12^						0.27	268	Söderlund et al., 1988
								0.07–0.52	159–454	
		12^^						0.63	428	
								0.19–2.15	324–716	
Southern Bothnian Sea	1986	5^						0.58	427	Söderlund et al., 1988
								0.21–0.99	308–604	
		5^^						0.54	677	
								0.49–0.59	457–877	
North Sea, German coast			26.0*			0.792*			32.7	Struck et al., 1997
	1994		0.2*	0.07–0.18						Ostapczuk et al., 1997a
Norwegian coast,	1973	1							55	Lande, 1977
Tronheimsfjord	1973	1							670	
UK estuaries and coasts	1976–80	5							85–1360	Bryan, 1983

UK coastal waters	1980–84		0.54±0.36		940±620	Langston, 1986
			0.16–1.26		210–1960	
UK estuaries			0.04–1.8		69–1740	Langston, 1986
Fucus serratus						
Öresund						
Area I	Pre-1973				122.8	Hägerhäll, 1973
					42.7–209.2	
Area II					169.4	
					39.2–330.5	
Fucus inflatus						
Öresund						
Area I	Pre-1973				95.2	Hägerhäll, 1973
					43.0–171.0	
Area II					122.5	
					45.1–212.6	
Ectocarpus siliculosus						
Gulf of Gdansk	1978			46	203	Szefer and Skwarzec, 1988
Pilayella littoralis						
Gulf of Gdansk	1987	12			120	Szefer et al., 1994a
					55–380	
Northern Baltic	1979–81					Struck et al., 1997
Tvärminne						
Laminaria saccharina						
Öresund						
Area I	Pre-1973				70.4	Hägerhäll, 1973
					29.5–83.8	
Area II					123.8	
					33.0–150.5	
Laminaria digitata						
Öresund						
Area I	Pre-1973				85.7	Hägerhäll, 1973
					63.5–108.2	
Area II					87.8	
					34.4–164.7	
Chorda filum						
Öresund						
Area I	Pre-1973				153.9	Hägerhäll, 1973
					50.6–334.3	
Area II					110.6	

Region	Sampling date	N	S	Se	Sn	Sr	Ti	V	Zn	References
									90.2–150.2	
Knotted wrack (*Ascophyllum nodosum*)										
Öresund										
Area I	Pre-1973								95.2	Hägerhäll, 1973
									30.4–285.8	
Area II									91.9	
									38.4–164.9	
Norwegian coast,	1972	14							199	Lande, 1977
Tronheimsfjord									59–446	
	1973	1							185	
Fal Estuary, UK		12							50–2081	Bryan et al., 1985
					BRYOPHYCEAE					
Fontinalis dalecarlica										
Southern Bothnian Sea	1986	3						4.57	226	Söderlund et al., 1988
								3.7–5.7	94–434	

* – mg g^{-1} dry wt.
** – Expressed as SO_4 in mg g^{-1} dry wt.
^ – Growing tips.
^ ^ – Old thallus.
[a] – Waters with trace elements concentrations that were normal for coastal areas.
[b] – Waters with high trace element concentrations.

TABLE 3.1. – continued

Region	Sampling date	N	Al	As	Ca	Cd	Co	Cr	Cu	Fe	References
					CHLOROPHYCEAE						
Enteromorpha sp.											
Southern Baltic											
Gulf of Gdansk and	1965–66	8			12.8*		0.35		12	930	Bojanowski, 1972
open waters					7.7–18.9		0.19–0.73		7.9–16.6	210–2740	
Gulf of Gdansk	1978–79	2	1.9*		0.56*	0.39	0.7		10.4	315	Szefer and Skwarzec, 1988
			1.2–2.6		0.49–0.66	0.33–0.45	0.60–0.80		5.6–15.2	260–370	
Enteromorpha intestinalis											
Öresund											
Area I[a]	Pre-1973					0.31		4.03	12.6		Hägerhäll, 1973
						ND–0.91		ND–5.15	3.65–27.4		
Area II[b]						6.22		8.11	21.6		
						1.41–37.0		0.81–15.1	6.95–35.5		
Gulf of Gdansk	1978	1	0.4*		0.78*	0.36	0.9		2.8	800	Szefer and Skwarzec, 1988
Enteromorpha crinita											
Southern Baltic, Polish coast	1978	1	1.2*		1.84*	0.37	0.6		2.6	300	Szefer and Skwarzec, 1988
Cladophora sp.											
Gulf of Gdansk and	1965–68	5			7.5*		0.53		9.9	1680	Bojanowski, 1972
open waters					6.3–8.9		0.28–1.87		6.2–13.3	380–3730	
Cladophora rupestris											
Southern Baltic, Polish coast	1978–79	2	3.85*		20.3*	0.7	1.6		3.5	4.75	Szefer and Skwarzec, 1988
			1.6–6.1			0.59–0.83	1.2–2.0		2.4–6.6	4.5–5.0	
Cladophora glomerata											
Öresund											
Area I	Pre-1973					0.54		2.4	7.1		Hägerhäll, 1973
						ND–0.85		050–1.60	3.10–14.3		
Area II						3.18		3.15	22.6		
						ND–4.55		0.65–3.96	22.6–24.4		
Northern Baltic, Tvärminne	1980	15				0.77			17	1770	Kangas and Autio, 1986
						0.2–1.4			12.0–25.0	200–5540	
Ulva lactuca											
Öresund											
Area I	Pre-1973					ND			9.48		Hägerhäll, 1973
									1.65–12.3		
Area II						ND			21.8		
									4.80–38.7		
Acrosiphonia centralis											
Öresund											
Area I	Pre-1973					0.34		0.4	13.1		Hägerhäll, 1973

Region	Sampling date	N	Al	As	Ca	Cd	Co	Cr	Cu	Fe	References
						ND–1.1		ND–1.60	10.8–16.1		
Area II						1.96		2.6	19		
						ND–8.50		0.50–3.00	7.10–22.5		
					RHODOPHYCEAE						
Furcellaria fastigiata											
Gulf of Gdansk and	1966–68	9			5.3*		1.82		12.1	820	Bojanowski, 1972
open waters					4.1–7.4		0.83–2.31		8.4–16.3	560–1020	
Öresund											
Area I	Pre-1973					0.45					Hägerhäll, 1973
						ND–0.90					
Area II						1.11					
						ND–6.44					
Ceramium rubrum											
Gulf of Gdansk and	1966–68	2			7.4*		3.98		23.1	1630	Bojanowski, 1972
open waters							2.34–5.61		20.1–26.0	1340–1920	
Öresund											
Area I	Pre-1973					ND					Hägerhäll, 1973
Area II						0.59					
						ND–4.20					
Ceramium tenuicorne											
Northern Baltic, Tvärminne	1980	12				0.4			20.8	840	Kangas and Autio, 1986
						0.20–0.60			17.0–24.0	215–1540	
Polysiphonia sp.											
Gulf of Gdansk and	1965–66	5			12.9*		3.15		18	2520	Bojanowski, 1972
open waters					6.4–22.4		1.68–5.60		13.5–22.4	830–3910	
Polysiphonia elongata											
Öresund											
Area I	Pre-1973					ND					Hägerhäll, 1973
Area II						2.11					
						ND–5.06					
Polysiphonia nigrescens											
Öresund											
Area I	Pre-1973					ND					Hägerhäll, 1973
Area II						1.7					
						ND–3.30					
Rhodomela confervoides											
Öresund											
Area I	Pre-1973					0.35					Hägerhäll, 1973
						ND–0.61					
Rhodomela subfusca											
Gulf of Gdansk and	1966–68	3			7.77*		4.87		25.8	1830	Bojanowski, 1972
open waters					4.8–10.6		2.84–7.01		19.8–36.8	1100–2800	

Phyllophora brodiaei									
Gulf of Gdansk and open waters	1968	1		7.3*		6.3	19.9	1190	Bojanowski, 1972
Öresund									
Area I	Pre-1973				ND				Hägerhäll, 1973
Area II					7.76				
					0.60–36.2				
Phyllophora membranifolia									
Öresund									
Area I	Pre-1973				ND				Hägerhäll, 1973
Area II					0.02				
					ND–0.10				
Dumontia incrassata									
Öresund									
Area I	Pre-1973				0.07				Hägerhäll, 1973
					ND–0.20				
Cystoclonium purpurascens									
Öresund									
Area I	Pre-1973				0.7				Hägerhäll, 1973
					ND–1.05				
Ahnfeltia plicata									
Area I	Pre-1973				0.35				Hägerhäll, 1973
Öresund					ND–1.0				
Membranoptera alata									
Öresund									
Area I	Pre-1973				ND				Hägerhäll, 1973
Area II					0.8				
					ND–1.70				
Phycodrys rubens									
Öresund									
Area I	Pre-1973				0.05				Hägerhäll, 1973
					ND–0.08				
Area II					11.4				
					0.30–48.1				
Rhodophyceae sp.	Pre-1986	6	3.5–6.1						Stoeppler et al., 1986

* – mg g^{-1} dry wt.
[a] – Waters with trace elements concentrations that were normal for coastal areas.
[b] – Waters with high trace element concentrations.

TABLE 3.1. – continued

Region	Sampling date	N	Hg	K	Mg	Mn	Na	Ni	P	References
					CHLOROPHYCEAE					
Enteromorpha sp.										
Southern Baltic										
Gulf of Gdansk and open waters	1965–66	8		18.2* 12.9–54.8	21.4* 13.4–31.0	100 50–220	34.9* 12.9–54.8	2.3 1.3–4.8	2.3* 0.72–4.13	Bojanowski, 1972
Gulf of Gdansk	1978–79	2		9.15* 8.7–9.6	10.5* 10.2–10.8	100 100–100	7.95* 5.30–10.6	1.85 1.2–2.5		Szefer and Skwarzec, 1988
Enteromorpha intestinalis										
Öresund										
Area I[a]	Pre-1973							11.6 8.00–14.4		Hägerhäll, 1973
Area II[b]								30.9 3.61–70.0		
Gulf of Gdansk	1978	1		10.2*	25.8*	600	3.5*	2.4		Szefer and Skwarzec, 1988
Enteromorpha crinita										
Southern Baltic, Polish coast		1		44.9*	2.3*	1700	3.7*	1.1		Szefer and Skwarzec, 1988
Cladophora sp.										
Gulf of Gdansk and open waters	1965–68	5		9.23* 1.7–24.5	7.80* 4.0–11.8	230 50–470	24* 12–52.9	3.3 1.6–5.2	0.57* 0.28–0.92	Bojanowski, 1972
Cladophora rupestris										
Southern Baltic, Polish coast	1978–79	2		13.1* 5.2–21.0	3.5* 2.2–4.8	1000 200–1800	6.5* 5.7–7.3	7.6 7.3–7.9		Szefer and Skwarzec, 1988
Cladophora glomerata										
Öresund										
Area I	Pre-1973							9.6 7.30–12.9		Hägerhäll, 1973
Area II								33.6 20.3–37.8		
Ulva lactuca										
Öresund										
Area I	Pre-1973							2.02 0.56–7.66		Hägerhäll, 1973
Area II								9.95 6.50–13.4		
Acrosiphonia centralis										
Öresund										
Area I	Pre-1973							2.56 0.55–8.30		Hägerhäll, 1973
Area II								15		

							0.90–67.9		
			RHODOPHYCEAE						
Furcellaria fastigiata									
Gulf of Gdansk and	1965–68	9	35.7*	9.2*	2820	14.2*	13.2	1.57*	Bojanowski, 1972
open waters			26.0–42.0	6.8–10.3	1100–4060	8.9–18.8	8.4–20.2	1.05–1.78	
Öresund									
Area I	Pre-1973						5.92		Hägerhäll, 1973
							5.50–6.35		
Area II							8.96		
							5.60–34.3		
Ceramium rubrum									
Gulf of Gdansk and	1966–68	2	27*	48.4*	3550	14.9*	17.3	0.86*	Bojanowski, 1972
open waters			15.9–38.1		3330–3770	13.4–16.3	13.6–21.0		
Öresund									
Area I	Pre-1973						ND		Hägerhäll, 1973
Area II							6.82		
Polysiphonia sp.									
Gulf of Gdansk and	1965–66	5	24.1*	51.7*	3620	16.1*	12	1.05*	Bojanowski, 1972
open waters			9.8–31.0	41.7–70	930–4880	10.1–22.3	7.9–15.8	0.29–1.42	
Polysiphonia elongata									
Öresund									
Area I	Pre-1973						1.76		Hägerhäll, 1973
							ND–6.61		
Area II							5.65		
							0.50–13.3		
Polysiphonia nigrescens									
Öresund									
Area I	Pre-1973						3.56		Hägerhäll, 1973
							ND–7.18		
Area II							7.5		
							ND–10.6		
Rhodomela confervoides									
Öresund									
Area I	Pre-1973						2.3		Hägerhäll, 1973
							160–4.00		
Rhodomela subfusca									
Gulf of Gdansk and	1966–68	3	28.4*	50.9*	4190	12.7*	14.4	1.4*	Bojanowski, 1972
open waters			24.3–32.4	38.6–63.1	3440–5230	11.7–13.6	12.3–16.8	1.34–1.45	
Phyllophora brodiaei									
Gulf of Gdansk and	1968	1	31.1*		4720	18.1*			Bojanowski, 1972
open waters									
Öresund									

Region	Sampling date	N	Hg	K	Mg	Mn	Na	Ni	P	References
Area I	Pre-1973							6.68		Hägerhäll, 1973
								0.06–9.18		
Area II								17.4		
								3.23–41.2		
Phyllophora membranifolia										
Öresund										
Area I	Pre-1973							5.54		Hägerhäll, 1973
								ND–10.0		
Area II								52.7		
								ND–69.8		
Dumontia incrassata										
Öresund										
Area I	Pre-1973							2.96		Hägerhäll, 1973
								ND–5.10		
Cystoclonium purpurascens										
Öresund										
Area I	Pre-1973							2		Hägerhäll, 1973
Ahnfeltia plicata										
Öresund										
Area I	Pre-1973							3.22		Hägerhäll, 1973
								ND–5.50		
Membranoptera alata										
Öresund										
Area I	Pre-1973							2.63		Hägerhäll, 1973
								ND–5.80		
Area II								5		
								ND–15.0		
Phycodrys rubens										
Öresund										
Area I	Pre-1973							12		Hägerhäll, 1973
								0.18–13.3		
Area II								19.2		
								0.60–56.2		
Ruppia maritima										
Gulf of Gdansk			0.011							Falandysz, 1994

* – mg g^{-1} dry wt.
[a] – Waters with trace elements concentrations that were normal for coastal areas.
[b] – Waters with high trace element concentrations.

TABLE 3.1. – continued

Region	Sampling date	N	Pb	S	Sr	Ti	Zn	References
				CHLOROPHYCEAE				
Enteromorpha sp.								
Southern Baltic								
Gulf of Gdansk and	1965–66	8		78.2**	0.135*		60	Bojanowski, 1972
open waters				66.2–95.7	0.095–0.18		35.0–90.0	
Gulf of Gdansk	1978–79	2	25			45.5	198	Szefer and Skwarzec, 1988
			20.0–30.0			26.0–65.0	140–256	
Enteromorpha intestinalis								
Öresund								
Area I[a]	Pre-1973		0.75				73.7	Hägerhäll, 1973
			ND–1.25				20.3–101	
Area II[b]			2.75				60.4	
			ND–22.0				24.1–325	
Gulf of Gdansk	1978	1	0.45			140	35	Szefer and Skwarzec, 1988
Enteromorpha crinita								
Southern Baltic, Polish coast		1	31				17	Szefer and Skwarzec, 1988
Cladophora sp.								
Gulf of Gdansk and	1965–68	5		50.4**	0.075*		80	Bojanowski, 1972
open waters				25.2–83.7	0.065–0.09		45–130	
Cladophora rupestris								
Southern Baltic, Polish coast	1978–79	2	17				122	Szefer and Skwarzec, 1988
			16.0–18.0				54–190	
Cladophora glomerata								
Öresund								
Area I	Pre-1973		4.84				41.5	Hägerhäll, 1973
			ND–6.21				24.2–61.6	
Area II			12.1				132	
			3.53–15.2				43.7–181	
Northern Baltic, Tvärminne	1980	15	6.1				71.8	Kangas and Autio, 1986
			1.5–9.9				44.0–102	
Ulva lactuca								
Öresund								
Area I	Pre-1973		ND					Hägerhäll, 1973
Area II			2.1				77.6	
			0.90–3.30				50.6–90.6	
Acrosiphonia centralis								
Öresund								
Area I	Pre-1973		0.86				43.3	Hägerhäll, 1973
			ND–3.70				23.3–78.6	
Area II			8.15				77.2	

Region	Sampling date	N	Pb	S	Sr	Ti	Zn	References
			ND–85.6				23.0–161	
				RHOD 'HYCEAE				
Furcellaria fastigiata								
Gulf of Gdansk and	1966–68	9		94.8**	0.09*		110	Bojanowski, 1972
open waters				87.2–101	0.06–1.45		60–190	
Öresund								
Area I	Pre-1973		ND				38.7	Hägerhäll, 1973
							18.5–58.8	
Area II			8.87				67.4	
			ND–42.7				18.2–167	
Ceramium rubrum								
Gulf of Gdansk and								
open waters	1966–68	2		6.2**	0.10*		325	Bojanowski, 1972
Öresund								
Area I	Pre-1973		1.7				67.8	Hägerhäll, 1973
			ND–2.0				63.6–71.9	
Area II			6.98				153	
			ND–43.6				75.8–218	
Ceramium tenuicorne								
Northern Baltic, Tvärminne	1980	12	6.05				113	Kangas and Autio, 1986
			2.2–9.9				97.0–179	
Polysiphonia sp.								
Gulf of Gdansk and	1965–66	5		8.74**	0.146*		206	Bojanowski, 1972
open waters				4.3–13.1	0.065–0.28		80–390	
Polysiphonia elongata								
Öresund								
Area I	Pre-1973		ND				60.7	Hägerhäll, 1973
							38.2–98.8	
Area II			24.2				274	
			ND–44.2				123–320	
Polysiphonia nigrescens								
Öresund								
Area I	Pre-1973		ND				90.9	Hägerhäll, 1973
							46.2–110	
Area II			26.6				169	
			ND–45.8				116–206	
Rhodomela confervoides								
Öresund								
Area I	Pre-1973		0.17				72.6	Hägerhäll, 1973
			ND–0.23				58.6–98.8	
Rhodomela subfusca								
Gulf of Gdansk and	1966–68	3		7.9**	0.098*		247	Bojanowski, 1972
open waters				6.8–10.1	0.07–0.11		130–440	

Phyllophora brodiaei							
Gulf of Gdansk and open waters	1968	1		10.0**	0.07*	260	Bojanowski, 1972
Öresund							
Area I	Pre-1973		0.05			68.2	Hägerhäll, 1973
			0.01–0.11			36.2–108	
Area II			35.7			321	
						50.2–501	
Phyllophora membranifolia							
Öresund							
Area I	Pre-1973		ND			45.7	Hägerhäll, 1973
Area II			24.1			218	
						37.5–492	
Dumontia incrassata							
Öresund							
Area I	Pre-1973		ND			52.5	Hägerhäll, 1973
						32.6–69.5	
Cystoclonium purpurascens							
Öresund							
Area I	Pre-1973		0.01			78.2	Hägerhäll, 1973
			ND–1.00			20.3–91.2	
Ahnfeltia plicata							
Öresund							
Area I	Pre-1973		0.05			24.7	Hägerhäll, 1973
			ND–0.09			17.6–32.7	
Membranoptera alata							
Öresund							
Area I	Pre-1973		ND			50.5	Hägerhäll, 1973
						43.2–96.2	
Area II			14.9			246	
			ND–25.4			51.9–428	
Phycodrys rubens							
Öresund							
Area I	Pre-1973		ND			29.4	Hägerhäll, 1973
						20.1–42.2	
Area II			30			220	
			15.0–79.8			31.6–486	

* – mg g^{-1} dry wt.
** – Expressed as SO_4 in mg g^{-1} dry wt.
[a] – Waters with trace elements concentrations that were normal for coastal areas.
[b] – Waters with high trace element concentrations.

TABLE 3.1. – continued

Region	Sampling date	Plant part	N	Al	Ca	Cd	Co	Cu	Fe	References
				SPERMATOPHYCEAE						
Eel grass (*Zostera marina*)										
Southern Baltic	1965–68		14		11.8*		1.91	15.2	480	Bojanowski, 1972
Gulf of Gdansk and open waters					8.8–21.8		0.27–6.80	8.0–33.5	120–1540	
Gulf of Gdansk	1978		1	0.20*	8.6*	0.38	0.6	1.6	400	Szefer and Skwarzec, 1988
	1988					1.1	0.7	6.1	1700	Szefer et al., 1994a
						1.1–1.2	0.6–0.7	5.5–7.2	1400–2000	
Danish waters, Limfjord	1980		40			0.1		4.06		Brix et al., 1983
						0.09–2.92		1.82–19.3		
Danish waters, Limfjord		A.G.				0.62		4.79		Brix et al., 1983
						0.09–2.92		1.86–16.6		
Danish waters, Limfjord		B.G.				0.3		3.33		
						0.13–0.92		1.82–19.3		
Danish waters										
Aalborg	1980	A.G.			0.7**				0.07**	Brix and Lyngby, 1983
		B.G.			0.8**				0.16**	
Nibe	1980	A.G.			1.0**				0.02**	
		B.G.			2.5**				0.26**	
Ronbjerg	1980	A.G.			1.6**				0.04**	
		B.G.			3.8**				0.27**	
Sago pondweed (*Potamogeton pectinatus*)										
Gulf of Gdansk and open waters	1965–66		11		14.4*		0.91	8.2	300	Bojanowski, 1972
					8.6–23.2		0.17–2.80	5.4–15.7	110–510	
Southern Baltic, Polish coast	1978		1	0.80*	11.1*	0.068		1.8	1300	Szefer and Skwarzec, 1988
Water thyme (*Elodea canadensis*)										
Southern Baltic, Polish coast	1978		1	1.20*	9.8*	0.63	0.8	1.6	1900	Szefer and Skwarzec, 1988
Sweet flag (*Acorus calamus*)										
Southern Baltic, Polish coast	1978		1	1.0*	14.6*	0.55	1.2	7.2	1400	Szefer and Skwarzec, 1988

* – mg g^{-1} dry wt.
** – g m^{-2}
A.G. – Aboveground parts.
B.G. – Belowground parts.

TABLE 3.1. – continued

Region	Sampling date	Plant part	N	K	Mg	Mn	Na	Ni	P	Pb	References
				SPERMATOPHYCEAE							
Eel grass (*Zostera marina*)											
Southern Baltic	1965–68		14	34.7*	9.9*	940	24.3*	4.6	2.52*		Bojanowski, 1972
Gulf of Gdansk and open waters				12.8–53.8	8.2–11.2	130–2270	10.0–34.8	1.3–11.8	1.82–3.84		
Gulf of Gdansk	1978		1	25.7*	8.5*	700	9.0*	2.6			Szefer and Skwarzec, 1988
	1988					300		1.6		3.5	Szefer et al., 1994a
						300–400		1.4–2.0		2.7–4.0	
Danish waters, Limfjord	1980		40							1.06	Brix et al., 1983
										0.35–37.5	
Danish waters, Limfjord		A.G.								1.07	Brix et al., 1983
										0.47–37.5	
Danish waters, Limfjord		B.G.								1.04	
										0.35–29.8	
Danish waters											
Aalborg	1980	A.G.		3.4**	0.7**	0.15**	3.2**				Brix and Lyngby, 1983
		B.G.		1.1**	0.5**	0.05**	2.5**				
Nibe	1980	A.G.		5.7**	1.0**	0.23**	3.6**				
		B.G.		2.1**	1.3**	0.01**	4.6**				
Ronbjerg	1980	A.G.		7.4**	1.6**	0.40**	6.4**				
		B.G.		6.2**	2.2**	0.03**	10.1**				
Sago pondweed (*Potamogeton pectinatus*)											
Gulf of Gdansk and open waters	1965–66		11	25.7*	10.4*	730	25.7*	4	2.29*		Bojanowski, 1972
				9.1–29.7	7.8–18.1	200–2100	18.9–32.2	2.5–7.1	1.42–3.30		
Southern Baltic, Polish coast	1978		1	13.7*	4.6*	1200	5.8*	2.9		34	Szefer and Skwarzec, 1988
Water thyme (*Elodea canadensis*)											
Southern Baltic, Polish coast	1978		1	29.2*	3.0*	1000	9.4*	5.3		33	Szefer and Skwarzec, 1988
Sweet flag (*Acorus calamus*)											
Southern Baltic, Polish coast	1978		1	56.1*	8.1*	3500	9.8*	4.6		30	Szefer and Skwarzec, 1988

* – mg g^{-1} dry wt.
** – g m^{-2}
A.G. – Aboveground parts.
B.G. – Belowground parts.

TABLE 3.1. – continued

Region	Sampling date	Plant part	N	S	Sr	Ti	Zn	References
				SPERMATOPHYCEAE				
Eel grass (*Zostera marina*)								
Southern Baltic	1965–68		14	12.5**	0.24*		300	Bojanowski, 1972
Gulf of Gdansk and open waters				10.4–15.5	0.155–0.355		80–820	
Gulf of Gdansk	1978		1			32	37	Szefer and Skwarzec, 1988
	1988						120	Szefer et al., 1994a
							84–149	
Danish waters, Limfjord	1980		40				66.5	Brix et al., 1983
							25.0–175	
Danish waters, Limfjord		A.G.					78	Brix et al., 1983
							41.0–175	
Danish waters, Limfjord		B.G.					55	
							25.0–125	
Sago pondweed (*Potamogeton pectinatus*)								
Gulf of Gdansk and open waters	1965–66		11	29.8**	0.165*		140	Bojanowski, 1972
				19.2–32.8	0.115–0.21		110–200	
Southern Baltic, Polish coast	1978		1			60	21	Szefer and Skwarzec, 1988
Water thyme (*Elodea canadensis*)								
Southern Baltic, Polish coast	1978		1			22	24	Szefer and Skwarzec, 1988
Sweet flag (*Acorus calamus*)								
Southern Baltic, Polish coast	1978		1			55	71	Szefer and Skwarzec, 1988

* – mg g^{-1} dry wt.
** – Expressed as SO_4 in mg g^{-1} dry wt.
A.G. – Aboveground parts.
B.G. – Belowground parts.

was not observed in the case of Cd, Cu and Pb. It is suggested that for those variations may be responsible factors such as the slow accumulation of trace elements or the higher dry weight of older parts (and therefore more numerous binding sites) and supposedly some contamination of the older parts with fine particles. Other reason for observed differences may be also epiphytes since they were, as filamentous algae, mainly confined to the older fragments of the *Fucus* (Bryan and Hummerstone, 1973c; Kangas et al., 1982; Forsberg et al., 1988).

Some reports concern the age-dependent morphological distribution of trace metals in seaweeds inhabited adjacent areas to the Baltic Sea. For instance, Brix and Lyngby (1982, 1983) investigated the distribution of trace elements in different parts (Fig. 3.2) of eelgrass (*Zostera marina*) collected in the Limfjord, Denmark. The tissue translocation of biomas (g dry wt. m^{-2}), Cd, Cu, Pb, Zn ($\mu g\ g^{-1}$), Fe, Mn, Ca, Mg, K and Na (%) is presented in Fig. 3.3. The concentrations of Ca, Cd, Cu, Fe, Mn, Pb and Zn were higher in the roots than in the rhizomes. In the aerial (above the sediment/water interface) parts two different age-dependent distribution patterns were detected. The concentrations of Ca, Cd, Fe, Mg, Mn, Na, Pb and Zn showed increasing trend with the age of leaves while the opposite relationship was observed for Cu and K. The roots contained the highest levels of Ca, Cd, Cu, Fe, Mn, Pb and Zn; the rhizomes were characterised by the highest levels of K, Mg and Na. It is reported (Brix and Lyngby, 1983) that the steam fraction *Z. marina* had highest levels of Fe, K, Mg, Na and Pb while the youngest leaves contained the most quantities of Ca, Cd, Mn and Zn (Fig. 3.3).

The accumulation of heavy metals in roots relative to rhizomes can be explained by the greater surface area per unit weight and also absorption capacity of the roots compared to the rhizomes (Lyngby et al., 1982; Brix and Lyngby,

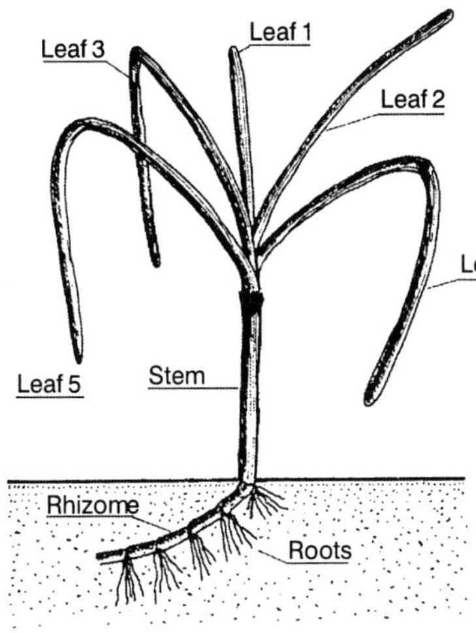

Fig. 3.2. Drawing of an eelgrass plant, showing the eight fractions into which eelgrass was divided. The age of the leaves increases from leaf 1 to leaf 5. The stem fraction is the plant portion from the rhizome to the leaf base. After Brix and Lyngby (1982); modified.

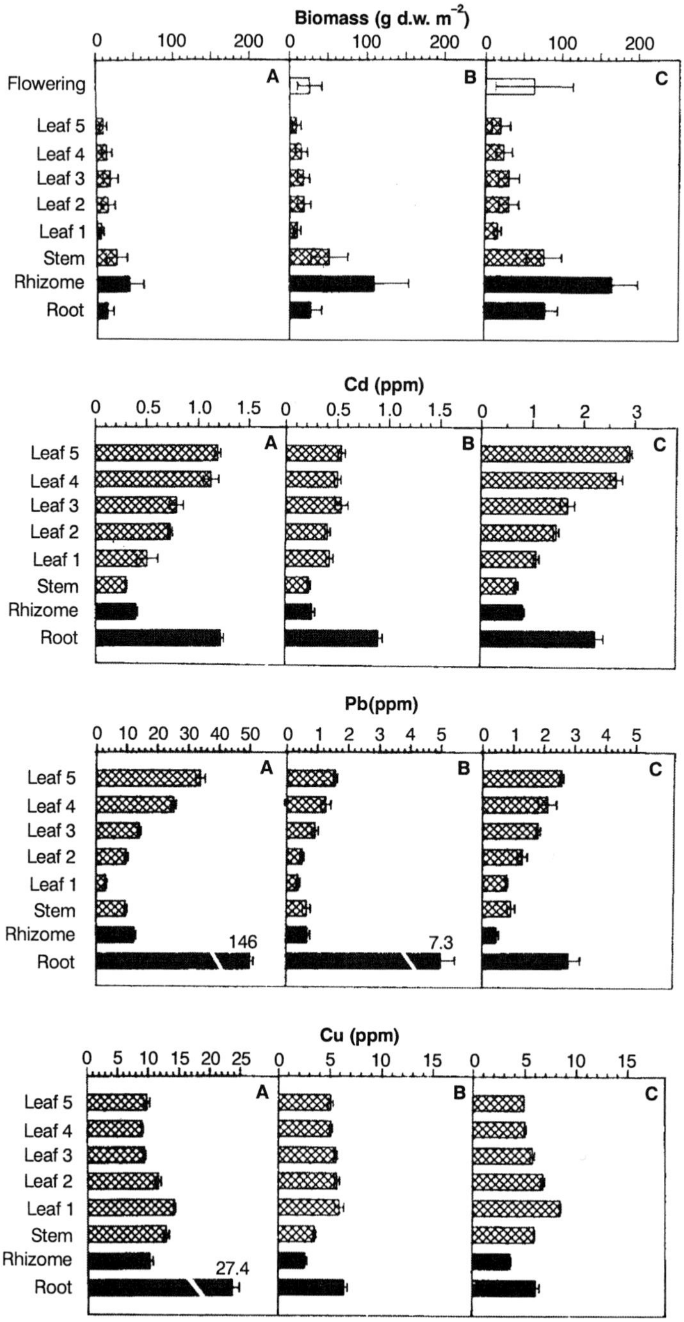

Fig. 3.3. The distribution of biomass and chemical elements in eelgrass (*Zostera marina* L.) at Aalborg (A), Nibe (B), and Rønbierg (C) in July, 1980. Bars indicate standard deviation of ten samples. After Brix and Lyngby (1982, 1983); modified.

Fig. 3.3. – continued.

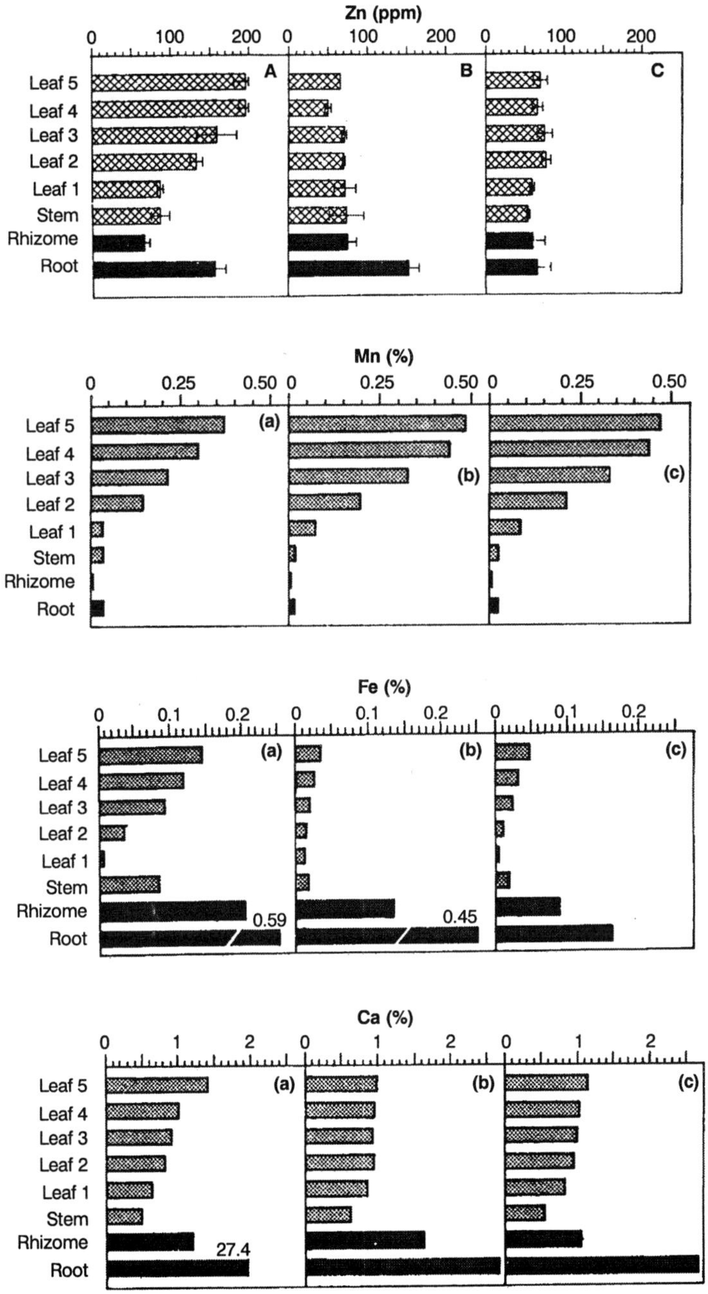

Fig. 3.3. – continued.

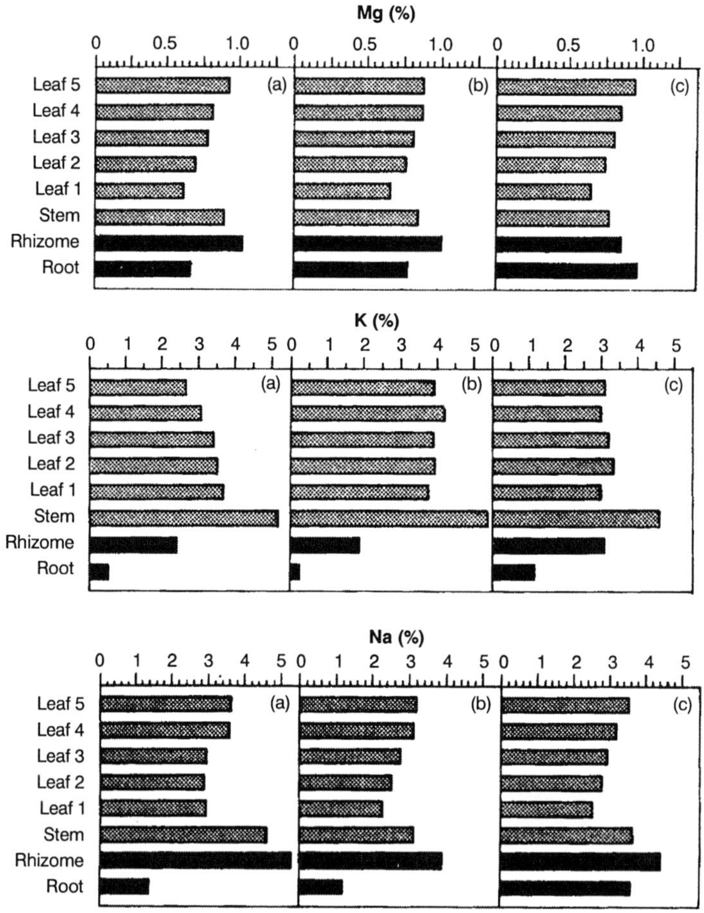

1983). The distribution pattern in the leaves may be attributed to an irreversible uptake or to occurrence of more binding sites in old tissues (Brix and Lyngby, 1982).

It is shown (Brix et al., 1983) that the concentrations of Cd, Cu and Zn in above-ground parts of *Z. marina* from the Limfjord were significantly greater than in the below-ground parts (Fig. 3.4). On the other hand, a significant correlation was found between heavy metal concentrations in above- and below-ground parts of *Z. marina* (Fig. 3.5) reflecting a relationship between trace element bioavailability in water and the adjacent sediment, respectively, or a transport within the plants.

Spatial trends

According to Phillips (1979) the distribution of Cd, Fe, Pb and Zn in growing tips of bladder wrack (*Fucus vesiculosus*) collected at nine locations of the Sound

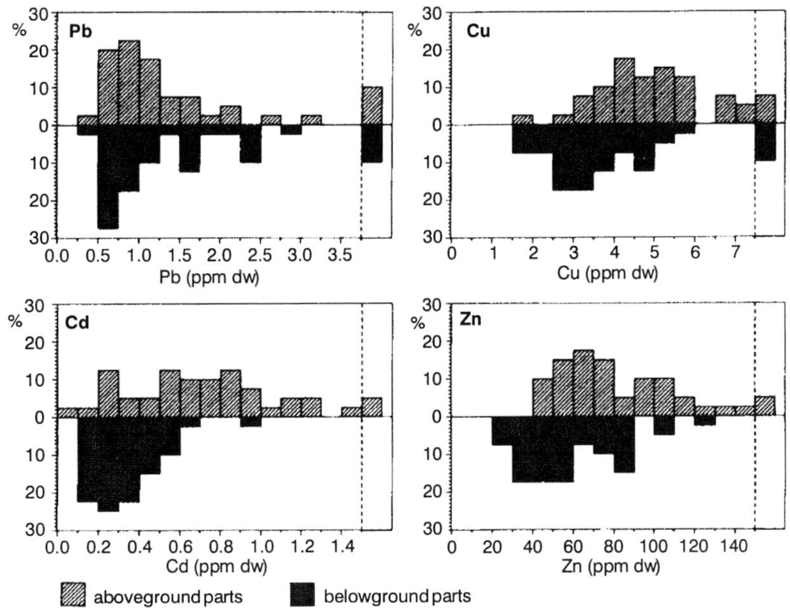

Fig. 3.4. The frequency distributions of trace metal concentrations (ppm dry weight) in aboveground parts (hatched columns) and belowground parts (black columns) of eelgrass *(Zostera marina* L.). Frequencies of concentrations higher than indicated by the dashed line are shown in the columns right of the dashed line. (A: lead; B: copper; C: cadmium; D: zinc). After Brix et al. (1983); modified.

(Öresund), Baltic Sea, showed some spatial variations in the trace metals contents. Pollution profile produced for selected metals corresponded to their profile in the alga studied reflecting the levels of these metals in surrounding waters of the Sound. Therefore *F. vesiculosus* appears to be responding exclusively to metals present in solution (Phillips, 1979; Bryan, 1983).

Brown alga *F. vesiculosus* inhabited along the coasts from the northern Baltic Proper into the Bothnian Sea indicated maximum concentrations of Cr and Ni when passing the outer Stockholm Archipelago and further increase of Zn levels up to the mouth of the Dalälven River and a continuos increase of Cd northwards in the Bothnian Sea (Söderlund et al., 1988). The same spatial trend was observed by Forsberg et al. (1988) who detected that the concentrations of metals in *Fucus* differed markedly in the direction leading from south to north of the outer zone of the Archipelago of Stockholm. Metals such as Al, Cd, Co, Cr, Cu, Fe, Mn, Ni, V and Zn showed similar tendencies with their elevated values in *F. vesiculosus* from the northern areas (Forsberg et al., 1988). According to Kangas and Autio (1986) the distribution of trace metals is dependent on *Fucus* habitation; the concentrations of Cu, Fe and Zn, in *F. vesiculosus* from the Tvärminne area, Northern Baltic Sea, were greater in coastal area than in outer one. Two annual filamentous algae *C. glomerata and P. littoralis* from the same area contained also higher levels of Pb at inner that at outer sampling sites (Kangas and Autio,

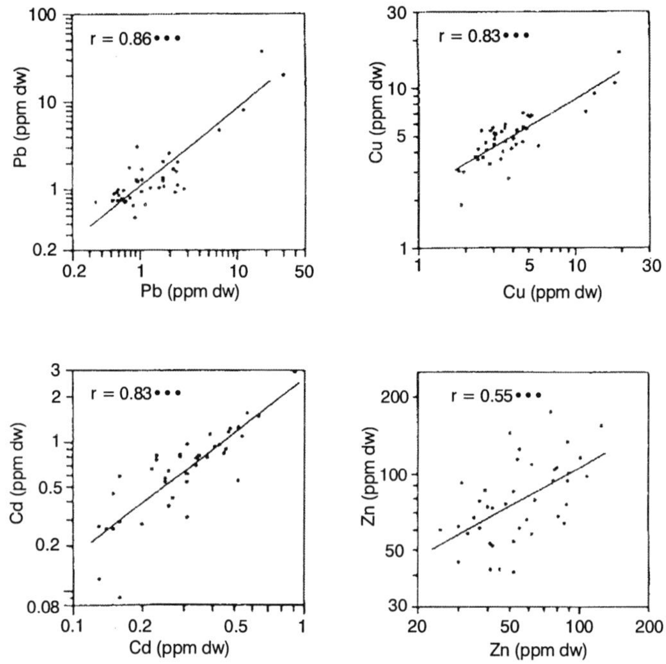

Fig. 3.5. Trace metal concentration (ppm dry weight) in aboveground parts of eelgrass *(Zostera marina* L.) (y-axis) plotted against concentration in below ground parts (x-axis) and the coefficients of correlation. Note double log-scale. (Significance level : ***p < 0.001). After Brix et al. (1983); modified.

1986). The concentrations of As, Cd, Co, Cu, Hg, Mn, Ni, Pb and Zn (trace elements) as well as Ca, Fe, K, Mg, Na, P and S (macroelements) were determined in this brown alga showing significant spatial differences between the Baltic Sea and North Sea (Struck et al., 1997). For example, increases of more than 50% of the North Sea mean concentration were observed in the Baltic Sea for Mn and Zn in *F. vesiculosus* (Fig. 3.6). The alga concentrations of As and Hg demonstrated decrease from the North Sea to the Baltic Sea which amounted to more than 50% of the North Sea mean concentrations. The concentrations of Cu decreased also in the same direction to the Baltic Sea (Fig. 3.6). These findings suggest that spatial trends of metal values are the most related to changes of salinity of the surrounding waters.

The southern Baltic seaweeds contained significantly larger amounts of Al, Fe, K and Zn and similar levels of Mn compared to plants taken from the Żarnowiec Lake (Szefer and Skwarzec, 1988).

Brix et al. (1983) reported data on the concentrations of several trace metals in above- and belowground fragments of *Z. marina* from the Limfjord, Denmark. The concentrations of Pb, Cu and Cd were significantly elevated in a restricted area at the cities of Aalborg (Cu and Pb) and at Struer (Cd) probably reflecting a significant discharge of the trace metals into the Limfjord in this area (Brix et al., 1983).

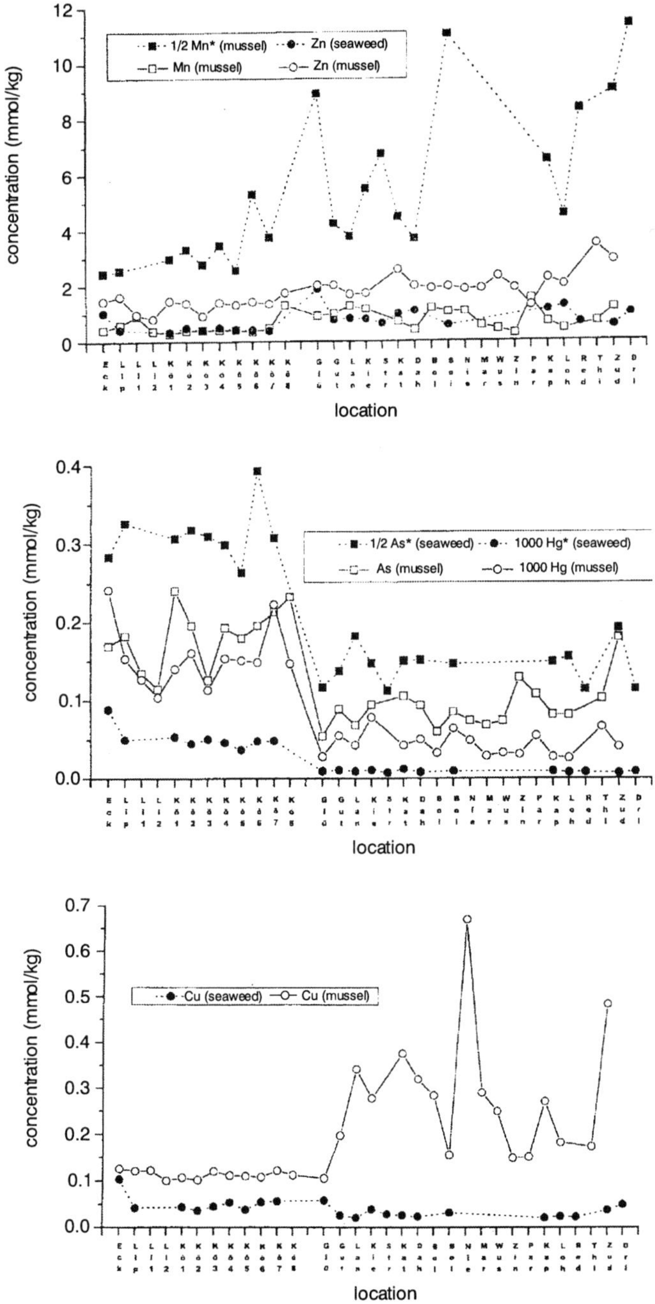

Fig. 3.6. Chemical element concentrations determined in seaweed and mussel from North and Baltic Sea locations. After Struck et al. (1997); modified.

The data reported above show that parameters such as input of trace metals from the land and inland waters, mostly anthropogenic in origin, as well as salinity variations of the ambient water may highly influence the metal levels in the alga tissues.

Temporal trends

A number of parameters influence the elemental composition of Baltic seaweeds. According to Forsberg et al. (1988) *F. vesiculosus* from the Swedish east coast showed pronounced differences in trace element concentrations with, for instance higher levels of Cu, Pb and V in 1983 and Co, Mn, Ni and Zn (old parts of thallus) in 1984. The elevated alga levels of Co and Ni may be derived principally from fossil-fuel burning in Sweden, particularly from oil-combustion. As regards the enhanced concentrations of Cu, Pb and V in seaweed collected in 1933; these may be explained by sulphide ore-mining activities at that time. This area is drained into the Dalälven River, which discharges into the Bothnian Sea (Forsberg, 1988). Seasonal changes in trace metal concentrations were also studied by Kangas and Autio (1986). Concentrations of Zn in both the tips and stipes of *F. vesiculosus* from Tvärminne area, Northern Baltic Sea, reached the highest values in mid summer and lowest in autumn. Similar trend was observed by Fuge and James (1974) in *Fucus* from the Bristol Channel. According to Stoeppler et al. (1986) total concentrations of As in *Fucus* from the Western Baltic ranged up to 40 $\mu g\ g^{-1}$ dry wt. and showed for the four locations studied significant seasonal variations for comparatively non polluted or non disturbed locations only.

Besides short-term seasonal dependent changes also long-term trends have been registered. For instance, samples of *F. vesiculosus* collected in the Baltic Sea and the North Sea during 1985–1994 were analysed for concentrations of As, Ba, Ca, Cd, Co, Cu, Fe, Hg, K, Mg, Mn, Na, Ni, P, Pb, S, Se, Sr, Tl and Zn (Ostapczuk et al., 1997a). The data indicated the occurrence of three groups of elements with respect to these ten-year long tendencies of their concentrations. The greatest differences between minimum and maximum concentrations with the sampling time were detected for Ni (more than 300%). The temporal trends were also noted for *Fucus* concentrations of Cd and As. The highest Cd levels occurred in 1988 and 1989. Between 1985 to 1994 the As concentration in *F. vesiculosus* has increased significantly indicating that the pollution of the Eckwarderhörne ecosystem with As or at least the bioavailable fraction of this element has increased during this decade.

A general seasonal variation patterns of Cd, Cu, Pb and Zn levels in different parts of *Z. marina* from the Limfjord, Denmark, were observed (Lyngby and Brix, 1984). The greatest concentrations were detected in late winter-early spring and the smallest concentrations in the autumn. Figure 3.7 well illustrates such seasonal relationships for Cu in above and below-ground tissues of *Z. marina* from the tree sampling sites of the Limfjord, Denmark. Such seasonal dependent varia-

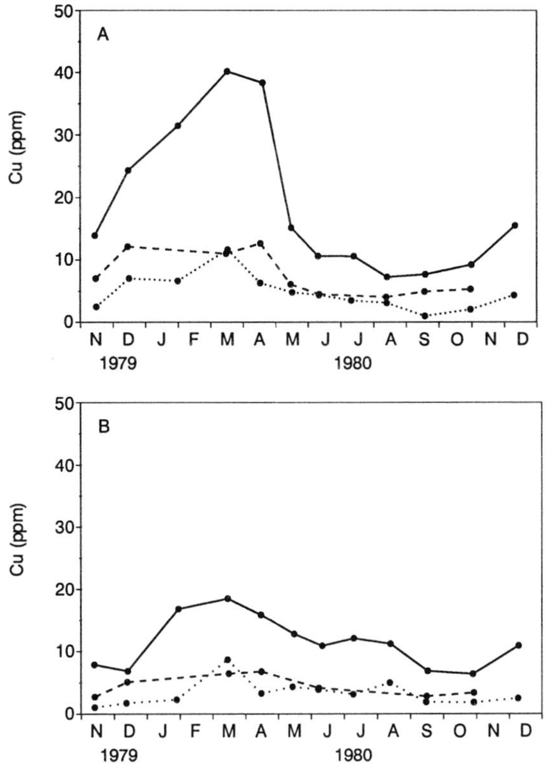

Fig. 3.7. Seasonal variation of copper (ppm dry weight) in above (A) and below-ground parts (B) of *Zostera marina* L. at Aalborg (solid line), Nibe (dotted line) and Rønbjerg (dashed line). Each point represents the mean of five replicates from a pooled sample. After Lyngby and Brix (1982); modified.

tion in trace metal content can be explained by the growth dynamic of *Z. marina* reflecting by very similar seasonal variation patterns. The greatest growth rate of the plant was observed in June and the lowest in the winter. Since maximum levels of heavy metals were recorded when the growth had ceased and distinct their decrease was noted at the beginning of the growth season it is suggested that some trace elements are irreversibly bound in *Z. marina* and these dilution effects may be caused by the increase in biomass (Brix and Lyngby, 1982; Lyngby and Brix, 1982).

(iii) Occurrence of Radionuclides in Seaweeds

The Chernobyl accident in 1986 provided a great opportunity to examine Baltic macroalgae as biomonitors of radionuclides, e.g. $^{239+240}$Pu and ^{210}Po. Such studies has been performed by several authors (Ilus et al., 1987, 1988, 1992; Carlson, 1990; Carlson and Holm, 1990; Dahlgaard and Boelskifte, 1992; Skwarzec and Bojanowski, 1992; Dahlgaard, 1994; Holm, 1995; Kanisch et al., 1995; Skwarzec,

1997; Christensen and Strålberg, 2000; Hou et al., 2000) although pre-Chernobyl accident works concerning Baltic seaweed concentrations of ^{110}Ag, ^{241}Am, ^{144}Ce, ^{60}Co, ^{58}Co, ^{137}Cs, ^{134}Cs, ^{131}I, ^{40}K, ^{54}Mn, ^{95}Nb, $^{239+240}$Pu, ^{106}Ru, ^{125}Sb, ^{90}Sr, ^{99}Tc and ^{65}Zn (Bojanowski and Pempkowiak, 1977; Ilus et al., 1981; Aarkrog et al., 1986; Christensen, 1986; Holm et al, 1986; Ilus et al, 1986; Jaworowski et al, 1986; Lazarev et al, 1986; Neumann et al., 1991) have been also made. According to Christensen and Strålberg (2000) the contribution from Sellafield now is negligible and main source of radiocaesium found in *Fucus* is the Chernobyl fallout being transported to the Baltic Sea by riverine runoff entering this Sea.

The concentrations of U (^{238}U, ^{235}U, ^{234}U) and Th (^{232}Th) were determined in several species of seaweeds collected mostly in east part of the southern Baltic, i.e. in the Gulf of Gdańsk (Szefer, 1987; Skwarzec, 1995). In Table 3.2 are collected concentration data of radioactive elements in seaweeds from the Baltic Sea. It can be seen that levels of some radionuclides vary depending on the distance and direction of the sampling site in respect to location of their emission source. According to Dahlgaard and Boelskifte (1992) *Fucus* can be used successfully as a semi-quantitative indicator for radioactive contaminants. Effects of biotic and abiotic factors on the accumulation of radionuclides (^{110}Ag, ^{241}Am, ^{60}Co, ^{58}Co, ^{137}Cs, ^{134}Cs, ^{40}K, ^{54}Mn, $^{239+240}$Pu, ^{106}Ru, ^{99}Tc and ^{65}Zn, ^{95}Zr) in *F. vesiculosus* from the Baltic Sea, Swedish coast were assayed by Carlson (1990). The levels of some radionuclides in Baltic algae strongly corresponded to their sampling sites affected by the deposition of the Chernobyl fallout (HELCOM, 1995). Maximum levels of Chernobyl-derived radiocaesium, ^{110m}Ag and ^{106}Ru in *F. vesiculosus* from Forsmark and Olkiluoto at the Bothnian Sea were observed in 1986 (Fig. 3.8). According to Holm (1995) the Chernobyl accident had no significant impact on plutonium concentration in *F. vesiculosus* along the Swedish coast. Similar results are reported for Gulf of Gdańsk seaweeds by Skwarzec and Bojanowski (1992) suggesting that the contribution of the Chernobyl-derived plutonium to Baltic plants was small. This finding was strongly supported by estimated the average ^{238}Pu/$^{239+240}$Pu activity ratio for that collection amounting to 0.032. This ratio is not very different from typical worldwide fallout and it significantly deviates from values of 0.47 reported for the Chernobyl fallout over Sweden (Holm et al., 1989). It is concluded that this Baltic ratio is comparable to values of 0.025 and 0.04 estimated for nuclear weapon test fallout and the SNAP-9A satellite accident in 1964, respectively (Perkins and Thomas, 1980).

The inter-tissue distribution of U and Th in *F. vesiculosus* showed a different character. Similarly to heavy metals, the highest levels of U and Th were observed in old thallus, while the lowest ones in younger off shoots (Szefer, 1987).

The concentrations of U in Baltic seaweeds were characterised by a great variability and, like heavy-metals, depended on the species and the sampling site at which specimens were collected. The average concentrations of U varied as follows (dry wt): 0.07–0.35 μg g^{-1} (Chlorophyta), 0.21–0.41 μg g^{-1} (Phaeophyta) and

TABLE 3.2.

Concentrations of radionuclides (Bq kg^{-1} dry wt.) in seaweeds of the Baltic Sea and other northern areas

Region	Sampling date	Plant part	N	Salinity (PSU)	110m-Ag	241-Am	140-Ba	7-Be	141-Ce	144-Ce	58-Co	60-Co	References
							PHAEOPHYCEAE						
Fucus vesiculosus													
Southern Baltic	1973	Y								110±20*			Bojanowski and Pempkowiak, 1977
Finnish coast													
Loviisa	0.5–08.86		2		180		ND–1900	ND–180	129	183	ND–10	40	Ilus et al., 1987
					130–230				8.1–250	35–330		35–44	
	1987		13	3.77	17.5					ND–11		8	Ilus et al., 1988
				2.44–4.10	11.0–29							ND–32	
	1989–90		18		5.41						1.09	10.2	Ilus et al., 1992
					ND–19.0						ND–2.1	0.80–34	
Olkiluoto	04–08.86		3		78		2850	58.5	265	166	ND–14	91.3	Ilus et al., 1987
					2.4–170		ND–4700	ND–74	1.7–700	7.1–360		70–110	
	1987		13	5.74	15.7					ND–17	ND–5.8	14	Ilus et al., 1988
				5.13–5.96	9.9–27							1.7–62	
	1989–90				1.09						6.23	48.4	Ilus et al., 1992
					ND–2.1						ND–21.0	1.9–170	
South coast of Finland	1987		25	5.63	11.8					ND–11		0–3.8	Ilus et al., 1988
				3.52–6.59	0.8–31								
*Fucus***													
Danish waters	1982		23	17.4–24.9		0.023–0.11						1.06–3.7	Aarkrog et al., 1986
	1983		158	14.1–29.7							0.38–1.04	0.36–8.3	
							CHLOROPHYCEAE						
Enteromorpha sp.													
Puck Bay	1973		2							110±40*			Bojanowski and Pempkowiak, 1977
Cladophora sp													
Southern Baltic	1973		3							180±30*			Bojanowski and Pempkowiak, 1977
							RHODOPHYCEAE						
Furcellaria fastigiata													
Gulf of Gdansk	1973									1220±140*			Bojanowski and Pempkowiak, 1977
Phyllophora brodiaei													
Gulf of Gdansk	1973									1790±150*			

Region	Sampling date	Plant part	N	Salinity (PSU)	110m-Ag	241-Am	140-Ba	7-Be	141-Ce	144-Ce	58-Co	60-Co	References
Ceramium diaphanum													
Gulf of Gdansk	1973									72–125*			Bojanowski and Pempkowiak, 1977
						SPERMATOPHYCEAE							
Zostera marina													
Gulf of Gdansk	1973	L								63±32*			Bojanowski and Pempkowiak, 1977
		R								120±34*			
		W								82±76*			
Southern Baltic	1979/88												
Potagometon pectinatus													
Southern Baltic	1973/79									140±70*			Bojanowski and Pempkowiak, 1977
Myriophyllum spicatum													
Southern Baltic	1973									440±30*			Bojanowski and Pempkowiak, 1977
Zarnowieckie Lake	1979												

* – pCi kg^{-1} dry wt.
** – *F. vesiculosus*, *F. serratus*, *F. spiralis*
Y – younger off-shoots; O – old thallus; R – receptacles; L – leaves; R – roots; W – whole plants

TABLE 3.2. – continued

Region	Sampling date	Plant part	N	Salinity (PSU)	51-Cr	134-Cs	136-Cs	137-Cs	131-I	40-K	140-La	References
						PHAEOPHYCEAE						
Fucus vesiculosus												
Southern Baltic	1973	Y						303±11*				Bojanowski and
		O						251±43*				Pempkowiak, 1977
		R						220±10*				
Finnish coast												
Loviisa	0.5–08.86		2			ND–210	1630	3000	ND–13000	1100	904	Ilus et al., 1987
							550–2700	1100–4900		1100–1100	7.1–1800	
	1987		13	3.77		204		510		880		Ilus et al., 1988
				2.44–4.10		150–260		370–670		790–100		
	1989–90		18			37.1		197		888		Ilus et al., 1992
						16–81		110–370		770–1100		
Olkiluoto	04–08.86		3			ND–150	286	535	9800	660	1500	Ilus et al., 1987
							8.3–710	26–1300	6.8–29000	590–690	17–3700	
	1987		13	5.74	ND–32	65.7		177		773		Ilus et al., 1988
				5.13–5.96		53–85		150–230		630–900		
	1989–90					24.6		135		704		Ilus et al., 1992
						16–48		110–220		540–870		
South coast of Finland	1987		25	5.63		81.7		217		830		Ilus et al., 1988
				3.52–6.59		30–170		80–440		480–1200		
*Fucus***												
Danish waters	1982		23	17.4–24.9		0.4		6.9–11.4		19–39^		Aarkrog et al., 1986
	1983		158	14.1–29.7		0.16–0.44		3.1–14.2	1.5–10.0	21–40^		
						CHLOROPHYCEAE						
Enteromorpha sp.												
Puck Bay	1973		2					116±6*				Bojanowski and Pempkowiak, 1977
Cladophora sp												
Southern Baltic	1973		3					146±17*				Bojanowski and Pempkowiak, 1977
						RHODOPHYCEAE						
Furcellaria fastigiata												
Gulf of Gdansk	1973							277±13*				Bojanowski and Pempkowiak, 1977

Region	Sampling date	Plant part	N	Salinity (PSU)	51-Cr	134-Cs	136-Cs	137-Cs	131-I	40-K	140-La	References
Phyllophora brodiaei												
Gulf of Gdansk	1973							146±11*				Bojanowski and Pempkowiak, 1977
Ceramium diaphanum												
Gulf of Gdansk	1973							365–1410*				Bojanowski and Pempkowiak, 1977
SPERMATOPHYCEAE												
Zostera marina												
Gulf of Gdansk	1973	L						33±2*				Bojanowski and
		R						152±6*				Pempkowiak, 1977
		W						31±4*				
Potagometon pectinatus												
Southern Baltic	1973/79							40±3*				Bojanowski and Pempkowiak, 1977
Myriophyllum spicatum												
Southern Baltic	1973							80±3*				Bojanowski and
Zarnowieckie Lake	1979											Pempkowiak, 1977

* – pCi kg^{-1} dry wt.
** – *F. vesiculosus*, *F. serratus*, *F. spiralis*
Y – younger off-shoots; O – old thallus; R – receptacles; L – leaves; R – roots; W – whole plants.
^ – g K kg^{-1} d.w.

TABLE 3.2. – continued

Region	Sampling date	Plant part	N	Salinity (PSU)	54-Mn	95-Nb	238-Pu	239+240-Pu (mBq/kg)	103-Ru	106-Ru	124-Sb	125-Sb	89-Sr	References
							PHAEOPHYCEAE							
Fucus vesiculosus														
Southern Baltic	1973	Y						118±15		29±12*				Bojanowski and
		O						30±7.0		12±10*				Pempkowiak, 1977
		R						115±30		32±8.0*				
Swedisch coast	1982		16					80.1						Holm, 1995
								21.0–193						
	1983		18					142						
								49–382						
	1986		31					68.3						
								26.0–268						
	1987		30					57.5						
								25.0–156						
	1991		29					48.3						
								23.0–85.0						
Belt Sea	1988–90							31–300						Kanisch et al., 1995
Finnish coast														
Loviisa	0.5–08.86		2						3100	1260	ND–16	60.5	380	Ilus et al., 1987
									260–5900	410–2100		24–97	53–710	
	1987		13	3.77			ND	ND–0.54		49.4		ND–8.3		Ilus et al., 1988
				2.44–4.10						3.5–90				
	1989–90		18							7.85	2.31			Ilus et al., 1992
										ND–15	ND–3.0			
Olkiluoto	04–08.86		3						840	254	1.49	14	ND–25	Ilus et al., 1987
									52–1900	61–590	ND–2.6	ND–24		
	1987		13	5.74			ND–011	ND–0.15		12.1		ND–5.6		Ilus et al., 1988
				5.13–5.96						7.3–22				
	1989–90									6.3				Ilus et al., 1992
										ND–10				
South coast of Finland	1987		25	5.63			ND–0.16	0.11		14.6		ND–14	16.9	Ilus et al., 1988
				3.52–6.59				0.036–0.16		4.0–38			13–22	
*Fucus***														
Danish waters	1982		23	17.4–24.9	0.4–0.95			0.23–0.62				1.0–1.47		Aarkrog et al., 1986
	1983		158	14.1–29.7	0.25–1.37	0.42–1.5				2.2–5.1		0.79–3.6		

Region	Sampling date	Plant part	N	Salinity (PSU)	54-Mn	95-Nb	238-Pu	239+240-Pu (mBq/kg)	103-Ru	106-Ru	124-Sb	125-Sb	89-Sr	References
Pilayella littoralis														
Gulf of Gdansk	1987		8					154±8.0						Skwarzec and Bojanowski, 1992
Laminaria saccharina														
Belt Sea	1987–88							31–63						Kanisch et al., 1995
CHLOROPHYCEAE														
Enteromorpha sp.														
Puck Bay	1973		2					37–80		31±11*				Bojanowski and Pempkowiak, 1977
Zarnowieckie Lake								13						
Enteromorpha intestinalis														
Gulf of Gdansk								26						Skwarzec and Bojanowski, 1992
Enteromorpha crinita														
Gulf of Gdansk								68						Skwarzec and Bojanowski, 1992
Enteromorpha compressa														
Gulf of Gdansk								28						Skwarzec and Bojanowski, 1992
Cladophora sp														
Southern Baltic	1973		3					67–489		49±32*				Bojanowski and Pempkowiak, 1977
RHODOPHYCEAE														
Furcellaria fastigiata														
Gulf of Gdansk	1973							1110±222		126±48*				Bojanowski and Pempkowiak, 1977
Phyllophora brodiaei														
Gulf of Gdansk	1973							592±74		900±66*				Bojanowski and Pempkowiak, 1977
Ceramium diaphanum														
Gulf of Gdansk	1973							81–211		23–76*				Bojanowski and Pempkowiak, 1977
SPERMATOPHYCEAE														
Zostera marina														
Gulf of Gdansk	1973	L						30.0±7.0		40±10*				Bojanowski and Pempkowiak, 1977
		R						63±18		24±10*				
		W						44±11		37±7*				

Southern Baltic	1979/88	23.3 9.0–44		Skwarzec and Bojanowski, 1992
Potagometon pectinatus				
Southern Baltic	1973/79	13–37	96±74*	Skwarzec and Bojanowski, 1992
Elodea canadensis				
Southern Baltic	1979	35		Skwarzec and Bojanowski, 1992
Acorus calamus				
Southern Baltic	1979	37		Skwarzec and Bojanowski, 1992
Rupia maritima				
Southern Baltic	1987	44±12		Skwarzec and Bojanowski, 1992
Zannichellia palustris				
Southern Baltic	1987	43±5		Skwarzec and Bojanowski, 1992
Myriophyllum spicatum				
Southern Baltic	1973	96±7	48.7*	Skwarzec and
Zarnowieckie Lake	1979	16		Bojanowski, 1992

* – pCi kg^{-1} dry wt.
** – *F. vesiculosus*, *F. serratus*, *F. spiralis*.
Y – younger off-shoots; O – old thallus; R – receptacles; L – leaves; R – roots; W – whole plants.

TABLE 3.2. – continued

Region	Sampling date	Plant part	N	Salinity (PSU)	90-Sr	99-Tc	129m-Te	132-Te	Th (tot.) ($\mu g\ g^{-1}$)	U (tot.) ($\mu g\ g^{-1}$)	Reference
					PHAEOPHYCEAE						
Fucus vesiculosus											
Southern Baltic	1973/78	Y			361±12*				0.25	0.26	Bojanowski and
		O			358±13*				0.33	0.41	Pempkowiak, 1977
		R			538±23*						Szefer, 1987
Finnish coast											
Loviisa	0.5–08.86		2		55.5		2920	ND			Ilus et al., 1987
					28–83		230–5600				
	1987		13	3.77	0–27						Ilus et al., 1988
				2.44–4.10							
	1989–90				18.6						Ilus et al., 1992
					15–22						
Olkiluoto	04–08.86		3		ND		960	400			Ilus et al., 1987
					16.3		22–2800	ND–500			
	1987		13	5.74	ND–23						Ilus et al., 1988
				5.13–5.96							
	1989–90				14–20						Ilus et al., 1992
South coast of Finland	1987		25		16.9						Ilus et al., 1988
					13–22						
*Fucus***											
Danish waters	1982		23	17.4–24.9	5.0–10.3	50–187					Aarkrog et al., 1986
	1983		158	14.1–29.7							
Ectocarpus siliculosus											
Southern Baltic	1978								0.09	0.21	Szefer, 1987
					CHLOROPHYCEAE						
Enteromorpha sp.											
Gulf of Gdansk	1973/78		2		128±6*				0.06–0.13	0.14–0.20	Bojanowski and Pempkowiak, 1977; Szefer, 1987
Zarnowieckie Lake											
Enteromorpha intestinalis											
Gulf of Gdansk	1978/79								0.05	0.07	Szefer, 1987
Enteromorpha crinita											
Southern Baltic	1978								0.23	0.09	Szefer, 1987

Region	Sampling date	Plant part	N	Salinity (PSU)	90-Sr	99-Tc	129m-Te	132-Te	Th (tot.) ($\mu g\ g^{-1}$)	U (tot.) ($\mu g\ g^{-1}$)	Reference
Cladophora sp.											
Southern Baltic	1973/86		3		67±16*				0.24–0.39	0.27–0.35	Bojanowski and Pempkowiak, 1977;Szefer, 1987
					RHODOPHYCEAE						
Furcellaria fastigiata											
Gulf of Gdansk	1973				49±6*						Bojanowski and Pempkowiak, 1977
Phyllophora brodiaei											
Gulf of Gdansk	1973				43±6*						Bojanowski and Pempkowiak, 1977
Ceramium diaphanum											
Gulf of Gdansk	1973				42–63*						Bojanowski and Pempkowiak, 1977
					SPERMATOPHYCEAE						
Zostera marina											
Gulf of Gdansk	1973	L			82±4*						Bojanowski and
		R			87±4*						Pempkowiak, 1977;
	1973/78	W							0.3	0.11–0.23	Szefer, 1987; Skwarzec, 1995
Potagometon pectinatus											
Southern Baltic	1978/79								0.17	0.16	Szefer, 1987
Elodea canadensis											
Southern Baltic	1978/79								0.21	0.24	Szefer, 1987
Acorus calamus											
Southern Baltic	1978/79								0.23	0.25	Szefer, 1987

* – pCi kg^{-1} dry wt.
** – *F. vesiculosus, F. serratus, F. spiralis*
Y – younger off-shoots; O – old thallus; R – receptacles; L – leaves; R – roots; W – whole plants.

TABLE 3.2. – continued

Region	Sampling date	Plant part	N	Salinity (PSU)	234-U	235-U	237-U	238-U	65-Zn	95-Zr	Reference
					PHAEOPHYCEAE						
Fucus vesiculosus											
Finnish coast											
Loviisa	0.5–08.86		2				ND		0–7.1	114	Ilus et al., 1987
										17–210	
	1987		13	3.77					2.45		Ilus et al., 1988
				2.44–4.10					ND–4.0		
Olkiluoto	04–08.86		3				ND–38		17.4	268	Ilus et al., 1987
									3.3–26	4.2–690	
	1987		13	5.74					3.93		Ilus et al., 1988
				5.13–5.96					1.4–8.4		
South coast of Finland	1987		25						2		Ilus et al., 1988
									ND–8.4		
*Fucus***											
Danish waters	1983		23	17.4–24.9					3.1		Aarkrog et al., 1986
Pilayella littoralis											
Gulf of Gdansk	1987		8		2.84–10.8	0.13–0.73		2.53–9.67			Skwarzec, 1995
					CHLOROPHYCEAE						
Enteromorpha sp.											
Gulf of Gdansk	1973/78		2		1.29±0.06	0.07±0.01		1.26±0.06			Skwarzec, 1995
Zarnowieckie Lake					0.80±0.07	0.04±0.01		0.64±0.06			
Enteromorpha intestinalis											
Gulf of Gdansk	1978/79				5.08±0.29	0.47±0.09		4.85±0.28			Skwarzec, 1995
Enteromorpha compressa											
Gulf of Gdansk	1986				11.9±0.28	0.40±0.05		9.74±0.26			Skwarzec, 1995
Cladophora sp.											
Southern Baltic	1973/86		3		4.08±0.11	0.16±0.02		3.37±0.10			Skwarzec, 1995

SPERMATOPHYCEAE

Zostera marina						
Gulf of Gdansk	1973/78	W	3.78±0.11	0.17±0.02	3.20±0.10	Skwarzec, 1995
Potagometon pectinatus						
Southern Baltic	1978/86		4.54±0.18	0.20±0.04	4.21±0.17	Skwarzec, 1995
Elodea canadensis						
Southern Baltic	1978/79		2.65±0.09	0.14±0.02	2.16±0.08	Skwarzec, 1995
Acorus calamus						
Southern Baltic	1978/86		1.80±0.07	0.09±0.01	1.50±0.06	Skwarzec, 1995
Rupia maritima						
Southern Baltic	1987		4.81±0.20	0.45±0.06	4.36±0.19	Skwarzec, 1995
Zannichellia palustris						
Southern Baltic	1987		4.16±0.11	0.22±0.03	3.71±0.11	Skwarzec, 1995
Myriophyllum spicatum						
Southern Baltic	1973					
Zarnowieckie Lake	1979		0.94±0.03	0.06±0.01	0.73±0.03	Skwarzec, 1995

* – pCi kg^{-1} dry wt.
** – *F. vesiculosus, F. serratus, F. spiralis*
Y – younger off-shoots; O – old thallus; R – receptacles; L – leaves; R – roots; W – whole plants.

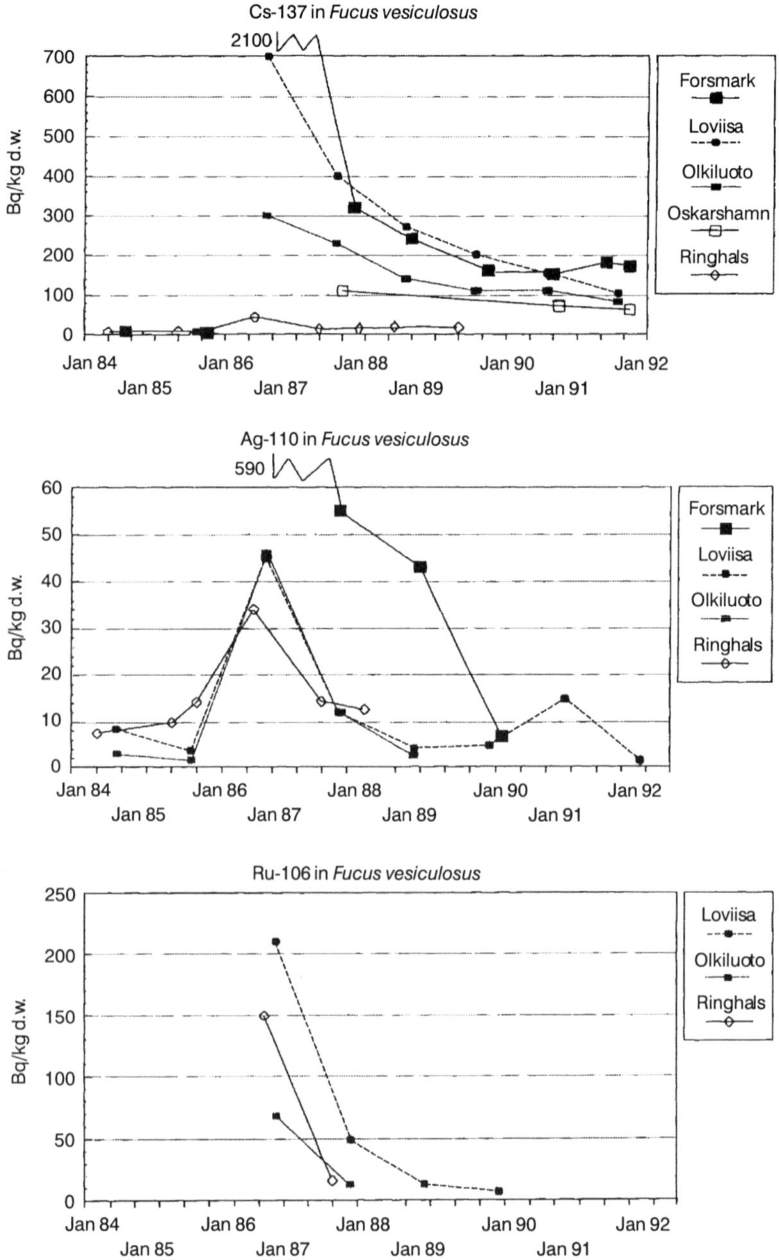

Fig. 3.8. Some radionuclides in *Fucus vesiculosus*. After Kanisch et al. (1995); modified.

0.11–0.30 μg g^{-1} (Spermatophyta). The following concentrations ranges were obtained for Th: 0.05–0.39 μg g^{-1} (Chlorophyta), 0.09–0.33 μg g^{-1} (Phaeophyta) and 0.17–0.60 μg g^{-1} (Spermatophyta). The average Th/U ratios ranged: 0.3–2.6 (Chlorophyta), 0.4–1.0 (Phaeophyta) and 0.9–2.0 (Spermatophyta) (Szefer, 1987).

B. PLANKTON

(i) Introduction

General Characteristics and Species Composition

In the phytoplankton from the Baltic Sea the most abundant are green algae Chlorophyceae, and diatoms Diatomophyceae, Bacillariophyceae and Pyrrophyceae. Phytoplankton composition changes during the year, i.e. between the three blooms in spring, summer and autumn. These differ in the dominance structure. In spring diatoms dominate, e.g. *Chaetoceros* sp., *Skeletonema costatum*, *Thalassiosira levanderi* and *Dinophysis* sp. In summer species diversity increases and *Flagellata* are also present, e.g. *Aphanizomenon* sp., *Eutreptiella* sp. and also *Prorocentrum* sp., *Gomphosphaeria* sp., *Nodularia spumigena.* In autumn, diatoms again dominate. Copepods such as *Acartia bifilosa, A. longiremis, Pseudocalanus minutus elongatus, Temora longicornis* dominate the zooplankton, while in the summer season Cladocera, e.g. *Bosmina coregoni maritima* become more abundant. Rotifera form also a major portion of mesozooplankton in summer. The macrozooplankton consists of a few species permanently present, e.g. *Aurelia aurita* – Mysidacea. Some species are only observed occasionally, when introduced with saline water inflows from the North Sea, e.g. *Pleurobranchia pileus*, *Cyanea capillata* and *Sagitta elegans* (Falandysz et al., 2000). In the mesozooplankton samples from the southern Baltic, 20 species or higher taxonomic units were identified. It can be seen (Szefer et al., 1985) that the species composition of mesozooplankton was similar in all the stations investigated and in the particular water layers. In particular, the differences in the abundance and in the percentage share of mesozooplankton components were noticed. The lowest values of the abundance of mesozooplankton were recorded in the 0–30 m layer (Szefer et al., 1985). The upper warm water layers created favourable conditions for typical summer components such as Rotatoria and Cladocera (Chojnacki, 1973; Hernroth and Ackefors, 1979; Kostriczkina et al., 1980; Koszteyn, 1982). *Temora longicornis* and *Acartia longiremis*, which prefer warm water, also appeared to be numerous. In the Baltic Proper the essential difference can be seen mainly in the percentage share of particular components, in spite of similar hydrological conditions appearing in surface layers. In the Słupsk Furrow, Rotatoria were not observed at all and *Bosmina coregoni-maritima*, *Evadne nordmanni*, nauplial stages of Copepoda and *Temora longicornis* were essentially scarce in comparison to their abundance in the Gdańsk Basin. The water of the Słupsk Furrow was dominanted by *Pseudocalanus elongatus* (about 64%) which in the area of the Gdańsk Basin constituted only 28–34% of all organisms. *Pseudocalanus elongatus* is a euryhaline species and its termic optimum is lower in comparison to other Copepoda (Ackefors and Hernroth, 1975; Hernroth and Ackefors, 1979; Kostriczkina et al., 1980; Koszteyn, 1983; Szefer et al., 1985). It is necessary to emphasise that this characteristics mainly concerns adult individuals. The younger stages of copepods, which were numerous, particularly in surface waters, have different requirements.

Overview of Worldwide Literature

Zooplankton play an important role in the cycling process of metals in the seas (Martin and Knauer, 1973; Greig et al., 1977; Li, 1981). Their particulate products, i.e. faecal pellets, immediately affect the chemical composition of pelagic sediments (Boström et al., 1974; Li, 1981; Fowler, 1977). Zooplankton are an important source of food for carnivorous animals such as chaetognaths and fish, therefore trace elements may be transported and biomagnified through the food-chain up to levels which can be dangerous for the organisms and for man (Knauer and Martin, 1972). The degree of bioaccumation of trace elements in plankton depends on various physicochemical parameters, e.g. temperature of water, salinity and depth of water, species composition etc. (Härdstedt-Roméo, 1982).

Plankton have been assayed in respect to them ability to concentrate of trace elements in the aquatic environments (Szabo, 1968; Martin, 1970; Martin and Knauer, 1972, 1973; Windom, 1972; Turekian et al., 1973; Martin and Broenkow, 1975; Martin et al., 1976a, 1976b; Knauer and Martin, 1972; Boström et al., 1974; Bohn and McElroy, 1976; Horowitz and Presley, 1977; Greig et al., 1977; Zafiropoulos and Grimanis, 1977; Brügmann, 1978; Davies, 1978; Demina and Fomina, 1978; Kosta et al., 1978; Leland et al., 1978; Moore and Boström, 1978; George and Kureishy, 1979; Phillips, 1980; Härdstedt-Roméo and Laumond, 1980; Patin et al., 1980; Sanders and Windom, 1980; Boyle, 1981; Li, 1981; Presley et al., 1981; Härdstedt-Roméo, 1982; Collier and Edmond, 1983; Knauss and Ku, 1983; Kureishy et al., 1983; Henning et al., 1985; Bryan et al., 1985; Fowler et al., 1985; Roméo et al, 1985; Fowler, 1986; Roméo et al., 1985; Roméo and Nicolas, 1986; Balogh, 1988; Diaz and Fernandez-Puelles, 1988; Witzel, 1989; Savenko, 1988; Pohl, 1992; Weber et al, 1992; Heyer et al., 1994; Ritterhoff and Zauke, 1997a, 1997b; Sydeman and Jarman, 1998; Zauke et al., 1996, 1998; Al-Majed and Preston, 2000; Fisher et al., 2000). The bioaccumulation and excretion kinetics of Se in the euphausiid *Meganyctiphanes norvegica* were examined (Fowler and Benayoun, 1977).

According to several authors (Phillips, 1980; Bryan et al., 1985; Brügmann and Hennings, 1994) plankton are not too much useful and effective organisms for biomonitoring of dissolved species of metals (see Chapter 7). Influence of leached trace metals from acidified areas on phytoplankton growth in coastal waters has been studied by Granéli and Haraldsson (1993).

Zooplankton fecal pellets have been also analysed for concentrations of selected metals (Fowler, 1977; Cherry and Higgo, 1978).

Marine plankton have been used as an bioindicator of low-level radionuclide contamination in the Southern Ocean (Marsch and Buddemeier, 1984). The abilities of marine plankton to accumulate transuranic elements such as ^{241}Am, ^{252}Cf, ^{235}Np and ^{237}Pu and their interaction were evaluated by Fisher et al. (1983a, 1983b).

(ii) Occurrence of Chemical Elements in Plankton

The concentration data in respect to heavy metals have been reported sporadically for plankton from the Baltic Sea (Brzezińska et al., 1984; Falandysz, 1984a; Szefer et al., 1985; Davidan and Savchuk, 1989; HELCOM, 1990; Brügmann and Hennings, 1994; Seisuma et al., 1995). Trace element concentrations in plankton from the River Odra mouth area have been reported by Protasowicki (1991a). Table 3.3 shows the concentration of chemical elements in plankton from the Baltic Sea and surrounding northern areas.

Interspecies and spatial trends

The regional variations of some metal levels in mesozooplankton from the southern Baltic were observed (Szefer et al., 1985). The mean levels of Mn, Zn, Pb, Cu and Fe (expressed on the dry biomass taken from the bottom to the surface) in mesozooplankton caught at the Słupsk Furrow region were higher in comparison to those in samples from the Gdańsk Deep. This may be the result of differences in the species structure of mesozooplankton inhabiting the Słupsk Furrow and the Gdańsk Basin regions. It can be seen in Figure 3.9 that the species composition of mesozooplankton was similar in all the stations investigated and in the particular water layers. In particular, the differences in the abundance and in the percentage share of mesozooplankton components were noticed.

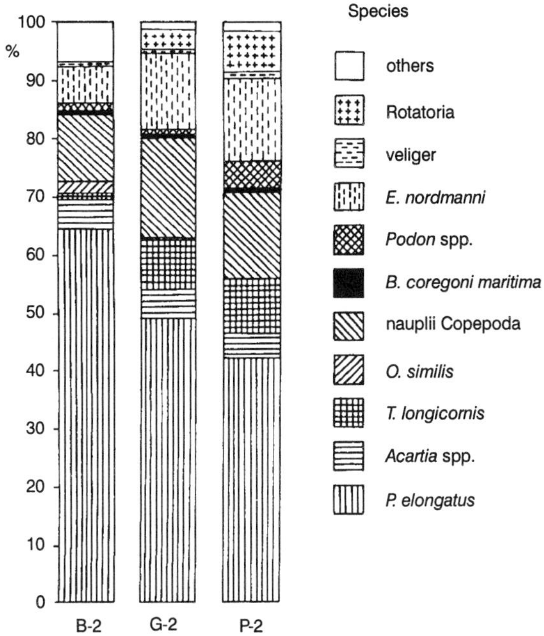

Fig. 3.9. The percentage share of fundamental components of mesozooplankton in the water of Słupsk Furrow (station B-2) and the Gdańsk Basin (stations G-2 and P-2). After Szefer et al. (1985).

TABLE 3.3.

Concentrations of chemical elements ($\mu g\ g^{-1}$ dry wt.) in plankton of the Baltic Sea and other northern areas

Region	Sampling date	Depth of water (m)	Mesh size (mm)	Dominant species	N	Ag	Al	Cd	Co	Cr	Cs	Cu	References
Southern Baltic													
Gdansk Bay	July 1980	0–80*	0.33	*Pseudocalanus elongatus* (63.8%)	2			< 0.9	< 0.5–0.8			49	Szefer et al., 1985
				nauplii Copepoda (8.40 %)								25–73	
				Evadne nordmanni (9.78 %)									
Gdansk Bay –	1979	20–30	0.2	Copepoda, Cladocera'''		0.62		1.81	0.25	5.5	0.05	233	Brzezińska et al., 1984
coastal zone				Copepoda, Cyanophyta''		0.75		2.06	0.45	17.1	0.23	475	Szefer et al., 1985
Gdansk Deep	July 1980	0–108*	0.33	*Pseudocalanus elongatus* (72.3%)	6			2.3	< 0.1–1.3			14.2	
				nauplii Copepoda (8.36 %)				0.9–3.7 (29.0)				5.0–33	
		0–108**	0.2	*Evadne nordmanni* (6.91 %)	1			< 2.2				17	
	1979		0.2	Copepoda, Cladocera'''		1.8		0.71	1	6.6	0.13	138	Brzezińska et al., 1984
Slupsk Furrow	July 1980	0–90*	0.33	*Pseudocalanus elongatus* (69.5%)	9			1.67	< 0.2–1.1			34.6	Szefer et al., 1985
				nauplii Copepoda (8.74 %)				< 0.7–4.6				7.0–78	
		0–90**	0.2	*Evadne nordmanni* (8.49 %)	2			< 0.2–0.20	< 0.2			5.0–31	
	1979		0.2	Copepoda, Cyanophyta''		1		0.86	0.71	4.8	0.09	307	Brzezińska et al., 1984
Bornholm Deep	1979		0.2	Copepoda, Cyanophyta''		2.5		0.07	0.31	3.1	0.05	151	Brzezińska et al., 1984
Gotland Deep	1979		0.2	Copepoda, Cladocera'''		0.24		0.4	0.08	0.9	0.03	175	
Pomeranian Bay	1979		0.2	Cyanophyta''		4.1		1.57	2.1	20	0.23	780	
Southern Baltic	Aug.–Sept. 1983			Mixed zooplankton	20			2.4				27	Falandysz, 1984a
								ND–7.2				6.1–200	
Baltic	Sept. 1980		0.2	Mixed zooplankton	13		180±67^	3.7±39^	0.61±37^	2.0±101^		23±58^	Brügmann and Hennings, 1994
	May–June 1981		0.2		18		200±95^	1.3±59^	0.86±25^	4.0±61^		15±30^	
	June–July 1983		0.2		20		50±60^	3.9±60^				13±37^	
	Nov.–Dec. 1984		0.2		21		600±162^	5.4±194^	0.34±71	8.3±11.6^		29±46^	
Gulf of Riga	1979–82		0.1–0.2	Mixed zooplankton	39			6.3				16	Davidan and Savchuk, 1989
Jaunkemeri	1987–91	10		*Acartia* sp. (87%)	4			0.26'				0.76'	Seisuma et al., 1995
				Eurytemora sp. (80%)				0.15–0.33				0.30–1.54	
				Synchaeta sp. (79%)									
				Evadne sp. (40%)									

Ragaciems	1988–91	20		*Eurytemora* sp. (83%)	3		0.26'			0.65'	
				Synchaeta sp. (72%)			0.18–0.31			0.59–0.75	
				Acartia sp. (38%)							
Roja	1987–91	10.0–20.0		*Synchaeta* sp. (97%)	8		0.23'			0.84'	
				Acartia sp. (81%)			0.10–0.31			0.26–2.22	
Open Baltic				*Synchaeta baltica* (81%)							
around Latvia	May–Sept. 1987	10.0–30.0		*Acartia* sp. (96%)	9		0.31'			2.43'	
				Synchaeta sp. (43%)			0.16–0.55			1.00–10.3	
				Temora sp. (41%)							
Baltic Proper	1979–82		0.1–0.2	Mixed zooplankton	233		1.4			21	Davidan and Savchuk, 1989
Gulf of Finland	1979–82		0.1–0.2		303		1.3			25	
Kattegat			0.2	Mixed zooplankton	5	70±43^	0.8±38^	0.80±7^	1.8±76^	9.0±31^	Brügmann and Hennings, 1994
					5	220±41^	2.0±14^			13±24^	
					6	600±47^	3.9±54^	0.59±26^	8.1±61^	73±116^	
North Sea	May–June1981		0.2	Mixed zooplankton	5	90±89^	2.0±72^	0.77±24^	3.9±90^	18±62^	Brügmann and Hennings, 1994
	June–July 1983				4	690±117^	1.8±15^			18±35^	
S. North Sea	1990–91	0–30	0.3	*Calanus finmarchicus/helgolandicus*			1.8			7.1	Zauke et al., 1996
Central North Sea							3.2			6.6	
German Bight				*Acartia* sp.			2.5			9.7	
				Mixed copepods			0.9			8.4	

* – Samples collected with Nansen net from three separate water layers
** – Samples collected with Hansen net (vertical hauls from the bottom to surface water).
' – Wet weight.
" – Copepoda: *Pseudocalanus elongatus, Acartia longiremis, A. bifilosa, Temora longicornis*, Cyanophyta: *Nodularia spumigena.*
''' – Cladocera: *Bosmina coregoni maritima, Evadne nordmanni, Podon intermedius, P. polyphemoides.*
^ – ±SD (%).

TABLE 3.3. – continued

Region	Sampling date	Depth of water (m)	Mesh size (mm)	Dominant species	N	Fe	Hg	K	Mn	Na	References
Southern Baltic											
Gdansk Bay	July 1980	0–80*	0.33	*Pseudocalanus elongatus* (63.8%)	2	2900		4300	11	4700	Szefer et al., 1985
				nauplii Copepoda (8.40%)		1100–4700		3200–5400	9.0–13.0	2400–7000	
				Evadne nordmanni (9.78%)							
Gdansk Bay –		20–30	0.2	Copepoda, Cladocera'''		640	0.76				Brzezińska et al., 1984
coastal zone	1979			Copepoda, Cyanophyta''		1300	1				
Gdansk Deep	July 1980	0–108*	0.33	*Pseudocalanus elongatus* (72.3%)	6	900		3450	15	5400	Szefer et al., 1985
				nauplii Copepoda (8.36%)		400–2100		1400–5000	11.0–22.0	2700–9200	
		0–108**	0.2	*Evadne nordmanni* (6.91%)	1			2800	5	2700	
	1979		0.2	Copepoda, Cladocera'''		8500	0.14				Brzezińska et al., 1984
Slupsk Furrow	July 1980	0–90*	0.33	*Pseudocalanus elongatus* (69.5%)	9	2000		3000	44.8	12800	Szefer et al., 1985
				nauplii Copepoda (8.74%)		600–6800		2100–6200	10.0–120	4500–37300	
		0–90**	0.2	*Evadne nordmanni* (8.49%)	2	300–1500		1500–1600	16.0–21.0	6600–9300	
	1979		0.2	Copepoda, Cyanophyta''		900	1.17				Brzezińska et al., 1984
Bornholm Deep	1979		0.2	Copepoda, Cyanophyta''		1120	0.17				Brzezińska et al., 1984
Gotland Deep	1979		0.2	Copepoda, Cladocera'''		160	0.16				
Pomeranian Bay	1979		0.2	Cyanophyta''		2300	2.14				
Southern Baltic	Aug.–Sept. 1983			Mixed zooplankton	20	720			31		Falandysz, 1984a
						60–4600			5.1–170		
Baltic	Sept. 1980		0.2	Mixed zooplankton	13	570±86^	0.19±49^		90±167^		Brügmann and Hennings, 1994
	May–June 1981		0.2		18	900±114^	0.06±32^		30±100^		
	June–July 1983		0.2		20	500±130^	0.05±29^		30±67^		
	Nov.–Dec. 1984		0.2		21	2640±95^	0.37±62^		40±75^		
Gulf of Riga											
Jaunkemeri	1987–91	10		*Acartia* sp. (87%)	4		0.012'		16.63'		Seisuma et al., 1995
				Eurytemora sp. (80%)			0.010–0.014		12.0–21.7		
				Synchaeta sp. (79%)							
				Evadne sp. (40%)							

Ragaciems	1988–91	*20*		*Eurytemora* sp. (83%)	3		0.011'	11.9'	
				Synchaeta sp. (72%)			0.010–0.011	4.6–17.4	
				Acartia sp. (38%)					
Roja	1987–91	10.0–20.0		*Synchaeta* sp. (97%)	8		0.013'	10.2'	
				Acartia sp. (81%)			0.011–0.018	1.5–23.6	
Open Baltic				*Synchaeta baltica* (81%)					
around Latvia	May–Sept. 1987	10.0–30.0		*Acartia* sp. (96%)	9		0.031'	7.49'	
				Synchaeta sp. (43%)			0.006–0.143	2.7–20.2	
				Temora sp. (41%)					
Baltic Proper	1979–82		0.1–0.2	Mixed zooplankton	233				Davidan and Savchuk, 1989
Gulf of Finland	1979–82		0.1–0.2		303				
Kattegat			0.2	Mixed zooplankton	5	320±94^	0.07±42^	20±50^	Brügmann and Hennings, 1994
					5	680±66^	0.05±40^	50±40^	
					6	1960±58^	0.65±27^	50±60^	
North Sea	May–June1981		0.2	Mixed zooplankton	5	1450±90^	0.06±36^	10±100^	Brügmann and Hennings, 1994
	June–July 1983				4	1760±81^	0.07±55^	50±120^	

* – Samples collected with Nansen net from three separate water layers.
** – Samples collected with Hansen net (vertical hauls from the bottom to surface water).
' – Wet weight.
" – Copepoda: *Pseudocalanus elongatus, Acartia longiremis, A. bifilosa, Temora longicornis*, Cyanophyta: *Nodularia spumigena.*
''' – Cladocera: *Bosmina coregoni maritima, Evadne nordmanni, Podon intermedius, P. polyphemoides.*
^ – ±SD (%).

TABLE 3.3. – continued

Region	Sampling date	Depth of water (m)	Mesh size (mm)	Dominant species	N	Ni	Pb	Sb	Se	Zn	References
Southern Baltic											
Gdansk Bay	July 1980	0–80*	0.33	*Pseudocalanus elongatus* (63.8%)	2	6.9	17			170	Szefer et al., 1985
				nauplii Copepoda (8.40%)		2.3–11.4	15.0–19.0			120–210	
				Evadne nordmanni (9.78%)							
Gdansk Bay –		20–30	0.2	Copepoda, Cladocera'''			133	0.7	2.5	476	Brzezińska et al., 1984
coastal zone	1979			Copepoda, Cyanophyta''			178	1.4	2.5	365	
Gdansk Deep	July 1980	0–108*	0.33	*Pseudocalanus elongatus* (72.3%)	6	7.82	22.2			185	Szefer et al., 1985
				nauplii Copepoda (8.36%)		3.0–16.4	3.0–65.0			120–290	
		0–108**	0.2	*Evadne nordmanni* (6.91%)	1	1					
	1979		0.2	Copepoda, Cladocera'''			74.7	1	2.3	284	Brzezińska et al., 1984
Slupsk Furrow	July 1980	0–90*	0.33	*Pseudocalanus elongatus* (69.5%)	9	9.66	23.3			2150	Szefer et al., 1985
				nauplii Copepoda (8.74%)		3.1–16.6	1.0–73 (450)			190–4800	
		0–90**	0.2	*Evadne nordmanni* (8.49%)	2	3.9–5.7	4.0–13.0			130–310	
	1979		0.2	Copepoda, Cyanophyta''			99.8	1.7	1.7	177	Brzezińska et al., 1984
Bornholm Deep	1979		0.2	Copepoda, Cyanophyta''		0.17	40.8	1.6	2.2	225	Brzezińska et al., 1984
Gotland Deep	1979		0.2	Copepoda, Cladocera'''		0.16	31.6	2.2	1.2	156	
Pomeranian Bay	1979		0.2	Cyanophyta''		2.14	319	5.5	1.3	780	
Southern Baltic	Aug.–Sept.1983			Mixed zooplankton	20	6.7	47			280	Falandysz, 1984a
						1.9–32.0	ND–780			96–1030	
Baltic	Sept. 1980		0.2	Mixed zooplankton	13	5.7±32^	8.3±96^		1.4*	42±38^	Brügmann and Hennings, 1994
	May–June 1981		0.2		18	5.7±35^	14.3±156^		0.2–3.2	94±95^	
	June–July 1983		0.2		20	4.6±26^	2.5±139^			320±53^	
	Nov.–Dec. 1984		0.2		21	10.8±68^	19.7±141^			573±62^	
Gulf of Riga	1979–82		0.1–0.2	Mixed zooplankton	39		8			130	Davidan and Savchuk, 1989
Jaunkemeri	1987–91	10		*Acartia* sp. (87%)	4		0.54'			28.7'	Seisuma et al., 1995
				Eurytemora sp. (80%)			0.31–1.01			10.2–80.5	
				Synchaeta sp. (79%)							
				Evadne sp. (40%)							
Ragaciems	1988–91	20		*Eurytemora* sp. (83%)	3		0.43'			13.4'	
				Synchaeta sp. (72%)			0.21–0.69			11.1–18.8	
				Acartia sp. (38%)							
Roja	1987–91	10.0–20.0		*Synchaeta* sp. (97%)	8		1.21'			25.6'	
				Acartia sp. (81%)			0.21–5.70			5.8–107	

Open Baltic				*Synchaeta baltica* (81%)						
around Latvia	May–Sept. 1987	10.0–30.0		*Acartia* sp. (96%)	9		1.63'		52'	
				Synchaeta sp. (43%)			0.42–3.70		17.5–96.3	
				Temora sp. (41%)						
Baltic Proper	1979–82		0.1–0.2	Mixed zooplankton	233		18		320	Davidan and Savchuk, 1989
Gulf of Finland	1979–82		0.1–0.2		303		17		340	
Kattegat			0.2	Mixed zooplankton	5	2.2±35^	4.2±61^	2.8[b]	36±24^	Brügmann and Hennings, 1994
					5	4.8±53^	4.1±57^	2.6–3.2	228±22^	
					6	4.8±34^	10.8±64^		673±67^	
North Sea	May–June1981		0.2	Mixed zooplankton	5	2.8±69	6.1±69	2	42±14	Brügmann and Hennings, 1994
	June–July 1983		0.2		4	6.0±42	9.4±54		408±45	
S. North Sea	1990–91	0–30	0.3	*Calanus finmarchicus/helgolandicus*			1		129	Zauke et al., 1996
Central North Sea							1		123	
German Bight				*Acartia* sp.			0.7		225	
				Mixed copepods			1		323	

* – Samples collected with Nansen net from three separate water layers.
** – Samples collected with Hansen net (vertical hauls from the bottom to surface water).
' – Wet weight.
" – Copepoda: *Pseudocalanus elongatus, Acartia longiremis, A. bifilosa, Temora longicornis*, Cyanophyta: *Nodularia spumigena.*
''' – Cladocera: *Bosmina coregoni maritima, Evadne nordmanni, Podon intermedius, P. polyphemoides.*
[a] – Average and range for 72 samples collected during 1980–84.
[b] – Average and range for 16 samples collected during 1981–84.
^ – ±SD (%).

Depth dependent trends

Besides the spatial, also the depth-dependent changes were observed for some trace elements in mesozooplankton from a southern Baltic (Szefer et al., 1985). The mesozooplankton caught at deeper waters of the Słupsk Furrow region generally contained more of some metals, with the exception of Pb and Co, than that from surface waters. One can notice different quantitative relations below 30 m depth. First of all, this water was characterised by a clearly visible dominance of *Pseudocalanus elongatus*. However, Rotatoria, Cladocera, nauplial stages of Copepoda, *Acartia* spp. and *Temora longicornis* were not observed at all or they were not numerous. *Oithona similis* and *Sagitta elegans*, which preferred a higher salinity, and colder, deeper water, were noticed (Ritter-Zahony, 1911; Różańska, 1971; Chojnacki, 1973; Siudziński, 1977; Hernroth and Ackefors, 1979). Such conditions appeared in deep water in the Słupsk Furrow and to a lower degree in the Gdańsk Deep. Generally, it was observed that the abundance of mesozooplankton was several times lower in deeper water layers in comparison to water up to 30 m depth which confirms the results of chemical analysis.

(iii) Occurrence of Radionuclides in Plankton

Baltic plankton have been studied for concentration of gamma emitting radionuclides, i.e. ^{110m}Ag, ^{140}Ba, ^{134}Cs, ^{136}Cs, ^{137}Cs, ^{141}Ce, ^{144}Ce, ^{131}I, ^{40}K, ^{140}La, ^{95}Nb, ^{103}Ru, ^{106}Ru, ^{125}Sb, ^{129m}Te and ^{95}Zr (Ilus et al., 1987; Bojanowski et al., 1995). The concentrations of alpha emitting radionuclides such as U (^{238}U, ^{235}U, ^{234}U) and Th (^{232}Th) were determined in phyto- and zooplankton collected mostly in east part of the southern Baltic, i.e. in the Gulf of Gdańsk (Szefer, 1987; Skwarzec, 1995).

The concentrations of the gamma and alpha emitting radionuclides in Baltic plankton are listed in Table 3.4. According to Bojanowski et al. (1995) the difference between two Pomeranian Bay plankton groups, consisted from predominant species belonged to zoo- and phytoplankon, was insignificant statistically and it did not seem to support a conclusion that the radiocaesium is preferentially accumulated in either phyto- or zooplankton.

Spatial variations in polonium concentrations in southern Baltic have been related to intense blue-green alga blooming in the Gdańsk Basin. The accumulation degree of this radionuclide estimated in relation to ambient seawater as a substrata increased as follows: phytoplankton < macrozooplankton < mesozooplankton (Skwarzec and Bojanowski, 1988). It is postulated that polonium is absorbed to a larger extent by organic matter consumers (zooplankton) than its producers (phytoplankton) (Skwarzec and Bojanowski, 1988; Skwarzec, 1999).

Seasonal variations in radionuclide concentrations in the Gulf of Finland have been reported. Phytoplankton inhabited the Loviisa, Gulf of Finland, was the most abundant in ^{103}Ru, ^{129m}Te and ^{137}Cs. Towards the autumn the levels of radionuclides decreased significantly (Ilus et al., 1987).

TABLE 3.4.

Concentrations of radionuclides (Bq kg^{-1} dry wt.) in phyto- and zooplankton of the Baltic Sea and other northern areas

Region	Sampling date	Species composition	Sample depth (m)	110m-Ag	140-Ba	134-Cs	136-Cs	137-Cs	141-Ce	131-I	40-K	210-Po (mBq kg^{-1})	References
Phytoplankton													
Gulf of Gdansk	1980/1985	Mixed sample*										33.9±3.2	Skwarzec, 1995
		Coscinodiscus granii										20.8±3.5	Skwarzec and Bojanowski, 1988
Słupsk Furrow		Mixed sample**										47.3±3.8	
		Dinobryon balticum										60.8±3.2	
Average												40.7±15.9	
Zooplankton													
Gdansk Basin		Macrozooplankton'	0–bottom									33±5	
		Mesozooplankton'	0–30									86±10	
			30–60									52±5	
			60–bottom									26±6	
Average												55±27	
Slupsk Furrow		Macrozooplankton'	0–90									58±23	
		Mesozooplankton'	0–30									351±95	
			30–60									150±50	
			60–90									142±33	
Average												214±116	
Mixed plankton													
Northern Baltic	June 1986	Diatom *Diatoma elongatus*		170	280	1350	140	2360	68	515	ND–39		Ilus et al., 1987
Hudöfjärden		Dinoflagellate *Gonyaulax catenata*		140–200		1200–1500		2200–2500		500–530			
Loviisa													
	July 1986	Blue green alga *Aphanizomenon flos-aquae*		170	ND	180	ND	365	ND	ND	ND–320		
						110–250		220–570					
	August 1986	Dinoflagellate *Dinophysis acuminata*, blue green alga *Gomphosphaeria lacustris*		1500	ND	500	ND	820	ND	ND	ND		
								790–850					
Pomeranian Bay	1993							2.0±1.5"				7.5±3.1"	Bojanowski et al., 1995

* – Blue-green alga (*Aphanizomenon flos-aquae, Nodularia spumigena, Nodularia herveyana, Anabaena flos-aquae, Anabaena spiroides, Anabaena affinis*) with admixture of green alga (*Pediastrum duplex, Oocystis* sp.).

** – *Microcystis aeruginosa* blue green alga and *Dinobryon balticum* chrysophycean.

' – *Pseudocalanus elongatus* (27.6–93.2%), nauplii Copepoda (0.4–24.1%), *Evadne nordmanni* (1.1–18.1%) and others (0.1–12%).

" – Wet weight base.

TABLE 3.4. – continued

Region	Sampling date	Species composition	140-La	95-Nb	103-Ru	106-Ru	125-Sb	129m-Te	U (tot.)[a]	234-U	235-U	238-U	95-Zr	References
Phytoplankton														
Gulf of Gdansk	1991	Mixed sample*							0.42±0.04	5.9±0.5	0.2±0.1	5.1±0.5		Skwarzec, 1995
Slupsk Furrow	1991	Mixed sample**							0.50±0.05	6.9±0.6	0.2±0.1	6.1±0.6		
Average									0.46±0.05	6.4±0.6	0.2±0.1	5.6±0.6		
Zooplankton														
Gulf of Gdansk	1988	Mixed sample'							0.11±0.02	1.5±0.2	0.1±0.1	1.3±0.2		Skwarzec, 1995
Mixed plankton														
Northern Baltic	June 1986	Diatom *Diatoma elngatus*	ND–430	ND–30 ND	2400	1055	ND	2350					ND–230	Ilus et al., 1987
Hudöfjärden		Dinoflagellate *Gonyaulax catenata*			2200–2600	910–1200		2200–2500						
Loviisa	July 1986	Blue green alga *Aphanizomenon flos-aquae*	ND	ND	ND–370	ND	ND	ND					ND	
	August 1986	Dinoflagellate *Dinophysis acuminata*, blue green alga *Gomphosphaeria lacustris*	ND	ND	180	ND	ND	ND					ND	

* – Mainly phytoplankton species *Coscinodiscus granii* (95%), *Chaetoceros* sp. (5 %) and admixture of zooplankton (*Pseudocalanus elongatus* and *Tintinopsis* sp.)

** – Mainly phytoplankton species *Coscinodiscus granii, Dinophysis acuminata, Aphanizomenon flos-aquae, Nodularia spumigena* and admixture of zooplankton (*Pseudocalanus elongatus, Evadne nordmanni, Tintinopsis* sp.).

' – Mainly zooplankton species *Pseudocalanus elongatus, Oithona similis* and *Evadne nordmanni.*

[a] – $\mu g\ g^{-1}$ d.w.

C. ZOOBENTHOS

1. MOLLUSCS

(i) Introduction

General Characteristics and Taxonomy

Bivalves dominate in macrozoobenthos of the southern Baltic: *Arctica islandica*, *Macoma balthica, Mya arenaria, Mytilus edulis,* and *Cardium glaucum*. The greatest species diversity is noted in the shallow littoral zone on sandy and sandy-muddy bottoms which offer habitat diversity. Macrofauna are much less diverse, as far as number of species is concerned, on the deep muddy bottom, where usually only a few species are present, e.g. *M. balthica.* At the same time, due to the great abundance of dominant bivalves – *M. balthica* on the muddy bottom, and *M. edulis* on the sandy and stony bottom (comprising in many cases almost 100% of the total macrofauna biomass) – the macrofauna biomass is relatively high, reaching up to 300–500 g wet weight per square meter. Decreasing species diversity, abundance and biomass of macrofauna with depth are a general pattern observed in the Bornholm, Gdańsk and Gotland Basins. The main factor responsible for this trend is oxygen deficits occurring in bottom waters of the Bornholm and Gotland Basins (Falandysz et al., 2000).

Taxonomy and description of habitat and food habits of particular species of molluscs inhabiting the Baltic Sea are given below.

Phylum: Mollusca
Class: Bivalvia
Family: Mytilidae

Species: Blue mussel, syn. Common mussel (*Mytilus edulis* Linnaeus, 1758)

Habitat and range: it lives in Atlantic waters of European coast; ranges from northern seas such as the Chukchi Sea, south-western areas of Kara Sea, White Sea, Barents Sea and Far East seas (Zatsepin et al., 1988) to the Mediterranean and Black Seas. It inhabits also the coastal waters of Island, southern part of the Greenland, Atlantic and Pacific coastal waters of the North America. In the Baltic Sea is distributed from the Kieler Bucht to the Bothnian Bay and Gulf of Finland (Gosling, 1992). *Mytilus* is wide-spread mollusc in all over the world, namely northern temperate latitudes, the Mediterranean Sea, the Pacific coast of North America, south-eastern and south-western coastal regions of South America, Australia, the New Zealand, the Kerguelen Islands, the Pacific coast of Asia (Goslin, 1992). *Food habits*: suspension (filter) feeder – planctivore – feeds on phytoplankton and bacteria (Miner, 1950; Mulicki, 1957; Ankar, 1977; Jagnow and Gosselck, 1987; Wiktor, 1990).

Class: Bivalvia
Family: Tellinidae

Species: Little Macoma (*Macoma balthica* Linnaeus, 1758)

Habitat and range: its population is distributed along the Atlantic coast (Zatsepin et al., 1988); as Atlantic-boreal species is observed from Arctic seas to Georgia, occurs in the Baltic Sea even to water depth of 100 m except some very low saline waters of the Bothnian Bay and Gulf of Finland, in the North Sea is typical representative of shallow-water fauna. Food habits: suspension (deposit) feeder, feeds mainly on bacteria, Protozoa, microalgae. (Miner, 1950; Mulicki, 1957; Ankar, 1977; Wiktor, 1985; Jagnow and Gosselck, 1987).

Class: Bivalvia
Family: Cardiidae

Species: Cockle shell (*Cardium glaucum* Brugiere), syn. (*Cerastoderma lamarcki* Reeve 1844)

Habitat and range: geographical distribution of *C. glaucum* is very wide; its population occurs along the European coasts from the Baltic Sea (except the Bothnian Bay and Gulf of Finland), western coasts of the Denmark and British gulfs across the Atlantic coast to the Mediterranean and Caspian Seas (Mars, 1951; Rygg, 1970; Labourg and Lasserre, 1980) and north African saline water bodies (Zaouali, 1977; Levy, 1985). *Food habits*: suspension feeder (Miner, 1950; Mulicki, 1957; Wiktor, 1985, Jagnow and Gosselck, 1987).

Class: Bivalvia
Family: Myidae

Species: Long clam (*Mya arenaria* Linnaeus, 1758)

Habitat and range: this Arctic-boreal species occurs in the Atlantic Ocean (Zatsepin et al., 1988); range: from Arctic seas to the North Carolina, very common in the Baltic Sea. *Food habits*: suspension (deposit) feeder (Miner, 1950; Mulicki, 1957; Ziegelmeier, 1957; Wiktor, 1985; Jagnow and Gosselck, 1987).

Class: Bivalvia
Family: Astartidae

Species: Northern Astarte (*Astarte* spp)

Habitat and range: Arctic Northern Astarte (*Astarte* spp.) occurs in all north seas, in the Baltic Sea lives *Astarte borealis* (Schumacher 1817) and *Astarte elliptica* (Brown 1827). *Astarte borealis* – at present is a relict species in the Baltic Sea originating from era corresponding to the Yoldia Sea; distributed in the Atlantic Ocean from the Greenland to the Massachusetts Bay. Range of its distribution in the Baltic Sea is limited to the Słupsk Furrow because of oxygen deficit below the halocline; food habits: deposit feeder? (Miner, 1950; Mulicki, 1957; Ziegelmeier, 1957; Jagnow and Gosselck, 1987). *Astarte elliptica* – Arctic species distributed in waters of the North Pole; distributed along European coasts of the Atlantic

Ocean to the Biscay, noted also along American coastal waters to the Massachusetts Bay. In the Pacific Ocean ranges to the British Columbia. From the North Sea enters the Kattegat, however it is not noted in the Danish Straits, its Baltic population occurs together with *A. borealis* in the Słupsk Furrow forming mixed populations. *Food habits*: deposit feeder? (Mulicki, 1957; Jagnow and Gosselck, 1987).

Class: Bivalvia
Family: Arcticidae

Species: Ocean quahog (*Arctica islandica*), syn. (*Cyprina islandica*)
Habitat and range: it is especially abundant in the northern part of its range, which extents from the Arctic Ocean to Cape Hatteras; occurs in western part of the Barents Sea and in some areas of the White Sea; in the Baltic Sea is observed from the Kiel Bay to deeper parts of the Arkona Basin. *Food habits*: filter feeder (Miner, 1950; Arntz and Weber, 1970; Zatsepin et al., 1988).

Class: Bivalvia
Family: Dreissenidae

Species: Zebra mussel (*Dreissena polymorpha*)
Habitat and range: inhabits riverine and brackish waters; observed in rivers, lakes and lagoons of Europe, e.g. the Netherlands, Belgium and Kiel and lagoons of the Black Sea and the Caspian Sea; in the lagoons of the Baltic Sea (Szczecin and Vistula lagoons) is observed sporadically, although more frequently in Szczecin Lagoon. *Food habits*: filter feeder (Wiktor, 1969; Zatsepin et al., 1988).

Class: Gastropoda
Family: Littorinidae

Species: Perwinkle, syn. Common winkle (*Littorina littorea*)
Habitat and range: distributed from Asturias (northern Spain) to northern Norway in the eastern Atlantic, and from New Jersey (USA) to Greenland in the western Atlantic. Occurs in the German North Sea and Baltic coast. It is shallow water species which inhabits rocky and sandy shores. In the Baltic Sea its geographical distribution range reaches eastern coast of the Bornholm and Rügen. Observed also in the White Sea. The bulk of the population occurs intertidally, however some specimens can be observed to a depth of 15 m. *Food habits*: feeds mainly on epilithic algae and vegetable detritus. *Fucus vesiculosus* is often eaten by perwinkle (Fretter and Graham, 1962; Nordsieck, 1968; Graham, 1988; Vlastov and Matekin, 1988; Taylor and Miller, 1989; Bauer et al., 1997).

Overview of Worlwide Literature

Most recently new directions for monitoring marine pollution and implications in estimation of metal bioavailability in Mussel Watch programmes are recommended (Soto et al., 2000; Szefer, 2000). Since the Minamata accident (Harada,

1995) many articles on Hg pollution in marine zoobenthos have been published. In many papers, emphasis has been placed on the ability of molluscs to concentrate of metallic pollutants in marine environments (Segar et al., 1971; Nickless et al., 1972; Greig et al., 1976; Boyden, 1977; Bryan and Hummerstone, 1977; Greig and Wenzloff, 1977; Lande, 1977; Phillips, 1977b, 1978, 1980; Luoma and Bryan, 1978, 1982; Windom and Kendall, 1979; Boyden and Phillips, 1981; Hung et al., 1981; Copper et al., 1982; McGreer, 1982; Bryan et al., 1983; Luoma, 1983; Wilson, 1983; Bryan, 1984, 1985; Martincic et al., 1984; Bryan et al., 1985; Luoma et al., 1985, 1990; Sunila and Lindström, 1985; Amiard et al., 1986; Ikuta, 1988; Phillips and Rainbow, 1988; Borchardt et al., 1989; Viarengo, 1989; Fowler, 1990; Savari et al., 1991; Viarengo and Canesi, 1991; Anderlini, 1992; Bordin et al, 1992; Bryan and Langston, 1992; Wilson and Elkaim, 1992; Fowler et al., 1993; Páez-Osuna et al., 1993, 1994; Phillips and Rainbow, 1993; Fujita, 1994; Sarkar et al., 1994; Andersen et al., 1996; Lee, 1996; Szefer et al., 1997c; Guns et al., 1999; Szefer et al., 1999a, 1999c, 1999d; Tedengren et al, 1999; De Wolf et al., 2000; Dietz et al., 2000a; Jeng et al., 2000; Ruiz and Saiz-Salinas, 2000; Ruelas-Inzunza and Páez-Osuna, 2000). Among others, especially organisms *Mytilus* spp. have been considered to be potential biomonitors of toxic metals in marine ecosystems (Fowler and Oregioni, 1976; Phillips, 1976a, 1976b, 1977b, 1978, 1985; Karbe et al., 1977; Goldberg et al., 1978, 1983; Schnier et al., 1978; Gordon et al., 1980; Phillips, 1980; Julshamn, 1981a, 1981c, 1981d, 1981e; Koide et al., 1982; Ritz et al., 1982; Bryan, 1983; Popham and D'Auria, 1983; Favretto and Favretto, 1984a, 1984b; Roesijadi et al., 1984; Bryan et al., 1985; Szefer and Szefer, 1985, 1990, 1991; Cossa, 1988, 1989; Fischer, 1988, 1989; Lobel et al., 1989; Knutzen and Skei, 1990; Marmolejo-Rivas and Páez-Osuna, 1990; Broman et al., 1991; Hamilton, 1991; Szefer, 1991; Lauenstein and Dolvin, 1992; Stronkhorst, 1992; Regoli and Orlando, 1993, 1994; Robinson et al., 1993; Fabris et al., 1994; Brown and Luoma, 1995; Julshamn and Grahl-Nielsen, 1996; Szefer et al., 1997a, 1997b, 1998a, 1998b; Beliaeff et al., 1998; Cantillo, 1998; Regoli, 1998; Giusti et al., 1999; Nicholson, 1999; Tedengren et al., 1999; Wright and Mason, 1999; Joiris et al., 2000b; Lee et al., 2000; Muñoz-Barbosa et al., 2000; Szefer and Nicholson, 2000; Wong et al., 2000). It is shown that the mussels Mytilidae from the coastal areas of the Kyushu Island (Japan), Korean waters, Scandinavian waters, as well as of northeast and estuarine waters of England are characterised by the highest concentrations of trace elements reported up to date (Phillips, 1978, 1979; Bryan et al., 1985; Julshamn and Grahl-Nielsen, 1996; Szefer et al., 1997b; Giusti et al., 1999; Lee et al., 2000). The concentrations and distribution of butyltin compounds have been recognised extensively under an International Scientific Research Program 1997–1999; butyltins residues were frequently detected in green mussel (*Perna viridis*) from Pacific coasts, indicating a widespread contamination along the coastal waters of Asian developing countries, i.e. Thailand, India, Philippines and Malaysia (Kan-atireklap et al., 1997, 1998; Prudente et al., 1999; Sudaryanto et al., 2000; Tanabe et al., 1998, 2000). Butyltin compounds have been also analysed in Mytilidae from other world areas, e.g. in *Mytilus edulis* from central-west Greenland (Jacobsen and Asmund, 2000).

The number of available articles on the distribution of trace elements in shells of the bivalves is scant (Stureson, 1976, 1978; Stureson and Reyment, 1971; Hamilton, 1980; Koide et al., 1982; Al-Dabbas et al., 1984; Bourgoin, 1990; Foster and Chacko, 1995; Soto et al., 1995; Puente et al., 1996).

Although most of papers are focused on investigations of soft tissue; however there are a few available data concerning concentrations of heavy metals in mussel byssus (Hamilton, 1980; Coombs and Keller, 1981; Koide et al., 1982; Ikuta, 1986a, 1986b; Szefer et al., 1997a). Relationships between selected metals in byssus and soft tissue of Mytilidae have been also reported (Ikuta, 1986b; Szefer et al., 1997a, 1997b, 1998a, 1998b, 1999a).

According to Phillips (1980) multivarious effects such as adsorption of metals to the shell surface influenced by e.g. salinity and metal interaction are main cause of difficulty in use of mussel shell as biomonitor of metallic pollutants in the marine environments. As it has been suggested by Koide et al. (1982), mussel shells as whole life integrator of metals may be better biomonitor of metallic pollutants than soft tissue. Other advantage in the use of this hard tissue as biomarker is also recommended; namely the samples must not be frozen and depurated before analysis. Moreover it is possible to make the comparison between the pollutant levels in recent shells and fossilised ones in order to estimate pre-civilisation background levels and to biomonitor the evolution of ecological parameters (Bourgoin, 1990; Puente et al., 1996).

Mytilus spp. attaches itself to bottom sediment by a network of threads, named byssi, which are secreted from a byssal gland in the foot. This material is composed of a protein component, collagen, which contains some potential metal binding sites, largely composed of glycine and proline amino acids' residues. The byssi have a significant contribution towards eliminating some elements from the mollusc's body; hence, metallic contaminants are transferred from the soft tissue to the byssus rather than adsorbed onto the surface of the byssus. *Mytilus* spp. byssus can concentrate a wide range of metals and in some instances to an astonishing degree (Coombs and Keller, 1981). The accumulation of metals in mussels with considering the mechanism of uptake, metabolism and detoxification has been reviewed by George (1980).

Accumulation of radionuclides by *Mytilus* soft tissue has been studied by several authors (see, e.g. Dahlgaard, 1981, 1991; Pentreath et al., 1979, Pentreath, 1981; Koide et al., 1982; Gouvea et al., 1987; Nolan and Dahlgaard, 1991). Additionally, shells and byssal threads of *Mytilus* have been analysed for actinides concentrations. From literature data clearly results that molluscs are useful bioindicators for radionuclides such as ^{106}Ru, $^{239+240}$Pu, ^{241}Am, ^{99}Tc and ^{137}Cs in the marine environments (Hamilton and Clifton, 1980; Charmasson et al., 1999). The concentrations of $^{239+240}$Pu and ^{137}Cs in soft tissue of six species of bivalve collected along the Japanese coast were within the values from 0.8 to 6.1 and from 47–77 mBq kg^{-1} wet wt. (Yamada et al., 1999). Long-term variations of artificial radionuclides concentrations in *M. edulis* from the French Mediterranean coast have been studied by Charmasson et al. (1999). The concentration and depuration of some radionuclides present in a chronically exposed population of *M. edulis* have been studied by Clifton et al. (1983). Experimental studies on the biokinetics of ^{241}Am and ^{237}Pu in the tissues of the molluscs *Tapes decussatus* and *Aporrhais pespelicani*, and the cephalopod *Octopus vulgaris* were performed by Grillo et al. (1983) and Guary and Fowler (1983). Szefer (1992) overviewed the distribution and bioaccumulation of U isotopes in marine biota including mussels.

(ii) Occurrence of Chemical Elements in Bivalvia

Different species of molluscs, especially *Mytilus* sp. from the Baltic Sea have been studied for tissue concentrations of heavy metals (Phillips, 1977a, 1978, 1979; Theede et al., 1979; Tervo et al., 1980; Möller et al., 1983; Szefer and Szefer, 1985; Szefer, 1986; Szefer and Szefer, 1990, 1991; Szefer and Wołowicz, 1993; Szefer et al., 1990a; Broman et al., 1991; Falandysz, 1994; Swaileh and Adelung, 1994; Seisuma et al., 1995; Perkowska and Protasowicki, 1996; Swaileh, 1996; Ostapczuk et al., 1997a, 1997b; Pempkowiak et al., 1999; Sokołowski et al., 1999; Rainbow et al., 2000; Szefer and Kusak, 2000; Szefer et al., 2000g). Protasowicki (1991a) has reported concentrations of Cd, Cu, Hg, Pb and Zn in the soft tissue of zebra mussel (*Dreissena polymorpha*) from the River mouth area. Bauer et al. (1997) analysed the soft tissue of gastropod mollusc, i.e. *Littorina littorea* from German cost of the North Sea and the Baltic Sea for concentration of TBT. Several metals have been also determined sporadically in mollusc shells (Szefer, 1986; Szefer and Szefer, 1985, 1990; Szefer et al., 2000g) and in byssal threads (Szefer et al., 2000g).

Metals in soft tissue

Inter-species trends

Concentration data listed in Table 3.5 show interspecies dependent changes in trace element concentrations. Soft tissue and shell of *Macoma balthica* were characterised by the highest levels of Zn and Cu while *Mya arenaria* from the same sampling sites contained the greatest quantities of Fe and Mn in the soft and hard parts. Cd and Ni were accumulated to the greatest extent by soft tissue of *Mytilus edulis* and *Cardium glaucum*, respectively. Bryan (1980) reported also significant bioavilability of these heavy metals to the same bivalve species from East Looe Estuary, contaminated with Ag and Pb. Therefore it is suggested that these molluscs as non-regulators incorporate quickly the trace metal levels from the environment because of their elevated biological tolerance and/or limited elimination with respect to the trace elements.

TABLE 3.5.

Concentrations of chemical elements ($\mu g\ g^{-1}$dry wt.) in soft tissue of mussels from the Baltic Sea and other northern areas

Region	Sampling date	Length (mm)	N	Ag	Al	As	Ba	Ca	Cd	Co	Cr	References
Blue = Common mussel (*Mytilus edulis trossulus*)												
Western Baltic	1973	40–60	13 (260) –	0.3		4.1	4.5	1.8*	4	0.41	3.3	Karbe et al., 1977,
			66 (1320)	0.06–1.2		3.6–5.0	2.0–9.2	1.09–2.98	2.2–7.7	0.10–1.2	0.83–21	Schnier et al., 1978
Western part of Baltic	1975–76	16.6–83.6	47 (429)						5.61			Theede et al., 1979
									1.40–34.1			
Western part of Baltic		30–60	19 (50)						3.65			Perkowska and Protasowicki, 1996
									0.84–6.06			
Western Baltic	1991							7.6*	1.91	0.71		Struck et al., 1997
Scandinavian waters	1976		54						2.67			Phillips, 1977a, 1978
									0.4–12.9			
Öresund												Phillips, 1979
Sweden	1977		9						2.4			
									0.6–7.6			
Denmark	1977		8						2.1			
									0.8–5.0			
Mid-Sound	1977		2						2.1			
									0.9–3.3			
Southern Baltic	1977–79								1.1–1.65			Szefer (unpublished data)
Gulf of Gdansk	1981	20–55	23 (627)					6.1*	6.2±0.7	1.5±0.3		Szefer and Szefer, 1985
								2.4–19.4	2.0–10.7	0.9–4.6		
Gulf of Gdansk	1987		6 (58)	< 0.5				2.19±0.93*	7.14±0.79	2.8±0.5		Szefer and Szefer, 1990
								1.27–3.11	4.97–9.54	1.7–4.4		
Gulf of Gdansk	1991		15	2.74±1.78	690±380			9.83±7.00*	9.03±3.53	2.49±0.78	6.96 5.78	Szefer and Kusak, 2000
				ND–4.90	140–1330			3.33–21.3	4.29–15.7	2.61–0.73	0.73–16.4	
Gulf of Gdansk	1997	17.7–39.5	64 (1300)	1.07±0.54					1.92±0.63	1.73±3.80	2.05 0.30	Szefer et al., 2000g
				0.21–2.18					1.18–3.50	ND–3.83	1.57–2.56	
	1998		61*						4.66			Rainbow et al., 2000
									2.89–8.33			
Pomeranian Bay	1997	16.4–41.6	94 (2110)	0.79±0.22					3.09±1.42	1.09±0.64	1.93 0.86	Szefer et al., 2000g
				0.47–1.43					ND–6.12	ND–2.33	ND–3.88	

Region	Sampling date	Length (mm)	N	Ag	Al	As	Ba	Ca	Cd	Co	Cr	References
Southern Baltic, middle part	1997	18.7–38.8	40 (826)						3.52±1.18	0.80±0.76	1.85±0.43	Szefer et al., 2000g
Northern Baltic Proper	1986		38						8.05			Broman et al., 1991
									5.0–11.5			
Southwestern Baltic	1979	40±2	32(640)	0.6		5.61		2.58*	1.13	0.63	1.12	Möller et al., 1983
				< 0.03–6.6		2.5–12.0		0.6–11.0	< 0.7–3.1	0.2–1.7	< 0.2–7.0	
Kiel Fjord	1979	10±2	(300)[a]	0.037		2.9		0.83*	1.4	0.23	0.78	Möller et al., 1983
		20±2	(70)[a]	0.067		3		1.80*	0.7	0.4	0.56	
		40±2	(20)[a]	0.063		2.5		0.59*	0.7	0.28	0.52	
		60±2	(20)[a]	0.089		6.7		6.50*	0.7	0.61	1	
North Sea	1973			0.114		9.74	5.84		1.83	0.56	1.84	Karbe et al., 1977,
				0.024–0.45		6.8–14.0	0.79–26.0		1.0–2.5	0.20–2.4	0.36–17.0	Schnier et al., 1978
North Sea	1975–76		11 (100)						2.48			Theede et al., 1979
									1.10–2.92			
North Sea, German coast	1991					13.5		4.8*	1.35	0.88		Struck et al., 1997
						12						
	1986–94					8.0–20						Ostapczuk et al., 1997a
Norwegian coast,									4.8–140			Stenner and Nickless, 1974
Hardangerfjord												
Norwegian coast,	1975							2.59*	33.6	0.46		Julshamn, 1981a, 1981e
Hardangerfjord								1.2–5.2	20.0–51.0	0.29–1.0		
Norwegian coast,	1972	40–50	8 (120)	2.43					2.5		19.9	Lande, 1977
Trondheimsfjord				1.0–6.0					< 1.0–5.0		4.0–49.0	
	1973	40–50	1 (15)	7							2	Lande, 1977
Western Norway	1993								0.35–1.4			Andersen et al., 1996

* – mg g^{-1} dry wt.

[a] – Weight adjusted.

TABLE 3.5. – continued

Region	Sampling date	Length (mm)	N	Cu	Fe	Hg	K	Mg	Mn	N	Na	Ni	References
Blue mussel (*Mytilus edulis trossulus*)													
Western Baltic	1973	40–60	13 (260) –		151	0.14						2.14	Karbe et al., 1977,
			66 (1320)		58–426	0.03–0.42						0.57–9.54	Schnier et al., 1978
Western part of Baltic		30–60	19 (50)	23.5									Perkowska and Protasowicki, 1996
				8.3–67.6									
Western Baltic	1991			17.8	692	0.008	9.36*	5.85*	47.2		38.4*	4.46	Struck et al., 1997
Scandinavian waters	1976		54		167				21.9				Phillips, 1978
					14–1367				4.9–91.7				
Öresund													Phillips, 1979
Sweden	1977		9		176								
					39–310								
Denmark	1977		8		131								
					61–346								
Mid-Sound	1977		2										
					79–182								
Southern Baltic	1988–89			15.6–23.2								2.5–6.6	Szefer (upublished data)
Gulf of Gdansk	1981	20–55	23 (627)	3.5	1200			2.7*	80±11			9.5±0.9	Szefer and Szefer, 1985
				2.8–5.2	280–1560			2.3–3.3	40–170			5.3–15.8	
Gulf of Gdansk	1987		6 (58)	3.3±1.1	1260±260		10.25±0.40*	2.28±0.19*	73.4±11.1		25.3±3.8*	5.1±0.6	Szefer and Szefer, 1990
							9.02–12.21	1.78–3.00	48.1–117		15.4–32.9	4.2–7.3	
Gulf of Gdansk	1991		15	13.1±6.97	560±330		17.04±7.39*	2.06±1.09*	41.4±16.9		16.39±8.13*	13.8±8.07	Szefer and Kusak, 2000
				5.86–26.3	210–1220		8.18–32.2	0.75–3.55	7.60–63.2		7.95–26.50	5.35–27.3	
Puck Bay	1987	15–30		8.9–39.9	140–940		7.0–12.4*	1.1–2.9*	6.6–99.6		2.1–32.3*	ND–26.1	Szefer et al., 1994a
Gulf of Gdansk	1988		9 (630)			0.11							Falandysz, 1994
				6.07–42.9	209–1620	0.13–0.15							
Gulf of Gdansk	1997	17.7–39.5	64 (1300)	7.14±1.21	558±385	0.10±0.06			33.8±14.8			1.41±0.59	Szefer et al., 2000g
				5.33–9.77	155–1660	0.04–0.23			10.7–61.3			0.56–2.23	
	1998		61[b]	10.9	809				32.9			5.65	Rainbow et al., 2000
				8.18–14.6	486–1881				19.0–61.5			3.01–12.6	
Pomeranian Bay	1997	16.4–41.6	94 (2110)	9.99±4.7	270±86.4	0.11±0.04			28.8±9.23			2.82±0.84	Szefer et al., 2000g

Region	Sampling date	Length (mm)	N	Cu	Fe	Hg	K	Mg	Mn	N	Na	Ni	References
				5.95–29.8	148–473	0.03–1.19			13.1–48.8			1.15–5.0	
Southern Baltic,	1997	18.7–38.8	40 (826)	8.38±1.19	322±133	0.08±0.04			48.8±56.8			2.81±0.40	Szefer et al., 2000g
middle part				6.89–10.9	220–650	0.03–0.13			11.1–169			2.19–3.38	
Northern Baltic Proper	1986		38										Broman et al., 1991
Southwestern Baltic	1979	40±2	32(640)		344	0.15	10.8*				11.6*	2.2	Möller et al., 1983
					77–817	0.03–0.44	< 4–42				3.0–43	< 1–11	
Kiel Fjord	1979	10±2	300		126	0.09[a]	< 4*				2.8*	< 1	Möller et al., 1983
		20±2	70		237	0.1[a]	< 4*				3.2*	< 1	
		40±2	20		266	0.09[a]	4.7*				3.2*	< 1	
		60±2	20		320	0.12[a]	9.3*				9.4*	2	
North Sea	1973				120	0.32						1.97	Karbe et al., 1977
					41–707	0.10–1.4						0.46–8.85	Schnier et al., 1978
North Sea, German coast	1991					0.028	10.2*	8.26*	29.7		57.9*	2.41	Struck et al., 1997
German coastal waters	1986–94					0.033–0.054							Ostapczuk, 1997a, 1997b
				6.99	532	0.006–0.01[a]							
Norwegian coast				3.0–22.0									Stenner and Nickless, 1974
Hardangerfjord													
Norwegian coast,	1975			7.49	98	0.93	5.79*	0.53*	11.2		3.67*		Julshamn, 1981a, 1981e
Hardangerfjord				5.2–10.0	71–130	0.38–2.0	2.7–9.1	0.34–0.85	6.0–19		2.0–7.4		
Norwegian coast,	1972	40–50	8 (120)	24.3	963							15.1	Lande, 1977
Trondheimsfjord				5.0–88.0	112–1620							6.0–43	
	1973	40–50	1 (15)	7	31	0.1						9	Lande, 1977
Western Norway	1993			4.5–18.0		0.045–0.29							Andersen et al., 1996

* – mg g^{-1} dry wt.
[a] – Me–Hg.
[b] – Weight adjusted.

TABLE 3.5 – continued

Region	Sampling date	Length (mm)	N	P	Pb	S	Se	Sn	Sr	Zn	References
Blue mussel (*Mytilus edulis trossulus*)											
Western Baltic	1973	40–60	13 (260)[a] –				2.8	4.9	71	169	Karbe et al., 1977,
			66 (1320)				1.3–5.1	2.4–7.1	28–100	52–575	Schnier et al., 1978
Western part of Baltic		30–60	19 (50)		21.2					211	Perkowska and Protasowicki, 1996
					7.0–45.8					31.8–367	
Western Baltic	1991			9.0*	2.07				52.6	139	Struck et al., 1997
Scandinavian waters	1976		54		54					104	Phillips, 1977a, 1978
					3.0–264					14–460	
Öresund											Phillips, 1979
Sweden	1977		9		92.3					175	
					34–202					45–396	
Denmark	1977		8		72					132	
					20–125					81–211	
Mid-Sound	1977		2		90.5					120	
					65–116					65–175	
Southern Baltic	1977–79				0.49–0.80					63.6–133	Szefer (unpublished data)
					< 2.0–4.5					79–255	
Gulf of Gdansk	1981	20–55	23 (627)		10±2					328±17	Szefer and Szefer, 1985
					4.0–20					210–600	
Gulf of Gdansk	1987		6 (58)		2.5±0.4					125±14	Szefer and Szefer, 1990
					1.3–3.5					91–167	
Gulf of Gdansk	1991		15		17.3±17.9					200±54.1	Szefer and Kusak, 2000
					4.07–56.2					126–267	
					ND–8.7						Szefer et al., 1994a
Gulf of Gdansk	1988		9 (630)							50–670	Falandysz, 1994
Gulf of Gdansk	1997	17.7–39.5	64 (1300)		1.16±0.87					126±20.2	Szefer et al., 2000g
					0.23–3.0					98.8–176	
	1998		61[c]		16.3					160	Rainbow et al., 2000
					7.63–36					76.8–276	
Pomeranian Bay	1997	16.4–41.6	94 (2110)		0.93±0.52					159±26.4	Szefer et al., 2000g
					ND–2.0					94.8–205	

Southern Baltic,	1997	18.7–38.8	40 (826)		0.91±0.61					140±25.2	Szefer et al., 2000g
middle part					0.18–2.35					100–189	
Northern Baltic Proper	1986		38							142	Broman et al., 1991
										119–174	
Southwestern Baltic	1979	40±2	32(640)				2.2	3.44(N=14)		161	Möller et al., 1983
							1.3–2.7	1.5–8.2		82–308	
Kiel Fjord	1979	10±2	(300)[a]				1.2	2.9		177	Möller et al., 1983
		20±2	(70)[a]				2	3.5		132	
		40±2	(20)[a]				1.6	3.2		117	
		60±2	(20)[a]				2.7	5		125	
North Sea	1973						3.88	3	46.6	89.8	Karbe et al., 1977
							1.5–9.9	1.3–5.0	18–98	38–200	Schnier et al., 1978
North Sea, German coast	1991			6.5*	2.07				61.3	87.6	Struck et al., 1997
German coastal waters	1986–94			10*		20*	2.8–3.8[b]				Ostapczuk et al., 1997
Norwegian coast					15–3100					170–2370	Stenner and Nickless, 1974
Hardangerfjord											
Norwegian coast,	1975				2–130					1300–3300	Julshamn, 1981a, 1981e
Hardangerfjord											
Norwegian coast,	1972	40–50	8 (120)							169	Lande, 1977
Trondheimsfjord										85–359	
	1973	40–50	1 (15)							22	Lande, 1977
Western Norway	1993				3.5–14.0					165–350	Andersen et al., 1996

* – mg g^{-1} dry wt.
[a] – No. of specimens in parentheses.
[b] – Recalculated from diagram.
[c] – Weight adjusted.

TABLE 3.5. – continued

Region	Sampling date	Length (mm)	N	Au	Br	Rb	Cs	Sb	Ta	Zr	References
Blue mussel (*Mytilus edulis*)											
Western Baltic	1973	40–60	13 (260) –	0.0063		6.6	0.013		0.053	6.2	Karbe et al., 1977,
			66 (1320)	0.002–0.0019		3.4–14.0	0.001–0.049		0.018–0.21	3.2–43.0	Schnier et al., 1978
Southwestern Baltic	1979	40±2	32(640)	0.0081	126	4.86	0.023	1.97 (N=12)	0.0034		Möller et al., 1983
				< 0.003–0.032	37–266	3.5–6.1	< 0.01–0.056	0.8–3.2	< 0.001–0.014		
Kiel Fjord	1979	10±2	(300)[a]	0.037		3.2	< 0.01				Möller et al., 1983
		20±2	(70)[a]	0.067		4.3	0.01				
		40±2	(20)[a]	0.063		3.8	< 0.01				
		60±2	(20)[a]	0.089		4.1	0.02				
North Sea	1973					5.84	0.017		0.021	3.9	Karbe et al., 1977
North Sea, German coast						3.5–12.0	0.004–0.13		0.004–0.33	0.62–42.0	

Region	Sampling date	Length (mm)	N	Sc	La	Hf	Eu	Ce	Tb	Yb	References
Blue mussel (*Mytilus edulis*)											
Western Baltic	1973	40–60	13 (260) –	0.021		0.018	0.0027		0.0034	0.015	Karbe et al., 1977
			66 (1320)	0.005–0.01		0.002–0.10	0.0007–0.11		0.0006–0.014	0.005–0.61	Schnier et al., 1978
Southwestern Baltic	1979	40±2	32(640)		0.27	0.071	0.0091	0.44	0.0055		Möller et al., 1983
					< 0.1–0.7	< 0.01–0.50	< 0.001–0.023	< 0.2–1.3	< 0.003–0.015		
Kiel Fjord	1979	10±2	(300)[a]	< 0.01	< 0.1		0.004	< 0.2	< 0.0018		Möller et al., 1983
		20±2	(70)[a]	< 0.01	< 0.1		0.003	< 0.2	0.0019		
		40±2	(20)[a]	0.2	< 0.1		< 0.001	< 0.2	< 0.0018		
		60±2	(20)[a]	0.2	< 0.1		0.004	0.5	< 0.0018		
North Sea	1973			0.014		0.014	0.0022		0.0031	0.011	Karbe et al., 1977
North Sea, German coast				0.001–0.076		0.0006–0.1	0.0005–0.015		0.0006–0.03	0.002–0.061	

[a] – No. of specimens in parentheses.

TABLE 3.5. – continued

Species Region	Sampling date	Length (mm)	N	Ag	Al	Ca	Cd	Co	Cr	Cu	Fe	References
Little macoma (*Macoma balthica*)												
Southern Baltic			20	3.1–12.9							0.46–6.0*	Szefer (unpublished)
Gulf of Gdansk	1981		11 (433)'			3.7±1.5*	0.40±0.08	1.4±0.4		17.9±4.9	0.45±0.018*	Szefer, 1986
						1.2–5.7	0.29–0.47	1.0–1.7		14.5–32.9	0.30–0.49	
Gulf of Gdansk	1987		24 (448)			2.35*	2.03	7.43		60.1	1.25*	Szefer and Szefer, 1990
				.5–5.2		0.89–5.25	0.97–2.98	1.5–25.7		144±79	0.49–3.40	
Gulf of Gdansk	1991		17	4.05±3.38	700±460	3.88±1.35*	1.76±0.78	7.84±4.02	13.2±11.0	54±217	1.60±0.69*	Szefer and Kusak, 2000
				0.31–7.86	245–1340	2.00–5.14	0.72–2.62	5.50–13.9	4.68–28.4	23.0–570	0.84–2.48	
Puck Bay	1987	32–40				5.1*	ND			110	0.27*	Szefer et al., 1994a
Slupsk Furrow	1993	8.65±2.72	6 (114)				0.54±0.33	1.03±0.46	4.74±2.29	19.4±6.42	0.52±0.14*	Ikuta and Szefer, 2000
Gulf of Finland	1979	7.2					0.7			44.9		Tervo et al., 1994
Helsinki	Pre-1991		63				0.98±0.47			160±65.2	0.60±0.24*	Ikuta and Szefer, 2000
Tvärminne	Pre-1991		59				3.37±0.91			297±101	1.28±0.53*	
Dutch coast, Westerschelde Estuary							0.19–1.13			16.8–32.1	0.51–1.98*	Bordin et al., 1992
UK estuarine areas				0.3–122	11.0–46.0		0.2–9.4	0.7–6.8	0.6–16.3	32–224	0.21–2.64*	Bryan et al., 1985
Long clam (*Mya arenaria*)												
Gulf of Gdansk	1981	10.0–55	15			10.2*	2.25	3.3		8.5	5.20*	Szefer, 1986
						7.5–14.6	1.0–5.1	ND–10.0		2.5–20.0	2.10–11.0	
Gulf of Gdansk	1987		5 (62)				2.28±0.28	10.1±2.9		23.8±2.0	12.6±1.30*	Szefer and Szefer, 1990
							2.01–3.70	3.7–20.7		20.5–31.8	10.3–17.5	
Gulf of Gdansk	1991		7		660±710	6.15±1.28*	2.40±0.67	4.79±3.80	3.74±1.35	16.3±5.72	1.70±1.28*	Szefer and Kusak, 2000
					51–1450	5.15–7.60	1.63–2.84	0.42–7.29	2.31–5.00	12.7–22.9	0.33–2.87	
Puck Bay	1987	25–50	3			14.7*	3.1			21.1	7.77*	Szefer et al., 1994a
						8.4–21.0	1.7–4.6			15.1–27.1	1.38–14.3	
Baltic, Polish coast			14				2.35±0.4	2.24±0.61	0.82±0.25	13.4±2.2	1.39±0.44*	Pempkowiak et al., 1999

Cockle shell (*Cardium glaucum*)												
Gulf of Gdansk	1981	8.0–23	6 (336)			19.0±3.0*	1.5±0.2	2.3±0.3		6.6±4.6	0.79±0.15*	Szefer and Szefer, 1985
						18–19	1.1–1.8	1.8–3.3		1.8–19.9	0.60–1.17	
Gulf of Gdansk	1986	10.0–20.0	12 (50)				2.54			8	0.73*	Szefer and Wołowicz, 1993
							1.4–3.7			6.7–12.6	0.52–0.92	
Gulf of Gdansk	1987		4 (45)				5.71±1.51	5.7±1.9		24.5±2.3	2.20±0.35*	Szefer and Szefer, 1990
							3.04–9.90	3.3–11.4		17.4–27.5	1.70–3.25	
Gulf of Gdansk	1981		4	2.49±1.24	800±560	6.15±1.70*	4.93±1.71	3.38±1.02	24.9±15.6	14.2±6.01	1.58±1.03*	Szefer and Kusak, 2000
				1.06–3.23	110–1460	4.25–7.87	2.70–6.81	2.62–4.54	3.67–40.7	10.7–23.1	0.99–3.12	
Cockle (*Cardium edule*)												
UK estuarine areas							0.4–16.9	1.9–12.0	0.5–20.0	9.0–174	0.43–5.52*	Bryan et al., 1985
Northern astarte (*Astarte borealis*)												
Slupsk Furrow	1987		14	ND		1.88*	30.9±2.7	20.5±4.1		54.2±15.4	3.96±0.23*	Szefer and Szefer, 1990
							25.8–40.9	11.6–31.5		34.7–116	3.36–4.76	
Slupsk Furrow	1993	6.37–16.1	60 (1153)				12.9±9.56	0.82±0.98	2.59±0.99	17.7±4.18	0.52±0.44*	Ikuta and Szefer, 2000
							11.9–14.0	0.56–1.16	2.44–2.67	17.7–18.0	0.42–0.66	
Ocean quahog (*Arctica islandica*)												
Western Baltic, Kiel Bay	1992–1993	30–60	–604				0.43–0.99			10.1–18.6		Swaileh, 1996
Zebra mussel (*Dreissena polymorpha*)												
Odra mouth	1986–88		8 (50)				1.69			19.2		Protasowicki, 1991a
							1.03–2.23			10.9–24.2		

[a] – Number of specimens in parentheses.
* – mg g^{-1} dry wt.

TABLE 3.5. – continued

Region	Sampling date	Length (mm)	N	Hg	K	Mg	Mn	Na	Ni	Pb	Se	Sn	Zn	References
Little macoma (*Macoma balthica*)														
Southern Baltic			20						2.5–7.2				368–1000	Szefer (unpublished)
Gulf of Gdansk	1981		11 (433)*				12.0±3.0		4.0±0.4	6.7±0.9			510±65	Szefer, 1986
							7.0–14.0		2.5–4.5	3.9–8.7			340–620	
Gulf of Gdansk	1987		24 (448)		6.63*	0.91*	47.7	9.73*	6.25	5.35			475	Szefer and Szefer, 1990
					1.61–8.96	0.33–1.46	17.5–124	5.0–24.9	2.2–16.1	1.6–12.2			174–900	
Puck Bay	1987	32–40			12.0*	2.7*	11	29.3*	5.1	2.5			600	
Gulf of Gdansk	1991		17		18.8±5.53*	2.09±1.17*	53.8±37.5	9.29±4.31*	15.1±7.84	10.6±6.26			790±360	Szefer and Kusak, 2000
					14.3–26.5	1.05–3.63	24.6–110	6.12–20.0	6.79–30.0	0.18–21.1			380–1550	
Slupsk Furrow	1993	8.65±2.72	6 (114)				553±167		1.53±0.62	2.18±1.26			313±81.2	Ikuta and Szefer, 2000
Gulf of Finland	1979	7.2		0.09						3.46			451	Tervo et al., 1994
Helsinki	Pre-1991		63				22.8±11.5						450±181	Ikuta and Szefer, 2000
Tvärminne	Pre-1991		59				40.6±23.2						1070±160	
Dutch coast, Westerschelde Estuary	1990–91												377–692	Bordin et al., 1992
UK estuarine areas				0.12–1.03			8.0–356		0.3–12.7	2.0–36.0	3	0.48–1.2	365–1510	Bryan et al., 1985
Long clam (*Mya arenaria*)														
Gulf of Gdansk	1981	10.0–55	15			3.05*	710		16	19			145	Szefer, 1986
						2.6–4.1	340–1800		5.6–49.0	13.0–40.0			110–170	
Gulf of Gdansk			5 (62)		6.77±0.47*	1.56±0.24*	245±64		9.9±2.0	6.2±2.0			212±43	Szefer and Szefer, 1990
					5.02–6.74	1.00–2.33	149–487		5.5–17.3	3.5–8.8			130–318	
Puck Bay	1987	25–50	3		8.32*	2.20*	273	12.8*	5.83	4.4			317	Szefer et al., 1994a
					6.0–9.5	1.7–2.8	48–410	5.3–26.3	5.2–6.4	2.1–8.1			70–470	
Gulf of Gdansk	1991		7		16.6±1.92*	1.95±0.29*	15.5±9.0	12.6±2.52*	7.43±2.81	5.5±2.7			112±36.7	Szefer and Kusak, 2000
					14.4–18.1	1.65–2.24	5.32–21.0	9.76–14.5	5.44–9.42	3.14–8.5			79.9–150	
Baltic, Polish coast			14				70.1±28.6		4.12±0.82	5.22±2.16			269±58	Pempkowiak et al., 1999

Cockle shell (*Cardium glaucum*)														
Gulf of Gdansk	1981	8.0–23	6 (336)			2.3±0.1*	32.0±3.0		60.0±9.0	12.5±1.4			210±160	Szefer and Szefer, 1985
						2.1–2.4	22.0–43.0		46–74	7.9–14.9			40–550	
Gulf of Gdansk	1986	1.0–2.0*	12 (50)				13.4		72.6				92.9	Szefer and Wołowicz, 1993
							8.2–23.5		60.6–105				80.3–120	
Gulf of Gdansk	1987		4 (45)			2.06±0.16*	60.0±5.3		39.6±7.0	7.8±0.1			98.8±8.0	Szefer and Szefer, 1990
						1.65–2.41	47.4–71.8		30.0–59.8	7.6–8.0			83.0–114	
Gulf of Gdansk	1991		4		8.10±4.53*	1.90±0.53*	60.2±23.3	13.6±7.74*	138±24.4	8.74±5.31			114±11.0	Szefer and Kusak, 2000
					1.59–11.3	1.10–2.42	38.5–86.5	7.74–24.9	110–162	5.29–16.5			107–130	
Cockle (*Cardium edule)*														
UK estuarine areas				0.26–0.86			5.0–317		22–174	0.4–371			46–309	Bryan et al., 1985
Northern astarte (*Astarte borealis)*														
Slupsk Furrow			14		12.2±3.78*	1.09±0.15*	19.5±0.21*		21.3±2.2	ND			128±7	Szefer and Szefer, 1990
					5.70–25.3	0.94–1.53	14.7–26.7		15.7–27.7				107–148	
Slupsk Furrow	1993	6.37–16.1	60 (1153)				709±310		5.58±2.64	2.66±2.06			126±78.3	Ikuta and Szefer, 2000
							640–781		4.95–6.57	1.92–3.40			105–150	
Ocean quahog (*Arctica islandica)*														
Western Baltic, Kiel Bay	1992–93	30–60	(604)[a]							0.91–2.70			104–232	Swaileh, 1996
Zebra mussel (*Dreissena polymorpha)*														
Odra mouth	1986–88		8 (50)	0.073						7.67			179	Protasowicki, 1991a
				0.038–0.099						3.40–16.5			109–290	

[a] – Number of specimens in parentheses
* – mg g^{-1} dry wt.

Inter-tissue trends

In order to identify the tissues and organs responsible for the accumulative abilities, their analyses for selected trace elements were performed using specimens of long clam, *Mya arenaria*, from the Gulf of Gdańsk (Szefer et al., 1990). It concerned especially elevated levels of Fe and Mn in soft tissue of this Baltic clam reported by several authors (Szefer, 1986; Szefer and Szefer, 1990; Szefer and Kusak, 2000a). The mantle and syphon of *M. arenaria*, comprising ca. 25 and 28% of the clam mass, respectively, contained 96% of the total Fe and Mn content. It should be stressed that other bivalve species such as little macoma, (*Macoma balthica*), blue mussel (*Mytilus edulis*) and cockle shell (*Cerastoderma glaucum*) were characterised by an order of magnitude lower tissue levels of Mn and Fe than *Mya arenaria* collected at the same sampling sites of the Gulf of Gdańsk (Szefer, 1986; Szefer and Szefer, 1990; Szefer and Kusak, 2000). The distribution of other elements, i.e. Cd, Co, Cu, Ni, and Pb in the syphon and mantle supports the finding about a key role of these tissues in accumulation of trace elements in *M. arenaria*. The digestive system appeared to be a little less important organ than syphon and mantle in respect to accumulation of Cd, i.e. it contained 24% of the total burden of Cd in the face of 31.1 and 24.6% of Cd contributions to the syphon and mantle, respectively.

According to Swaileh and Adelung (1994) different organs of *Arctica islandica* from Kiel Bay display different capacities for accumulating selected metals. Highest metal levels occurred in the gills, followed by the kidney, digestive gland, mantle, foot, anterior adductor muscle and finally posterior adductor muscle. Metal levels appear to be associated with organ function. The gills are responsible for the water flow and are exposed to a large water volume and hence are expected to have high metal levels. The kidney, digestive gland and the mantle play a key role in filtration, digestion and secretion of the shell material, respectively, and thus contain elevated metal levels. Greater concentrations of metals detected in the foot muscle than in the anterior and posterior adductor muscles are possibly associated with contact of the foot muscle with the sediment particles (Swaileh and Adelung, 1994).

Age-dependent trends

The variations of trace element concentrations with weight or shell length in bivalvia from the coastal areas of Baltic Sea have been reported by several authors (Szefer and Szefer, 1985; Brix and Lyngby, 1985; Swaileh and Adelung, 1994; Szefer and Kusak, 2000). The effect of mussel size upon both the concentrations and contents of Cd, Cr, Cu, Hg, Pb and Zn in the soft tissues of *M. edulis* in the Limfjord, Denmark was investigated in detail by Brix and Lyngby (1985). All the above heavy metals significantly correlated with the mussel size (Fig. 3.10). The concentrations of tissue Hg and Pb increased significantly with size. The levels of Cd, Cu and Zn in the soft tissue were independent of mussel size.

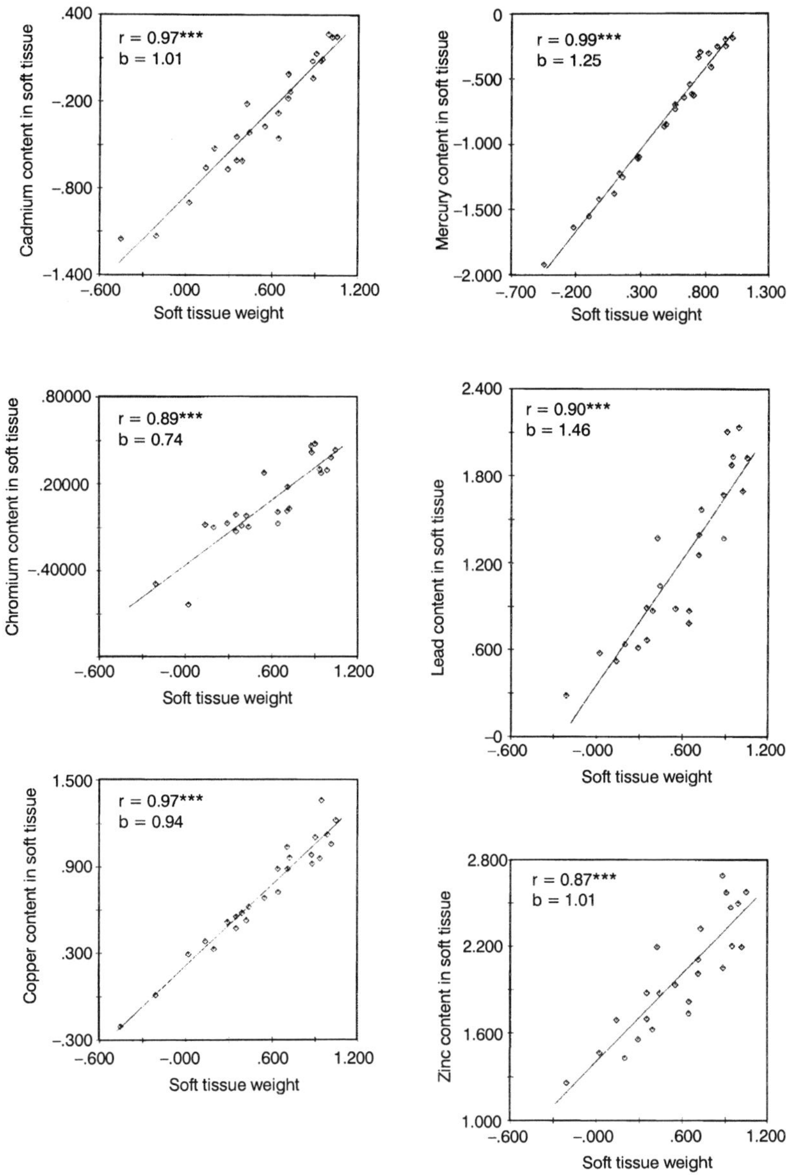

Fig. 3.10. Relationships between trace elements in soft tissue of *Mytilus edulis* from the Limfjord, Denmark. After Brix and Lyngby (1985); modified.

According to Brix and Lyngby (1985) a positive correlation between metal concentrations and size may be assigned to:

- growth dilution, i.e. when tissue increment is faster than metal accumulation. This is frequently observed for small individuals when the growth rate of younger specimens nearly always exceeds that of older ones,

- a net uptake (bioaccumulation) of trace elements throughout the life-time of the mussel.

A negative correlation between metal concentrations and size is then detected when the trace element uptake by smaller individuals is more rapid than uptake by large individuals. An example of such relationship is the distribution pattern of Cr in the soft tissue of *M. edulis* from the Limfjord, Denmark, showing negative trend with the size (Brix and Lyngby, 1985).

Accumulated trace elements in *Mytilus* may be stored as low-molecular weight protein metalloproteins, i.e. metallothioneine or in membrane–limited vesicles and thereby be detoxified. These processes are induced by exposure to high levels and may be attributed to the accumulation of toxic elements with increasing age (George, 1980; Brix and Lyngby, 1985; Roesijadi et al., 1982).

According to Swaileh and Adelung (1994) smaller individuals of *Arctica islandica* from Kiel Bay have higher levels of Cu and Zn while larger individuals have higher levels of Cd and Pb. Based on statistical data it is suggested that Cd and Cu appear to be affected by maturation.

Spatial trends

To characterise the region-dependent variations, the concentration data were compared for molluscs *M. edulis*, *Macoma balthica*, *Mya arenaria* and *Cardium glaucum* taken in the same period but from different sampling sites of the southern Baltic (Phillips, 1977a, 1978, 1979; Theede et al., 1979; Broman et al., 1991; Szefer and Kusak, 2000; Szefer et al., 2000g). Tissue levels of several trace metals depend on various environmental parameters such as, e.g. salinity and temperature of water, contents of organic matter, geochemical composition of suspended matter and bottom sediments as well as on anthropogenic impact.

It is well documented that salinity of waters is an important factor influencing concentrations of selected metals in the soft tissue of *M. edulis* in the Baltic Sea. According to Karbe et al. (1977) and Struck et al. (1997) the levels of Hg and As are lower in brackish waters of the Baltic Sea as compared to those in more saline waters of the North Sea. In contrast, the Ag and Zn levels are higher in the Baltic Sea. Statistical multivariate analysis; e.g. factor analysis, cluster analysis and discriminant analysis (Struck et al., 1997) demonstrated influence of salinity on the uptake of the trace metals and macroelements in *M. edulis*. According to Broman et al. (1991) bioavailabilty of Cd to the soft tissue of this mussel is dependent on salinity of the adjacent water. Tissue Cd levels in *M. edulis* inhabited the southern coastal waters of Sweden were up to an order of magnitude lower as compared to those detected in the northern area characterised by low salinity. However such relationship between Zn levels and salinity was not observed (Broman et al., 1991). Variations of content of Cd and Zn of *M edulis* as a function of the distance from Hornslandet (Cd) and Gävlebukten (Zn) are illustrated in Figure 3.11. Phillips (1977a) noted greater tissue concentrations of both Cd and Zn in Baltic *M. edulis* from low salinity waters (the Gulf of Finland, Southern Both-

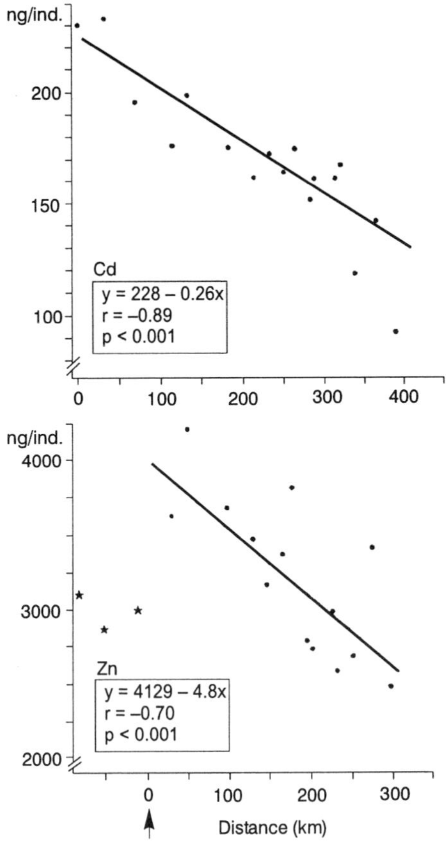

Fig.3.11. Metal content (ng per individual) of *Mytilus edulis*, as a function of the distance from Hornslandet (Cd) and Gävlebukten (Zn), respectively. The relation between Zn and the distance from Gävlebukten was tested in two directions, one north and one south. The equations of the regression lines, the correlation coefficients (r) and the probability values (p) for both Cd and Zn are given for the southern directions. After Broman et al. (1991); modified.

nian Sea, Baltic Proper) as compared to those from high-salinity regions. The lowest levels of Cd, Pb, Zn and Fe were observed in mussels taken from relatively high-salinity waters (Catgut, Eastern Skagerrak, the Oslofjord), especially from the Sound and Great Belt, which are areas of rapid salinity change were mixing of Baltic water with water from Kattegat/Skagerrak origin has place (Phillips, 1977a). Other example of the dependence of metal levels in molluscs from environmental parameters is nature of their potential food. According to Phillips (1979) the metal gradients in *M. edulis* must have been generated by two main factors, i.e. a greater availability of metals from inorganic particulates in waters of the Baltic Sea as compared to those of Kattegat and a greater metals' availabilty from phytoplankton in the Baltic waters than in Kattegat waters.

Chemical composition of mollusc inhabited specific marine environments could be extremely different from that effected by typical ambient bottom sediment as a substrata. Spectacular example of such action is unusual abilities of *Astarte borealis* in an accumulation of several trace metals in the Słupsk Furrow, southern Baltic Sea (Szefer and Szefer, 1990; Ikuta and Szefer, 2000). The elevated levels of tissue Cd, Co, Mn and Fe in this species originated from ferromanganese concretions, associated with surface sediments. Similarly enhanced tissue concentrations were observed for Mn in *Macoma balthica* from this region which were much greater than those reported for other Baltic subareas (Table 3.5).

Influence of anthropogenic factors on trace metals concentrations in the soft tissue of *M. edulis* in some areas of the Baltic Sea and surrounding areas was reported by several authors (Phillips, 1977a, 1977b, 1979; Theede et al., 1979; Ostapczuk et al., 1997a, 1997b). For example, the mussels from polluted areas of the Baltic Sea are characterised by elevated tissue levels of Cd (20–40 $\mu g\ g^{-1}$ dry wt. in the innermost part of the Kiel Fjord in the Kiel Bay), Pb and Fe (210–264 μg Pb g^{-1} dry wt. and 510–1367 μg Fe g^{-1} dry wt. in industrial areas of Rånö and Oxelösund in the vicinity of ironworks). Elevated concentrations of Ba, Fe, Hg, Mn, Pb and Se in *M. edulis* from Eckwarderhörne, in comparison with the mollusc from Königshafen, may reflect the industrial pollution of the River Weser basin and the coastal region of Wilhelmshaven (Ostapczuk et al., 1997a).

Szefer et al. (2000g) have reported spatial differences in concentrations of Ag, Cd, Co, Cr, Cu, Fe, Hg, Mn, Ni, Pb and Zn in *M. edulis* from the southern Baltic. From Table 3.5 results that soft tissue of *M. edulis trossulus* from the Pomeranian Bay contained the highest levels of Zn, Ni, Cu and Hg. Specimens inhabited the Gulf of Gdańsk and middle part of a southern Baltic were characterised by the greatest concentrations of Ag, Co, Cr, Fe, Pb, and Cd and Mn, respectively.

Based on the results obtained in the present study and earlier published data for other geographical regions (Bryan, 1984; Bryan and Langston, 1992; Bryan et al., 1985; Cossa, 1988; Fowler, 1990; Anderlini, 1992; Szefer et al., 1997a, Szefer et al., 1998b) including numerous studies have been published in the U.S., e.g., both NOAA's and EPA's mussel watch programmes (NOAA, 1989; Lauenstein and Dolvin, 1992; Lauenstein et al., 1990), it can be concluded that southern Baltic is not one of the extremely polluted areas reported to date.

Tributyltin (TBT) results from the German Survey in 1994/1995 comprising 19 sampling sites clearly exhibited the highest TBT levels in a snail perwinkle, *Littorina littorea*, from the marina at Kiel/Schilksee (above 2.8 μg TBT–Sn g^{-1} dry wt.) on the Baltic coast (Bauer et al., 1997). This area was one of the most contaminated areas investigated in the survey.

Temporal trends

Although the statistical analysis was applied in an assessment of temporal trends of several trace metals in the soft tissue of *M. edulis*, changes of Hg levels have been only discussed (Harms, 1996). It is reported that tissue Hg and As concentrations in this species collected during 1980–94 from the Kattegat and the Swedish west coast and between 1986 and 1994 from the German Bay were approximately constant (Harms, 1996; Ostapczuk et al., 1997a). Seasonal variations in the levels of Cd, Cu, Pb and Zn in soft tissue of *Arctica islandica* from Kiel Bay, Western Baltic were studied by Swaileh (1996). Copper and Zn exhibited maximum values during the summer months, when the dry soft tissue weight was reaching its the highest values. The opposite trend was observed for toxic metals, i.e. Cd and Pb showing their maximum tissue values in the winter when the dry soft tissue weight of Baltic *A. islandica* had minimum values (Fig. 3.12).

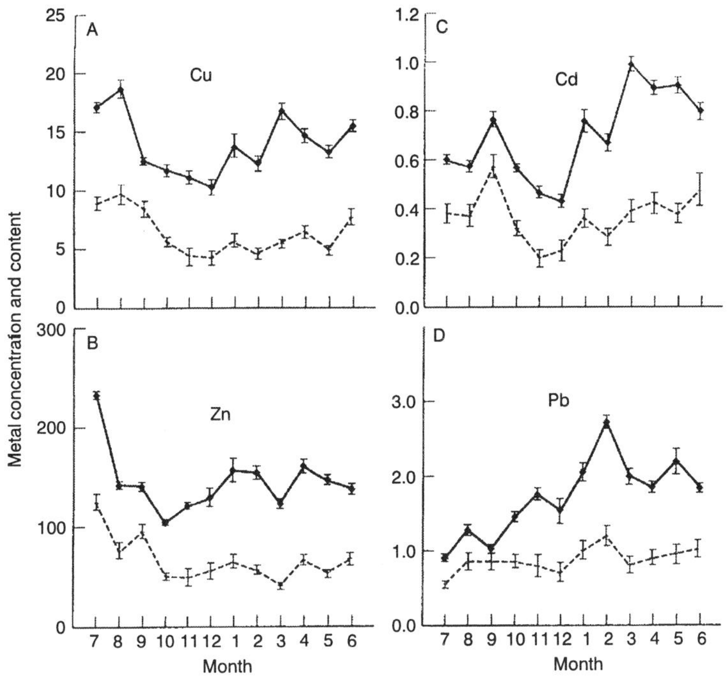

Fig. 3.12. Monthly profiles for the concentrations in $\mu g\ g^{-1}$ (solid lines) and contents in μg (broken lines) of Cu, Cd, Pb and Zn in *Arctica islandica* samples (shell length 30-60 mm) collected from Kiel Bay from July, 1992 to June, 1993. After Swaileh (1996); modified.

Metals in shells

Several metals have been sporadically determined in shells of Baltic mollusc (Brix and Lyngby, 1985; Szefer and Szefer, 1985, 1990; Szefer, 1986; Ikuta and Szefer, 2000; Szefer et al., 2000g). Shells of *M. edulis trossulus* from the Gulf of Gdańsk (Table 3.6) contained the highest levels of Ag (up to 3.34 $\mu g\ g^{-1}$), Mn (69.4±31 $\mu g\ g^{-1}$) and Fe (81.6±65.4 $\mu g\ g^{-1}$) which are suspected to be natural in origin. Since there is the lack of available information on shell metals for the adjacent regions to the south-western Baltic, concentration data related to even remote geographical zones are also listed in Table 3.6. It should be emphasised that this data matrix is unique and valuable from analytical point of view because it is free from any potential errors connected with using different procedures. Specimens of molluscs Mytilidae from the southern Baltic as well as from other temperate, subtropical and tropical areas were sampled, processed and analysed for metal concentrations with the participation of the same person and using the same apparatuses served by the same scientific staff (Szefer et al., 1997a, 1997b, 1998a, 1998b, 1999a, 2000g; Szefer and Nicholson, 2000). As can be seen in Table 3.6 shell concentrations of Fe and Mn in *Mytilus* from the Sepetiba Bay (Rio de Janeiro) and Öxelosund (Swedish coast) as well as shell concentrations of Cu in *Mytilus* from Saganoseki area (Japan) are characterised by maximum values. Elevated shell levels were also detected for Zn in *Mytilus* from the Gulf of Gdańsk (up to 20 $\mu g\ g^{-1}$), Dutch estuaries (up to 31.5 $\mu g\ g^{-1}$) and the Sepetiba Bay (Rio de Janeiro) (up to 13.1 $\mu g\ g^{-1}$) possibly corresponding to its anthropogenic provenance (Table 3.6).

Unique phenomenon was observed in the case of shell trace metals in *A. borealis* inhabited the Słupsk Furrow area where numerous ferromanganese nodules accompanied this species (Szefer and Szefer, 1990; Ikuta and Szefer, 2000). Significantly enhanced levels of shell Co, Fe, Ni, Pb, Zn and especially Mn were found in *A. borealis* collected in this region (Table 3.6). Special attention was paid to isolate shell coatings for further analysis which chemical composition was similar to that of ferromanganese nodules. Additionally periostracum was assayed in order to determine microdistribution of the metals studied in shells of *A. borealis* (Ikuta and Szefer, 2000). It is noteworthy that this organic microlayer was highly enriched in Co, Cr, Cu and Pb in respect to whole shell material while Zn levels were very low, i.e. below the limit of the method used. It is important to note that ferromanganese coatings occurred on the shells of living specimens *of A. borealis* which were generally 0.1 mm thick or less and thicker on the posterior halves of the shell than on the anterior halves. Having in mind that *A. borealis* belongs to one of hemi-endobenthos, the anterior halves are embedded vertically in the bottom sediments as a substrata whereas the posterior halves are exposed to seawater. Outer surfaces of *M. balthica* shells, on the other hand, were completely free of ferromanganese coatings since these shells are totally embedded in the substrata (Szefer et al., 1998c).

According to Brix and Lyngby (1985) the concentrations of Cd, Cr, Hg and Zn in the shells of *M. edulis* from the Limfjord, Danish waters, decreased significantly with shell weight while their contents as well as those of Cu and Pb exhibited positive trends with increasing of size, similarly to the distribution patterns of all these metals in the soft tissue.

Metals in byssal threads

There is very poor knowledge about distribution of trace metals in mussel byssus in the Baltic Sea except data reported by Szefer et al. (2000g). Mean levels of Ag, Cr and Fe in byssi threads reached maximum values in the Gulf of Gdańsk;

TABLE 3.6.

Concentrations of chemical elements ($\mu g\ g^{-1}$ dry wt.) in shells of molluscs from the Baltic Sea and other areas

Species Region	Sampling date	Length (mm)	N	Ag	Al	As	Ca	Cd	Co	Cr	Cu	Fe	References
Blue mussel (*Mytilus edulis*)													
Southern Baltic Sea													
Gulf of Gdansk	1981						350*	0.017±0.008	0.37±0.008			2800±40	Szefer and Szefer, 1985
							310–380	0.002–0.047	< 0.006–0.92			430–8100	
								< 0.10	< 0.10		3.2±0.2	50.0±10.0	Szefer and Szefer, 1990
											1.9–4.7	20.0–80.0	
	1991		15	1.34±1.28	62±42		490±113*	1.06±0.47	5.57±1.61	2.09±2.97	7.07±12.1	289±200	Szefer and Kusak, 2000
				< 0.1–3.34	40–134		287–570	0.53–1.62	3.65–7.55	< 0.10–6.49	1.04–34.5	16.4–540	
	1997		21								6.62±0.82	81.6	Szefer et al., 2000g
											5.05–8.25	12.0–317	
Pomeranian Bay	1997		36								6.53±0.73	18.3±9.95	
											5.25–7.90	2.60–38.0	
Slupsk Bank	1997		15								6.08±0.64	13.8±7.45	
											4.70–6.95	2.1–31.8	
Northern Baltic													
Asko	1996										14.2	24.1	Szefer et al., 2000g
Oxelosund	1996										14.4	> 440	
Norwegian coast	1972		5 (75)	3.5						1.8	5.4	44.6	Lande, 1977
Trondheimsfjord				2.0–4.0				< 1.0–3.0		1.0–2.0	4.0–8.0	17–55	
Dutch estuaries	1985–90	25–70	67			1.04–1.09		0.10–0.60		0.34–0.43	1.62–2.11		Stronkhorst, 1992
Oosterschelde	1996										11.8	14.9	Szefer et al., 2000i
Dutch coast											11.3–12.2	12.6–17.4	
Japan coast													
Urashiro	1994										11.9	2.73	Szefer et al., 2000i
											11.3–12.5	2.35–3.15	
Akamizu	1994										13.8	6.04	
											12.6–14.7	3.88–8.88	
Saganoseki	1994										26.1	6.82	
											23.2–30.0	4.77–8.43	

Species Region	Sampling date	Length (mm)	N	Ag	Al	As	Ca	Cd	Co	Cr	Cu	Fe	References
Mytilus sp.													
Brazilian coast													
Sepetiba Bay	1996										9.61	102	
											6.44–12.2	63.8–142	
U.S.A. Coast				0.012				0.029			0.96		Koide et al., 1982
				0.002–0.09				0.008–0.074			0.43–2.39		
Mytilus galloprovincialis													
Mediterranean Sea, Nice	1998										12.91	2.98	Szefer et al., 2000i
											11.85–14.23	1.85–5.54	
Spanish coast													
Vigo	1996										11.7	4	Szefer et al., 2000i
											11.70–11.70	2.46–5.54	
Pontevedra	1996										10.7	1.97	
											10.4–11.0	1.85–2.08	
Little macoma (*Macoma balthica*)													
Southern Baltic													
Gulf of Gdansk	1981							0.024±0.01			0.44±0.14	1280±110	Szefer, 1986
								0.020–0.030			0.35–0.78	910–1820	
	1987							< 0.5	< 0.5		16.6	440	Szefer and Szefer, 1990
											10.0–30.4	60–1000	
	1991		9	0.93±0.65	153		253±117*	1.45±0.35	5.64±0.92	1.5±1.14	18.0±11.0	480±320	Szefer and Kusak, 2000
				0.16–2.00	67–325		147–520	1.05–2.07	3.65–6.80	< 0.1–3.10	7.40–44.0	63.0–1200	
Slupsk Furrow	1993	4.02–11.9	6 (114)					6.03±2.41	0.23±0.17	0.17±0.06	5.86±0.76	130±32.9	Ikuta and Szefer, 2000
Long clam (*Mya arenaria*)													
Southern Baltic													
Gulf of Gdansk	1981	10.0–55	15				330*	0.12	< 0.50		< 0.30–0.50		Szefer, 1986
							320–340	0.02–0.25					
	1987		5 (62)								2.5±0.7	2210±250	Szefer and Szefer, 1990
											0.6–4.4	1300–2720	
	1991		7	0.41±0.04	68.0±7.0		220±61.6*	0.57±0.31	4.15±2.61	1.38±0.42	1.47±0.61	170±150	Szefer and Kusak, 2000
				0.38–0.43	62.9–73		171–291	0.28–0.89	2.02–7.06	1.08–1.67	1.1–2.17	57.5–340	

Cockle shell (*Cerastoderma glaucum*)													
Southern Baltic													
Gulf of Gdansk	1981	8.0–23	4 (224)				400±10*	0.015±0.005	.08			660±40	Szefer and Szefer, 1985
							390–410	0.009–0.020				410–960	
	1987		7 (45)								2.7±0.2	880±70	Szefer and Szefer, 1990
											1.9–3.5	670–1180	
	1991		7	1.17±0.70	378±210		257±77.8*	0.92±0.46	5.19±2.81	1.85±1.12	1.31±0.09	329	Szefer and Kusak, 2000
				0.31–1.62	229–395		155–287	0.46–0.82	1.27–6.17	0.57–3.31	1.22–1.42	35.0–1180	
Northern astarte (*Astarte borealis*)													
Southern Baltic													
Slupsk Furrow	1987		8 (84)					0.44±0.06	14.8±2.8		5.4±0.6	8100±700	Szefer and Szefer, 1990
								0.15–0.60	8.3–19.8		3.4–7.8	588–11900	
	1993	6.37–16.1	60 (1153)					0.38±0.09	4.99±3.90	0.61±0.15	6.33±1.38	740±215	Ikuta and Szefer, 2000
								0.37–0.39	4.07–6.24	0.56–0.64	5.91–6.73	586–905	
								0.07±0.06[a]	15.6±9.90[a]	20.2 6.66[a]	45.6±26.0[a]	1620±506[a]	
								0.04–0.11	11.4–22.9	14.9–23.4	25.0–67.7	1090–1950	
								2.76±0.41[b]	171±21.6[b]	12.8±1.20[b]	131±37.4[b]	25500±1530[b]	
Patella vulgata													
Norwegian coast	1972	40–50	7 (105)	4				1.57		1.71	3.43	48	Lande, 1977
Trondheimsfjord				4.0–5.0				< 1.0–2.0		1.0–2.0	1.0–6.0	39–64	

* – mg g^{-1}
[a] – Concentration in periostracum.
[b] – Concentration in ferromanganese deposit in the shell

TABLE 3.6. – continued

Region	Sampling date	Length (mm)	N	Hg	K	Mg	Mn	Na	Ni	Pb	Zn	References
Blue mussel (*Mytilus edulis*)												
Southern Baltic												
Gulf of Gdansk	1981					1.6*	115±13		13.3	19.9	9.7±2.6	Szefer and Szefer, 1985
						1.2–1.9	80–200		6.6–19.3	13.1–31.0	3.3–20.0	
							99.4±7.2			1.0±0.2	9.5±1.2	Szefer and Szefer, 1990
							65.0–136			< 0.5–2.0	5.1–17.7	
	1991		15		500±230	760±160	75.0±66.3	2.50±0.69*	15.3±2.40	5.30±2.69	9.9	Szefer and Kusak, 2000
					87–790	475–930	5.25–150	1.22–3.33	10.7–18.5	3.57–11.2	0.84–15.2	
Gulf of Gdansk	1997		21				64.9±31				3.76±1.04	Szefer et al., 2000g
							27.9–119				1.22–5.07	
Pomeranian Bay	1997		36				39.4±12.8				5.16±1.7	Szefer et al., 2000g
							7.49–65.1				2.49–9.46	
Slupsk Bank region	1997		15				25.5±14.6				4.78±2.29	Szefer et al., 2000g
							15.26–55.0				2.18–9.75	
Northern Baltic												
Asko	1996		1				43.9				2.71	Szefer et al., 2000i
Oxelosund	1996		1				233				6.15	
Norwegian coast	1972		5 (75)						6.6		7	Lande, 1977
Trondheimsfjord									6.0–8.0		4.0–12.0	
Dutch estuaries	1985–90	25–70	67	0.035–0.038						0.30–0.49	18.2–31.5	Stronkhorst, 1992
Japan coast												
Urashiro	1994		3				7.71				0.6	Szefer et al., 2000i
							7.4–8.11				0.43–0.85	
Akamizu	1994		4				8.83				0.99	
							8.14–9.70				0.70–1.32	
Saganoseki	1994		3				24.2				1.81	
							14.0–35.3				1.59–2.07	
Mytilus sp.												
Brazilian coast												
Sepetiba Bay	1996		3				42.1				9.47	
							27.6–70.2				4.36–13.1	
U.S.A. coast			31						0.19	0.85		Koide et al., 1982
									0.04–0.68	0.07–4.65		
Mytilus galloprovincialis												
Mediterranean Sea, Nice	1998		4				8.79				0.84	
							8.10–9.96				0.67–1.11	
Spanish coast												
Vigo	1996		2				7.81				0.65	
							7.63–7.99				0.64–0.66	
Pontevedra	1996		2				7.94				0.84	

						7.76–8.11				0.45–1.22	
Little macoma (*Macoma balthica*)											
Southern Baltic											
Gulf of Gdansk	1981	8.0–22	7			13.0±4.0		18.5±5	27.0±3.0	8.0±0.9	Szefer, 1986
						8.0–20.0		15.0–20.0	22.0–29.0	7.8–8.2	
	1987		38			21.5		ND	1.7	18.3	Szefer and Szefer, 1990
						7.5–46.8			< 0.5–4.3	6.6–36.0	
	1991		9			19.6±14.2		20.4±12.2	4.1±1.18	21±20	Szefer and Kusak, 2000
						3.40–48.1		9.70–51.5	1.89–5.30	6.0–70.1	
Slupsk Furrow	1993	4.02–11.9	6 (114)			104±36.9		1.07±0.52	0.13±0.09	8.90±2.11	Ikuta and Szefer, 2000
Long clam (*Mya arenaria*)											
Southern Baltic											
Gulf of Gdansk	1981	10.0–55	15			31		24	35.5	22.9	Szefer, 1986
						16–53		19–28	26–48	13–40	
	1987		5 (62)			52.5±7.5			1.9±0.6	13.9±1.5	Szefer and Szefer, 1990
						32.6–76.7			< 0.5–3.8	8.8–17.4	
	1991		7	280±65	330±94	36.3±31.0	3.61±1.34*	12.3±0.36	5.77±2.5	9.55±4.99	Szefer and Kusak, 2000
				230–350	250–440	5.61–67.6	2.73–5.15	12.1–12.7	3.83–8.59	3.90–13.4	
Cockle shell (*Cerastoderma* = *Cardium glaucum*)											
Southern Baltic											
Gulf of Gdansk	1981	8.0–23	4 (224)		280±100	26.0±1.0		12.8±1.5	17.7±2.7	9.9±3.8	Szefer and Szefer, 1985
					250–290	23.0–31.0		8.2–17.4	10.5–24.8	3.2–18.9	
	1987		7 (45)			30.2±2.4			0.9±0.2	9.0±0.8	Szefer and Szefer, 1990
						24.4–40.0			0.5–2.1	6.5–11.9	
			7	305±174	378±121	37.2±24.0	4.06±0.39*	15.0±6.0	4.35±0.52	5.25±3.87	Szefer and Kusak, 2000
				105–380	229–395	16.2–58.0	3.64–4.36	6.7–21.0	3.76–4.69	0.8–7.11	
Northern astarte (*Astarte borealis*)											
Southern Baltic											
Slupsk Furrow	1987		8 (84)			34100±3400		12.7±1.5	9.9±1.2	97.0±9.7	Szefer and Szefer, 1990
						18200–45000		8.4–17.4	5.3–13.4	56–131	
	1993	6.37–16.1	60 (1153)			10900±1160		14.6±3.29	2.68±0.73	47.3±8.56	Ikuta and Szefer, 2000
						8370–10900		11.7–16.0	2.42–3.01	43.3–52.7	
						332±694[a]		4.03±2.40[a]	8.61±8.87[a]	ND	
						24.5–793		ND–4.21	5.93–11.9		
						233000±7000[b]		12.6±5.14[b]	37.4±6.66[b]	967±20.5[b]	
Patella vulgata											
Norwegian coast	1972	40–50	7 (105)					4.9		9.3	Lande, 1977
Trondheimsfjord								3.0–7.0		3.0–18.0	

* – mg g^{-1}

[a] – Concentration in periostracum

[b] – Concentration in ferromanganese deposit in the shell

specimens inhabited the Słupsk Bank region contained the greatest byssi amounts of Ni, Pb and Cd (Table 3.7). The data obtained for this Baltic area (Szefer et al., 2000g) are compared to that reported for other, even remote areas of all over the world (Hamilton, 1980; Coombs and Keller, 1981; Szefer et al., 1997a, 1997b, 1998a, 1998b, 1999a, 2000i, 2000j) because of the lack in the available literature of byssi data for the Baltic Sea and the adjacent areas (Table 3.7). The levels of Ag are the highest among those reported for all of the world up to day (Table 3.7). For such elevated byssal concentrations of this element are probably responsible anthropogenic factors and/or specifically higher background levels of Ag in bottom sediments of the Polish EEZ. According to Szefer et al. (1998c) Ag is characterised by concentrations much higher than even elevated background levels in the $< 2\ \mu m$ fraction of the deeper Baltic sediments. These findings reflect the fact that Poland has the highest abundance of silver deposits per unit area of any country (Singer, 1995).

The mean Baltic values of Mn, and Fe, and Cr (Pomeranian Bay and Gulf of Gdańsk) are comparable to their highest levels observed in *Mytilus* byssus from Saganoseki (Japan) and the Sepetiba Bay, Rio de Janeiro (Brazil), respectively, which are known to be very much industrialised coastal areas (de Lacerda et al., 1983; Magalhães and Pfeiffer, 1995; Szefer et al., 1997a, 1997b, 1998a, 1998b, 1999a). It is supposed that elevated levels of byssal Mn and Fe correspond to specific geochemical composition of southern Baltic bottom sediments. According to Szefer et al. (1995a) amorphic oxyhydroxides of Fe and Mn are precipitated at the hydrological front of the Gulf of Gdańsk where mixing of Vistula River waters and brackish bay waters has place. As a result of this process, adjacent sediments, frequently inhabited by specimens of mussels, are enriched in Fe and Mn compounds which abundance could be reflected by their higher levels in byssus of *M. edulis trossulus*. In contrast to mainly natural origin of Fe and Mn in the southern Baltic samples studied, for elevated levels of these metals in byssus from Saganoseki and Rio de Janeiro (Sepetiba Bay) are exclusively responsible anthropogenic factors (Szefer et al., 1997b, 1998a). It is interesting to note that extremely great concentrations of byssi Cu (1870 $\mu g\ g^{-1}$ dry wt.) and Pb (182 $\mu g\ g^{-1}$ dry wt.) were found in Saganoseki area while *Mytilus* from Dee Aberdeen and the Sepetiba Bay, Rio de Janeiro, concentrated the greatest amounts of byssal Zn amounting to 1230 and 670 $\mu g\ g^{-1}$ dry wt., respectively (Table 3.7). Great concentrations of these metals are also suspected to be connected with industrial activity of man.

Partition of metals between the soft tissue, shells and byssus

In order to evaluate the relation between both the shell (byssus) and tissue concentrations of the metals studied, the ratio of metal concentrations in these parts of *M. edulis trossulus* were computed. The mean ratios of both the byssus metal (BTR) and shell metal (STR) to the tissue metal are presented by Szefer et al. (2000g). Coefficients of partition between shell and tissue concentrations of metals exhibit significant fluctuations depending on metal and sampling site.

TABLE 3.7.

Concentrations of chemical elements ($\mu g\ g^{-1}$ dry wt.) in byssus of *Mytilus* from the Baltic Sea and other areas

Region	Sampling date	Ag	Al	As	Au	Ba	Ca	Cd	References
Blue mussel (*Mytilus edulis trossulus*)									
Southern Baltic									
Gulf of Pomerania	1997	2.30±1.44						1.00±0.82	Szefer et al., 2000g
		0.34–5.37						0.28–3.25	
Slupsk Bank	1997	2.88±1.34						1.92±1.25	
		1.30–4.37						0.62–3.65	
Gulf of Gdansk	1997	4.29±2.50						0.81±0.35	
		1.01–8.39						0.30–1.20	
Mytilus galloprovincialis									
Mediteranean Sea, Carteau	1995							0.75±0.02	Szefer et al., 1998a
Mediteranean Sea, Nice	1998	ND–3.14						ND–0.58	Szefer et al., 1998a
Vigo, Spain coast	1996							0.13	Szefer et al., 1998a
								0.12–0.14	
Pontevedra Rias, Spain coast	1996							ND–0.20	
M. edulis									
Japan coast of Kyushu Island									
Urashiro, Japan coast	1994	< 1.36						0.40±0.26	Szefer et al., 1997b, 1999a
Akamizu	1994	1.44±0.26						0.16±0.08	
Saganoseki	1994	2.87±0.55						0.64±0.34	
Mytilus sp.									
Sepetiba Bay, Brazilian coast	1996							1.07±0.07	Szefer et al., 1998a
M. edulis									
Ravenglass Cumbria	1977	0.4	> 300	2		3.6			Hamilton, 1980
Trebarwith Cornwall	1977	0.01	47	0.1		0.3			Hamilton, 1980
Dee Aberdeen	Pre-1981	ND					1020	3.7	Coombs and Keller, 1981
Mytilus californianus									
Seattle Washington, USA	Pre-1981	0.8			0.8		837		Coombs and Keller, 1981

TABLE 3.7. – continued

Region	Sampling date	Ce	Co	Cr	Cu	F	Fe	Hg	I	References
Blue mussel (*Mytilus edulis trossulus*)										
Southern Baltic										
Gulf of Pomerania	1997		4.52±2.02	4.38±1.68	25.9±10.7		1290±730	0.094±0.08		Szefer et al., 2000g
			ND–6.37	ND–7.48	17.9–58.5		430–3270	0.031–0.30		
Slupsk Bank	1997		4.20±2.61	2.92±2.43	21.3±7.94		847±476	0.095±0.04		
			ND–7.94	ND–4.87	11.0–32.6		240–1250	0.032–0.14		
Gulf of Gdansk	1997		4.20±3.81	7.15±1.47	24.5±5.94		4690±3020	0.081±0.03		
			0.59–11.25	5.54–9.82	17.27–33.6		1260–10310	0.038–0.125		
Mytilus galloprovincialis										
Mediteranean Sea, Carteau	1995		4.12±0.14	2.58±0.47	17.6±1.88		860±94.9			Szefer et al., 1998a
Mediteranean Sea, Nice	1998		ND–2.69	ND–1.98	31.7–55.0		450–610	0.030–0.042		Szefer et al., 1998a
Vigo, Spain coast	1996		1.02	5.19	14.25		336			Szefer et al., 1998a
			0.75–1.28	4.44–5.93	13.7–14.8		284–387			
Pontevedra Rias, Spain coast	1996		0.61	3.31	10.4		126			Szefer et al., 1998a
			0.57–0.65	2.84–3.78	10.3–10.5		122–130			
M. edulis										
Japan coast of Kyushu Island										
Urashiro, Japan coast	1994		1.26±0.46	0.85±0.02	22.9±0.96		203±29.1			Szefer et al., 1997b, 1999a
Akamizu	1994		3.13±0.56	2.24±0.79	876±94.4		891±94.4			
Saganoseki	1994		12.0±0.96	5.93±1.88	1870±91.1		943±11.1			
Mytilus sp.										
Sepetiba Bay, Brazilian coast	1996		3.77±0.54	11.1±0.08	6.66±0.29		7080±530			Szefer et al., 1998a
M. edulis										
Ravenglass Cumbria	1977	0.4	0.3	2.9	104	2			6.1	Hamilton, 1980
Trebarwith Cornwall	1977	0.02	0.06	0.1	3	0.3			1.7	Hamilton, 1980
Dee Aberdeen	Pre-1981		2.8		31		61			Coombs and Keller, 1981
Mytilus californianus										
Seattle Washington, USA	Pre-1981	7.5			15					Coombs and Keller, 1981

TABLE 3.7. – continued

Region	Sampling date	La	Mn	Mo	Nb	Nd	Ni	Pb	References
Blue mussel (*Mytilus edulis trossulus*)									
Southern Baltic									
Gulf of Pomerania	1997		476±227				16.8±3.72	2.31±1.69	Szefer et al., 2000g
			139–941				8.66–22.3	0.50–5.43	
Slupsk Bank	1997		181±141				19.0±3.98	7.17±5.52	
			73.2–426				14.1–25.2	ND–12.0	
Gulf of Gdansk	1997		434±169				11.8±6.63	2.77±2.33	
			118–602				1.50–19.6	0.35–5.77	
Mytilus galloprovincialis									
Mediteranean Sea, Carteau	1995		198±12.3				12.7±1.2	4.11±0.21	Szefer et al., 1998a
Mediteranean Sea, Nice	1998		48.5–114				11.1–24.7	ND–4.76	Szefer et al., 1998a
Vigo, Spain coast	1996		29.6				4.18	4.65	Szefer et al., 1998a
			21.2–38.0				3.13–5.23	2.69–6.60	
Pontevedra Rias, Spain coast	1996		20				2.91	3.22	
			14.6–25.4				2.63–3.19	1.85–4.59	
M. edulis									
Japan coast of Kyushu Island									
Urashiro, Japan coast	1994		46.2±4.89				5.71±0.95	3.96±0.77	Szefer et al., 1997b, 1999a
Akamizu	1994		341±144				5.56±0.19	22.2±2.2	
Saganoseki	1994		997±44				22.4±0.75	182±18.7	
Mytilus sp.									
Sepetiba Bay, Brazilian coast	1996		155±10.1				5.16±1.68	7.04±0.98	Szefer et al., 1998a
M. edulis									
Ravenglass Cumbria	1977	0.3	93	0.5	0.1	0.5	66	0.1	Hamilton, 1980
Trebarwith Cornwall	1977	< 0.01	2.6	0.2	0.01		12	0.6	Hamilton, 1980
Dee Aberdeen	Pre-1981		9.7				16	5.3	Coombs and Keller, 1981
Mytilus californianus									
Seattle Washington, USA	Pre-1981			ND				7.9	Coombs and Keller, 1981

TABLE 3.7. – continued

Region	Sampling date	Se	Sn	Sr	Ti	V	W	Zn	References
Blue mussel (*Mytilus edulis trossulus)*									
Southern Baltic									
Gulf of Pomerania	1997							218±73.9	Szefer et al., 2000g
								87.9–396	
Slupsk Bank	1997							170±61.1	
								120–264	
Gulf of Gdansk	1997							163±43.8	
								121–226	
Mytilus galloprovincialis									
Mediteranean Sea, Carteau	1995							190±23.0	Szefer et al., 1998a
Mediteranean Sea, Nice	1998							55.5–95.7	Szefer et al., 1998a
Vigo, Spain coast	1996							103	Szefer et al., 1998a
Pontevedra Rias, Spain coast	1996							85.9	
								80.6–91.1	
M. edulis									
Japan coast of Kyushu Island									
Urashiro, Japan coast	1994							98.4±13.1	Szefer et al., 1997b
Akamizu	1994							243±13.0	
Saganoseki	1994							297±8.67	
Mytilus sp.									
Sepetiba Bay, Brazilian coast								668±67.5	Szefer et al., 1998a
M. edulis									
Ravenglass Cumbria	1977		2	17		0.2			Hamilton, 1980
Trebarwith Cornwall	1977		0.3	0.8		0.06			Hamilton, 1980
Dee Aberdeen	Pre-1981							1230	Coombs and Keller, 1981
Mytilus californianus									
Seattle Washington, USA	Pre-1981	1.8			173		1.1		Coombs and Keller, 1981

The STR values lower than unity show that Zn and Fe concentrations are generally greater in the dried soft tissue as compared with those in the shells, while for Cu and Mn their mean ratios approximate to unity. However, the ratios of metal content in shell to metal content in soft tissue were mostly greater than unity for Cu and Mn and smaller than unity for Zn and Fe. Koide et al. (1982) reported also the ratio < 1 for Zn in *M. edulis* inhabited the West and East Coast of U.S.A.

As for the BTR values, the highest values are noted for Ni and Mn amounting up to 8.37 and 16.5, respectively, while its lowest values are observed for Hg (0.80–1.22), and especially for Cd (0.32–0.55). It means that concentrations of Ni and Mn are significantly greater in byssi threads than in soft tissue of the specimens studied.

Inter-Elemental Relationships

Figures 3.13 displays the Hg vs. the Se and Eu vs Fe concentration relationships in *M. edulis* from the Baltic Sea and surrounding areas. It is clearly shown

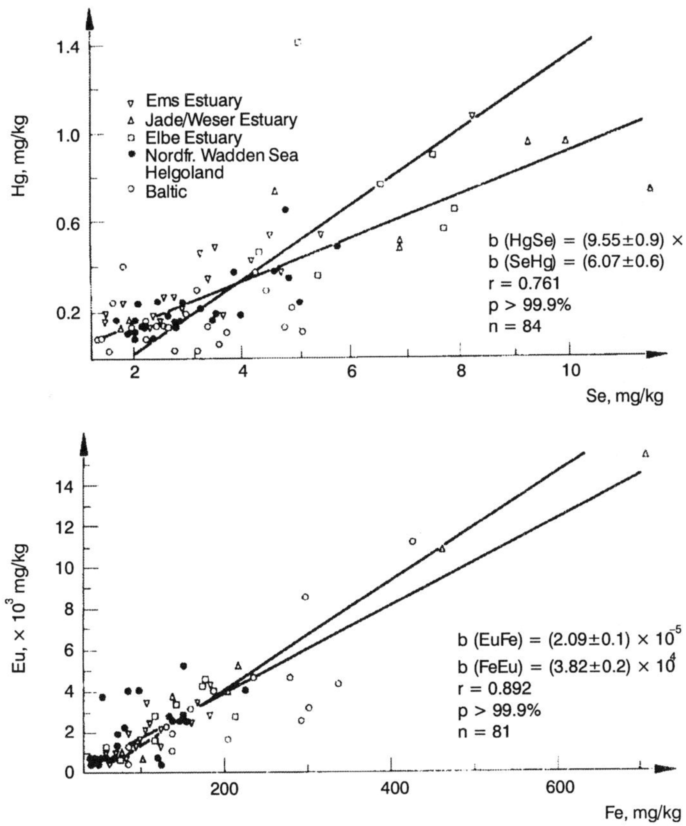

Fig. 3.13. Correlation between Hg and Se, and Eu and Fe contents in mussels. After Karbe et al. (1977); modified.

that the Baltic results are located below the regression line. This suggests that the ratios of tissue Hg to Se concentrations are smaller in the Baltic mollusc than in the surrounding areas (Karbe et al., 1977). This difference well identifies Baltic specimens of *M. edulis*. Figure 3.14 presents comparison of trace element ratios (in logarithmic scale) computed for the German coastal waters of the Baltic Sea and the North Sea. It can be seen that the ratios of Ag, Cd, Cr and Zn to other elements are generally higher for the Baltic mussel than those for the North Sea mussel. Inverse trends are observed for the ratios of As, Co, Cs, Hg and Se. There are exceptions from the increased values for the North Sea for ratios Co/As, Co/Hg, Se/As, Se/Hg, Cs/As and Cs/Hg. According to Karbe et al. (1977) this may be attributed to the high concentration differences of the North Sea and the Baltic mussels (As, Hg Baltic < As, Hg North Sea).

Figures 3.15 and 3.16 illustrate significant relationships ($p < 0.01$ or $p < 0.05$) between concentrations of some metals in the soft tissue (Cd–Mn, Cu–Mn,

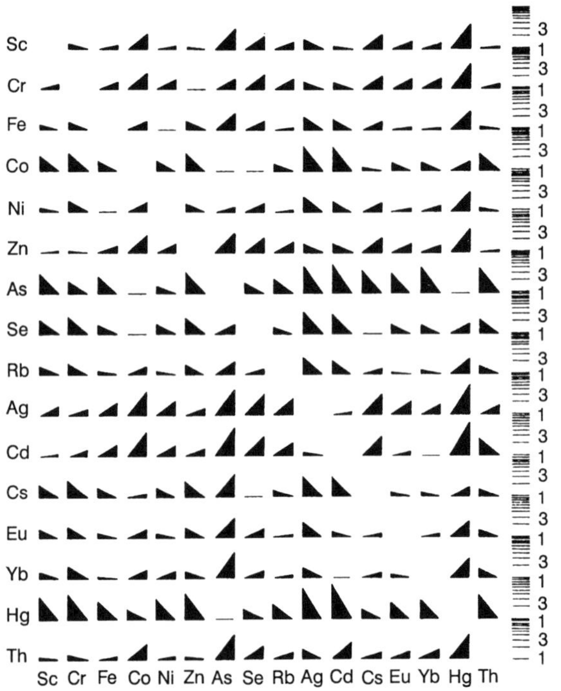

Fig. 3.14. Comparison of the trace element ratios calculated for German coastal waters of the North Sea and the Baltic (logarithmic scale). Ratios are calculated by dividing the concentrations of elements in the left column by those in the bottom line. Each triangle shows the increase or decrease of the ratios comparing the North Sea (left) and Baltic (right). For example the triangle on the left side of the bottom gives a Hg/Cr ratio for the North Sea 5.1 times greater than for the Baltic. The whole figure shows regional differences in the multielement ratios. For example arsenic has higher values in the North Sea compared to all other elements shown in the figure. On the other hand silver has in all cases higher values in the Baltic. After Karbe et al. (1977).

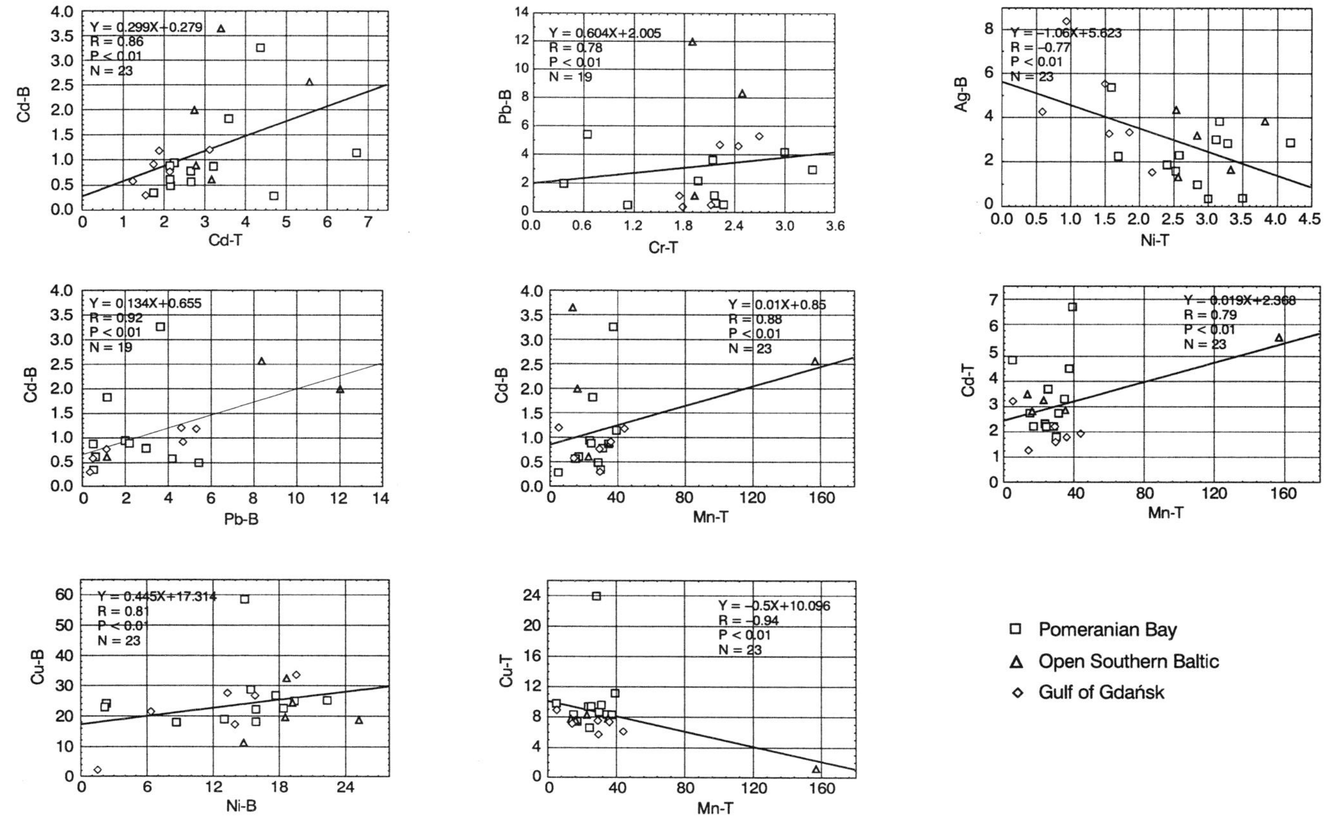

Fig. 3.15. Relationships between concentrations of trace elements in the soft tissue (T) and byssus (B) of *Mytilus edulis trossulus* from the three regions of the Southern Baltic Sea. After Szefer et al. (2000g).

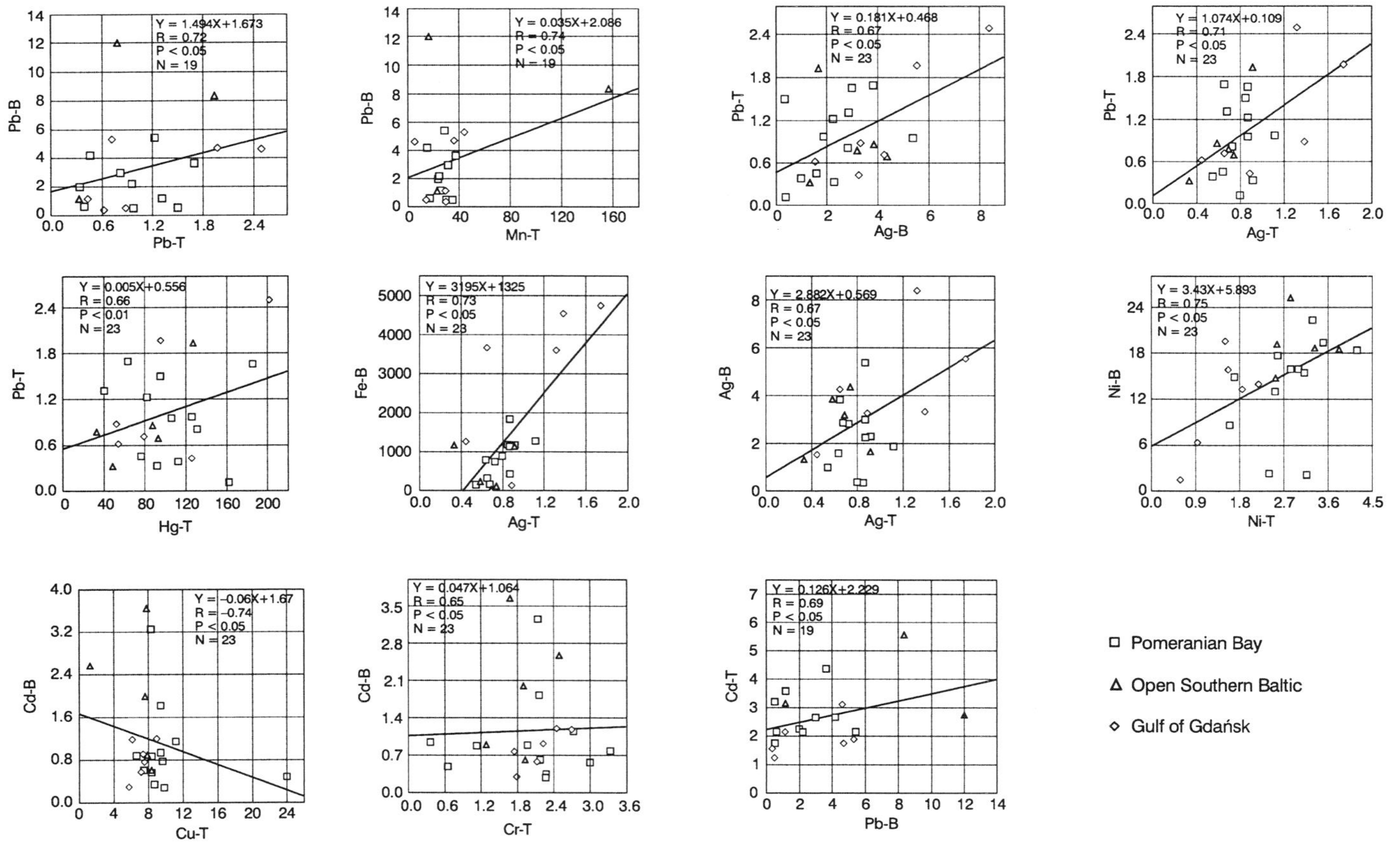

Fig. 3.16. Relationships between concentrations of trace elements in the soft tissue (T) and byssus (B) of *Mytilus edulis trossulus* from the three regions of the Southern Baltic Sea. After Szefer et al. (2000g).

Pb–Ag, Pb–Hg) and byssi (Cd–Pb, Cu–Ni). There is also strong covariance between concentrations of Cd, Pb, Ag and Ni in the soft tissue and byssus.

(iii) Occurrence of Radionuclides in Bivalvia

Dahlgaard (1986, 1991) studied variation in radionuclide (^{110m}Ag, ^{241}Am, ^{144}Ce, ^{60}Co, ^{244}Cm, ^{134}Cs, ^{155}Eu, ^{54}Mn, ^{237}Np, ^{238}Pu, ^{239}Pu, ^{95m}Tc, ^{99m}Tc and ^{65}Zn) loss rates from Baltic *Mytilus edulis.* Extensive studies of Baltic molluscs (*M. edulis*, *M. balthica*) for concentrations of ^{241}Am, ^{137}Cs, $^{239+240}$Pu and ^{90}Sr before the Chernobyl accident have been performed (Holm et al., 1986; Tuomainen et al, 1986). After the Chernobyl accident several authors have made more intensive studies of Baltic molluscs for $^{239+240}$Pu, ^{210}Po and ^{137}Cs (Skwarzec and Falkowski, 1988; Skwarzec and Bojanowski, 1992; Bojanowski et al., 1995; Kanisch et al., 1995; Skwarzec, 1995, 1997; Stepnowski and Skwarzec, 2000b). The concentrations of tissue U (^{238}U, ^{235}U, ^{234}U) and Th (^{232}Th) were determined in several species of molluscs from the Gulf of Gdańsk, southern Baltic (Szefer and Wenne, 1987; Skwarzec, 1995, 1997). Shell concentrations of U and Th are reported in Szefer and Wenne (1987).

The concentrations of selected radionuclides in molluscs from the Baltic Sea are listed in Table 3.8. The levels of radiocaesium (^{137}Cs) in whole body of mussels *Mytilus edulis* and *Macoma balthica* showed maximum values for specimens collected in Bothnian Sea in year period 1986–1987, i.e. after the Chernobyl accident (Fig. 3.17). Additional distinct radiocaesium maximum as a function of time was observed for *M. edulis* (whole body) from Forsmark area in 1989 (Fig. 3.17). The levels of ^{60}Co in *Mytilus edulis* and *M. balthica* from Forsmark indicated the distribution pattern (Figs. 3.17) which was similar to temporal trends observed for radiocaesium levels in the mussels from the same subarea (HELCOM, 1995). Most probably contribution of this radioisotope to the total radioactivity of mussels after 1986–1987 corresponded to ^{60}Co emission from the nuclear power plants located at Forsmark. Mussels collected at other sites close to the nuclear power plants indicated also rather regular temporal changes of relatively small ^{60}Co activities (HELCOM, 1995).

In contrast to ^{137}Cs and ^{60}Co, the distribution of ^{90}Sr in mussels was irregular because of its global fallout origin (HELCOM, 1995).

Concentrations of plutonium isotopes (^{238}Pu, $^{239+240}$Pu) have been sporadically reported for Baltic mussels (Skwarzec and Bojanowski, 1992; HELCOM, 1995). Small differences in plutonium concentrations in mussels before and after the Chernobyl accident were observed, namely the levels were a little higher in mussels collected in 1986 than in the following two years (Skwarzec and Bojanowski, 1992). The Chernobyl-derived plutonium in molluscs is supported by ^{238}Pu/$^{239+240}$Pu ratio amounting on the average to 0.092 (0.075–0.093) (Skwarzec and Bojanowski, 1992) which was slighty higher than value of 0.06 being representative for the global fallout (HELCOM, 1995).

TABLE 3.8.

Concentrations of radionuclides in molluscs of the Baltic Sea and other northern areas

Region	Sampling date	Body part	N	241-Am (Bq kg^{-1} d.w.)	60-Co (Bq kg^{-1} w.w.)	137-Cs (Bq kg^{-1} w.w.)	210-Po (Bq kg^{-1}d.w.)	239+240-Pu (Bq kg^{-1} d.w.)	90-Sr (Bq kg^{-1}d.w.)	References
Blue mussel (*Mytilus edulis*)										
Swedish coast	Pre-1986	Soft tissue	4	0.05		3.5*		0.22		Holm et al., 1986
				0.02–0.11		2.2–4.8		0.007–0.41		
		Shell	5	0.02		0.37*		0.024		
				0.01–0.03		0.37–0.37		0.007–0.048		
Finnish coast	1982	Soft tissue	4			2.02±0.10		0.07	5.9	Tuomainen et al., 1986
		Shell	4			< 0.16		0.038	26±2	
		Whole body	4	0.016±0.006		< 0.55		0.028±0.013	21±2	
Gulf of Gdansk	1985–88	Soft tissue	3				82.7–164.3	0.018–0.226 (N=7)		Skwarzec and Falkowski, 1988
		Shell	3				4.8–6.1			Skwarzec, 1995
		Whole body	3				17.6–26.1			Skwarzec and Bojanowski, 1992
Southern Baltic	1996–97	Soft tissue					272±27.6			Stepnowski and Skwarzec, 2000b
		Byssal threads					30.0±1.7			
		Shell					0.9±0.1			
Pomeranian Bay	1993	Soft tissue				3.8				Kanisch et al., 1995
Kattegat	1984–91	Soft tissue	9		4.0–56	1.5–9.5				
Bornholm Sea	1984–91	Soft tissue	4			20–30				
Bothnian Sea	1984–91	Whole body	4			10–160				
Forsmark area	1984–91	Whole body	6		5–215					
Belt Sea	1991	Soft tissue						19		Kanisch et al., 1995
Belt Sea	1991	Whole body						30		
Little macoma (*Macoma balthica*)										
Finnish coast	1982–83	Whole body	8	0.036±0.016		1.5±1.1*		0.090±0.055	34±8.2	Tuomainen et al., 1986
Gulf of Gdansk	1985–88	Soft tissue					237±5.3	0.031–0.086 (N=6)		Skwarzec and Falkowski, 1988
		Shell					21.5±1.4			Skwarzec and Bojanowski, 1992
		Whole body					68.4±1.6			Skwarzec, 1995
Slupsk Furrow	1985	Soft tissue					170±4.6			
		Shell					12.1±1.3			
		Whole body					32.1±1.3			
Bothnian Sea	1984–91	Whole body	8		4.0–42	2–120		14		Kanisch et al., 1995

Gotland West	1984–91	Whole body	4	3.0–15				
Forsmark area	1984–91	Whole body	5	5.0–41	2–120			
Long clam (*Mya arenaria*)								
Gulf of Gdańsk	1985–88	Soft tissue				146±4.0	0.024–0.217 (N=2)	Skwarzec and Falkowski, 1988
		Shell				5.8±0.7		Skwarzec and Bojanowski, 1992
		Whole body				33.9±1.0		Skwarzec, 1995
	1996–97	Soft tissue				10.1±1.7		Stepnowski and Skwarzec, 2000
		Shell				0.4±0.1		
Cockle shell (*Cardium glaucum*)								
Gulf of Gdansk	1985–88	Soft tissue				170±7.1	0.040±0.017	Skwarzec and Falkowski, 1988
		Shell				4.4±0.3		Skwarzec and Bojanowski, 1992
		Whole body				12.7±0.4		Skwarzec, 1995
Northern astarte (*Astarte borealis*)								
Slupsk Furrow	1985	Soft tissue				88.6±6.7		Skwarzec and Falkowski, 1988
		Shell				7.6±0.6		Skwarzec and Bojanowski, 1992
		Whole body				11.3±0.7		Skwarzec, 1995
Ocean quahog (*Arctica islandica*)								
Belt Sea	1988	Soft tissue				76		Kanisch et al., 1995
Arkona Sea	1991	Soft tissue				36		

* – Dry wt.

TABLE 3.8. – continued

Region	Sampling date	Body part	Shell length (mm)	N	Th (tot.) ($\mu g\ g^{-1}$ d.w.)	U (tot.) ($\mu g\ g^{-1}$ d.w.)	234-U ($Bq\ kg^{-1}$ d.w.)	235-U ($Bq\ kg^{-1}$ d.w.)	238-U ($Bq\ kg^{-1}$ d.w.)	References
Blue mussel (*Mytilus edulis*)										
Gulf of Gdansk	1981/88	Soft tissue		18(475)[a]	0.18±0.02	0.19±0.02				Szefer and Wenne, 1987
					0.08–0.40	0.13–0.27				
							3.28–4.93	0.19–0.42	2.96–4.12	Skwarzec, 1995
		Shell		18(475)	0.016±0.002	0.023±0.004				
					0.007–0.027	0.006–0.037				
Western Baltic	1973	Soft tissue			0.026					Karbe et al., 1977
					0.004–0.26					
Southwestern Baltic	1979	Soft tissue	40±2	32(640)	0.061	0.18				Möller et al., 1983
					< 0.01–0.22	< 0.1–0.6				
Kiel Fjord	1979	Soft tissue	10±2	(300)[a]	< 0.01	< 0.1				Möller et al., 1983
			20±2	(70)[a]	< 0.01	< 0.1				
			40±2	(20)[a]	< 0.01	< 0.1				
			60±2	(20)[a]	0.03	0.17				
Little macoma (*Macoma balthica*)										
Gulf of Gdansk	1981/85	Soft tissue		7(433)	0.10±0.01	0.35±0.04	3.20–7.16	0.10–0.51	2.93–6.32	Skwarzec, 1995
					0.05–0.17	0.20–0.41				
		Shell		7(433)	0.032±0.002	0.019±0.003				Szefer and Wenne, 1987
					0.010–0.042	0.011–0.024				
Long clam (*Mya arenaria*)										
Gulf of Gdańsk	1985	Soft tissue		9(295)	0.23±0.04	0.15±0.02	5.64±0.17	0.18±0.03	5.06±0.17	Skwarzec, 1995
					0.15–0.32	0.11–0.18				
		Shell		9(295)	0.043±0.015	0.020±0.008				Szefer and Wenne, 1987
Cockle shell (*Cardium glaucum*)										
Gulf of Gdansk	1985	Soft tissue		4(228)	0.14±0.03	0.27±0.04	6.89±0.30	0.39±0.07	6.08±0.28	Skwarzec, 1995
					0.12–0.16	0.21–0.33				
		Shell			0.061±0.010	0.021±0.003				Szefer and Wenne, 1987
					0.051–0.071	0.012–0.030				

[a] – No. of specimens in parentheses

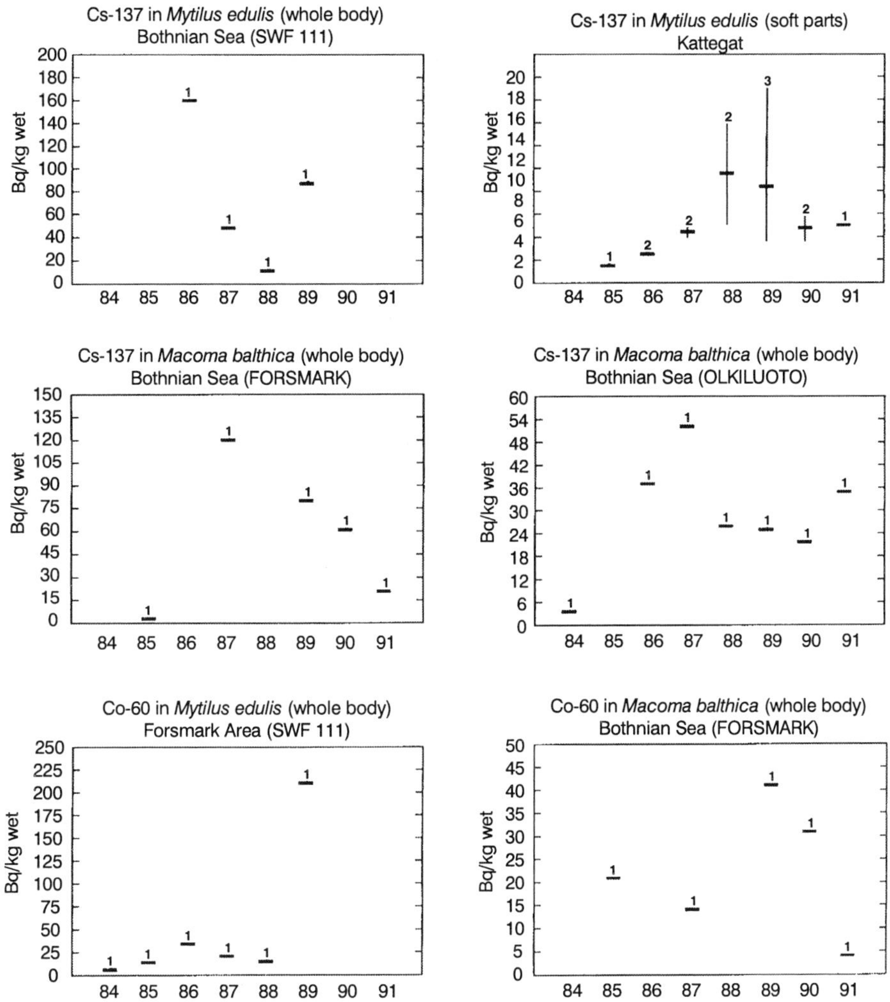

Fig. 3.17. Activities of some radionuclides in molluscs from different Baltic Sea subareas. After Kanisch et al. (1995); modified.

The concentrations of ^{210}Po in soft tissue of *M. edulis* from the Gulf of Gdańsk and Danish water were similar and amounted to 124 and 149 Bq/kg dry wt., respectively (Skwarzec and Bojanowski, 1992; Dahlgaard, 1996). Intertissue variations in polonium concentrations in Baltic *M. edulis* and *Mya arenaria* have been reported by Skwarzec and Falkowski (1988) and Stepnowski and Skwarzec (2000b). The highest polonium concentrations were found in the hepatopancreas followed by alimentary tract, gill and muscle.

2. CRUSTACEANS

(i) Introduction

Crustaceans *Saduria entomon, Pontoporeia femorata* and *Monoporeia affinis* are typical species inhabiting the Baltic Sea. On the deep muddy bottom such species as *S. entomon* and *P. femorata* are usually present. The assessment of the state of the macrozobenthos was based on samples collected at the Gulf of Finland. Both abundance and biomass values, excluding the occasionally occurring big isopod *Saduria entomon*, were significantly higher during 1989–93 as compared to period 1984–88 (Andersin et al., 1996). The mean abundance value for period 1990–93 exceeded by an order of magnitude that estimated for 1984–88. During 1989–93, the abundance was strongly dominated by the amphipods *Pontoporeia femorata* and *Monoporeia affinis*. Biological characteristics and taxonomy of crustaceans are given by several authors (Miner, 1950; Mulicki, 1957, Birshteyn and Pasternak, 1988a, 1988b; Köhn and Gosselck, 1989; Hill and Elmgren, 1992).

General Characteristics and Taxonomy

Phylum: Arthropoda
Class: Crustacea
Order: Isopoda
Family: Idoteidae Milne Edwards, 1840
Species: *Saduria* (syn. *Mesidothea*) *entomon* Linnaeus, 1758

Habitat and range: it is relict form of Arctic origin, lives in brine and fresh waters; ranges in the Baltic Sea from the Danish Straits to the Bothnian Bay, noted in the Caspian Sea and White Sea. *Food habits*: predator, carnivore (scavenger) – feeds on *Harmothoe sarsi, Pontoporeia* spp. (Mulicki, 1957, Birshteyn and Pasternak, 1988b; Köhn and Gosselck, 1989; Hill and Elmgren, 1992).

Family: Idoteidae
Species: *Idotea balthica* (Pallas, 1722), *Idotea chelipes* (Pallas, 1766), *Idotea granulose* (Rathke, 1843)

Habitat and range: Idotea balthica – widely distributed (cosmopolitan), euryhaline species; occurs at eastern coast of the North, coastal waters of Brazil, New Zealand and Java as well as in the Mediterranean Sea, Red Sea and the North Sea, noted also in the Skagerrak and Kattegat. In the Baltic Sea its distribution reaches both the Bothnian Bay and Gulf of Finland. *Idotea chelipes* – prefers brine waters, distributed along Atlantic coastal waters from Murmansk to France, enters western part of the Mediterranean Sea and North Sea. Its distribution in the Baltic Sea reaches as far as both the Bothnian Bay and Gulf of Finland. *Idotea granulosa* – occurs along coastal waters of France, Holland, British Isles and Denmark. On the north occurs in entering sector of the White Sea, coastal waters of the Island and the North Sea. In the Baltic Sea reaches entrance to the

Gulf of Finland (isohaline 6‰). *Food habits*: herbivores (Miner, 1950; Mulicki, 1957; Wiktor, 1985; Köhn and Gosselck, 1989).

Family: Balanidae
Species: Barnacle *Balanus improvisus* Darwin

Habitat and range: Atlantic-boreal species, euryhaline; in the coastal waters of America ranges from the Nova Scotia to Patagonia. In the Baltic Sea reaches the Aland Islands and enters the Gulf of Finland. *Food habits*: suspension feeder (Köhn and Gosselck, 1989; Miner, 1950; Mulicki, 1957; Wiktor, 1985).

Order: Amphipoda
Family: Pontoporeiidae
Species: *Monoporeia* (syn. *Pontoporeia*) *affinis* Bousfield, 1989

Habitat and range: prefers brine and fresh waters, being in the Baltic Sea a relict species of the Ancylus Lake era; occurs in Arctic estuaries of the Eurasia and Canada. Observed also in lakes of north Europe and the North America. Its eastern range reaches the White Sea. *Food habits*: deposit feeder – feeds on bacteria, microalgae, meiofauna and also young molluscs (Mulicki, 1957; Wiktor, 1985; Elmgren et al., 1986; Köhn and Gosselck, 1989; Jażdżewski and Konopacka, 1995; Moor, 1977).

Order: Amphipoda
Family: Gammaridae Leach, 1813
Gammarus sp.

Habitat and range: *Gammarus* sp. is distributed in almost all surficial waters of the Arctic Ocean, prefers shallow waters. *Food habits*: deposit feeder (Miner, 1950; Jażdżewski, 1975; Köhn and Gosselck, 1989; Jażdżewski and Konopacka, 1995).

Order: Amphipoda
Family: Talitridae
Species: Sandhopper (*Talitrus saltator* Montagu, 1808)

Habitat and range: this Mediterranean-boreal species is distributed along European coasts from western part of the Mediterranean Sea to southern part of Norway and the Baltic Sea (Jażdżewski and Konopacka, 1995), inhabits sandy beaches among decaying macroalgae and detritus, it lives buried beneath the strandline (Rainbow et al., 1998). *Food habits*: deposit feeder – feeds on small carrion and macroalgae on beach.

Order: Decapoda
Suborder: Natantia
Family: Crangonidae
Species: Common shrimp (*Crangon crangon* Linnaeus, 1758)

Habitat and range: distributed along Atlantic coasts from the White Sea to the Mediterranean Sea. In the Baltic Sea reaches the Gulf of Finland. *Food habits*: as predator feeds on Amphipoda, Mysidacea, Polychaeta, small fish and carrion (Köhn and Gosselck, 1989).

Suborder: Natantia
Species: *Neomysis vulgaris*

Habitat and range: distributed along Atlantic coasts from the White Sea to the Mediterranean Sea. In the Baltic Sea reaches the Gulf of Finland. *Food habits*: predator – feeds on Amphipoda, Mysidacea, Polychaeta and also small fish and carrion (Köhn and Gosselck, 1989).

Suborder: Reptantia
Species: *Eriocheir sinensis* H. Milne Edwards

Habitat and range: this euryhaline species originates from China, distributed along coastal and estuarine waters of the North Sea and Baltic Sea. Enters rivers such as Elba. *Food habits*: predator, scavenger (Schellenberg, 1928; Arndt, 1969).

Species: Green (Shore) crab (*Carcinus maenas* Linnaeus, 1758) syn. *Carcinides maenas*

Habitat and range: Atlantic species, common in the Baltic Sea to the Danish Straits. On the east recorded sporadically. *Food habits*: predator – feeds on benthal and epibenthal invertebrates (Schellenberg, 1928; Miner, 1950; Arndt, 1969; Köhn and Gosselck, 1989).

Family: Pilumnidae
Species: Mud crab (*Rhitropanopeus harrisi* Gould, 1841)

Habitat and range: occurs along western coastal waters of the North America. In the Baltic Sea observed in gulfs (Kiel Bucht, Gulf of Gdańsk – Dead Vistula) and the Vistula Lagoon. *Food habits*: predator, mature specimens feed on Mysidacea (Miner, 1950; Köhn and Gosselck, 1989).

Order: Cumacea
Species: Cumacean (*Diastylis rathkei* Kröyer)

Habitat and range: occurs in the Western Baltic and in the German Bight, North Sea. *Food habits*: deposit feeder in muddy sand bottoms but it can survive as an epistratum feeder on coarser sediment (Forsman, 1938; Habermehl et al., 1990). It is one of the major benthic producers (e.g. 1500 t yr^{-1} in the Kiel Bay) and the most important food item of demersal fish (dab, cod, flounder) in the Western Baltic (Arntz, 1971, 1974, 1977a, 1977b; Rachor et al., 1982; Swaileh and Adelung, 1995).

Overview of Worldwide Literature

Marine crustaceans have different abilities to bioconcentrate some heavy metals in their body from the environment (Bryan, 1968; Dethlefsen, 1977; Amiard et al., 1980; Phillips 1980; Anil and Wagh, 1988; Rainbow, 1989, 1993, 1995a, 1995b, 1997, 1998; Rainbow and Moore, 1990; Rainbow et al., 1989a, 1989b, 1990; Moore et al., 1991; Phillips and Rainbow 1993; Rainbow and Phillips, 1993; Ismail et al., 1995; Rainbow, 1995b; Hockett et al., 1997; Scott-Fordsmand and Depledge, 1997; Kress et al., 1998; Abdennour et al., 2000; Jewett and Naidu, 2000;

Roast et al., 2000). They are very interesting benthic organisms as potential biomonitors because of them widespread geographical distribution (Rainbow and Phillips, 1993; Rainbow, 1995a, 1995b, 1996). According to many authors (Ireland, 1974; Walker et al., 1975a, 1975b; Walker and Foster, 1979; Rainbow et al, 1980; White and Walker, 1981; Rainbow, 1985, 1987; Chan et al., 1986; Phillips and Rainbow, 1988, 1993; Rainbow and White, 1989; Powell and White, 1990; Rainbow, 1995a; Watson et al., 1995; Blackmoore et al., 1998; Fialkowski and Newman, 1998; Blackmoore, 1999) amongst crustaceans, barnacles appear to be most effective biomonitors of metallic pollutants. They are the most sedentary as compared to other crustaceans, and are relatively easy to age in temperate areas (Rainbow, 1995a).

Some barnacles, e.g. *Balanus amphitrite*, are well known as fouling crustaceans on shipping; moreover they closely attached to rocky bed or to manmade structures such as piers. The distribution of trace elements has been studied extensively in barnacles inhabited the Indo-Pacific, from southern Japan and Korea to the Gulf of Thailand and Bombay (Rainbow, 1995a), Central and South America (Birshteyn and Pasternak, 1988b), the Atlantic (the Azores), the Adriatic Sea and temperate environments such as the Baltic Sea, Black Sea and Azov Sea (Barbaro et al., 1978; Birshteyn and Pasternak, 1988b; Weeks et al., 1995). These organisms appear to be potential cosmopolitan biomonitors in tropical and subtropical zones such as coastal waters of Hong Kong (Chan et al., 1986; Phillips and Rainbow, 1988; Blackmore, 1999), China (Rainbow et al., 1993b; Blackmore et al., 1998) and Malaysian mangroves (Rainbow et al., 1989a). Barnacles were also recognised as biomonitors of metallic pollutants in the Atlantic, the Azores (Weeks et al., 1995), North Adriatic lagoons (Barbaro et al., 1978) and in the subtropical Pacific coast of Mexico (Páez-Osuna et al., 1999). Talitrid amphipods appeared to be promising bioaccumulators of trace metals in coastal marine environments (Moore et al., 1991; Weeks and Rainbow, 1991, 1993; Rainbow et al., 1989b, 1993a, 1998).

The biological availability to marine crustaceans of transuranium and other long-lived nuclides has been reported extensively by Pentreath (1981). $^{239+240}Pu$ concentrations in crustaceans *Trackypenaeus curvirostris* and *Ovalipes punctatus* from the Japanese coast were 5.0 and 2.5 mBq kg^{-1} kg wet wt., respectively while ^{137}Cs concentrations amounted to 140 and 36 mBq kg^{-1} kg wet wt., showing distinct interspecies differences (Yamada et al., 1999).

(ii) Occurrence of Chemical Elements in Crustaceans

Different species of Baltic crustaceans have been analysed for concentration of selected metals to recognise actual pollution status of the sea, e.g. *Saduria entomon* (Lithner, 1974; Niemi, 1977; Häkkilä, 1980; Kauppinen, 1980; Tervo et al., 1980; Sandler, 1984, 1986; Skwarzec et al., 1984; Kulikova et al., 1985; Szefer, 1986; Szefer et al., 1990a; Voloz et al. 1990; Falandysz, 1994; Pynnönen, 1996; Voipio et al., 1997; Szefer and Kusak, 2000), *Crangon crangon* (Szefer, 1986; Falandysz, 1994; Szefer and Kusak, 2000), *Balanus improvisus* (Szefer, 1986; Rainbow et al., 2000; Szefer et al., 2000b), *Idotea* sp. (Szefer and Kusak, 2000), *Diastylis rathkei* (Swaileh and Adelung, 1995) and *Talitrus saltator* (Rainbow et al., 1998).

Inter-species trends

Since isopod *Saduria entomom* constitutes a representative part of macrobenthic community with wide distribution and relatively long life period of ca, 8–9 years, this crustacean is included in the Baltic monitoring programme. From Table 3.9 results that *S. entomon* from the Gulf of Gdańsk, Baltic Sea, contained in its whole body higher levels of Cu than barnacle *Balanus improvisus* and talitrid amphipod crustacean *Talitrus saltator* from the same region (Szefer, 1986; Rainbow et al., 1998; Szefer and Kusak, 2000; Szefer et al., 2000b). The latter species was characterised by the greatest concentration of Cd while *B. improvisus* from the same location accumulated the highest amounts of Mn and Zn. Inter-species changes in metals contents are also well marked for two benthic crustaceans inhabited the Gulf of Riga (Kulikova et al., 1985). Specimens of *S. entomon* contained higher levels of Ca, Cd, Co, Cr, Cu, Fe, Hg, Mg, Mn, Ni, Pb, Sr and Zn as compared to those in *Neomysis vulgaris* from the Gulf (Table 3.9).

Inter-tissue trends

Analyses of trace element contents of *Saduria entomon* = *Mesidothea entomon* have shown that Cd, Co, Cu, Mn, Ni, Pb and Zn are non-uniformly distributed within the crustacean (Szefer et al., 1990a). The hepatopancreas of *Saduria entomon*, representing only 2.4% of the whole body, contained 56% of the total burden of Cu; it indicates the dominant role of the digestive organ in accumulation of Cu in these invertebrates. According to Skwarzec (unpublished data) haemolymphe isolated from hepatopancreas of *S. entomon* contained from 250 to 440 μg Cu g^{-1} dry wt. Kulikova et al. (1985) supposed that Cu is bound by haemocyanin, the respiratory pigment in the blood of crustaceans such as *S. entomon*. Enhanced levels of Cu, covering the range from 64 to 349 $\mu g\ g^{-1}$ dry wt., were found in this crustacean from open areas of the Bothnian Sea and the Gulf of Bothnia (Sandler, 1984; Tervo et al., 1980). As can be seen in Table 3.9 these values are an order of magnitude higher as compared to those reported for the Gulf of Gdańsk.

Inter-tissue distribution of trace elements in *B. improvisus* from the Gulf of Gdańsk studied by Szefer et al. (2000b) clearly showed that concentrations of Fe and Zn in the soft tissue were smaller than those in the exoskeleton which contained higher levels of Cd, Cu and especially Mn.

Spatial trends

From the data reported clearly results (HELCOM, 1993; Sandler, 1984; Szefer, 1986; Szefer and Kusak, 2000) that there was a distinctive geographical variance in trace metal contents of isopoda *Saduria entomon* inhabited the Baltic Sea.

TABLE 3.9.

Concentrations of chemical elements ($\mu g\ g^{-1}$ dry wt.) in crustaceans from the Baltic Sea and other northern areas

Region	Sampling date	Length (mm)	N	Al	Ag	Ca	Cd	Co	Cr	Cu	Fe	References
Saduria (Mesidothea) entomon												
Gulf of Gdansk	1981	20–70	16 (604)			102±18*	0.36±0.09	1.48±0.35		15.3±2.2	5150±440	Szefer, 1986
						77–120	0.26–0.49	0.79–1.79		9.0–21.1	3170–6650	
Gulf of Gdansk			6	1090±325	1.84±1.07	272±164*	1.78±1.10	5.91±1.31	2.89±2.19	110±34.3	3370±1910	Szefer and Kusak, 2000
				620–1400	0.22–2.63	108–503	0.32–3.29	3.94–6.90	0.54–5.62	72.9–151	680–6300	
Gulf of Bothnia	1979	40–92	3				0.87			199		Tervo et al., 1980
							0.66–1.15			164–258		
Gulf of Riga	Pre-1985					26290[b]	0.97[b]	3.9[b]	1.5[b]	35.0[b]	542[b]	Kulikova et al., 1985
Bothnian Sea, open area		36–85	10 (66)							153.7		Sandler, 1984
										64.0–349		
Bothnia Sea, open area										122–240		Voipio et al., 1977
Bothnia Sea, off Pori										65–206		Hakkila, 1980
Bothnia Bay, off Kokkola										194–268		Niemi, 1977
The Quark off Vaasa										159–170		Kauppinen, 1980
Neomysis vulgaris												
Gulf of Riga	Pre-1985					4300[b]	0.12[b]	0.4[b]	0.3[b]	4.8[b]	108[b]	Kulikova et al., 1985
Pontoporeia affinis												
Bothnian Sea		15								97.8		Sandler, 1984
Open sea										90–130		
Bothnian Bay										120–195		Lithner, 1974
Bay of Skelleften												
Barnacle (*Balanus improvisus*)												
Gulf of Gdansk	1981		2 (1350)			330*	0.06	< 1.2			1420	Szefer, 1986
							0.03–0.09				1340–1500	
	1994	1.0–16	28 (558)				1.36			9.34	270	Szefer et al., 2000b
							0.02–2.3			4.34–17.7	10.6–990	

Region	Sampling date	Length (mm)	N	Al	Ag	Ca	Cd	Co	Cr	Cu	Fe	References
	1995	1.0–16	32 (924)				1.35			8.43	310	
							0.29–2.08			2.8–10.2	79.2–630	
	1997	2.0–9					0.18			3.9	550	
							0.07–0.33			3.56–4.63	205–910	
	1998		22 (32)				21.3[a]			66.2[a]	13.31[a,b]	Rainbow et al., 2000
							18.7–23.6			42.5–86.2	6.74–22.23	
Common shrimp (*Crangon crangon*)												
Gulf of Gdansk	1981	35–50	1 (5)			79*	1.01	< 2.1		4.3	200	Szefer, 1986
			12	450±465	2.46±1.29	58.8±23.7*	1.44±0.52	2.28±0.52	2.65±1.81	63.9±26.8	190	Szefer and Kusak, 2000
				114–1820	0.62–4.97	32.3–126	0.64–2.32	1.51–2.95	0.56–6.82	33.8–104	280–450	
Puck Bay			2			65.4*	3.4			38.5	67.1	Szefer et al., 1994a
						61.8–68.9	2.7–4.1			37.2–39.8	66.7–67.4	
Idotea sp.												
Gulf of Gdansk			3	390±36	1.63±1.42	71.6±29.9*	1.70±0.59	5.27±1.00	1.27±0.62	79.1±4.10	365±76	Szefer and Kusak, 2000
				350–420	0.62–2.63	39.3–98.3	1.04–2.19	4.36–6.34	0.57–1.51	75.3–83.4	278–420	
Sandhopper (*Talitrus saltator*)												
Gulf of Gdansk	1996		242		2.46		20.8			59.2	294	Rainbow at al., 1998
					1.70–3.40		8.04–35.1			49.7–70.4	161–492	
Gammarus sp.												
Gulf of Gdansk	1981	5	1 (98)				0.53	< 1.85		12.4	360	Szefer, 1986
Gulf of Gdansk			5	860±730		153±112*	1.22±0.33	4.45±1.88	17.7±14.5	40.1±20.9	430±313	Szefer and Kusak, 2000
				340–1370		72.0–346	0.85–1.58	2.54–7.56	4.15–34.4	7.52–59.2	82–760	
Bothnian Sea		5.0–48	1 (50)							45		Sandler, 1984

* – mg g^{-1}
[a] – Weight adjusted mean concentration.
[b] – Wet weight.

TABLE 3.9. – continued

Region	Sampling date	Length (mm)	N	Hg	K	Mg	Mn	Na	Ni	Pb	Sr	Zn	References
Saduria (Mesidothea) entomon													
Gulf of Gdansk	1981	20–70	16 (604)			7.3±0.9*	232±19		14.0±1.0	31.0±3.0		63.7±7.0	Szefer, 1986
						6.9–8.5	140–329					26–90	
Gulf of Gdansk			6		13.02±6.77*	1.67±1.40*	247±73	15.58±5.74*	16.6±4.39	6.01±2.31		101±7.7	Szefer and Kusak, 2000
					4.17–22.07	0.39–4.18	130–329	8.50–25.7	12.5–24.4	3.99–9.46		70.9–119	
Gulf of Riga	Pre-1985			0.053[b]		1670[b]	286[b]		4.7[b]	6.1[b]	266[b]	32[b]	Kulikova et al., 1985
	1988	30–79	32 (66)	0.057									Falandysz, 1994
				0.033–0.077									
Gulf of Bothnia	1979	40–92	3							0.97		75	Tervo et al., 1980
										0.74–1.36		62.3–92.8	
Bothnian Sea, open area		36–85	10 (66)									76.1	Sandler, 1984
												42–108	
Bothnia Sea, open area												69–114	Voipio et al., 1977
Bothnia Sea, off Pori												108–154	Hakkila, 1980
Bothnia Bay, off Kokkola												104–125	Niemi, 1977
The Quark off Vaasa												87–97	Kauppinen, 1980
Pontoporeia affinis													
Bothnian Sea		15										74	Sandler, 1984
Open sea												56–137	
Bothnian Bay												50–100	Lithner, 1974
Bay of Skelleften													
Neomysis vulgaris													
Gulf of Riga	Pre-1985			0.012[b]		675[b]	10[b]	400[b]	0.9[b]	1.1[b]	38[b]	16[b]	Kulikova et al., 1985
Barnacle (*Balanus improvisus*)													
Gulf of Gdansk	1981		2 (1350)			4.0*	720		18	28		47	Szefer, 1986
						3.3–4.7	580–860		14.8–21.1	22–34		42–51	
	1994	1.0–16	28 (558)				540					220	Szefer et al., 2000b
							24–1200					65–990	

Region	Sampling date	Length (mm)	N	Hg	K	Mg	Mn	Na	Ni	Pb	Sr	Zn	References
	1995	1.0–16	32 (924)				520					250	
							26.7–1200					69.5–1800	
	1997	2.0–9					14.9					1650	
							7.98–25.8					125–3999	
	1998		22 (32)				228[a]		53.1[a]	106[a]		6.92[a]*	Rainbow et al., 2000
							187–307		39.1–59.4	92.0–130		4.96–11.06	
	1988		1 (50)	0.006									Falandysz, 1994
Common shrimp (*Crangon crangon*)													
Gulf of Gdansk	1981	35–50	1 (5)			4.1*	26		18	48			Szefer, 1986
			12		12.5±7.53*	1.10±0.53*	20.4±9.9	12.48±5.21*	5.99±1.59	10.5±8.17		123±13.3	Szefer and Kusak, 2000
					3.04–26.9	0.48–2.20	10.0–39.4	3.68–23.55	3.94–9.35	1.57–26.9		97.5–145	
	1988		1 (200)	0.14									Falandysz, 1994
Puck Bay	1987		2		8.9*	2.45*	10	8.75*	7.75	ND–0.5		146	Szefer et al., 1994a
					3.1–14.7	2.4–2.5	9.0–11	7.3–10.2	5.6–9.9			82–210	
Idotea sp.													
Gulf of Gdansk			3		8.76±6.31*	5.27±6.52*	97.8±17.5	22.4±5.03*	26.5±23.5	5.88±0.28		75.3±6.7	Szefer and Kusak, 2000
					2.29–14.9	0.98–12.77	78.2–112	17.03–27.01	12.4–53.6	5.68–6.07		67.7–80.4	
Sandhopper (*Talitrus saltator*)													
	1996		242						5.18	27.9		211	Rainbow at al., 1998
									2.07–8.91	20.9–32.3		162–262	
Gammarus sp.													
Gulf of Gdansk	1981	5	1 (98)			8.6*	57		17.7	20		2700	Szefer, 1986
Gulf of Gdansk			5		20.08±13.5*	5.18±4.94*	51.7±33.3	37.97±42.47*	15.7±11.2	11.1±11.4		86.0±10.9	Szefer and Kusak, 2000
					7.63–42.9	1.47–12.7	15.3–102	10.2–111	7.1–35.2	4.12–30.9		70.6–98.8	
Bothnian Sea		5.0–48	1 (50)									85	Sandler 1984

* – mg g^{-1}
[a] – Weight adjusted mean concentration.
[b] – Wet weight.

The concentrations of Cu in whole body of this isopoda were significantly greater in specimens from the Bothnia Sea than from the Gulf of Finland, Gulf of Gdańsk and the Gulf of Riga (Kulikova et al., 1985, Szefer, 1986; HELCOM, 1993). An inverse tendency was observed for Pb which concentration reached the highest values in *S. entomon* from the Gulf of Gdańsk and the Gulf of Riga (Table 3.9).

Extensive studies of talitrid amphipod *T. saltator* collected from the strandline of sites around the Gulf of Gdańsk, southern Baltic, were performed to determine the concentrations of Ag, Cd, Cu, Fe, Mn, Ni, Pb and Zn (Rainbow et al., 1998). Significant geographical differences in metal levels were detected depending on outflows from the Vistula River (Cd, Fe, Mn, Zn) or from local sources around the Gulf of Gdańsk (Cu, Pb).

Temporal trends

Positive temporal trends were observed for Zn and Fe in whole body of *B. improvisus* from the Gulf of Gdańsk (Southern Baltic) while negative temporal pattern was registered for Cd, Cu and Mn during 1994–1997 (Szefer et al., 2000b). Statistically significant ($p < 0.0001$) seasonal variations in the concentrations of Cd, Cu, Pb and Zn in the cumacean, *Diastylis rathkei*, from Kiel Bay (Western Baltic) were observed (Swaileh and Adelung, 1995). In general, high levels of the four metals were detected during the summer months (May–August) (Fig. 3.18) corresponded to the main growth period of this crustacean. Growth could lead to dilution of metals if tissue assimilation exceeds metal accumulation, however this is not that case in *D. rathkei* since it feeds on detritus enriched in heavy metals (Rainbow, 1990) as well as its moulting is attributed to a temporary increase in the concentration of metals inside bodies of crustaceans (White and Rainbow, 1984). The lowest monthly average levels of Cd and Zn occurred in August and those of Cu and Pb in December (Fig. 3.18). The ratio between the seasonal average maximum and minimum concentrations was the highest for Pb (factor 2.5) and the lowest for Zn (factor 1.4) (Swaileh and Adelung, 1995).

(iii) Occurrence of Radionuclides in Crustaceans

Among crustaceans *Saduria entomom* has been most extensively analysed for concentrations of selected radionuclides in the Baltic Sea (Szefer and Wenne, 1987; Skwarzec, 1995, 1997; Skwarzec and Falkowski, 1988; Skwarzec and Bojanowski, 1992; Bojanowski et al., 1995; HELCOM, 1995; Stepnowski and Skwarzec, 2000a). Table 3.10 lists concentration data of radionuclides in crustaceans from the Baltic and other northern areas. The levels of radiocaesium (^{137}Cs) radiostrontium (^{90}Sr) and other radioisotopes (^{60}Co, ^{110m}Ag) in whole body of *S. entomom* from Gulf of Finland (Loviisa) were characterised by similar distribution pattern relative to year of sampling indicating maximum values during 1986–1987 (HELCOM, 1995). As can be seen in Fig. 3.19 concentration of radiocaesium in *S. entomom* from the Gulf of Finland reaching maximum value of 550 Bq kg^{-1} dry

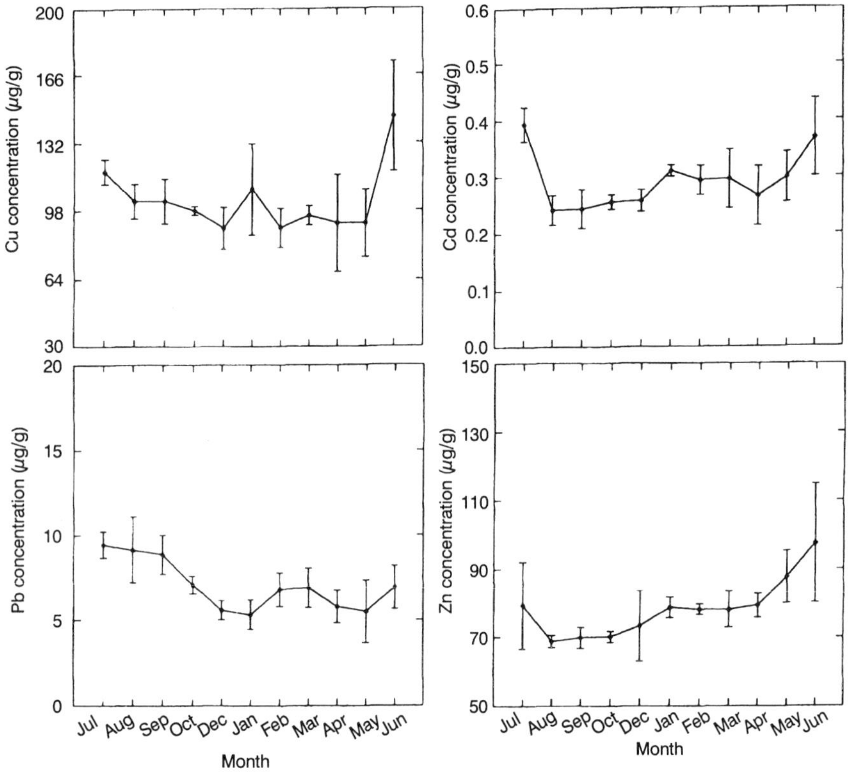

Fig. 3.18. Seasonal profiles for the concentrations (μg g^{-1} dry wt. ± SE) of Cu, Cd, Pb and Zn in *Diastylis rathkei* from Kiel Bay in the period from July 1992 to June 1993, n=4-8 pooled samples per month ($p < 0.0001$). After Swaileh and Adelung (1995); modified.

wt. after the Chernobyl accident (Kanisch et al., 1995) was ca. fifteen-times greater than that from the Pomeranian Bay (Bojanowski et al., 1995). Elevated concentrations of ^{90}Sr and ^{110m}Ag in this crustacean from the Gulf of Finland in 1986 and 1987, respectively (Fig. 3.19) were attributable to an influence by the Chernobyl deposition. According to Agnedal (1988) the highest level of ^{110m}Ag (620 Bq kg^{-1} dry wt.) in this crustacean from the south western part of the Bothnian Sea originated from the Chernobyl accident emission.

Several authors (Skwarzec, 1995; Skwarzec and Bojanowski, 1988; HELCOM, 1995) reported concentration data for plutonium ($^{239+240}$Pu) in Baltic crustaceans such as *Saduria entomon*, *Balanus improvisus* and *Gammarus* sp. The concentrations of plutonium in the latter species from the Gulf of Gdańsk were somewhat greater in 1986 than in 1987 and 1988 (Skwarzec and Bojanowski, 1988). This trend is attributed to the Chernobyl deposition in 1986.

The concentrations of polonium (^{210}Po) in whole body of *Saduria entomon* from the Gulf of Gdańsk and the Pomeranian Bay were within the range of 29.5–54 Bq kg^{-1} dry wt. (Skwarzec, 1995; Bojanowski et al., 1995; Stepnowski and

TABLE 3.10.
Concentrations of radionuclides (Bq kg^{-1}) in crustaceans of the Baltic Sea and other northern areas

Region	Sampling date	Body part	N	110m-Ag (dry wt.)	60-Co (wet wt.)	134-Cs (wet wt.)	137-Cs (wet wt.)	54-K (dry wt.)	54-Mn (dry wt.)	210-Po (dry wt.)	239+240-Pu (dry wt.)	References
Saduria entomon												
Baltic Sea	1983–84	Whole body					4.0*					Agnedal, 1988
	1986	Whole body		136		46.8	85.4*		18.8			Agnedal, 1988
Gulf of Gdansk	1985	Whole body								48.5–54.0	0.025–0.089	Skwarzec, 1995
Southern Baltic	1996	Chitinous shell								0.6±0.1		Stepnowski and Skwarzec, 2000
		Whole body								29.5±0.8		
Pomeranian Bay	1993	Whole body					3.8			15		Bojanowski et al., 1995
Gulf of Finland	1984–91	Whole body	8		2.5–19		< 20–550					Kanisch et al., 1995
	1986	Whole body		59	ND	25	58	260	1.6			Ilus et al., 1987
	1989–90	Whole body	5								0.084	Kanisch et al., 1995
	1989–90	Whole body		16–20	7.2–7.8*	11.0–12.0*	69–92*	230–250	ND			Ilus et al., 1992
Bothnian Sea	1986	Whole body	2	49–180	ND	26–74*	49–150*	220–300	ND–8.8			Ilus et al., 1987
Common shrimp (*Crangon crangon*)												
Gulf of Gdansk	1985	Whole body								78.9±2.7		Skwarzec, 1995
Pomeranian Bay	1993		5				7.1±2.0			40±6		Bojanowski et al., 1995
Gammarus sp.												
Gulf of Gdansk	1985	Whole body								60.2±3.0	60±8	Skwarzec, 1995 Skwarzec and Bojanowski, 1992
Pontoporeia affinis												
Baltic Sea	1986	Whole body		590 412–772		82.3 74–91	193* 157–229					Agnedal, 1988
Barnacle (*Balanus improvisus*)												
Gulf of Gdansk	1985	Whole body									8±1	Skwarzec and Bojanowski, 1992

* – Dry wt.

TABLE 3.10. – continued

Region	Sampling date	Body part	N	103-Ru (Bq kg^{-1} d.w.)	90-Sr (Bq kg^{-1} d.w.)	Th (tot.) (μg g^{-1} d.w.)	U (tot.) (μg g^{-1} d.w.)	234-U (Bq kg^{-1} d.w.)	235-U (Bq kg^{-1} d.w.)	238-U (Bq kg^{-1} d.w.)	References
Saduria entomon											
Baltic Sea	1986	Whole body		26.2							Agnedal, 1988
				8.3–44							
Gulf of Gdansk	1985	Whole body	15(604)*			0.33±0.02	0.032±0.004				Szefer and Wenne, 1987
						0.16–0.70	0.027–0.044				
								1.04–1.66	0.04–0.08	0.75–1.46	Skwarzec, 1995
Gulf of Finland	1986	Whole body		6							Ilus et al., 1987
	1989–90	Whole body	2		26–28						Ilus et al., 1992
Bothnian Sea	1986		2	ND–22							Ilus et al., 1987
Common shrimp (*Crangon crangon*)											
Gulf of Gdansk	1985	Whole body	1(5)			< 0.10	< 0.10				Szefer and Wenne, 1987
								2.86±0.09	0.12±0.02	2.48±0.08	Skwarzec, 1995
Gammarus sp.											
Gulf of Gdansk	1985	Whole body	1(98)			0.32	0.76				Szefer and Wenne, 1987
Barnacle (*Balanus improvisus*)											
Gulf of Gdansk	1985	Whole body	2(1350)			0.04	0.05				Szefer and Wenne, 1987
						0.01–0.06	0.02–0.07				

* – No. of specimens in parentheses

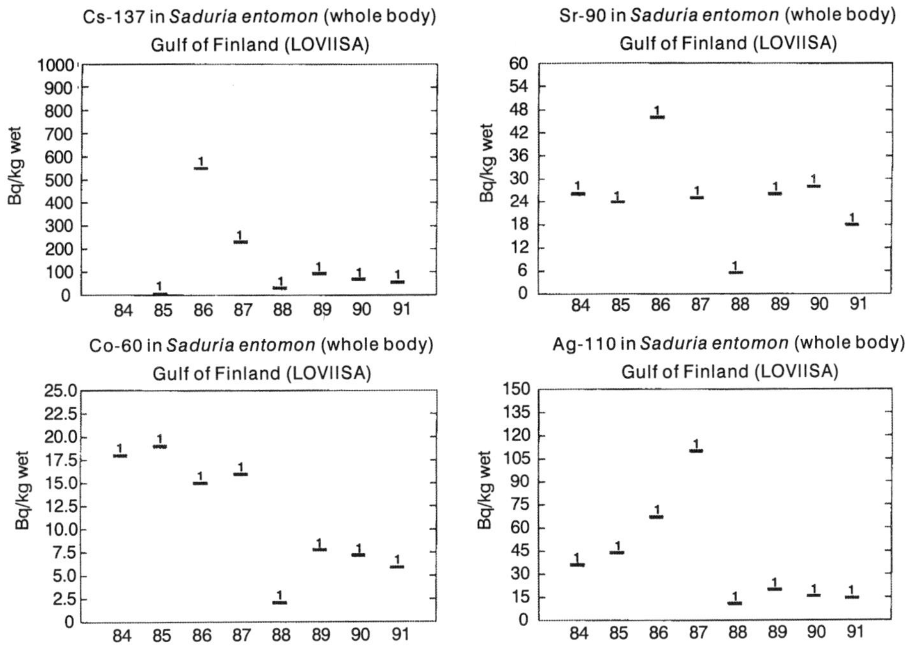

Fig. 3.19. Activities of some radionuclides in *Saduria entomon* from the Loviisa area, Gulf of Finland. After Kanisch et al. (1995); modified.

Skwarzec, 2000a). Intertissue studies of this crustacean indicated that the highest values of polonium were accumulated in the hepatopancreas (Skwarzec and Falkowski, 1988; Stepnowski and Skwarzec, 1999, 2000a).

3. ZOOBENTHAL WORMS AND ASTEROIDS

(i) Introduction

General Characteristics and Taxonomy

Phylum: Echinodermata
Class: Asteroidea
Species: Common sea star, syn. Starfish (*Asterias rubens* L.)

Habitat and range: distributed along coastal waters of the north-eastern Atlantic Ocean; inhabits shallow waters of the Barents, White and Baltic Seas; in the Baltic Sea reaches its western part to salinity of 8‰; i.e. near the Rügen (Bielyaev, 1988). *Food habits*: predator – feeds mainly on small snails (*Hydrobia ulvae*), mussels (*Mytilus edulis* and *Macoma balthica*) and their spawn (Arndt, 1969; Anger et al., 1977).

Phylum: Annelida
Class: Polychaeta
Family: Polynoidae
Species: *Harmothoe sarsi* (Kinberg, 1985), Syn. *Antinöella sarsi*

Habitat and range: Arctic-boreal species distributed along American and European coasts of north Atlantic from the Arctic Ocean to the North Sea; range: occurs in whole the Baltic Sea including both the Bothnian Bay and Gulf of Finland. In the Landsort Deep observed to water depth of 450 m. It is nectobenthic organism, i.e. lives on the bottom as well as in pelagic zone. (Mulicki, 1957; Bick and Gosselck, 1985). *Food habits*: predator (carnivore) – feeds on zooplankton, meiofauna, Oligochaeta, juvenile stadium of *Pontoporeia* spp, *Macoma balthica*, *Harmothoe sarsi* (Abrams et al., 1990; Hill et al., 1990; Ankar, 1997).

Class: Polychaeta
Family: Nereidae
Species: Ragworm (*Nereis diversicolor* O.F. Müller, 1776)

Habitat and range: Atlantic-boreal, eurythermal and euryhaline species. Inhabits mainly brine and estuarine waters at both European and American coast of north Atlantic and adjacent. Occurs in whole the Baltic Sea. *Food habits*: Omnivores (carnivore, deposit and suspension feeder) – feeds on young specimens of *C. volutator* and mussels (*Cardium*, *Macoma*) (Mulicki, 1957; Reise, 1979; Goerke, 1971; Bick and Gosselck, 1985; Svieshnikov, 1987; Bick and Arlt, 1993).

Phylum: Priapulidae
Class: Priapulida
Species: *Halicryptus spinulosus* Siebold

Habitat and range: Arctic species distributed in northern seas; occurs in the Barents, White, Kara, Laptiev Seas and in waters of the Greenland. In the Baltic Sea is a relict species from the Yoldia Sea era and ranges from the Danish Straits to the Aland Islands. *Food habits*: predator – feeds on *Hydrozoa*, *Halicryptus spinulosus*, *Harmothoe sarsi*, *Pygospio elegans*, *Naididae*, *Peloscolex benedeni*, *Pontoporeia affinis*) (Miner, 1950; Mulicki, 1957; Sarvala, 1971; Ankar, 1977; Wiktor, 1985; Ioffe, 1987).

Overview of Worldwide Literature

Analyses of the estuarine benthal worms have been leaded mainly by Bryan and Hummerstone (1971, 1973a, 1973b, 1977), Renfro (1973), Bryan (1974, 1976, 1980), Bryan and Gibbs (1980a, 1980b, 1987), Klöckner (1979), Gibbs and Bryan (1980a, 1980b), Langston (1980, 1986), Bryan et al. (1985), Packer et al. (1980), Ray et al. (1980), Gibbs et al. (1981, 1983), Luoma and Bryan (1982), Bryan and Gibbs (1983, 1987), Luoma (1983), Howard and Brown (1983), Amiard et al. (1987); Jenner and Bowmer (1990), Everaarts and Saraladevi (1996), Ali et al. (1997), Saizsalinas and Franceszubillaga (1997) and Bernds et al. (1998). The distribution of trace elements has been also analysed in marine asteroid *Asterias rubens* (Temara et al., 1997).

Accumulation, tissue distribution and loss of ^{237}Pu, ^{241}Am and ^{242}Cm were examined with the tissues of the polychaete *Hermione hystrix*, the echinoderms *Stichopus regalis* and *Ophiura texturata* (Grillo et al., 1983). Pentreath (1981) has presented an overview on the biological availability to Polychaeta *Nereis diversicolor* and seastar *Asterias forbesi* of transuranium and other long-lived nuclides. Concentrations of selected radionuclides have been analysed sporadically in some organisms, e.g. seastar (Galey et al., 1983).

(ii) Occurrence of Chemical Elements in Benthal Worms and Asteroids

Other representative species of Baltic zoobenthic community, i.e. Polychaeta, Priapulida, Asteroidea have been also studied in this respect (Lithner, 1974; Sandler, 1984; Brügmann and Lange, 1988; Szefer, 1986, Szefer and Kusak, 2000).

As in the case of crustaceans, pronounced interspecies pattern was observed for selected trace elements in some Baltic species belonging to the Polychaeta class (Table 3.11). For instance, the levels of Co, Cr, Cu, Fe and Zn in *Gammarus* sp. from the Gulf of Gdańsk were significantly higher than those in *Nereis diversicolor* from the Gulf (Szefer and Kusak, 2000).

As can be seen in Table 3.11 within burrowing Polychaeta class, *Nereis diversicolor* from the Gulf of Gdańsk concentrated smaller amounts of Co, Cr, Cu, Fe, Mn and Zn as compared with that inhabited adjacent region such as UK coastal areas. Opposite spatial tendency was observed for Cr, Mn and Ni indicating their higher values in Baltic *Nereis* (Bryan et al., 1985; Langston, 1986; Szefer and Kusak, 2000). This spatial pattern may be caused by different state of metal contamination as well as their various biological availability in both the Polish and British coastal zones.

Distinct spatial differences were detected for content of Cu, Fe, Mn, Se, Zn and especially Cu in *Asterias rubens* and these are attributed to different hydrographic conditions and to the composition of the bottom sediments acting as a substrate for their prey, i.e. mussels and snails (Brügmann and Lange, 1988).

Parallel analyses of *Asterias rubens* arms and the central discs showed that Cu, Fe, Hg and Zn levels were from 16 to 30% higher and Ca Mg, Mn, Pb and Se levels were from 4 to 9% higher in the arms. Concentrations of Cd were 20% greater in the central discs as compared to the arms (Brügmann and Lange, 1988).

(iii) Occurrence of Radionuclides in Benthal Worms

There are only a few data for selected radionuclides ($^{239+240}$Pu, ^{210}Po) in Baltic zoobenthos other than crustaceans, i.e. *Halicryptus spinulosus*, *Antinöella sarsi*, *Nereis diversicolor* and *Asterias rubens*. The concentrations of selected radionuclides in zoobenthal worms from the Baltic Sea are listed in Table 3.12. The

TABLE 3.11.

Concentrations of chemical elements ($\mu g\ g^{-1}$ dry wt.) in Priapulida, Polychaeta and other zoobentic organisms from the Baltic Sea and other northern areas

Region	Sampling date	Length (mm)	N	Ag	Al	As	Ca	Cd	Co	Cr	Cu	Fe	References
							POLYCHAETA						
Harmothoe sarsi													
Gulf of Gdansk	1981	< 25–>30	2 (205)				4.5*	0.61	1.93		1.8	5700	Szefer, 1986
							3.5–5.5	0.41–0.82	1.42–2.43		0.3–3.2		
Ragworm (*Nereis diversicolor)*													
Gulf of Gdansk			2	< 0.5	2690±2540		137±53.2*	1.63±2.14	0.75±0.67	6.69±4.67	14.8±7.45	725±320	Szefer and Kusak, 2000
					890–4480		101–176	0.12–3.14	0.27–1.22	3.38–10.0	9.48–20.0	497–954	
UK southwest areas				1.5±0.8		19.9±4.7		0.7±1.0		0.6±0.4	46±22	448±163	Langston, 1986
UK estuarine waters				0.4–3.1		14.3–29.8		0.06–3.8		0.07–1.6	19.0–97.0	265–966	
				0.1–18.0		8.0–84.0		0.14–5.0	5.1–14.2	.3–0.6	19.0–1430	349–739	Bryan et al., 1985
Pontoporeia affinis													
Bothnian Sea (open sea)			15								97.8		Sandler, 1984
Bothnian Bay											90–130		
Bay of Skelleften											120–195		Lithner, 1974
							PRIAPULIDA						
Halicryptus spinulosus													
Gulf of Gdansk	1981	5.0–10	1 (51)				320*	0.67	< 1.20		2.5	5640	Szefer, 1986
Bothnian Sea			1 (7)								45		Sandler, 1984
							ASTEROIDEA						
Common sea star (*Asterias rubens)*													
Western Baltic	1984	65–100	104					0.42±0.18	0.45±0.26		7.1±5.5	25±13	Brügmann and Lange, 1988
								0.10–1.12	0.12–1.32		1.8–33.7	2.0–61.0	

* – mg g^{-1} dry wt.

TABLE 3.11. – continued

Region	Sampling date	Length (mm)	N	Hg	K	Mg	Mn	Na	Ni	Pb	Sn	Zn	References
							POLYCHAETA						
Harmothoe sarsi													
Gulf of Gdansk	1981	< 25–>30	2 (205)			3.5*	37		6.8	14		157	Szefer, 1986
						3.2–3.8	34–39		6.3–7.3	12.0–16.0		73–240	
Ragworm (*Nereis diversicolor)*													
Gulf of Gdansk			3		19.35±5.72*	2.08±0.58*	51±33.5	13.96±2.45*	20.6±14.9	25.9±29.9		340±210	Szefer and Kusak, 2000
					15.30–23.40	1.67–2.49	27.3–74.7	12.22–15.69	10.0–31.1	4.73–47.1		194–490	
UK southwest areas				0.05–2.5			5.7–14.1		2.3–13.3	2.0–685	0.08–1.30	163–470	Langston, 1986
UK estuarine waters				0.91±0.62			26.0±22.0		4.8±1.8	9.5±4.2	0.55±0.45	196±45	Bryan et al., 1985
				0.26–2.8			9.0–123		1.8–9.0	3.2–21.3	0.12–1.76	130–294	
Pontoporeia affinis													
Bothnian Sea (open sea)			15									74	Sandler, 1984
												56–137	
Bothnian Bay												50–100	Lithner, 1974
Bay of Skelleften													
							PRIAPULIDA						
Halicryptus spinulosus													
Gulf of Gdansk	1981	5.0–10	1 (51)			2.3*	23		7.9	10		310	Szefer, 1986
Bothnian Sea			1									213	Sandler, 1984
							ASTEROIDEA						
Common sea star (*Asterias rubens)*													
Western Baltic	1984	65–100	104	0.06±0.022			29±8		9.6±2.3	0.45±0.26		268±61	Brügmann and Lange 1988
				0.017–0.163			13–51		4.2–18.5	0.12–1.32		158–460	

* – mg g^{-1} dry wt.

TABLE 3.12.

Concentrations of radionuclides in Priapulida, Polychaeta and Asteroidea of the Baltic Sea and other northern areas

Region	Sampling date	N	210-Po (Bq kg^{-1} d.w.)	239+240-Pu (Bq kg^{-1} d.w.)	90-Sr (Bq kg^{-1} w.w.)	Th (tot.) (μg g^{-1} d.w.)	U (tot.) (μg g^{-1} d.w.)	References
				PRIAPULIDA				
Halicryptus spinulosus								
Gulf of Gdansk	1982–85		53.1±2.1					Skwarzec and Falkowski, 1988
	1987			0.957±0.070				Skwarzec and Bojanowski, 1992
	1981	1(51)				< 0.05	< 0.05	Szefer and Wenne, 1987
				POLYCHAETA				
Antinöella sarsi								
Gulf of Gdansk	1982–85		72.5±7.4					Skwarzec and Falkowski, 1988
	1988			0.169±0.045				Skwarzec and Bojanowski, 1992
	1981	2(205)				0.11	0.28 0.24–0.32	Szefer and Wenne, 1987
	1985				0.11(N=1)			
Ragworm (*Nereis diversicolor)*								
Gulf of Gdansk	1985		57.3					Skwarzec and Falkowski, 1988
				ASTEROIDEA				
Starfish (*Asterias rubens)*								
Belt Sea	1989			22				Kanisch et al., 1995
	1990			28				

higher levels of polonium are related to polychaeta, priapulida and malacostraca and lower to molluscs (Skwarzec and Falkowski, 1988). It should be emphasised that level of plutonium in *Halicryptus spinulosus* was an order magnitude higher than that in *Antinöella sarsi* from the Gulf of Gdańsk (Skwarzec and Bojanowski, 1992).

D. FISH

(i) Introduction

The number of Baltic marine fish species decreases from 57 in Arkona Basin up to 22 in the Gulf of Finland. Cod, herring, sprat, plaice and brill are typical, marine species spawning in the Baltic Proper. Also other marine species occur here sporadically, e.g.: anchovy, whiting, horse mackerel and mackerel. These species, however, do not spawn in the Baltic Sea. Some authors consider Baltic herring and sprat to be a separate subspecies, typical for this water body. A high variability in marine fish species number is observed, as a consequence of different intensity of saline water inflows from the North Sea. It is especially true for species which embryo stages incubate in pelagic zone, and thus, water salinity (density) and oxygen conditions determine their embryos survival. Characteristic feature of this region is occurrence of many fresh-water species, e.g. perch *Perca fluviatilis* being very abundant in coastal waters (Falandysz et al., 2000).

General Characteristics and Taxonomy

Order: Gadiformes
Suborder: Gadoidei
Family: Gadidae
Species: Cod (*Gadus morhua*)

Habitat and range: this bottom fish lives in the North Atlantic and neighbouring seas (Rutkowicz, 1982); breeds in southern Baltic and the waters of Island (March–June), the North Sea (April–July), the water of Newfoundland (December–March) (Rutkowicz, 1982). Baltic cod (*G. morhua callarisa*) and White Sea cod (*G. morhua maris-albi*) are typical for the Baltic and White Seas, respectively (Marti, 1983). *Food habits*: its diet consists mainly from fish (herring, mackerel, capelin) and it feeds also on crustaceans, mussels and squids (Cięglewicz et al., 1972; Rutkowicz, 1982; Marti, 1983).

Species: Whiting (*Merlangus merlangus*)

Habitat and range: this fish occurs mainly at water depth of 30–100 m in the Atlantic and Mediterranean coasts (Rutkowicz, 1982); it is distributed in western part of the Mediterranean Sea, waters of the North Sea, Irish Sea and the waters of Island; observed also in the Black Sea and south-western waters of the Barents Sea (Rutkowicz, 1982; Marti, 1983). *Food habits*: young specimens feed on plankton, older fish are eaters of fish, e.g. herring, as well as crustaceans (Rutkowicz, 1982; Marti, 1983).

Species: Fourbearded rockling (*Enchelyopus cimbrius*)

Habitat and range: this bottom fish occurs in shelf waters of the north-western Atlantic, e.g. the North Sea and Norwegian Sea; prefers mainly silty bottom at water depth of 20–270 m (Rutkowicz, 1982; Marti, 1983); recorded in European waters from the Bay of Biscay to western part of the Baltic Sea, the waters of Island and south-western part of the Barents Sea. It is observed in shelf waters of the North America from the North Carolina to the Gulf of Saint Lawrence (Rutkowicz, 1982; Marti, 1983). *Food habits*: feeds on crustaceans, molluscs and small fish (Rutkowicz, 1982; Marti, 1983).

Species: Haddock (*Gadus aeglefinus*)

Habitat and range: it occurs in shelf waters of the North Atlantic at water depth of ca. 300 m (Rutkowicz, 1982); its range very similar to recorded for Cod (*Gadus morhua*) (Rutkowicz, 1982; Marti, 1983). *Food habits*: feeds on crustaceans, molluscs, bottom worms and fish (Rutkowicz, 1982).

Suborder: Macruroidei
Family: Macruridae
Species: Grenadier (*Macrurus rupestris*)

Habitat and range: occurs in subarctic and boreal waters of the North Atlantic (Rutkowicz, 1982; Sazonov, 1983); recorded from the White Sea to Island, the Greenland to Labrador, Newfoundland and Nova Scotia (Rutkowicz, 1982). *Food habits*: feeds on batypelagic crustaceans and small zoobenhal organisms (Sazonov, 1983).

Order: Clupeiformes
Suborder: Clupeoidei
Family: Clupeidae
Species: Herring (*Clupea harengus*)

Habitat and range: this pelagic fish lives in the North Atlantic and neighbouring seas (Rutkowicz, 1982; Rass, 1983a); its distribution is very similar to that of Cod (*G. morhua*). *Food habits*: it feeds mainly on zooplankton.

Species: Sprat (*Sprattus sprattus*)

Habitat and range: this pelagic fish occurs in shelf seas of the (Rutkowicz, 1982; Rass, 1983a); it is frequently observed in the Black and Adriatic Seas. In breeding season its shoal is found in the Baltic Sea (May–July) and in the North Sea (January–July). *Food habits*: its main food is plankton (Rutkowicz, 1982; Rass, 1983a).

Suborder: Salmonidei
Family: Salmonidae
Species: Sea trout (*Salmo trutta*)

Habitat and range: it occurs in coastal waters of the North Atlantic (Rutkowicz, 1982; Savvaitova and Miednikov, 1983); migrates to European rivers, from the Iberian Peninsula to the Pechora Sea; observed in the waters of Island, the White and Baltic Seas as well as in the Black and Aral Seas (Savvaitova and Miednikov, 1983). *Food habits*: feeds mainly on small fish, e.g. Sand eel, young Herring, Smelt sparling, Stickleback, and also small crustaceans (Savvaitova and Miednikov, 1983).

Species: Atlantic salmon (*Salmo salar*)

Habitat and range: lives in shelf waters of the North Atlantic and adjacent seas (Rutkowicz, 1982; Savvaitova and Miednikov, 1983); it may cover very long distance, enters European rivers as far as to their source, from Portugal to the White and Barents Seas; observed among other in the Baltic Sea and North Sea. In contrast to Scandinavian rivers, Salmon enters the Vistula and other Polish rivers sporadically because of their pollution (Rutkowicz, 1982); moreover inhabits coastal waters of the North America, from Connecticut to the Greenland (Savvaitova and Miednikov, 1983). *Food habits*: feeds mainly on small fish and small crustaceans (Rutkowicz, 1982).

Species: Rainbow trout (*Salmo gairdneri*)

Habitat and range: occurs in coastal waters of the North-Eastern part of Pacific and rivers entering the ocean (Rutkowicz, 1982; Savvaitova and Miednikov, 1983); enters rivers in the California and Alaska (Savvaitova and Miednikov, 1983). *Food habits*: feeds mainly on fish and also insects, crustaceans and squids (Rutkowicz, 1982; Savvaitova and Miednikov, 1983).

Suborder: Salmonoidei
Family: Coregonidae
Species: Vendace (*Coregonus albula*)

Habitat and range: occurs in lakes of north-eastern Europe (Savvaitova and Miednikov, 1983); inhabits lakes of the Baltic countries, Murmansk district, lakes in up watershed of the Volga River, the Gulf of Finland. It enters the Neva River to breed in the Lake Ladoga (Savvaitova and Miednikov, 1983). *Food habits*: feeds mainly on plankton (Savvaitova and Miednikov, 1983).

Order: Pleuronectiformes
Suborder: Pleuronectoidei
Family: Pleuronectidae
Species: Flounder (*Platichthys flesus*)

Habitat and range: this bottom fish occurs in European shelf waters from the Barents Sea to the Mediterranean and Black Seas (Rutkowicz, 1982; Ostroumova, 1983); breeds in the Baltic Sea and North Sea during February–June and January–May at the water depth ranging of 20–50 m; occurs also in the White, Black and Azov Seas; visits estuarine and adjacent river waters. *Food habits*: its main food is invertebrates and small bottom fish (Rutkowicz, 1982; Ostroumova, 1983).

Species: Plaice (*Pleuronectes platessa*)

Habitat and range: this bottom fish is recorded in shelf waters of western Europe to the Mediterranean and Black Seas; occurs at water depth of ca. 250 m on sea bottom (Rutkowicz, 1982; Ostroumova, 1983); ranges from south France and Portugal to the Barents and White Seas, neighbourhood of the Island waters, south Greenland and western areas of the Mediterranean Sea; it breeds in the Baltic Sea from May to July, and in the North Sea from January to June (Rutkowicz, 1982; Ostroumova, 1983). *Food habits*: feeds mainly invertebrates, e.g. molluscs, Polychaeta, and small bottom fish (Rutkowicz, 1982; Ostroumova, 1983).

Species: Dab (*Limanda limanda*)

Habitat and range: recorded in the waters of western and northern; occurs at water depth of ca. 20–300 m usually on sandy or muddy sea bottom. (Rutkowicz, 1982; Ostroumova, 1983); ranges from the Biscay Bay to Cheshskoj Deep, frequently recorded in the White Sea, it breeds in the Baltic Sea from April to August, and in the North Sea from February to July (Rutkowicz, 1982; Ostroumova, 1983). *Food habits*: feeds mainly invertebrates, e.g. molluscs and crustaceans (Rutkowicz, 1982).

Species: Witch (*Glyptocephalus cynoglossus*)

Habitat and range: occurs in shelf waters of the North Atlantic (Rutkowicz, 1982); recorded from waters of France to the Barents Sea and from Nova Scotia to Labrador and Greenland. *Food habits*: feeds mainly on small bottom invertebrates (Rutkowicz, 1982).

Suborder: Pleuronectoidei
Family: Bothidae
Species: Turbot (*Psetta maxima*)

Habitat and range: it occurs in European waters of the Atlantic and neighbouring seas (Rutkowicz, 1982); observed in shelf waters of north-western Europe, the North Sea, Baltic Sea, the Mediterranean and Black Seas (Rutkowicz, 1982; Ostroumova, 1983). *Food habits*: feeds mainly on small bottom invertebrates and fish (Rutkowicz, 1982).

Order: Petromyzoniformes
Family: Petromyzonidae
Species: Lampern (*Lampetra fluviatilis*)

Habitat and range: lives in European seas, in western coasts of the North America and southern coastal waters of the Greenland and Island (Rutkowicz, 1982); distributed in shelf waters of north-western Atlantic, the North Sea, the Baltic and Mediterranean Seas; it lives in lakes and ponds (Rutkowicz, 1982; Abakumov, 1983). *Food habits*: feeds on marine carrion, small representatives of bottom fauna, especially crustaceans (Rutkowicz, 1982).

Order: Anguilliformes
Suborder: Anguilloidei
Family: Anguillidae
Species: Eel (*Anguilla anguilla*)

Habitat and range: occurs in central and north-eastern waters of the Atlantic and adjacent seas (Rutkowicz, 1982); lives also in European rivers and lakes from Pechora to rivers entering the Black Sea; recorded in coastal areas of the North Sea, the Baltic and the Mediterranean Seas, waters of the Canary Islands, Azores, Madeira Islands, the Great Britain, Ireland and Island; breeds in the Sargasso Sea (Rutkowicz, 1982; Miednikov, 1983). *Food habits*: feeds on invertebrates and small fish (Rutkowicz, 1982).

Order: Beloniformes
Family: Belonidae
Species: Garfish (*Belone belone*)

Habitat and range: this boreal-Mediterranean species occurs in moderately warm waters of south-western coasts of Europe and north Africa. (Rutkowicz, 1982); in summer it lives in coastal waters entering sometimes estuaries while in winter season is distributed in open sea waters; occurs from Cape Verde to Island

and Norway, the Mediterranean and Black Seas. It breeds in the North Sea and the Baltic Sea from April to September (Rutkowicz, 1982; Parin, 1983). *Food habits*: its main food is fish and crustaceans (Rutkowicz, 1982).

Order: Perciformes
Suborder: Ammodytoidei
Family: Ammodytidae
Species: Sand eel (*Ammodytes tobianus*)

Habitat and range: occurs in shelf waters of north-western Europe (Rutkowicz, 1982; Rass, 1983b); lives in the waters of Island, Greenland, the North Sea and the Baltic Sea; prefers bottom waters (Rutkowicz, 1982). *Food habits*: its main food consists of planktonic crustaceans and benthos (Rutkowicz, 1982).

Suborder: Zoarcoidei
Family: Zoarcidae
Species: Eel-pout (*Zoarces viviparus*)

Habitat and range: lives in shelf waters of north-western Europe (Rutkowicz, 1982); ranges from western waters of the British Isles, Orkney and Shetland Islands to the White Sea; numerous in coastal waters of the Baltic Sea, the North Sea, Norway and Denmark; enters sometimes estuarine waters and ponds (Rutkowicz, 1982; Makuszok, 1983; Muus and Dahlström, 1985); prefers sea bottom (Rutkowicz, 1982). *Food habits*: feeds mainly on bottom invertebrates, e.g. molluscs and crustaceans (Rutkowicz, 1982; Makuszok, 1983).

Suborder: Percoidei
Family: Percidae
Species: Perch (*Perca fluviatilis*)

Habitat and range: occurs in Europe, except Island, Italy and north Scandinavian (Spanovskaja, 1983); observed from Ireland, France, the Netherlands, Denmark and Baltic countries to north Asia, (Spanovskaja, 1983); inhabits lakes, rivers and ponds. *Food habits*: feeds on zooplankton and insects larvas (Spanovskaya, 1983).

Order: Gasterosteiformes
Family: Gasterosteidae
Species: Stickleback (*Gasterosteus aculeatus*)

Habitat and range: occurs in coastal waters of north-western Europe, the North America and Pacific Ocean (Rutkowicz, 1982); ranges from coastal waters of the Black and Mediterranean Seas to the Baltic Sea, Faeroe Islands, Island, Greenland and coastal waters of the North America; recorded in the Pacific Ocean from the Bering Sea to Korea and California; lives also in coastal waters of Murmansk district; inhabits river and pond waters (Rutkowicz, 1982; Rutenberg, 1983). *Food habits*: feeds on plankton, roe and larvas of other fish (Rutkowicz, 1982; Rutenberg, 1983).

Overview of Worldwide Literature

The pollution of fish with heavy metals is still actual problem from both hygenic and ecotoxicological points of view. In several countries, especially participating in International Council for the Exploration of the Sea (ICES) monitoring programmes, a great attention has been paid to determine chemical pollutantssuch as heavy metals in fish. Two main approaches are considered in this respect, namely evaluation of toxic metals in edible tissues (muscle, liver) in relation to human health and assaying of metallic pollutants in estuarine and coastal areas using fish tissues as biomonitors (Phillips, 1980; Cossa et al., 1992). Muscles of fish are recommended to be sensitive and selective biomonitors of Hg pollution of the aquatic ecosystems (Olsson, 1976; Köhler et al., 1986; Cossa et al., 1992; Stronkhorst, 1992; Akagi et al., 1995; Joiris et al., 1995, 1997, 1999, 2000a; Malm et al., 1995a. 1995b; Julshamn and Grahl-Nielsen, 1996). Many pollution studies have been performed concerning distribution of selected trace elements in the muscle and/or liver of different species of fish from all over the world (Eisler and LaRoche, 1972; Nishigaki et al., 1974; Mackay et al., 1975; Tamura et al., 1975; Bebbington et al., 1977; Grimanis et al., 1978; Kobayashi et al., 1979; Plaskett and Potter, 1979; Katsuki et al., 1980; Pinder and Giesy, 1981; Von Westernhagen et al, 1981; Yamamoto and Takizawa, 1982; Greig et al., 1983; Honda et al., 1983b; Jothy et al., 1983; Jacobsen et al., 1986; Norton and Murray, 1983; Windom et al., 1973, 1987; Moriarty et al., 1984; Jensen and Cheng, 1987; Macdonald and Sprague 1988; Steimle et al., 1990; Saiki, 1990; Saiki and Palawski, 1990; Benemariya et al., 1991; Szefer et al., 1993a, 1993b; Chan, 1995; Gibbs and Miskiewicz, 1995; Joiris et al., 1995, 1997, 1999, 2000a; Mathieson and McLusky, 1995; Collings et al., 1996; Dietz et al., 1996; Hellou et al., 1996; Andersen and Depledge, 1997; Prudente et al., 1997; Riget et al., 1997; Cronin et al., 1998; Meador et al., 1998; Catsiki and Strogyloudi, 1999; Roméo et al., 1999; Zauke et al., 1999; Al-Majed and Preston, 2000; Al-Yousuf et al., 2000; Alonso et al., 2000; Andres et al., 2000; Elgethun et al., 2000). Retention of some trace elements in the liver cod has been studied by Lie et al. (1989). Various chemical elements occur in otoliths and scales. In recent years there has been a growing research interest in the distribution of chemical elements of the otoliths of teleosts because of their potential use for distinguish populations of a species, determine migration routes, detect anadromy and for reconstructing the environmental history of individual fish (Edmonds et al., 1991; Thresher et al., 1994; Halden et al., 1995; Secor et al., 1995; Townsend et al., 1995; Thorrold et al., 1997). Fish otoliths are known to be effective accumulators of some heavy metals (Edmonds et al., 1989, 1991, 1992; Mugiya et al., 1991). Metals incorporated into both otoliths and scales were Al, Ba and Sr, and Ba, Sr and Zn, respectively (Mugiya et al., 1991).

Fish have been analysed for concentration of several radionuclides including the subcellular distribution studies. For instance, Durand et al. (1999) investigated the subcellular distribution of ^{210}Po in the liver of the Atlantic mackerel *Scomber scombrus*. The majority of the ^{210}Po was found in the cytosol of the liver cells and ca. 30% of this naturally occurring radionuclide was bound to ferritin and ca. 28% to metallothioneins. Pentreath (1981) has presented an overview on the biological availability of plutonium to place *Pleuronectes platessa*.

(ii) Occurrence of Chemical Elements in Fish

Among fish from the Baltic Sea and the surrounding areas the following species have been studied for heavy metal levels: cod (*Gadus morhua*), herring (*Clupea harengus*), spratt (*Sprattus sprattus*), flounder (*Platichthys flesus*) and sea trout (*Salmo trutta*) (Dąbrowski et al., 1967; Kuźma, 1971; ICES, 1977; Harms, 1975; Enoberg, 1976; Gajewska and Nabrzyski, 1977, 1978; Stoeppler and Nurnberg, 1979; Nuurtamo et al., 1980; Tervo et al., 1980; Protasowicki and Chodyniecki, 1980; Westernhagen et al., 1981; Perttilä et al., 1982a, 1982b; Protasowicki, 1982, 1986a, 1986b, 1989, 1991, 1992; Szefer et al., 1982, 1990a, 1990b; Perttilä et al., 1982a; Protasowicki et al., 1983; Brzezińska et al., 1984; Falandysz and Lorenc-Biała, 1984; Falandysz, 1985, 1986a, 1986b, 1986c; 1992a; Szefer and Falandysz, 1985; Hellou et al., 1992; Vuorinen et al., 1994, 1998; Gajewska et al., 2000; Harms and Kanisch, 2000; Szefer et al., 2000a). Less extensive pollution studies have been performed using other Baltic species such as whiting (*Merlangus merlangus*), fourbearded rockling (*Enchelyopus cimbrius*), flounder (*Pltichthys flesus*), plaice (*Pleuronectes platessa*), turbot (*Psetta maxima*), sea trout (*Salmo trutta*), Atlantic salmon (*Salmo salar*), vendace *Coregonus albula*), whitefish (*Coregonus* sp.), lampern (*Lampetra fluviatilis*), eel (*Anguilla anguilla*), garfish (*Belone belone*), sand eel (*Ammodytes tobianus*), eelpont (*Zoarces viviparus*), stickleback (*Gasterosteus aculeatus*) and perch (*Perca fluviatilis*) (Falandysz and Lorenc-Biała, 1984; Szefer and Falandysz, 1985; Falandysz and Falandysz, 1986; Falandysz and Centkowska, 1986; Falandysz, 1992b; Falandysz et al, 1992; Falandysz and Kowalewska, 1993; Schladot et al., 1997). Butyltin compounds have been analysed in Baltic fish by Kannan and Falandysz (1997a, 1997b) and Senthilkumar et al. (1999). Polemic articles presenting interesting discussions concerning the concentration data reported have been published in Marine Pollution Bulletin (Kannan and Falandysz, 1997b; Robinson et al., 1999).

Protasowicki and Kosior (1987, 1988) reported concentration data for otoliths of cod from southern Baltic. Protasowicki (1989) and Szefer et al. (1990a) determined concentrations of selected metals in particular tissues and organs of Baltic cod.

Metals in soft tissues

Interspecies trends

Tables 3.13 and 3.14 present the concentrations data of selected chemical elements in muscle and liver of fish, respectively. From data listed clearly results that muscle concentrations of Cd, Cu and Zn in sprat and herring are generally

TABLE 3.13.

Concentrations of trace elements ($\mu g\ g^{-1}$ wet wt.) and Ca, Mg, K, Na, N, P and S ($mg\ g^{-1}$ wet wt.) in muscle of fish from the Baltic Sea and other northern areas

Region	Sampling date	Length (cm)	N	Ag	Al	As	B	Br	Ca	Cd	Co	Cr	Cu	F	Fe	References
										GADIDAE						
Cod *(Gadus morhua)*																
Baltic Proper																
Southern Baltic	1964		1										0.5		16.4	Dąbrowski et al., 1967
	1971		3 (9)								0.002		0.31		5.7	Kuźma, 1971
											0.001–0.003		0.20–0.40		3.0–8.3	
	1979		3							0.053			1.3			Brzezińska et al., 1984
										0.008–0.124			0.6–2.3			
	1977–80		160							0.101			0.27			Protasowicki et al., 1983
	1981		97							0.003±0.001			0.21±0.03		2.9±0.5	Falandysz and Lorenc-Biala, 1984
										ND–0.011			0.02–0.53		0.8–9.5	
	1981		70							0.003±0.001	< 0.005–0.014		0.18±0.05		4.0±0.6	Szefer and Falandysz, 1985
										< 0.001–0.045			0.01–1.06		1.1–12.9	
	1983		201							0.005			0.15		3.7	Falandysz, 1986c
										ND–0.057			0.01–1.0		0.73–14.0	
	1973–75	20–85	10							0.035						Gajewska and Nabrzyski, 1977
										< 0.01–0.092						
	1974–77		24[a]							0.023			0.363			Gajewska and Nabrzyski, 1978
										0.01–0.084			0.151–0.552			
Gulf of Gdansk	1983								0.1							Szefer (unpublished data)
Western Baltic																
	1973		15										0.27			ICES, 1977
													0.19–0.95			
	1974–75		21 (119)							0.013			0.51			ICES, 1977
										0.006–0.024			0.08–1.1			
	1975		30							0.003			0.17			ICES, 1977
										0.003–0.007			0.10–0.22			
	1987		60							0.052±0.031			0.30±0.08			Protasowicki, 1991

Northern Baltic																
Gulf of Finland																
Hanko			27							0.056			6.5			Tervo et al., 1980
										0.015–0.163			2.2–20.2			
Kotka			21							0.065			11.5			
										0.019–0.193			1.83–53.3			
Gulf of Bothnia																
Vaasa			22							0.016			4			Tervo et al., 1980
										0.006–0.0022			1.3–8.1			
Pori			4							0.012			7			
										0.007–0.016			5.1–84			
Gulf of Finland,	Pre-1980		3		< 2	0.5	< 0.2	3	0.28	< 0.005	< 0.01	0.01	0.33	1.6	1.6	Nuurtamo et al., 1980
Gulf of Bothnia									0.35–0.31		0.002–0.010	0.002–0.01	0.23–0.52	1.3–2.2	1.3–1.8	
Off Danish coast	1976	40–53	4							< 0.001			0.137			Stoeppler and Nürnberg, 1979
North Sea													0.104–0.164			
Norwegian coast	1973	48–53	6	0.2						0.04		0.3	0.7		6	Lande, 1977
Trondheimsfjord										0.003–0.004		0.2–0.4	0.6–0.8		5.6–6.0	
Whiting *(Merlangus merlangus)*																
Southern Baltic	1973		2 (3)							0.005	< 0.005		0.14–0.20		3.2–3.8	Szefer and Falandysz, 1985 Falandysz and Lorenc-Biała, 1984
Norwegian coast	1973	32	2							0.02		0.3			5.5	Lande, 1977
												0.2–0.4			4.6–6.4	
Haddock *(Gadus aeglefinus)*																
	1973	46–49	3									1.5			27.5	Lande, 1977
												1.0–2.0			23.0–32.0	
Fourbearded rockling *(Enchelyopus cimbrius)*																
Southern Baltic	1981	26–38	3							0.002–0.004	< 0.005		0.07–0.35		4.2–6.2	Szefer and Falandysz, 1985 Falandysz and Lorenc-Biała, 1984

[a] – Fish caught in the southern Baltic and the North Sea.

TABLE 3.13. – continued

Region	Sampling date	Length (cm)	N	Hg	K	Mg	Mo	Mn	N	Na	Ni	Pb	Rb	S	Se	Si	Sn	Zn	References
									GADIDAE										
Cod (*Gadus morhua*)																			
Baltic Proper																			
Southern Baltic	1964		1					0.2											Dąbrowski et al., 1967
	1971		3(9)					0.37 0.30–0.48			0.14							5.3 4.3–6.3	Kuźma, 1971
	1979		3									0.06 0.04–0.10						7 3.0–21.3	Brzezińska et al., 1984
	1977–80		160									1.07						4.5	Protasowicki et al., 1983
	1981		97					0.42±0.09 0.09–1.30				0.13±0.02 0.05–0.41						5.4±0.7 0.8–12.0	Falandysz and Lorenc-Biała, 1984
	1981		70					0.22±0.05 0.05–0.78			0.27±0.08 0.02–1.03	0.06 < 0.01–0.34						3.0±0.5 0.9–8.8	Szefer and Falandysz, 1985
	1983		201								0.081 ND–0.71	0.086 ND–0.45						3.9 1.2–8.4	Falandysz, 1986c
	1973–75	20–85	10	0.046 0.016–0.24								0.17 0.016–0.6							Gajewska and Nabrzyski, 1977
	1974–77		19–26[a]	0.027 0.016–0.056								0.139 0.02–0.23						5.58 1.69–10.9	Gajewska and Nabrzyski, 1978
Gulf of Gdansk	1983				4.17	0.28				0.57									Szefer (unpublished data)
	1990	33–36	2 (3)														0.019 0.014–0.024		Kannan and Falandysz, 1997
Western Baltic																			
	1973		15									0.08 0.04–0.27						4.5 3.7–5.6	ICES, 1977
	1974–75		21(119)									0.08 0.04–0.14						4.6 1.2–9.2	ICES, 1977
	1975		30									0.05 0.03–0.09						3.5 2.3–4.1	ICES, 1977
	1987		60									0.43±0.18						4.65±1.13	Protasowicki, 1991

Northern Baltic																	
Gulf of Finland	1979																
Hanko			27	0.022							0.07					13.1	Tervo et al.,
				0.014–0.042							0.01–0.41					6.6–20.9	1980
Kotka			21	0.076							0.07					14.5	
				0.017–0.173							0.01–0.26					7.0–33.2	
Gulf of Bothnia	1979																
Vaasa			22	0.025							0.03					8	Tervo et al.,
				0.015–0.042							0.02–0.05					4.6–11.8	1980
Pori			4	0.025							0.04					6.7	
				0.022–0.031							0.02–0.09					5.7–7.3	
Gulf of Finland,	Pre-1980		3	0.17	3.4	0.21	< 0.1	0.25	26	0.02	0.03	0.67	2.6	0.27	< 5.0	4.2	Nuurtamo et al., 1980
Gulf of Bothnia				0.099–0.23	2.7–3.8	0.14–0.25		0.22–0.27	25–26	0.005–0.05	0.01–0.04	0.61–0.70	1.8–2.2			3.4–4.9	
Off Danish coast	1976	40–53	4	0.12							< 0.01						Stoeppler and Nürnberg, 1979
North Sea				0.04–0.26													
Norwegian coast	1973	48–53	6	0.04						0.9						8.6	Lande, 1977
Trondheimsfjord										0.8–1.0						7.6–9.6	
Whiting (*Merlangus merlangus*)																	
Southern Baltic	1973		2 (3)	0.005				0.28		0.42	0.05–0.18					1.3–2.1	Szefer and Falandysz, 1985; Falandysz and Lorenc-Biała, 1984
Norwegian coast	1973	32	2	< 0.02						1	0.5					5.8	Lande, 1977
Haddock (*Gadus aeglefinus*)																	
	1973	46–49	3	0.1–0.1						4.5						39	Lande, 1977
										3.0–6.0						30–48	
Fourbearded rockling (*Enchelyopus cimbrius*)																	
Southern Baltic	1981	26–38	3					0.37–2.6		1.2	0.04–0.13					1.4–14.0	Szefer and Falandysz, 1985 Falandysz and Lorenc-Biała, 1984

[a] – Fish caught in the southern Baltic and the North Sea.

TABLE 3.13. – continued

Region	Sampling date	Length (cm)	N	Al	As	B	Br	Ca	Cd	Co	Cr	Cu	F	Fe	References
									CLUPEIDAE						
Herring (*Clupea harengus*)															
Baltic Proper															
Southern Baltic	1964		1									0.69		21	Dąbrowski et al., 1967
	1971		13							0.005		0.58		7.9	Kuźma, 1971
										0.003–0.006		0.41–0.69		7.1–9.2	
	1973–75	19–28	10						0.04						Gajewska and Nabrzyski, 1977
									0.01–0.10						
	1974–77		6–13*						0.022			0.85			Gajewska and Nabrzyski, 1978
									0.01–0.05			0.46–1.1			
	1977		25									0.69			Protasowicki and Chodyniecki, 1980
												0.33–1.1			
	1979		1 (5)						0.047			1.3			Brzezińska et al., 1984
	1977–80		200						0.13			0.88			Protasowicki et al., 1983
									0.06–0.20			0.70–0.99			
	1981		42 (315)						0.02			0.63		8.8	Falandysz and Lorenc-Biała, 1984
									0.007–0.042			0.25–1.4		2.0–13.0	
	1981		34 (285)						0.013±0.003	0.008		0.39±0.10		1	Szefer and Falandysz, 1985
									0.003–0.048	< 0.005–0.019		0.10–0.71		6.7–15.3	
Gulf of Gdansk	1983		187						0.009			0.49		9.9	Falandysz, 1986a
									ND–0.074			0.18–1.0		3.9–17.0	
	Pre-1991		53–60 (365)**						0.062±0.026			0.91±0.22			Protasowicki, 1991
Western Baltic															
	1973		9 (35)									1			ICES, 1977
												0.7–1.4			
	1975		30						0.004			0.47			ICES, 1977
									0.002–0.009			0.31–0.70			
	1976	25–28	10						< 0.002			0.421 0.35–0.55			Stoeppler and Nürnberg, 1979

Northern Baltic															
	1974		10 (102)						0.023			0.39			ICES, 1977
									0.007–0.06			0.30–0.54			
	1975		85						0.04			1.5			ICES, 1977
									0.012–0.06			1.0–1.9			
	1978								0.006			0.54			Perttilä et al., 1982a
Gulf of Finland	1979		4 (80)						0.007			0.5			Tervo et al., 1980
									0.005–0.007			0.45–0.57			
Gulf of Bothnia	1979		2 (40)						0.007			0.67			Tervo et al., 1980
									0.006–0.008			0.66–0.67			
Gulf of Finland and	Pre-1980		3	< 2	0.51	< 0.2	2	0.71	0.005	0.01	0.02	0.6	1	5.9	Nuurtamo et al., 1980
Gulf of Bothnia						0.1–0.2		0.66–0.77	0.002–0.010	0.002–0.01	0.01–0.02	0.44–0.75	0.7–1.1	5.3–6.6	
	1981		5 (70)						0.003			0.45			Perttilä et al., 1982a
									0.002–0.006			0.39–0.54			
Gulf of Bothnia	1979–92	16–62*													Vuorinen et al., 1998
Bothnian Bay									0.002–0.01						
Bothnian Sea									0.002–0.01						
Norwegian coast Trondheimsfjord	1973	32	2						0.02			0.8		14.4	Lande, 1977

* – Weight (g).
** – At least 365 (5–10 individuals in eatch batch).
[a] – Fish caught in the southern Baltic and the North Sea.

TABLE 3.13. – continued

Region	Sampling date	Length (cm)	N	Hg	K	Mg	Mn	Mo	N	Ni	P	References
						CLUPEIDAE						
Herring (*Clupea harengus*)												
Baltic Proper												
Southern Baltic	1964		1				0.24					Dąbrowski et al., 1967
	1971		13				0.63					Kuźma, 1971
							0.44–0.78					
	1973–75	19–28	10	0.03								Gajewska and Nabrzyski, 1977
				0.008–0.08								
	1974–77		7–33[a]	0.024								Gajewska and Nabrzyski, 1978
				0.015–0.053								
	1977		25									Protasowicki and Chodyniecki, 1980
	1979	22–24	1 (5)	0.02								Brzezińska et al., 1984
	1981		42(315)				0.54					Falandysz and Lorenc-Biała, 1984
							0.18–1.8					
	1981		34(285)				0.3			0.39		Szefer and Falandysz, 1985
							0.14–0.72			0.04–1.07		
Gulf of Gdansk	1983		187				0.19			0.1		Falandysz, 1986a
							0.08–0.57			< 0.04–0.47		
	Pre-1991		53–73(365)**	0.019±0.01								Protasowicki, 1991
Firth of Vistula	1997	20–23	6									
Western Baltic	1976	25–28	10	0.063								Stoeppler and Nürnberg, 1979
				0.055–0.071								
Northern Baltic	1979–81	2[b]		0.021±0.009								Perttilä et al., 1982a

Gulf of Finland	1979	15–17	4(80)	0.026								Tervo et al., 1980
				0.011–0.042								
Gulf of Bothnia	1979	14–17	2(40)	0.019								Tervo et al., 1980
				0.019–0.019								
Gulf of Finland and	Pre-1980		3	0.037	2.7	0.22	0.59	< 0.1	10	0.02	2	Nuurtamo et al., 1980
Gulf of Bothnia				0.026–0.044	2.2–3.1	0.17–0.24	0.46–0.70		8.6–11		1.7–2.2	
	Pre-1981		5(70)									Perttilä et al., 1982b
Gulf of Bothnia	1979–92	16–62*										Vuorinen et al., 1998
Bothnian Bay				0.016								
				0.012–0.04								
Bothnian Sea				0.022								
				0.010–0.032								
Norwegian coast	1973	32	2	0.02						0.8		Lande, 1977
Trondheimsfjord												

* – Weight (g).
** – At least 365 (5–10 individuals in eatch batch).
[a] – Fish caught in the southern Baltic and the North Sea.
[b] – Age: 2-year-old

TABLE 3.13. – continued

Region	Sampling date	Length (cm)	N	Pb	Rb	S	Se	Si	Sn	Zn	References
						CLUPEIDAE					
Herring (*Clupea harengus*)											
Baltic Proper											
Southern Baltic	1971		13							5.8	Kuźma, 1971
										5.7–5.9	
	1973–75	19–28	10	0.24							Gajewska and Nabrzyski, 1977
				0.033–1.3							
	1974–77		7–33[a]	0.24						11.2	Gajewska and Nabrzyski, 1978
				0.024–1.0						9.2–12.8	
	1977		25							6.6	Protasowicki and Chodyniecki, 1980
										4.8–10.0	
	1979		1 (5)	0.05						4.7	Brzezińska et al., 1984
	1977–80		200	1.04						8	Protasowicki et al., 1983
				0.01–1.9						7.0–9.5	
	1981		42(315)	0.15						12.6	Falandysz and Lorenc-Biała, 1984
				0.07–0.41						4.9–27	
	1981		34(285)	0.09						8.5	Szefer and Falandysz, 1985
				0.01–0.71						3.9–14.1	
Gulf of Gdansk	1983		187	0.079						9.2	Falandysz, 1986a
				0.006–0.50						4.7–20	
	Pre-1991		53–73(365)*	0.48±0.20						9.0±3.0	Protasowicki, 1991
	1990	25–26	1(6)						0.04		Kannan and Falandysz, 1997
Firth of Vistula	1997	20–23	6						0.078"		
Western Baltic											
	1973		9(35)	0.1						11	ICES, 1977

				0.06–0.20						8.5–14.0	
	1975		30	0.06						4.6	ICES, 1977
				0.03–0.11						3.4–7.3	
	1976	25–28	10	< 0.01							Stoeppler and Nürnberg, 1979
Northern Baltic											
	1974		10(102)	0.12						6.4	ICES, 1977
				0.01–0.85						5.0–8.2	
	1975		85	0.85						25	ICES, 1977
				0.15–1.40						20–32	
	1979–81	2[b]		0.05						3.4	Perttilä et al., 1982a
Gulf of Finland	1979	15–17	4(80)	0.07						4.3	Tervo et al., 1980
				0.06–0.07						3.9–4.9	
Gulf of Bothnia	1979	14–17	2(40)	0.05						5.4	Tervo et al., 1980
				0.05–0.05						5.4–5.4	
Gulf of Finland and	Pre-1980		3	0.04	0.57	1.5	0.18	< 5		17	Nuurtamo et al., 1980
Gulf of Bothnia				0.02–0.07	0.48–0.66	1.2–1.7	0.11–0.22			16.0–18.0	
	Pre-1981		5(70)	0.04						2.7	Perttilä et al., 1982b
				0.04–0.06						2.1–3.0	
Norwegian coast	1973	32	2							16	Lande, 1977
Trondheimsfjord											

* – At least 365 (5–10 individuals in eatch batch).
" – Concentration converted to butyltin ion.
[a] – Fish caught in the southern Baltic and the North Sea.
[b] – age: 2-year-old.

TABLE 3.13. – continued

Region	Sampling date	Length (cm)	N	Al	As	B	Br	Ca	Cd	Co	Cr	Cu	F	Fe	Hg	References
									CLUPEIDAE							
Sprat (*Sprattus sprattus*)																
Baltic Proper																
Southern Baltic	1964		1									1.06		25.1		Dąbrowski et al., 1967
	1971		3 (60)							0.002–0.005		0.8		35		Kuźma, 1971
												0.77–1.02		20.9–48.4		
	1973–75		3						0.062						0.022	Gajewska and
									0.05–0.076						0.007–0.04	Nabrzyski, 1977
	1974–77		4–5[*]						0.151			0.515			0.042	Gajewska and
									0.135–0.175			0.375–0.65			0.033–0.053	Nabrzyski, 1978
	1977		20									0.87				Protasowicki and
												0.56–1.30				Chodyniecki, 1980
	1979		1 (5)						0.026			2.5				Brzezińska et al., 1984
	1977–80		70						0.09			0.8				Protasowicki et al., 1983
	1981		6 (90)						0.026			0.65		9.8		Falandysz and
									0.016–0.034			0.50–0.79		3.8–15.0		Lorenc-Biała, 1984
	1981		5 (75)						0.019±0.005	0.008±0.001		0.40±0.07		13.5		Szefer and Falan-
									0.01–0.036	< 0.006–0.01		0.15–0.53		9.5–19.0		dysz, 1985
Gulf of Gdansk	1983	13	208						0.031			0.59		12		Falandysz, 1986b
									ND–0.13			0.14–1.2		3.4–30.0		
Western Baltic	1977		5									0.77				Protasowicki and Chodyniecki, 1980
Northern Baltic	1974		7 (67)						0.021			0.42				ICES, 1977
									0.01–0.027			0.36–0.50				
	1975		5						0.081			1.2				ICES, 1977
									0.021–0.20			1.0–1.6				
	Pre-1980		<2		0.7	< 0.2	2	1.7	0.01	0.01	0.01	0.48	1.7	6.7	0.56	Nuurtamo et al.,
						0.07–0.2		1.7–1.9	0.01–0.02	0.002–0.01	0.002–0.01	0.46–0.51	1.5–1.8	5.5–7.9	0.40–0.73	1980

[*] – Fish caught in the southern Baltic and the North Sea.

TABLE 3.13. – continued

Region	Sampling date	Length (cm)	N	K	Mg	Mn	Mo	N	Ni	Pb	Rb	S	Se	Si	Zn	References
								CLUPEIDAE								
Sprat (*Sprattus sprattus*)																
Baltic Proper																
Southern Baltic	1964		1			1.3										Dąbrowski et al., 1967
	1971		3 (60)												14	Kuźma, 1971
															11.8–16.7	
	1973–75		3							0.15						Gajewska and Nabrzyski, 1977
										0.083–0.37						
	1974–77		4–5[a]							0.17					13.5	Gajewska and Nabrzyski, 1978
										0.12–0.24					11.0–15.0	
	1977		20												25	Protasowicki and Chodyniecki, 1980
															21.0–35.0	
	1979		1 (5)							0.12					4	Brzezińska et al., 1984
	1977–80		70							1.7					26	Protasowicki et al., 1983
	1981		6 (90)			0.45				0.27					15	Falandysz and Lorenc-Biala, 1984
						0.20–0.91				0.07–0.63					9.6–22.0	
	1981		5 (75)			0.40±0.05			0.33±0.13	0.09±0.02					12.6±0.6	Szefer and Falandysz, 1985
						0.24–0.52			0.08–0.67	0.03–0.15					11.4–13.8	
Gulf of Gdansk	1983	13	208			0.26			0.23	0.1					14	Falandysz, 1986b
						0.02–1.0			ND–1.8	0.029–0.31					3.2–26.0	
Western Baltic	1977		5												25	Protasowicki and Chodyniecki, 1980
Northern Baltic	1974		7 (67)							0.045					10	ICES, 1977
										0.01–0.10					7.0–14.0	
	1975		5							0.71					19	ICES, 1977
										0.11–1.9					17.0–20.0	
	Pre-1980		2	2.3	0.2	1.3	< 0.1	18	0.02	< 0.04	0.51	1.2	0.1	< 5.0	16	Nuurtamo et al., 1980
				2.2–2.4	0.19–0.2			17–18		0.03–0.04	0.46–0.55	1.2–1.3	0.09–0.11			

[a] – Fish caught in the southern Baltic and the North Sea

TABLE 3.13. – continued

Region	Sampling date	Length (cm)	N	Ag	Al	As	B	Br	Ca	Cd	Co	Cr	Cu	F	Fe	References
									PLEURONECTIDAE							
Flounder (*Platichthys flesus*)																
Baltic Proper																
Southern Baltic	1971		3 (12)										0.48		4.5	Kuźma, 1971
													0.47–0.48		3.3–5.3	
	1973–75		63 (116)							0.012			0.38			ICES, 1977
										0.003–0.03			0.13–0.89			
	1973–75									0005–0.023			0.29–0.52			Enoberg, 1976
	1975–75		9							0.051						Gajewska and
										0.01–0.15						Nabrzyski, 1977
	1979									0.14			1			Brzezińska et al., 1980;
																unpublished data
	1987–88		19							0.01			0.4		5.4	Falandysz and Lorenc-
										ND–0.019			0.18–0.47		6.4–7.3	Biała, 1984
	1981		2 (17)							0.003	0.006		0.24		6.8	Szefer and Falandysz,
										0.001–0.004	0.005–0.007		0.23–0.25		4.3–9.2	1985
	1983		103							0.013			0.18		3.4	Falandysz, 1985
										ND–0.11			0.048–0.82		0.57–22.0	
Gulf of Gdansk	1983		1						0.06							Szefer (unpublished
																data)
Northern Baltic	Pre-1980		2		< 2	0.95	0.2	1	0.59	< 0.005	< 0.01	0.02	0.47	1.5	2.8	Nuurtamo et al., 1980
									0.49–0.68	0.002–0.005	0.002–0.01	0.004–0.03	0.46–0.48	0.6–2.5	2.5–3.0	
Plaice (*Pleuronectes platessa*)																
Southern Baltic	1971		3 (12)										0.22		3.7	Kuźma, 1971
													0.17–0.27		2.9–4.1	
	1975		4 (21)							0.064			0.54			ICES, 1977
										0.014–0.11			0.45–1.2			
	1983		27							0.009			0.14		1.9	
										ND–0.037			0.014–0.33		0.46–5.8	
German Bight	1976	19–23	16							< 0.001			0.17			Stoeppler and Nürn-
													0.12–0.22			berg, 1979
Off Dutch coast	1976	34–37	4							< 0.001			0.133			
													0.12–0.142			
Witch (*Glyptocephalus cynoglossus*)																
Norwegian coast	1973	72	1	0.02						0.002		0.2	0.2		6.4	Lande, 1977
									BOTHIDAE							
Turbot (*Psetta maxima*)																
Southern Baltic	1981												0.42			Falandysz, 1985
													0.37–0.46			
	1983									0.012			0.15		2.2	Falandysz, 1985
										ND–0.051			0.031–1.1		0.59–1.0	

TABLE 3.13. – continued

Region	Sampling date	Length (cm)	N	Hg	K	Mg	Mn	Mo	N	Ni	References
						PLEURONECTIDAE					
Flounder (*Platichthys flesus*)											
Baltic Proper											
Southern Baltic	1971		3 (12)				0.26				Kuźma, 1971
							0.22–0.29				
	1973–75		63 (116)								ICES, 1977
	1975–75		9	0.07							Gajewska and Nabrzyski, 1977
				0.018–0.215							
	1979										Brzezińska et al., 1980
	1987–88		19				0.38				Falandysz and Lorenc-Biała, 1984
							0.29–0.46				
	1981		2 (17)				0.19			0.12	Szefer and Falandysz, 1985
							0.08–0.30			0.02–0.23	
	1983		103				0.14			0.14	Falandysz, 1985
							0.015–0.80			0.008–0.81	
Gulf of Gdańsk	1987–88	8.5–37.5	65 (85)	0.052							Falandysz, 1992a
				0.019–0.14							
Northern Baltic	Pre-1980		2	0.12	2.7	0.21	0.39	< 0.1	10	0.03	Nuurtamo et al., 1980
				0.11–0.14	2.3–3.0	0.19–0.23	0.36–0.42		8.8–11	0.01–0.06	
Dutch Estuaries	1985–90		258	0.106–0.107							Stronkhorst, 1992
Plaice (*Pleuronectes platessa*)											
Southern Baltic	1971		3 (12)				0.35				Kuźma, 1971
							0.17–0.58				
	1983		27				0.087				Falandysz, 1985
							0.015–0.24				
German Bight	1976	19–23	16	0.058							Stoeppler and Nürnberg, 1979
				0.039–0.074							
Off Dutch coast	1976	34–37	4	0.025							
				0.01–0.045							
Witch (*Glyptocephalus cynoglossus*)											
Norwegian coast	1973	72	1	0.02						1	Lande, 1977
						BOTHIDAE					
Turbot (*Psetta maxima*)											
Southern Baltic		1983					0.14			0.24	Falandysz, 1985
							0.047–0.45			0.011–1.0	

TABLE 3.13. – continued

Region	Sampling date	Length (cm)	N	P	Pb	Rb	S	Se	Si	Sn	Zn	References
						PLEURONECTIDAE						
Flounder (*Platichthys flesus*)												
Baltic Proper												
Southern Baltic	1971		3 (12)								4.7	Kuźma, 1971
											4.6–5.2	
	1973–75		63 (116)		0.09						5.7	ICES, 1977
					0.02–0.35						3.5–11.0	
	1973–1975				0.04–0.10						7.0–10.0	Enoberg, 1976
	1975–75		9		0.2							Gajewska and Nabrzyski, 1977
					0.043–0.63							
	1979				0.06						1.6	Brzezińska et al., 1980
	1987–88		19		0.2						1	Falandysz and Lorenc-Biała, 1984
					0.10–0.25						3.1–13.0	
	1981		2 (17)		0.08						4	Szefer and Falandysz, 1985
					0.07–0.09						2.5–5.5	
	1983		103		0.12						4.4	Falandysz, 1985
					ND–0.83						1.6–8.5	
Gulf of Gdańsk												
	1990	32–37	1 (3)							0.316		Kannan and Falandysz, 1997
	1997	19.8–25.0	6							0.083"		Senthilkumar et al., 1999
Northern Baltic	Pre-1980		2	1.8	0.05	0.86	2	0.24	10		5	Nuurtamo et al., 1980
				1.7–1.9	0.04–0.06	0.70–1.0	1.7–3.3	0.19–0.29			4.9–5.1	
Plaice (*Pleuronectes platessa*)												
Southern Baltic	1971		3 (12)								3.3	Kuźma, 1971
											2.9–3.8	
	1975		4 (21)		0.12						5.5	ICES, 1977
					0.05–1.1						4.7–10.0	
	1983		27		0.17						4.5	Falandysz, 1985
					ND–0.79						1.1–12.0	
German Bight	1976	19–23	16		< 0.01							Stoeppler and Nürnberg, 1979
Off Dutch coast	1976	34–37	4		< 0.01							
Witch (*Glyptocephalus cynoglossus*)												
Norwegian coast	1973	72	1								7.2	Lande, 1977
							BOTHIDAE					
Turbot (*Psetta maxima*)												
Southern Baltic		1981			0.14						11	Falandysz, 1985
					0.11–0.17						8.0–16.0	
		1983			0.16						5	Falandysz, 1985
					ND–0.50						2.1–12.0	
Gulf of Gdansk	1990	32–35	1 (3)							0.039		Kannan and Falandysz, 1997
	1997	17.5–21.5	6							0.11"		Senthilkumar et al., 1999

" – butyltin ion

TABLE 3.13. – continued

Region	Sampling date	Length (cm)	N	Ag	Al	As	B	Br	Ca	Cd	Co	Cr	Cu	F	Fe	References
										AMMODYTIDAE						
Greater sand eel (*Hyperoplus lanceolatus*)																
Southern Baltic	1981	17	3 (15)							0.035	< 0.003–0.006		0.14		45	Szefer and Falandysz, 1985
										0.012–0.057			0.03–0.32		22–58	
										BELONIDAE						
Garfish (*Belone belone*)																
Southern Baltic	1981	63–66	4							0.011	0.013		0.24		6.85	Szefer and Falandysz, 1985
										0.004–0.031	< 0.005–0.02		0.21–0.29		5.2–8.6	
										SALMONIDAE						
Atlantic salmon (*Salmo salmo*)																
Southern Baltic	1981		2							0.003–0.012	0.007		0.09–0.55		3.0–4.9	Szefer and Falandysz, 1985
																Falandysz and Lorenc-Biała, 1984
Vistula estuary	1975	65–66	4							0.093						Gajewska and Nabrzyski, 1977
										0.043–0.14						
Northern Baltic	Pre-1980		2		4	1.2	< 0.2	2	0.16	0.005	0.01	< 0.01	0.58	0.3	4	Nuurtamo et al., 1980
							0.1–0.2		0.57–0.23	0.002–0.005	0.003–0.01		0.54–0.61	0.1–0.7	3.7–4.1	
Sea trout (*Salmo trutta*)																
Gulf of Bothnia	1990–91	53–73	25							< 0.01		0.05	0.31			Vuorinen et al., 1994
												0.04–0.07	0.25–0.35			
Rainbow trout (*Salmo gairdnerii*)																
Northern Baltic	Pre-1980		5		3	0.27	< 0.2	3	1.3	0.005	0.01	0.01	0.53	0.3	3.8	Nuurtamo et al., 1980
					1.0–7.0				0.10–5.4	0.002–0.005	0.003–0.01	0.003–0.03	0.10–0.75	0.1–0.7	1.8–5.8	
										COREGONIDAE						
Whitefish (*Coregonus* sp.)																
Northern Baltic	Pre-1980		5		< 2	0.016	< 0.2	1	0.6	0.005	0.01	0.02	0.5	1	4.8	Nuurtamo et al., 1980
							0.08–0.2		0.18–1.3	0.001–0.02	0.003–0.03	0.008–0.03	0.42–0.62	0.4–2.3	3.4–7.5	
Vendace (*Coregonus albula*)																
Northern Baltic	Pre-1980		3		< 2	0.29	< 0.2		1.9	0.01	< 0.01	0.01	0.68	1	7.6	Nuurtamo et al., 1980
							0.07–0.2		1.6–2.2	0.005–0.02		0.002–0.02	0.51–0.70	0.4–1.0	6.8–8.4	

Region	Sampling date	Length (cm)	N	Ag	Al	As	B	Br	Ca	Cd	Co	Cr	Cu	F	Fe	References
										ANGUILLIDAE						
Eel (*Anguilla anguilla*)																
Southern Baltic	1981		2							0.004–0.007	0.01		2.03–2.9		4.5	Szefer and Falandysz, 1985
																Falandysz and Lorenc-Biała, 1984
Gulf of Gdansk	1982	36–75	64							0.017		0.22	0.46		10.9	Falandysz and
										0.004–0.041		ND–0.70	0.23–0.87		3.4–27.0	Lorenc-Biała, 1987
	1983	38–70	40[a]							0.019		0.023	0.27		10.8	Falandysz and Falan-
		95–545*								ND–0.037		ND–0.11	0.08–0.67		3.3–33	dysz, 1986
Puck Bay	1983	33–84	59[b]							0.013		0.038	0.39		14.9	Falandysz and Cent-
		50–1590*								0.007–0.079		ND–0.12	0.14–1.7		3.7–42	kowska, 1986
Northern Baltic	Pre-1980		2		2		0.3		4.3	< 0.02	0.01	0.14	0.4		10	Nuurtamo et al., 1980
											0.004–0.01	0.04–0.23				
										PETROMYZONIDAE						
Lampern (*Lampetra fluviatilis*)																
Southern Baltic	1981		1							0.025			0.87			Falandysz and Lorenc-Biała, 1984
										GASTEROSTEIDAE						
Stickleback (*Gasterosteus aculeatus*)																
Southern Baltic	1981		1 (16)							0.042	< 0.005		0.09		127	Szefer and Falandysz, 1985
										ZOARCIDAE						
Eelpont (*Zoarces viviparus*)																
Gulf of Gdansk	1986–89	21–36											0.55		13.4	Falandysz et al., 1992
													0.10–0.92		2.1–49.0	
																Schladot at al., 1997
Western Baltic																
Darßer Ort	1994					2.9[d]							1.0[d]		7[d]	Schladot et al., 1997
										PERCIDAE						
Perch (*Perca fluviatilis*)																
Gulf of Gdańsk	1987–89	20–403*	34										0.26±0.15		2.9±1.4	Falandysz, 1992b
Swina estuary	Autumn 96'	15–32	15							0.02±0.017			0.2±0.03			Szefer et al., 2000a
		1.0–3.0[c]								0.007–0.043			0.2–0.3			
	Spring 97'	14–29	17							0.01±0.00			0.4±0.03			

		1.0–2.0[c]								0.006–0.011			0.3–0.4			
	Summer 96'	16–24	21							0.004±0.000			0.3±0.07			
		1.0–2.0[c]								0.003–0.004			0.2–0.4			
	Winter 96/97'	20–35	20							0.016±0.006			0.2±0.04			
		2.0–3.0[c]								0.012–0.022			0.2–0.3			
Pomeranian Bay	Summer 97'	19–27	10							0.003±0.001			0.14±0.043			
		2[c]								0.002–0.003			0.13–0.15			
Northern Baltic	Pre-1980		5		< 2	0.56	< 0.2	3	1.1	< 0.005	< 0.01	0.01	0.45	1	2.8	Nuurtamo et al., 1980
							0.06–0.2		0.38–2.4				0.32–0.66	0.2–2.4	2.1–4.4	
Pikeperch (*Lucioperca lucioperca*)																
Northern Baltic	Pre-1980		2		< 2		< 0.2		5.4	< 0.005	0.01	0.01	0.34	0.4	2.9	Nuurtamo et al., 1980
									3.7–7.0		0.002–0.02	0.004–0.01	0.31–0.37	0.2–0.7	1.7–4.2	
										MACRURIDAE						
Grenadier (*Macrurus rupestris*)																
Norwegian coast	1973	60–65	2	0						0.004		0.2	0.4		10.2	Lande, 1977
										CYPRINIDAE						
Roach (*Rutilus rutilus*)																
Northern Baltic	Pre-1980		5		< 2	0.25	< 0.2	2	0.7	0.005	< 0.01	0.01	0.68	0.4	5.5	Nuurtamo et al., 1980
							0.06–0.2		0.36–1.1	0.001–0.01		0.002–0.02	0.55–1.0	0.2–0.6	4.8–6.1	
										CARCHARHINIDAE						
Galeus melastomus																
Norwegian coast	1973	55–60	2	0						0.006		0.2	0.6		17.8	Lande, 1977
										SQUALIDAE						
Etmopterus spinax																
Norwegian coast	1973	30–36	2	0						0.004		0.2	1		17.7	Lande, 1977
										ARGENTINIDAE						
Argentina silus																
Norwegian coast	1973	32–36	2	0						0.004		0.2	0.2		6	Lande, 1977

* – weight (g).
[a] – Without any pathological symptoms on skin.
[b] – With some pathological symptoms on skin.
[c] – Age (year).
[d] – dry wt.

TABLE 3.13. – continued

Region	Sampling date	Length (cm)	N	Hg	K	Mg	Mn	Mo	P	N	Na	Ni	References
							AMMODYTIDAE						
Greater sand eel (*Hyperoplus lanceolatus*)													
Southern Baltic	1981	17	3 (15)				3.5					0.13	Szefer and Falandysz, 1985
							3.3–3.7					0.10–0.15	
							BELONIDAE						
Garfish (*Belone belone*)													
Southern Baltic	1981	63–66	4				0.14					0.67	Szefer and Falandysz, 1985
							0.07–0.21					0.44–1.0	
							SALMONIDAE						
Atlantic salmon (*Salmo salmo*)													
Southern Baltic	1981		2				0.13–0.33					006	Szefer and Falandysz, 1985
													Falandysz and Lorenc-Biała, 1984
Vistula estuary	1975	65–66	4	0.034									Gajewska and Nabrzyski, 1977
				0.03–0.04									
Northern Baltic	Pre-1980		2	0.01	3.9	0.25	0.14	< 0.10	2.4	32		0.02	Nuurtamo et al., 1980
				0.078–0.136	3.7–4.1	0.22–0.29	0.10–0.17		2.2–2.6	29–34		0.01–0.02	
Sea trout (*Salmo trutta*)													
Gulf of Bothnia	1990–91	53–73	25	0.10–0.19									Vuorinen et al., 1994
Gulf of Bothnia	1979–92	16–62**											Vuorinen et al., 1998
Bothnian Bay				0.016									
				0.012–0.040									
Bothnian Sea				0.022									
				0.010–0.032									
Rainbow trout (*Salmo gairdnerii*)													
Northern Baltic	Pre-1980		5	0.13	4.2	0.25	0.18	< 0.10	2.6	35		0.02	Nuurtamo et al., 1980
				0.11–0.15	3.3–4.8	0.06–0.33	0.03–0.27		1.7–3.2	31–38		0.01–0.03	
							COREGONIDAE						
Vendace (*Coregonus albula*)													
Gulf of Bothnia	1990	19–67*	40	0.10–0.10									Vuorinen et al., 1998
Northern Baltic	Pre-1980		3	0.14	3.3	0.25	0.87	< 0.10	2.9	28		0.02	Nuurtamo et al., 1980
				0.073–0.20	3.1–3.5	0.22–0.26	0.70–1.1		2.9–3.1	26–30		0.007–0.02	

Whitefish (*Coregonus* sp.)													
	1990	77–1278*	39	0.1									Vuorinen et al., 1998
Northern Baltic	Pre-1980		5	0.13	3.8	0.3	0.49	< 0.10	2.9	33		0.02	Nuurtamo et al., 1980
				0.047–0.39	3.6–4.4	0.29–0.31	0.23–1.3		2.6–3.4	30–36		0.008–0.03	
							ANGUILLIDAE						
Eel (*Anguilla anguilla*)													
Southern Baltic	1981		2				0.3					0.15	Szefer and Falandysz, 1985
													Falandysz and Lorenc-Biała, 1984
Gulf of Gdansk	1982	36–75	64				0.37					0.07	Falandysz and Lorenc-Biała, 1987
							0.17–1.2					ND–0.21	
	1983	38–70	40[a]				0.16					0.04	Falandysz and Falandysz, 1986
		95–545*					0.08–0.34					ND–0.38	
Puck Bay	1983	33–84	59[b]				0.22					0.058	Falandysz and Centkowska, 1986
		50–1590*					0.09–0.38					0.006–0.42	
Northern Baltic	Pre-1980		2	0.088	2.8	0.26	3		3.9	28		0.03	Nuurtamo et al., 1980
				0.082–0.094	2.8–2.9	0.25–0.26	2.9–3.1			27–30		0.01–0.04	
							GASTEROSTEIDAE						
Stickleback (*Gasterosteus aculeatus*)													
Southern Baltic	1981		1 (16)*				11.7					0.73	Szefer and Falandysz, 1985
Gulf of Gdańsk	1988–89	1.0–7.9	124 (837)*	0.01–0.11									Falandysz and Kowalewska, 1993
							ZOARCIDAE						
Eelpont (*Zoarces viviparus*)													
Gulf of Gdańsk	1986–89	21–36					0.32						Falandysz et al., 1992b
							0.09–0.92						
Western Baltic													
Darßer Ort	1994			0.02[d]	18[d]	1.5[d]			10[d]		1.4[d]		Schladot et al., 1997
							PERCIDAE						
Perch (*Perca fluviatilis*)													
Gulf of Gdańsk	1987–89	20–403*	34				0.31±0.16						Falandysz, 1992b
Pomeranian Bay Swina estuary	Autumn 96'	15–32	15	0.05±0.02									Szefer et al., 2000a
		1.0–3.0[c]		0.028–0.067									
	Spring 97'	14–29	17	0.073±0.03									
		1.0–2.0[c]		0.043–0.104									
	Summer 96'	16–24	21	0.07±0.03									
		1.0–2.0[c]		0.042–0.101									

Region	Sampling date	Length (cm)	N	Hg	K	Mg	Mn	Mo	P	N	Na	Ni	References
	Winter 96/97[f]	20–35	20	0.12±0.04									
		2.0–3.0[c]		0.112–0.120									
Szczecin Lagoon	Summer 97[f]	19–27	10	0.046±0.015									
		2[c]		0.041–0.051									
Northern Baltic	Pre-1980		5	0.17	3.3	0.26	0.62	< 0.10	2.4	28		0.03	Nuurtamo et al., 1980
				0.086–0.26	2.8–3.8	0.24–0.26	0.22–1.9		2.2–2.8	28–29		0.006–0.08	
Pikeperch (*Stizostedion lucipera*)													
Gulf of Gdansk	1990	44–56	1 (3)										Kannan and Falandysz, 1997
Northern Baltic	Pre-1980		2	0.12	3.4	0.29	0.28		2.2	32		0.03	Nuurtamo et al., 1980
				0.097–0.14	3.3–3.5		0.24–0.31			31–33		0.02–0.04	
							CYPRINIDAE						
Roach (*Rutilus rutilus*)													
Vistula River	1997	11.5–16.0	4										
Northern Baltic	Pre-1980		5	0.29	3.6	0.26	0.53	< 0.10	2.3	28		0.02	Nuurtamo et al., 1980
				0.074–1.01	3.2–3.8	0.25–0.27	0.21–0.95		2.1–2.5	26–29		0.006–0.04	
							MACRURIDAE						
Grenadier (*Macrurus rupestris*)													
Norwegian coast	1973	60–65	2	0.02								0.8	Lande, 1977
							CARCHARHINIDAE						
Galeus melastomus													
Norwegian coast	1973	55–60	2	0.14								0.8	Lande, 1977
							SQUALIDAE						
Etmopterus spinax													
Norwegian coast	1973	30–36	2	0.08								1.2	Lande, 1977
							ARGENTINIDAE						
Argentina silus													
Norwegian coast	1973	32–36	2	0.04								0.6	Lande, 1977

* – Weight (g).
[a] – Without any pathological symptoms on skin.
[b] – With some pathological symptoms on skin.
[c] – Age (year).
[d] – Dry wt.

TABLE 3.13. – continued

Region	Sampling date	Length (cm)	N	Pb	Rb	S	Se	Si	Sn	Tl	V	Zn	References
							AMMODYTIDAE						
Greater sand eel (*Hyperoplus lanceolatus*)													
Southern Baltic	1981	17	3 (15)	0.23								18.4	Szefer and Falandysz, 1985
				0.15–0.27								17.0–19.3	
							BELONIDAE						
Garfish (*Belone belone*)													
Southern Baltic	1981	63–66	4	0.045								16.2	Szefer and Falandysz, 1985
				0.01–0.10								11.8–22.7	
							SALMONIDAE						
Atlantic salmon (*Salmo salmo*)													
Southern Baltic	1981		2	0.01–0.08								11.2–19.0	Szefer and Falandysz, 1985
													Falandysz and Lorenc-Biała, 1984
Vistula estuary	1975	65–66	4	0.48									Gajewska and Nabrzyski, 1977
				0.13–0.77									
Northern Baltic	Pre-1980		2	0.06	0.84	2.1	0.26	5				4.3	Nuurtamo et al., 1980
				0.03–0.10	0.78–0.88	1.9–2.3	0.20–0.34					3.3–5.1	
Sea trout (*Salmo trutta*)													
Gulf of Gdansk	1990	73–87	2 (9)						0.051				Kannan and Falandysz, 1997
									0.045–0.057				
	1997	17	1						0.078"				Senthilkumar et al., 1999
Gulf of Bothnia	1990–91	53–73	25	0.03–0.09							0.03–0.03	4.34–8.02	Vuorinen et al., 1994
Rainbow trout (*Salmo gairdnerii*)													
Northern Baltic	Pre-1980		5	0.03	3.8	2.4	0.26	5				4.8	Nuurtamo et al., 1980
				0.01–0.05	0.75–7.5	2.0–2.7	0.17–0.34					1.3–6.5	
							COREGONIDAE						
Vendace (*Coregonus albula*)													
Northern Baltic	Pre-1980		3	0.04	3.9	1.9	0.26	< 5				40	Nuurtamo et al., 1980
				0.02–0.07	1.5–6.4	1.7–2.1	0.22–0.33					35–43	
Whitefish (*Coregonus* sp.)													
Northern Baltic	Pre-1980		5	0.07	2.8	2.3	0.37	< 5				12	Nuurtamo et al., 1980
				0.03–0.13	1.0–5.7	2.1–2.5	0.16–0.65					5.2–31	
							ANGUILLIDAE						
Eel (*Anguilla anguilla*)													
Southern Baltic	1981		2	0.05–0.11								13.6–17.0	Szefer and Falandysz, 1985

Region	Sampling date	Length (cm)	N	Pb	Rb	S	Se	Si	Sn	Tl	V	Zn	References
													Falandysz and Lorenc-Biała, 1984
Gulf of Gdansk	1982	36–75	64	0.16								22.3	Falandysz and Lorenc-Biała, 1987
				0.03–0.48								16.0–31.0	
	1983	38–70	40[a]	0.18								18.7	Falandysz and Falandysz, 1986
		95–545*		0.02–0.92								6.3–27	
	1990	28–33	1 (20)						0.13				Kannan and Falandysz, 1997
Puck Bay	1983	33–84	59[b]	0.16								18	Falandysz and Centkowska, 1986
		50–1590*		ND–0.37								9.6–33	
Northern Baltic	Pre-1980		2	.02	4.7	1.8	0.57					27	Nuurtamo et al., 1980
							0.51–0.62					26–27	
						PETROMYZONIDAE							
Lampern (*Lampetra fluviatilis*)													
Southern Baltic	1981		1	0.14								26	Falandysz and Lorenc-Biała, 1984
						GASTEROSTEIDAE							
Stickleback (*Gasterosteus aculeatus*)													
Southern Baltic	1981		1 (16)*	0.85								49	Szefer and Falandysz, 1985
Gulf of Gdańsk	1988–89	1.0–7.9	124										Falandysz and Kowalewska, 1993
			(837)*										
						ZOARCIDAE							
Eelpont (*Zoarces viviparus*)													
Gulf of Gdańsk	1986–89	21–36										21.5	Falandysz et al., 1992b
												13.0–35.0	
Western Baltic													
Darßer Ort	1994			0.19[d]		13[d]	1.0[d]			0.002[d]		50[d]	Schladot et al., 1997
						PERCIDAE							
Perch (*Perca fluviatilis*)													
Gulf of Gdańsk	1987–89	20–403*	34									7.7±2.4	Falandysz, 1992b
Pomeranian Bay Swina estuary	Autumn 96'	15–32	15	0.02±0.007								5.8±1.6	Szefer et al., 2000a
		1.0–3.0[c]		0.007–0.024								3.7–7.6	
	Spring 97'	14–29	17	0.02±0.003								5.1±1.0	
		1.0–2.0[c]		0.019–0.025								4.0–5.7	
	Summer 96'	16–24	21	0.011±0.002								4.9±1.3	
		1.0–2.0[c]		0.009–0.011								3.8–6.2	
	Winter 96/97'	20–35	20	0.03±0.005								4.3±0.5	
		2.0–3.0[c]		0.023–0.033								3.8–4.7	
Szczecin Lagoon	Summer 97'	19–27	10	0.013±0.006								5.84±1.16	

		2[c]		0.013–0.014						5.78–5.90	
Northern Baltic	Pre-1980		5	0.03	4.4	2.3	0.28	< 5		8.1	Nuurtamo et al., 1980
				0.02–0.06	1.3–13	2.2–2.4	0.20–0.30			6.0–9.6	
Pikeperch (*Stizostedion lucipera*)											
Gulf of Gdansk	1990	44–56	1 (3)						0.455		Kannan and Falandysz, 1997
Northern Baltic	Pre-1980		2	0.02	1.4	2.6	0.33			5.7	Nuurtamo et al., 1980
				0.02–0.03						4.8–6.6	
Ruff (*Acerina cernua*)											
Firth of Vistula	1997	13.3–14.0	6						0.044"		Senthilkumar et al., 1999
						SCOMBRIDAE					
Mackerel (*Scomber scombrus*)											
Gulf of Gdansk	1990	36–37	2 (5)						0.027		Kannan and Falandysz, 1997
						CYPRINIDAE					
Roach (*Rutilus rutilus*)											
Puck Bay	1997	14.5	1						3.3"		Senthilkumar et al., 1999
Vistula River	1997	11.5–16.0	4						0.1"		
Northern Baltic	Pre-1980		5	0.03	4	2.1	0.24	< 5		12	Nuurtamo et al., 1980
				0.01–0.04	1.3–7.6	2.0–2.1	0.15–0.53			6.9–19	
						MACRURIDAE					
Grenadier (*Macrurus rupestris*)											Lande, 1977
Norwegian coast	1973	60–65	2							11	
						CARCHARHINIDAE					
Galeus melastomus											
Norwegian coast	1973	55–60	2							9.8	
						SQUALIDAE					
Etmopterus spinax											
Norwegian coast	1973	30–36	2							12.6	
						ARGENTINIDAE					
Argentina silus											
Norwegian coast	1973	32–36	2							6.2	

* – Weight (g).
[a] – Without any pathological symptoms on skin.
[b] – With some pathological symptoms on skin.
[c] – Age (year).
[d] – Dry wt.
" – Concentration is converted to butyltin ion.

TABLE 3.14.

Concentrations of trace elements ($\mu g\ g^{-1}$ wet wt.) in liver of fish from the Baltic Sea and other northern areas

Region	Sampling date	Length (cm)	N	Cd	Cr	Cu	Fe	References
				GADIDAE				
Cod (*Gadus morhua)*								
Gulf of Bothnia								
Vaasa	1974–pre-1991	25–35	22	0.016		4		Tervo et al., 1980
				0.006–0.022		1.3–8.1		
Peri	1974–pre-1991	25–35	4	0.012		7		
				0.007–0.016		5.1–8.4		
Gulf of Finland								
Hanka	1974–pre-1991	25–35	27	0.056		6.5		Tervo et al., 1980
				0.015–0.163		2.2–20.2		
Kotka	1974–pre-1991	25–35	21	0.065		11.5		
				0.019–0.193		1.83–53.5		
Southern Baltic	1974–pre-1991		29–33 (> 150)	0.05±0.03		8.83±3.85		Protasowicki, 1991
				CLUPEIDAE				
Herring (*Clupea harengus)*								
Southern Baltic	1974–pre-1991		31–32 (> 160)	0.62±0.29		5.35±1.99		Protasowicki, 1991
				ANGUILLIDAE				
Eel (*Anguilla anguilla)*								
Gulf of Gdansk	1982	< 45 –> 61	48 (208)[a]	0.16	1.6	7.1	190	Falandysz and Lorenc-Biała, 1987
				0.089–0.37	0.11–4.8	3.1–19	79–550	
Gulf of Gdansk	1983	39–70	27 (36)[a]	0.11	0.31	5.17	133	Falandysz and Falandysz, 1986
		100–545*		0.04–0.28	ND–0.78	0.24–16.0	46–240	
Puck Bay	1983	36–81	56 (72)[b]	0.11	0.11	11.2	147	Falandysz and Centkowska, 1986
		58–1000*		0.02–0.81	ND–0.30	0.81–33.0	31–360	

PERCIDAE						
Perch (*Perca fluviatilis*)						
Gulf of Gdańsk	1987–89	20–403*	14		5.4±2.6	Falandysz, 1992b
					1.3–10.0	
Pomeranian Bay	Autumn 96'	15–32	15	0.030±0.008	2.6±1.9	Szefer et al., 2000a
Swina estuary		1.0–3.0[c]		0.021–0.041	0.8–5.2	
	Spring 97'	14–29	17	0.039±0.01	3.1±0.2	
		1.0–2.0[c]		0.031–0.047	2.9–3.2	
	Summer 96'	16–24	21	0.030±0.003	4.2±1.3	
		1.0–2.0[c]		0.022–0.028	2.5–5.4	
	Winter 96/97'	20–35	20	0.058±0.004	4.4±1.9	
		2.0–3.0[c]		0.054–0.062	3.3–6.6	
Szczecin Lagoon	Summer 97'	19–27	10	0.032±0.015	6.48±2.10	
		2[c]		0.031–0.032	5.53–7.43	

* – Weight (g).
[a] – Without any pathological symptoms on skin.
[b] – With some pathological symptoms on skin.
[c] – Age (year).

TABLE 3.14. – continued

Region	Sampling date	Length (cm)	N	Hg	Mn	Ni	Pb	Sn	Zn	References
					GADIDAE					
Cod (*Gadus morhua)*										
Gulf of Bothnia										
Vaasa	1974–pre-91	25–35	22				0.03		8	Tervo et al., 1980
							0.02–0.05		4.6–11.8	
Peri	1974–pre-91	25–35	4				0.04		6.7	
							0.02–0.09		5.7–7.3	
Gulf of Finland										
Hanka	1974–pre-91	25–35	27				0.07		13.1	Tervo et al., 1980
							0.01–0.41		6.6–20.9	
Kotka	1974–pre-91	25–35	21				0.07		14.5	
							0.01–0.26		7.0–33.2	
Southern Baltic	1974–pre-91		29–33 (> 150)	0.021±0.018			0.26±0.13		16.8±5.22	Protasowicki, 1991
Burbot (*Lota lota)*										
Vistula River	1997	21.5–25.0	3					0.032"		Senthilkumar et al., 1999
					CLUPEIDAE					
Herring (*Clupea harengus)*										
Southern Baltic	1974–pre-91		31–32 (> 160)	0.037±0.029			0.70±0.60		35.3±20.32	Protasowicki, 1991
Firth of Vistula	1997	20–23	6					4.8"		Senthilkumar et al., 1999
					PLEURONECTIDAE					
Flounder (*Platichthys flesus)*										
Gulf of Gdańsk	1987–88	8.5–37.5	59 (63)	0.037						Falandysz, 1992a
				0.011–0.080						

				ANGUILLIDAE					
Eel (*Anguilla anguilla*)									
Gulf of Gdansk	1982	< 45–> 61	48 (208)[a]	0.9	0.06	0.21		33.3	Falandysz and Lorenc-Biała, 1987
				0.54–3.2	ND–0.25	0.06–0.90		23–60	
Gulf of Gdansk	1983	39–70	27 (36)[a]	1.34	0.23	0.67		33	Falandysz and Falandysz, 1986
		100–545*		0.11–1.8	0.05–2.2	0.11–20		19–66	
Puck Bay	1983	36–81	56 (72)[b]	1.01	0.15	0.45		48.3	Falandysz and Centkowska, 1986
		58–1000*		0.23–1.7	0.02–1.4	0.03–2.10		12–230	
				PERCIDAE					
Perch (*Perca fluviatilis*)									
Gulf of Gdańsk	1987–89	20–403*	14	3.9±2.9				19±5	Falandysz, 1992b
				1.4–9.3				12.0–30.0	
Pomeranian Bay	Autumn 96'	15–32	15			0.04±0.017		24.8±2.4	Szefer et al., 2000a
Swina estuary		1.0–3.0[c]				0.026–0.067		21.9–27.1	
	Spring 97'	14–29	17			0.05±0.01		27.9±4.0	
		1.0–2.0[c]				0.046–0.056		23.3–30.5	
	Summer 96'	16–24	21			0.023±0.010		26.9±3.3	
		1.0–2.0[c]				0.013–0.036		23.4–30.7	
	Winter 96/97'	20–35	20			0.066±0.003		19.8±2.5	
		2.0–3.0[c]				0.063–0.069		17.5–22.5	
Szczecin Lagoon	Summer 97'	19–27	10			0.031±0.001		24.4±1.87	
		2[c]				0.029–0.033		23.7–25.1	
Vistula River	1997	13.5–15.5	7				0.41"		Senthilkumar et al., 1999

* – Weight (g).
[a] – Without any pathological symptoms on skin.
[b] – With some pathological symptoms on skin.
[c] – Age (year).
" – Concentration is converted to butyltin ion.

greater than those in cod (Table 3.13). Hepatic levels of Cd in herring from the southern Baltic are also higher as compared to those reported for cod from the same area (Table 3.14). This finding has been supported by Swedish monitoring data (Harms, 1996) which indicated that concentrations of Cd in liver of herring are mostly greater than those in liver of cod from southeastern part of Gotland and from the Kattegat. Such interspecies specific difference is supposedly due to the different lipid concentrations in liver of cod and herring. Bearing in mind that cod liver contains generally greater amounts of lipids as compared to herring liver, the accumulative abilities of cod liver in respect to Cd may generally be less effective than those of herring liver. This is in accordance with negative correlation between concentrations of metals and lipids in liver tissue (Harms, 1996).

Intertissue trends

Concentrations of several trace elements were generally greater in liver than in muscle of different species of fish. Szefer et al. (1990a) reported data on intertissue distribution of Cd, Co, Cu, Fe, Mn, Ni, Pb and Zn in cod from the Gulf of Gdańsk. High levels of Cd were observed in kidney and the pyloric caeca although gills contained its highest levels as well as other trace elements, i.e. Co, Ni and Pb. This finding may be attributed to the presence of adsorbed suspended matter on the gills rather than to active biological uptake of these trace metals (Szefer et al., 1990a). Baltic values are highly comparable to enhanced renal Cd values as well as intestine levels of Cd, Co and Pb reported for cod from the coastal region of Norway and from the Barents Sea (Julshamn et al., 1978). According to Protasowicki (1989) among particular tissues and organs of cod, herring and flounder from a southern Baltic, the liver was characterised by maximum levels of Cu; otoliths of cod and flounder accumulated maximum amounts of Cd and Pb.

The concentrations of Cd, Cu, Hg, Pb and Zn were determined in the liver, kidney, gills and muscle of healthy and diseased dab *Limanda limanda* from the German Bight transect (Protasowicki, 1992). A two-way analysis of variance showed that for only about 20% of the cases were observed statistically significant variations between healthy and diseased fish. For instance, higher levels of Cu and Zn were found in livers of healthy specimens while their kidney generally contained less Zn.

Inter-age trends

The influence of the age of perch on hepatic and muscle levels of selected metals was studied by Szefer et al. (2000a). The hepatic and muscle data were treated separately and applied for each season, age class and by station of sampling. The inter-age differences concerned Cd and Hg in muscle (Szefer et al., 2000a).

Inter-sex trends

Protasowicki (1986a) reported sex dependent changes in trace metals concentrations in some organs of Baltic fish. Males of cod, herring and perch contained

higher hepatic levels of Zn than females of these species. In contrast, gills of females were characterised by ca. 5 times greater concentrations of Cu and Zn as compared to those of males. It confirms the importance of these essential elements in fish embryonic development. The reverse distribution pattern for liver of male and female showed that during the female gonad development the essential elements are taken up from the liver. Toxic metals such as Cd and Pb were accumulated more distinguishably in organs of males suggesting that a mechanism of physiological protection against intoxication plays a more important role in females as potential reproductive specimens (Protasowicki, 1986a).

Spatial trends

From Table 3.14 results that concentration of Cu in cod liver was significantly greater in specimens caught at the Gulf of Finland that those from the southern Baltic. The levels of muscle Hg in herring caught in the Bothnian Bay were slightly higher than those in herring from the Baltic Proper and Kattegat (Table 3.13).

According to Perttilä at al. (1982a) both herring (muscle) and cod (liver) exhibited in most cases considerably higher concentrations of trace metals in the Danish sea areas than in the Gulf of Finland and in the Gulf of Bothnia. In the Baltic Sea area, the mean values of the trace metal contents in herring muscle did not differ significantly from one to another. Cod liver, however, exhibited spatial trends which, with the exception of Pb, followed the areal differences of metal concentrations (Zn, Cu, Cd and Hg) in seawater of the northern Gulf of Bothnia with the lowest concentrations, and the eastern Gulf of Finland with the highest ones.

Significantly lower levels of Hg were observed in muscle of *Zoarces viviparus* from Darßer Ort, Baltic Sea, than in that from Meldorf Bay, North Sea, (Schladot et al., 1997).

Hellou et al. (1992) reported concentration data for numerous elements (Ag, As, Ca, Cd, Co, Cs, Cu, Fe, Mg, Mn, Mo, Ni, Rb, Se, Sr, Zn) in the muscle, liver and ovaries of cod from the northwest Atlantic, the levels of Cd, Hg and Pb in muscle and liver were similar or lower than those reported for cod from the Baltic Sea, the North Sea and the North Atlantic.

Some tendency in spatial distribution is observed for muscle Cu which reached maximum values in the Pomeranian Bay in all the age-groups of perch (*Perca fluviatilis*) and during all the seasons of their capture (Szefer et al., 2000a).

Temporal trends

The temporal changes of Pb levels in cod liver caught south-easterly of Gotland and from the Kattegat indicated negative trend of ca. 5% yr^{-1} during 1981–94 (Fig. 3.20). However time series of hepatic Pb in herring from the Bothnian Bay, Bothnian Sea, Baltic Proper and Kattegat as well as data on cod from Polish zone of the Southern Baltic showed insignificant trends in respect to the statistical assessment. The data obtained for flounder from the Belt Sea and the

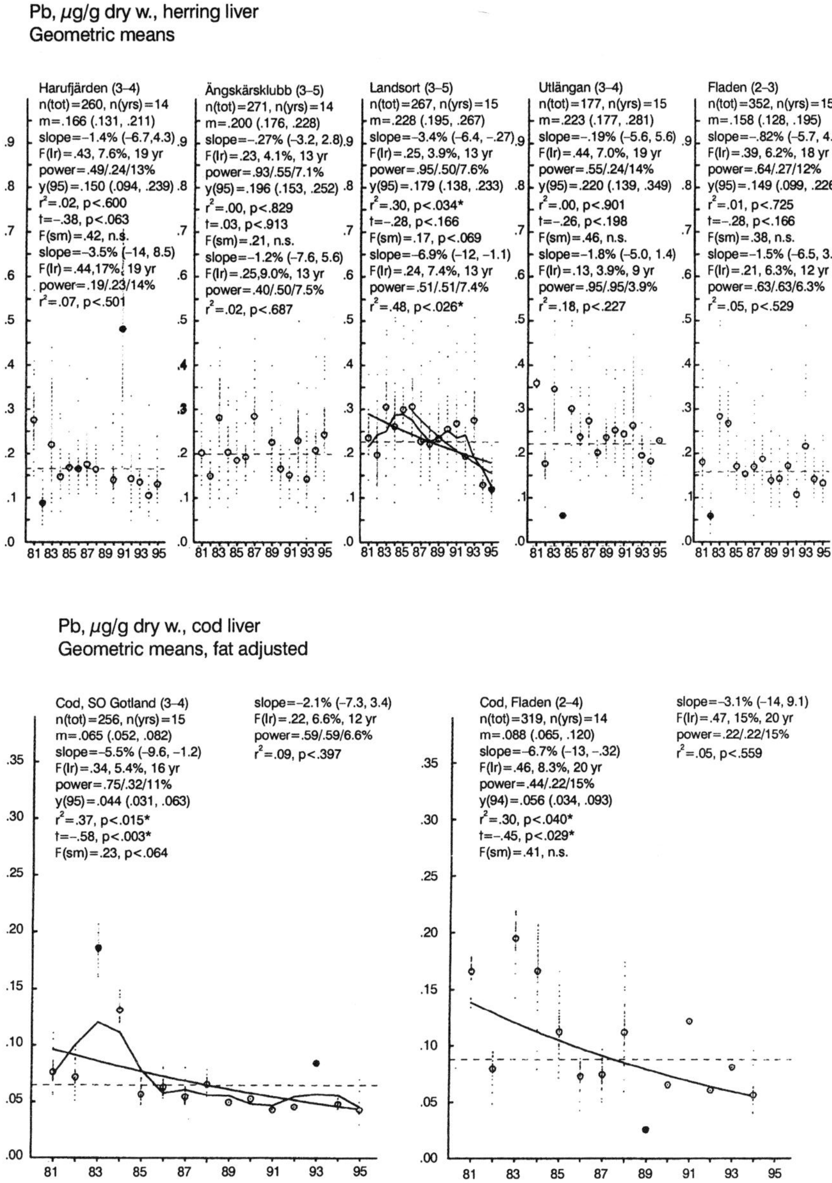

Fig. 3.20. Temporal trends of Pb in liver tissue of herring collected from the Bothnian Bay (Harufjärden), Bothnian Sea (Ängskärsklubb), Northern Baltic Proper (Landsort), Southern Baltic Proper (Utlängan), Kattegat (Fladen) and south-easterly of Gotland. After Harms (1996); modified.

Sound as well as data reported for the Gulf of Finland and the Bothnian Sea were excluded from the data matrix because less than five years of monitoring studies were performed (Harms, 1996).

The decline in atmospheric Pb detected over a time scale of 10 years, namely from 1979 to 1989, caused by reduction of leaded gasoline in western European countries, should be reflected in decreasing Pb levels in surficial water and biota of the Baltic Sea. The temporal negative trend for cod registered under the Swedish monitoring programme (Fig. 3.20) seems to support this argumentation.

In the case of hepatic Cd, the significant temporal upward trend for herring from the Bothnian Sea, Northern and Southern Baltic Proper was detected (Fig. 3.21). It is suggested that the continuous decrease of salinity in the Baltic Proper and Bothnian Bay may have created favourable conditions for a gradual enhancement of Cd species which were strongly adsorbed by plankton. If trace elements associated with plankton are consumed by fish, their bioaccumulation via uptake into the digestive canal will be a predominant route of exposure (Harms, 1996). In contrast to cod eating mainly bottom invertebrates and smaller fish, herring mainly feeds on plankton; hence respective variations of Cd levels in water and in consequence in plankton could not be reflected in cod but in herring. It is suggested that spatial variations of Cd levels at top levels of marine food chain are probably related to regional differences in water salinity. The temporal positive trend for Cd in herring from the Bothnian Sea and the Baltic Proper supports these earlier findings because it is associated with the long-term decreasing trend of salinity in the sub-areas mentioned. Owing to the lack of major inflows of high saline North Sea waters during the last two decades until 1993 a continuos decrease of salinity in almost all subareas since ca. the mid-1970s has been noted (Harms, 1996). The long-term changes of Cd in food (plankton) of herring could be an explanation for the increase its content in herring over the corresponding time.

The concentrations of Hg in herring muscle from Ängskärsklubb, the Bothnian Sea, showed a significant decreasing temporal trend. However, it could not be accepted as representative for the whole the Bothnian Sea because this sampling location might be influenced by local input of pollutants from the Gävleån River. No systematic temporal changes have been detected for muscle Hg levels in herring collected in the Northern Baltic Proper although upward trends of ca. 4% yr^{-1} (Fig. 3.22) were identified for herring from the Baltic Proper and the Kattegat. Dab from the Kattegat showed approximately constant levels of muscle Hg covering the period of 1979–94 while its values in cod from the same subarea changed irregularly, i.e. decreased during 1980–84, followed by a rather constant level up to 1990 and finally decreased thereafter (Harms, 1996).

It seems that temporal changes of muscle Hg in Baltic fish are associated with emission sources of this element and its chemical speciation. The formation and fate of organic species of Hg in marine ecosystems are dependent upon several parameters, e.g. surface water temperature, nutrient supply and on phytoplankton abundances and its species composition. Although methyl Hg constitutes a main contribution to the total Hg content in muscle fish, factors controlling the in situ alkylation of inorganic Hg are only partly understood (Harms, 1996).

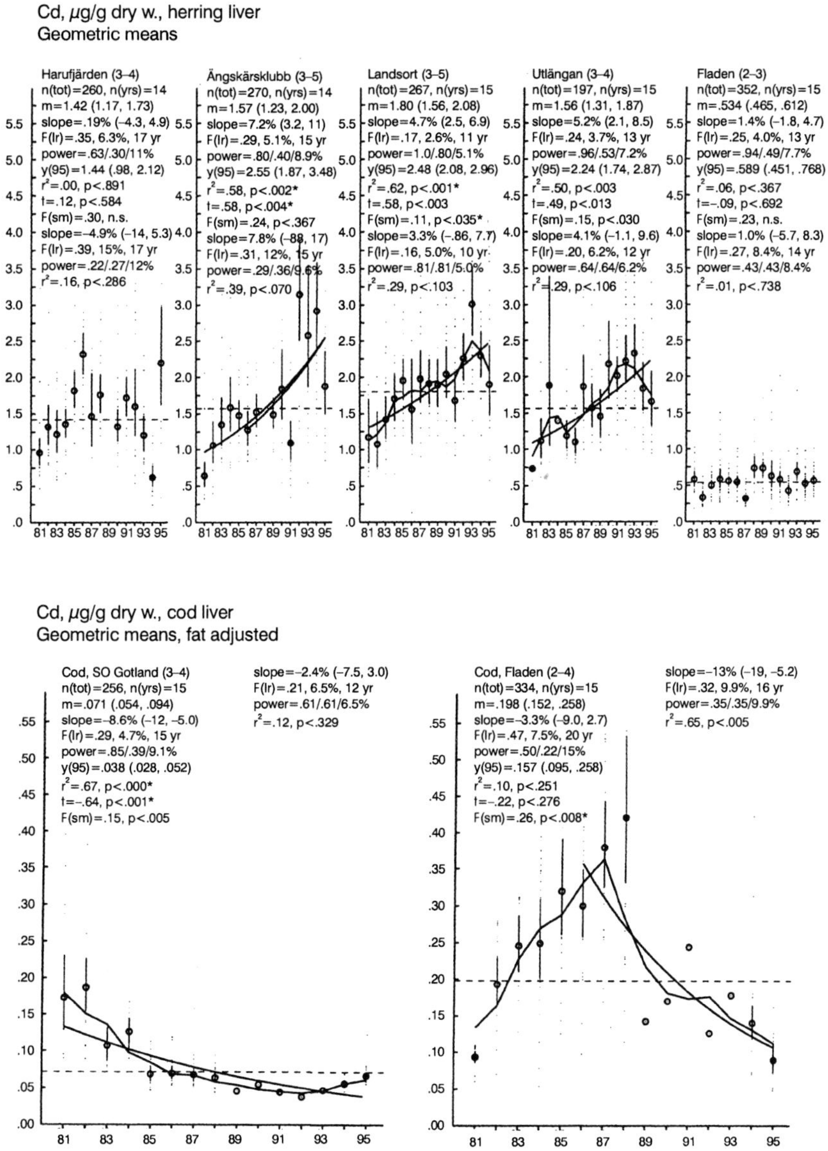

Fig. 3.21. Temporal trends of Cd in liver tissues of herring and cod collected from the Bothnian Bay (Harufjärden), Bothnian Sea (Ängskärsklubb), Northern Baltic Proper (Landsort), Southern Baltic Proper (Utlängan), Kattegat (Fladen) and south-easterly of Gotland. After Harms (1996); modified.

The temporal trends for Cu and Zn levels in cod and herring from the Kattegat and in flounder from the Sound are almost constant over the period studied. Herring collected from other subareas such as the Baltic Proper and the Bothnian Sea showed also approximately unchangable levels of these trace ele-

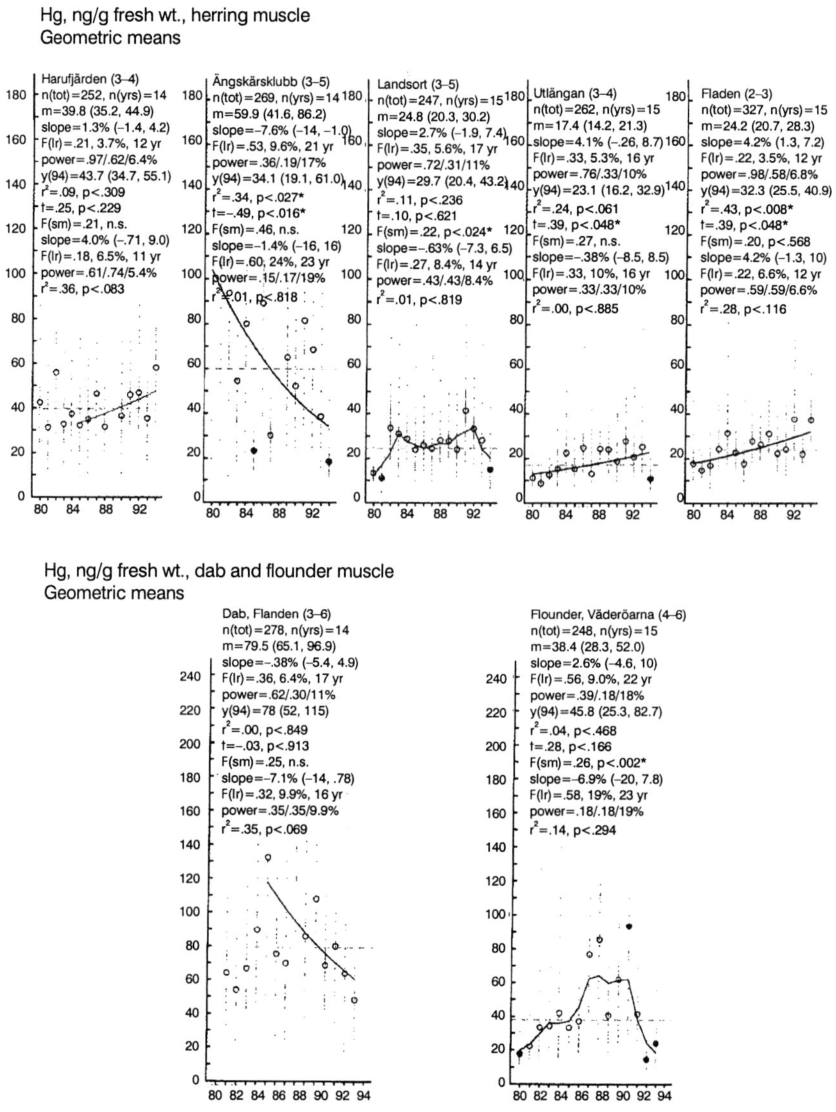

Fig. 3.22. Temporal trends of Hg in muscle tissue of herring, dab and flounder collected from the Bothnian Bay (Harufjärden), Bothnian Sea (Ängskärsklubb), Northern Baltic Proper (Landsort), Southern Baltic Proper (Utlängan), Kattegat (Fladen) and the Swedish west coast (Väderöarna). After Harms (1996); modified.

ments with sampling time. On the other side, significant upward trends were detected for Cu in herring caught in the Bothnian Bay (Harms, 1996).

In spite of somewhat confusing trends making an interpretation of the data difficult it should be taken into account that Cu and Zn as essential trace elements, in contrast to toxic elements Cd, Hg and Pb, are involved in various metabolic processes and play specific physiological roles. Bearing in mind physiological

function of both the elements, it is clear that temporal trends for their concentrations in fish may be more associated with natural factors including homeostasis than with anthropogenic impact.

The comparison of muscle concentrations of Hg and hepatic concentrations of Pb, Cd, Cu and Zn in perch was assessed within the same region (Pomeranian Bay) but between the three seasons (spring, summer and autumn) using two-way variance analysis ANOVA (Szefer et al., 2000a). The data obtained clearly demonstrate that in the Pomeranian Bay there are significant ($p < 0.001, p < 0.05$) seasonal variations of muscle Pb and Cd, and hepatic Cd, Pb, and Cu concentrations. For instance, the liver of perch during summer season contained the highest levels of Zn and Cd, while muscle Hg reached the highest levels in winter season. The observed seasonal variations in selected metals in perch are reflected by different metal bioavailability depending on the ligands present in the biotopes and the chemical speciations between two the dissolved and particulate phases (Andres, 2000). Moreover fish metabolism may be dependent on the abiotic conditions, food supply and the stage of the cycle reproduction (Kock et al., 1996; Olsson et al., 1996; Andres et al., 2000).

Metals in otoliths

According to Protasowicki and Kosior (1987, 1988) otoliths of Baltic cod *Gadus morhua* caught during 1969–1985 have a great abilities to accumulate trace metals such as Cd, Cu, and Pb and Cr in particular. The levels of Cd, Cr, Cu and Zn in otoliths decreased with age, length, weight of the fish, while the levels of Pb increased. The levels of these trace elements in cod otoliths showed some fluctuations during the 17 years considered. According to Protasowicki and Kosior (1988) otoliths of cod from the southern Baltic exhibited significantly higher levels of Cd, Cu, Pb and especially Cr than those in the liver. In contrast to these trace elements, Zn showed no strong ability to bioaccumulate in the cod otoliths.

Interrelationships between selected metals

Statistically significant ($p < 0.01$, $p < 0.05$) correlations between length-weight-age of particular specimens and metal concentrations as well as between their hepatic and muscle concentrations were obtained. Muscle concentrations of Hg exhibit positive relationships with length-size-age of the specimens studied. This correlation is probably due to the specific bioaffinity for organic matter of CH_3Hg (with high biological halftime) which constitutes the dominant pool of the total Hg in the fish muscle (Cossa et al., 1992). Significant coassociations between the muscle concentrations of Hg and length of perch (Szefer et al., 2000a) are in an agreement with those have been reported by several authors (Olsson, 1976; Cossa et al., 1992; Stronkhorst, 1992) for flounder and pike from different areas. In contrast to Hg, the levels of muscle Cd are relatively low and in an agreement with those reported for perch from Lot River, France, extremely polluted area by Cd and Zn (Andres et al., 2000). It can be explained partly by the lack of bio-

magnification of Cd in muscle of perch. As for Zn, unlike to other metals, its concentrations varied insignificantly reflecting strong abilities of the fish in Zn regulation. Similar tendency has been observed by Andres et al. (2000).

Fish Parasites

The concentration data have been reported for Cd, Co, Cu, Mn, Ni, Pb and Zn for parasites such as copepods *Thersitina gasterostei* and cestodes *Schistocephalus solidus* and in their host tissues, i.e. muscle and gills of stickleback *Gasterosteus aculeatus* from the Gulf of Gdańsk, Baltic Sea (Morozińska-Gogol et al., 1998). There were insignificant differences between metal concentrations in the parasites and their host tissues. Sures et al. (1997) reported also rather low levels of Pb (0.06–0.09 $\mu g\ g^{-1}$ wet wt.) and Cd (0.07–0.12 $\mu g\ g^{-1}$ wet wt.) in cestode parasite *Bothriocephallus scorpii* from its final host turbot (*Scopthalmus maximus*) caught at the Gulf of Gdańsk, southern Baltic. The metals are predominantly present in the posterior parts of the strobila. Levels of Pb and Cd were significantly higher in the gravid proglottides as compared to the anterior parts of the worm. Since the gravid posterior parts of the cestodes contain many eggs, the shell may detoxify or deposit these elements. It can be also possible that the quantities of the heavy metals merely depend on the age the proglottides. Posterior parts were older than the anterior ones, hence were exposed to heavy metals for a longer time and as a result should have higher levels of the metals. The positive correlation between Pb and Cd, and the weight of *B. scorpii* seems to support this hypothesis bearing in mind that the weight of the parasite should depend on its age (Sures et al., 1997).

(iii) Occurrence of Radionuclides in Fish

After the Chernobyl accident several authors have made more intensive studies of Baltic fish for ^{137}Cs, ^{210}Po, $^{239+240}Pu$, ^{90}Sr and other radionuclides (Ilus et al., 1987, 1992, 1993; Skwarzec, 1988, 1995, 1997; Grzybowska, 1989; Holm, 1994; Skwarzec et al., 1994, 2000; Bojanowski et al., 1995; Kanisch et al., 1995; Dahlgaard, 1996; Herrmann, 2000; Rissanen and Ikäheimonen, 2000). Pre-Chernobyl studies have been performed by Aarkrog et al. (1986), Holm et al. (1986), Jaworowski et al. (1986) and Weiss and Moldenhawer (1986). The concentrations of radionuclides in Baltic fish have been extensively studied in respect to their significance as a marine food. Environmental measurements of ^{137}Cs and ^{90}Sr in the Baltic Sea were compiled from 1961 to 1995 focussing on fish and seawater as the sources most relevant to doses to man in Baltic countries (Herrmann, 2000).

The concentration data are listed in Table 3.15. Elevated levels of Chernobyl radiocaesium (^{137}Cs) in Baltic subareas such as the Bothnian Sea, the Gulf of Finland, the Åland and the Archipelago Seas corresponded to maximum levels of this radioisotope in fish in 1986 and 1987 with tendency to their decrease during the following years (HELCOM, 1995). As can be seen in Fig. 3.23 this pattern is

supported well for herring from the Bothnian Sea, the Gulf of Finland and the Northern Baltic Proper as well as for cod from the Bothnian Sea. Radiocaesium activity in fillet of pike caught in Vaasa (Bothnian Sea), in contrast to herring from the same area, was characterised by slower increase after the Chernobyl deposition, reaching maximum level from 1988 to 1990 (Fig. 3.23). It can be explained by specific position of pike as a top predator in food chain, reaching maximum radioactivity later after eating of fish containing in their muscle the highest levels of radiocaesium accumulated earlier (HELCOM, 1995). Herring caught in the Bothnian Bay during 1988–1989 had in its edible parts maximum levels of radiocaesium (Fig. 3.23). It could be a result of the riverine flux of caesium from drainage areas and of a slow its transport in more radiated waters

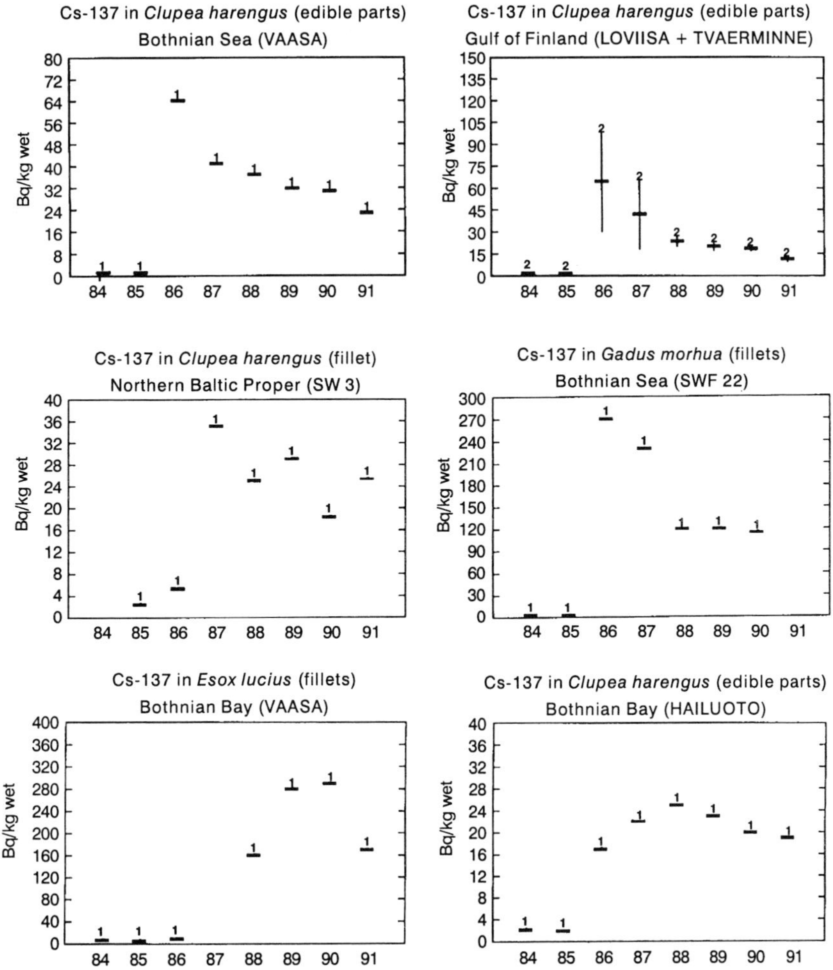

Fig. 3.23. Activities of some radionuclides in fish from different Baltic Sea subareas. After Kanisch et al. (1995); modified.

Fig. 3.23. – continued.

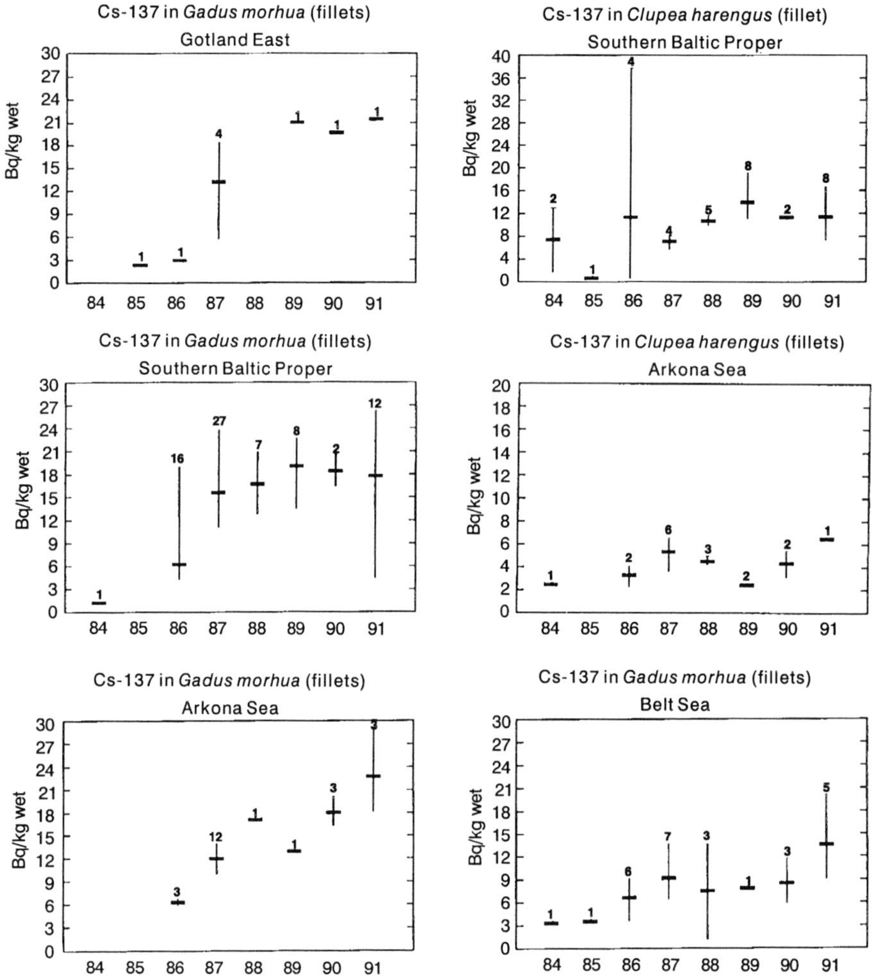

from the south (HELCOM, 1995). Fish from subareas such as East Gotland and southern Baltic Proper accumulated in their muscle less Chernobyl radiocaesium as compared to those from the Archipelago Sea and adjacent subareas (HELCOM, 1995). For example, a distinct radiocaesium maximum as a function of time is not registered for herring and cod from southern Baltic Proper (Fig. 3.23). Radiocaesium levels in fillets of fish from the most southern subareas (Bornholm, Arkona and Belt Seas) indicated an increasing trend after 1986, although their values being still relatively low (Fig. 3.23). This pattern may be explained by fact that the above mentioned southern subareas are affected by the counter clockwise movements of surficial Baltic waters and their outflow to the North Sea via the Danish straits. Therefore fish from this latter region, influenced by a relatively low

Fig. 3.23. – continued.

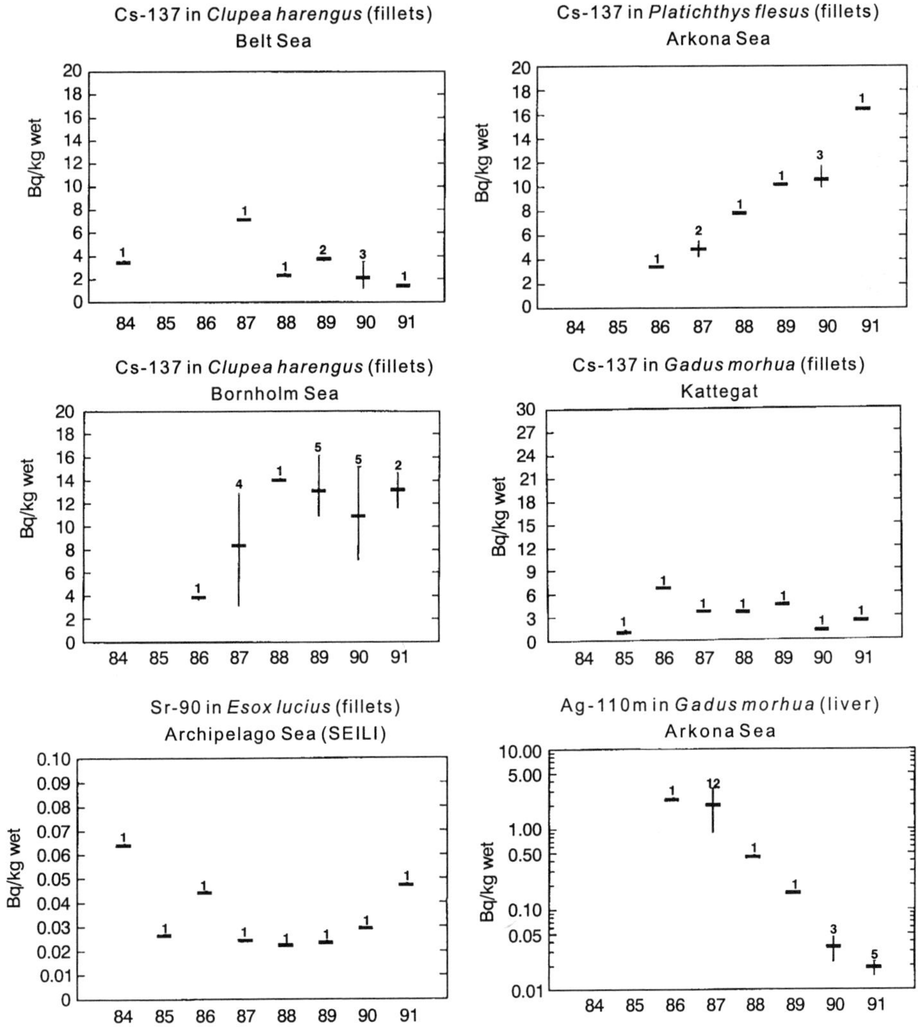

direct Chernobyl fallout, were characterised by much smaller activity of radiocaesium than those from other subareas (Fig. 3.23). This resulted in transport of the radionuclide with a southward directed stream along the Swedish east coast from the areas most affected by the Chernobyl accident to the southern subareas (HELCOM, 1995). The concentrations of muscle radiocaesium in fish from the Pomeranian Bay (Bojanowski et al., 1995) are in good agreement with the values reported for fish caught around Bornholm in approximately the same sampling time (Anon. 1994, 1995). According to Rissanen and Ikäheimonen (2000) salmon from the Gulf of Bothnia contained up to 1.3 Bq ^{137}Cs kg^{-1} originating from the Chernobyl accident.

TABLE 3.15.

Concentrations of radionuclides in fish (wet wt.) of the Baltic Sea and other northern areas

Region	Sampling date	Body part	N	110m-Ag (Bq kg^{-1})	241-Am (Bq kg^{-1})	60-Co (Bq kg^{-1})	137-Cs (Bq kg^{-1})	134-Cs (Bq kg^{-1})	131-I (Bq kg^{-1})	40-K (Bq kg^{-1})	54-Mn (Bq kg^{-1})	63-Ni (mBq kg^{-1})	210-Po (Bq kg^{-1})	239+240-Pu (mBq kg^{-1})	106-Ru (pCi kg^{-1})	References
Cod (*Gadus morhua*)																
Swedish waters	Pre-1986	Meat			0.007*											Holm et al., 1986
Gulf of Gdansk	1983/92	Lateral muscle											2±1			Skwarzec, 1995
		Liver											138±11			Skwarzec, 1988
		Whole body										345	18±2	3.0±0.3		Skwarzec et al., 1994
	1997	Liver												0.40±0.09		Skwarzec et al., 2000
		Muscle												0.29±0.06		
		Whole body												2.35±0.11		
Belt Sea	1984–91	Fillet	27				3.5–14									Kanisch et al., 1995
Bothnian Sea	1984–91	Fillet	7				5.0–270									
Gotland East	1984–91	Fillet	9				2.0–22									
Arkona Sea	1984–91	Fillet	25				6.0–23									
Bornholm Sea	1984–91	Fillet	33				5.0–22									
Southern Baltic Proper	1984–91	Fillet	73				1.0–19									
Kattegat	1984–91	Fillet	7				1.0–7.0									
Nordic area																
Fladen	1992	Muscle	1				13±1.1*						3.0±1.4(6)*			Holm, 1994
SE Gotland	1992	Muscle	1				89±4*	6.1±0.5*					3.9±0.6 (6)*			
Herring (*Clupea harengus*)																
Gulf of Gdansk	1983/92	Lateral muscle											5±1			Skwarzec, 1995
		Liver											9±1			Skwarzec, 1988
		Kidney											11±2			Skwarzec et al., 1994
		Whole body										41	6±1	2.2±0.3		
	1997	Liver												9.36±1.13		Skwarzec et al., 2000
		Muscle												0.24±0.05		
		Whole body												2.22±0.11		
Belt Sea	1984–91	Fillet	9				3.5–14									Kanisch et al., 1995
Bothnian Bay	1984–91	Fillet	5				4.0–32									
Bothnian Bay		Edible parts	8				2.0–25									
Bothnian Sea	1984–91	Fillet	5				60–270									
		Edible parts	8				1.0–64									
	1989–90	Flesh and bones	8	ND		ND	34.9 20–41	6.4 4.0–8.3		124 120–130	ND					Ilus et al., 1992
Archipelago Sea	1984–91	Edible parts	8				1.0–22									
Northern Baltic Proper		Fillet	7				2.0–35									

Region	Sampling date	Body part	N	110m-Ag ($Bq\ kg^{-1}$)	241-Am ($Bq\ kg^{-1}$)	60-Co ($Bq\ kg^{-1}$)	137-Cs ($Bq\ kg^{-1}$)	134-Cs ($Bq\ kg^{-1}$)	131-I ($Bq\ kg^{-1}$)	40-K ($Bq\ kg^{-1}$)	54-Mn ($Bq\ kg^{-1}$)	63-Ni ($mBq\ kg^{-1}$)	210-Po ($Bq\ kg^{-1}$)	239+240-Pu ($mBq\ kg^{-1}$)	106-Ru ($pCi\ kg^{-1}$)	References
Southern Baltic Proper		Fillet	34				1.0–14									
Gulf of Finland	1984–91	Edible parts	16				2.0–63									
	1986	Edible parts	4				51	24.7	ND–13	115						Ilus et al., 1987
							11.0–91	5.3–44		110–120						
	1989–90	Flesh and bones	8	ND–0.08		ND	19.3	3.41		114	ND					Ilus et al., 1992
							15–23	2.3–4.4		110–120						
	1993	Muscle	1				63±3*	3.7±0.3*					4.8±1.1*			Holm, 1994
Bornholm Sea		Fillet	20				2.8–14									
Arkona Sea		Fillet	17				2.5–6.3									
Kattegat		Fillet	8				3.0–8.0									
Gotland-Bornholm	1991–92	Muscle	6				90.7*	6.63*					4.14*			Holm, 1994
							59–152	4.1–10.0					1.6–9.6			
Fladen	1991/92	Muscle	2				6.15*						3.8*			
							4.8–7.5						3.2–4.4			
Sprat (*Sprattus sprattus*)																
Gulf of Gdansk	1991	Whole body										2.63		4.4±0.5		Skwarzec, 1995
	1997	Muscle												0.16±0.03		Skwarzec et al., 2000
		Whole body												0.33±0.04		
Flounder (*Pleuronectes flesus*)																
Gulf of Gdansk	1983/92	Lateral muscle											3±1			Skwarzec, 1995
		Liver											13±1			Skwarzec, 1988
		Kidney											9±1			Skwarzec et al., 1994
		Whole body											3±1			
	1997	Liver												1.51±0.19		Skwarzec et al., 2000
		Muscle												0.13±0.02		
		Whole body												0.94±0.05		
Plaice (*Pleuronectes platessa*)																
Danish waters	1991–94	Meat	14										0.96			Dahlgaard, 1996
Arkona Sea	1984–91	Fillet	9				3.5–16.5									Kanisch et al., 1995
Atlantic salmon (*Salmo salar*)																
Gulf of Finland	1989–90		3[b]				2.8	0.37		107						Ilus et al., 1992
							2.2–3.7	0.25–0.60		100–110						
Rainbow trout (*Salmo gairdneri*)																
Gulf of Finland	1989–90		9[b]				3.1	0.45		110						Ilus et al., 1992
							1.2–4.7	0.13–0.70		77–130						

Pike (*Esox lucius*)													
Bothnian Bay	1984–91	Fillet	7			6.0–54							Kanisch et al., 1995
	1993	Muscle	1			599±24*	35±1.6*				1.7±0.3*		Holm, 1994
Bothnian Sea		Fillet	7			5.0–290							Kanisch et al., 1995
	1989–90	Flesh	8	ND	ND	87.3	16.1		115	ND			Ilus et al., 1992
						51–140	7.3–28		100–120				
Archipelago Sea		Fillet	8			2.5–38							
Gulf of Finland		Fillet	16			4.0–100							
	1986	Edible parts	7			26.6	12.4	ND–12	122			ND–1.4[a]	Ilus et al., 1987
						7.3–5.6	2.0–26		120–130				
	1989–90	Flesh	8	ND	ND	87.3	16.1		115	ND			Ilus et al., 1992
						51–140	7.3–28		100–120				
Perch (*Perca fluviatilis*)													
Gulf of Finland	1986	Edible parts	7			134	68.7	ND–13	134			0–1.6[a]	Ilus et al., 1987
						10–190	3.3–96		98–120				
	1989	Flesh and bones	8	ND		109	20.9		106				Ilus et al., 1992
						63–230	8.8–50		93–150				
	1993	Muscle	1			248*	14*				0.2±0.1*		Holm, 1994
Bothnian Bay	1993	Muscle	1			781*	45*				0.2±0.1*		Holm, 1994
	1989	Flesh and bones	8	ND	ND–0.1	113	21		96	ND			Ilus et al., 1992
						100–140	15–31		88–100				
Pike perch (*Lucioperca lucioperca*)													
Gulf of Finland	1986	Edible parts	1			15	5.9	ND	120			ND	Ilus et al., 1987
	1993	Muscle	1			153±6*	8.6±0.5*				3.8±0.7*		Holm, 1994
Roach (*Rutilus rutilus*)													
Gulf of Finland	1986	Edible parts	8			47	19.9	ND–15	117			ND–1.0[a]	Ilus et al., 1987
						13–50	6.4–26		100–150				
	1989–90	Flesh and bones	8	ND–0.10	ND–0.18	23	4.2		99	ND–0.15			Ilus et al., 1992
						19–30	2.7–6.4		92–110				
Whitefish (*Coregonus lavaretus*)													
Bothnian Bay	1993	Muscle	1			148±6*	8.3±0.6*				0.8±0.1*		Holm, 1994
Gulf of Finland			1			107±4*	5.9±0.5*				2.9±0.2*		

* – Dry wt.
[a] – Bq kg^{-1} fresh wt.
[b] – Organisms caught in the Loviisa fish farm.

TABLE 3.15. – continued

Region	Sampling date	Body part	N	90-Sr ($Bq\ kg^{-1}$)	Th (tot.) ($ng\ g^{-1}$)	U (tot.) ($ng\ g^{-1}$)	234-U ($mBq\ kg^{-1}$)	235-U ($Bq\ kg^{-1}$)	238-U ($mBq\ kg^{-1}$)	65-Zn ($Bq\ kg^{-1}$)	References
Cod (*Gadus morhua)*											
Southern Baltic	1981	Muscle	26–55(70)		0.41±0.06	0.30±0.05					Szefer et al., 1990b
		Whole body					14.9±1.1	0.38±0.04	13.4±1.0		Skwarzec, 1995
Belt Sea	1984–91	Fillet	27	0.002–0.005							Kanisch et al., 1995
Bornholm Sea	1984–91	Fillet	33	0.004–0.005							
Herring (*Clupea harengus)*											
Southern Baltic	1981	Muscle	14–30(235)		0.38±0.08	< 0.15–0.65					Szefer et al., 1990b
		Whole body					16.1±1.5	0.40±0.05	14.5±1.3		Skwarzec, 1995
Bothnian Bay		Edible parts	8	0.084–0.160							
Bothnian Sea		Edible parts	8	0.057–0.120							
	1989–90	Flesh and bones	8							ND	Ilus et al., 1992
Archipelago Sea	1984–91	Edible parts	8	0.046–0.150							
Gulf of Finland	1984–91	Edible parts	16	0.038–0.230							
Sprat (*Sprattus sprattus)*											
Southern Baltic	1981	Muscle	4 (75)		< 0.30–0.82	< 0.30–0.44					Szefer et al., 1990b
Gulf of Gdansk	1991	Whole body					5.9±0.6	0.14±0.03	5.2±0.05		Skwarzec, 1995
Flounder (*Platichthys flesus)*											
Southern Baltic	1981	Muscle	2 (17)		0.43–0.53	0.22					Szefer et al., 1990
Perch (*Perca fluviatilis)*											
Gulf of Finland	1989	Flesh and bones	8							ND	Ilus et al., 1992
	1989–90	Flesh and bones	8							ND	
Pike (*Esox lucius)*											
Bothnian Bay	1984–91	Fillet	7	0.036–0.250							Kanisch et al., 1995
Bothnian Sea		Fillet	7	0.034–0.083							
Archipelago Sea		Fillet	8	0.022–0.064							

Gulf of Finland		Fillet	16	0.017–0.059				
	1989–90	Flesh	8				ND	Ilus et al., 1992
Roach (*Rutilus rutilus*)								
Gulf of Finland	1989–90	Flesh and bones	8				ND–0.32	Ilus et al., 1992
Greater sand eel (*Hyperoplus lanceolatus*)								
Southern Baltic	1981	Muscle	1–2 (10)		1.2	0.44		Szefer et al., 1990b
Garfish (*Belone belone*)								
Southern Baltic	1981	Muscle	2–4 (4)		0.63–1.30	0.30–0.42		Szefer et al., 1990b
Atlantic salmon (*Salmo salmo*)								
Southern Baltic	1981	Muscle	1 (1)		0.50			Szefer et al., 1990b
Eel (*Anguilla anguilla*)								
Southern Baltic	1981	Muscle	1 (1)		0.66			Szefer et al., 1990b
Whiting (*Merlangus merlangus*)								
Southern Baltic	1981	Muscle	1 (1)		1.6			Szefer et al., 1990b
Stickleback (*Gasterosteus aculeatus*)								
Southern Baltic	1981	Whole specimens	1 (16)		1.0	2.6		Szefer et al., 1990b

Concentrations of radiostrontium (^{90}Sr) in Baltic fish do not indicate any significant temporal trends in relation to a Chernobyl indications (HELCOM, 1995) except its small enhancement in fillets of pike observed in 1986 (Fig. 3.23). This species inhabits waters with low salinity, for which CF (concentration factor) for flesh radiostrontium is suspected to be significantly higher than in fish from marine waters.

In spite of substantial improvements of radioanalytical methodology since 1990, most of the plutonium ($^{239+240}Pu$) data concerning fish fillets was below the detection limit. Chernobyl plutonium in Gulf of Gdansk fish was not observed in 1987–1991. The data published recently (Skwarzec et al., 2000) report significant contribution of Chernobyl-derived plutonium to its total concentrations in fish from this region. Whole body and muscle of cod and herring from the Gulf of Gdansk collected in 1997 contained higher plutonium levels as compared to those of sprat from the same region.

Activity concentrations of radiosilver (^{110m}Ag) in liver of Baltic cod were surprisingly high reaching maximum levels of 2.0–2.5 Bq kg^{-1} wet wt. around the years 1986 and 1987. After this period the levels of the hepatic radionuclide gradually declined to minimum value amounting on 0.02 Bq kg^{-1} wet wt. (Fig. 3.23). It is evidently known that silver isotope ^{110m}Ag originated from the Chernobyl accident, although the source of ^{108m}Ag is not clarly identified still up to date (HELCOM, 1995).

The activity concentrations of ^{210}Po in meat of cod and herring show significant interspatial variability; specimens from the Gulf of Gdansk (Skwarzec, 1988) contained by an order of magnitude higher levels of this radionuclide than those from Nordic especially Danish waters (Holm, 1994; Dahlgaard, 1996). Bojanowski et al. (1995) reported levels of muscle ^{210}Po in fish from the Pomeranian Bay which were similar to Dahlgaard's data. Significant interspecies differences in polonium levels were observed for fish caught at the latter region, i.e. place concentrated in its meat ca. 2- and 3-times more ^{210}Po than herring and cod, respectively (Dahlgaard, 1996). As polonium is accumulated in fish from the diet, the great levels in plaice are supposedly a result of relatively high concentrations of ^{210}Po in mussels constituting its preferable food (Dahlgaard, 1996). Intertissue variations in polonium distribution in Baltic fish have been studied by Skwarzec (1988). It is shown that this radionuclide is non-uniformly distributed within cod, herring and flounder with its highest levels being found in the digestive organs, especially within the intestine. This finding supports the previous note about the food provenience of polonium in fish. Isotope of nickel ^{63}Ni similarly to polonium is distributed non-uniformly within Baltic fish. It is suggested, that mechanism for uptake of this radionuclide in cod and herring is by passive adsorption onto the surface of skin and scale mucus (Skwarzec et al., 1994).

The concentrations of tissue U (^{238}U, ^{235}U, ^{234}U) and Th (^{232}Th) and other radionuclides were determined in several species of fish from southern Baltic (Szefer et al., 1990b; Anon., 1994, 1995; Dahlgaard, 1996; Skwarzec, 1995, 1997;

Rissanen and Ikäheimonen, 2000). Although the stable isotopes of U and Th were determined (Szefer et al., 1990b), their concentrations practically correspond to their radioactive counterparts, i.e. ^{238}U and ^{232}Th. The mean levels of U and Th in the muscle tissue of cod, herring and sprat from the southern Baltic were ranged between 0.29–0.44 and 0.38–0.64 ng g^{-1} wet wt., respectively (Szefer et al., 1990b).

(iv) Environmental Implications and Future Recommendations

The assessment of temporal changes in the Baltic Sea as a whole is not possible because fish sampling sites covered only the Baltic Basin in part. Since no temporal trends were detectable for these sites it is not possible to pointed out whether this may be due to low statistical 'power' of the time series or because of the absence of this trend. Upward trends of hepatic Cd were registered in herring from the Baltic Proper and the Bothnian Sea. This finding allows to present a hypothesis on interrelationship between the bioavailability of Cd and long-term changes of salinity in the Baltic Sea. Further research, however, is needed in order to verify this hypothesis (Harms, 1996).

E. WATERFOWLS

(i) Introduction

General Characteristics and Taxonomy

Various water birds (especially Anseriformes) gather in quite large numbers on shallow coastal waters and coastal lagoons during their migrations. The most important Baltic wintering sites are located in the Baltic Proper: shallow lagoons and estuaries along the German – Polish coast (Vorpommern and Odra lagoons), Pomeranian Bay between Denmark, Germany and Poland, and the Gulf of Riga. Each of these sites gathers around 1 million of birds. The most abundant wintering species are: tufted duck *Aythya fuligula*, pochard *Aythya ferina*, scaup *Aythya marila*, goosander *Mergus merganser*, long-tailed duck *Clangula hyemalis*, velvet scoter *Melanitta fusca*, common scoter *Melanitta nigra*, black-throated diver *Gavia arctica*, red-throated diver *Gavia stellata* and slavonian grebe *Podiceps auritus*. Shores along river mouths are important stop-over sites for migratory waders (Charadrii). Flocks of birds gathering there account up to maximum a few hundred individuals. However birds stay only for a short period of time, and thus the exchange of individuals is very frequent. An intensive migration of passerines (Passeriformes) takes place along the southern coast of the Baltic. During favorable weather conditions, flocs of birds accounting up to thousands of small birds may be observed. Birds are migratory vertebrates, e.g. ducks wintering in the Baltic Sea, and they can transfer pollutants in their tissues from one to another area. Therefore it is important to know their worldwide distribution as well as food habits. The data provided in this chapter were taken from Sokołowski (1958, 1965), Brehm (1962), Bourne (1974, 1980); Summers et al. (1977), Wilson et al. (1980), Greenwood (1984, 1986), Mikheev (1986a, 1986b, 1986c), Mikheev and Kuroczkin (1986); Mikheev et al. (1986a, 1986b, 1986c), Blomqvist et al. (1987), Jonsson, 1992; Andell et al. (1994), Hudec (1993), The Illustrated Encyclopaedia of Birds (1993) and National Audubon Society, Field Guide to North American Birds (1996), Stokes and Stokes (1996a, 1996b).

Order: Anseriformes
Family: Ducks (Anatidae)
Species: Long-tailed duck, syn. Oldsquaw (*Clangula hyemalis*)

Habitat and range: breeds on tundra, winters on open bays and inshore waters; common in the Greenland and Island and northern Norway. In winter it visits the Baltic Sea, coastal regions of Denmark, Sweden, the Netherlands and England. *Food habits*: its food consists mainly of mussels, crabs, shrimp, and other crustaceans

Species: Scaup duck, syn. Greater scaup (*Aythya marila*)

Habitat and range: lakes, bays, and ponds; in winter, often on salt water; breeds in the Scotland, Scandinavia and on tundra of the Northern Asia and the North America. Scaup duck visits coastal areas of the Baltic Sea in September; winters in the western coasts of Europe and the Mediterranean Sea. *Food habits*: because it dives for molluscs and other animals and is not as much of a vegetarian as the Readhead or the Canvasback, the Greater scaup is not considered as choice a game bird, although it is still shot in large numbers annually.

Species: Common goldeneye, syn. Goldeneye (*Bucephala clangula*)

Habitat and range: breeds on wooded lakes and ponds; winters mainly on coastal bays and estuaries; breeds on Alaska and across Canada to Newfoundland and Maritime Provinces, south to mountains of Montana and Great Lakes. Winters in much of United States, wherever water is open, also in Eurasia. *Food habits*: wintering in small groups or large flocks, goldeneyes feed mainly on mollusks, in summer, their diet shifts to aquatic plants and insects.

Species: Black scoter, syn. Common scoter (*Melanitta nigra*)

Habitat and range: breeds on ponds in boreal forests; winters on ocean and in large salt bays; breeds in western Alaska and in Labrador and Newfoundland. Winters along coasts from Alaska south to California and from Newfoundland south to Carolinas, as well as on portions of Gulf Coasts and on Great Lakes. Also in Eurasia. *Food habits*: like the other scoters, it preys on mussels and other molluscs, and also tears barnacles, chitons, and limpets from submerged rocks and reefs. Newly hatched young feed on fresh water mussels and the larvae of aquatic insects before moving to salt water.

Species: Tufted duck (*Aythya fuligula*)
Habitat and range: wooded lakes, streams, and marshes; in winter, often in estuaries and shallow coastal bays; breeds in northern Eurasia, very rare in North America. Most often seen near urban area. *Food habits*: feeds on small bivalvia and less frequently shoots of aquatic plants.

Species: White-winged scoter, syn. Velvet scoter (*Melanitta fusca*)
Habitat and range: breeds on large lakes, winters mainly on the ocean and on large coastal bays, but a few remain on lakes in the interior; breeds in Alaska and much of western and central Canada. Winters along coasts, from Alaska south to California and from Newfoundland south to Carolinas, rarely to Florida and Texas. Noted also in Eurasia. *Food habits*: this species feeds chiefly on molluscs, which it collects from mussel beds at depths of 15 to 40 feet (5 to 12 meters). These birds also feed on crabs, starfish, sea urchins, and some fish.

Species: Red-breasted merganser (*Mergus serrator*).
Habitat and range: breeds on wooded lakes and tundra ponds; winters mainly on salt water; breeds in Alaska and across northern Canada to Newfoundland and south to Great Lakes. Winters chiefly along coasts from Alaska south to California, from Maritime provinces south to Florida, and along Gulf Coast. Observed also in Eurasia. *Food habits*: feeds mainly on fish, which it captures in swift underwater dives, aided by its long pointed bill lined with sharp, tooth-like projections.

Species: Common merganser, syn. Goosander (*Mergus merganser*)
Habitat and range: breeds on wooded rivers and ponds; winters mainly on lakes and rivers, occasionally on salt water; breeds across Canada from eastern Alaska, Manitoba, and Newfoundland south in mountains to California, northern New Mexico, Great Lakes, and northern New England. Winters south to northern Mexico and Georgia (rarely farther), also in Eurasia. *Food habits*: often called "Sawbills", mergansers have fine, tooth-like serrations along the sides of their bills that help in grasping slippery fish.

Species: Common eider (*Somateria mollissima*)
Habitat and range: rocky coasts and coastal tundra; breeds on Arctic coasts of Alaska and Canada south to Massachusetts. Winters along Alaska coast south to Washington and along Atlantic Coast south to Long Island, occasionally farther, also northern Asia. *Food habits*: their principal foods are mussels and other shellfish.

Order: Colymbi = Gaviiformes
Family: Loons (Colymbidae = Gaviidae)
Species: Black-throated loon (*Gavia arctica*)
Habitat and range: coastal and tundra ponds during summer; large lakes, bays, estuaries, and ocean during migration and winter; breeds from Aleutian Islands, Alaska, also in northern Asia. *Food habits*: feeds mainly on fish, also frogs and aquatic insects.

Species: Red-throated loon (*Gavia stellata*)

Habitat and range: islands and continents in the vicinity of the North Pole, in Europe its distribution reaches Latvia and observed in Scandinavia. *Food habits*: as excellent swimmers and divers they prey mainly on fish.

Order: Podicipediformes
Family: Grebes (Podicipedidae)
Species: Great crested grebe (*Podiceps cristatus*)

Habitat and range: breeds on lakes, ponds, and marshes; breeds in whole Europe, its maximum reach – southern Sweden and Finland to north and the Nile estuary and central Asia to south. *Food habits*: Grebes are expert divers, propelling themselves with their lobed toes as they pursue fish, crustaceans, and aquatic insects.

Order: Charadriiformes
Suborder: Charadrii
Family: Waders (Scolopacidae)
Species: Dunlin (*Calidris alpina alpina* L.) and Sandpiper (*Calidris ferruginea* Pont.)

Habitat and range: mudy and sandy estuarine and coastal areas, also observed on rocky shores with seaweed accumulation and inland along shallow rivers and lakes; *Calidris* is distributed in the Old and the New World, breeds mainly in the Taiga, both sub-Arctic and Arctic areas. *Food habits*: feeds on insects and other small organisms.

Suborder: Alcae
Family: Murres (Alcidae)
Species: Common murre (*Uria aalge*)

Habitat and range: rocky coasts; range: breeds on islands along coasts of northern and western Europe from islands located near the North Pole to France, Spain and Portugal, also in northern Asia. *Food habits*: they dive for their food, primarily small fish.

Species: Black guillemot (*Cepphus grylle*)

Habitat and range: rocky coasts, breeds from Arctic Alaska and Canada south along Atlantic Coast to Maine. Winters south to the Bering Sea and Long Island (rarely). Also in northern Europe, Scandinavia and Alaska. *Food habits*: its diet consists primarily of small fish, crustaceans, molluscs, and marine worms.

Species: Razorbil (*Alca torda*)

Habitat and range: Inhabits rocky islands of the northern seas, e.g. on Swedish islands and along Baltic coast (Swedish); occurs along coasts of Denmark, the Netherlands, Norway, Great Britain, Island, Greenland and the North America. *Food habits*: it feeds on small fish, mostly specimens of stickleback and sand eel.

Suborder: Lari
Family: Sternidae
Species: Little tern (*Sterna albifrons*)

Habitat and range: distributed all over the world except the South America. *Food habits*: its diet consists primarily of small fish, to a less extent of small crustaceans and molluscs

Order: Pelecaniformes
Family: Cormorants (Phalacrocoracidae)
Species: Black cormorant (*Phalacrocorax carbo*)

Habitat and range: Wide-spread all over the world except northern Asia and the South America. *Food habits*: feeds on different species of fish.

Order: Accipitres
Family: Eagles (Accipitridae)
Species: Sea eagle (*Haliaeëtus albicilla*)

Habitat and range: sea coasts, forest areas neighbouring lakes and rivers, lives in northern Asia, Europe and Greenland, observed sporadically in Germany and Poland. *Food habits*: it preys on waterfowls (ducks, geese, loons etc), fish, mammals.

Overview of Worldwide Literature

There are numerous available reports on concentrations of heavy metals in different tissues of waterfowl, e.g. ducks, loons, grebes, murres, waders, cormorants, ibises and eagles (Johnels et al., 1968; Jensen et al., 1972; Lande, 1977; Greichus et al., 1977, 1978; Frank and Borg, 1979; Koranda et al., 1979; Vermeer and Peakall, 1979; Scanlon et al., 1980; Fleming, 1981; Nicholson, 1981; Di Giulio and Scanlon, 1984a, 1984b, 1984c; Goede, 1985; Goede and de Bruin, 1985a, 1985b; Goede and de Voogt, 1985; Honda et al., 1985a, 1985b, 1986a, 1990; Norheim, 1987; Muirhead and Furness, 1988; Lee et al., 1989; Thompson and Furness, 1989; Dietz et al., 1990, 1996; Furness et al., 1990; Thompson et al., 1990; Elliott et al., 1992; Lock et al., 1992; Szefer et al., 1993a, 1993b; Custer and Hohman, 1994; Goede and Wolterbeek, 1994; Franson et al., 1995; Kalisińska and Szuberla, 1996; Kim et al., 1996a, 1996b, 1996c, 1998b, 1999; Möller, 1996; Dańczak et al., 1997; Debacker et al., 1997; Donaldson et al., 1997; Furness and Camphuysen, 1997; Klekowski et al., 1999; Monteiro et al., 1999; Custer et al., 2000; Saeki et al., 2000).

Residues of eggs or feathers have been also analysed for metallic pollutants (Wiemeyer et al., 1980, 1984; Goede and de Bruin, 1984a, 1984b; Furness et al., 1990; Goede, 1993; Wenzel et al., 1996; Fasola et al., 1998; Sydeman and Jarman, 1998; Bearhop et al., 2000a, 2000b; Burger and Gochfeld, 2000; Malinga and Szefer, 2000a, 2000b; Movalli, 2000; Goutner et al., 2001). The chronic toxicity effect of acidification on the availability of trace elements to wild birds has been extensively overviewed by Scheuhammer (1987, 1991).

(ii) Occurrence of Chemical Elements in Waterfowls

Despite extensive worldwide studies of waterfowl for heavy metals, relatively less data in this respect are reported for birds collected in the Baltic region (Henriksson et al., 1966; Falandysz and Szefer, 1983; Szefer and Falandysz, 1983a, 1986, 1987; Falandysz, 1984b, 1986d; Appelquist et al., 1984; Blomqvist et al., 1987; Falandysz et al., 1988; Falandysz and Szefer, 1993; Dańczak et al., 1997; Kannan and Falandysz, 1997a; Falandysz et al., 2000a; Szefer et al., 2000c; Thyen et al., 2000). Besides soft tissues, feathers of birds from area of the Baltic Sea have been also analysed for trace element concentrations (Berg et al., 1966; Szefer and Falandysz, 1983a, 1987; Goede and Bruin, 1985a, 1985b; Falandysz et al., 1988; Goede et al., 1989). Within the framework of temporal trends studies, Berg et al. (1966) reported data of Hg levels in feathers of Swedish birds since 1834. Intercomparison studies of As, Hg and Se in feather and kidney of waders from the Dutch Wadden Sea, northern part of Norwegian coast and the Baltic Sea (Scandinavia) have been performed by Goede and Bruin (1985b), Goede et al. (1989) and Goede and Wolterbeek (1994).

Inter-species trends

Tables 3.16–3.20 present concentration data of selected chemical elements in soft and hard parts of waterfowls from the Baltic Sea and surrounding areas. Inter-species variations in trace element concentrations in birds are due to different factors including their food habits. The uptake of several metals from food ingested by birds is significantly more important than the uptake through respiratory surfaces or skin, although some metals, e.g. Ni, are much more readily accumulated via respiratory surfaces than via the intestine (Outridge and Scheuhammer, 1993; Furness, 1996). An example of influence of chemical composition of food on metal levels in tissues and organs of marine birds is greater concentration of Cd in ducks than in grebes and murres. This finding can be explained by occurrence of higher levels of this element in molluscs constituting an important food for ducks. Fish being extremely poor in muscle Cd are however consumed mainly by grebes and murres which contain also lower levels of hepatic Cd (Szefer et al., 2000c). Falandysz et al (2000a) analysed particulate tissues and organs of white-tailed sea eagle collected dead in Poland for concentrations of numerous elements, i.e. As, Ag, Bi, Ca, Cd, Co, Cr, Cs, Cu, Fe, Hg, In, K, Mg, Mn, Ni, P, Pb, Rb, S, Sr, Tl, V, Y, Zn, REE (La, Ce, Pr, Nd, Sm, Eu, Gd, Tb, Dy, Ho, Er, Tm, Yb, Lu), Th and U. The data obtained revealed that metal of risk is Pb, and to some extent also Hg. An intoxication of white-tailed sea eagle with Pb is probably caused by ingestion of lead pellets from its prey, i.e. the waterfowl injured or killed by the hunters. For elevated levels of Hg in eagle tissues and organs is responsible food (waterfowl and fish) originating from a coast of the Baltic Sea and the Firth of Szczecin (Falandysz et al., 2000a).

TABLE 3.16.

Concentrations of chemical elements ($\mu g\ g^{-1}$ wet wt.) in liver of waterfowl from the Baltic Sea and other northern areas

Region	Sampling date	Sex	Age	N	Ca	Cd	Co	Cr	Cu	Fe	References
						DUCKS – ANATIDAE					
Long tailed duck (*Clangula hyemalis*)											
Gulf of Gdansk, S. Baltic	1980–81	M		10 (50)		0.76±0.06	0.09±0.02		1.50±0.2	750±110	Szefer and Falandysz, 1983
						0.60–0.98	0.06–0.22		1.10–66.0	370–1200	
		F		8 (40)		0.59±0.10	0.06±0.008		1.30±0.10	670±130	
						0.30–1.1	0.013–0.075		0.81–1.80	110–1000	
Gulf of Pomerania	Pre-1997	M		14		1.30±0.56			7.40±2.3		Dańczak et al., 1997
						0.51 2.40			3.23–14.6		
		F		14		1.40±0.55			7.8±2.3		
						0.50–2.29			4.89–14.3		
Scaup duck (*Aythya marila*)											
Gulf of Gdansk	1980–84	M		6 (30)	110±40	0.41±0.03	0.10±0.01		14.0±2.9	340±20.0	Szefer and Falandysz, 1987
					50–290	0.20–0.54	0.08–0.11		7.9–25.0	290–450	Szefer, 1989, unpublished data
		F		7 (35)	110±40	0.41	0.10±0.01		17.0±4.3	360±20.0	
					70–220	0.15–0.65	0.09–0.13		8.7–43.0	300–430	
Mallard (*Anas platyrhynchos*)											
Warta and Oder		M		10		1.55±1.24			29.7±13.5		Dańczak et al., 1997
flood waters						0.00–4.35			11.6–61.9		
		F		4		1.71±2.75			18.6±9.2		
						0.04–5.78			5.16–25.0		
Greylag geese (*Anser anser*)											
Warta and Oder		M		3		0.52±0.30			31.3±4.2		Dańczak et al., 1997
flood waters						0.32–0.86			27.1–35.6		
		F		6		1.37±1.37			25.5±7.4		
						0.22–3.10			13.5–34.1		
Tufted duck (*Nyroca fuligula*)											
Gulf of Gdansk	1980–84	M		18 (27)		0.43		0.13	19		Szefer et al., 2000c
						0.17–0.64		0.01–0.22	13.3–27.2		
		F		9 (11)		0.21		0.16	15.5		
						< 0.01–0.42		0.11–0.21	12.8–17.1		
Goldeneye (*Bucephala clangula*)											
Gulf of Gdansk	1980–84	M		3 (3)		0.71		0.35	8.25		Szefer et al., 2000c
Velvet scoter (*Oidemia fusca*)											
Gulf of Gdansk	1980–84	M		3 (2)		1.2		0.19	7.11		
		F		6 (3)		0.92		0.1	7.65		Szefer et al., 2000c
Common eider (*Somateria mollissima*)											
Gulf of Gdansk	1980–84	M		3 (4)		1.03		0.19	7.64		Szefer et al., 2000
Norwegian coast Trondheimsfjord	1973	F		6		2.6		0.2	73		Lande, 1977

Region	Sampling date	Sex	Age	N	Ca	Cd	Co	Cr	Cu	Fe	References
Goosander (*Mergus merganser)*											
Gulf of Gdansk	1980–84	F		3 (1)		0.21		0.02	3.58		Szefer et al., 2000c
Merganser (*Mergus serrator)*											
Gulf of Gdansk	1980–84	M		3 (4)		0.12		0.11	6.72		Szefer et al., 2000c
		F		6 (5)		0.12		0.12	4.58		
GREBES – PODICIPEDIAE											
Great crested grabe (*Podiceps cristatus)*											
Gulf of Gdansk,	1980–84	M		3 (5)		0.02		0.2	5.8		Szefer et al., 2000c
		F		6 (6)		0.01		< 0.01	3.79		
MURRES – ALCIDAE											
Common murre (*Uria aalge)*											
Gulf of Gdansk	1980–84	F		3 (1)		0.1		0.18	4.41		Szefer et al., 2000c
Belgian beaches	1989–95	M+F		143		2.4±1.6*			52±17*		Debacker et al., 1997
Black guillemot (*Cepphus grylle)*											
Gulf of Gdansk	1980–84	F		6 (3)		0.16		0.11	5.02		Szefer et al., 2000c
WADERS – SCOLOPACIDAE											
Migrating dunlin (*Calidris alpina)*											
Baltic Sea	1981–83		O	70	74±211	0.26±0.20	0.04±0.03		4.69±1.23	480±129	Blomqvist et al., 1987
Swedish island of Öland					24–1800	< 0.06–1.01	< 0.02–0.14		2.57–9.10	208–805	
			J	17	57±12	0.06±0.06	0.03±0.03		5.42±0.82	543±177	
					39–78	< 0.06–0.13	< 0.02–0.08		3.89–6.90	315–805	
			A	26	24±70	0.32±0.18	0.04±0.03		4.24±0.95	456±113	
					24–1800	0.13–1.01	< 0.02–0.14		2.57–5.87	231–715	
Curlew sandpiper (*Calidris ferruginea)*											
Baltic Sea	1981–83		O	28	96±246	0.76±1.14	0.05±0.02		5.83±1.62	565±131	Blomqvist et al., 1987
Swedish island of Öland					28–1350	< 0.06–4.29	< 0.02–0.10		3.68–9.40	340–830	
			J	12	53±12	0.10±0.11	0.05±0.01		6.44±1.82	652±111	
					38–76	< 0.06–0.32	0.03–0.07		3.68–9.40	433–830	
			A	16	128±326	1.26±1.31	0.06±0.03		5.37±1.34	500±106	
					28–1350	0.19–4.29	< 0.02–0.10		3.70–8.70	340–705	
EAGLES – ACCIPITRIDAE											
White-tailed eagle (*Halliaeëtus albicilla)**											
Western Polish coast	1981–87	F	A	4		0.04–0.07	0.335	0.2	0.62–13.0	240–800	Falandysz, 1984; Falandysz and Szefer, 1983;
	1991–94	M+F	A	7		< 0.05–0.35			5.66–38.7		Szefer et al., 2000c
South of the Polish cost	1987	F	A	2		0.07–0.17			3.3–8.4	700–1300	Falandysz et al., 1988
	1994	M	N	2		< 0.05			11.2–14.4		Szefer et al., 2000c

* – Dry wt.
A – Adult, J – Juvenile, N – Nestling, O – Overall, M – Male, F – Female.

TABLE 3.16. – continued

Region	Sampling date	Sex	Age	N	Hg	Mg	Mn	Na	Ni	Pb	Sn	Zn	References
							DUCKS – ANATIDAE						
Long tailed duck (*Clangula hyemalis*)													
Gulf of Gdansk, S. Baltic	1980–81	M		10 (50)			4.2±0.8		0.014±0.005	0.13±0.02		15±0.9	Szefer and Falandysz, 1983
							0.80–6.40		.005–0.04	0.06–0.19		11.0–18.0	
		F		8 (40)			4.9±0.7		0.021±0.006	0.14±0.03		14±0.7	
							2.60–8.00		0.005–0.047	0.06–0.20		12.0–17.0	
	1989	F+M		2							2.44		Kannan and Falandysz, 1997
											0.28–4.6		
Gulf of Pomerania		M		14						0.16±0.07		24±4.9	Dańczak et al., 1997
										0.03–0.30		8.20–28.3	
		F		14						0.18±0.09		27.2±2.9	
										0.05–0.36		23.4–34.4	
Scaup duck (*Aythya marila*)													
Gulf of Gdansk	1980–84	M		6 (30)		190±10	3.50±0.20	1070±30	0.08±0.01	0.23±0.01		31±1.0	Szefer and Falandysz, 1987
						160–210	1.60–5.50	1000–1100	0.06–0.11	0.18–0.28		27.0–34.0	Szefer, 1989
		F		7 (35)		190±10	2.80±0.20	1020±120	0.08±0.01	0.21±0.00		33.0±4.0	Szefer (unpublished data)
						170–230	1.80– 3.90	760–1470	0.06–0.12	0.16–0.26		24.0–52.0	
Mallard (*Anas platyrhynchos*)													
Warta and Oder		M		10						0.65±0.84		43.6±4.8	Dańczak et al., 1997
flood waters										0.00–2.09		37.0–50.3	
		F		4						0.05±0.06		53.9±7.0	
										0.00–0.11		44.5–61.6	
Greylag geese (*Anser anser*)													
Warta and Oder		M		3						0.09±0.03		45.0±11.2	Dańczak et al., 1997
flood waters										0.07–0.12		37.2–57.9	
		F		6						0.12±0.10		45.5±13.6	
										0.04–0.27		32.1–68.8	
Tufted duck (*Nyroca fuligula*)													
Gulf of Gdansk	1980–84	M		18 (27)			2.94			< 0.01–0.03			Szefer et al., 2000c
							1.94–3.89						
		F		9 (11)			2.52						
							2.0–2.96						
Goldeneye (*Bucephala clangula*)													
Gulf of Gdansk	1980–84	M		3 (3)			3.95						Szefer et al., 2000c

Region	Sampling date	Sex	Age	N	Hg	Mg	Mn	Na	Ni	Pb	Sn	Zn	References
Velvet scoter (*Oidemia fusca*)													
Gulf of Gdansk	1980–84	M		3 (2)			2.5						Szefer et al., 2000c
		F		6 (3)			3.24						
Common eider (*Somateria mollissima*)													
Gulf of Gdansk	1980–84	M		3 (4)			2.76						Szefer et al., 2000c
Norwegian coast Trondheimsfjord	1973	F		6	0.12				0.2			41	Lande, 1977
Goosander (*Mergus merganser)*													
Gulf of Gdansk	1980–84	F		3 (1)			3.44						Szefer et al., 2000c
Merganser (*Mergus serrator)*													
Gulf of Gdansk,	1980–84	M		3 (4)			1.88						Szefer et al., 2000c
Southern Baltic		F		6 (5)			2.18						
Red-throuted loon (*Gavia stellata)*													
Gulf of Gdansk	1989	M		1							0.61		Kannan and Falandysz, 1997
GREBES – PODICIPEDIAE													
Great crested grabe (*Podiceps cristatus)*													
Gulf of Gdansk	1980–84	M		3 (5)			1.75						Szefer et al., 2000c
		F		6 (6)			2.61						
	1989	M		1							0.54		Kannan and Falandysz, 1997
MURRES – ALCIDAE													
Common murre (*Uria aalge)*													
Gulf of Gdansk	1980–84	F		3 (1)			1.1						Szefer et al., 2000c
	1989	M		3							0.5		Kannan and Falandysz, 1997
Belgian beaches	1989–95	M+F		143	6.1±3.4*							145±39*	Debacker et al., 1997
Black guillemot (*Cepphus grylle)*													
Gulf of Gdansk		F		6 (3)			1.27		ND	ND			Szefer et al., 2000c
Razorbil (*Alca torda)*													
Gdansk Bay	1989	M		3							0.33 0.26–0.38		Kannan and Falandysz, 1997

						CORMORANTS – PHALACROCORACIDAE						
Black cormorant (*Phalacrocorax carbo*)												
Gdansk Bay	1989	M		1						0.87		Kannan and Falandysz, 1997
						WADERS – SCOLOPACIDAE						
Migrating dunlin (*Calidris alpina*)												
Baltic Sea	1981–83			O	70	218±43	2.51±0.047		0.04±0.06		25±3	Blomqvist et al., 1987
Swedish island of Öland						161–365	1.38–3.50		< 0.06–1.0		18–34	
				J	17	225±52	2.82±0.42		0.06±0.06		27±3	
						177–365	1.88–3.50		< 0.06–0.24		21–32	
				A	26	206±40	2.40±0.5		0.02±0.04		24±4	
						161–295	1.38–3.23		< 0.06–0.20		18–34	
Curlew sandpiper (*Calidris ferruginea*)												
Baltic Sea	1981–83			O	28	247±76	2.73±0.61		0.06±0.07		25±5	Blomqvist et al., 1987
Swedish island of Öland						175–435	1.92–4.40		< 0.06–0.23		17–41	
				J	12	256±87	2.75±0.67		0.08±0.06		25±3	
						175–385	1.92–4.40		< 0.06–0.19		21–29	
				A	16	240±70	2.70±0.58		0.05±0.07		25±6	
						177–435	1.97–4.40		< 0.06–0.23		17–41	
						EAGLES – ACCIPITRIDAE						
White-tailed eagle (*Halliaeëtus albicilla*)*												
Western Polish coast	1981–87	F	A	4	5.1–33		2.10–4.50	0.015–0.18	0.06–0.47		70.4–100	Falandysz, 1984; Falandysz and Szefer, 1983;
	1991–94	M+F	A	7			4.72–14.1		ND–29.3			Szefer et al., 2000c
South of the Polish cost	1987	F	A	2	0.86–1.8		1.80–3.30		18–28		38.0–39.0	Falandysz et al., 1988
	1994	M	N	2			3.74–10.1		ND			Szefer et al., 2000c
Gorzów Province	1989	F		1						0.035		Kannan and Falandysz, 1997

* – Dry wt.
A – Adult, N – Nestling,
M – Male, F – Female.

TABLE 3.17.
Concentrations of chemical elements ($\mu g\ g^{-1}$ wet wt.) in kidney of waterfowl from the Baltic Sea and other northern areas

Region	Sampling date	Sex	Age	N	Ag	As	Ca	Cd	Co	Cr	Cu	Fe	References
							DUCKS – ANATIDAE						
Long tailed duck (*Clangula hyemalis*)													
Gulf of Pomerania	1980–81	M		14				4.22±1.75			3.69±0.41		Dańczak et al., 1997
								1.96–6.91			3.01–4.68		
		F		14				5.26±2.26			3.92±0.52		
								2.11–10.6			3.19–5.10		
Scaup duck (*Aythya marila*)													
Gulf of Gdansk	1980–84	M		4 (20)			270±70	1.29±0.14	0.14±0.01		3.7±0.1	230±10	Szefer and Falandysz, 1987
							150–500	0.63–1.90	0.13–0.15		3.4–4.0	200–270	Szefer, 1989
		F		5 (15)			300±20	1.10±0.36	0.14±0.01		5.0±0.2	240±20	Szefer (unpublished data)
							290–330	0.35–1.94	0.13–0.15		4.3–5.3	210–280	
Mallard (*Anas platyrhynchos*)													
Warta and Oder		M		10				14.2±14.4			4.43±1.33		Dańczak et al., 1997
flood waters								0.02–43.9			2.21–6.52		
		F		4				20±31.9			6.35±2.19		
								0.16–76.1			3.76–9.46		
Greylag geese (*Anser anser*)													
Warta and Oder		M		3				3.41±2.94			2.98±0.53		Dańczak et al., 1997
flood waters								1.04–6.69			2.39–3.39		
		F		6				6.24±6.26			2.94±0.79		
								0.74–13.7			2.00–4.22		
Tufted duck (*Nyroca fuligula*)													
Gulf of Gdansk	1980–84	M		18 (27)				1.6		0.11	4.15		Szefer et al., 2000c
								0.24–2.58		< 0.05–0.19	3.29–5.86		
		F		6 (9)				2.06		0.18	4.85		
								1.9–2.2					
Velvet scoter (*Oidemia fusca*)													
Gulf of Gdansk	1980–84	M		3 (2)				< 0.01		0.08	2.28		Szefer et al., 2000c
Common eider (*Somateria mollissima*)													
Gulf of Gdansk	1980–84	M		3 (4)				1.03±0.07		0.19±0.00	76.4±7.55		Szefer et al., 2000c
		F		12 (11)				3.24		0.14	8.61		
								1.32–5.13		< 0.05–0.20	7.19–11.5		
Norwegian coast	1973	F		6	1.4			5		0.2	8.6		Lande, 1977
Trondheimsfjord													

Goosander (*Mergus merganser*)											
Gulf of Gdansk	1980–84	F		3 (1)		1.7		0.3	3.18		Szefer et al., 2000c
GREBES – PODICIPEDIAE											
Great crested grabe (*Podiceps cristatus*)											
Gulf of Gdansk	1980–84	M		9 (11)		0.21		0.09	2.38		Szefer et al., 2000c
						0.05–0.50		0.07–0.14	2.22–2.51		
		F		6 (6)		0.12		0.05	2.23		
						0.02–0.23		< 0.05–0.06			
LOONS – GAVIIDAE (COLYMBIDAE)											
Black-throated loon (*Colymbus arcticus*)											
Gulf of Gdansk	1980–84	M		3 (2)		0.65		0.19	1.85		Szefer et al., 2000c
		F		3 (1)		0.36		0.1	1.94		
Red-throated loon (*Colymbus stellatus*)											
Gulf of Gdansk	1980–84	M		3 (4)		0.17		0.1	2.36		Szefer et al., 2000c
MURRES – ALCIDAE											
Common murre (*Uria aalge*)											
Gulf of Gdansk	1980–84	M		3 (2)		0.15		0.1	3.82		Szefer et al., 2000c
		F		5 (3)		0.15		0.1	2.22		
						0.10–0.19		0.09–0.12			
Belgian beaches	1989–95	M+F		143		7.8±6.6*			28±12*		Debacker et al., 1997
Black guillemot (*Cepphus grylle*)											
Gulf of Gdansk	1980–84	M		3 (3)		0.88		0.05	3.9		Szefer et al., 2000c
		F		6 (3)		0.33		0.25	4.12		
						0.05–0.60		0.05–0.42			
WADERS – SCOLOPACIDAE											
Migrating dunlin (*Calidris alpina*)											
Baltic Sea											
Swedish Island of Öland	1981–83		O	69	202±510	0.76±0.77	0.04±0.02		3.54±0.81	184±30	Blomqvist et al., 1987
					32–2700	< 0.07–4.08	< 0.02–0.07		2.38–6.50	114–254	
			J	17	273±627	0.07±0.10	0.03±0.03		4.31±0.85	176±38	
					48–2460	< 0.07–0.30	< 0.02–0.07		3.41–6.50	114–254	

Region	Sampling date	Sex	Age	N	Ag	As	Ca	Cd	Co	Cr	Cu	Fe	References
			A	26			150±339	1.23±0.92	0.05±0.01		3.32±0.55	193±27	
							32–1550	0.14–4.08	< 0.02–0.07		2.50–4.50	145–254	
Ottenby, S. Sweden	1982		J	5		< 0.4*							Goede et al., 1989
			A	5		< 0.3*							
Curlew sandpiper (*Calidris ferruginea*)				28									
Baltic Sea													
Swedish Island of Öland	1981–83		O	28			149±329	2.21±3.37	0.09±0.03		4.42±0.82	176±30	Blomqvist et al., 1987
							40–1500	0.06–14.1	0.04–0.20		3.06–6.10	116–240	
			J	12			156±295	0.13±0.05	0.09±0.02		4.92±0.68	188±28	
							47–1090	0.06–0.22	0.05–0.12		3.97–6.10	145–240	
			A	16			144±362	3.76±3.30	0.09±0.04		4.05±0.72	167±28	
							40–1500	0.40–14.1	0.04–0.20		3.06–5.80	116–210	
EAGLES – ACCIPITRIDAE													
White-tailed eagle (*Halliaeëtus albicilla*)*													
Western Polish coast	1981–87	F	A	4				44.0–56.0	0.33	ND	2.0–11.0	40.0–230	Falandysz, 1984b; Falandysz and Szefer, 1983;
	1991–94	M+F	A	7				0.05–4.10			8.54–34.7		Szefer et al., 2000c
South of the Polish cost	1987	F	A	2				2.5–8.8			3.2–4.9	170–230	Falandysz et al., 1988
	1994	M	N	2				< 0.05–0.49			11.4–18.9		Szefer et al., 2000c

* – Dry wt.
A – Adult, N – Nestling, O – Overall.
M – Male, F – Female.

TABLE 3.17. – continued

Region	Sampling date	Sex	Age	N	Hg	Mg	Mn	Na	Ni	Pb	Se	Zn	References
						DUCKS – ANATIDAE							
Long tailed duck (*Clangula hyemalis*)													
Gulf of Pomerania	Pre-1977	M		14						0.11		22.6	Dańczak et al., 1997
										0.04–0.27		16.7–22.1	
		F		14						0.09		22.9	
										0.04–0.20		19.3–27.2	
Scaup duck (*Aythya marila*)													
Gulf of Gdansk	1980–84	M		4 (20)		200±10	1.80±0.20	1400±10	0.13±0.01	0.27±0.03		22.0±1.0	Szefer and Falandysz, 1987
						190–219	1.30–2.20	380–1400	0.10–0.15	0.23–0.35		21.0–23.0	Szefer, 1989
		F		5 (15)		180±10	2.20±0.30	1020±240	0.15±0.03	0.23±0.03		24.0±3.0	Szefer (unpublished data)
						170–200	1.70–2.80	540–1350	0.11–0.23	0.19–0.28		22.0–72.0	
Mallard (*Anas platyrhynchos*)													
Warta and Oder		M		10						9.16		26	Dańczak et al., 1997
flood waters										0.00–103		20.432.3	
		F		4						0.25		29.2	
										0.00–0.81		24.7–35.4	
Greylag geese (*Anser anser*)													
Warta and Oder		M		3						0.3		20.8	Dańczak et al., 1997
flood waters										0.14–0.55		19.6–22.3	
		F		6						0.5		19.4	
Tufted duck (*Nyroca fuligula*)													
Gulf of Gdansk	1980–84	M		18 (27)			1.75			< 0.05			Szefer et al., 2000c
		F		6 (9)			1.98			< 0.05			
Velvet scoter (*Oidemia fusca*)													
Gulf of Gdansk	1980–84	M		3 (2)			0.84			< 0.05			Szefer et al., 2000c
Common eider (*Somateria mollissima*)													
Gulf of Gdansk	1980–84	M		3 (4)			1.62			< 0.05			Szefer et al., 2000c
Norwegian coast	1973	F		6					0.4			23.4	Lande, 1977
Trondheimsfjord													
Goosander (*Mergus merganser*)													
Gulf of Gdansk	1980–84	F		3 (1)			2.06			< 0.05			Szefer et al., 2000c
						GREBES – PODICIPEDIAE							
Great crested grabe (*Podiceps cristatus*)													
Gulf of Gdansk	1980–84	M		9 (11)			1.42			< 0.05			Szefer et al., 2000c
		F		6 (6)			1.3			< 0.05			
						LOONS – GAVIIDAE (COLYMBIDAE)							
Black-throated loon (*Colymbus arcticus*)													
Gulf of Gdansk	1980–84	M		3 (2)			1.15			< 0.05			Szefer et al., 2000c

Region	Sampling date	Sex	Age	N	Hg	Mg	Mn	Na	Ni	Pb	Se	Zn	References
		F		3 (1)			0.74			< 0.05			
Red-throated loon (*Colymbus stellatus*)													
Gulf of Gdansk	1980–84	M		3 (4)			1.07			< 0.05			Szefer et al., 2000c
MURRES – ALCIDAE													
Common murre (*Uria aalge*)													
Gulf of Gdansk	1980–84	M		3 (2)			1.01			< 0.05			Szefer et al., 2000c
		F		5 (3)			0.79			< 0.05			
Belgian beaches	1989–95	M+F		143	4.6±2.9*							169±41*	Debacker et al., 1997
Black guillemot (*Cepphus grylle*)													
Gulf of Gdansk	1980–84	M		3 (3)			1.18			< 0.05			Szefer et al., 2000c
		F		6 (3)			1.12			< 0.05			
WADERS – SCOPOLACIDAE													
Migrating dunlin (*Calidris alpina*)													
Baltic Sea													
Swedish Island of Öland	1981–83		O	69		220±35	2.13±0.37			0.10±0.11		20.0±2.0	Blomqvist et al., 1987
						176–365	1.41–3.30			0.07–0.63		16.0–25.0	
			J	17		232±43	2.42±0.38			0.12±0.13		21.0±2.0	
						201–365	1.84–3.30			< 0.07–0.42		18.0–25.0	
			A	26		217±33	2.09±0.30			0.10±0.08		20.0±2.0	
						176–295	1.50–2.60			< 0.07–0.38		16.0–25.0	
Ottenby, S. Sweden	1982		J	5	1.9±2.2*						6±1*		Goede et al., 1989
			A	5	2.1±0.8*						8±3*		
	1982		J	5							5.8±1.2*		Goede and Bruin, 1985
Curlew sandpiper (*Calidris ferruginea*)													
Swedish island of Öland	1981–83		O	28		227±33	2.04±0.26			0.06±0.06		21.0±2.0	Blomqvist et al., 1987
Baltic						193–305	1.60–2.45			< 0.07–0.20		18.0–25.0	
			J	12		239±42	2.02±0.25			0.11±0.05		21.0±2.0	
						196–305	1.60–2.40			< 0.07–0.20		18.0–24.0	
			A	16		218±21	2.06±0.26			0.03±0.04		21.0±2.0	
						193–270	1.60–2.45			< 0.07–0.12		18.0–25.0	
EAGLES – ACCIPITRIDAE													
White-tailed eagle (*Halliaeëtus albicilla*)*													
Western Polish coast	1981–87	F	A	4	44.0–56.0		0.67–2.40			0.20–0.73			Falandysz, 1984b;
	1991–94	M+F	A	7			2.52–5.65			ND–26.4			Szefer et al., 2000c
South of the Polish cost	1987	F	A	2	2.50–8.80		0.98–1.30			9.7–13.0			Falandysz et al., 1988
	1994	M	N	2			2.12–3.80			ND			Szefer et al., 2000c

* – Dry wt.
A – Adult, J – Juvenile, N – Nestling, O – Overall.
M – Male, F – Female.

TABLE 3.18.

Concentrations of chemical elements ($\mu g\ g^{-1}$ wet wt.) in muscle of waterfowl from the Baltic Sea and other northern areas

Region	Sampling date	Sex	Tissue or age	N	Ag	Cd	Co	Cr	Cu	Fe	References
					DUCKS – ANATIDAE						
Long tailed duck (*Clangula hyemalis*)											
Gulf of Gdansk	1980–81	M	B.M.	10 (50)		0.01±0.003	0.013±0.003		0.86±0.04	87±10	Szefer and Falandysz, 1983a
						0.002–0.032	0.002–0.025		0.05–1.2	30.0–140	
			L.M.	3 (30)		0.015±0.012	0.042±0.004		0.59±0.10	100±5	
						0.001–0.028	0.039–0.047		0.44–0.79	91–110	
		F	B.M.	8 (40)		0.005±0.001	0.015±0.003		0.86±0.13	92±11	
						0.001–0.011	0.007–0.03		0.22–1.2	63.0–130	
			L.M.	3 (30)		0.020±0.002	0.061±0.015		0.96±0.05	150±20	
						0.018–0.023	0.044–0.091		0.87–1.0	120–190	
Scaup duck (*Aythya marila*)											
Gulf of Gdansk	1980–84	M	B.M.	12 (42)		0.03±0.00	0.04±0.01		5.5±0.4	80±10	Szefer and Falandysz, 1987
						0.02–0.06	0.01–0.07		3.4–7.5	50.0–140	
			L.M.	9 (22)		0.04±0.01	0.06±0.02		3.7±0.4	80±10	
						0.02–0.06	0.03–0.09		2.0–5.6	50.0–110	
		F	B.M.	17 (40)		0.03±0.00	0.04±0.01		5.1±0.4	80±10	
						0.01–0.05	0.02–0.07		1.6–7.6	50.0–110	
			L.M.	6 (35)		0.04±0.01	0.06±0.01		3.7±0.3	70±10	
						0.02–0.06	0.04–0.09		2.1–4.2	40.0–80	
Tufted duck (*Nyroca fuligula*)											
Gulf of Gdansk	1980–84	F	B.M.	12 (12)		< 0.05		0.12–0.21	2.79–10.9		Szefer et al., 2000c
		M	L.M.	6 (8)		< 0.05		0.02–0.11	3.73–15.8		
		F	L.M.	3 (5)		< 0.05		0.11	4.38		
Goldeneye (*Bucephala clangula*)											
Gulf of Gdansk	1980–84	F	B.M.	3 (2)		< 0.05		0.11	6.39		Szefer et al., 2000c
Velvet scoter (*Oidemia fusca*)											
Gulf of Gdansk	1980–84	F	B.M.	3 (1)		< 0.05		0.15	4.75		Szefer et al., 2000c
Common scoter (*Oidemia nigra*)											
		M	B.M.	3 (2)				0.12	6.65		
		M	L.M.	3 (2)				0.17	2.86		
Common eider (*Somateria mollissima*)											
Gulf of Gdansk	1980–84	M	B.M.	3 (2)		< 0.05		0.35	9.62		Szefer et al., 2000c
		F	B.M.	6 (3)		< 0.05		0.07–0.18	6.07–16.4		
		F	L.M.	6 (5)		< 0.05–0.08		< 0.01–0.12	3.05–3.35		
Norwegian coast Trondheimsfjord	1973	F		6	0.4			0.2	6.5	52	Lande, 1977
Goosander (*Mergus merganser*)											
Gulf of Gdansk	1980–84	F	L.M.	3 (1)		< 0.05		0.21	2.32		Szefer et al., 2000c

Region	Sampling date	Sex	Tissue or age	N	Ag	Cd	Co	Cr	Cu	Fe	References
Merganser (*Mergus serrator*)											
Gulf of Gdansk	1980–84	F	B.M.	3 (1)		0.91		0.13	6.26		Szefer et al., 2000c
		M	L.M.	3 (3)		< 0.05		0.2	2.66		
		F	L.M.	3 (2)		< 0.05		0.09	2.9		
GREBES – PODICIPEDIAE											
Great crested grabe (*Podiceps cristatus*)											
Gulf of Gdansk	1980–84	M	B.M.	6 (9)		< 0.05		0.08–0.23	6.40–9.71		Szefer et al., 2000c
		M	L.M.	6 (3)		< 0.05		0.06–0.20	1.94–2.64		
LOONS – GAVIIDAE (COLYMBIDAE)											
Black-throated loon (*Colymbus arcticus*)											
Gulf of Gdansk	1980–84	M	H.M.	1 (1)		< 0.05		0.17	2.01		Szefer et al., 2000c
Red-throated loon (*Colymbus stellatus*)											
Gulf of Gdansk	1980–84	N.I.	B.M.	6 (4)		< 0.05		0.14–0.16	5.79–6.75		Szefer et al., 2000c
		N.I.	L.M.	6 (4)		< 0.05–0.10		< 0.05–0.13	1.36–36.2		
MURRES – ALCIDAE											
Common murre (*Uria aalge*)											
Gulf of Gdansk	1980–84	F	B.M.	3 (1)		< 0.05		0.07	3.29		Szefer et al., 2000c
		M	L.M.	3 (1)		< 0.05		< 0.05	2.06		
		F	L.M.	6 (4)		< 0.05		010–0.14	1.63–1.84		
Belgian beaches	1989–95	M+F		143		ND			18±6.0*		Debacker et al., 1997
Black guillemot (*Cepphus grylle*)											
Gulf of Gdansk		F	B.M.	1		< 0.05		0.21	3.48		Szefer et al., 2000c
EAGLES – ACCIPITRIDAE											
White-tailed eagle (*Halliaeëtus albicilla*)*											
Western Polish coast	1981–87	F	A	4							Falandysz, 1984b;
						0.006–0.021	0.03	0.005	1.26–23.0		Falandysz and Szefer, 1983;
	1991–94	M+F	A	7							Szefer et al., 2000c
						ND			5.94–16.5		
South of the Polish cost	1987	F	A	2							Falandysz et al., 1988
						0.019–0.058			3.8–4.1		
	1994	M	N	2							Szefer et al., 2000c
						ND			4.7–8.3		

* – Dry wt.
M – Male, F – Female.
B.M. – Brest muscle.
L.M. – Leg muscle.
H.M. – Heart.
A – Adult, N – Nestling.

TABLE 3.18. – continued

Region	Sampling date	Sex	Tissue or age	N	Hg	Mn	Ni	Pb	Zn	References
					DUCKS – ANATIDAE					
Long tailed duck (*Clangula hyemalis*)										
Gulf of Gdansk	1980–81	M	B.M.	10 (50)		0.19±0.03	0.005±0.001	0.03±0.02	3.6±0.4	Szefer and Falandysz, 1983a
						0.05–0.39	< 0.004–0.008	0.02–0.04	2.3–4.7	
			L.M.	3 (30)		0.68±0.03	0.025±0.008	0.065±0.034	12±0.2	
						0.63–0.73	< 0.009–0.035	0.021–0.11	11.0–12	
		F	B.M.	8 (40)		0.30±0.05	0.019±0.003	0.04±0.05	3.8±0.4	
						0.14–0.50	< 0.004–0.027	0.01–0.11	2.6–5.6	
			L.M.	3 (30)		1.0±0.3	0.039±0.034	0.13±0.05	13±3	
						0.4–1.4	0.019–0.064	0.074–0.23	7.6–18	
Scaup duck (*Aythya marila*)										
Gulf of Gdansk	1980–84	M	B.M.	12 (42)		0.44±0.04	0.05±0.02	0.10±0.01	11±1	Szefer and Falandysz, 1987
						0.22–0.64	0.03–0.09	0.05–0.14	7.0–19.0	
			L.M.	9 (22)		0.74±0.13	0.08±0.01	0.12±0.03	23±3	
						0.26–1.40	0.04–0.11	0.02–0.28	15–35	
		F	B.M.	17 (40)		0.43±0.02	0.04±0.01	0.09±0.01	11±1	
						0.31–0.55	0.02–0.06	0.02–0.15	6.0–14	
			L.M.	6 (35)		0.67±0.11	0.07±0.02	0.10±0.07	19±2	
						0.25–0.99	0.02–0.13	0.03–0.38	17–23	
Tufted duck (*Nyroca fuligula*)										
Gulf of Gdansk	1980–84	F	B.M.	12 (12)			< 0.05	< 0.05		Szefer et al., 2000c
		M	L.M.	6 (8)			< 0.05	< 0.05		
		F	L.M.	3 (5)			< 0.05	< 0.05		
Goldeneye (*Bucephala clangula*)										
Gulf of Gdansk	1980–84	F	B.M.	3 (2)			< 0.05	< 0.05		Szefer et al., 2000c
Velvet scoter (*Oidemia fusca*)										
Gulf of Gdansk	1980–84	F	B.M.	3 (1)			< 0.05	< 0.05		Szefer et al., 2000c
Common scoter (*Oidemia nigra*)										
	1980–84	M	B.M.	3 (2)			< 0.05	< 0.05		
		M	L.M.	3 (2)			< 0.05	< 0.05		

Region	Sampling date	Sex	Tissue or age	N	Hg	Mn	Ni	Pb	Zn	References
Common eider (*Somateria mollissima)*										
Gulf of Gdansk	1980–84	M	B.M.	3 (2)			< 0.05	< 0.05		Szefer et al., 2000c
		F	B.M.	6 (3)			< 0.05	< 0.05		
		F	L.M.	6 (5)			< 0.05	< 0.05		
Norwegian coast	1973	F		6	0.04		0.4		6.6	Lande, 1977
Trondheimsfjord										
Goosander (*Mergus merganser)*										
Gulf of Gdansk	1980–84	F	L.M.	3 (1)			< 0.05	< 0.05		Szefer et al., 2000c
Merganser (*Mergus serrator)*										
Gulf of Gdansk	1980–84	F	B.M.	3 (1)			< 0.05	< 0.05		Szefer et al., 2000c
		M	L.M.	3 (3)			< 0.05	< 0.05		
		F	L.M.	3 (2)			< 0.05	< 0.05		
GREBES – PODICIPEDIAE										
Great crested grabe (*Podiceps cristatus)*										
Gulf of Gdansk	1980–84	M	B.M.	6 (9)			< 0.05	< 0.05		Szefer et al., 2000c
		M	L.M.	6 (3)						
LOONS – GAVIIDAE (COLYMBIDAE)										
Black-throated loon (*Colymbus arcticus)*										
Gulf of Gdansk	1980–84	M	H.M.	1 (1)			< 0.05	< 0.05		Szefer et al., 2000c
Red-throated loon (*Colymbus stellatus)*										
Gulf of Gdansk	1980–84	N.I.	B.M.	6 (4)			< 0.05	< 0.05		Szefer et al., 2000c
		N.I.	L.M.	6 (4)			< 0.05	< 0.05		
MURRES – ALCIDAE										
Common murre (*Uria aalge)*										
Gulf of Gdansk	1980–84	F	B.M.	3 (1)			< 0.05	< 0.05		Szefer et al., 2000c
		M	L.M.	3 (1)			< 0.05	< 0.05		
		F	L.M.	6 (4)			< 0.05	< 0.05		
Belgian beaches	1989–95	M+F		143	2.1±1.2*				60±14*	Debacker et al., 1997

Black guillemot (*Cepphus grylle*)										
Gulf of Gdansk	1980–84	F	B.M.	1			< 0.05	< 0.05		Szefer et al., 2000c
				EAGLES – ACCIPITRIDAE						
White-tailed eagle (*Halliaeëtus albicilla*)*										
Western Polish coast	1981–87	F	A	4			< 0.05	0.01–0.06		Falandysz, 1984b;
					1.2–6.2	0.27–0.83			22.8–86.0	Falandysz and Szefer, 1983;
	1991–94	M+F	A	7			ND	ND		Szefer et al., 2000c
						0.80–4.43				
South of the Polish cost	1987	F	A	2						Falandysz et al., 1988
					0.48–0.83	0.47–0.52			43.0–54.0	
	1994	M	N	2						Szefer et al., 2000c
						0.84–2.07				
				TERNS – STERNIDAE						
Little tern (*Sterna albifrons*)										
Baltic coast of Scleswig-	1995–96		N.D.E.	17	1.64±1.00					Thyen et al., 2000
Holstein			D.E.	15	2.01±0.73					
			C.	12	2.24±1.18					
			D.	22	8.36±4.65					

* – Dry wt.
M – Male, F – Female.
B.M. – Brest muscle.
L.M. – Leg muscle.
H.M. – Heart.
A – Adult, N – Nestling.
N.D.E. – Non-developed egg, D.E. – Developed egg, C. – Chick, D. – Down.

TABLE 3.19.

Concentrations of trace elements ($\mu g\ g^{-1}$ dry wt.) in bones of waterfowl from the Baltic Sea

Region Bone part	Sampling date	Sex	Age	N	Ag	Cd	Co	Cr	Cu	Fe	References
					DUCKS – ANATIDAE						
Scaup duck (*Aythya marila*)											
Gulf of Gdansk	1982/83	M+F									Szefer and Falandysz, 1986
Sternum				19 (38)	< 0.05–1.2	0.07	< 0.20–0.30	< 1.0	4.4	270	
						0.02–0.21			2.2–7.0	160–370	
Skull				8 (32)	< 0.05	0.05	< 0.20	< 1.0	5.6	340	
						0.03–0.11			3.9–8.0	240–410	
Wing bones				14 (28)	< 0.05–0.20	0.02	< 0.10–0.20	< 1.0	1.3	130	
						0.01–0.07			0.4–3.1	80–290	
Backbone				6 (24)	0.05–0.10	0.05	< 0.20	< 1.0	3.9	230	
neck sector						0.02–0.09			2.5–6.7	70–370	
Leg bones				16 (32)	< 0.05–0.15	0.09	< 0.20–0.50	< 1.0	2.2	160	
upper part						0.01–0.34			1.3–4.0	130–400	
Leg bones				16 (32)	< 0.05–0.40	0.06	< 0.20	< 1.0	1.2	220	
lower part						0.01–0.16			0.2–2.6	70–420	
Beak				9 (36)	< 0.05	0.04	< 0.20	< 1.0	3.4	320	
						0.03–0.06			2.5–4.6	160–450	
Trachea				5 (25)	< 0.05–0.25	0.09	< 0.20–0.40	< 1.0	4	555	
						0.06–0.76			3.5–5.7	370–630	
					WADERS – SCOPOLACIDAE						
Dunlin (*Calidris alpina*)											
Baltic Sea											
Ottenby, S. Sweden	1982										
Leg with claw			A	5		0.013					Goede et al., 1989
						0.009–0.021					
			J	5		0.020					
						0.007–0.033					
					EAGLES – ACCIPITRIDAE						
White-tailed eagle (*Haliaeëtus albicilla*)											
Coast of southern Baltic											
and Province Gorzów	1960/80										
Wing bone			A	1		0.52			1.7	110	Falandysz, 1984, 1986;
			I	1		0.21			1	86	
Leg bone			A			ND			0.57	31	
			I	1		0.5			1.7	84	
Femur			A	4		< 0.05–0.34			1.39	49.5	Falandysz et al., 1988
									0.56–1.9	30–120	

A – Adult, J –Juvenile, I – Immature.

TABLE 3.19. – continued

Region Bone part	Sampling date	Sex	Age	N	Mn	Ni	Pb	Zn	References
				DUCKS – ANATIDAE					
Scaup duck (*Aythya marila*)									
Gulf of Gdansk									Szefer and Falandysz, 1986
Sternum	1982/83	M+F		19 (38)	5.7	0.21	3.2	125	
					2.8–9.6	< 0.10–0.34	0.7–10.5	100–200	
Skull				8 (32)	4.4	0.55	4.3	130	
					2.6–5.5	< 0.10–1.1	1.7–7.8	100–270	
Wing bones				14 (28)	3.7	0.21	1.2	140	
					1.4–6.9	< 0.10–0.43	0.03–4.1	80–210	
Backbone				6 (24)	3.9	0.4	3.1	165	
neck sector					2.0–5.0	< 0.10–0.63	1.2–6.0	120–180	
Leg bones				16 (32)	7.3	< 0.10–0.43	6.3	240	
upper part					5.9–8.9		0.03–19.9	120–550	
Leg bones				16 (32)	6.6	< 0.10–2.0	5.1	175	
lower part					2.3–10.6		0.5–17.1	90–390	
Beak				9 (36)	5.7	0.32	3.8	135	
					4.7–6.6	< 0.10–0.80	0.6–8.5	100–230	
Trachea				5 (25)	10.5	0.45	19.9	195	
					9.2–13.4	< 0.30–1.30	15.2–24.2	170–200	
				EAGLES – ACCIPITRIDAE					
White-tailed eagle (*Haliaeëtus albicilla*)									
Coast of southern Baltic									
and Province Gorzów	1960/80								
Wing bone			A	1	3.9	6.7	3.7	280	Falandysz, 1984, 1986;
			I	1	2.7	8.1	3.5	86	
Leg bone			A		2.5	0.74	1.9	140	
			I	1	2.7	7.8	3.5	140	
Femur			A	4	2.95		10.4	284	Falandysz et al., 1988
					1.8–5.9		4.7–19	175–390	

A – Adult, I – Immature.

TABLE 3.20.

Concentrations of trace elements ($\mu g\ g^{-1}$ dry wt.) in feathers of waterfowl from the Baltic Sea and other northern areas

Region	Sampling date	Sex	Age	N[a]	As	Ca	Cd	Co	Cu	Fe	References
					DUCKS – ANATIDAE						
Long tailed duck (*Clangula hyemalis*)											
Gulf of Gdansk	1980–81	M		7 (28)			0.10±0.02	0.19±0.05	6.3±0.9	380±63	Szefer and Falandysz, 1983a
							0.08–0.17	0.10–0.40	3.4–8.9	290–610	
		M		5 (15)[b]			0.05±0.005	0.18±0.04	8.4±0.4	91±27	
							0.04–0.055	0.12–0.22	7.6–8.9	59–140	
		F		2 (6)			0.10±0.01	0.13±0.03	6.3±0.7	290±20	
							0.09–0.12	0.09–0.16	5.6–7.0	270–300	
Scaup duck (*Aythya marila*)											
Gulf of Gdansk	1980–84	M		4 (8)		490±40	0.30±0.05	0.51±0.07	10.6±1.4	1700±200	Szefer and Falandysz, 1987;
						400–580	0.22–0.38	0.35–0.70	6.9–13.6	1300–2300	Szefer (unpublished data)
		F		6 (12)		490±180	0.38±0.08	0.40±0.04	7.9±1.5	900±300	
						200–1400	0.15–1.10	0.25–0.45	3.4–12.5	200–1900	
					WADERS – SCOLOPACIDAE						
Knot (*Calidris canutus*)											
Dutch Wadden Sea	1979–82		J	33 (Vane)	< 0.21–1.20		< 0.14–0.84				Goede and de Bruin, 1984
			J	33 (Shaft)	< 0.05		< 0.05–0.12				
Dunlin (*Calidris alpina*)											
Baltic Sea											
Ottenby, S. Sweden	1982		A	4 (Vane)	1.0±0.5*						Goede et al., 1989
				5 (Vane)	1.1±0.6**						
					EAGLES – ACCIPITRIDAE						
White-tailed eagle (*Haliaeëtus albicilla*)											
Wolin and Uznam Islands, S.W. Baltic	1987	F	A	2			1.2"		16.0"	120"	Falandysz et al., 1988

[a] – Number of pooled samples; the total number of specimens analysed in parentheses.
[b] – Oiled.
F – Female; M – Male; J – Juvenile; A – Adult.
" – Tetrices.
* – Summer plumage.
** – Primary feathers.

TABLE 3.20. – continued

Region	Sampling date	Sex	Age	N[a]	Hg	Mg	Mn	Na	Ni	Pb	Se	Zn	References
							DUCKS – ANATIDAE						
Long tailed duck (*Clangula hyemalis*)													
Gulf of Gdansk,	1980–81	M		28			7.9±1.8		0.44±0.15	5.2±0.8		110±10.0	Szefer and Falandysz, 1983a
Southern Baltic							4.2–14.0		0.19–1.0	3.2–7.0		81.0–160	
		M		15[a]			2.1±0.7		0.69±0.25	1.2±0.2		95.0±3.0	
							1.5–3.5		0.50–0.84	0.7–1.5		90.0–100	
		F		15			3.3±0.6		0.21±0.05	4.4±0.7		110±8.0	
							2.8–3.9		0.16–0.26	3.7–5.1		100–120	
Scaup duck (*Aythya marila*)													
Gulf of Gdansk,	1980–84	M		8		1.18±0.05	32.0±1.1	4.03±0.25	2.90±0.10	0.51±0.07		130±10.0	Szefer and Falandysz, 1987;
Southern Baltic						1.11–1.30	32.0–75.0	3.30–4.46	2.70–3.10	0.35–0.70		110–140	Szefer (unpublished data)
		F		12		0.88±0.04	18.0±6.0	4.48±0.37	1.80±0.50	0.40±0.04		120±30.0	
						0.73–0.98	4.1–43.0	3.61–5.33	0.90–3.90	0.25–0.45		6.0–230	
							WADERS – SCOLOPACIDAE						
Knot (*Calidris canutus*)													
Dutch Wadden Sea	1979–82		J	33 (Vane)	< 0.80–1.95					2.19–16.4	< 1.95–13.8	198–232	Goede and de Bruin, 1984
			J	33 (Shaft)	< 0.40–1.04					< 0.10–0.66	< 0.70–1.50	88.6–91.0	
Dunlin (*Calidris alpina*)													
Baltic Sea													
Ottenby, S. Sweden	1982		J	5 (Vane)							< 1.9–2.7		Goede and Bruin, 1985
	1982		A	5 (Vane)	5.5±2.7*								Goede et al., 1989
			A	4 (Vane)	4.1±0.7**								
							EAGLES – ACCIPITRIDAE						
White-tailed eagle (*Haliaeëtus albicilla*)													
Wolin and Uznam	1987	F	A	2	18'								Falandysz et al., 1988
Islands, S.W. Baltic					54"		7.5"			2.7"		88"	
					33"'								

F – Female; M –Male; J – Juvenile; A – Adult.
' – Flight.
" – Tetrices.
"' – Fluff.
* – Winter plumage.
** – Summer plumage.
[a] – Oiled.

Inter-tissue trends

Significant inter-tissue variations in metal concentration in *Clangula hyemalis* wintering in the Gulf of Gdańsk were observed (Szefer and Falandysz, 1983a). There were no significant differences in the concentrations of selected metals between male and female of the ducks. For such result could be responsible the unknowing of age of the specimens studied. Among the soft tissues and organs, the liver contained the highest levels of Fe and Cd. Feathers were characterised by greater concentrations of Zn, Cu, Pb, Co and Ni as compared with the soft tissues. Several authors (Szefer and Falandysz, 1986, 1987) found distinct differences in concentrations of some metals in particular tissues and organs of *Aythya marila* from the Gulf of Gdańsk. Kidney contained the maximum concentrations of Cd, while liver had the highest levels of Cu and Mn. Both the liver and stomach were characterised by greatest concentrations of Zn; Fe was detected, sometimes at highly eleveted levels, in lung and liver. Feathers contained maximum levels of Fe, Zn, Mn, Pb, Cd, Co and Ni (Table 3.20). There were also significant differences in the concentrations of most metals for different bones (Szefer and Falandysz, 1986). It should be stressed that bones are generally enriched in Pb (Connors et al., 1975; Finley et al., 1976; White and Stendell, 1977; Fleming, 1981; Szefer and Falandysz, 1986) as compared to soft tissues (Szefer and Falandysz, 1987). The highest concentrations of Fe were found in skull and the lowest in wing bones (Table 3.19). The distribution of Cu was similar to that of Fe. Leg bones contained, on average, the highest amounts of Zn, Mn and Pb, while wing bones were characterised by the lowest levels of these metals.

Inter-age trends

The age-related concentration variations of hepatic and renal Cd and Cu in dunlin and curlew sandpiper staging at the Baltic Sea have been well documented by Blomqvist et al. (1987). The accumulation was already significant in first-year bird. A subsequent positive age-related trends were observed for renal Cd, i.e. 1.7-fold increase is statistically significant ($p < 0.02$). Such age-related relationship between first-year birds and adults was not detectable in the liver of dunlins, However distinct age-dependent trends were observed for hepatic and renal Cd in both dunlin and curlew sandpiper; adults contained maximum metal levels being several times higher than those in juveniles (Fig. 3.24). The distribution pattern of Cu changed with age of the birds showing negative age-related tendency (Fig. 3.24). According to Blomqvist et al. (1987) a pronounced age-related organ-specific increase is highest for the kidney tissues, with an average 17.6 and 28.9-fold increase ($p < 0.001$) between two age classes, i.e. juveniles and adults of dunlin and curlew sandpiper, respectively. The relative accumulation was higher in the latter species. The levels of Cd in the kidney vs. the liver of both the species were significantly correlated (Fig. 3.25).

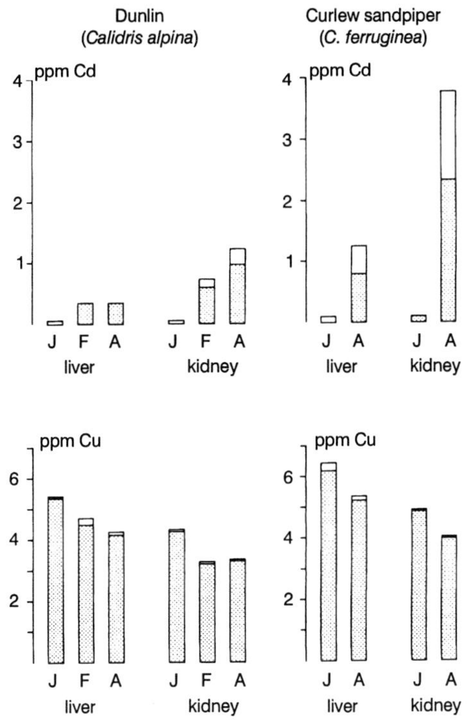

Fig. 3.24. *Calidris alpina* and *C. ferruginea.* Age-related concentration changes of cadmium and copper in liver and kidney tissues (fresh wt) of dunlin and curlew sandpiper, expressed as geometric mean (stippled part of columns) and arithmetic mean (open part). Age classes: J = juveniles, F = first-year birds, A = adults. Note different scales of axes. After Blomqvist et al. (1987); modified.

(iii) Occurrence of Radionuclides in Waterfowls

There are extremely few available data on concentration of radionuclides in birds from the Baltic Sea area (Szefer and Falandysz, 1983b). The U concentration in tissues of long tailed duck wintering in the Baltic Sea ranged from 0.037 to 0.42 ng g^{-1} wet wt., the Th levels were comparable to the U values and lay within the range of 0.024 and 0.79 ng g^{-1} wet wt. (Szefer and Falandysz, 1983b).

(iv) Recommendations for Future Studies

It is sometimes found for animals, especially birds, a large discrepancy between Pb determination in various samples of the same organ; hence contamination by this metal was suspected. For solution of this finding, Frank (1986) analysed the liver and kidney of shot eider ducks (*Somateria mollissima*) and long-tailed ducks (*Clangula hyemalis*) from Sweden for environmental pollutants obtaining for Pb irreproducible results. The author concluded that these birds shot

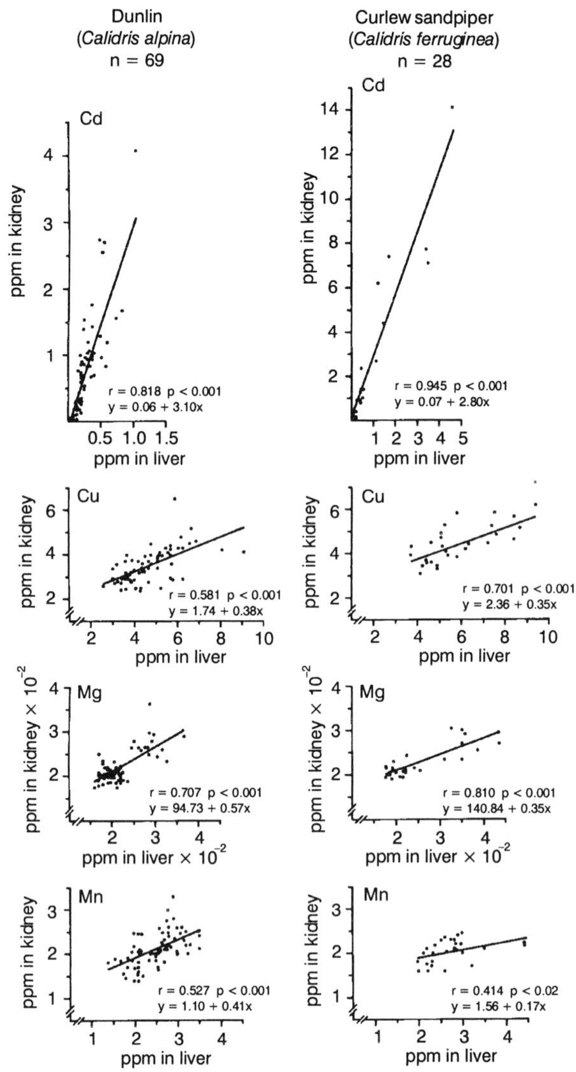

Fig. 3.25. *Calidris alpina* and *C. ferruginea.* Linear regressions of cadmium, copper, magnesium and manganese concentrations in kidney versus liver tissues (fresh wt) of dunlin and curlew sandpiper. Symbol r indicates Pearson product-moment correlation coefficient. Note different scales of the axes. After Blomqvist et al. (1987); modified.

with lead pellets should not be used for Pb determination unless careful X-ray studies are performed prior to the chemical analysis. Determinations are recommended to be made on at least two various samples of the tissue. This finding appears to be applicable to tissues of other shot animals, e.g. marine mammals (seals, dolphins and porpoises). Other Pb sources may by lead shots deposited after hunting in shallow waters and next ingested by birds, e.g. the swans (Frank and Borg, 1979).

According to Bryan and Langston (1992), data on geographical variations in exposed seabirds to heavy metals recorded before 1990 were rather limited and of doubtful toxicological significance. More work is needed on metals and organometals in estuarine birds, particularly in respect to their metabolism and their effects on individuals affected by stresses such as starvation (Bryan and Langston, 1992). It seems that this situation still prevails because since the beginning of 90s a state of this research has not been significantly improved. Studies on the microdistribution of toxic metals and radionuclides in organisms (at the cellular level) and on the toxicokinetics of organic pollutants in organisms in respect to ecotoxicological risk assessment are also necessary. As an example, Reijnders (1994), in his overview paper, presents data on the toxicokinetics of chlorobiphenyls and associated physiological responses in marine mammals, with particular reference to their potential for ecotoxicological risk assessment. Great attention should be paid to study of biogeochemical interactions affecting trace metals levels in the tissue of organisms. For example, hepatic concentrations of Se, Mn and Fe in aquatic birds indicate the induction of enzymes in response to oxidative stress (Möller, 1996).

F. MARINE MAMMALS

(i) Introduction

General Characteristics and Taxonomy

Marine mammals occur mainly in the northern and western parts of the Baltic. These are noted only occasionally in the Baltic Proper. The most commonly occurring species is the largest Baltic seal – grey seal (*Halichoerus grypus*) which migrates mainly from Estonia, the nearest hatching area. The ringed seal *Pusa hispida* is very seldom observed. The common seal *Phoca vitulina* occurs mainly in the Kattegat and Skagerrak. There is not much information on the Baltic porpoise *Phocoena phocoena,* which occurs mainly in the western Baltic, although it is sometimes found in fish nets elsewhere in the Baltic (Segerstrale, 1957; Jansson, 1972; Augustowski et al., 1987; Andell et al., 1994; Rheinheimer, 1998; Falandysz et al., 2000).

Order: Cetacea
Suborder: Odontoceti
Family: Phocoenidae
Species: Harbour porpoise (*Phocoena phocoena*, Linnaeus, 1758).

Habitat and range: it is a generally coastal species, inhabits relatively shallow, cold temperate and subarctic waters of the Northern Hemisphere. This sea mammal is found in bays, estuaries, tidal rivers and channels, and adjacent offshore shallows (Klinowska, 1991); distributed in the eastern North Atlantic and ranged from the Kara and Barents Seas to Senegal and Dakar in West Africa, including the Faeroe, Cape Verde and Azores Islands; occurred in the Baltic and White Seas. In the western North Atlantic it is recorded from southern Greenland and southern Baffin Island to North Carolina. Harbour porpoise may occasionally be found in the eastern Pacific Ocean. This mammalian species is present near islands in the Bering Sea and is found along the western Pacific coasts (the northern Japanese islands). Isolated population is noted in the Black Sea (Tomilin, 1989; Klinowska, 1991; Jefferson et al., 1993); In the Baltic Sea, especially in the Danish Lille Belt catches (for use as food but also for oil, meat, meal and other commercial products) have taken place since the 14th century. This species appeared frequently in the southern Baltic up to 1940, a sharp decline in its population being observed thereafter. About 1,600 and 770 specimens were caught for human consumption during the World War I and World War II, respectively (Klinowska, 1991). According to Andersen (1982) only ca. 20 animals were registered in the same area from 1969 to 1970. However, since the 80's harbour porpoises have been observed rather more often, especially in Gdańsk Bay, although it is still a extremely rare species in Polish zone of the Baltic Sea (Skóra et al, 1988; Skóra, 1991). Although there is a good evidence for major population declined in some Baltic Sea areas, especially such as Polish and Danish waters, the situation elsewhere is unclear (Klinowska, 1991). Clarke et al. (1998) estimated the bycatch mortality of harbour porpoise in the Baltic Sea. According to Otterlind (1976) this mammalian species has been endangered in Swedish waters. *Food habits*: its diet consists mainly from fish and it feeds less frequently on bivalvia and squids. The Baltic porpoises' stomachs contained mainly bottom fish (Gobiidae, *Gadus morhua, Zoarces viviparus*), pelagic fish (*Clupea harengus, Sprattus sprattus*) and semi-pelagic fish (*Ammodytes tobianus*) (Szefer et al., 1995b, Szefer et al., 2000f). Invertebrates were observed in the stomach contents sporadically. Diet of porpoises from the Danish Baltic is very similar to that of southern Baltic porpoises. Composition of stomach contents of the Greenland porpoises was as follows: mainly fish (*Mallotus villosus*, *Boreogadus saida*, *Reinhardtius hippoglossoides*), squids and Euphausids (Paludan-Müller et al., 1993; Heide-Jørgensen and Lockyer, 1999).

Order: Cetacea
Suborder: Odontoceti
Family: Delphinidae
Species: White-beaked dolphin (*Lagenorhynchus albirostris*).

Habitat and range: distributed in temperate waters of North Atlantic; range: frequent in the waters of Newfoundland and the Labrador Sea during the summer months (Lien et al., 1984), it may spend winter in the Gulf of Lawrence (Muir et al., 1988), in the Russia observed at Murmansk coast and in the Baltic Sea reaches Finnish Bay (Tomilin, 1989; Klinowska, 1991; Jefferson et al., 1993), sporadically noted in southern Baltic. *Food habits*: feeds mainly on fish, its diet shifts sporadically to squids and crustaceans (Tomilin, 1989).

Species: Striped dolphin (*Stenella coeruleoalba*)

Habitat and range: distributed in tropical, subtropical and temperate waters of Pacific Ocean; observed in northern waters, e.g. Southern Greenland, Shetland and Orkney Islands, Japan, Kuril Islands and British Columbia, south to La Plata, South Africa and New Zealand (Tomilin, 1989; Klinowska, 1991; Jefferson et al., 1993), recently very sporadically visits southern Baltic. *Food habits*: feed mainly on fish, squids, shrimps and myctophid fish (Honda et al., 1982; Tomilin, 1989).

Beluga whale (*Delphinapterus leucas*)

Habitat and range: occurred in waters of the Arctic and surrounding basins – the Bering and Okhotsk Seas, in very frost winters observed in coastal areas of Japan, Massachusetts Bay, England and visits also the Baltic Sea (Tomilin, 1989; Klinowska, 1991; Jefferson et al., 1993. *Food habits*: feed mainly on fish, crustaceans and squids (Tomilin, 1989).

Order: Pinnipedia
Suborder: Phocids
Family: Phocidae
Species: Grey seal (*Halichoerus grypus*)

Habitat and range: it occurs in temperate and subarctic waters of the North Atlantic – Gulf of St. Lawrence, Iceland, British and Norwegian coasts, the Baltic Sea (King, 1983; Naumov, 1989). Breeding colonies are found on islands in ice-free waters and on fasty ice and islands closely connected with fast ice. In Canadian waters Grey seal is more widespread in summer and may be observed from Cape Chidley on the Labrador coast to Nova Scotia. It is found on Islands in the Gulf of Lawrence, in the Bay of Fundy and on the shores of Nova Scotia as well as in the estuary of the St Lawrence River (King, 1983; Naumov, 1989). *Food habits*: feed mainly on fish, some crustaceans and molluscs (King, 1983; Naumov, 1989).

Species: Harbour (Common) seal (*Phoca vitulina*)

Habitat and range: it occurs on the coast of North Atlantic and North Pacific (King, 1983); Eastern Atlantic Harbour seal (*P. vitulina vitulina*) occurs in the Wash and breeding colonies are found on the English east coast. It inhabits the eastern Scottish coast, Shetland Islands, the Outer and Inner Hebrides, the shallow waters of the eastern and north eastern coasts of Ireland. It is found also on the sandy shores of the Netherlands and the adjacent areas of the German and Danish coasts; small numbers of seals extend into the Baltic Sea as far as Stockholm. Moreover, it is presented along all the Norwegian coast (King, 1983). Western Atlantic Harbour seal (*P. vitulina concolor*) occurs on the Greenland coast, and across to the southern and western shores of Baffin Island, the shores of Hudson Bay, less frequently observed north of Southampton Island. Its main region of distribution is from Labrador to New Brunswick, on the coasts of Newfoundland and Nova Scotia as well as in some areas of the Gulf of St Lawrence. Sometimes occurs on the coasts of Virginia and Carolina. Food habits: Eastern Atlantic Harbour seal (*P. vitulina vitulina*) is a fish eater, e.g. flounder, sole, herring, eel, cod and whiting; feeds also on shrimps, whelks, crabs and molluscs. Western Atlantic Harbour seal (*P. vitulina concolor*) feeds mainly on fish, e.g. herring and flounder (King, 1983; Naumov, 1989).

Species: Ringed seal (*Phoca hispida*)

Habitat and range: it occupies the circumpolar Arctic, even as far as the North Pole (King, 1983; Naumov, 1989); it is seen on the northerly coast of Iceland during the winter, but does not breed there and its range may extend down the Labrador coast as far as Newfoundland. *P. hispida hispida* occurs at coasts of Russia, Europe, Canada and Alaska. *P. hispida krascheninikovi* is seen in the northern parts of the Bering Sea, the Okhotsk Sea and western coast of Kamchatka, along the Pacific Japanese coasts and Korea. *P. hispida botnica* Gmelin inhabits the Baltic Sea, including the Gulfs of Finland and Bothnia. Freshwater Ringed seals *P. hispida saimensis* and *P. hispida ladogensis* are observed in the Finnish Lake Saimaa and the Russian Lake Ladoga, respectively (King, 1983; Naumov, 1989). *Food habits*: feeds on small pelagic amphipods, e.g. *Parathemisto*, euphausians and other crustaceans, e.g. *Mysis*, and small fish, e.g. *Boreogadus saida*, and molluscs (King, 1983; Naumov, 1989).

Overview of Worldwide Literature

Recently heavy metal influx to some marine coastal and estuarine areas has been greatly increased. It is well known that marine mammals are affected by exposure to metallic pollutants. There are numerous papers on heavy metals in marine mammals such as whales (Stoneburner, 1978; Hyvärinen and Sipilä, 1984; Honda et al., 1987; Muir et al., 1988; Morris et al., 1989; Hansen et al., 1990; Wagemann et al., 1991; Meador et al., 1993; Sanpera et al., 1993; Dietz et al., 1996, 2000a; Law et al., 1996, 1997a, 1997b; Holsbeek et al., 1999; Parsons et al., 1999; Frodello et al., 2000; Hernández et al., 2000; Nielsen et al., 2000), belugas (Hansen et al., 1990; Wagemann et al., 1991; Muir et al., 1999), narwhals (Wagemann et al., 1983; Hansen et al., 1990; Dietz et al., 1996; Muir et al., 1999), walruses (Taylor et al., 1989; Muir et al., 1999), dugongs (Denton et al., 1980), seals (Gaskin et al., 1973; Heppleston and French, 1973; Drescher et al., 1977; Roberts et al., 1976; Duinker et al., 1979; McKie et al., 1980; Reijnders, 1980; Helle, 1981; McClurg, 1984; Ronald et al., 1984; Tohyama et al., 1986; Yamamoto et al., 1987; Wagemann et al., 1988; Wagemann, 1989; Morris et al., 1989; Skaare et al., 1990, 1994; Law et al., 1991, 1992; Szefer et al., 1993a, 1994c; Malcolm et al., 1994; Noda et al., 1995; Watanabe et al., 1998; Muir et al., 1999; Yeats et al., 1999; Dietz et al., 2000a; Julshamn and Grahl-Nielsen, 2000), porpoises (Gaskin et al., 1972; Falconer et al., 1983; Fujise et al., 1988; Joiris and Bossicart, 1989; Morris et al., 1989; Joiris et al., 1991; Law et al., 1991, 1992; Paludan-Müller et al., 1993;

Teigen et al., 1993; Mackey et al., 1995; Dietz et al., 1996; Tibury et al., 1997; Bennett et al., 2001), dolphins (Honda et al., 1982; Honda and Tatsukawa, 1983; Muir et al., 1988; Morris et al., 1989; Law et al., 1991, 1992; Kuehl et al., 1994; Mackey et al., 1995; Holsbeek et al., 1998; Cardellicchio et al., 2000; Das et al., 2000), sea lions (Martin et al., 1976b; Sydeman and Jarman, 1998) and polar bears (Norstrom et al., 1986; Norheim et al., 1992; Dietz et al., 1996, 2000a, 2000b). Kemper et al. (1994) presented a review of heavy metal levels in marine mammals of Australia. An overview and evaluation of contaminants in baleen whales based on data for ca. 1000 individuals in 10 species from all over the world have been reported by O'Shea and Brownell Jr. (1994). Wagemann and Muir (1984) evaluated in their extensive review the concentration data of trace elements in marine mammals of northern waters.

In recent years a great attention has been paid to the distribution of butyltins in marine mammals. Although many studies of these compounds pollution have been conducted, but to date only a few have included determinations in the tissues and organs of marine mammals. According to several authors (Law et al., 1999) butyltins can accumulate in these animals and this has been reported for seals, a range of cetaceans and Steller sea lions from the North Pacific Ocean and Japanese coastal waters (Iwata et al., 1994a, 1994b; Kim et al., 1996d), Dall's porpoises, 3 species of dolphins and 4 species of whales from the North Pacific Ocean and Asian coastal waters (Tanabe et al., 1999), bottlenose dolphins from the Italian coast (Kannan et al, 1996) and from the U.S. Atlantic and Gulf coasts (Kannan et al., 1997b), pelagic cetaceans, harbour porpoises and grey seals from the coastal waters of England and Wales (Law et al., 1998, 1999) and harbour porpoise from the Baltic Sea (Kannan and Falandysz, 1997a). Accumulation patterns of butyltins in Ganges river dolphins have been also studied by Kannan et al. (1997a). Kim et al. (1998a) described *in vitro* inhibition of hepatic cytochrome P450 and enzyme activity by butyltin compounds in marine mammals. The butyltin environmental problem reported has been of particular concern, attracting a great attention and simulating to scientific hot discussion and debate. An example of such great interest and polemics are comments on butyltins' articles and author's responses to these comments (Robinson et al., 1999; Kannan and Falandysz, 1997b; Kannan and Tanabe, 1997). Most recently, Tanabe (1999) summarised and discussed extensively world data concerning butyltin pollution in marine mammals.

The concentrations of trace elements in soft tissues and organs of marine mammals have been extensively studied, however such investigations of bones have been conducted sporadically (Honda et al., 1984a, 1984b, 1986b).

There are a few available data of radionuclide concentrations in marine mammals (Anderson et al., 1990; Berrow et al., 1998; Watson et al., 1999).

(ii) Occurrence of Chemical Elements in Marine Mammals

Data on distribution trace metals in porpoises from the Baltic Sea and adjacent areas especially from its southern region have been published by several authors (Gaskin et al., 1972; Andersen and Rebsdorff, 1976; Falconer et al., 1983; Harms et al., 1977/1978; Clausen and Andersen, 1988; Joiris et al., 1991; Szefer et al., 1994b, 1995b; Kannan and Falandysz, 1997a); however dolphins have been studied less frequently in this respect (Harms et al., 1977/1978; Szefer et al., 2000d). Baltic seals have been analysed for metallic pollutants by several authors (Herva and Häsänen, 1972; Kari and Kauranen, 1978; Helle, 1981; Perttilä et al., 1986; Frank et al., 1992; Szefer et al., 2000d; Fant et al., 2001). Spatial investigations of harbour porpoises from the Baltic Sea, Danish and Greenland waters have been performed by Szefer et al. (1999b, 2000f, 2000h). The concentration and distribution of radionuclides in marine mammals from the southern Baltic have been studied by Szefer et al. (2000e).

Marine mammals are known to be effective bioaccumulators of metallic pollutants from their food. A special attention should be paid to Hg and Cd because of their very toxic effect to organisms. Some metals, especially Hg can biomagnify along the trophic levels in the marine environment reaching very high hepatic and renal levels in pinnipeds and cetaceans. It should be emphasised that diseases of cetaceans are associated with elevated levels of some metallic toxicants, e.g. Hg (Viale, 1994). It is reported that Se protects against Hg toxicity and both the elements are present generally in a 1:1 molar ratio in the liver of sea mammals and humans (Koeman et al., 1973, 1975; Nielsen and Dietz, 1990; Skaare et al., 1990, 1994), therefore there is important to know relationships between concentrations of both these elements in Baltic mammals. Seals from fresh and brackish waters of Finland have been also analysed for Hg and Se concentrations (Kari and Kauranen, 1978; Hyvärinen and Sipilä, 1984; Hyvärinen et al., 1998). Data of Hg and/or Se have been published for seals from the Baltic Sea and adjacent areas (Perttilä et al., 1986; Morris et al., 1989; Law et al., 1991, 1992; Frank et al., 1992; Skaare et al., 1994). There are only a few available data concerning concentrations of Hg in harbour porpoise and dolphins from German coastal waters (Harms et al., 1977/1978), southern Baltic (Szefer et al., 1994b, 1995b, 2000h); more data are related to adjacent areas (Andersen and Rebsdorff, 1976; Falconer et al., 1983; Clausen and Andersen, 1988; Joiris and Bossicart, 1989; Morris et al., 1989; Joiris et al., 1991; Law et al., 1991, 1992; Teigen et al., 1993; Tibury et al., 1997; Das et al., 2000). As for porpoise Se, respective Baltic data are reported by Szefer et al. (2000h).

Representatives of Baltic mammals such as harbour porpoise (*Phocoena phocoena*), striped dolphin (*Stenella coeruleoalba*), white-beaked dolphin (*Lagenorhynchus albirostris*) as well as ringed seal (*Phoca hispida*), harbour (common) seal (*Phoca vitulina*) and grey seal (*Halichoerus grypus*) have been analysed for trace element concentrations (Table 3.21–3.25).

TABLE 3.21.

Concentrations of chemical elements ($\mu g\ g^{-1}$ wet wt.) in liver of marine mammals from the Baltic Sea and other northern areas

Region	Sampling date	Length (cm)	Age (year)	N	Ag	Cd	Co	Cr	Cu	Fe	References
						PHOCOENIDAE					
Harbour porpoise (*Phocoena phocoena*)											
Southern Baltic	1989–93		J	36*	5.78±7.3	0.29±0.16			19.5±6.2		Szefer et al., 1995b
					0.19–24.1	0.02–0.65			10.6–35.8		
Southern Baltic	1990–96		< 1	8*		0.07±0.07	< 0.5	0.60±0.05	14.2±3.74	660±290	Szefer et al., 2000f
						0.02–0.2		0.55–0.68	7.32–21.2	270–1200	
			1.0–2.0	10*		0.05	< 0.5	0.58±0.07	14.4±3.75	660±250	
						0.03–0.08		0.47–0.73	8.50–20.3	220–960	
			3.0–6.0	16*		0.06	< 0.5	0.60±0.15	14.8±4.71	630±260	
						0.02–0.1		0.43–0.88	7.61–20.9	330–980	
German coastal waters	Pre-1978	80		1		0.023			15		Harms et al., 1977/78
		126		1		0.025			4		
		153		1		0.19			6		
Danish waters	1972–73								4.5		Andersen and Rebsdorf, 1976
									2.6–8.3		
Danish waters	1993–96		< 1.0	6*		< 0.02		0.64±0.09	100±61	690±250	Szefer et al., 2000f
								0.52–0.79	30.7–160	280–890	
			1.0–2.0	5*		0.17±0.15		0.69±0.13	19.4±3.65	960±250	
						0.03–0.38		0.53–0.84	18.1–24.0	680–1200	
			> 2.0	1*		0.43		0.41	23.2	550	
Waters around the British Isles											
Cadigan Bay	1988–89					< 0.06			13.2		Law et al., 1991
									6.6–30		
	Pre-1989		Y–J			< 0.06		< 0.5	7.65		Morris et al., 1989
									6.6–8.7		
Scotland	1988–89					0.15			58		Law et al., 1991
						0.07–0.24			7.4–160		
North British Isles	1988–89					0.16			10		Law et al., 1991
						0.07–0.23			6.6–16		
Welsh coast	1988–91		Y–J			0.16			19.8		Law et al., 1992
						< 0.07–0.41			2.8–120		

Region	Sampling date	Length (cm)	Age (year)	N	Ag	Cd	Co	Cr	Cu	Fe	References
Western Greenland	1988–89		0–>6.0	44		3.25			12		Paludan-Müller et al., 1993
						0.05–11.7			5.03–50.3		
Central West part of	Pre-1995		2.0–>7	34		4.1					Dietz et al., 1996
Southern Greenland						3.18–5.0					
Southwest Greenland	1995		< 1->6	43*		10.7		0.61	22.5	510	Szefer et al., 2000f
						0.03–31.7		0.49–0.78	8.61–52.0	160–1100	
						DELPHINIDAE					
White beaked dolphin (*Legenorhynchus albirostris)*											
Southern Baltic	1989–95	119–229		3*		0.13		0.51	15	450	Szefer et al., 2000d
						0.06–0.23		0.50–0.52	13.8–16.5	257–760	
Waters of the British Isles	1989		A	1		0.11		< 0.6	6.8		Law et al., 1991, 1992
Welsh coast and Irish Sea											
Striped dolphin (*Stenella coeruleoalba*)											
Southern Baltic	1998/99	187		2*		4.28		1.15	17.05	404	Szefer et al., 2000d
						3.62–4.94		0.99–1.30	16.8–17.3	403–404	
Waters of the British Isles	1990		J+NI	2		9.7		< 0.5	11		Law et al., 1991
						8.4–11.0			10.0–12.0		
Welsh coast and Irish Sea	1990/91		A	3		7.03			10.1		Law et al., 1992
						1.7–11.0		< 0.5	8.4–12.0		
Beluga whale (*Delphinapterus leucas*)											
Baltic	Pre-1978	271		1		0.9			20.4		Harms et al., 1977/78

* – Dry wt.
J – Juvenile, Y–J – young/juvenile, N – Neonate, A – Adult.
NI – No information

TABLE 3.21. – continued

Region	Sampling date	Sex	Length (cm)	Age (year)	N	Ag	Al	Ca	Cd	Co	Cr	Cu	Fe	References
PHOCIDAE														
Ringed seal (*Phoca hispida*)														
Gulf of Finland	1976–82			7.3	11				0.17			3.9		Perttilä et al., 1986
				0.5–18					0.04–0.41			1.9–6.8		
Baltic Sea	1988				10		0.96	43	0.11	0.02	0.085	7.2		Frank et al., 1992
							0.35–1.36	37–59	0.07–2.51	0.008–0.026	0.005–0.156	3.2–18	320	
German coastal waters	Pre-1978		84		1				0.31			2.1	196–939	Harms et al., 1977/78
Grey seal (*Halichoerus grypus*)														
Gulf of Finland	1976–82			10.9	9				0.04			10.7		Perttilä et al., 1986
				0.10–40					ND–0.16			2.4–17.3		
Baltic Sea	1988				10		0.83	55	0.1	0.021	0.054	15	560	Frank et al., 1992
							0.36–12	48–79	< 0.02–0.17	0.017–0.035	0.047–0.068	7.4–24	303–681	
E. England	1988								0.06			22		Law et al., 1991
N. Ireland	1988–89								0.85			13		
									0.06–2.9			7.8–17.0		
Welsh coast and Irish Sea	1988–91				21				0.46		< 0.5–2.0	14.8		Law et al., 1992
									< 0.06–2.9			2.2–39.0		
Cadigan Bay, W. Wales				< 1.0	2				2.1		< 0.5	17.6		Morris et al., 1989
									1.7–2.5			9.1–26.0		
Southern Baltic	1999				2*	ND			ND		1.07	27.5	1220	Szefer et al., 2000d
											0.93–1.21	17.4–37.5	690–1750	
Harbour seal (*Phoca vitulina*)														
German coastal waters	Pre-1978			2					0.01–0.21			2.6–17.0		Harms et al., 1977/78
Skagerrak	1988				10		1	57	0.04	< 0.002	0.025	9.3	369	Frank et al., 1992
							< 0.02–3.83	44–91	< 0.02–0.10	< 0.002–0.035	0.017–0.035	5.0–16.0	248–642	
Kattegat	1988				10		0.65	58	0.04	0.019	0.032	12	319	
							< 0.02–1.38	48–69	< 0.02–0.06	< 0.002–0.025	0.023–0.058	8.1–20	204–668	
Kalmarsund	1988				10		1.88	64	0.02	0.008	0.138	4	350	
							0.23–5.64	49–91	< 0.02–0.06	< 0.002–0.022	0.107–0.157	2.2–9.2	188–855	

Region	Sampling date	Sex	Length (cm)	Age (year)	N	Ag	Al	Ca	Cd	Co	Cr	Cu	Fe	References
German coast, North Sea	1974–76			0–6**	27				0.018			6.1		Drescher et al., 1977
									0.01–0.042			2.8–13.5		
				7.0–12.0**	11				0.024			7		
									0.018–0.03			3.3–10.5		
				13–18**	11				0.028			5.7		
									0.01–0.063			2.6–10.8		
				19–24**	4				0.032			10.6		
									0.025–0.038			6.0–17.0		
				3.0–5.0	1				0.042			2.8		
				6.0–8.0	4				0.135			10.7		
									0.05–0.20			8.5–14.8		
Dutch Wadden Sea	Pre-1979	M		3 weeks	8				0.03–0.21			2.0–20.0	28.0–600	Duinker et al., 1979
Norwegian waters	1988				71				< 0.1–1.0			7.68		Skaare et al., 1990
E. England	1988	M			16				0.08			8.1		Law et al., 1991
									< 0.06–0.15			6.2–12.0		
		F			12				0.2			7.8		
									< 0.07–0.50			5.2–16.0		
N. Ireland	1988–89								0.26					
									< 0.06–0.94			14		
												7.8–21.0		
Welsh coast and Irish Sea	1988–91				13						< 0.5–1.0	13.5		Law et al., 1992
												7.2–21.0		

* – Dry wt.
** – Months.

TABLE 3.21. – continued

Region	Sampling date	Length (cm)	Age (year)	N	Hg	Mn	Ni	Pb	Se	Sn	Zn	References
						PHOCOENIDAE						
Harbour porpoise (*Phocoena phocoena*)												
Southern Baltic	1991									0.023*		Kannan and Falandysz, 1997a
										0.018–0.027		
Southern Baltic	1989–93		J	36*	19.6±15.8	17.7±4.0		0.76±0.31			114±26.7	Szefer et al., 1995b
					2.12–52.3	8.30–34.5		0.28–1.26			78–181	
Southern Baltic	1990–96		< 1	8*	10.8±10.0	9.53±3.89	< 0.5	< 0.5	10.7±4.4		93.4±47.7	Szefer et al., 2000f, 2000h
					3.02–23.4	0.60–13.3			7.6–15.7		54.9–110	
			1.0–2.0	10*		9.14±2.26	< 0.5	< 0.5			100±73.9	
						5.81–12.7					64.2–310	
			2.0–4.0	5*	19.4±13.1				20.8±19.4			
					2.22–37.9				3.75±53.2			
			3.0–6.0	16*		8.98±1.65	< 0.5	< 0.5			90.4±22.6	
						6.74–11.0					60.9–120	
			> 5	2*	48.9±59.3				52.2±59.8			
					7.02–90.9				9.92–94.5			
German coastal waters	Pre-1978	80		1	0.7			0.17			50	Harms et al., 1977/78
		126		1	2.5			0.43			34	
		153		1	28			0.35			49	
Danish waters	1972–73				22			3.5			59	Andersen and Rebsdorf, 1976
					1.5–69			1.9–5.3			45–72	
Danish waters	1993–96		< 1.0	6*		6.16±3.0		< 0.5			180±130	Szefer et al., 2000f, 2000h
						3.11–10.9					81.8–420	
			1.0–2.0	5*	2.8	12.3±2.35		< 0.5	4.95		120±37.6	
					2.4–3.1	9.30–15.5			3.37–6.53		92.9–190	
			11	1*	90.7	11		< 0.5	47		130	
Kattegat and	1987–90	67–172		12	1.99							Joiris et al., 1991
Belgian water					0.35–132.0							
Norwegian coast												Teigen et al., 1993
Oslofjord	1989–90			16	4.0±4.5				3.5±3.4			
Western coast				10	4.5±3.7				4.4±3.1			
Waters around the British Isles												
Cadigan Bay	1988–89				1.3			< 0.7			58	Law et al., 1991
					0.5–2.8						25–87	
	Pre-1989		Y–J		0.62		< 0.5	< 0.6			37	Morris et al., 1989
					0.61–0.63						25–49	
Scotland	1988–89				2.2			< 0.7			39	Law et al., 1991
					0.7–4.9						34–49	
North Island	1988–89				14			< 0.7			16	Law et al., 1991
					1.4–30						8.8–28	

Region	Sampling date	Length (cm)	Age (year)	N	Hg	Mn	Ni	Pb	Se	Sn	Zn	References
Welsh coast	1988–91		Y–J	36	20 0.5–190		< 0.5–1.8	0.08–4.3			64 20–150	Law et al., 1991
Coastal waters of England and Wales	1992–1996		0–8**	29						0.214ª 0.05–0.64		Law et al., 1998
Belgium	1987–88			3	0.7 1.36–5.0							Joiris and Bossicart, 1989
Western Greenland	1988–89		0–>6.0	44	4.17 0.48–20.7				2.77 0.58–8.95		50.6 36.2–92.3	Paludan-Müller et al., 1993
Central West part of Southern Greenland	Pre-1995		2.0–>7	34	5.53 2.38–8.21				3.07 1.99–4.34			Dietz et al., 1996
Southwest Greenland	1995		< 1–>6	43*	52.6 1.08–150	13.3 9.50–16.9	0.27 ND–0.43	ND	20.1 5.7–52.9		130 87.4–250	Szefer et al., 2000f, 2000h
						DELPHINIDAE						
White beaked dolphin (*Legenorhynchus albirostris*)												
Southern Baltic	1989–95	119–229		3*		5.28 4.23–6.25	ND–0.21	0.99 0.66–1.35			78 74.8–82.5	Szefer et al., 2000d
Coastal waters of England and Wales	1994–1998	215–257	5–>9	3						0.113ª 0.03–0.17		Law et al., 1999
Waters of the British Isles	1989		A	1	27		< 0.6	3.8			27	Law et al., 1991
Striped dolphin (*Stenella coeruleoalba*)												
Southern Baltic	1998/99	187		2*		8.56 7.98–9.13		ND			173 163–183	Szefer et al., 2000d
Coastal waters of England and Wales	1996	219–219	17–20	2						0.237ª 0.16–0.31		Law et al., 1999
Waters of the British Isles Welsh coast and	1990		J+NI	2	10.5 10.0–11.0		< 0.6	< 0.7			55.5 46–65	Law et al., 1991, 1992
Beluga whale (*Delphinapterus leucas*)												
Baltic	Pre-1978	271		1	4.4			0.36			32	Harms et al., 1977/78

* – Dry wt.
ND – Not detected.
A – Adult, J – Juvenile, Y–J – young/juvenile.
** – Age not determined in 13 specimens.
ª – concentration of total butyltins.
NI – No information.

TABLE 3.21. – continued

Region	Sampling date	Sex	Length (cm)	Age (year)	N	Hg	Mg	Mn	Ni	Pb	Se	Sn	V	Zn	References
									PHOCIDAE						
Ringed seal (*Phoca hispida*)															
Gulf of Finland	1976–82			7,3	11	11,9				0,11					Perttilä et al., 1986
				0.5–18		1.3–26.0				0.04–0.22					
Baltic Sea	1988				10	1,99	186	3,4	0,018	0,1	1,35		0,043	36	Frank et al., 1992
						0.83–9.74	132–235	2.7–4.6	< 0.006–0.044	0.05–0.21	1.10–5.25		0.023–0.117	29–50	
German coastal waters	Pre-1978		84		1	0,64				0,24				40	Harms et al., 1977/78
Norwegian waters	1989–90	F		J	3	9,23		5,5							
						3.75–19.9		2.32–11.4							
		M		J		2,93		2,06							
						2.08–3.79		1.79–2.33							
		M		A		18,16		11,35							
						13.1–23.3		6.75–16.0							
Grey seal (*Halichoerus grypus*)															
Gulf of Finland	1976–82			10,9	9	99,7				0,17	26,8			54,8	Perttilä et al., 1986
				0.10–40		2.0–493				0.07–0.17	0.2–106			39.3–65.0	
Baltic Sea	1988				10	15	189	3,3	0,009	0,25	6,31		0,033	46	Frank et al., 1992
						7.03–37	182–212	2.6–4.1	< 0.006–0.015	0.16–0.43	3.42–12		0.024–0.077	40–57	
Southern Baltic	1999	M			2*			6,98						127	Szefer et al., 2000d
								6.37–7.58		ND–2.29				106–147	
German coastal waters	Pre-1978			2		19,5				0,31				61	
E. England	1988					110				< 0.6				25	Law et al., 1991
N. Ireland						28				< 0.6				60	
						3.50–95.0								36.0–79.0	
Welsh coast and Irish Sea	1988–91				21	108				< 0.60–1.9				60	Law et al., 1992
						1.5–430								22.0–110	
Cadigan Bay, W. Wales				< 1.0	2	2,1				< 0.7				84	Morris et al., 1989
						1.7–2.5								80.0–88.0	
Coastal waters of England and Wales	1993–1995	M+F		ND	5							0.016*			Law et al., 1998
												ND–0.022			
Harbour seal (*Phoca vitulina*)															
German coastal waters	Pre-1978					1.5–160				0.09–0.74				27.0–60.0	Harms et al., 1977/78
Skagerrak	1988				10	3,56	156	4,1	0,017	0,12	2,04		0,045	36	Frank et al., 1992
						0.72–7.69	135–186	2.4–5.1	< 0.006–0.028	0.09–0.25	1.17–4.88		0.018–0.173	25–46	

Region	Sampling date	Sex	Length (cm)	Age (year)	N	Hg	Mg	Mn	Ni	Pb	Se	Sn	V	Zn	References
Kattegat	1988				10	2,42	179	4,7	0,02	0,08	2,07		0,042	35	
						1.44–5.29	147–202	4.1–5.0	0.008–0.033	0.03–0.91	1.42–3.58		0.022–0.077	32–43	
Kalmarsund	1988				10	0,44	174	3,7	< 0.006	0,1	1,02		0,024	28	
						0.20–0.85	143–238	1.4–6.2	< 0.006–0.010	0.04–0.22	0.69–1.42		0.015–0.056	22–40	
German coast, North Sea	1974–76			0–6**	27	3,5				0,23				39,4	Drescher et al., 1977
						1.5–8.9				0.10–0.57				27.0–56.0	
				7.0–12.0**	11	6,7				0,23				39,6	
						1.6–26.5				0.11–0.55				27.5–50.5	
				13–18**	11	15,3				0,17				42,2	
						1.8–53.0				0.10–0.28				34.0–49.0	
				19–24**	4	11,2				0,23				37,3	
						7.7–14.7				0.14–0.30				29.0–50.0	
				3.0–5.0	1	29,9				0,25				27,5	
				6.0–8.0	4	96				0,46				36	
						59–160				0.39–0.53				29.0–41.0	
Dutch Wadden Sea	Pre-1979	M		3 weeks	8			2.0–6.0		< 0.05–2.3				16.0–64.0	Duinker et al., 1979
Norwegian waters	1988				71	5,21					6,3			52	Skaare et al., 1990
						0.1–89					0.5–49			19–99	
E. England	1988	M			16	30				< 0.70				42	Law et al., 1991
						1.40–110								39.0–49.0	
		F			12	56				< 0.80				49	
						6.4–100								32.0–66.0	
N. Ireland	1988–89					51				< 0.60				51	
						0.98–170								29.0–72.0	
Welsh coast and Irish Sea	1988–91				13	50,6									Law et al., 1992
						1.0–170				< 0.6–0.18				50,5	

* – Dry wt.
** – Months.
J – Juvenile, A – Adult.

TABLE 3.22.

Concentrations of chemical elements ($\mu g\ g^{-1}$ wet wt.) in kidney of marine mammals from the Baltic Sea and other northern areas

Region	Sampling date	Length (cm)	Age (year)	N	Ag	Cd	Co	Cr	Cu	Fe	References
					PHOCOENIDAE						
Harbour porpoise (*Phocoena phocoena*)											
Southern Baltic	1989–93		J	33*	< 0.62	1.59±1.39			12.4±2.6		Szefer et al., 1995b
						0.21–3.85			9.4–18.7		
Southern Baltic	1990–96		< 1	6*		0.58±0.45	< 0.5	0.48±0.05	10.9±5.93	420±260	Szefer et al., 2000f
						0.06–1.27		0.40–0.54	7.46–24.2	210–580	
			1.0–2.0	12*		0.63±0.31	< 0.5	0.52±0.05	8.33±1.70	410±140	
						0.22–1.26		0.44–0.61	6.61–10.1	190–700	
			3.0–6.0	6*		0.56±0.42	< 0.5	0.52±0.04	8.17±1.47	390±150	
						0.45–1.29		0.47–0.56	6.38–10.3	180–530	
German coastal waters	Pre-1978	126		1		0.077			3.1		Harms et al., 1977/78
		153		1		0.95			3.2		
Danish waters	1980–81		0–1.0	27		0.1					Clausen and Andersen, 1988
						0.01–0.8					
			2.0–3.0	18		0.3					
						0.01–1.3					
			4.0–12	29		0.5					
						0.01–1.6					
Danish waters	1993–96		< 1.0	6*		0.14±0.03	< 0.5	0.69±0.26	15.5±4.88	470±100	Szefer et al., 2000f
						0.10–0.17		0.42–1.10	12.0–25.5	330–580	
			1.0–2.0	6*		1.21±1.14	< 0.5	0.58±0.17	11.6±0.85	510±82	
						0.10–2.29		0.42–0.91	10.5–12.7	430–650	
			> 2.0	1*		4.15	< 0.5	0.92	10.4	340	
Western Greenland	1988–89		0–6.0	44		13.2			5.53		Paludan-Müller et al., 1993
						0.11–72.5			3.69–7.95		
Central West part of Southern Greenland	Pre-1995		4.0–7.0	20		18.4					Dietz et al., 1996
						16.5–20.3					
Sothwest Greenland	1995		< 1–12	42*		55.3		0.68	13.2	250	Szefer et al., 2000f
						0.32–210		0.20–1.60	5.77–20.6	92.8–640	
					DELPHINIDAE						
White beaked dolphin (*Legenorhynchus albirostris*)											
Southern Baltic	1989–95	119–229		3*		1.19		0.56	5.78	510	Szefer et al., 2000d
						0.57–1.55		0.43–0.69	5.26–6.49	425–625	
Striped dolphin (*Stenella coeruleoalba*)											
Southern Baltic	1998/99	187		2*							Szefer et al., 2000d
						30.6–54.7		1.73–6.27	18.4–36.7	240–470	
Beluga whale (*Delphinapterus leucas*)											
Baltic	Pre-1978	271		1		1.9			3.1		Harms et al., 1977/78

* – Dry wt.
J – Juvenile.

TABLE 3.22. – continued

Region	Sampling date	Sex	Age (year)	N	Ag	Al	As	Ca	Cd	Co	Cr	Cu	Fe	References
								PHOCIDAE						
Ringed seal (*Phoca hispida*)														
Gulf of Finland	1976–82		7.3	11					1.19			2.9		Perttilä et al., 1986
			0.5–18						0.04–3.9			1.9–3.6		
Baltic Sea	1988			10		0.45	2.6	72	0.78	0.033	0.05	3.9	209	Frank et al., 1992
						0.22–1.77	2.0–3.3	53–101	0.45–13	0.013–0.047	0.031–0.077	2.7–5.1	157–314	
Grey seal (*Halichoerus grypus*)														
Gulf of Finland	1976–82		10.9	8					0.63			2.41		Perttilä et al., 1986
			0.10–40						ND–2.8			1.8–2.8		
Baltic Sea	1988			10		0.32	3	78	0.93	0.024	0.059	2.8	187	Frank et al., 1992
						0.17–0.58	1.2–3.9	64–107	0.25–1.93	0.017–0.048	0.029–0.329	2.6–3.1	140–265	
Southern Baltic	1999	M		2*	ND				ND–0.32		1.15–1.30	10.3–11.7	520–770	Szefer et al., 2000d
Harbour seal (*Phoca vitulina*)														
German coast	Pre-1978								0.06–1.0			2.3–4.0		Harms et al., 1977/78
Skagerrak	1988			10		0.41	1.6	65	0.21	0.022	0.07	3.5	169	Frank et al., 1992
						0.10–0.60	1.1–2.5	59–78	0.07–0.44	0.015–0.025	0.056–0.11	2.6–5.7	118–274	
Kattegat	1988			10		0.29	2.3	64	0.23	0.018	0.044	3.6	155	
						0.18–1.75	1.4–3.4	53–82	0.12–0.57	< 0.002–0.025	0.020–0.14	2.6–4.1	139–193	
Kalmarsund	1988			10		0.59	0.83	69	0.1	0.017	0.139	3.3	150	
						0.17–2.08	0.3–1.7	61–82	< 0.02–0.24	0.005–0.036	0.069–0.15	2.8–4.0	115–237	
Dutch waters, Wadden Sea	Pre-1979			2					0.15–0.17		0.15–0.59	4.8–5.1	31.0–66.0	Duinker et al., 1979
German coast, North Sea	1974–76		7.0–12.0**	9					0.1			3.2		
									0.06–0.15			2.5–3.8		Drescher et al., 1977
			19–24**	4					0.22			3.3		
									0.19–0.29			3.0–4.0		
			6.0–8.0	3					0.31			2.5		
									0.26–0.38			2.3–2.8		

* – Dry wt.
** – Months.

TABLE 3.22. – continued

Region	Sampling date	Length (cm)	Age (year)	N	Hg	Mn	Ni	Pb	Se	Zn	References
					PHOCOENIDAE						
Harbour porpoise (*Phocoena phocoena*)											
Southern Baltic	1989–93		J	36*	5.15±3.30	2.63±0.45		< 0.5–7.48		113±17.6	Szefer et al., 1995b
					2.14–13.2	1.91–3.50				84–136	
Southern Baltic	1990–96		< 1	8*	2.93±1.84	2.26±0.53	< 0.5	< 0.5	10.6±3.26	77.4±25.7	Szefer et al., 2000f, 2000h
					1.69–5.05	1.57–3.15			7.66–14.4	51.7–120	
			1.0–2.0	12*		1.95±0.33	< 0.5	< 0.5		75.3±15.3	
						1.37–2.68				58.1–105	
			2.0–4.0	5*	6.23±2.82				16.9±11.9		
					1.77–8.78				4.13–36.5		
			3.0–6.0	6*		1.97±0.52	< 0.5	< 0.5		78.2±21.4	
						1.42–2.87				57.2–110	
			> 5	2*	12.5±4.58				20.1±0.64		
					9.02–15.5				19.6–20.5		
German coastal waters	Pre-1978	126		1				0.15		20	Harms et al., 1977/78
		153		1				0.17		24	
Danish waters	1980–81		0–1.0	28	0.4						Clausen and Andersen, 1988
			2.0–3.0	21	0.7						
			4.0–12	40	1.3						
Danish waters	1993–96		< 1.0	6*	1.09	2.50±0.55	< 0.5	< 0.5	21.5	93.2±21.7	Szefer et al., 2000f, 2000h
					0.82–1.35	1.76–3.02			9.4–33.5	72.1–130	
			1.0–2.0	6*		3.60±2.01	< 0.5	< 0.5		77.9±10.7	
						2.06–7.12				67.4–98.4	
			11	1*	13.02	4.39	< 0.5	< 0.5	39.9	84.5	
Kattegat and	1987–90	67–172		12	1.24						Joiris et al., 1991
Belgian water					0.41–8.30						
Norwegian coast	1989–90		2.1 1.6	16	0.86±0.84				2.8±1.4		Teigen et al., 1993
			4.1 3.0	10	1.2±0.81				3.5±2.2		
			2.3 2.0	21	0.70±0.29				3.6±1.3		
			2.4 1.3	23	0.80±0.66				4.6±1.4		

Region	Sampling date	Length (cm)	Age (year)	N	Hg	Mn	Ni	Pb	Se	Zn	References
			4.1 2.8	13	0.77±0.55				3.6±0.7		
Western Greenland	1988–89		0–>6.0	44	0.92				5.79	34.5	Paludan-Müller et al., 1993
					0.19–2.51				1.96–11.9	26.0–43.8	
Central West part of	Pre-1995		4.0–>7.0	20	1.15				6		Dietz et al., 1996
Southern Greenland					1.02–1.29				5.77–6.30		
Southwest Greenland	1995		< 1.0–>6	7*	4.33	2.87	ND	ND	29.7	120	Szefer et al., 2000f, 2000h
					0.60–12.7	2.41–5.92			15.3–63.5	81.4–190	
					DELPHINIDAE						
White beaked dolphin (*Legenorhynchus albirostris*)											
Southern Baltic	1989–95	119–229		3*		2.31	ND			65	Szefer et al., 2000d
						2.23–2.45		ND–0.68		58.8–73.4	
Striped dolphin (*Stenella coeruleoalba*)											
Southern Baltic	1998/99	187		2*							Szefer et al., 2000d
						1.97–2.89		ND		104–168	
Beluga whale (*Delphinapterus leucas*)											
Baltic	Pre-1978	271		1				0.13		29.5	Harms et al., 1977/78

* – Dry wt.
J – Juvenile.

TABLE 3.22. – continued

Region	Sampling date	Sex	Age (year)	N	Hg	Mg	Mn	Ni	Pb	Se	V	Zn	References
							PHOCIDAE						
Ringed seal (*Phoca hispida*)													
Gulf of Finland	1976–82		7.3	11	2.07				0.11			35.6	Perttilä et al., 1986
			0.5–18		0.9–4.4				0.01–0.33			22.5–68.0	
Baltic Sea	1988			10		130	1	0.022	0.08		0.015	21	Frank et al., 1992
						118–142	0.7–1.5	0.012–0.046	0.03–0.13		0.009–0.043	19–35	
Norwegian waters	1989–90	F	J	3	1.66±0.55					2.56±0.32			Skaare et al., 1994
					1.05–2.12					2.20–2.81			
		M	J	2	0.85					2.67			
					0.62–1.07					2.47–2.86			
		M	A	2	1.96					3.4			
					1.89–2.02					3.34–3.46			
Grey seal (*Halichoerus grypus*)													
Gulf of Finland	1976–82		10.9	8	9.5				0.1			34.1	Perttilä et al., 1986
			0.10–40		0.8–28.0				0.03–0.15			18.0–48.0	
Baltic Sea	1988			10		142	0.9	0.02	0.15		0.012	22	Frank et al., 1992
						125–166	0.6–1.1	< 0.006–0.053	0.09–0.23		0.006–0.033	19–28	
Southern Baltic	1999	M		2*									Szefer et al., 2000d
							2.07–2.20		ND			72.2–97.7	
Harbour seal (*Phoca vitulina*)													
German coast	Pre-1978								0.08–0.60			15.5–34.0	Harms et al., 1977/78
Skagerrak	1988			10		149	0.9	< 0.006	0.04		0.018	19	Frank et al., 1992
						125–171	0.7–1.1	< 0.006–0.014	.02–0.07		0.011–0.040	15–27	
Kattegat	1988			10		149	0.9	0.014	0.04		0.015	21	
						138–171	0.7–1.2	0.008–0.029	.02–0.07		0.006–0.026	19–22	
Kalmarsund	1988			10		163	0.9	< 0.006	0.07		0.018	21	
						139–187	0.7–1.3	< 0.006–0.027	0.03–0.21		0.010–0.066	19–47	
Dutch waters, Wadden Sea	Pre-1979			2			1.9–3.4		0.16–0.23			15.0–25.0	Duinker et al., 1979
German coast, North Sea	1974–76		7.0–12.0**	9	3.3				0.32			22	
					1.6–5.9				0.18–0.48			18.8–26.5	Drescher et al., 1977
			19–24**	4	3.9				0.38			26.5	
					3.1–4.7				0.14–0.55			23.3–32.0	
			6.0–8.0	3	7.9				0.46			18.8	
					3.9–12.5				0.40–0.51			16.3–20.0	

* – Dry wt.
** – Months.
J – Juvenile, A – Adult, M – Male, F – Female.

TABLE 3.23.

Concentrations of chemical elements ($\mu g\ g^{-1}$ wet wt.) in muscle of marine mammals from the Baltic Sea and other northern areas

Region	Sampling date	Length (cm)	Age (year)	N	Ag	Cd	Co	Cr	Cu	Fe	References
					PHOCOENIDAE						
Harbour porpoise (*Phocoena phocoena*)											
Southern Baltic	1989–93		J	24*	0.15±0.06	< 0.01–0.23			6.36±1.08		Szefer et al., 1995b
					0.05–0.21				4.81–7.91		
Southern Baltic	1990–96		< 1	7*		< 0.01			7.6±1.6	700±200	Szefer et al., 2000f
									6.51–11.0	490–1100	
			1.0–2.0	9*		< 0.01			6.98±1.14	730±110	
									5.60–8.92	570–910	
			3.0–6.0	4*		< 0.01			6.52±0.20	730±97	
									5.58–6.82	610–820	
German coastal waters	Pre-1978	80		1		0.002			2.7		Harms et al., 1977/78
		126		1		0.002			1.8		
		153		1		0.002			2.1		
Danish waters	1993–96		< 1.0	6*		< 0.01	< 0.5	0.36±0.10	8.32±1.90	570±150	Szefer et al., 2000f
								0.26–0.62	6.04–11.7	350–810	
			1.0–2.0	6*		< 0.01	< 0.5	0.32±0.05	5.53±2.19	780±150	
								0.26–0.40	2.95–9.22	590–950	
			> 2.0	1*		< 0.01	< 0.5	0.59	3.31	640	
Cadigan Bay	Pre-1989		Y–J	2		< 0.06		< 0.5	2.26		Morris et al., 1989
									1.5–3.0		
Western Greenland	1988–89		0–>6.0	77	0.05				1.97		Paludan-Müller et al., 1993
					< 0.02–0.33				1.08–5.37		
Central West part of Southern Greenland	Pre-1995		4.0–>7.0	55	0.06						Dietz et al., 1996
					< 0.02–0.11						
Southwest Greenland	1995		< 1.0–>6	43*	0.19	2.87		0.57	7.7	720	Szefer et al., 2000f
					0.03–0.45	2.41–5.92		0.30–8.38	4.27–12.8	430–970	
					DELPHINIDAE						
White beaked dolphin (*Legenorhynchus albirostris*)											
Southern Baltic	1989–95	119–229		3*		ND		0.3	6.67	470	Szefer et al., 2000d
								0.26–0.33	5.79–7.15	420–520	
Striped dolphin (*Stenella coeruleoalba*)											
Southern Baltic	1998/99	187		2*							Szefer et al., 2000d
						ND–0.06		0.63–1.40	4.46–17.6	400–1030	
Beluga wale (*Delphinapterus leucas*)											
German coast, Baltic	Pre-1978	271		1		0.007			1.1		Harms et al., 1977/78

* – Dry wt.
Y–J – Young/juvenile.

TABLE 3.23. – continued

Region	Sampling date	Sex	Age (year)	N	Ag	Cd	Co	Cr	Cu	Fe	References
						PHOCIDAE					
Ringed seal (*Phoca hispida*)											
Gulf of Finland	1976–82		7.3	11		0.01			1.2		Perttilä et al., 1986
			0.5–18			ND–0.03			0.8–1.6		
Grey seal (*Halichoerus grypus*)											
Gulf of Finland	1976–82		10.9	8		0.004			1.2		Perttilä et al., 1986
			0.10–40			0.001–0.010			1.0–1.7		
Southern Baltic	1999	M	1–3 months	2*	ND	ND	ND	1.24	5	400	Szefer et al., 2000d
								1.08–1.40	4.83–5.06	390–405	
Harbour seal (*Phoca vitulina*)						0.002–0.08			0.05–2.0		
German coast	Pre-1978										Harms et al., 1977/78

* - Dry wt.
M - Male

TABLE 3.23. – continued

Region	Sampling date	Length (cm)	Age (year)	N	Hg	Mn	Ni	Pb	Se	Zn	References
					PHOCOENIDAE						
Harbour porpoise (*Phocoena phocoena*)											
Southern Baltic	1989–93		J	36*	2.65±0.94	0.90±0.30		0.19±0.08		35.3±6.4	Szefer et al., 1995b
					1.20–4.24	0.60–1.46		0.08–0.30		2.41–55.9	
Southern Baltic	1990–96		< 1	8*	20.8±37.4	0.94±0.20		< 0.5	4.28±0.46	86.7±21.1	Szefer et al., 2000f, 2000h
					2.05–56.6	0.68–1.39			3.75–4.60	58.6–98.4	
			1.0–2.0	9*		0.63±0.30		< 0.5		78.2±14.7	
						0.10–1.05				61.2–110	
			2.0–4.0	5*	4.41±2.45				3.14±0.27		
					1.98–7.48				2.72–3.48		
			3.0–6.0	4*		0.62±0.10		< 0.5		75.6±12.0	
						0.48–0.71				60.3–87.7	
			> 5	2*	7.80±0.02				3.75±0.15		
					7.79–7.82				3.64–3.85		
German coastal waters	Pre-1978	80		1				0.03		21	Harms et al., 1977/78
		126		1				0.07		12.4	
		153		1				0.05		14.4	
Danish waters	1993–96		< 1.0	6*		0.89±0.20	< 0.5	< 0.5		190±43.6	Szefer et al., 2000f, 2000h
						0.60–0.97				120–220	
			1.0–2.0	6*	0.97	0.87±0.60	< 0.5	< 0.5	1.67	120±54.8	
					0.68–1.27	0.26–2.09			1.65–1.69	67.5–210	
			11	1*	9.83	0.8	< 0.5	< 0.5	2.27	98.9	
Kattegat and Belgian coast	1987–90	67–172		12	0.95						Joiris et al., 1991
					0.33–6.5						
Cadigan Bay	Pre-1989		Y–J	2	0.66			< 0.7		23	Morris et al., 1989
					0.22–1.1					22–23	
Belgium	1987–88			2	0.67						Joiris and Bossicart, 1989
					0.33–1.0						

Western Greenland	1988–89		0–>6.0	77	0.49				0.54	17.7	Paludan-Müller et al., 1993
					0.07–1.10				0.23–1.37	10.3–33.1	
Central West part of	Pre-1995		4.0–>7.0	55	0.48				0.52		Dietz et al., 1996
Southern Greenland					0.32–0.66				0.50–0.55		
Southwest Greenland	1995		< 1.0–>6	43*	1.71	0.61	NA	NA	4.09	60.9	Szefer et al., 2000f, 2000h
					0.42–3.46	0.34–1.18			1.78–6.19	33.3–125	
					DELPHINIDAE						
White beaked dolphin (*Legenorhynchus albirostris)*						0.54				59.8	
Southern Baltic	1989–95	119–229		3*		0.49–0.60				52.2–69.6	Szefer et al., 2000d
Striped dolphin (*Stenella coeruleoalba)*											
Southern Baltic	1998/99	187		2*		0.53–1.19		ND		38.7–74.0	Szefer et al., 2000d
Beluga wale (*Delphinapterus leucas)*											
German coast, Baltic	Pre 1978	271		1	1.6			0.08		20	Harms et al., 1977/78

* – Dry wt.
Y–J – Young/juvenile, Y – Young.
NA – Not analysed.

TABLE 3.23. – continued

Region	Sampling date	Sex	Age (year)	N	Hg	Mn	Ni	Pb	Zn	References
					PHOCIDAE					
Ringed seal (*Phoca hispida*)										
Gulf of Finland	1976–82		7.3	11	0.64			0.08	38.2	Perttilä et al., 1986
			0.5–18		0.31–1.03			0.01–0.26	21.3–64.1	
Grey seal (*Halichoerus grypus*)										
Gulf of Finland	1976–82		10.9	8	1.45			0.07	35	Perttilä et al., 1986
			0.10–40		0.2–4.9			0.02–0.14	20.9–48.6	
Southern Baltic	1999	M	1–3 months	2*		0.77	ND	ND	138	Szefer et al., 2000d
						0.67–0.86			119–156	
Harbour seal (*Phoca vitulina*)										
German coast	Pre-1978				1.0–10.0			0.03–0.10	15.0–36.0	Harms et al., 1977/78

* – Dry wt.
M – Male

TABLE 3.24.

Concentrations of trace elements ($\mu g\ g^{-1}$ wet wt.) in bluber of marine mammals from the Baltic Sea and other northern areas

Region	Sampling date	Age (year)	N	As	Cd	Cr	Cu	Fe	References
Ringed seal (*Phoca hispida*)									
Gulf of Finland	1976–82	7.3	11		0.0025		0.22		Perttilä et al., 1986
		0.5–18			0.001–0.009		0.10–0.80		
Baltic Sea	1988		10	2.6					Frank et al., 1992
				2.0–3.3					
Grey seal (*Halichoerus grypus*)									
Gulf of Finland	1976–82	10.9	9		0.0017		0.32		Perttilä et al., 1986
		0.10–40			0.001–0.006		0.10–1.80		
Baltic Sea	1988		10	3					Frank et al., 1992
				1.2–3.9					
Cadigan Bay, West Wales	1988		1		< 0.06	< 0.5	< 0.1		Morris et al., 1989
Harbour seal (*Phoca vitulina*)									
Skagerrak	1988		10	1.6					Frank et al., 1992
				1.1–2.5					
Kattegat	1988		10	2.3					
				1.4–3.4					
Kalmarsund	1988		10	0.83					
				0.3–1.7					
Dutch coast, Wadden Sea	Pre-1979		3		< 0.01–0.02	0.49	0.90–3.00	27.–75.0	Duinker et al., 1979
Harbour porpoise (*Phocoena phocoena*)									
Cadigan Bay, West Wales	1988		4		< 0.07	< 0.6	0.62		Morris et al., 1989
							0.21–1.70		
Striped dolphin (*Stenella coeruleoalba*)									
Cadigan Bay, West Wales	1988		2		< 0.08	< 0.06	0.52		Morris et al., 1989
							0.33–0.70		

TABLE 3.24. – continued

Region	Sampling date	Age (year)	N	Hg	Mn	Ni	Pb	Se	Zn	References
Ringed seal (*Phoca hispida*)										
Gulf of Finland	1976–82	7.3	11	0.04			0.05	0.13	3.52	Perttilä et al., 1986
		0.5–18		ND–0.25			0.01–0.17	0.10–0.20	0.50–10.9	
Grey seal (*Halichoerus grypus*)										
Gulf of Finland	1976–82	10.9	9	0.15			0.11	0.11	3.93	Perttilä et al., 1986
		0.10–40		0.02–0.75			0.01–0.70	0.10–0.20	0.10–17.7	
Cadigan Bay, West Wales	1988		1	0.05			< 0.6		1.8	Morris et al., 1989
Harbour seal (*Phoca vitulina*)										
Dutch coast, Wadden Sea	Pre-1979		3		< 0.04–2.70		< 0.05–1.0		3.0–14.0	Duinker et al., 1979
Harbour porpoise (*Phocoena phocoena*)										
Cadigan Bay, West Wales	1988		4	0.06		< 0.60	< 0.70		3.86	Morris et al., 1989
				0.01–0.18					1.80–5.50	
Belgian coast	1989		1	0.57						Joiris and Bossicart, 1989

TABLE 3.25.

Concentrations of trace elements ($\mu g\ g^{-1}$ dry wt.) in bones of marine mammals from the Baltic Sea

Region	Part analysed	Sampling date	Length (cm)	Age	N	Ag	Cd	Co	Cr	Cu	Fe	References
Striped dolphin (*Stenella coeruleoalba*)												
Southern Baltic	Rib	1998/99	187–187		2	ND	1.2	ND	3.75	3.09	111	Szefer et al., 2000d
							1.07–1.33		0.86–6.63	2.65–3.53	66–155	
Grey seal (*Halichoerus grypus*)												
Southern Baltic	Rib	1996/99		1–3 months	2	ND	ND–0.98	ND	5.86	2.91	218	Szefer et al., 2000d
									5.48–6.24	2.76–3.05	198–238	

ND – Not detected

TABLE 3.25 – continued

Region	Sampling date	Length (cm)	Age	N	Mn	Ni	Pb	Zn	References
Striped dolphin (*Stenella coeruleoalba)*									
Southern Baltic	1998/99	187		2	3.13	ND	3.49	430	Szefer et al., 2000d
					2.94–3.32		3.24–3.74	388–478	
Grey seal (*Halichoerus grypus)*									
Southern Baltic	Rib	1996/99	1–3 months	2	2.66	ND	3.41	114	Szefer et al., 2000d
					2.64–2.68		1.69–5.13	101–126	

ND – Not detected

Intertissue trends

There are significant variations in concentrations of chemical elements in particular tissues and organs of Baltic mammals.

The concentrations of several trace elements have been determined in nine tissues and organs of harbour porpoise (*Phocoena phocoena*) from the southern Baltic and Danish waters (Szefer et al., 2000f, 2000h). Higher levels of Zn were observed in the liver, spleen and digestive tract; Cu was occurred in greater quantities in the liver and heart; Fe was more concentrated in the lungs, liver and spleen. The kidney and liver, and spleen were the target organs for Cd and Mn, respectively while the higher concentrations of Pb and Cr were found in the spleen (Szefer et al., 1994b, 1995b, 2000d). Besides the inter-tissue variations, intra-tissue changes in the concentrations of selected metals were also observed (Szefer et al., 2000f, 2000h).

Fourteen tissues and organs of white-beaked dolphin (*Lagenorhynchus albirostris*) from the southern Baltic were recognised in respect to their burden of selected trace elements. The liver appeared to be the most enriched organ in Pb, Mn and Fe; the kidney was characterised by the highest abundance of Cd and the diaphragma contained higher levels of Zn. The highest levels of Cu occurred in heart, liver and brain while Cr was highly concentrated in both the kidney and liver (Szefer et al., 2000d).

The distribution of trace elements in thirteen tissues and organs of striped dolphin (*Stenella coeruleoalba*) from southern Baltic has been analysed (Szefer at al., 2000d). The liver was characterised by elevated concentrations of Zn and Mn, kidney accumulated more Cd, Cu and Cr and significant tissue burden of Fe was observed in lung. The levels of Pb were generally below the method detection.

Fourteen different tissues and organs of grey seal (*Halichoerus grypus*) have been investigated in respect to their abundance in trace metal concentrations (Szefer at al., 2000d). Liver concentrated the higher levels of Cu and Mn, diaphragma and eye balls had greater concentrations of Pb and Zn, respectively while Fe was highly concentrated in lung, spleen and liver. It is interesting to note that and Cd levels were very low in all the tissues, i.e. below the limit of the method used. Cr was distributed in all samples rather uniformly.

Spatial trends

Since it is suspected that southern Baltic porpoises constitute part of the North Sea population and had migrated into the Baltic Sea recently (Kannan et al., 1993), the Baltic data are compared to those corresponding to the North Sea and adjacent area as Greenland area (Tables 3.21–3.23). Relatively small the hepatic and renal concentrations of Cd in both the Baltic and Danish harbour porpoises are similar to those reported earlier for porpoises from German, Danish and British waters (Harms et al., 1977/1978; Clausen et al., 1988; Law et al., 1991, 1992). It reflects low rates of Cd exposure, supposedly an alimentary origin, for porpoises from the temperate marine ecosystems. For example, the food composition of the Baltic porpoises, consisted mainly of fish such as cod *(Gadus morhua)*, herring (*Clupea harengus*) and Gobiidae, is characterised by very small levels of Cd (Szefer and Falandysz, 1985; Falandysz, 1985, 1986a, 1986b; Szefer et al., 1995b). Apparently, this species from other NW European areas feeds also on fish, mainly or exclusively. The Cd concentrations in the kidney, liver and muscle of the Greenland porpoises (Tables 3.21–3.23) are much greater than those from Polish and Danish waters as well as other NW European regions. Greenland's values are similar to those reported for harbour porpoise from SW Greenland (Paludan-Müller et al., 1993) and Dall's porpoise (*Phocoenoides dalli*) from the north-western Pacific, feeding mainly on squids (Fujise et al., 1988). These values converted to a dry weight basis, are in agreement with concentration data obtained for the Greenland porpoises and 2–3 orders of magnitude higher than data registered for both the Baltic and Danish porpoises (Tables 3.21–3.23). It is concluded that various Cd content in porpoise diet (prey) is responsible for such great geographical difference of the hepatic and renal concentrations. Important food component of Dall's porpoise is squid. It should be emphasised that the levels of Cd in polar cod (*Boreogadus saida*), shorthorn sculpin (*Anarhichas minor*) and Greenland halibut (*Reinhardtius hippoglossoides*) from Greenland waters (Dietz et al., 1996; Riget et al., 1997) were higher than those in cod (*Gadus morhua*), herring (*Clupea harengus*) and flounder (*Platichthys flesus*) from Baltic Sea and other north European waters (Szefer and Falandysz, 1995; Falandysz, 1985, 1986a, 1986b; Paludan-Müller et al., 1993). Stomach content of Greenland porpoises contained up to 37% of squid (Heide-Jørgensen and Lockyer, 1999) characterised by elevated hepatic concentrations of Cd, amounting up to 200 $\mu g\ g^{-1}$ dry wt. (Honda and Tatsukawa, 1983). Therefore significantly higher levels of Cd are observed in the liver and kidney of porpoises inhabiting the Greenland and north-western Pacific areas as compared with other geographical zones. Since there is insignificant difference in Hg concentrations, in contrast to Cd, in potential food of porpoises from their various habitations therefore observed spatial variations in tissue Hg levels in porpoises can be due to their different expose to Hg (Szefer et al., 1995b, 2000h).

The results obtained for harbour porpoise from the Baltic Sea are generally the same order of magnitude of those reported for other northern areas such as the North Sea. Exceptionally high levels of hepatic Cu (up to 160 g g^{-1} wet wt.) have been observed in porpoise calves from waters around the British Isles (Table 3.21).

The extremely high levels of Cd were detected in Arctic marine seals (Dietz et al., 1996, 1998; Wagemann et al., 1996; Szefer et al., 2000f; Fant et al., 2001) compared to the Baltic seals. It can be explained by a large accumulation of Cd in hyperiid amphipod, i.e. *Parathemisto libellula* which makes up a significant part of the diet of various Arctic vertebrates (Macdonald and Sprague, 1988; Dietz et al., 1996; Fant et al., 2001). This amphipod is not present, however, in the Baltic Sea. Moreover, it has been also postulated that the elevated levels of Cd in Arctic seals may be caused by slower their growth rates in the Arctic (AMAP, 1998; Fant et al., 2001). Spatial differences in metal concentrations in ringed seals (*Phoca hispida*) have been reported by Fant et al. (2001). The levels of Hg and Se were considerably higher in the Baltic ringed seals, but the Cd levels lower as compared to those in the Svalbard ringed seals (Fig. 3.26). There were insignificant geographical variations of Pb concentrations.

It should be emphasised that observed spatial trends for metals in harbour porpoise (Szefer et al., 2000f) and ringed seal (Fant et al., 2001) are based on

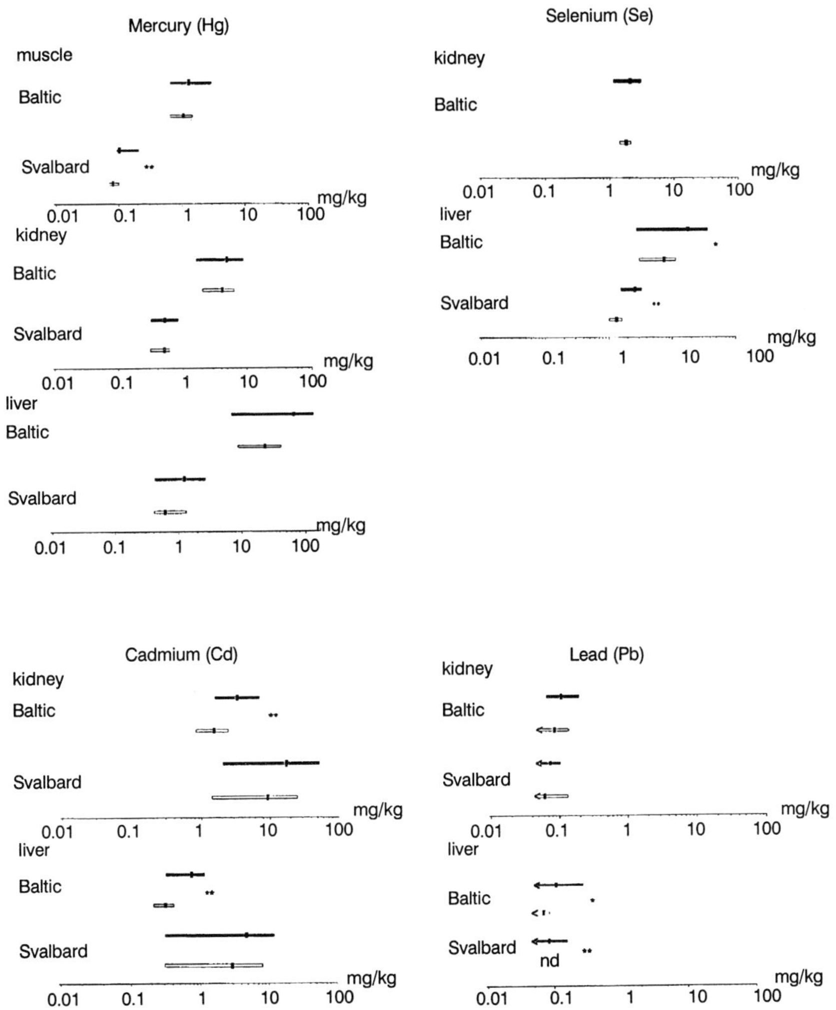

Fig. 3.26. Concentrations of Hg, Se, Cd and Pb (minimum, mean and maximum value; $\mu g\ g^{-1}$ wet wt.) in ringed seals from the Baltic Sea and Svalbard. The white bars represent seals ≤ 5 years old and the dark bars represent seals > 5 years old. Significant differences between the two age groups are indicated by two asterisks for $p \leq 0.01$ and one asterisk for $p \leq 0.05$. After Fant et al. (2001); modified.

concentration data obtained within the studies where the same analytical techniques have been used for each other as well as data on age structure of the mammals have been known. Therefore other than geographically-dependent factors should not influence the validity of the spatial trends observed.

Temporal trends

According to Fant et al. (2001) the use of various analytical methods, insufficient information on age structure and individual variations in metal levels make the temporal estimate complicated. Nevertheless such intercomparison approach based on available literature data has been made by Fant et al. (2001). As can be seen in Fig. 3.27, the concentrations of Hg, Se, Cd and Pb have been approximately constant since the 1980s. Surprisingly large variations of Hg levels were observed during the 1970s. Possibly incidental Hg pollution in these years were responsible for such irregular temporal trend pattern.

Age and sex trends

Age dependent concentrations are well documented in Tables 3.21 and 3.22. It is important to note that levels of both the hepatic and renal Cd as well as renal Zn in Danish, especially in Baltic porpoises seem to be unchangeable at 11 and 6

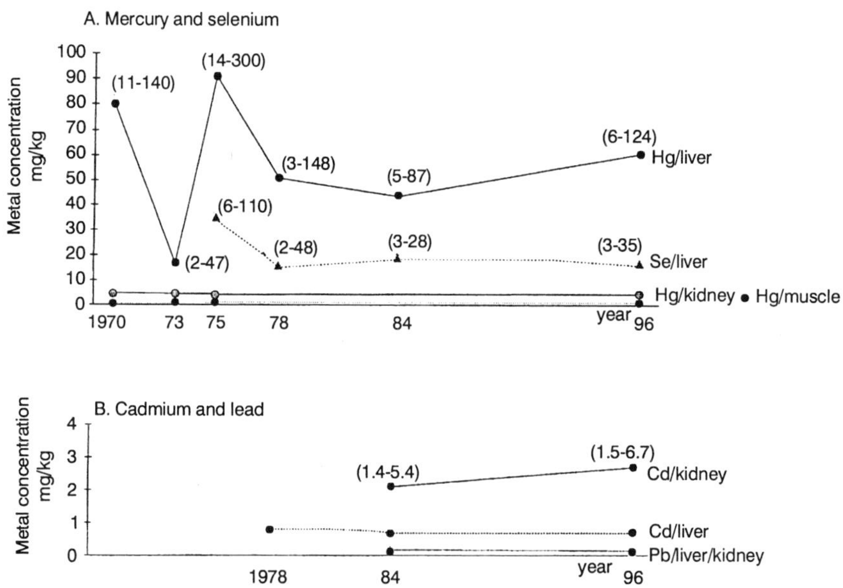

Fig. 3.27. Temporal trends of (A) Hg and Se levels and (B) Cd and Pb levels in ringed seals from the Gulf of Bothnia. The metal concentrations ($\mu g\ g^{-1}$ fresh wt.) are presented as mean values and (range), if nothing else is indicated. The sampling years refer to the following studies (mean age, n, sample size): 1970 – adults, n = 4; 1973 – age unknown, n = 10; 1975 – age unknown, n = 12 for liver, n = 2 for kidney, n = 8 for muscle; 1978 – 16.5 years, n = 22; 1984 – adults, n = 3 for liver, n = 4 for kidney; 1996 – 14.7 years, n = 15. After Fant et al. (2001); modified.

years of age, respectively. The concentrations of Cd in the liver and the concentrations of Cd and Zn in the kidney were well correlated with age in porpoises from the Greenland. No significant variations in mean levels of Hg and Se were observed in males and females of *Phocoena phocoena* within the age groups studied. This note is in an agreement with data reported for hepatic and renal metal associations in West Greenland and Norwegian porpoises (Paludan-Müller et al., 1993; Teigen et al., 1993). The concentrations of Hg in the liver and Se in the kidney significantly correlated with age of the specimens studied (Fig. 3.28). No significant sex variations were observed for selected metals within the age groups studied (Szefer et al., 2000h).

The average concentrations of Hg, Se, Cd and Pb were greater in older ringed seals than in young ones in both populations, i.e from the Baltic Sea and the Svalbard (Fig. 3.26). However, there was insignificant relationship between metal concentrations and age structure in the Baltic seals. It is well marked in Fig. 3.26 that the distribution pattern of Hg in the muscle, kidney and liver varies with age. Both populations were characterised by a larger bioaccumulation of hepatic Hg in old seals, especially in Baltic ones, than in young seals.

Toxicological aspects

There is insufficient available information on the toxicological effects of metal burdens in marine mammals attributed mostly to a lack of toxicological data (Hyvärinen and Sipilä, 1984; Law, 1996; Bennet et al., 2001; Fant et al., 2001). According to Fant et al. (2001) among the metals analysed only the Hg burden in Baltic ringed seals can be considered high while levels of Cd and Pb were below reported effect threshold levels in marine mammals (AMAP, 1998; Fant et al., 2001). Almost half of the Baltic seals exhibited hepatic Hg burdens higher than value of 60 $\mu g\ g^{-1}$ which is a recommended threshold level for liver damage in marine mammals (AMAP, 1998; Fant et al., 2001). It is concluded that marine mammals may have adapted to tolerate high levels of metals, e.g. in respect to naturally enhanced levels of Cd in a diet such as squids and others. In the Baltic seals the molar ratio of Hg/Se was close to unity but a slight exceeded of Hg in liver was observed. This could be explained by limitations in the detoxification process. Enhanced levels of Hg in the kidney could have negative toxic effects since the kidney is highly sensitive to inorganic Hg. However, the Baltic seals did not exhibit any symptoms of renal damage. Moreover, having very low muscular levels of Hg, these mammals indicated no signs of insufficient transfer of Hg to the liver.

Inter-elemental relationships

Significant positive correlations were found for some metal concentrations in the organs and tissue of *Phocoena phocoena* from the Baltic Sea (Figs 3.28 and 3.29). There was a strong correlation ($p < 0.01$; $p < 0.05$) for assemblage Cd–Zn–Cu in the kidney of both the Baltic and Greenland porpoises (Fig. 3.29). A similarly strong correlation between Cd and Zn has been found in the kidney

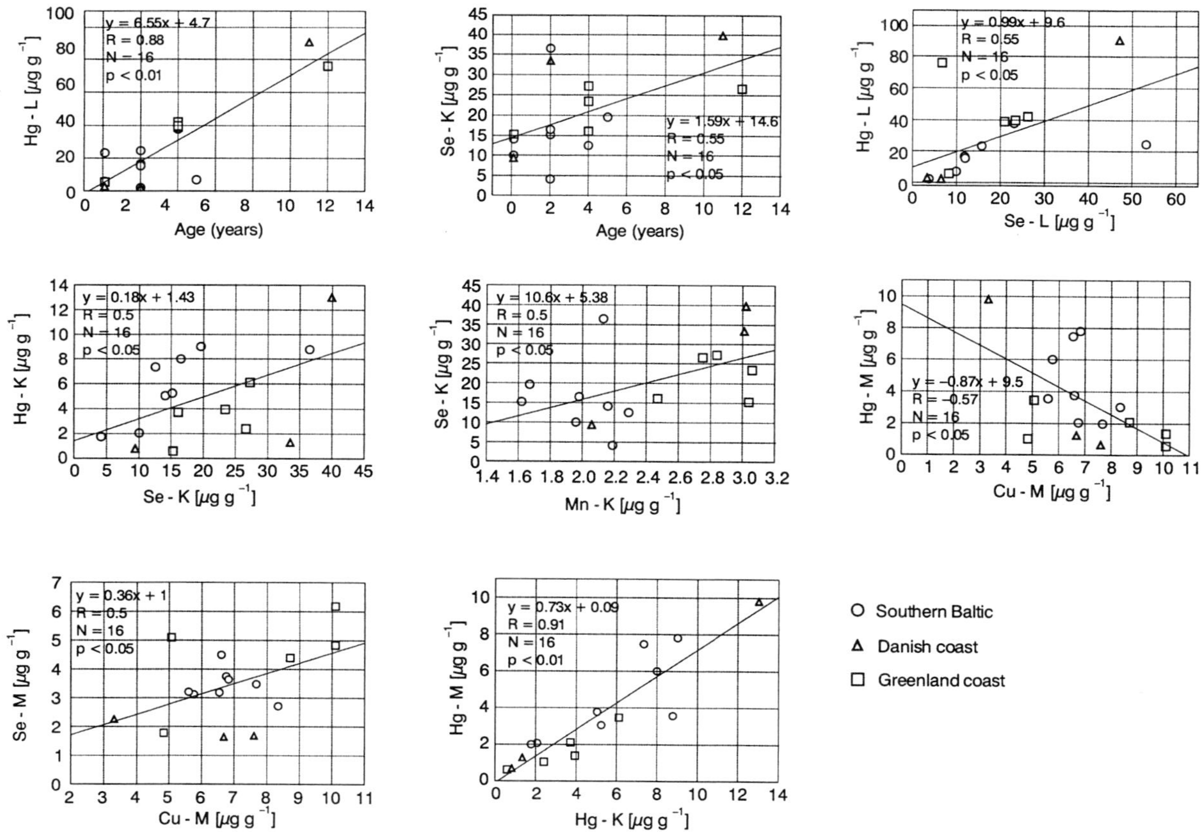

Fig. 3.28. Relationships between Hg and Se and age as well as between Hg, Se and other trace elements in harbour porpoise, *Phocoena phocoena,* from the three geographical regions reflecting significant ($p < 0.01$, $p < 0.05$) intra- and inter-tissue differentiation of their concentrations. After Szefer et al. (2000h).

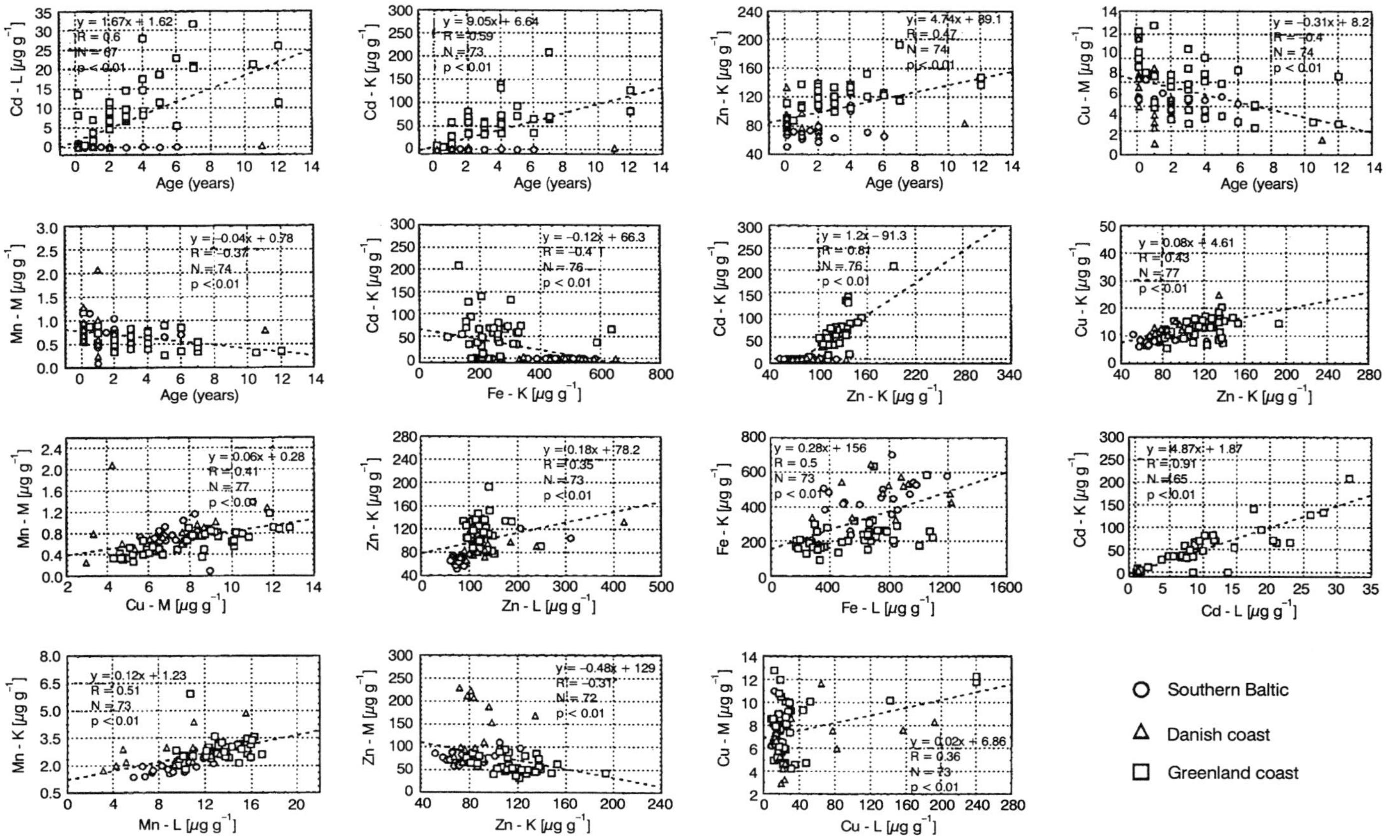

Fig. 3.29. Relationships between concentrations of trace elements in the liver, kidney and muscle of harbour porpoise, *Phocoena phocoena,* from the three geographical regions reflecting significant ($p < 0.01$, $p < 0.05$) differentiation of their concentrations. After Szefer et al. (2000f).

of harbour porpoises from West Greenland (Paludan-Müller et al., 1993) and some other marine mammals such as beluga (*Delphinapterus leucas*), narwhal (*Monodon monoceros*) from West Greenland (Hansen et al., 1990) and striped dolphin (*Stenella coeruleoalba*) from the Pacific coast (Honda and Tatsukawa, 1983; Honda et al., 1983a). This relationships can be explained by similar physical/chemical properties of these two metals; Cd may displace Zn from metallothioneins because of stronger affinity of Cd for this low molecular weight metal-binding proteins involved in metal homeostasis and detoxification (Mochizuki et al., 1985; Wagemann and Hobden, 1986; Tohyama et al., 1986; Bremner, 1987). There is no correlation between Cd and Zn in the liver what is in agreement with the hepatic inter-elemental assemblage reported for West Greenland porpoises, minke wales, belugas and narwhals (Hansen et al., 1990; Paludan-Müller et al., 1993). In contrast, significant co-association between Cd and Zn is well documented for liver of some species of seals and whales (Honda and Tatsukawa, 1983; Honda et al., 1987; Szefer et al., 1994c). There is significant correlation between concentrations of Hg and Se in the liver and kidney (Fig. 3.28), similar to those reported for harbour porpoises from coastal Norwegian waters (Teigen et al., 1993) and West Greenland (Paludan-Müller et al., 1993). Renal Se, Cd and Zn as well as hepatic Hg and Cd correlated significantly with age of harbour porpoise (Figs. 3.28 and 3.29).

(iii) Occurrence of Radionuclides in Marine Mammals

The concentration data of radionuclides in marine mammals from the Baltic Sea and from British coasts are listed in Tables 3.26–3.28. The levels of ^{137}Cs in the liver, kidney and muscle of Baltic mammals (harbour porpoise and seals) are comparable to those (liver and kidney) recorded for harbour porpoise and seals from waters around British Isles (Berrow et al., 1998; Watson et al., 1999). It has been suggested that radiocaesium in Baltic mammals are originating from the Chernobyl accident (Szefer et al., 2000e). Anderson et al. (1990) concluded that most of the ^{137}Cs burden (66–78%) in grey seals from North Rona and from the Isle of May originated from Sellafield discharges while remainder was attributable to the Chernobyl accident. The radiocaesium contamination in porpoise and seals from coastal British waters decreased with distance from processing plant at Sellafield indicating that this plant was major source of this radionuclide (Watson et al., 1999).

(iv) Porpoise parasites

There were significant differences in concentrations of selected metals such as Cd, Cu, Zn, Cr, Ni, Co, Mn and Fe in lung nematode *Pseudalius inflexus* and its host organ lung of porpoise. It is well documented that concentrations of Cu, Cr, Fe, Zn and especially Mn were significantly greater than those in the hosts lung (Fig. 3.30). Concentration of Co in all and Ni in some cases in both the nema-

TABLE 3.26.

Concentrations of radionuclides (mBq/g wet wt.) in liver of marine mammals from the Baltic Sea and other northern areas

Region	Sampling date	Length (cm)	Age [Weight (kg)]	N	134-Cs	137-Cs	40-K	210-Po	239+240-Pu	238-Pu	References
Harbour porpoise (*Phocoena phocoena*)											
Southern Baltic	1999/00			4		15.4	83.8	64.9			Szefer et al., 2000e
						11.7–19.5	78.7–87.3	26.3–110			
Waters around the British	1988–95	127	I [27]		0.1*	2.7*	85.1*		0.007	< 0.006	Watson et al., 1999
Isles					ND–6.2	ND–30.5	41–116				
Striped dolphin (*Stenella coeruleoalba*)											
Southern Baltic	1998			2		7.3	83.2	120			Szefer et al., 2000e
						3.81–10.8	76.7–89.7	19–210			
Grey seal (*Halichoerus grypus*)											
Southern Baltic	1996/99			2		17.4	59	ND			Szefer et al., 2000e
						16.7–18.1	53.7–64.4				
Waters around the British		108	I [20]		0.15*	2.2*	79.4*		0.027	0.005	Watson et al., 1999
Isles					ND–2.0	ND–22.2	19–104		0.015–0.037	0.004–0.008	
Ringed seal (*Phoca hispida*)											
Southern Baltic	1999			1		4.08	66	22.4			Szefer et al., 2000e
Harbour seal (*Phoca vitulina*)											
Waters around the British		110	I [34]		0.1*	3.4*	77*		0.009		Watson et al., 1999
Isles					ND–6.1	ND–17.5	43–107		0.006–0.014	< 0.004–0.006	

* – Median.
I – Immature.

TABLE 3.27.
Concentrations of radionuclides (mBq/g wet wt.) in kidney of marine mammals from the Baltic Sea

Region	Sampling date	N	137-Cs	40-K	210-Po	References
Harbour porpoise (*Phocoena phocoena*)						
Southern Baltic	1999/00	4	13.8	81	72.9	Szefer et al., 2000e
			13.2–14.3	77.0–86.3	41.2–110	
White beaked dolphin (*Legenorhynchus albirostris*)						
Southern Baltic						Szefer et al., 2000e
Striped dolphin (*Stenella coeruleoalba*)						
Southern Baltic	1998	2	25.5	89.7	284	Szefer et al., 2000e
			12.4–38.7	88.0–91.4	41.4–523	
Grey seal (*Halichoerus grypus*)						
Southern Baltic		1			0.28	Szefer et al., 2000e
Ringed seal (*Phoca hispida*)						
Southern Baltic	1999	1			11.9	Szefer et al., 2000e

TABLE 3.28.

Concentrations of radionuclides (mBq/g wet wt.) in muscle of marine mammals from the Baltic Sea and other northern areas

Region	Sampling date	Length (cm)	Age [Weight (kg)]	N	134-Cs	137-Cs	40-K	210-Po	References
Harbour porpoise (*Phocoena phocoena*)									
Southern Baltic	1999/00			4		18.7	110	29.2	Szefer et al., 2000e
						16.5–23.0	105–113	17.1–41.7	
Waters around the British	1988–95	127	I [27]		0.2*	6.9*	89.3*		Watson et al., 1999
Isles					ND–1.1	ND–66.6	69–116		
White beaked dolphin (*Legenorhynchus albirostris*)									
Southern Baltic									
Striped dolphin (*Stenella coeruleoalba*)									
Southern Baltic	1998			2		37.6	90	104	Szefer et al., 2000e
						12.7–52.5	79.5–100	5.24–203	
Grey seal (*Halichoerus grypus*)									
Southern Baltic	1996/99			2		27	87	ND	Szefer et al., 2000e
						25.2–28.7	72.9–91.1		
Waters around the British		108	I [20]		0.2*	2.7*	71.4*		Watson et al., 1999
Isles					ND–2.9	ND–57.2	26–116		
Ringed seal (*Phoca hispida*)									
Southern Baltic	1999			1		5.02	82.9	4.73	Szefer et al., 2000e
Harbour seal (*Phoca vitulina*)									
Southern Baltic									
Waters around the British		110	I [34]		0.2*	3.6*	102.6*		Watson et al., 1999
Isles					ND–2.3	ND–13.3	29–123		

I – Immature
* – Median

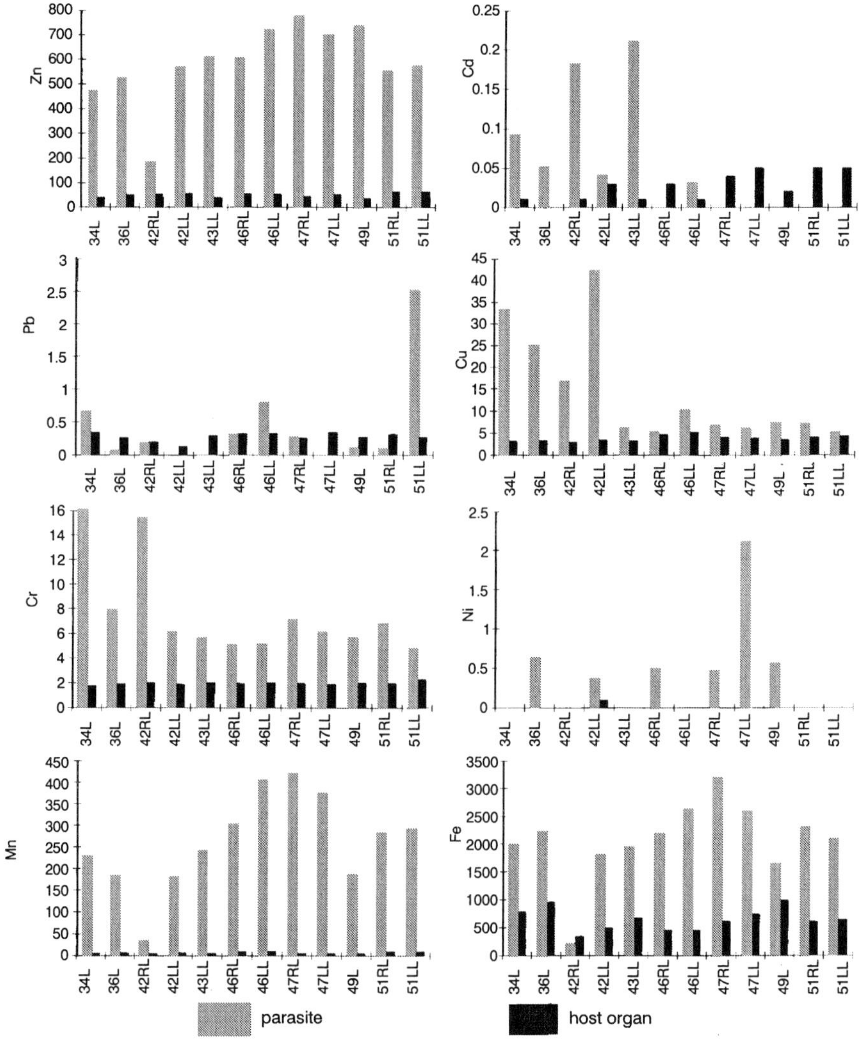

Fig. 3.30. Inter-specimen variations of the selected metals ($\mu g\ g^{-1}$ dry wt.) analyzed in *Pseudalius inflexus* in relation to both parts of lung (L), left lung (LL) and right lung (RL) of eight specimens of *Phocoena phocoena* (coded as follows: Nos. 34, 36, 42, 43, 46, 47, 49 and 51). After Szefer et al. (1998d).

todes and lungs samples were below the detection limit of the method used. Inter-specimen differences in metal concentrations were also observed, mainly concerning parasite Cd, Cr, Cu, Fe, Ni, Mn and Pb (Szefer et al., 1998d).

(v) General Remarks and Recommendations

Fant et al. (2001) have attempted to study possible health effects of metal burdens in ringed seals. Some haematological and biochemical parameters differed insignificantly between the Baltic and the Swalbard populations resulted supposedly from e.g. differences in diet and fasting. The major limitation of the sufficient interpretation of the data obtained was a lack of basic information on normal physiological parameters in ringed seals as well as a small sample size restricted the analyses. Therefore, according to Fant et al. (2001) for future studies the normal ranges of physiological parameters would be needed to know toxic effects of trace elements in ringed seals. It appears also that future work is needed to assess the potential additive or synergistic effects of toxic metals, e.g. Hg and Se and organochlorines on the health status of Baltic mammals. Bennet et al. (2001) suggested that such investigations using complete data for key variables would allow to apply "more robust and powerful statistical tests of the hypothesis that exposure to toxic contaminants influences susceptibility to infectious disease in harbour porpoises". These authors reported that 37 porpoises from England and Wales died of infectious diseases caused by parasitic, bacterial, fungal and viral pathogens and it was found that average hepatic levels of Hg, Se and Zn were significantly higher in the porpoise that died of infectious disease than in healthy porpoises died from physical trauma.

References

Aarkrog, A., H. Dahlgaard and S. Boelskifte, 1986. Transfer of radiocesium and ^{90}Sr from Sellafield to the Danish Straits, in: Study of Radioactive materials in the Baltic Sea. (International Atomic Energy Agency, Vienna). Report (IAEA-TECDOC-362) of the Final Research Co-ordination Meeting on the Study of Radioactive Materials in the Baltic Sea organized by the IAEA and held in Helsinki, Finland 24–28 September, 1984, pp. 32–51.

Aarkrog, A., S. Boelskifte, H. Dahlgaard, S. Duniec, L. Hallstadius, E. Holm and J.N. Smith, 1987. Technetium-99 and caesium-134 as long distance tracers in Arctic waters. Estuar. Coast. Shelf Sci. **24**, 637–647.

Abakumov, W.A., 1983. Podklass Bieschejjustnyie (*Agnatha*), in: Zhizn' Zhivotnyh – Lancetniki – Kruglorotnyie – Chrashchebyie Ryby – Kostnyie Ryby (The Life of Animals – Fish), ed. T.S. Rass (Prosveshchenie, Izdatelstvo, Moscow), **Vol. 4**, 11–20 (in Russian).

Abdennour, C., B.D. Smith, M.S. Boulakoud, B. Samraoui and P.S. Rainbow, 2000. Trace metals in marine, brackish and freshwater prawns (Crustacea, Decapoda) from northeast Algeria. Hydrobiologia **432**, 217–227.

Abrams, P.A., C. Hill and R. Elmgren, 1990. The functional response of the predatory polychaete, *Harmothoe sarsi*, to the amphipod, *Pontoporeia affinis*. OIKOS **59**, 261–269.

Ackefors, H., and L. Hernroth, L., 1975. The distribution and biomass of zooplankton in the Baltic proper in 1972. Medd. Havsfiskelab. Lysekil, 184.

Agadi, V.V., N.B. Bhosle and A.G. Untawale, 1978. Metal concentration in some seaweeds of Goa (India). Bot. Mar. **21**, 247–250.

Agnedal, P.O., 1988. Cs-137 and other radionuclides in the benthic fauna in the Baltic Sea before and after the Chernobyl accident, in: Radionuclide: A Tool for Oceanography, eds. J.C. Guarry, P. Gueguenjat and R.J. Pentreath (Elsevier Applied Science, London, New York) pp. 240–249.

Akagi, H., O. Malm, Y. Kinjo, M. Harada, F.J.P. Branches, W.C. Pfeiffer and H. Kato, 1995. Methylmercury pollution in the Amazon, Brazil. Sci. Total Environ. **175**: 85–95.

Al-Dabbas, M.A.M., F.H. Hubbard and J. McManus, 1984. The shell of *Mytilus* as an indicator of zonal variations of water quality within an estuary. Estuar. Coast. Shelf Sci. **18**, 263–270.

Ali, I.B., C.R. Joiris and L. Holsbeek, 1997. Total and organic mercury in the starfish *Ctenodiscus crispatus* and the polychaete *Maldanes sarsi* from the Barents Sea. Sci. Total Environ. **201**, 189–194.

Al-Majed, N.B., and M.R. Preston, 2000. An assessment of the total and methyl mercury content of zooplankton and fish tissue collected from Kuwait territorial waters. Mar. Pollut. Bull. **40**, 298–307.

Alonso, D., P. Pineda, J. Olivero, H. González and N. Campos, 2000. Mercury levels in muscle of two fish species and sediments from the Cartagena Bay and the Ciénaga Grande de Santa Marta, Colombia. Sci. Total Environ. **109**, 157–163.

Al-Yousuf, M.H., M.S. El-Shahawi and S.M. Al-Ghais, 2000. Trace metals in liver, skin and muscle of *Lethrinus lentjan* fish species in relation to body length and sex. Sci. Total Environ. **256**, 87–94.

AMAP, 1998. AMAP (Arctic Monitoring and Assessment Programme) Assessment Report: Arctic Pollution Issues (Oslo, Norway), **Chapt. 7**.

Amiard, J.-C., C. Amiard-Triquet, C. Metayer, J. Marchand and R. Ferre, 1980. Etude du transfert de Cd, Pb, Cu et Zn dans les chaines trophiques neritiques et estuariennes – I. Etat dans l'estuaire interne de la Loire (France) au cours de l'ete 1978. Water Res. **14**, 665–673.

Amiard, J.C., C. Amiard-Triquet, B, Berthet and C. Metayer, 1986. Contibution to the ecotoxicological study of cadmum, lead, copper and zinc in the mussel *Mytilus edulis*. I. Field study. Mar. Biol. **90**, 425–431.

Amiard, J.C., C. Amiard-Triquet, B. Berthet and C. Metayer, 1987. Comparative study of the patterns of bioaccumulation of essential (Cu, Zn) and non-essential (Cd, Pb) trace metals in various estuarine and coastal organisms. J. Exp. Mar. Biol. Ecol. **106**, 73–89.

Andell, P., J. Dürinck and H. Skov, 1994. Baltic marine areas of outstanding importance for wintering seabirds. WWF Bulletin **5**, 8 pp.

Anderlini, V.C., 1992. The effect of seawage on trace metal concentrations and scope for growth in *Mytilus edulis aoteanus* and *Perna canaliculus* from Wellington Harbour, New Zealand. Sci. Total Environ. **125**, 263–288.

Andersen, S.H., 1982. Change in occurrence of the harbour porpoise, *Phocoena phocoena*, in Danish waters as illustrated by catch statistics from 1834 to 1970. Mammals in the Seas. **Vol. 4**, FAO Fisheries 5, 131–143.

Andersen, S.H., and A. Rebsdorff, 1976. Polychlorinated hydrocarbons and heavy metals in harbour porpoise (*Phocoena phocoena*) and white-beaked dolphin (*Lagenorhynchus albirostris*) from Danish waters. Aquat. Mamm. **4**, 14–20.

Andersen, V., A. Maage and P.J. Johannessen, 1996. Heavy metals in blue mussels (*Mytilus edulis*) in the Bergen Harbor area, western Norway. Bull. Environ. Contam. Toxicol. **57**, 589–596.

Andersen, J.L., and M.H. Depledge, 1997. A survey of total mercury and methylmercury in edible fish and invertebrates from Azorean waters. Mar. Environ. Res. **44**, 331–350.

Andersin, A.-B., I. Kotta and A. Maksimov, 1996. Macrozoobenthos, in: HELCOM, Third periodic assessment of the state of the marine enviroment of the Baltic Sea, 1989–1993. Baltic Sea Environment Proceedings (Baltic Marine Environment Protection Commission, HELCOM), No. **64 B**.

Anderson, S.S., F.R. Livens and D.L. Singleton, 1990. Radionuclides in grey seals. Mar. Pollut. Bull. **21**, 343–345.

Andres, S., F. Ribeyre, J.-N. Tourencq and A. Boudou, 2000. Interspecific comparison of cadmium and zinc contamination in the organs of four fish species along a polymetallic polution gradient (Lot River, France). Sci. Total Environ. **248**, 11–25.

Anger, K., U. Rogal, G. Schriever and C. Valentin, 1977. In situ investigations on the echinoderm *Asterias rubens* as a predator of soft-bottom communities in the western Baltic. Helgoländ. Meeresunters. **29**, 439–459.

Anil, A.C., and A.B. Wagh, 1988. Accumulation of copper and zinc by *Balanus amphitrite* in a tropical estuary. Mar. Pollut. Bull. **19**, 177–180.

Ankar, S., 1977. The soft bottom ecosystem of the Northern Baltic Proper with special reference to the macrofauna. Contr. Askö Lab. **No. 19**, 20–26.

Anon., 1994. Sources of radioactivity in the marine environment and their relative contributions to overall dose assessment from the marine radioactivity, in: Final Report of the CRP, ed. P.P. Povinec, IAEA-MEL-R2/94, June 1994.

Anon., 1995. Environmental radioactivity in Denmark in 1992 and 1993. Risø-R-756 (EN).

Appelquist, H., S. Asbirk and I. Drabćk, 1984. Mercury monitoring: mercury stability in bird feathers. Mar. Pollut. Bull. **15**, 22–24.

Arndt, E.A., 1969. Zwischen Düne und Meeresgrund (Urania–Verlag , Leipzig–Jena–Berlin) 376 pp.

Arntz, W.E., 1971. Die Nahrung der Kliesche (*Limanda limanda* L.) in der Kieler Bucht. Ber. dt. wiss. Kommn. Meeresforsch. **22**, 129–183.

Arntz, W.E., 1974. A conribution of the feeding of juvenile cod (*Gadus morhua* L.) in the Western Baltic. Rapp. P.-v. Reun. Cons. Perm. Int. Explor. Mer. **166**, 13–19.

Arntz, W.E., 1977a. The food of adult cod (*Gadus morhua* L.). Meeresforsch. **26**, 60–69.

Arntz, W.E., 1977b. Predation on benthos by flounder, *Platichthys flesus* (L) in deeper parts of Kiel Bay. Meeresforsch. **26**, 70–78.

Arntz, W.E., and W. Weber, 1970. *Cyprina islandica* L. (Mollusca, Bivalvia) als Nahrung von Dorsch und Kliesche in der Kieler Bucht. Ber. dt. wiss. Kommn. Meeresforsch. **21**, 193–209.

Augustowski, B., 1987. Southern Baltic (Polish Academy of Sciences, Wrocław). (in Polish)..

Balogh, K., 1988. Comparison of mussels and crustacean plankton to monitor heavy metal pollution. Water Air Soil Pollut. **37**, 281–292.

Barbaro, A., A. Francescon, B. Polo and M. Bilio, 1978. *Balanus amphirite* (Cirripedia: Thoracica) – a potential indicator of fluoride, copper, lead, chromium and mercury in North Adriatic lagoons. Mar. Biol. **46**, 247–257.

Barnett, B.E., and C.R. Ashcroft, 1985. Heavy metals in *Fucus vesiculosus* in the Humber Estuary. Environ. Pollut. **9B**, 193–213.

Bauer, B., P. Fioroni, U. Schulte-Oehlmann, J. Oehlmann and W. Kalbfus, 1997. The use of *Littorina littorea* for tributyltin (TBT) effect monitoring – results from the German TBT survey 1994/1995 and laboratory experiments. Environ. Pollut. **96**, 299–309.

Bäck, S., J. Ekebom, C. Johansson, P. Kangas, H. Kautsky, D. Krause-Jensen, A. Mäkinen and K. Nielesen, 1998. Operative methods for mapping and monitoring phytobenthic zone biodiversity in the Baltic Sea, in: II Report on the Phytobios Project.TemaNord **568**, 70 pp.

Bearhop, S., G.D. Ruxton and R. Furness, 2000a. Dynamics of mercury in blood and feathers of great skua. Environ. Toxicol. Chem. **19**, 1638–1643.

Bearhop, S., S. Waldron, D. Thompson and R. Furness, 2000b. Bioamplification of mercury in great skua *Catharacta skua* chicks: the influence of trophic status as determined by stable isotope signatures of blood and feathers. Mar. Pollut. Bull. **40**, 181–185.

Bebbington, G.N., N.J. Mackay, R. Chvojka, R.J. Williams, A. Dunn and E.A. Auty, 1977. Heavy metals, selenium and arsenic in nine species of Australian commercial fish. Aust. J. Mar. Freshwater Res. **28**, 277–286.

Beliaeff, B., T.P. O'Connor and D. Claisse, 1998. Comparison of chemical concentrations in mussels and oysters from the United States and France. Environ. Monit. Assess. **49**, 87–95.

Benemariya, H., H. Robberecht and H. Deelstra, 1991. Atomic absorption spectrometric determination of zinc, copper, and selenium in fish from Lake Tanganyika, Burundi, Africa. Sci. Total Environ. **105**, 73–85.

Bennett, P.M., P.D. Jepson, R.J. Law, B.R. Jones, T. Kuiken, J.R. Baker, E. Rogan and J.K. Kirkwood, 2001. Exposure to heavy metals and infectious disease mortality in harbour porpoises from England and Wales. Environ. Pollut. **112**, 33–40.

Berg, W., A. Johnels, B. Sjöstrand and T. Westermark, 1966. Mercury content in feathers of Swedish birds from the past 100 years. Oikos **17**, 71–83.

Bernds, D., D. Wübben and G.-P. Zauke, 1998. Bioaccumulation of trace metals in polychaetes from the German Wadden Sea: Evaluation and verification of toxicokinetic models. Chemosphere **37**, 2573–2587.

Berrow, S.D., S.C. Long, A.T. McGarry, D. Pollard, E. Rogan and C. Lockyer, 1998. Radionuclides (^{137}Cs and ^{40}K) in harbour porpoises *Phocoena phocoena* from British and Irish coastal waters. Mar. Pollut. Bull. **36**, 569–576.

Bick, A. and F. Gosselck, 1985. Arbeitsschlüssel zur Bestimmung der Polychaeten der Ostsee. Mitt. Zool. Mus. Berl. **61**, 272 pp.

Bick, A. and G. Arlt, 1993. The influence of *Hediste diversicolor* (O.F. Müller, 17776) on the macro- and meiozoobenthos of a shallow water area of Mecklenburg Bay (Western Baltic Sea). Rostock. Meeresbiolog. Beitr. **1**, 9–23.

Bielyaev, G.M., 1988. Klass Morskije Zwiezdy (*Asteroidea*), in: Zhizn' Zhivotnyh – Molljuski – Igłokożie – Chlenistonogie (The life of Animals – Molluscs – Crustaceans), ed. R.K. Pasternak (Prosveshchenie, Izdatelstvo, Moscow), **Vol. 2**, 200–221 (in Russian).
Birshteyn, Ya.A., and R.K. Pasternak, 1988a. Podklass Maksillopody (Maxillopoda), in: Zhizn' Zhivotnyh (The Life of Animals – Molluscs – Crustaceans), ed. R.K. Pasternak (Prosveshchenie, Izdatelstvo, Moscow), **Vol. 2**, 314–341 (in Russian).
Birshteyn, Ya.A., and R.K. Pasternak, 1988b. Ravnonogie rakoobraznyie (*Isopoda*), in: Zhizn' Zhivotnyh – Molljuski – Igłokożie – Chlenistonogie (The Life of Animals – Molluscs – Crustaceans), ed. R.K. Pasternak (Prosveshchenie, Izdatelstvo, Moscow), **Vol. 2**, 365–377 (in Russian).
Black, W.A.P., and R.L. Mitchell, 1952. Trace elements in the common brown algae and in sea water. J. Mar. Biol. Assoc. U.K. **30**, 575–584.
Blackmoore, G., 1999. Temporal and spatial biomonitoring of heavy metals in Hong Kong coastal waters using *Tetraclita squamosa*. Environ. Pollut. **106**, 273–283.
Blackmoore, G., B. Morton and Z.G. Huang, 1998. Heavy metals in *Balanus amphitrite* and *Tetraclita squamosa* (Crustacea: Cirripedia) collected from the coastal waters of Xiamen, China. Mar. Polut. Bull. **36**, 32–40.
Blomqvist, S., A. Frank and L.R. Petersson, 1987. Metals in liver and kidney tissues of autumn migrating dunlin *Calidris alpina* and curlew sandpiper *Calidris ferruginea* staging at the Baltic Sea. Mar. Ecol. Prog. Ser. **35**, 1–13.
Bohn, A., 1979. Trace metals in Fucoid algae and purple sea urchins near high arctic lead/zinc ore deposit. Mar. Pollut. Bull. **10**, 325–327.
Bohn, A., and R.O. McElroy, 1976. Trace metals (As, Cd, Cu, Fe and Zn) in Arctic cod, *Boreogadus saida*, and selected zooplankton from Strathcona Sound, northern Baffin Island. J. Fish. Res. Board Can. **33**, 2836–2840.
Bojanowski, R. 1972. The occurrence of major and minor chemical elements in the more common Baltic seaweed. Oceanologia **2**, 81–152.
Bojanowski, R., and J. Pempkowiak, 1977. Akumulacja ^{90}Sr, ^{137}Cs, ^{106}Ru, ^{144}Ce i ^{239,240}Pu w roślinach wodnych południowego Bałtyku. Oceanologia **7**, 89–104.
Bojanowski, R., Radecki, Z. and Knapinska-Skiba, D. (1995). The distribution of ^{137}Cs, $^{239+240}$Pu and ^{210}Po in the Pomeranian Bay (southern Baltic) ecosystem. Biul. Mors. Inst. Ryb. **3**, 15–24.
Bok, C.S., and W.M. Keong, 1976. Heavy metals in marine biota from coastal waters around Singapore. J. Singapure Nat. Acad. Sci. **5**, 47–53.
Borchardt, T., S. Burchert, L. Karbe and R. Zeitner, 1989. Enhanced heavy metal concentrations in *Mytilus edulis* from the central North Sea. Scientia Mar. **53**, 725–728.
Bordin, G., J. McCourt and A. Rodriguez, 1992. Trace metals in the marine bivalve *Macoma balthica* in the Westerschelde Estuary (The Netherlands). Part 1. Analysis of total copper, cadmium, zinc and iron concentrations-locational and seasonal variations. Sci. Total Environ. **127**, 255–280.
Boström, K., O. Joensuu and I. Brohm, 1974. Plankton: its chemical composition and its significance as a source of pelagic sediments. Chem. Geol. **14**, 255–271.
Bourgoin B.P., 1990. *Mytilus edulis* shell as a bioindicator of lead pollution: considerations on bioavailability and variability. Mar. Ecol. Prog. Ser. **61**, 253–262.
Bourne, W.R.P., 1974. Birds of the North Atlantic Ocean. Proc. 16th Intern. Ornithol. Congress, August 1974 (Australian Academy of Sciences, Canberra), 705–715.
Bourne, W.R.P., 1980. The habitats, distribution, and numbers of northern seabirds. Trans. Linnaean Soc. New York **IX**, 1–14.
Boyden, C.R., 1977. Effects of size upon metal content in shellfish. J. Mar. Biol. Ass. UK **57**, 675–714.
Boyden, C.R., and D.J.H. Phillips, 1981. Seasonal variations and inherent variability of trace elements in oysters and their implication for indicator studies. Mar. Ecol. Prog. Ser. **5**, 29–40.
Boyle, E.A., 1981. Cadmium, zinc, copper, and barium in foraminifera tests. Earth Planet. Sci. Letters **53**, 11–35.
Brehm, A., 1962. The Life of Animals – Birds; title of original – Brehms Tierleben, Dritter Band, Vögel, Urania – Verlag 1956. (Państwowe Wydawnictwo Naukowe, Warsaw) 568 pp. (in Polish).
Bremner, I., 1987. Interactions between metallothionein and trace elements. Prog. Food Nutr. Sci. **11**, 1–39.

Brix, H., and J.E. Lyngby, 1982. The distribution of cadmium, copper, lead, and zinc in eelgrass (*Zostera marina* L.). Sci. Total Environ. **24**, 51–63.

Brix, H., and J.E. Lyngby, 1983. The distribution of some metallic elements in eelgrass (*Zostera marina* L.) and in sediment in the Limfjord, Denmark. Estuar. Coast. Shelf Sci. **16**, 455–467.

Brix, H., and J.E. Lyngby, 1985. The influence of size upon the concentrations of Cd, Cr, Cu, Hg, Pb and Zn in the common mussel (*Mytilus edulis* L.). Symposia Biologia Hungarica **29**, 253–271.

Brix, H., J.E. Lyngby and H.-H. Schierup, 1983. Eelgrass (*Zostera marina* L.) as an indicator organism of trace metals in the Limfjord, Denmark. Mar. Environ. Res. **8**, 165–181.

Broman, D., L. Lindquist and I. Lundbergh, 1991. Cadmium and zinc in *Mytilus edulis* L. from the Bothnian Sea and the northern Baltic proper. Environ. Pollut. **74**, 227–244.

Brown, C.L., and Luoma S.N., 1995. Use of the euryhaline bivalve *Potamocorbula amurensis* as a biosentinel species to assess trace metal contamination in San Francisco Bay. Mar. Ecol. Prog. Ser. **124**, 129–142.

Brown, M.T., W.M. Hodgkinson and C.L. Hurd, 1999. Spatial and temporal variations in the copper and zinc concentrations of two green seaweeds from Otago Harbour, New Zealand. Mar. Environ. Res. **47**, 175–184.

Brügmann, L., 1978. Zum Gehalt eniger Schwermetalle in marinen Organismen aus dem Auftriebsgebiet vor NW-Afrika. Fisherei-Forchung **16**, 53–58.

Brügmann, L., and D. Lange, 1988. Trace metal studies on the starfish *Asterias rubens* L. from Western Baltic Sea. Chem. Ecol. **3**, 295–311.

Brügmann, L., and U. Hennings, 1994. Metals in zooplankton from the Baltic Sea, 1980–84. Chem. Ecol. **9**, 87–103.

Bryan, G.W., 1968. Concentrations of zinc and copper in the tissues of decapod crustaceans. J. Mar. Biol. Assoc. UK **48**, 308–321.

Bryan, G.W., 1969. The absorption of zinc and other metals by the brown seaweed *Laminaria digitata*. J. Mar. Biol. Assoc. UK **49**, 225–243.

Bryan, G.W., 1974. Adaptation of an estuarine polychaete to sediments containing high concentrations of heavy metals, in: Pollution and Physiology of Marine Organisms, eds. F.J. Vernberg and W.B. Vernberg (Academic Pres, New York), 123–135.

Bryan G.W., 1976. Heavy metal contamination in the sea, in: Marine Pollution, ed. R. Johnston (Academic Press, London, New York), pp. 185–302.

Bryan, G.W., 1980. Recent trends in research on heavy-metal contamination in the sea. Helgoländer Meeresunters. **33**, 6–25.

Bryan, G.W., 1983. Brown seaweed, *Fucus vesiculosus*, and the gastropod, *Littorina littoralis*, as indicators of trace-metal availability in estuaries. Sci. Total Environ. **28**, 91–104.

Bryan, G.W., 1984. Pollution due to heavy metals and their compounds, in: Marine Ecology, ed. O. Kinne (John Wiley & Sons Ltd, Chichester) **Vol. 5**, Part 3, 1289–1431.

Bryan, G.W., 1985. Bioavailability and effects of heavy metals in marine deposits. Wastes in the oceans, in: Nearshore Waste Disposal, eds. B.H. Ketchum, J.M. Capuzzo, W.V. Burt, I.W. Duedall, P.K. Park and D.R. Kester (John Wiley & Sons Ltd, New York) **Vol. 6**, 42–79.

Bryan, G.W. and L.G. Hummerstone, 1971. Adaptation of the polychaete *Nereis diversicolor* to estuarine sediments containing high concentrations of heavy metals. I. General observations and adaptation to copper. J. Mar. Biol. Assoc. UK **51**, 845–863.

Bryan, G.W. and L.G. Hummerstone, 1973a. Adaptation of the polychaete *Nereis diversicolor* to manganese in estuarine sediments. J. Mar. Biol. Assoc. UK **53**, 859–872.

Bryan, G.W. and L.G. Hummerstone, 1973b. Adaptation of the polychaete *Nereis diversicolor* to estuarine sediments containing high concentrations of zinc and cadmium. J. Mar. Biol. Assoc. UK **53**, 839–857.

Bryan, G.W., and L.G. Hummerstone, 1973c. Brown seaweed as an indicator of heavy metals in estuaries in south-west England. J. Mar. Biol. Assoc. U.K. **53**, 705–720.

Bryan, G.W., and L.G. Hummerstone, 1977. Indicators of heavy metal contamination in the Looe Estuary (Cornwall) with particular regard to silver and lead. J. Mar. Biol. Assoc. U.K. **57**, 75–92.

Bryan, G.W. and P.E. Gibbs, 1980a. Metals in nereid polychaetes: the contribution of metals in the jaws to the total body burden. J. Mar. Biol. Assoc. UK **60**, 641–654.

Bryan, G.W. and P.E. Gibbs, 1980b. Zinc – a major inorganic componenent of nereid polychaete jaws. J. Mar. Biol. Assoc. UK **59**, 969–973.

Bryan, G.W., and P.E. Gibbs, 1983. Heavy metals in the Fal Estuary, Cornwall: a study of long-term contamination by mining waste and its effects on estuarine organisms. Mar. Biol. Assoc. UK, Occasional Publication **No. 2**, 112 pp.

Bryan, G.W. and P.E. Gibbs, 1987. Polychaetes as indicators of heavy-metal availability in marine deposits, in: Oceanic Processes in Marine Pollution. Vol. 1. Biological Processes and Wastes in the Ocean, ed. J.M. Capuzzo and D.R. Kester (Krieger Publishing Co. Inc. Melbourne, FL, USA), 37–49.

Bryan, G.W. and W.J. Langston, 1992. Bioavailability, accumulation and effects of heavy metals in sediments with special reference to United Kingdom estuaries: a review. Environ. Pollut. **76**, 89–131.

Bryan, G.W., W.J. Langston, L.G. Hummerstone, G.R. Burt and Y.B. Ho, 1983. An assessment of the gastropod, *Littorina littorea*, as an indicator of heavy-metal contamination in United Kingdom estuaries. J. Mar. Biol. Assoc. UK **63**, 327–345.

Bryan, G.W., W.J. Langston, L.G. Hummerstone and G.R. Burt, 1985. A guide to the assessment of heavy-metal contamination in estuaries using biological indicators. Mar. Biol. Assoc. U.K., Occasional Publication (The Laboratory, Citadel Hill, Plymouth, Devon, England) **No. 4**, 92 pp.

Brzezińska, A., A. Trzosińska, W. Żmijewska and L. Wódkiewicz, 1984. Trace metals in some organisms from the southern Baltic. Oceanologia **18**, 79–94.

Burdon-Jones, C., G.R.W. Denton, G.B. Jones and K.A. McPhie, 1982. Regional and seasonal variations of trace metals in tropical Phaeophyceae from north Queensland. Mar. Environ. Res. **7**, 13–30.

Burger, J. and M. Gochfeld, 2000. Metal level in feathers of 12 species of seabirds from Midway Atoll in the northern Pacific Ocean. Sci. Total Environ. **257**, 37–52.

Butterworth, J., P. Lester and G. Nickless, 1972. Distribution of heavy metals in Severn estuary. Mar.Pollut. Bull. **3**, 72–74.

Caçador, I., C. Vale and F. Catarino, 2000. Seasonal variation of Zn, Pb, Cu and Cd concentrations in the root-sediment system of *Spartina maritima* and *Halimione portulacoides* from Tagus estuary salt marshes. Mar. Environ. Res. **49**, 279–290.

Campanella, L., M.E. Conti, F. Cubadda and C. Sucupane, 2001. Trace metals in seagrass, algae and molluscs from an uncontaminated area in the Mediterranean. Environ. Pollut. **111**, 117–126.

Cantillo, A.Y., 1998. Comparison of results of Mussel Watch programs of the United States and France with worldwide Mussel Watch studies. Mar. Pollut. Bull. **36**, 712–717.

Cardellicchio, N., S. Giandomenico, P. Ragone and A. Di Leo, 2000. Tissue distribution of metals in striped dolphins (*Stenella coeruleoalba*) from the Apulian coasts, Southern Italy. Mar. Environ. Res. **49**, 55–66.

Carlson, L., 1990. Effects of biotic and abiotic factors on the accumulation of radionuclides in *Fucus vesiculosus* L. Dissertation (Department of Ecology, Marine Ecology, Lund University, Sweden), 111 pp.

Carlson, L., and E. Holm, 1990. Radioactivity in the Baltic Sea following the Chernobyl accident. SSI project P392-86 (University of Lund, Sweden), 21 pp.

Catsiki, V.-A., and E. Strogyloudi, 1999. Survey of metal levels in common fish species from Greek waters. Sci. Total Environ. **237/238**, 387–400.

Chan, K.M., 1995. Concentrations of copper, zinc, cadmium and lead in rabbitfish (*Siganus oramin*) collected in Victoria Harbour, Hong Kong. Mar. Pollut. Bull. **31**, 277–280.

Chan, H.M., P.S. Rainbow and D. J.H. Phillips, 1986. Barnacles and mussels as monitors of trace metal bio-availability in Hong Kong waters, in: The Marine Flora and Fauna of Hong Kong and Southern China, Proc. Second Int. Mar. Biol. Workshop (ed. B. Morton), Hong Kong, 1986 (Hong Kong University Press) 1239–1268.

Charmasson, S., E. Barker, D. Calmet, A.-S. Pruchon and H. Thébault, 1999. Long-term variations of man-made radionuclide concentrations in a bio-indicator *Mytilus galloprovincialis* from the French Mediterranean coast. Sci. Total Environ. **237/238**, 93–103.

Cherry, R.D., and J.J.W. Higgo, 1978. Zooplankton fecal pellets and element residence times in the ocean. Nature **274**, 246–248.

Chojnacki, J., 1973. The zooplankton of the southern Baltic in 1960. Zesz. Nauk. Akad. Roln. Szczecinie **40**, 11–24 (in Polish).

Christensen, G.C., 1986. Radioactivity in *Fucus vesiculosus* along the Norwegian North Sea and Skagerrak coast 1980–1983, in: Study of Radioactive materials in the Baltic Sea. (International Atomic Energy Agency, Vienna). Report (IAEA-TECDOC-362) of the Final Research Coordination Meeting on the Study of Radioactive Materials in the Baltic Sea organized by the IAEA and held in Helsinki, Finland 24–28 September, 1984, pp. 133–140.

Christensen, C.Ch., and E. Strålberg, 2000. Can the outflow of radiocaesium from the Baltic Sea be detected in brown algae along the Norwegian coast?, in: The Radiological Exposure of the Population of the European Community to Radioactivity in the Baltic Sea. Marina-Balt Project, ed. S.P. Nielsen. Proceedings of a Seminar held at Hasseludden Conference Centre, Stockholm, 9–11 June 1998, European Commission, Directorate-General Environment, EUR 19200 EN (European Communities, 2000, Belgium), pp. 433–438.

Cięglewicz, W., W. Załachowski, J. Szypuła, S. Krzykawski and I. Krzykawska, 1972. Food habits of commercial fish of the southern Baltic, in: Marine Ecosystems – Biology, ed. K. Siudziński (Morski Instytut Rybacki, Gdynia), **Vol. 2**, pp. 363–450 (in Polish).

Clarke, E.D., L. Hilby and S.T. Buckland, 1998. The estimation of the bycatch mortality of harbour porpoise in the Baltic Sea. ASCOBANS/ADV./COM/5/DOC. 15, p. 41.

Clausen, B., and S. Andersen,1988. Evaluation of bycatch and health status of the harbour porpoise (*Phocoena phocoena*) in Danish waters. Dan. Rev. Game Biol. **13**, 1–20.

Clifton, R.J., H.E. Stevens and E.I. Hamilton, 1983. Concentration and depuration of some radionuclides present in a chronically exposed population of mussels (*Mytilus edulis*). Mar. Ecol. Prog. Ser. **11**, 245–256.

Collier, R.W. and J.M. Edmond, 1983. Plankton compositions and trace element fluxes from the surface ocean. NATO Conf. Ser., 4, , Trace Metals in Sea Water **9**, 789–809.

Collings, S.E., M.S. Johnson and R.T. Leah, 1996. Metal contamination of angler-caught fish from the Mersey Estuary. Mar. Environ. Res. **41**, 281–297.

Connors, P.G., V.C. Anderlini, R.W. Risebrough, M. Gilbertson and H. Hays, 1975. Investigations of heavy metals in common tern populations. Can. Field Nat. **89**, 157–162.

Coombs, T.L., and P.J. Keller, 1981. *Mytilus* byssal threads as an environmental marker for metals. Aquatic Toxicol. **1**, 291–300.

Copper, R.J., D. Langlois and J. Olley, 1982. Heavy metals in Tasmanian shellfish. I. Monitoring heavy metal contamination in the Dervent estuary. Use of oysters and mussels. J. Appl. Toxicol. **2**, 99–109.

Cossa, D., 1988. Cadmium in *Mytilus* spp.: Worldwide survey and relationship between seawater and mussel content. Mar. Environ. Res. **26**, 265–284.

Cossa, D., 1989. A review of the use of *Mytilus* spp. as quantitative indicators of cadmium and mercury contamination in coastal waters. Oceanol. Acta **12**, 417–432.

Cossa, D., D. Auger, B. Averty, M. Lucon, P. Masselin and J. Nol, 1992. Flounder (*Platichthys flesus*) muscle as an indicator of metal and organochlorine contamination of French Atlantic coastal waters. Ambio **21**, 176–182.

Cronin, M., I.M. Davies, A. Newton, J.M. Pirie, G. Topping and S. Swan, 1998. Trace metal concentrations in deep sea fish from the North Atlantic. Mar. Environ. Res. **45**, 225–238.

Custer, T.W., and W.L. Hohman, 1994. Trace elements in canvasbacks (*Aythya valisineria*) wintering in Louisiana, USA, 1987–1988. Environ. Pollut. **84**, 253–259.

Custer, T.W., C.M. Custer, R.K. Hines and D.W. Sparks, 2000. Trace elements, organochlorines, polycyclic aromatic hydrocarbons, dioxins, and furns in lesser scaup wintering on the Indiana Harbor Canal. Environ. Pollut. **110**, 469–482.

Dahlgaard, H., 1981. Loss of ^{51}Cr, ^{54}Mn, ^{57}Co, ^{59}Fe, ^{65}Zn and ^{134}Cs by the mussel *Mytilus edulis*. Impacts of Radionuclide Releases into the Marine Environment (IAEA, Vienna), pp. 361–370.

Dahlgaard, H., 1986. Effects of season and temperture on long-term *in situ* loss rates of Pu, Am, Np, Eu, Ce, Ag, Tc, Zn, Co and Mn in a Baltic *Mytilus edulis* population. Mar. Ecol. Prog. Ser. **33**, 157–165.

Dahlgaard, H., 1991. Variation in radionuclide loss rates from Baltic *Mytilus edulis*, in: Trace Metals in the Environment 1: Heavy Metals in the Environment, ed. J.-P. Vernet (Elsevier, Amsterdam), pp. 261–271.

Dahlgaard, H., 1994. Marine radioecology, in: Radioecology: Lectures in Environmental Radioactivity, ed. E. Holm (World Scientific Singapore, Singapore, New Jersey, London, Hong Kong).

Dahlgaard, H., 1996. Polonium-210 in mussels and fish from the Baltic-North Sea estuary. J. Environ. Radioactivity **32**, 91–96.

Dahlgaard, H., and S. Boelskifte, 1992. "Sensi": A model describing the accumulation and time-integration of radioactive discharges in the bioindicator *Fucus vesiculosus*. J. Environ. Radioactivity **16**, 49–63.

Dahlgaard, H., T.D.S. Bergan and G.C. Christensen, 1986. Technetium-99 and caesium-137 time series at the Norwegian coast monitored by the brown alga *Fucus vesiculosus*. Proceedings Part 1 of International Symposium on Radionuclides in the Ocean – RADOC 96–97, Cherbourg-Octeville, France, 7–11 October 1996. Rdioprotection – Colloques, **Vol. 32**, C2, 353–358.

Dańczak, A., M. Ligocki and E. Kalisińska, 1997. Heavy metals in the organs of anseriform birds. Polish J. Environ. Studies **6**, 39–42.

Das, K., G. Lepoint, V. Loizeau, V. Debacker, P. Dauby and J.M. Bouquegnau, 2000. Tuna and dolphin associations in the North-east Atlantic; Evidence of different ecological niches from stable isotope and heavy metal measurements. Mar. Pollut. Bull. **40**, 102–109.

Davidan, I.N, and O.P. Savchuk, 1989. Problems in studies and mathematical modelling of the Baltic Sea ecosystem. General tendencies in the evolution of the ecosystem. International Project "Baltica" (Leningrad, Gidrometeoisdat) Chap. **4**, 179–187 (in Russian).

Davies, A.G., 1978. Pollution studies with marine plankton. Part II. Heavy metals. Adv. Mar. Biol. **15**, 381–508.

Dąbrowski, T., J. Zamojski and J. Skup, 1967. Oznaczanie mikroelementów w mięsie niektórych ryb morskich i słodkowodnych metodą analizy spektralnej (Determination of microelements in muscle of some marine and freshwater fish by method of spectral analysis). Prace MIR **XIVB**, 105–114 (in Polish).

De Lacerda, L.D., N.R.W. Lima, W.C. Pfeiffer and M. Fiszman, 1983. Size and metal concentration in the mangrove mussel *Mytella guyanensis* (Mollusca: Bivalvia) from Baia de Sepetiba, Brazil. Rev. Biol.Trop **31**, 333–335.

De Wolf, H., T. Backeljau and R. Blust, 2000. Heavy metal accumulation in the periwinkle *Littorina littorea*, along a pollution gradient in the Scheldt estuary. Sci. Total Environ. **262**, 111–121.

Debacker, V., L. Holsbeek, G. Tapia, S. Gobert, C.R. Joiris, T. Jauniaux, F. Coignoul and J.-M. Bouquegneau, 1997. Ecotoxicological and pathological studies of common guillemots *Uria aalge* beached on the Belgian coast during six successive wintering periods (1989–90 to 1994–95). Dis. Aquat. Org. **29**, 159–168.

Demina, L.L., and L.S Fomina, 1978. On the occurrence forms of Fe, Mn, Zn and Cu in surficial suspension of the Pacific Ocean. Geokhimya **11**, 1710–1726 (in Russian, with an English abstract).

Denton, G.R.W., H. Marsh, G.E. Heinsohn and C. Burdon-Jones, 1980. The unusual metal status of the dugong, *Dugong dugong*. Mar. Biol. **57**, 201–219.

Dethlefsen, V., 1977. Uptake, retention and loss of cadmium by brown shrimp (*Crangon crangon*). Berichte der Deutschen Wissenschaftlichen Kommission für Meeresforschung **26**, 137–152.

Di Giulio, R.T., and P.F. Scanlon, 1984a. Sublethal effects of cadmium ingestion on mallard ducks. Arch. Environ. Contam. Toxicol. **13**, 765–771.

Di Giulio, R.T., and P.F. Scanlon, 1984b. Heavy metals in tissues of waterfowl from the Chesapeake Bay, USA. Environ. Pollut. **35A**, 29–48.

Di Giulio, R.T., and P.F. Scanlon, 1984c. Effects of cadmium and lead ingestion on tissue concentrations of cadmium, lead, copper and zinc in mallard ducks. Sci. Total Environ. **39**, 103–110.

Di Giulio, R.T., and P.F. Scanlon, 1985. Heavy metals in aquatic plants, clams, and sediments from the Chesapeake Bay, U.S.A. Implication for waterfowl. Sci. Total Environ. **41**, 259–274.

Diaz, C., and M.L. Fernandez-Puelles, 1988. Contents of heavy metals in zooplankton of Canary Islands. Biol. Instit. Espanol. Oceanogr. **5**, 57–61 (in Spanish).

Dietz, R., C.O. Nielsen, M.M. Hansen and C.T. Hansen, 1990. Organic mercury in Greenland birds and mammals. Sci. Total Environ. **95**, 41–51.

Dietz, R., F. Riget and P. Johansen, 1996. Lead, cadmium, mercury and selenium in Greenland marine animals. Sci. Total Environ. **186**, 67–93.

Dietz, R., J. Norgaard and C.J. Hansen, 1998. Have Arctic marine mammals adapted to high cadmium levels? Mar. Pollut. Bull. **36**, 490–492.

Dietz, R., F. Riget and E.W. Born, 2000a. An assessment of selenium to mercury in Greenland marine mammals. Sci. Total Environ. **245**, 15–24.

Dietz, R., F. Riget and E.W. Born, 2000b. Geographical differences of zinc, cadmium, mercury and selenium in polar bears (*Ursus maritimus*) from Greenland. Sci. Total Environ. **245**, 25–47.

Donaldson, G.M., B.M. Braune, A.J. Gaston and D.G. Noble, 1997. Organochlorine and heavy metal residues in breast muscle of known-age thick-milled murres (*Uria lomvia*) from the Canadian Arctic. Arch. Environ. Contam. Toxicol. **33**, 430–435.

Drescher, H.E., U. Harms and E. Huschenbeth, 1977. Organochlorines and heavy metals in the harbour seal *Phoca vitulina* from the German North Sea coast. Mar. Biol. **41**, 99–106.

Drifmeyer, J.E., G.W. Thayer, F.A. Cross and J.C. Zieman, 1980. Cycling of Mn, Fe, Cu and Zn by eelgrass, *Zostera marina* L. Am. J. Bot. **67**, 1089–1096.

Druehl L., M. Cackette and J.M. D'Auria, 1988. Temporal and spatial distribution of iodine-131 in the brown seaweed *Fucus* subsequent to the Chernobyl incident. J. Mar. Biol. **98**, 125–129.

Duinker, J.C., M.T.J. Hillebrand and R.F. Nolting, 1979. Organochlorines and metals in harbour seals (Dutch Wadden Sea). Mar. Pollut. Bull. **10**, 360–364.

Durand, J.P., F.P. Carvalho, F. Goudard, J. Pieri, S.W. Fowler and O. Cotret, 1999. ^{210}Po binding to metallothioneins and feritin in the liver of teleost marine fish. Mar. Ecol. Prog. Ser. **177**, 189–196.

Edmonds, J.S., M.J. Moran, N. Caputi and M. Morita, 1989. Trace element analysis of fish sagittae as an aid to stock identification: pink snapper (*Chrysophrys auratus*) in Western Australian waters. Can. J. Fisher. Aquatic Sci. **46**, 50–54.

Edmonds, J.S., N. Caputi and M. Morita, 1991. Stock discrimination by trace-element analysis of otoliths of orange roughy (*Hoplostethus atlanticus*), a deeper-water marine teleost. Aust. J. Mar. Freshwater Res. **42**, 383–389.

Edmonds, J.S., R.C.J. Lenanton, N. Caputi and M. Morita, 1992. Trace elements in otoliths of yellow-eye mullet (*Aldrichetta forsteri*) as an aid to stock identification. Fish. Res. **13**, 39–51.

Eide, I., S. Myklestad and S. Melsom, 1980. Long-term uptake and release of heavy metals by *Ascophyllum nodosum* (L.) Le Jol. (Phaeophyceae) *in situ*. Environ. Pollut. Ser. A **23**, 19–28.

Eisler, R., and G. LaRoche, 1972. Elemental composition of the estuarine teleost *Fundulus heteroclitus* (L.). J. Exp. Mar. Biol. Ecol. **9**, 29–42.

Elgethun, K., C. Neumann and P. Blake, 2000. Butyltins in shellfish, finfish, water and sediment from the Coos Bay estuary (Oregon, USA). Chemosphere **41**, 953–964.

Elliott, J.E., A.M. Scheuhammer, F.A. Leighton and P.A. Pearce, 1992. Heavy metal and methallothionein concentrations in Atlantic Canadian seabirds. Arch. Environ. Contam. Toxicol. **22**, 63–73.

Elmgren, R., S. Ankar, B. Marteleur and G. Ejdung, 1986. Adult interference with postlarvae in oft sediments: the *Pontoporeia – Macoma* example. Ecology **67**, 827–836.

Enoberg, A., 1976. Statens Levnedsmiddelinstitut Pub. 33, after Falandysz, J. (1985). Z. Lebensm. Unters. Forsch. **181**, 117–120.

Everaarts, J.M., and K. Saraladevi, 1996. Cadmium distribution in sediment and the lugworm *Arenicola marina* in a low concentration exposure experiment. Bull. Environ. Contam. Toxicol. **57**, 771–778.

Fabris, J.G., B.J. Richardson, J.E. O'Sullivan and F.C. Brown, 1994. Estimation of cadmium, lead, and mercury concentrations in estuarine waters using the mussel *Mytilus edulis planulatus* L. Environ. Toxicol. Water Quality: An International Journal **9**, 183–192.

Falandysz, J., 1984a. Trace metals and organochlorines in plankton from the southern Baltic. Mar. Pollut. Bull. **15**, 416–418.

Falandysz, J., 1984b. Metals and organochlorines in a female white-tailed eagle from Uznam Island, southwestern Baltic Sea. Environ. Conservation **11**, 262–263.

Falandysz, J., 1985. Trace metals in flatfish from the southern Baltic, 1983. Z. Lebensm. Unters. Forsh. **181**, 117–120.

Falandysz, J., 1986a. Trace metals in herring from the southern Baltic, 1983. Z. Lebensm. Unters. Forsh. **182**, 36–39.

Falandysz, J., 1986b. Trace metals in sprats from the southern Baltic, 1983. Z. Lebensm. Unters. Forsh. **182**, 40–43.

Falandysz, J., 1986c. Trace metals in cod from the southern Baltic, 1983. Z. Lebensm. Unters. Forsh. **182**, 228–231.

Falandysz, J., 1986d. Metals and organochlorines in adult and immature males of white-tailed eagle. Environ. Conservation **13**, 69–70.

Falandysz, J., 1992a. Zawartość rtęci w mięśniach i wątrobie storni (*Platichthys flesus*) złowionych w części zachodniej Zatoki Gdańskiej (The content of mercury in muscles and liver of flounder (*Platichthys flesus*) caught in the western part of the Gulf of Gdańsk). Stud. Mater. Oceanol. **No. 62**, 13–18 (in Polish, with English abstract).

Falandysz, J., 1992b. Metale w mięsnach i wątrobie okoni *Perca fluviatilis* z Zatoki Gdańskiej (Metals content in the muscular tissue and liver of perch *Perca fluviatilis* from the Gulf of Gdańsk). Bromat. Chem. Toksykol. **XXV**, 333–335 (in Polish, with English abstract).

Falandysz, J., 1994. Mercury concentrations in benthic animals and plants inhabiting the Gulf of Gdańsk, Baltic Sea. Sci. Total Environ. **141**, 45–49.

Falandysz, J., and P. Szefer, 1983. Metals and organochlorines in a specimen of white-tailed eagle. Environ. Conservation **10**, 256–258.

Falandysz, J., and H. Lorenc-Biała, 1984. Trace metals in fish from the southern Baltic. Meeresforsch. **30**, 111–119.

Falandysz, J., and D. Centkowska, 1986. Metale i związki polichlorowe w węgorzach z Zatoki Gdańskiej i wód jej przyległych (Metals and organochlorine compounds in eels from the Gulf of Gdańsk and the adjacent waters). Roczn. PZH 1987, **XXXVIII**, 124–131 (in Polish, with English abstract).

Falandysz, J., and J. Falandysz, 1986. Metale i związki polichlorowe w węgorzach z Zatoki Gdańskiej i wód jej przyległych (Metals and organochlorine compounds in eels from the Gulf of Gdańsk and the adjacent waters). Roczn. PZH **XXXVII**, 414–421 (in Polish, with English abstract).

Falandysz, J., and M. Kowalewska, 1993. Mercury concentration of stickleback *Gasterosteus aculeatus* from the Gulf of Gdańsk. Bull. Environ. Contam. Toxicol. **51**, 710–715.

Falandysz, J., B. Jakuczun and T. Mizera, 1988. Metals and organochlorines in four female white-tailed eagles. Mar. Pollut. Bull. **19**, 521–526.

Falandysz, J., K. Ośmiałowski and B. Sicińska, 1992. Manganese, copper, iron, and zinc content in eel-pout muscles *Zoarces viviparus*. Bromat. Chem. Toksykol. **XXV**, 73–76.

Falandysz, J., A. Trzosińska, P. Szefer, J. Warzocha and B. Draganik, 2000. The Baltic Sea, in: Seas at the Millennium: An Environmental Evaluation, Vol. I: Europe, The Americas and West Africa, ed. C.R.C. Sheppard (Pergamon, Elsevier), **Chapt. 7**, pp. 99–120.

Falandysz, J., H. Ichihashi, T. Mizera and S. Yamasaki, 2000a. Mineral composition of selected tissues and organs of white-tailed sea eagle. Roczn. Państw. Zakł. Hig. **51**, 1–5.

Falconer, C.R., I.M. Davies and G. Topping, 1983. Selected trace metals in porpoises (*Phocoena phocoena*) from the North East Coast of Scotland. Mar. Environ. Res. **8**, 119–125.

Fant, M.L., M. Nyman, E. Helle and E. Rudbäck, 2001. Mercury, cadmium, lead and selenium in ringed seals (*Phoca hispida*) from the Baltic Sea and from Svalbard. Environ. Pollut. **111**, 493–501.

Fasola, M., P.A. Movalli and C. Gandini, 1998. Heavy metal, organochlorine pesticide, and PCB residues in eggs and feathers of herons breeding in Northern Italy. Arch. Environ. Contam. Toxicol. **34**, 87–93.

Favretto, L., and L.G. Favretto, 1984a. Multivariate data analysis of some xenobiotic trace metals in mussels from the Gulf of Trieste. Z. Lebensm. Unters. Forsch. **179**, 201–204.

Favretto, L., and L.G. Favretto, 1984b. Principal component analysis as a tool for studying interedependences among trace metals in edible mussels from the Gulf of Trieste. Z. Lebensm. Unters. Forsch. **179**, 377–380.

Fialkowski, W., and W.A. Newman, 1998. A pilot study of heavy metal accumulations in barnacle from the Salton, Southern California. Mar. Pollut. Bull. **36**, 138–143.

Filho, G.M.A., L.R. Andrade, C.S. Karez, M. Farina and W.C. Pfeiffer, 1999. Brown algae species as biomonitors of Zn and Cd at Sepetiba Bay, Rio de Janeiro, Brazil. Mar. Environ. Res. **48**, 213–224.

Finley, M.T., M.P. Dieter and L.N. Locke, 1976. Lead in tissues of mallard ducks dosed with two types of lead shot. Bull. Environ. Contam. Toxicol. **16**, 261–269.

Fischer, H., 1988. *Mytilus edulis* as a quantitative indicator of dossolved Cd. Final study and synthesis. Mar. Ecol. Prog. Ser. **48**, 159–169.

Fischer, H., 1989. Cadmium in seawater recorded by mussels: regional decline established. Mar. Ecol. Prog. Ser. **55**, 163–174.

Fisher, N.S., P. Bjerregaard and S.W. Fowler, 1983a. Interactions of marine plankton with transuranic elements. 1. Biokinetics of neptunium, plutonium, americium, and californium in phytoplankton. Limnol. Oceanogr. **28**, 432–447.

Fisher, N.S., P. Bjerregaard and S.W. Fowler, 1983b. Interactions of marine plankton with transuranic elements. 3. Biokinetics of americium in euphausiids. Mar. Biol. **75**, 261–268.

Fisher, N.S., I. Stupakoff, S. Sañudo-Wilhelmy, W.-X. Wang, J.-L. Teyssié, S.W. Fowler and J. Crusius, 2000. Trace metals in marine copepods: a field test of a bioaccumulation model coupled to laboratory uptake kinetics data. Mar. Ecol. Prog. Ser. **194**, 211–218.

Fleming, W.J., 1981. Environmental metal residues in tissues of canvasbacks. J. Wildl. Manage **45**, 508–511.

Forsberg, A., S. Söderlund, A. Frank, L.R. Petersson and M. Pedersen, 1988. Studies on element content in seaweed, *Fucus vesiculosus*, from Archipelago of Stockholm. Environ. Pollut. **49**, 245–263.

Forsman, B., 1938. Untersuchungen über die Cumaceen des Skageraks. Zool. Bidr. Upps. **18**, 161 pp.

Foster, P., 1976. Concentrations and concentration factors of heavy metals in brown algae. Environ. Pollut. **10**, 45–53.

Foster, P., and J. Chacko, 1995. Minor and trace elements in the shell of *Patella vulgata* (L.). Mar. Environ. Res. **40**, 55–76.

Fowler, S.W., 1977. Trace elements in zooplankton particulate products. Nature **269**, 51–53.

Fowler, S.W., 1986. Trace metal monitoring of pelagic organisms from the open Mediterranean Sea. Environ. Monitor. Assess. **7**, 59–78.

Fowler, S.W., 1990. Critical review of selected heavy metal and chlorinated hydrocarbon concentrations in the marine environment. Mar. Environ. Res. **29**, 1–64.

Fowler, S.W., and B. Oregioni, 1976. Trace metals in mussels from the N. W. Mediterranean. Mar. Pollut. Bull. **7**, 26–29.

Fowler, S.W., and G. Benayoun, 1977. Selenium kinetics in marine zooplankton, in Radioactivity in the Sea (International Atomic Energy Agency, Vienna), **No. 63**, 23 pp. (Reprinted from *Marine Science Communication* **2**, 43–67 (1976)).

Fowler, S.W., C. Papadopoulou and D. Zafiropoulos, 1985. Trace elements in selected species of zooplankton and nekton from the open Mediterranean Sea, in: Heavy Metals in the Environment, ed. T.D. Lekkas (CEP Consultants Ltd, Edinburgh, UK), 670–672.

Fowler, S.W., J.W. Readman, B. Oregioni, J.-P. Villeneuve and K. McKay, 1993. Petroleum hydrocarbons and trace metals in nearshore gulf sediments and biota before and after the 1991 war: an assessment of temporal and spatial trends. Mar. Pollut. Bull. **27**, 171–182.

Frank, A., 1986. Lead fragments in tissues from wild birds: A cause of misleading analytical results. Sci. Total Environ. **54**, 275–281.

Frank, A. and K. Borg, 1979. Heavy metals in tissues of the mute swan (*Cygnus olor*). Acta Vet. Scand. **20**, 447–465.

Frank, A., V. Galgan, A. Roos, M. Olsson, L.R. Petersson and A. Bignert, 1992. Metal concentrations in seals from Swedish waters. Ambio **21**, 529–538.

Franson, J.C., P.S. Koehl, D.V. Derksen, T.C. Rothe, C.M. Bunck and J.F. Moore, 1995. Heavy metals in seaducks and mussels from Misty Fjords National Monument in Southeast Alaska. Environ. Monit. Assess. **36**, 149–167.

Fretter V., and A. Graham, 1962. British Prosobranch Molluscs, their Functional Anatomy and Ecology (Ray Society, London), 755 pp.

Frodello, J.P., M. Roméo and D. Viale, 2000. Distribution of mercury in the organs and tissues of five toothed-whale species of the Mediterranean. Environ. Pollut. **108**, 447–452.

Fuge, R., and K.H. James, 1973. Trace metal concentrations in brown seaweeds, Cadigan Bay, Wales. Mar. Chem. **1**, 281–293.

Fuge, R., and K.H. James, 1974. Trace metal concentrations in *Fucus* from the Bristol Channel. Mar. Pollut. Bull. **5**, 9–12.

Fujise, Y., K. Honda, R. Tatsukawa and S. Mishima, 1988. Tissue distribution of heavy metals in Dall's porpoise in the northwestern Pacific. Mar. Pollut. Bull. **19**, 226–230.

Fujita, Y., 1994. Specific characteristics of heavy metal accumulation by marine bivalves, gastropods and polyplacophores, MSc Thesis, Miyazaki University, Miyazaki, Japan.

Furness, R.W., 1996. Transfer of heavy metals from contaminated land to top predators; implications for birds of growing trees for soil amelioration, in: Heavy Metals and Trees, ed. F. Glimmerveen (Institute of Chartered Forests, Edinburgh), 107–122.

Furness, R.W., and C.J. Camphuysen, 1997. Seabirds as monitors of the marine environment. ICES J. Mar. Sci. **54**, 726–737.

Furness, R.W., S.A. Lewis and J.A. Mills, 1990. Mercury levels in the plumage of red-billed gulls *Larus novaehollandiae scopulinus* of known sex and age. Environ. Pollut. **63**, 33–39.

Gajewska, R. and M. Nabrzyski, 1977. Mercury, cadmium and lead content in sea and fresh-water fish). Roczn. PZH **XXVIII**, 215–224 (in Polish, with English abstract).

Gajewska, R., and M. Nabrzyski, 1978. Distribution of some trace metals in marine and freshwater fish). 1st National Conference on Effects of Trace Elements Pollutants on Agricultural Environment Quality, Puławy, May 4–6, 1978, pp. 135–143.

Gajewska, R., E. Malinowska, M. Nabrzyski and Z. Ganowiak, 2000. Porównanie zawartości rtęci w rybach bałtyckich poławianych w latach 1971–1997 (A comparative study on total mercury content of the Baltic fish 1971–1997). Bromat. Chem. Toksykol. **XXXIII**, 233–236 (in Polish, with English summary).

Galey, J., F. Goudard, J. Pieri, S.W. Fowler and F.P. Carvalho, 1983. Tissue and subcellular distribution of ^{252}Cf and ^{241}Am in the seastar *Marthasterias glacialis*. Mar. Biol. **75**, 253–259.

Gaskin, D.E., K. Ishida and R. Frank, 1972. Mercury in harbour porpoises (*Phocoena phocoena*) from the Bay of Fundy region. J. Fish. Res. Bd. Can. **29**, 1644–1646.

Gaskin, D.E., R. Frank, M. Holdrinet, K. Ishida, C.J. Walton and M. Smith, 1973. Mercury, DDT, PCBs in harbour seals (*Phoca vitulina*) from the Bay of Fundy and Gulf of Maine. J. Fish. Res. Bd. Can. **30**, 471–475.

George, S.G., 1980. Correlation of metal accumulation in mussels with the mechanism of uptake, metabolism and detoxification: a review. Thallasia Jugoslavica **16**, 347–365.

George, M.D. and T.W. Kureishy, 1979. Trace metals in zooplankton from the Bay of Bengal. Ind. J. Mar. Sci. **8**, 190–192.

Gibbs, P.E., and G.W. Bryan, 1980a. A note of the elemental composition of the jaws of *Goniada maculata* (Polychaeta: Goniadidae). J. Mar. Biol. Assoc. UK **60**, 541–542.

Gibbs, P.E., and G.W. Bryan, 1980b. Copper – the major metal component of glycerid polychaete jaws. J. Mar. Biol. Assoc. UK **60**, 205–214.

Gibbs, P.E., G.W. Bryan and K.P. Ryan, 1981. Copper accumulation by the polychaete *Melinna palmata*: an antipredation mechanism? J. Mar. Biol. Assoc. UK **61**, 707–722.

Gibbs, P.E., W.J. Langston, G.R. Burt and P.L. Pascoe, 1983. *Tharyx marioni* (Polychaeta), a remarkable accumulator of arsenic. J. Mar. Biol. Assoc. UK **63**, 313–325.

Gibbs, P.J., and A.G. Miskiewicz, 1995. Heavy metals in fish near a major primary treatment sewage plant outfall. Mar. Pollut. Bull. **30**, 667–674.

Giusti, L., A.C. Williamson and A. Mistry, 1999. Biologically available trace metals in *Mytilus edulis* from the coast of Northeast England. Environ. Intern. **25**, 969–981.

Gnassia-Barelli, M., R. Lemee, D. Pesando and M. Romeo, 1995. Heavy metal distribution in *Caulerpa taxifolia* from the north-western Mediterranean. Mar. Pollut. Bull. **30**, 749–755.

Goede, A.A. (1985). Selenium in Marine Waders. Ph.D. Thesis (Delft University of Technology, Delft, The Netherlands) 159 pp.

Goede, A.A. (1993). Mercury, selenium, arsenic and zinc in waders from the Dutch Wadden Sea. Environ. Pollut. **37A**, 287–309.

Goede, A.A., and M. de Bruin, 1984a. The use of bird feather parts as a monitor for metal pollution. Environ. Pollut. **8B**, 281–298.

Goede, A.A., and M. de Bruin, 1984b. The use of bird feather for indicating heavy metal pollution. (Paper presented at the Fifth International Conference on Nuclear Methods in Environmental and Energy Research, Mayaguez, Puerto Rico, April 2–6, 1984). IRI-reportnr. 133-84-25 (Interuniversitair Reactor Instituut, Delft), 10 pp.

Goede, A.A., and M. de Bruin, 1985a. Arsenic in the dunlin (*Calidris alpina*) from the Dutch Waddenzee. Bull. Environ. Contam. Toxicol. **34**, 617–622.

Goede, A.A., and M. de Bruin, 1985b. Selenium in a shore bird, the dunlin, from the Dutch Waddenzee. Mar. Pollut. Bull. **16**, 115–117.

Goede, A.A., and P. de Voogt, 1985. Lead and cadmium in waders from the Dutch Wadden Sea. Environ. Pollut. **37A**, 311–322.

Goede, A.A., and H.Th. Wolterbeek, 1994. Have high selenium concentrations in wading birds their origin in mercury. Sci. Total Environ. **144**, 247–253.

Goede, A.A., T. Nygard, M. de Bruin and E. Steinnes, 1989. Selenium, mercury, arsenic and cadmium in the lifecycle of the dunlin, *Calidris alpina*, a migrant wader. Sci. Total Environ. **78**, 205–218.

Goerke, H., 1971. De Ernährungsweise der Nereis – Arten (Polychaeta, Nereidae) der Deutschen Küsten. Verroffentlichungen des Instituts für Meeresforschung in Bremerhaven **13**, 50 pp.

Goldberg, E.D., V.T. Bowen, J.W. Farrington, G. Harvey, J.H. Martin, P.L. Parker, R.W. Risebrough, W. Robertson, E. Schneider and E. Gamble, 1978. The Mussel Watch. Environ. Conserv. **5**, 101–125.

Goldberg, E.D., M. Koide, V. Hodge, A.R. Flegal and J. Martin, 1983. U. S. Mussel Watch: 1977–1978 results on trace metals and radionuclides. Estuar. Coast. Shelf Sci. **16**, 69–93.

Gordon, M., G.A. Knauer and J.H. Martin, 1980. *Mytilus californianus* as a bioindicator of trace metal pollution: variability and statistical considerations. Mar. Pollut. Bull. **11**, 195–198.

Gosling, E.M., 1992. Systematics and geographic distribution of *Mytilus*, in: The mussel *Mytilus*: ecology, physiology, genetics and culture, ed. E. Gosling (Elsevier, Amsterdam), 1–20.

Goutner, V., R.W. Furness and G. Papakostas, 2001. Mercury in feathers of Squacco Heron (*Ardeola ralloides*) chicks in relation to age, hatching order, growth, and sampling dates. Environ. Pollut. **111**, 107–115.

Gouvea, R.C., P.L. Santos and V.A. Gouvea, 1987. Contribution to the study of radioactivity in marine organisms; dosage of ^{210}Po in *Perna perna*, L. Sci. Total Environ. **61**, 117–120.

Graham, A., 1988. Molluscs: Prosobranch and Pyramidellid Gastropods (Academic Press, London, New York), 112 pp.

Granéli, E., and C. Haraldsson, 1993. Can increased leaching of trace metals from acidified areas influence phytoplankton growth in coastal waters? Ambio **22**, 308–311.

Greenwood, J.G., 1984. Migration of dunlin *Calidris alpina*: a worldwide overview. Ring. Migr. **5**, 35–39.

Greenwood, J.G., 1986. Geographical variation and taxonomy of the dunlin *Calidris alpina* (L.). Bull. Brit. Orn. Cl. 106, 43–56.

Greichus, Y.A., A. Greichus, B.D. Amman, D.J. Call, D.C.D. Hamman and R.M. Pott, 1977. Insecticides, polychlorinated biphenyls and metals in African lake ecosystems. I. Hartbeespoort Dam, Transvaal and Voëlvlei Dam, Cape Province, Republic of South Africa. Archiv. Environ. Contam.Toxicol. **6**, 371–383.

Greichus, Y.A., A. Greichus, H.A. Draayer and B. Marshall, 1978. Insecticides, polichlorinated biphenyls and metals in African lake ecosystems. II. Lake Mc Ilwaine, Rhodesia. Bull. Environ. Contam. Toxicol. **7**, 444–453.

Greig, R., and D. Wenzloff, 1977. Trace metals in finfish from the New York Bight and Long Island Sound. Mar. Pollut. Bull. **8**, 198–200.

Greig, R., D. Wenzloff and J. Pearce, 1976. Distribution and abundance of heavy metals in finfish, invertebrates, and sediments collected at a deepwater disposal. Mar. Pollut. Bull. **7**, 185–187.

Greig, R.A., A. Adams and D.R. Wenzloff, 1977. Trace metal content of plankton and zooplankton collected from the New York Bight and Long Island Sound. Bull. Environ. Contam. Toxicol. **18**, 3–8.

Greig, R.A., S. Schuman, J. Pereira and P. Naples, 1983. Metals and PCB concentrations in windowpane flounder from Long Island Sound. Bull. Environ. Contam. Toxicol. **31**, 257–262.

Grillo, M.C., J.-C. Guary and S.W. Fowler, 1983. Comparative studies on transuranium nuclide biokinetics in sediment-dwelling invertebrates, in Radioactivity in the Sea (International Atomic Energy Agency, Vienna), **No. 70**, 273–291. (Reprinted from *Proc. Symp. Vienna*, 1980, IAEA, Vienna, 1981, 273–291).

Grimanis, A.P., D. Zafiropoulos and M. Vassilaki-Grimani, 1978. Trace elements in the flesh and liver of two fish species from polluted and unpolluted areas of the Aegean Sea. Environ. Sci. Technol. **12**, 723–726.

Grzybowska, D., 1989. Concentrations of ^{137}Cs and ^{90}Sr in marine fish from the southern Baltic Sea. Acta Hydrobiol. **31**, 139–147.

Guary, J.-C., and S.W. Fowler, 1983. Experimental studies on the biokinetics of plutonium and americium in the cephalopod *Octopus vulgaris*, in Radioactivity in the Sea (International Atomic Energy Agency, Vienna), **No. 72**, 9 pp. (Reprinted from *Mar. Ecol. Progr. Ser.* **7**, 327–335 (1982)).

Guns, M., P. Van Hoeyweghen, W. Vyncke and H. Hillewaert, 1999. Trace metals in selected benthic invertebrates from Belgian coastal waters (1981–1996). Mar. Pollut. Bull. **38**, 1184–1193.

Habermehl, M., A. Jarre and D. Adelung, 1990. Field and laboratory studies on the vertical migration of *Diastylis rathkei* (Crustacea: Cumacea) in Kiel Bay, Western Baltic. Meeresforsch. **32**, 295–305.

Halden, N.M., J.A. Babaluk, J.L. Campbell and W.J. Teesdale, 1995. Scanning microprobe analysis of strontium in an arctic char, *Salvelinus alpinus*, otolith: implications for the interpretation of anadromy. Environ. Biol. Fish. **43**, 333–339.

Hamilton, E. I, 1980. Concentration and distribution of uranium in *Mytilus edulis* and associated materials. Mar. Ecol. Prog. Ser. **2**, 61–73.

Hamilton, E. I, 1991. Metals in *Mytilus edulis* from the northern coast of Portugal. Mar. Pollut. Bull. **22**, 249–253.

Hamilton, E.I., and R.J. Clifton, 1980. Concentration and distribution of the transuranium radionuclides $^{239+240}Pu$, ^{238}Pu and ^{241}Am in *Mytilus edulis*, *Fucus vesiculosus* and surface sediment of Esk estuary (1980). Mar. Ecol. Prog. Ser. **3**, 267–277.

Hansen, C.T., C.O. Nielsen, R. Dietz and M.M. Hansen, 1990. Zinc, cadmium, mercury and selenium in minke whales, belugas and narwhals from West Greenland. Polar. Biol. **10**, 529–539.

Harada, M., 1995. Minamata disease: methylmercury poisoning in Japan caused by environmental pollution. Crit. Rev. Toxicol. **25**, 1–24.

Harms, U., 1975. The levels of heavy metals (Mn, Fe, Co, Ni, Cu, Zn, Cd, Pb, Hg) in fish from onshore and offshore waters of the German Bight. Z. Lebensm. Unters. Forsh. **157**, 125–132.

Harms, U., 1996. Biota. Third Periodic Assessment of the State of the Marine Environment of the Baltic Sea, 1989–93; Background Document. Baltic Sea Environment Proc. (Baltic Marine Environment Protection Commission, Helsinki), HELCOM **No. 64B**, pp. 149–153.

Harms, U., and G. Kanisch, 2000. Intra-individual variability of levels of lead detected in the liver of Baltic cod (*Gadus morhua*). Mar. Pollut. Bull. **40**, 710–712.

Harms, U., H.E. Drescher and E. Huschenbeth, 1977/1978. Further data on heavy metals and organochlorines in marine mammals from German coastal waters. Meeresforschung. **26**, 153–161.

Haug, A., S. Melsom and S. Omang, 1974. Estimation of heavy metal pollution in two Norwegian fjord areas by analysis of the brown alga *Ascophyllum nodosum*. Environ. Pollut. **7**, 179–192.

Hägerhäll, B., 1973. Marine botanical-hydrographical trace element studies in the Öresund area. Botanica Mar. **16**, 53–64.

Häkkilä, K., 1980. Pohjasedimenttien ja pohjaelinten raskasmetalleista Porin edustan merialueella. Nat. Board of Waters, Rep. **190**, 1–39 (in Finnish in an English abstract).

Härdstedt-Roméo, M., 1982. Some aspects of the chemical composition of plankton from the North-Western Mediterranean Sea. Mar. Biol. **70**, 229–236.

Härdstedt-Roméo, M., and F. Laumond, 1980. Zinc, copper and cadmium in zooplankton from the N.W. Mediterranean. Mar. Pollut. Bull. **11**, 133–138.

Heide-Jørgensen M.P., and C. Lockyer, 1999. The biology of harbour porpoises from West Greenland, in: Proceedings of the Twelfth Annual Conference of the European Cetaceans Society Monaco 20–24.01.1998 (eds. P.G.H. Evans and E.C.M. Parsons), 1998 (Published in Valencia, February 1999), 111.

HELCOM, 1990. Second Periodic Assessment of the State of the Marine Environment of the Baltic Sea, 1984–1988; Background Document. Baltic Sea Environment Proceedings (Baltic Marine Environment Protection Commission, HELCOM), **No. 35B**.

HELCOM, 1993. First Assessment of the State of the coastal waters of the Baltic Sea. Baltic Sea Environment Proceedings (Baltic Marine Environment Protection Commission, HELCOM), **No. 54**.

HELCOM, 1995. Radioactivity in the Baltic Sea 1984–1991. Baltic Sea Environment Proceedings (Baltic Marine Environment Protection Commission, HELCOM) **No. 61**.

Helle, E., 1981. Reproductive trends and occurrence of organochlorines and heavy metals in the Baltic seal populations. ICES (CM papers and reports), **E:37**, 13 pp. (Mimeo.).

Hellou, J., W.G. Warren, J.F. Payne, S. Belkhode and P. Lobel, 1992. Heavy metals and other elements in three tissues of cod, *Gadus morhua* from the northwest Atlantic. Mar. Pollut. Bull. **24**, 452–458.

Hellou, J., V. Zitko, J. Friel and T. Alkanani, 1996. Distribution of elements in tissues of yellowtail flounder *Pleuronectes ferruginea*. Sci. Total Environ. **181**, 137–146.

Henning H.F.-K.O., G.A. Eagle, C.D. McQuaid and L.H. Rickett, 1985. Metal concentrations in Antarctic zooplankton species, in: Antarctic Nutrient Cycles and Food Webs, eds. W.R. Siedfrid, P.R. Condy and R.M. Laws (Springer-Verlag Berlin Heidelberg), pp. 656–661.

Henriksson, K., E. Karppanen and M. Helminen, 1966. High residues of mercury in Finnish White-tailed eagle. Ornis Fennica **43**, 38–45.

Heppleston, P.B., and M.C. French, 1973. Mercury and other metals in British seals. Nature (Lond.) **243**, 302–304.

Hernández, F., R. Serrano, A.F. Roig-Navarro, Y. Martinez-Bravo and F.J. López, 2000. Persistent organochlorines and organophosphorus compounds and heavy elements in common whale (*Balaenoptera physalus*) from the Western Mediterranean Sea. Mar. Pollut. Bull. **40**, 426–433.

Hernroth, L., and H. Ackefors, 1979. The zooplankton of Baltic proper. Fishery Board of Sweden. Rep. **No. 2**.

Herrmann, J., 2000. Levels of radioactivity, in: The Radiological Exposure of the Population of the European Community to Radioactivity in the Baltic Sea. Marina-Balt Project, ed. S.P. Nielsen. Proceedings of a Seminar held at Hasseludden Conference Centre, Stockholm, 9–11 June 1998, European Commission, Directorate-General Environment, EUR 19200 EN (European Communities, 2000, Belgium), pp. 77–129.

Herva, E., and E. Häsänen, 1972. Mercury in seals of the Gulf of Bothnia. Suomen Eläinlääkärilehti **78**, 445–448 (in Finnish).

Heyer, K., M. Engel, U.H. Brockmann, H.-J. Rick, C.-D. Dürselen, H. Hühnerfuss, U. Kammann, H. Steinhart, W. Kienz, M. Krause, L. Karbe, A. Faubel and S. Regier, 1994. Local studies in the German Bight during Winter/Spring 1988/89, in: Circulation and Contaminant Fluxes in the North Sea, ed. J. Sündermann (Springer, Berlin), pp. 190–249.

Hill, C.R. and R. Elmgren, 1992. Predation by the isopod *Saduria entomon* on the amphipods *Monoporeia affinis* and *Pontoporeia femorata*: experiments on prey vulnerability. Oecologia **91**, 153–156.

Hill, C., R. Elmgren and P. Abrams, 1990. Predation by the polychaete *Harmothoe sarsi* on different size classes of the amphipod *Pontoporeia affinis*. Proc. 24th Europ. Mar. Biol. Symp. 468–477.

Hockett, D., P. Ingram and A. LeFurgey, 1997. Strontium and manganese uptake in the barnacle shell: electron probe microanalysis imaging to attain fine temporal resolution of biomineralization activity. Mar. Environ. Res. **43**, 131–143.

Hoek van den, C.D.G. Mann and H.M. Jahns, 1995. Algae. An Introduction to Phycology (Cambridge University Press), 627 pp.

Holm, E., 1994. Polonium-210 and radiocaesium in muscle tissue of fish from different Nordic marine areas, in: Nordic Radioecology, The Transfer of Radionuclides through Nordic Ecosystems to Man, ed. H. Dahlgaard (Elsevier Science, Amsterdam), pp. 119–126.

Holm, E., 1995. Plutonium in the Baltic Sea. Appl. Radiat. Isot. **46**, 1225–1229.

Holm, E., S. Duniec, L. Hallstadius, S. Mattsson, J. Rioseco and C. Samuelsson, 1986. Releases distribution and pathways for radionuclides in the Baltic Sea, in: Study of Radioactive materials in the Baltic Sea. (International Atomic Energy Agency, Vienna). Report (IAEA-TECDOC-362) of the Final Research Co-ordination Meeting on the Study of Radioactive Materials in the Baltic Sea organized by the IAEA and held in Helsinki, Finland 24–28 September, 1984, pp.152–161.

Holm, E., A. Aarkrog, S. Ballestra and J.J. Lopez, 1989. Fallout deposition of actinides in Monaco and Denmark after the Chernobyl accident, in: Proc. of the Impact of Nuclear Origin Accidents on the Environment (Centre d'Etudes Nucleaires de Cadarache, France), 22 pp.

Holsbeek, L., U. Siebert and C.R. Joiris, 1998. Heavy metals in dolphins stranded on the French Atlantic coast. Sci. Total Environ. **217**, 241–249.

Holsbeek, L., C.R. Joiris, W. Debacker, I.B. Ali, P. Roose, J.-P. Nellissen, S. Gobert, J.-M. Bouquegneau and M. Bossicart, 1999. Heavy metals, organochlorines and polycyclic aromatic hydrocarbons in sperm whales stranded in the southern North Sea during the 1994/1995 winter. Mar. Pollut. Bull. **38**, 304–313.

Honda, K, and R. Tatsukawa, 1983. Distribution of cadmium and zinc in tissues and organs, and their age-related changes in striped dolphins, *Stenella coeruleoalba*. Arch. Environ. Contam. Toxicol. **12**, 543–550.

Honda, K., R. Tatsukawa and T. Fujiyama, 1982. Distribution characteristics of heavy metals in the organs and tissues of striped dolphin, *Stenella coeruleoalba*. Agric. Biol. Chem. **46**, 3011–3021.

Honda, K., R. Tatsukawa, K. Itano, N. Miyazaki and T. Fujiyama, 1983a. Heavy metal concentrations in muscle, liver and kidney tissue of striped dolphin, *Stenella coeruleoalba*, and their variations with body length, weight, age and sex. Agric. Biol. Chem. **47**, 1219–1228.

Honda, K., M. Sahrul, H. Hidaka and R. Tatsukawa, 1983b. Organ and tissue distribution of heavy metals, their growth-related changes in Antarctic fish, *Pagothenia borchgrevinki*. Agric. Biol. Chem. **47**, 2521–2532.

Honda, K., Y. Fujise, K. Itano and R. Tatsukawa, 1984a. Composition of chemical components in bone of striped dolphin, *Stenella coeruleoalba*: Distribution characteristics of heavy metals in various bones. Agric. Biol. Chem. **48**, 677–683.

Honda, K., Y. Fujise, R. Tatsukawa and N. Miyazaki, 1984b. Composition of chemical components in bone of striped dolphin, *Stenella coeruleoalba*: Distribution characteristics of major inorganic and organic components in various bones, and their age-related changes. Agric. Biol. Chem. **48**, 409–418.

Honda, K., B.Y. Min and R. Tatsukawa, 1985a. Heavy metal distribution in organs and tissues of eastern great white egret *Egretta alba modesta*. Bull. Environ. Contam. Toxicol. **35**, 781–789.

Honda, K., B.Y. Min and R. Tatsukawa, 1985b. Distribution of heavy metals and their age-related changes in the eastern great white egret, *Egretta alba modesta*. Arch. Environ. Contam. Toxicol. **15**, 185–197.

Honda, K., Y. Yamamoto, H. Hidaka and R. Tatsukawa, 1986a. Heavy metal accumulations in Adelie penguins, *Pygoscelis adeliae*, and their variations with the reproductive processes. Mem. Natl Inst. Polar Res., Spec. Issue **40**, 443–453.

Honda, K., Y. Fujise, R. Tatsukawa, K. Itano, and N. Miyazaki, 1986b. Age-related accumulation of heavy metals in bone of the striped dolphin, *Stenella coeruleoalba*. Mar. Environ. Res. **20**, 143–160.

Honda, K., Y. Yamamoto, H. Kato and R. Tatsukawa, 1987. Heavy metal accumulations and their recent changes in southern minke whales *Balaenoptera acutorostrata*. Arch. Environ. Contam. Toxicol. **16**, 209–216.

Honda, K., J.E. Marcovecchio, S. Kan, R. Tatsukawa and H. Ogi, 1990. Metal concentrations in pelagic seabirds from the North Pacific Ocean. Arch. Environ. Contam. Toxicol. **19**, 704–711.

Hornung, H., D. Raviv and B.S. Krumgalz, 1981. The occurrence of mercury in marine algae and some gastropod molluscs of the Mediterranean shoreline of Israel. Mar. Pollut. Bull. **12**, 397–340.

Horowitz, A., and B.J. Presley, 1977. Trace metal concentrations and partitioning in zooplankton, neuston, and benthos from the south Texas Outer Continental Shelf. Arch. Environ. Contam. Toxicol. **5**, 241–255.

Hou, X., H. Dahlgaard and S.P. Nielsen , 2000. Iodine-129 time series in Danish, Norwegian and Northwest Greenland coast and the Baltic Sea by seaweed. Estuar. Coast. Shelf Sci. **51**, 571–584.

Howard, L.S., and B.E. Brown, 1983. Natural variation in tissue concentration of copper, zinc and iron in the polychaete *Nereis diversicolor.* Mar. Biol. **78**, 87–97.

Hudec, K., 1993. Przewodnik Ptaki (A Guide to Birds). ©Copyright for the Polish edition by MULTICO sp. z.o.o. joint venture. Warszawa, 295 pp.

Hung, T.-Ch., Ch.-Y. Kuo and M.H. Chen, 1981. Mussel watch in Taiwan, Republic of China. (1) Bio-accumulative factors of heavy metals. Acta Oceanographica Taiwanica **12**, 67–83.

Hyvärinen, H., and T. Sipilä, 1984. Heavy metals and high pup mortality in the Saimaa ringed seal population in eastern Finland. Mar. Pollut. Bull. **15**, 335–337.

Hyvärinen, H., T. Sipilä, M. Kunnasranta and J.T. Koskela, 1998. Mercury pollution and the Saimaa ringed seal (*Phoca hispida saimensis*). Mar. Pollut. Bull. **36**, 76–81.

ICES, 1977. Coop. Res. Rep. 63.

Ikuta, K., 1986a. Metal concentrations in byssuses and soft bodies of bivalves. Bull. Fac. Agric., Miyazaki Univ. **33**, 255–264.

Ikuta, K., 1986b. Correlations between ratios of metal concentrations in byssuses to those in soft bodies and metal concentrations in soft bodies of bivalves. Bull. Fac. Agric., Miyazaki Univ. **33**, 265–273.

Ikuta, K., 1988. Inherent differences in some heavy metal contents among ostreids, mytilids and acmaeids. Nippon Suisan Gakk. **54**, 811–816.

Ikuta, K., and P. Szefer, 2000. Distribution of heavy metals in soft tissue and shell of *Astarte borealis* and *Macoma balthica* from Slupsk Furrow, a southern Baltic (in preparation).

Ilus, E., J. Ojala, K.-L. Sjöblom and K. Toumainen, 1981. *Fucus vesiculosus* as bioindicator of radioactivity in Finnish coastal waters. 1. Gulf of Finland. ISBN 951-46-5408-0 (Inst. Rad. Prot.).

Ilus, E., R. Saxén and T.K. Taipale, 1986. Accumulation of certain long-lived radionuclides by fish, in: Study of Radioactive materials in the Baltic Sea. (International Atomic Energy Agency, Vienna). Report (IAEA-TECDOC-362) of the Final Research Co-ordination Meeting on the Study of Radioactive Materials in the Baltic Sea organized by the IAEA and held in Helsinki, Finland 24–28 September, 1984, pp. 63–78.

Ilus, E., K.-L. Sjöblom, R. Saxén, H. Aaltonen and K. Taipake, 1987. Finnish studies in the Baltic Sea after the Chernobyl accident in 1986. Suppl. 11 to Annual Rep. STUK-A55. STUK-A66, April 1987 (Finnish Centre for Radiation and Nuclear Safety, Helsinki, Finland) 40 pp.

Ilus, E., S. Klemola, K.-L. Sjöblom and T. Ikäheimonen, 1988. Radioactivity of *Fucus vesiculosus* along the Finnish coast in 1987. Suppl. 9 to Annual Rep. 1987 (STUK-A74). STUK-A83, May 1988 (Finnish Centre for Radiation and Nuclear Safety, Helsinki, Finland) 36 pp.

Ilus, E., K.-L. Sjöblom, S. Klemola and H. Arvela, 1992. Monitoring of radionuclides in the environs of Finnish nuclear power plants in 1989–1990. Suppl. 9 to Annual Rep. STUK-A89). STUK-A102, January 1992 (Finnish Centre for Radiation and Nuclear Safety, Helsinki, Finland) 91 pp.

Ilus, E., K.-L. Sjöblom, T.K. Ikäheimonen, R. Saxén and S. Klemola, 1993. Monitoring of radionuclides in the Baltic Sea in 1989–1990. Suppl. 10 to Annual Rep. STUK-A89. STUK-A103, February 1993 (Finnish Centre for Radiation and Nuclear Safety, Helsinki, Finland) 35 pp.

Ioffe, B.I., 1987. Klass Priapulidy (Priapulida), in: Zhizn' Zhivotnyh – Prostiejshie – Kishechnopolostnyie – Chervy (The Life of Animals – Worms), ed. Yu.I. Polyanskij (Prosveshchenie, Izdatelstvo, Moscow), **Vol. 1**, 350–355 (in Russian).

Ireland, M.P., 1974. Variations in the zinc, copper, manganese and lead content of *Balanus balanoides* in Cadigan Bay, Wales. Environ. Pollution **7**, 65–75.

Ismail, A., N.R. Jusoh and I.A. Ghani, 1995. Trace metal concentrations in marine prawns off the Malaysian coast. Mar. Pollut. Bull. **31**, 108–110.

Iwata, H., S. Tanabe, N. Miyazaki and R. Tatasukawa, 1994a. Detection of butyltin compound residues in the blubber of marine mammals. Mar. Pollut. Bull. **28**, 607–612.

Iwata, H., S. Tanabe, T. Mizuno and R. Tatasukawa, 1994b. High accumulation of toxic butyltins in marine mammals from Japanese coastal waters. Environ. Sci, Techn. **29**, 2959–2962.

Jacobsen, J.A., and G. Asmund, 2000. TBT in the marine sediments and blue mussels (*Mytilus edulis*) from central-west Greenland. Sci. Total Environ. **245**, 131–136.

Jacobsen, A., E. Neumann and G. Thoresson, 1986. The viviparous blenny as an indicator of environmental effects of harmful substances. Ambio **15**, 236–238.

Jagnow, B., and F. Gosselck, 1987. Bestimmundsschlüssel für die Gehäuseschnecken und Muschekln der Ostsee. Mitt. Zool. Mus. Berl. **63**, 191–268.

Jahnke, A., T. Shimmen, H. Koyama-Ito and T. Yamazaki, 1981. PIXE trace element analysis of an aquatic plant as indicator of heavy metal pollution. Chemosphere **10**, 303–312.

Jankovski, H., H. Kukk and Y. Volozh, 1988. Accumulation of heavy metals in the Baltic Sea algae. Proc. Acad. Sciences, Biology (Est. SSR) **37**, 233–241.

Jansson, B.O., 1972. Ecosystem approach to the Baltic problem. Bull. Ecol. Res. Comm. NFR **16**, 82.

Jaworowski, Z., A. Barański, D. Grzybowska, A. Pawlak, J. Tomczak, A. Pietruszewski, M. Suplińska, D. Skiba and T. Wardaszko, 1986. Monitoring of the radioactive materials in the Baltic Sea, in: Study of Radioactive materials in the Baltic Sea. (International Atomic Energy Agency, Vienna). Report (IAEA-TECDOC-362) of the Final Research Co-ordination Meeting on the Study of Radioactive Materials in the Baltic Sea organized by the IAEA and held in Helsinki, Finland 24–28 September, 1984, pp. 141–151.

Jayasekera, R., and M. Rossbach, 1996. Use of seaweeds for monitoring trace elements in coastal waters. Environ. Geochem. Health **18**, 63–68.

Jażdżewski, K., 1975. Morfologia, taksonomia i występowanie w Polsce kiełży z rodzajów Gammarus Fabr. I Chaetogammarus Mart. (Crustacea, Amphipoda). Wyd. UŁ. Łódź, 185 pp. (in Polish).

Jażdżewski, K. and A. Konopacka, 1995. Katalog fauny Polski. Część XIII, Pancerzowce (Malacostraca) (Wyd. Muz. Inst. Zoologii PAN, Warszawa) **Vol. 1**, 165 pp. (in Polish).

Jefferson, T.A., S. Leatherwood and M.A. Webber, 1993. FAO Species Identification Guide. Marine Mammals of the World. (United Environment Programme, Food and Agriculture Organization of the United Nations, Rome), 320 pp. + 587 Figs.

Jeng, M.-S., W.-L. Jeng, T.-Ch. Hung, Ch.-Y. Yeh, R.-J. Tseng, P.-J. Meng and B.-Ch. Han, 2000. Mussel Watch: a review of Cu and other metals in various marine organisms in Taiwan, 1991–98. Environ. Pollut. **110**, 207–215.

Jenner, H.A., and T. Bowmer, 1990. The accumulation of metals and their toxicity in the marine intertidal invertebrates *Cerastoderma edule*, *Macoma balthica*, *Arenicola marina* exposed to pulverised fuel ash mesocosms. Environ. Pollut. **66**, 139–156.

Jensen, A., and Z. Cheng, 1987. Statistical analysis of trend monitoring data on heavy metals in flounder (*Platichthys flesus*). Mar. Pollut. Bull. **18**, 230–238.

Jensen, S., A.G. Johnels, M. Olsson and T. Westermark, 1972. The avifauna of Sweden as indicators of environmental contamination with mercury and chlorinated hydrocarbons. Proc. XVth Intern. Ornithol. Congress (Leiden, E.J. Brill), 455–465.

Jewett, S.C., and A.S. Naidu, 2000. Assessment of heavy metals in red king crabs following offshore placer gold mining. Mar. Pollut. Bull. **40**, 478–490.

Johnels, A.G., M. Olsson and T. Westermark, 1968. *Esox lucius* and some other organisms as indicators of mercury contamination in Swedish lakes and rivers. Bull. Off. Int. Epiz. **69**, 1439–1452.

Joiris, C.R., and M. Bossicart, 1989. Mercury contamination of Belgian and Peruvian Cetacea. ICES 1989, C.M. 1989/N:14 (Abstract), 3 pp.

Joiris, C.R., L. Holsbeek, J.M. Bouquegneau and M. Bossicart, 1991. Mercury contamination of the harbour porpoise (*Phocoena phocoena*) and other cetaceans from the North Sea and the Kattegat. Water Air Soil Pollut. **56**, 283–293.

Joiris, C.R., I.B. Ali, L. H. Holsbeek, M. Bossicart and G. Tapia, 1995. Total and organic mercury in Barents Sea pelagic fish. Bull. Environ. Contam. Toxicol. **55**, 674–681.

Joiris, C.R., I.B. Ali, L. H. Holsbeek, M. Kanuya-Kinokti and Y. Telele-Michael, 1997. Total and organic mercury in Greenland and Barents seas demersal fish. Bull. Environ. Contam. Toxicol. **58**, 101–107.

Joiris, C.R., L. Holsbeek and N.L. Moatemri, 1999. Total and methylmercury in sardines *Sardinella aurita* and *Sardina pilchardus* from Tunisia. Mar. Pollut. Bull. **38**, 188–192.

Joiris, C.R., K. H. Das and L. Holsbeek, 2000a. Mercury accumulation and speciation in marine fish from Bangladesh. Mar. Pollut. Bull. **40**, 454–457.

Joiris, C.R., L. Holsbeek and F.A. Otchere, 2000b. Mercury in the bivalves *Crassostrea tulipa* and *Perna perna* from Ghana. Mar. Pollut. Bull. **40**, 457–460.

Jonsson, L., 1992. Birds of Europe with North Africa and the Middle East (Christopher Helm Limited) 559 pp.

Jothy, A.A., E. Huschenbeth and U. Harms, 1983. On the detection of heavy metals, organochlorine pesticides and polychlorinated biphenyls in fish and shellfish from the coastal waters od Peninsular Malaysia. Arch. Fisch. Wiss. **33**, 161–206.

Julshamn, K., 1981a. Studies on major and minor elements in molluscs in western Norway. I. Geographical variations in the contents of 10 elements in oyster (*Ostrea edulis*), common mussel (*Mytilus edulis*) and brown seaweed (*Ascophyllum nodosum*) from three oyster farms. Fisk. Dir. Skr., Ser. Ernæring **1**, 161–182.

Julshamn, K., 1981b. Studies on major and minor elements in molluscs in western Norway. III. Effects of size and age on the contents of 10 elements in oyster (*Ostrea edulis*) taken from unpolluted waters. Fisk. Dir. Skr., Ser. Ernæring **1**, 199–214.

Julshamn, K., 1981c. Studies on major and minor elements in molluscs in western Norway. V. Protein binding of zinc, cadmium and copper in tissues of oyster (*Ostrea edulis*), common mussel (*Mytilus edulis*) and horse mussel (*Modiolus modiolus*) taken from unpolluted waters. Fisk. Dir. Skr., Ser. Ernæring **1**, 235–245.

Julshamn, K., 1981d. Studies on major and minor elements in molluscs in western Norway. VI. Accumulation and depletion of cadmium and lead and 5 further elements in tissues of oyster (*Ostrea edulis*), and common mussel (*Mytilus edulis*) by transfer between waters of highly different heavy metal loads. Fisk. Dir. Skr., Ser. Ernæring **1**, 247–265.

Julshamn, K., 1981e. Studies on major and minor elements in molluscs in western Norway. VII. The contents of 12 elements, including copper, zinc, cadmium and lead in common mussel (*Mytilus edulis*) and brown seaweed (*Ascophyllum nodosum*) relative to the distance from the industrial sites in Sorfjorden, inner Hardangerfjord. Fisk. Dir. Skr., Ser. Ernæring **1**, 267–287.

Julshamn, K., and O. Grahl-Nielsen, 1996. Distribution of trace elements from industrial discharges in the Hardangerfjord, Norway: A mulivariate data analysis of saithe, flounder and blue mussel as sentinel organisms. Mar. Pollut. Bull. **32**, 564–571.

Julshamn, K., and O. Grahl-Nielsen, 2000. Trace element levels in harp seal (*Pagophilus groenlandicus*) nd hooded seal (*Cystophora cristata*) from the Greenland Sea. A multivariate approach. Sci. Total Environ. **250**, 123–133.

Julshamn, K., J. Haugsnes and J. Eriksen, 1978. Major and minor element (mineral) levels in products and offal from the fishing industries. Fisk. Dir. Skr., Ser. Ernæring **1**, 137–144.

Kalisińska, E., and U. Szuberla, 1996. Heavy metals in the brain of long-tailed duck (*Clangula hyemalis*) wintering in the Pomeranian Bay, Poland. Biol. Trace Elem. Res. **55**, 191–197.

Kan-atireklap, S., S. Tanabe, J. Sanguansin, M.S. Tabucanon and M. Hungspreugs, 1997. Contamination by butyltin compounds and organochlorine residues in green mussel (*Perna viridis*, L.) from Thailand coastal waters. Environ. Pollut. **97**, 79–89.

Kan-atireklap, S., N.T.H. Yen, S. Tanabe and A.N. Subramanian, 1998. Butyltin compounds and organochlorine residues in green mussel (*Perna viridis* L.) from India. Toxicol. Environ. Chem. **67**, 409–424.

Kangas, P., and H. Autio, 1986. Macroalgae as indicators of heavy metal pollution. Publication of the Water Research Institute, National Board of Waters, Finland **No. 68**, 183–189.

Kangas, P., H. Autio, G. Hällfors, H. Luther, Å. Niemi and H. Salemaa, 1982. A general model of the decline of *Fucus vesiculosus* at Tvärminne, south coast of Finland in 1977–81. Acta Bot. Fennica **118**, 1–27.

Kanisch, G., G. Neumann and E. Ilus, 1995. Radionuclides in biota, in: Radioactivity in the Baltic Sea 1984–1991. Baltic Sea Environment Proceedings HELCOM **No. 61**, 93–120.

Kannan, K., and J. Falandysz, 1997a. Butyltin residues in sediment, fish, fish-eating birds, harbour porpoise and human tissues from the Polish coast of the Baltic Sea. Mar. Pollut. Bull. **34**, 203–207.

Kannan, K., and J. Falandysz, 1997b. Response to the comment on: Butyltin residues in sediment, fish, fish-eating birds, harbour porpoise and human tissues from the Polish coast of the Baltic Sea. Mar. Pollut. Bull. **38**, 61–63.

Kannan, K., and S. Tanabe, 1997. Response to the comment on: Response to comment on "Elevated accumulation of tributyltin and its breakdown products in bottlenose dolphins (*Tursiops truncatus*) found stranded along the U.S. Atlantic and Gulf coasts". Environ. Sci. Techn. **31**, 3035–3036.

Kannan, K., J. Falandysz, S. Tanabe and R. Tatsukawa, 1993. Persistent organochlorines in harbour porpoises from Puck Bay, Poland. Mar. Pollut. Bull. **26**, 162–165.

Kannan, K., S. Corsolini, S. Focardi, S. Tanabe, and R. Tatsukawa, 1996. Accumulation pattern of butyltin compounds in dolphin, tuna, and shark collected from Italian coastal waters. Arch. Environ. Contam. Toxicol. **31**, 19–23.

Kannan, K., K. Senthilkumar and R.K. Sinha, 1997a. Sources and accumulation of butyltin compounds in Ganges river dolphin *Platanista gangetica*. Appl. Organometal. Chem. **11**, 223–230.

Kannan, K., K. Senthilkumar, B.G. Loganathan, S. Takahashi, D.K. Odell and S. Tanabe, 1997b. Elevated accumulation of tributyltin and its breakdown products in bottlenose dolphins (*Tursiops truncatus*) found stranded along the U.S. Atlantic and Gulf coasts. Environ. Sci. Techn. **31**, 296–301.

Karbe, L., Ch. Schnier and H.O. Siewers, 1977. Trace elements in mussels (*Mytilus edulis*) from coastal areas of the North Sea and the Baltic. Multielement analyses using instrumental neutron activation analysis. J. Radioanal. Chem. **37**, 927–943.

Kari, T., and P. Kauranen, 1978. Mercury and selenium contents of seals from fresh and brackish waters in Finland. Bull. Environ. Contam. Toxicol. **19**, 273–280.

Katsuki, Y, H. Ishii, A. Itoh, Y. Naoi and Y. Kimura, 1980. Study on amounts of trace elements in marine fishes (V) Distribution of heavy metals and arsenic in flatfish, graymullet and gizzard-shad tissue. Ann. Rep. Tokyo Metr. Res. Lab. P.H. **31**, 174–179.

Khan, A.H., M. Ali, S.K. Biaswas and D.A. Hadi, 1987. Trace elements in marine fish from the Bay of Bengal. Sci. Total Environ. **61**, 121–130.

Kauppinen, V., 1980. Vaasan edustan merialueen kalataloustarkkailu– ja pohjasedimenttitutkimukset vuonna 1980. Pohjois-Suomen vesitutkimustoimisto, 1–42 (in Finnish).

Kemper, C., P. Gibbs, D. Obendorf, S. Marvanek and C. Lenghaus, 1994. A review of heavy metal and organochlorine levels in marine mammals in Australia. Sci. Total Environ. **154**, 129–139.

Khristoforova N.K., and N.N. Bogdanova, 1980. Mineral composition of seaweeds from coral islands of the Pacific Ocean as a function of environmental conditions. Mar. Ecol. Prog. Ser. **3**, 25–29.

Kim, E.Y., T. Murakami, K. Saeki and R. Tatsukawa, 1996a. Mercury levels and its chemical forms in tissues and organs of seabirds. Arch. Environ. Contam. Toxicol. **30**, 259–266.

Kim, E.Y., K. Saeki, S. Tanabe, H. Tanaka and R. Tatsukawa, 1996b. Specific accumulation of mercury and selenium in seabirds. Environ. Pollut. **94**, 261–265.

Kim, E.-Y., H. Ichihashi, K. Saeki, G. Atrashkevich, S. Tanabe and R. Tatsukawa, 1996c. Metal accumulation in tissues of seabirds from Chaun, Northeast Siberia, Russia. Environ. Pollut. **92**, 247–252.

Kim, G.B., J.S. Lee, S. Tanabe, H. Iwata, R. Tatsukawa and K. Shimazaki, 1996d. Specific accumulation and distribution of butyltin compounds in various organs and tissues of the Steller sea lion (*Eumetopias jubatus*): comparison with organochlorine accumulation pattern. Mar. Pollut. Bull. **32**, 558–563.

Kim, G.B., H. Nakata and S. Tanabe, 1998a. *In vitro* inhibition of hepatic cytochrome P450 and enzyme activity by butyltin compounds in marine mammals. Environ. Pollut. **99**, 255–261.

Kim, E.-Y., R. Goto, S. Tanabe, H. Tanaka and R. Tatsukawa, 1998b. Distribution of 14 elements in tissues and organs of oceanic seabirds. Arch. Environ. Contam. Toxicol. **35**, 638–645.

Kim, E.-Y., R. Goto, H. Iwata, Y. Masuda, S. Tanabe and S. Fujita, 1999. Preliminary survey of lead poisoning of Steller's sea eagle (*Haliaeetus pelagicus*) and white-tailed sea eagle (*Haliaeetus albicilla*) in Hokkaido, Japan. Environ. Toxicol. Chem. **18**, 448–451.

King, J.E., 1983. Seals of the World. British Museum (Natural History) (Oxford University Press), 240 pp.

Klekowski, E.J., S.A. Temple, A.M. Siung-Chang and K. Kumarsingh, 1999. An association of mangrove mutation, scarlet ibis, and mercury contamination in Trinidad, West Indies. Environ. Pollut. **105**, 185–189.

Klinowska, M., 1991. Dolphins, Porpoises and Whales of the World (The IUCN Red Data Book, IUCN, Gland, Switzerland and Cambridge), viii + 429 pp.

Klöckner, K., 1979. Uptake and accumulation of cadmium by *Ophryotrocha diadema* (Polychaeta). Mar. Ecol. Prog. Ser. **1**, 71–76.

Knauer, G.A., and J.H. Martin, 1972. Mercury in a marine pelagic food chain. Limnol. Oceanogr. **17**, 868–876.

Knauss, K., and T.-L. Ku, 1983. The chemical composition and decay-series radionuclide content of plankton from the East Pacific. Chem. Geol. **39**, 125–145.

Knutzen, J., and J. Skei, 1990. Quality criteria for micropollutants in water, sediments and organisms and preliminary proposals for classification of environmental quality. NIVA-Rep., l. nr. 2540 (NIVA, Oslo, Norway) pp.139.

Kobayashi, R., E. Hirata, K. Shiomi, H. Yamanaka and T. Kikuchi, 1979. Heavy metal contents in deep-sea fishes. Bull. Japan. Soc. Scient. Fish. **45**, 493–497.

Kock, G., M. Triendl and R. Hofer, 1996. Seasonal pattern of metal accumulation in Arctic char (*Salvelinus alpinus*) from an oligotrophic Alpine like related to temperture. Can. J. Fish. Aquat. Sci. **53**, 780–786.

Koeman, J.H., W.H.M. Peeters, C.H.M. Koudstaal-Hoi, P.S. Tijoe and J.J.M. de Goeij, 1973. Mercury-selenium correlations in marine mammals. Nature (Lond.) **245**, 285–286.

Koeman, J.H., W.S.M. van der Ven, J.J.M. de Goeij, P.S. Tijoe, and J.L. van Haaften, 1975. Mercury and selenium in marine mammals and birds. Sci. Total Environ. **3**, 279–287.

Koide, M., D.S. Lee and E.D. Goldberg, 1982. Metals and transuranic records in mussel shells, byssal threads and tissues. Estuar. Coast. Shelf Sci. **15**, 679–695.

Koranda, J.J., M. Stuart, S. Thompson and C. Conrado, 1979. Biogeochemical studies of wintering waterfowl in the Imperial and Sacramento Valleys. Lawrence Livermore Laboratory, Livermore, CA, UCID-18288, October, 1–119.

Kosta, L., V. Ravnik, A.R. Byrne, J. Štirn, M. Dermelj and P. Stegnar, 1978. Some trace elements in the waters, marine organisms and sediments of the Adriatic by neutron activation analysis. J. Radioanal. Chem. **44**, 317–332.

Kostriczkina, E.M., R.J. Line, L.L. Sidrevic and M.T. Witinja, 1980. The ecological complexes of zooplankton of the Baltic Sea. Ekosistemy Baltiki (Morski Instytut Rybacki, Gdynia) 157–177 (in Russian).

Koszteyn, J., 1982. Seasonal changes in the vertical distribution and composition of mesozooplankton in the Gulf of Gdańk. Int. Counc. Explor. Sea, C.M. **L57**, 1–14.

Koszteyn, J., 1983. Differences in the species composition of mesozooplanktonc communities in the southern Baltic in July 1981. Int. Counc. Explor. Sea, C.M. **L11**, 1–9.

Köhler, A., U. Harms and B. Lucas, 1986. Accumulation of organochlorines and mercury in flounder – an approach to pollution assessments. Helgoländer Meeresuntersuch. **40**, 431–440.

Köhn, J., and F. Gosselck, 1989. Bestimmungsschlüssel der Malacostraken der Ostsee. Mitt. Zool. Mus. Berl. **65**, 114 pp.

Kress, N., H. Hornung and B. Herut, 1998. Concentrations of Hg, Cd, Cu, Zn, Fe and Mn in deep sea benthic fauna from the southeastern Mediterranean Sea: a comparison styudy between fauna collected at a pristine area and at two waste disposal sites. Mar. Pollut. Bull. **36**, 911–921.

Kuehl, D.W., R. Haebler and C. Potter, 1994. Coplanar PCB and metal residues in dolphins from the U.S. Atlantic coast including bottlenose obtained during the 1987/88 mass mortality. Chemosphere **28**, 1245–1253.

Kulikova, I. Z. Seisuma and M. Legzdina, 1985. Heavy metals in marine organisms, in: Heavy Metals in Water Organisms, ed. J. Salánki (Akadémiai Kiadó Budapest). Symposia Biologia Hungarica **29**, 141–151.

Kureishy, T.W., S. Sanzgiry, M.D. George and A. Braganca, 1983. Mercury, cadmium and lead in different tissues of fishes and in zooplankton from the Andaman Sea. Ind. J. Mar. Sci. **12**, 60–63.

Kuźma, W., 1971. Oznaczanie zawartości żelaza, miedzi, cynku, manganu, kobaltu i niklu w mięsie ryb (Determination of concentration of iron, copper, zinc, manganese, cobalt and nickel in fish muscle). Prace MIR **XVIB**, 205–219 (in Polish).

Labourg, P.J., and G. Lasserre, 1980. Dynamique des populations de *Cerastoderma glaucum* dans une lagune aurenagee de la region d'Arcachon. Mar. Biol. **60**, 147–157.

Lande, E., 1977. Heavy metal pollution in Trondheimsfjorden, Norway, and the recorded effects on the fauna and flora. Environ. Pollut. **12**, 187–198.

Langston, W.J., 1980. Arsenic in U.K. estuarine sediments and its availability to benthic organisms. J. Mar. Biol. Ass. UK **60**, 868–881.

Langston, W.J., 1986. Metals in sediments and benthic organisms in the Mersey estuary. Estuar. Coast. Shelf Sci. **23**, 239–261.

Lauenstein, G.G., and S.S. Dolvin, 1992. Mollusk monitoring of United States coastal and estuarine environments. Analusis Magazine **20**, M23-M26.

Lauenstein, G.G., A. Robertson and T.P. O'Connor, 1990. Comparison of trace metal data in mussel and oysters from a Mussel Watch Program of the 1970s with those from a 1980s Program. Mar. Pollut. Bull. **21**, 440–447.

Law, R., 1996. Metals in marine mammals, in: Environmental Contaminants in Wildlife, eds. N. Beyer, G. Heinz, and A. Redmond-Nordwood. SETAC Special Publications Series (CRC Press, Boca Raton, FL, USA), pp. 357–376.

Law, R.J., C.F. Fileman, A.D. Hopkins, J.R. Barker, J. Harwood, D.B. Jackson, S. Kennedy, S.R. Martin and R.J. Morris, 1991. Concentrations of trace metals in the livers of marine mammals (seals, porpoises and dolphins) from waters around the British Isles. Mar. Pollut. Bull. **22**, 183–191.

Law, R.J., B.R. Jones, J.R. Baker, S. Kennedy, R. Milne and R.J. Morris, 1992. Trace metals in the livers of marine mammals from the Welsh coast and Irish Sea. Mar. Pollut. Bull. **24**, 396–304.

Law, R.J., R.L. Stringer, C.R. Allchin and B.R. Jones, 1996. Metals and organochlorines in sperm whales (*Physeter macrocephalus*) stranded around the North Sea during the 1994/1995 winter. Mar. Pollut. Bull. **32**, 72–77.

Law, R.J., C.R. Allchin, B.R. Jones, P.D. Jepson, J.R. Baker and C.J.H. Spurrier, 1997a. Metals and organochlorines in tissues of a Blainville's beaked whale (*Mesoplodon densirostris*) and a killer whale (*Orcinus orca*) stranded in the United Kingdom. Mar. Pollut. Bull. **34**, 208–212.

Law, R.J., R.J. Morris, C.R. Allchin and B.R. Jones, 1997b. Metals and chlorobiphenyls in tissues of sperm whales (*Physeter macrocephalus*) and other cetacean species exploiting similar diets. Bull. Kon. Belg. Inst. Natuurwet. **67**, 79–89.

Law, R.J., S.J. Blake, B.R. Jones and E. Rogan, 1998. Organotin compounds in liver tissues of harbour porpoises (*Phocoena phocoena*) and grey seals (*Halichoerus grypus*) from the coastal waters of England and Wales. Mar. Pollut. Bull. **36**, 241–247.

Law, R.J., S.J. Blake and C.J.H. Spurrier, 1999. Butyltin compounds in liver tissues of pelagic cetaceans stranded on the coasts of England and Wales. Mar. Pollut. Bull. **38**, 1258–1261.

Lazarev, L.N., Yu.V. Kuznetsov, L.I. Gedeonov, D.I. Gusev, V.V. Anisimov, V.A. Blinov, L.M. Ivanova, T.E. Orlova, A.V. Stepanov, N.A. Tishkova, V.P. Tishkov, V.M. Flegontov, S.M. Vakulovskiy, A.I. Nikitin, I.Yu. Katrich, V.B. Chumichev, O.A. Pavlovskiy and V.D. Stepanova, 1986. Investigation of the radioactive contamination of the Baltic Sea in 1978–83, in: Study of Radioactive materials in the Baltic Sea. (International Atomic Energy Agency, Vienna). Report (IAEA-TECDOC-362) of the Final Research Co-ordination Meeting on the Study of Radioactive Materials in the Baltic Sea organized by the IAEA and held in Helsinki, Finland 24–28 September, 1984, pp. 162–175.

Lee, C.-L., 1996. Use of oyster, *Crassostrea gigas*, and ambient water to assess metal pollution status of the Charting coastal area, Taiwan, after the 1986 green oyster incident. Chemosphere **33**, 2505–2532.

Lee, D.P., K. Honda K., R. Tatsukawa and P.-O. Won, 1989. Distribution of residual level of mercury, cadmium and lead in Korean birds. Bull. Environ. Contam. Toxicol. **43**, 550–555.

Lee, R.E., 1999. Phycology (Cambridge University Press), pp. 614.

Lee, Ch.-B., B.S. Kim and P. Szefer, 2000. Distribution of mercury and other trace elements in soft tissue and byssus of *Mytilus* from Korean waters (in preparation).

Leland, H.V., S.N. Luoma, J.F. Elder and D.J. Wilkes, 1978. Heavy metals and related trace elements. J. Water Pollut. Control Fed., 1469–1514.

Levy, A., 1985. Une nouvelle conception de l'origiene enigmatique des *Cerastoderma glaucum* quaternaires du Sahara. C.R. Acad. Sc. Paris. **301**, 437–442.

Li Y.-H., 1981. Geochemical cycles of elements and human perturbation. Geochim. Cosmochm. Acta **45**, 2073–2084.
Lie, O., K. Julshamn, E. Lied and G. Lambertsen, 1989. Growth and feed conversion in cod (*Gadus morhua*) on different feeds, retention of some trace elements in the liver. Fisk. Dir. Skr., Ser. Ernaering **11**, 235–244.
Lien, J., L. Fawcett and S. Staniforth, 1984. Whales and Seals of Newfoundland and Labrador. Breakwater Books, St. John's Newfoundland, 136 pp.
Lithner, G., 1974. Rönnskärsunderskningen 1973. Statens naturvdrdsverk PM **497**, 1–65.
Lobel P.B., S.P. Belkhode, S.E. Jackson and H.P. Longerich, 1989. A universal method for quantifying and comparing the residual variability of elemenent concentrations in biological tissues using 25 elements in the mussel *Mytilus edulis* as a model. Mar. Biol. **102**, 513–518.
Lock, J.W., D.R. Thompson, R.W. Furness and J.A. Bartle, 1992. Metal concentrations in seabirds of the New Zealand region. Environ. Pollut. **75**, 289–300.
Lunde, G., 1970. Analysis of trace elements in seaweeds. J. Sci. Fd. Agric. **21**, 416–418.
Luoma, S.N., 1983. Bioavailability of trace metals to aquatic organisms – a review. Sci. Total Environ. **28**, 1–22.
Luoma, S.N., and G.W. Bryan, 1978. Factors controlling the availability of sediment-bound lead to the estuarine bivalve *Scrobicularia plana*. J. Mar. Biol. Assoc. U.K. **58**, 793–802.
Luoma, S.N., and G.W. Bryan, 1982. A statistical study of environmental factors controlling concentrations of heavy metals in the burrowing bivalve *Scrobicularia plana* and the polychaete *Nereis diversicolor*. Estuar. Coast. Shelf Sci. **15**, 95–108.
Luoma, S.N., G.W. Bryan and W.J. Langston, 1982. Scavenging of heavy metals from particulates by brown seaweed. Mar. Pollut. Bull. **13**, 394–396.
Luoma, S.N., D.J. Cain and C. Johansson, 1985. Temporal fluctuations of silver, copper and zinc in the bivalve *Macoma balthica* at five stations in South San Francisco Bay. Hydrobiologia **129**, 109–120.
Luoma, S.N., R. Dagovitz and E. Axtmann, 1990. Temporally intensive study of trace metals in sediments and bivalves from a large river-estuarine system: Suisun Bay/Delta in San Francisco Bay. Sci. Total Environ. **97/98**, 685–712.
Lüning, K., 1990. Seaweeds. Their Environment, Biogeography, and Ecophysiology (John Wiley & Sons, Inc., New York), pp. 527.
Lyngby, J.E., and H. Brix, 1982. Seasonal and environmental variation in cadmium, copper, lead and zinc concentrations in eelgrass (*Zostera marina* L.) in the Limfjord, Denmark. Aquat. Bot. **14**, 59–74.
Lyngby, J.E., and H. Brix, 1984. The uptake of heavy metals in eelgrass *Zostera marina* and their effect on growth. Ecol. Bull. (Stockholm) **36**, 81–89.
Lyngby, J.E., and H. Brix, 1987. Monitoring of heavy metal contamination in the Limfjord, Denmark, using biological indicators and sediment. Sci. Total Environ. **64**, 239–252.
Lyngby, J.E., H. Brix and H.-H. Schierup, 1982. Absorption and translocation of zinc in eelgrass (*Zostera marina* L.). J. Exp. Mar. Biol. Ecol. **58**, 259–270.
Macdonald, C.R., and J.B. Sprague, 1988. Cadmium in marine invertebrates and arctic cod in the Canadian Arctic. Distribution and ecological implications. Mar. Ecol. Prog. Ser. **47**, 17–30.
Mackay, N., M.N. Kazacos, R.J. Williams and M.I. Leedow, 1975. Selenium and heavy metals in black marlin. Mar. Pollut. Bull. **6**, 57–61.
Mackey, E.A., R. Demiralp, P.R. Becker, R.R. Greenberg, B.J. Koster and S.A. Wise, 1995. Trace element concentrations in cetacean liver tissues archived in the National Marine Mammal Tissue Bank. Sci. Total Environ. **175**, 25–41.
Magalhães, V.F. and W.C. Pfeiffer, 1995. Arsenic concentration in sediments near a metallurgical plant (Sepetiba Bay, Rio de Janeiro, Brazil). J. Geochem. Explor. **52**, 175–181.
Makuszok, W.M., 1983. Siemiejstvo Troiepieryie (*Tripterygiidae*), in: Zhizn' Zhivotnyh – Lancetniki – Krugłorotnyie – Chrashchebyie Ryby – Kostnyie Ryby (The Life of Animals – Fish), ed. T.S. Rass (Prosveshchenie, Izdatelstvo, Moscow), **Vol. 4**, 436–443 (in Russian).
Malcolm, H.M., I.L. Boyd, D. Osborn, M.C. French and P. Freestone, 1994. Trace metals in Antarctic fur seal (*Arctocephalus gazella*) livers from Bird Island, South Georgia. Mar. Pollut. Bull. **28**, 375–380.

Malinga, M., and P. Szefer, 2000a. Distribution and relationships between heavy metals in Glaucous Gull (*Larus hyperboreus*) from Bear and Jan Mayen Islands, the Arctic (in preparation).

Malinga, M., and P. Szefer, 2000b. Biomagnification of selected heavy metals in Glaucous Gull (*Larus hyperboreus*) relative to lower trophic levels of the Arctic food chain (in preparation).

Malm O., F.J.P. Branches, H. Akagi, M.B. Castro, W.C. Pfeiffer, M. Harada, W.R. Bastos and H. Kato, 1995a. Mercury and methylmercury in fish and human hair from Tapajos river basin, Brazil. Sci. Total Environ.**175**, 141–150.

Malm, O., M.B. Castro, W.R. Bastos, F.J.P. Branches, J.R.D. Guimarães, C.E. Zuffo and W.C. Pfeiffer, 1995b. An assessment of Hg pollution in different goldmining areas, Amazon Brazil. Sci. Total Environ. **175**, 127–140.

Marmolejo-Rivas, C. and F. Paez-Osuna, 1990. Trace metals in tropical coastal lagoon bivalves, *Mytella strigata*. Bull. Environ. Contam. Toxicol. **45**, 545–551.

Mars, P., 1951. Essai d'interpretation des formes generalement groupees sous le nom de *Cardium edule* Linne. Bull. Mus. Hist. Nat. Marseille. **11**, 1–31.

Marsch, K.V., and R.W. Buddemeier, 1984. Marine plankton as an indicator of low-level radionuclide contaminaton in the Southern Ocean. Rep. UCRL-53549 (Lawrence Livermore Laboratory, University of California) 1–39.

Marti, Yu.Yu., 1983. Smiejstvo Treskovyie (*Gadidae*), in: Zhizn' Zhivotnyh – Lancetniki – Krugłorotnyie – Chrashchebyie Ryby – Kostnyie Ryby (The Life of Animals – Fish), ed. T.S. Rass (Prosveshchenie, Izdatelstvo, Moscow), **Vol. 4**, 310–321 (in Russian).

Martin, J.H., 1970. The possible transport of trace metals via moulted copepod exoskeletons. Limnol. Oceanogr. **15**, 756–761.

Martin, J.H., and G.A. Knauer, 1972. A comparison of inshore vs. offshore levels of 21 trace and major elements in marine plankton, in: Baseline Studies of Pollutants in the Marine Environment, ed. E.D. Goldberg (N.S.F./I.D.O.E. Brookhaven, New York), p. 35.

Martin, J.H., and G.A. Knauer, 1973. The elemental composition of plankton. Geochim. Cosmochim. Acta **37**, 1639–1653.

Martin, J.H., and W.W. Broenkow, 1975. Cadmium in plankton: Elevated concentrations off Baja California. Science **190**, 884–885.

Martin, J.H., K.W. Bruland and W.W. Broenkow, 1976a. Cadmium transport in the California Current, Marine Pollutant Transfer, eds. H.L. Windom and R.A., Duce (Lexington Books, D.C. Heat and Company, Lexington, MA), 159–184.

Martin, J.H., P.D. Elliott, V.C. Anderlini, D. Girvin, S.A. Jacobs, R.W. Risebrough, R.L. Delong and W.G. Gilmartin, 1976b. Mercury-selenium-bromine imbalance in premature parturient California sea lions. Mar. Biol. **35**, 91–104.

Martincic, D., H.W. Nurnberg, M. Stoeppler and M. Branica, 1984. Bioaccumulation of heavy metals by bivalves from Lim Fjord (North Adriatic Sea). Mar. Biol. **81**, 177–188.

Mathieson, S., and D.S. McLusky, 1995. Inter-species variation of mercury in skeletal muscle of five fish species from inshore waters of the Firth of Clyde, Scotland. Mar. Pollut. Bull. **30**, 283–286.

McClurg, T.P., 1984. Trace metals and chlorinated hydrocarbons in Ross seals from Antarctica. Mar. Pollut. Bull. **15**, 384–389.

McGreer, E.R., 1982. Factors affecting the distribution of the bivalve, *Macoma balthica* (L.) on a mudflat receiving sewage effluent, Fraser River estuary, British Columbia. Mar. Environ. Res. **7**, 131–149.

McKie, J.C., I.M. Davies and G. Topping, 1980. Heavy metals in grey seals (*Halichoerus grypus*) from the east coast of Scotland. ICES CM 1980/E:41, 13 pp.

Meador, J.P., U. Varanasi, P.A. Robisch and S.-L. Chan, 1993. Toxic metals in pilot whales (*Globicephala melaena*) from strandings in 1986 and 1990 on Cape Cod, Massachusetts. Can. J. Fish. Aquat. Sci. **50**, 2698–2706.

Meador, J.P., P.A. Robisch, R.C. Clark Jr and D.W. Ernest, 1998. Elements in fish and sediment from the Pacific coast of the United States: results from the National Benthic Surveillance Project. Mar. Pollut. Bull. **37**, 56–66.

Melhuus, A., K.L. Seip, H.M. Seip and S. Myklestad, 1978. A preliminary study of the use of benthic algae as biological indicators of heavy metal pollution in Sorfjorden, Norway. Environ. Pollut. **15**, 101–107.

Miednikov, B.M, 1983. Nadotrjad Angvilloidnyie (*Anguillomorpha*), in: Zhizn' Zhivotnyh – Lancetniki – Kruglorotnyie – Chrashchebyie Ryby – Kostnyie Ryby (The Life of Animals – Fish), ed. T.S. Rass (Prosveshchenie, Izdatelstvo, Moscow), **Vol. 4**, 205–218 (in Russian).

Mikheev, A.V.,1986a. Order Arseniformes, in: The Life of Animals – Birds, eds. V.D. Ilichev and A.V. Mikheev (Prosveshchenie, Izdatelstvo, Moscow), **Vol. 6**, 65–112 (in Russian).

Mikheev, A.V., and E.N. Kuroczkin, 1986b. Suborder Lari, in: The Life of Animals – Birds, eds. V.D. Ilichev and A.V. Mikheev (Prosveshchenie, Izdatelstvo, Moscow), **Vol. 6**, 245–260 (in Russian).

Mikheev, A.V., E.N. Kuroczkin and A.K. Rustamov, 1986c. Order caniformes and Steganopodes, in: The Life of Animals – Birds, eds. V.D. Ilichev and A.V. Mikheev (Prosveshchenie, Izdatelstvo, Moscow) **Vol. 6**, 52–63 (in Russian).

Miner, R.W., 1950. Field Book of Seashore Life. (G.P. Putnam's Sons. New York), 888 pp.

Mochizuki, Y., K.T. Suzuki, H. Sunaga, T, Kobayashi and R. Doi, 1985. Separation and characterization of metallothionein in two species of seals by high performance liquid chromatography – atomic absorption spectrophotometry. Comp. Biochem. Physiol. **82C**, 249–254.

Monteiro, L.R., J.P. Granadeiro, R.W. Furness and P. Oliveira, 1999. Contemporary patterns of mercury contamination in the Portuguese Atlantic inferred from mercury concentrations in seabird tissues. Mar. Environ. Res. **47**, 137–156.

Moor, J.W., 1977. Importance of algae in the diet of subarctic populations of *Gammarus lacustris* and *Pontoporeia affinis*. Canadian J. Zool. **55**, 637–641.

Moore, C., and K. Boström, 1978. The elemental compositions of lower marine organisms. Chem. Geol. **23**, 1–9.

Moore, P.G., P.S. Rainbow and E. Hayes, 1991. The beach-hopper *Orchestia gammarellus* (Crustace: Amphipoda) as a biomonitor for copper and zinc: North Sea. Sci. Total Environ. **106**, 221–238.

Moriarty, F., H.M. Hanson and P. Freestone, 1984. Limitations of body burdens as an index of environmental comtamination: heavy metals in fish *Cottus gobio* L. from the River Ecclesbourne, Derbyshire. Environ. Pollut. Ser. A **34**, 297–320.

Morozińska-Gogol, J., J. Rokicki, K. Frelek and P. Szefer, 1998. Trace metals in *Gasterosteus aculeatus* L. and its parasites from the Gulf of Gdańsk. Oceanol. Studies **27**, 71–75.

Morris, A.W., and A.J. Bale, 1975. The concentration of cadmium, copper, manganese and zinc by *Fucus vesiculosus* in the Bristol Channel. Estuar. Coast. Mar. Sci. **3**, 153–163.

Morris, R.J., R.J. Law, C.R. Allchin, C.A. Kelly and C.F. Fileman, 1989. Metals and organochlorines in dolphins and porpoises of Cardigan Bay, West Wales. Mar. Pollut. Bull. **20**, 512–523.

Movalli, P.A., 2000. Heavy metal and other residues in feathers of laggar falcon *Falco biarmicus jugger* from six districts of Pakistan. Environ. Pollut. **109**, 267–275.

Möller G., 1996. Biogeochemical interactions affecting hepatic trace element levels in aquatic birds. Environ. Toxicol. Chem. **15**, 1025–1033.

Möller, H., R. Schneider and Ch. Schnier, 1983. Trace metal and PCB content of mussels (*Mytilus edulis*) from the southwestern Baltic Sea. Int. Revue ges. Hydrobiol. **68**, 633–647.

Mugiya, Y., T. Hakomori and K. Hatsutori, 1991. Trace metal incorporation into otoliths and scales in the goldfish, *Carassius auratus*. Comp. Biochem. Physiol. **99C**, 327–331.

Muir, D.C.G., R. Wagemann, N.P. Grift, R.J. Norstrom, M. Simon and J. Lien, 1988. Organochlorine chemical and heavy metal contaminants in white-beaked dolphins (*Lagenorhynchus albirostris*) and pilot whales (*Globicephala melaena*) from the coast of Newfounland, Canada. Arch. Environ. Contam. Toxicol. **17**, 613–629.

Muir, D., B. Braune, B. DeMarch, R. Norstrom, R. Wagemann, L. Lockhart, B. Hargrave, D. Bright, R. Addison, J. Payne and K. Reimer, 1999. Spatial and temporal trends and effects of contaminants in the Canadian Arctic marine ecosystem: a review. Sci. Total Environ. **230**, 83–144.

Muirhead, S.J., and R.W. Furness, 1988. Heavy metal concentrations in the tissues of seabirds from Gough Island, South Atlantic Ocean. Mar. Pollut. Bull. **19**, 278–283.

Mulicki, Z., 1957. Ekologia ważniejszych bezkręgowców dennych Bałtyku. Prace. Morsk. Inst. Ryb., Gdynia, **No. 9**, 379 pp. (in Polish).

Munda, I.M., 1978. Trace metal concentrations in some Icelandic seaweeds. Bot. Mar. **21**, 261–263.

Munda, I.M., 1984. Salinity dependent accumulation of Zn, Co and Mn in *Scytosiphon lomentaria* (Lyngb.) and *Enteromorpha intestinalis* (L.) from the Adriatic Sea. Bot. Mar. **27**, 371–376.

Muñoz-Barbosa, A., E.A. Gutiérrez-Galindo and G. Flores-Muñoz, 2000. *Mytilus californianus* as an indicator of heavy metals on the nortwest coast of Baja California, Mexico. Mar. Environ. Res. **49**, 123–144.

Muse, J.O., J.D. Stripeikis, F.M. Fernandez, L. d'Huicque, M.B. Tudino, C.N. Carducci and O.E. Troccoli, 1999. Seaweeds in the assessment of heavy metal pollution in the Gulf San Jorge, Argentina. Environ. Pollut. **104**, 315–322.

Muus, B.J., and P. Dahlström, 1985. Marine fish and fishing in Northwest Europe. Nordisk Rotorgravyr, Stockholm.

Myklestad, S., and I. Eide, 1978. Exchange of heavy metals in *Ascophyllum nodosum* (L.) Le jol. *in situ* by means of transplanting experiments. Environ. Pollut. **16**, 277–284.

National Audubon Society, Field Guide to North American Birds. Revised Edition – John Bull & John Farrand, Jr. Revised by John Farrand, Jr. Visual Key by Amanda Wilson and Lori Hogan. Alfred A. Knopf (New York Copyright© 1994, 1996 by Chanticleer Press Inc.), 800 pp.

Naumov, S.P., 1989. Otriad Lastonogie (*Pinnipedia*), in: Zhizn' Zhivotnyh – Mlekopitayushchie (The Life of Animals – Mammals), ed. W.E. Sokolow (Prosveshchenie, Izdatelstvo, Moscow), **Vol. 7**, 326–337 (in Russian).

Neumann, G., M. Notter and H. Dahlgaard, 1991. Bladder-wrack (*Fucus vesiculosus* L.) as an indicator for radionuclides in the environment of Swedish nuclear power plants. Report 3931 (Swedish Environmental Protection Agency, Solna, Sweden), pp. 35.

Nicholson, J.K., 1981. The comparative distribution of zinc, cadmium and mercury in selected tissues of the herring gull (*Larus argentatus*). Comp. Biochem. Physiol. **68C** 91–94.

Nicholson, S., 1999. Cytological and physiological biomarker responses from green mussels, *Perna viridis* (L.) transplanted to contaminated sites in Hong Kong coastal waters. Mar. Pollut. Bull. **39**, 261–268.

Nickless G., R. Stenner and N. Terrille, 1972. Distribution of cadmium, lead and zinc in the Bristol Channel. Mar. Pollut. Bull. **3**, 188–190.

Nicolaidou, A., and J.A. Nott, 1998. Metals in sediment, seagrass and gastropods near a nickel smelter in Greece: possible interactions. Mar. Pollut. Bull. **36**, 360–365.

Nielsen, C.O., and R. Dietz, 1990. Distribution pattern of zinc, cadmium, mercury, and selenium in livers of Hooded Seal (*Cystophora cristata*). Biol. Trace Element Res. **24**, 61–71.

Nielsen, J.B., F. Nielsen, P.-J. Jørgensen and P. Grandjean, 2000. Toxic metals and selenium in blood from pilot whales (*Globicephala melas*) and sperm whales (*Physeter catodon*). Mar. Pollut. Bull. **40**, 348–351.

Niemi, A., 1977. Avustavan virkamiehen lausunto jtevesien vaikutuksesta Kokkolan merialueen tilaan ja kalatalouteen. Statement, Nat. Board Waters, 1–27 (in Finnish).

Nishigaki, S., Y. Tamura, T. Maki, H. Yamada, Y. Shimamura, S. Ochiai, and Y. Kimura, 1974. Studies on behaviour of accumulation of trace elements in fishes (I) Mercury- selenium correlations in connection with body weight in muscle of seafish. Ann. Rep. Tokyo Metr. Res. Lab. P.H. **25**, 235–239.

NOAA, 1989. National Status and Trends Program for the Marine Environmental Quality progress report: a summary of data on tissue contamination from the first three years (1986–89) of the Mussel Watch Project. NOAA Technical Memorandum NOS OMA 49, 22 p + appendinces. Rockville, MD.

Noda K., H. Ichihashi, T.R. Loughlin, N. Baba, M. Kiyoto and R. Tatsukawa, 1995. Distribution of heavy metals in muscle, liver, and kidney of northern fur seal (*Callorhinus ursinus*) caught off Sanriku, Japan and from the Pribulof Islands, Alaska. Environ. Pollut. **90**, 51–59.

Nolan, C., and H. Dahlgaard, 1991. Accumulation of metal radiotracers by *Mytilus edulis*. Mar. Ecol. Prog. Ser. **70**, 165–174.

Nordsieck, F., 1968. Die europäischen Meeres-Gehäuseschnecken (Prosobranchia). Vom Eismeer bis Kapverden und Mittelmeer (Fischer, Stuttgart).

Norheim, G.,1987. Levels and interactions of heavy metals in sea birds from Svalbard and the Antarctic. Environ. Pollut. **47**, 83–94.

Norheim, G., J.U.Skaare and O. Wiig, 1992. Some heavy metals, essential elements, and chlorinated hydrocarbons in polar bear (*Ursus maritimus*) at Svalbard. Environ. Pollut. **77**, 51–57.

Norstrom, R.J., R.E. Schweinsberg and B.T. Collins, 1986. Heavy metals and essential elements in livers of the polar bear (*Ursus maritimus*) in the Canadian Arctic. Sci. Total Environ. **48**, 195–212.

Norton, M.G., and A.J. Murray, 1983. The metal content of fish and shellfish in Liverpool Bay. Chem. Ecol. **1**, 159–172.

Nuurtamo, M., P. Varo, E. Saari and P. Koivistoinen, 1980. Mineral element composition of Finnish foods, in: Minerak Element Composition of Finnish Foods: N, K, Ca, Mg, P, S, Fe, Cu, Mn, Zn, Mo, Co, Ni, Cr, F, Se, Si, Rb, Al, B, Br, Hg, As, Cd, Pb and Ash, ed. P. Koivistoinen (Scand. Assoc. Agric. Scient. Royal Swed. Acad. Agric. Forest., Stockholm), Acta Agric. Scand. Suppl. **22**, 77–87.

O'Shea, T.J., and R.L. Brownell Jr., 1994. Organochlorine and metal contaminants in baleen whales: a review and evaluation of conservation implications. Sci. Total Environ. **154**, 179–200.

Olsson, M., 1976. Mercury level as a function of size and age in northern pike, one and five years after the mercury ban in Sweden. Ambio **5**, 73–76.

Olsson, P.E., A. Larsson and C. Haux, 1996. Influence of seasonal changes in water temperature on cadmium inducibility of hepatic and renal metallothionein in rainbow trout. Mar. Environ. Res. **42**, 41–44.

Ostapczuk, P., M. Burrow, K. May, C. Mohl, M. Froning, B. Süßenbach, E. Waidmann and H. Emons, 1997a. Mussels and algae as bioindicators for long-term tendencies of element pollution in marine ecosystems. Chemosphere **34**, 2049–2058.

Ostapczuk, P., J.D. Schladot, H. Emons, K. Oxynos, K.-W. Schramm, G. Grimmer and J. Jacob, 1997b. Environmental monitoring and banking of marine pollutants by using common mussels. Chemosphere **34**, 2143–2151.

Ostroumova, T.A., 1983. Otrjad Kambaloobraznyie (*Pleuronectiformes*), in: Zhizn' Zhivotnyh – Lancetniki – Krugłorotnyie – Chrashchebyie Ryby – Kostnyie Ryby (The Life of Animals – Fish), ed. T.S. Rass (Prosveshchenie, Izdatelstvo, Moscow), **Vol. 4**, 494–502 (in Russian).

Otterlind, G., 1976 The harbour porpoise (*Phocoena phocoena*) endangered in Sweden waters. ICES CM 1976/N16. International Council for the Exploration of the Seas, Copenhagen.

Outridge, P.M., and A.M. Scheuhammer, 1993. Bioaccumulation and toxicology of nickel, implications for wild mammals and birds. Environ. Rev. **1**, 172–197.

Packer, D.M., M.P. Ireland and R.J. Wootton, 1980. Cadmium, copper, lead, zinc and manganese in the polychaete *Arenicola marina* from sediments around the coast of Wales. Environ. Pollut. **A22**, 309–321.

Páez-Osuna, F., J.I. Osuna-Lopez, G. Izaguirre-Fierro and H.M. Zazueta-Padilla, 1993. Heavy metals in clams from a subtropical coastal lagoon associated with an agricultural drainage basin. Bull. Environ. Contam. Toxicol. **50**, 915–921.

Páez-Osuna, F., J.l. Osuna Lopez, G. Izaguirre-Fierro and H.M. Zazueta-Padilla, 1994. Trace metals in mussels from the Ensenada del Pabellon lagoon Mexico. Mar. Pollut. Bull. **28**, 124–126.

Páez-Osuna, F., H. Bojorquez-Leyva and J. Ruelas-Inzunza, 1999. Regional variations of heavy metal concentrations in tisues of barnacles from the subtropical Pacific coast of Mexico. Environ. Intern. **25**, 647–654.

Páez-Osuna, F., M.J. Ochoa-Izaguirre, H. Bojórquez-Leyva and I.L. Michel-Reynoso, 2000. Macroalgae as biomonitors of heavy metal availability in coastal lagoons from the subtropical Pacific of Mexico. Bull. Environ. Contamin. Toxicol. **64** (in press).

Paludan-Müller, P., C.T. Agger, R. Dietz and C.Ch. Kinze, 1993. Mercury, cadmium, zinc, copper and selenium in harbour porpoise (*Phocoena phocoena*) from West Greenland. Polar. Biol. **13**, 311–320.

Parin N.V., 1983. Siemiejstvo Połuryłowyie (*Hemirhamphidae*), in: Zhizn' Zhivotnyh – Lancetniki – Krugłorotnyie – Chrashchebyie Ryby – Kostnyie Ryby (The Life of Animals – Fish), ed. T.S. Rass (Prosveshchenie, Izdatelstvo, Moscow), **Vol. 4**, 300–308 (in Russian).

Parsons, E.C.M., H.M. Chan and R. Kinoshita, 1999. Trace metal and organochlorine concentrations in a pygmy Bryde's whale (*Balaenoptera edeni*) from the South China Sea. Mar. Pollut. Bull. **38**, 51–55.

Patin, S.A., N.P. Morozov, A.S. Romanteeva, R.M. Melnikova and G.S. Borisenko, 1980. Trace metals in ecosystem of the Japan Sea. Geokhimya **3**, 423–429 (in Russian, with an English abstract).

Pempkowiak, J., A. Sikora and E. Biernacka, 1999. Speciation of heavy metals in marine sediments vs their bioaccumulation by mussels. Chemosphere **39**, 313–321.

Pentreath, R.J., 1981. The biological availability to marine organisms of transuranium and other long-lived nuclides, in: Impacts of Radionuclide Releases into the Marine Environment (International Atomic Energy Agency, Vienna), pp. 241–272.

Pentreath, R.J., D.F. Jefferies, M.B. Lovett and D.M. Nelson, 1979. The behaviour of transuranic and other long-lived radionuclides in the Irish Sea and its relevance to the deep sea disposal of radioactive wastes. Proc. 3rd NEA Seminar on Marine Radioecology (Nuclear Agency Organisation for Economic Co-operation and Development), pp. 203–221.

Pergent-Martini, Ch., 1998. *Posidonia oceanica*: a biological indicator of past and present mercury contamination in the Mediterranean Sea. Mar. Environ. Res. **45**, 101–111.

Perkins, R.W., and C.W. Thomas, 1980. World wide fallout, in: Transuranic Elements in the Environment, ed. W.C. Hanson (National Technical Information Service, Springfield, VA) DOE/TIC-22800.

Perkowska, A., and M. Protasowicki, 1996. Heavy metal contents in sediments and the blue mussel *Mytilus edulis* from three regions of the southern Baltic Sea. Oceanol. Studies **No. 4**, 55–63.

Perttilä, M., V. Tervo and R. Parmanne, 1982a. Heavy metals in Baltic herring and cod. Mar. Pollut. Bull. **13**, 391–393.

Perttilä, M., V. Tervo and R. Parmanne, 1982b. Age dependence on the concentration of harmful substances in Baltic herring (*Clupea harengus*). Chemosphere **11**, 1019–1026.

Perttilä, M., O. Stenman, H. Pyysalo and K. Wickström, 1986. Heavy metals and organochlorine compounds in seals in the Gulf of Finland. Mar. Environ. Res. **18**, 43–59.

Phillips, D.J.H., 1976a. The common mussel *Mytilus edulis* as an indicator of pollution by zinc, cadmium, lead and copper. I. Effects of environmental variables on uptake of metals. Mar. Biol. **38**, 59–69.

Phillips, D.J.H., 1976b. The common mussel *Mytilus edulis* as an indicator of pollution by zinc, cadmium, lead and copper. II. Relationship of metals in the mussel to those discharged by industry. Mar. Biol. **38**, 71–80.

Phillips, D.J.H., 1977a. The common mussel *Mytilus edulis* as an indicator of trace metals in Scandinavian waters. I. Zinc and cadmium. Mar. Biol. **43**, 283–291.

Phillips, D.J.H., 1977b. Effects of salinity on the net uptake of zinc by the common mussel *Mytilus edulis*. Mar. Biol. **41**, 79–88.

Phillips D.J.H., 1977c. The use of biological indicator organisms to monitor trace metal pollution in marine and estuarine environments – A review. Environ. Pollut. **13**, 281–317.

Phillips, D.J.H., 1978. The common mussel *Mytilus edulis* as an indicator of trace metals in Scandinavian waters. II. Lead, iron and manganese. Mar. Biol. **46**, 147–156.

Phillips, D.J.H., 1979. Trace metals in the common mussel, *Mytilus edulis* (L.), and in the alga *Fucus vesiculosus* (L.) from the region of the Sound (Öresund). Environ. Pollut. **18**, 31–43.

Phillips, D.J.H., 1980. Quantitative Aquatic Biological Indicators (Applied Science Publishers Ltd, London) 488 pp.

Phillips, D.J.H., 1985. Organochlorines and trace metals in green-lipped mussels *Perna viridis* from Hong Kong waters: a test of indicator ability. Mar. Ecol. Prog. Ser. **21**, 251–258.

Phillips, D.J.H., and P.S. Rainbow, 1988. Barnacles and mussels as biomonitors of trace elements: a comparative study. Mar. Ecol. Prog. Ser. **49**, 83–93.

Phillips, D.J.H., and P.S. Rainbow, 1993. Biomonitoring of Trace Aquatic Contaminants. Elsevier Science Publishers Ltd., London, 371 pp.

Pinder, J.E., III and J.P. Giesy, 1981. Frequency distribution of the concentrations of essential and nonessential elements in largemouth bass, *Micropterus salmoides*. Ecol. **62**, 456–468.

Plaskett, D., and I.C. Potter, 1979. Heavy metal concentrations in the muscle tissue of 12 species of teleost from Cockburn Sound, Western Australia. Aust. J. Mar. Freshwater Res. **30**, 607–616.

Podbielkowski, Z., and H. Tomaszewicz, 1979. Zarys Hydrobotaniki (PWN, Warszawa), 531 pp. (in Polish).

Pohl, C., 1992. Correlation between trace metal concentrations (Cd, Cu, Pb, Zn) in seawater and zooplankton organisms (*Copepoda*) of the Arctic and Atlantic Ocean. Berichte Polarforsch. **101**, 198 pp.

Popham, J.D., and J.M. D'Auria, 1983. Statistical approach for deciding if mussels (*Mytilus edulis*) have been collected from a water body pollyted with trace metals. Environ. Sci. Technol. **17**, 576–582.

Powell, M.I., and K.N. White, 1990. Heavy metal accumulation by barnacles and its implications to their use as biological monitors. Mar. Environ. Res. **30**, 91–118.

Presley, B.J., J.S. Schofield and J. Trefry, 1981. Waste material behaviour and inorganic geochemistry at the Puerto Rico waste dumpsite, in: Ocean Dumping of Industrial Wastes, eds. H.B. Ketchum, D.R. Kester and P.K. Park (Plenum, New York), 233–246.

Preston, A., D.F. Jefferies, J.W.R. Dutton, B.R. Harvey and A.K. Steele, 1972. British Isles coastal waters: the concentrations of selected heavy metals in sea water, suspended matter and biological indicators – a pilot survey. Environ. Pollut. **3**, 69–82.

Protasowicki, M., 1982. Changes in contents of selected heavy metals in fish from the Baltic and the River Odra mouth area. Zesz. Nauk. Akad. Roln. w Szczecinie **No. 93**, 213–221 (in Polish, with English abstract).

Protasowicki, M., 1986a. Sex effects on Cd, Pb, Cu and Zn contents in selected fish organs. Baltic Sea Environment Proceedings of Baltic Sea Monitoring Symposium, Tallinn, USSR, 10–15 March 1986 (Baltic Marine Environment Protection Commission, Helsinki Commission), **No. 19**, 433–441.

Protasowicki, M., 1986b. The long-term observations on heavy metals content of fish in the southern Baltic. I. Mercury. Baltic Sea Environment Proceedings of Baltic Sea Monitoring Symposium, Tallinn, USSR, 10–15 March 1986 (Baltic Marine Environment Protection Commission, Helsinki Commission), **No. 19**, 62–75.

Protasowicki, M., 1989. Bioaccumulation and distribution of heavy metals in fish organs. Proc. of the 21st EMBS, Gdańsk, 14–19 September 1986 (Polish Academy of Sciences – Institute of Oceanology, Sopot) pp. 609–614.

Protasowicki, M., 1991. Heavy metals in southern Baltic fish: Present situation and future trends. Acta Ichtyol. (Suppl.) **21**, 291–300.

Protasowicki, M., 1991a. Long-term studies on heavy metals in aquatic organisms from the River Odra mouth area. Acta Ichthyol. Piscat. **XXI Suppl**. 301–309.

Protasowicki, M., 1992. Preliminary studies on selected elements in organs of dab *Limanda limanda* and their relation to fish disease state. Mar. Ecol. Prog. Ser. **91**, 57–60.

Protasowicki, M., and A. Chodyniecki, 1980. Fish as an indicator of pollution of the Baltic Sea waters with some heavy metals. Oceanologia **13**, 71–75.

Protasowicki, M. and M. Kosior, 1987. Long-term observations of selected heavy metals contained in otoliths of cod from Southern Baltic. Mar. Environ. Qual. Comm. (ref. J. Baltic Fish Comm.) C.M., **E:5**, 15pp.

Protasowicki, M. and M. Kosior, 1988. Long-term observations of selected heavy metals contained in otoliths of cod from Southern Baltic. Kieler Meeresforsch., Sonderh. **6**, 424–431.

Protasowicki, M., A. Chodyniecki and A. Ociepa, 1983. Metale ciężkie w rybach przemysłowych poławianych w latach 1976–1980 (Heavy metals in the fish caught in 1976–1980). Zesz. Nauk. AR w Szczecinie No. **103**, 181–198 (in Polish, with English abstract).

Prudente, M., E.-Y. Kim, S. Tanabe, and R. Tatsukawa, 1997. Metal levels in some commercial fish species from Manila Bay, the Philippines. Mar. Pollut. Bull. **34**, 671–674.

Prudente, M., H. Ichihashi, S. Kan-atireklap, I. Watanabe and S. Tanabe, 1999. Butyltins, organochlorines and metal levels in green mussel, *Perna viridis* L. from the coastal waters of the Philippines. Fish. Sci. **65**, 441–447.

Puente, X., R. Villares, E. Carral and A. Carballeira, 1996. Nacreous shell of *Mytilus galloprovincialis* as a biomonitor of heavy metal pollution in Galiza (NW Spain). Sci. Total Environ. **183**, 205–211.

Pynnönen, K., 1996. Heavy metal-induced changes in the feeding and burrowing behaviour of a Baltic isopod, *Saduria* (*Mesidothea*) *entomon* L. Mar. Environ. Res. **41**, 145–156.

Rachor, E., W.E. Arntz, H. Rumohr and K.H. Mantau, 1982. Seasonal and long-term population fluctuations in *Diastylis rathkei* (Crustacea: Cumacea) of Kiel Bay and German Bight. Neth. J. Sea Res. **16**, 141–150.

Rainbow, P.S., 1985. Accumulation of Zn, Cu and Cd by crabs and barnacles. Estuar. Coast. Shelf Sci. **21**, 669–686.

Rainbow, P.S., 1987. Heavy metals in barnacles, in: Biology of Barnacles, ed. A.J. Southward (A.A. Balkema, Rotterdam), 405–417.

Rainbow, P.S., 1989. Copper, cadmium and zinc concentrations in oceanic amphipod and euphausiid crustaceans, as a source of heavy metals to pelagic seabirds. Mar. Biol. **103**, 503–518.

Rainbow, P.S., 1990. Heavy metal levels in marine invertebrates, in: Heavy Metals in the Marine Environment, eds. R.W. Furness and P.S. Rainbow (CRC Press, Boca Raton, USA), pp. 67–79.

Rainbow, P.S., 1993. The significance of trace metal concentrations in marine invertebrates, in: Ecotoxicology of Metals in Invertebrates, eds. R. Dallinger and P.S. Rainbow (Lewis Publishers, Boca Raton), **Chapt. 1**, 3–23.

Rainbow, P.S., 1995a. Biomonitoring of heavy metal availability in the marine environment. Mar. Pollut. Bull. **31**, 183–192.

Rainbow, P.S., 1995b. Physiology, physicochemistry and metal uptake – a crustacean perspective. Mar. Pollut. Bull. **31**, 55–59.

Rainbow, P.S., 1996. Heavy metals in aquatic invertebrates, in: Environmental Contaminants in Wildlife. Interpreting Tissue Concentrations, eds. W.N. Beyer, G.H. Heinz and A.W. Redmon-Norwood (CRC Lewis Publishers, Boca Raton, Florida), pp. 405–425.

Rainbow, P.S., 1997. Trace metal accumulation in marine invertebrates: marine biology or marine chemistry ? J. Mar. Biol. Assoc. U.K. **77**, 195–210.

Rainbow, P.S., 1998. Phylogeny of trace metal accumulation in crustaceans, in: Metal Metabolism in Aquatic Environments, eds. W.J. Langston and M.J. Bebianno (Chapman & Hall, London), 285–319.

Rainbow, P.S., and S.L. White, 1989. Comparative strategies of heavy metal accumulation by crustaceans: Zinc, copper and cadmium in a decapod, an amphipod and a barnacle. Hydrobiology **174**, 245–262.

Rainbow, P.S., and P.G. Moore, 1990. Seasonal variation in copper and zinc concentrations in three talitrid amphipods (Crustacea). Hydrobiologia **196**, 65–72.

Rainbow, P.S., and D.J.H. Phillips, 1993. Cosmopolitan biomonitors of trace metals. Mar. Polut. Bull. **26**, 593–601.

Rainbow, P.S., A.G. Scott, E.A. Wiggins and R.W. Jackson, 1980. Effect of chelating agents on the accumulation of cadmium by the barnacle *Semibalanus balanoides*, and complexation of soluble Cd, Zn and Cu. Mar. Ecol.-Progr. Ser. **2**, 143–152.

Rainbow, P.S., J. Green and P. Denny, 1989a. A note on the intertidal ecology of two little known barnacles, *Balanus kondakovi* (Tarasov and Zevina) and *Balanus patelliformis* (Bruguière, 1789) from a Malaysian mangrove swamp (Cirripedia, Balanomorpha). Crustaceana **57**, 104–107.

Rainbow, P.S., P.G. Moore and D. Watson, 1989b. Talitrid amphipods (Crustacea) as biomonitors for copper and zinc. Estuar. Coast. Shelf Sci. **28**, 567–582.

Rainbow, P.S., D.J.H. Phillips and M.H. Depledge, 1990. The significance of trace metal concentrations in marine invertebrates. A need for laboratory investigation of accumulation strategies. Mar. Polut. Bull. **21**, 321–324.

Rainbow, P.S., R.H. Emson, B.D. Smith, P.G. Moore and P.V. Mladenov, 1993a. Talitrid amphipods as biomonitors of trace metals near Dunedin, New Zealand. New Zealand J. Mar. Freshwat. Res. **27**, 201–210.

Rainbow, P.S., H. Zongguo, Y. Songkai and B.D. Smith, 1993b. Barnacles as biomonitors of trace metals in the coastal waters near Xiamen, China. Asian Mar. Biol. **10**, 109–121.

Rainbow, P.S., W. Fialkowski and B.D. Smith, 1998. The sandhopper *Talitrus saltator* as a trace metal biomonitor in the Gulf of Gdansk, Poland. Mar. Pollut. Bull. **36**, 193–200.

Rainbow, P.S., M. Wolowicz, W. Fialkowski, B.D. Smith and A. Sokolowski, 2000. Biomonitoring of trace metals in the Gulf of Gdansk using mussels (*Mytilus trossulus*) and barnacles (*Balanus improvisus*). Wat. Res. **34**, 1823–1829.

Ramirez, M., H. Gonzalez, N. Ablanedo and I. Torres, 1990. Heavy metals in macroalgae of Havana's Northern Littoral, Cuba. Chem. Ecol. **4**, 49–55.
Rass T.S., 1983a. Otriad Sel'deobraznyie (*Clupeiformes*), in: Zhizn' Zhivotnyh – Lancetniki – Kruglorotnyie – Chrashchebyie Ryby – Kostnyie Ryby (The Life of Animals – Fish), ed. T.S. Rass (Prosveshchenie, Izdatelstvo, Moscow), **Vol. 4**, 106–131 (in Russian).
Rass T.S., 1983b. Podotriad Pieschankovidnyie (*Ammodytoidei*), in: Zhizn' Zhivotnyh – Lancetniki – Kruglorotnyie – Chrashchebyie Ryby – Kostnyie Ryby (The Life of Animals – Fish), ed. T.S. Rass (Prosveshchenie, Izdatelstvo, Moscow), **Vol. 4**, 444–445 (in Russian).
Ray, S., D. McLeese and D. Pezzack, 1980. Accumulation of cadmium by *Nereis virens*. Arch. Environ. Contam. Toxicol. **9**, 1–8.
Regoli, F., 1998. Trace metals and antioxidant enzymes in gills and digestive gland of the Mediterranean mussel *Mytilus galloprovincialis*. Arch. Environ. Contam. Toxicol. **34**, 48–63.
Regoli, F., and E. Orlando, 1993. *Mytilus galloprovincialis* as a bioindicator of lead pollution: biological variables and cellular responses. Sci. Total Environ. **Suppl. 2**, 1283–1292.
Regoli, F., and E. Orlando, 1994. Seasonal variation of trace metal concentrations (Cu, Fe, Mn, Pb, Zn) in the digestive gland of the Mediterranean mussel *Mytilus galloprovincialis*: comparison between a polluted and a non polluted site. Arch. Environ. Contam. Toxicol. **27**, 36–43.
Reijnders, P.J.H., 1980. Organochlorine and heavy metal residues in harbour seals from the Wadden Sea and their possible effects on reproduction. Neth. J. Sea Res. **14**, 30–65.
Reijnders, P.J.H., 1994. Toxicokinetics of chlorobiphenyls and associated physiological responses in marine mammals, with particular reference to their potential for ecotoxicological risk assessment. Sci. Total Environ. **154**, 229–236.
Reise, K., 1979. Spatial configurations generated by mobile benthic polychaetes. Helgoländer Meeresunters. **32**, 55–72.
Renfro, W.C., 1973. Transfer of Zn-65 from sediments by marine polychaete worms. Mar. Biol. **21**, 305–316.
Rheinheimer, G., 1998. Pollution in the Baltic Sea. Naturwissenschaften **85**, 318–329.
Riget, F., P. Johansen and G. Asmund, 1995. Natural seasonal variation of cadmium, copper, lead and zinc in brown seaweed (*Fucus vesiculosus*). Mar. Pollut. Bull. **30**, 409–413.
Riget, F., R. Dietz and P. Johansen, 1997. Zinc, cadmium, mercury and selenium in Greenland fish. Meddelelser om Gronland, Bioscience **48**, 1–29.
Rissanen, K., and T.K., Ikäheimonen, 2000. Caesium and plutonium concentrations in salmon caught in River Teno (Norwegian Sea) and in River Tornionjoki (Gulf of Bothnia), in: The Radiological Exposure of the Population of the European Community to Radioactivity in the Baltic Sea. Marina-Balt Project, ed. S.P. Nielsen. Proceedings of a Seminar held at Hasseludden Conference Centre, Stockholm, 9–11 June 1998, European Commission, Directorate-General Environment, EUR 19200 EN (European Communities, 2000, Belgium), pp. 439–448.
Ritterhoff, J. and G.-P. Zauke, 1997a. Trace metals in field samples of zooplankton from the Fram Strait and the Greenland Sea. Sci. Total Environ. **199**, 255–270.
Ritterhoff, J. and G.-P. Zauke, 1997b. Influence of body length, life-history status and sex on trace metal concentrations in selected zooplankton collectives from the Greenland Sea. Mar. Pollut. Bull. **34**, 614–621.
Ritter-Zahony, R., 1911. Chaetognathi. das Tierreich (Berlin) 29.
Ritz, D.A., R. Swain and N.G. Elliott, 1982. Use of the mussel *Mytilus edulis planulatus* (Lamarck) in monitoring heavy metal levels in seawater. Australian J. Mar. Freshwater Res. **33**, 415–421.
Roast, S.D., J. Widdows and M.B. Jones, 2000. Mysids and trace metals: disruption of swimming as a behavioural indicator of environmental contamination. Mar. Environ. Res. **50**, 107–112.
Roberts, T.M., P.B. Heppleston and R.D. Roberts, 1976. Distribution of heavy metals in tissues of the common seal. Mar. Pollut. Bull. **7**, 194–196.
Robinson, W.E., D.K. Ryan and G.T. Wallace, 1993. Gut contents: a significant contaminant of *Mytilus edulis* whole body metal concentrations. Arch. Environ. Contam. Toxicol. **25**, 36–43.
Robinson, S., J. Volosin, J. Keithly and R. Cardwell, 1999. Comment on: Butyltin residues in sediment, fish, fish-eating birds, harbour porpoise and human tissues from the Polish coast of the Baltic Sea (Kannan and Falandysz 1997). Mar. Pollut. Bull. **38**, 57–61.

Roesijadi, G., A.S. Drum, J.M. Thomas and G.W. Fellingham, 1982. Enhanced mercury tolerance in marine mussels and relationship to low molecular weight, mercury binding proteins. Mar. Pollut. Bull. **13**, 250–253.

Roesijadi, G., J.S. Young, A.S. Drum and J.M. Gurtisen, 1984. Behaviour of trace metals in *Mytilus edulis* during a reciprocal transplant field experiment. Mar. Ecol.-Progr. Ser. **18**, 155–170.

Roméo, M., and E. Nicolas, 1986. Cadmium, copper, lead and zinc in three species of planktonic crustaceans from the east coast of Corsica. Mar. Chem. **18**, 359–367.

Roméo, M., M. Gnassia-Barelli and E. Nicolas, 1985. Concentrations en plomb du plancton de la Mer Ligure (Mediterrannee Nord-Occidentale). Chemoshere **14**, 1423–1431.

Roméo, M., Y. Siau, Z. Sidoumou and M. Gnassia-Barelli, 1999. Heavy metal distribution in different fish species from the Mauritania coast. Sci. Total Environ. **232**, 169–175.

Romeril, M.G., 1977. Heavy metal accumulation in the vicinity of a desalination plant. Mar. Pollut. Bull. **8**, 84–87.

Ronald, K., R.J. Frank, J. Dougan, R. Frank and H.E. Braun, 1984. Pollutants in harp seals (*Phoca groenlandica*) II. Heavy metals and selenium. Sci. Total Environ. **38**, 153–166.

Różańska, Z., 1971. The biological and ecological studies over chaetognatha in the Baltic Sea (WSR Olsztyn) (in Polish).

Ruelas-Inzunza, J.R., and F. Páez-Osuna, 2000. Comparative bioavailability of trace metals using three filter-feeder organisms in a subtropical coastal environment (Southeast Gulf of California). Environ. Pollut. **107**, 437–444.

Ruiz, J.M., and J.I. Saiz-Salinas, 2000. Extreme variation in the concentration of trace metals in sediments and bivalves from the Bilbao estuary (Spain) caused by the 1989–90 drought. Mar. Environ. Res. **49**, 307–317.

Rutenberg E.P., 1983. Otrjad Koljuszkoobraznyie (*Gasterosteiformes*), in: Zhizn' Zhivotnyh – Lancetniki – Krugłorotnyie – Chrashchebyie Ryby – Kostnyie Ryby (The Life of Animals – Fish), ed. T.S. Rass (Prosveshchenie, Izdatelstvo, Moscow), **Vol. 4**, 348–356 (in Russian).

Rutkowicz, S., 1982. Encyklopedia Ryb Morskich (Wydawnictwo Morskie, Gdańsk), 663 pp.

Rygg, B., 1970. Studies on *Cerastoderma edule* (L.) and *Cerastoderma glaucum* (Poiret). Sarsia **43**, 65–50.

Saeki, K., Y. Okabe, E.-Y. Kim, S. Tanabe, M. Fukuda and R. Tatsukawa, 2000. Mercury and cadmium in common cormorants (*Phalacrocorax carbo*). Environ. Pollut. **108**, 249–255.

Saenko, G.N., M.D. Koryakova, V.F. Makienko and I.G.Dobrosmyslova, 1976. Concentration of polyvalent metals by seaweeds in Vostok Bay, Sea of Japan. Mar. Biol. **34**, 169–176.

Saiki, M.K., 1990. Elemental concentrations in fishes from the Salton Sea, southeastern California. Water Air Soil Pollut. **52**, 41–56.

Saiki, M.K. and D.U. Palawski, 1990. Selenium and other elements in juvenil striped bass from the San Joaquin Valley and San Francisco Estuary, California. Arch. Environ. Contam. Toxicol. **19**, 717–730.

Saizsalinas, J.I., and G. Franceszubillaga, 1997. *Nereis diversicolor*: an unreliable biomonitor of metal contamination in the 'Ria de Bilbao' (Spain). Mar. Ecol. **18**, 113–125.

Sanders, J.G., and H.L. Windom, 1980. The uptake and reduction of arsenic species by marine algae. Estuar. Coast. Mar. Sci. **10**, 555–567.

Sandler, H., 1984. Zinc and copper concentrations in benthic invertebrates considered in relation to concentrations in sediments and water in the Bothnian Sea (northern Baltic). Finn. Mar. Res. **No. 250**, 19–32.

Sandler, H., 1986. Heavy metals in benthic crustaceans and mysids in the Bothnian Sea. Publ. Water Res. Inst. **68**, 205–210.

Sanpera, C., R. Capelli, V. Minganti and L. Jover, 1993. Total and organic mercury in North Atlantic fin whales: distribution pattern and biological related changes. Mar. Pollut. Bull. **26**, 135–139.

Sarkar, S.K., B. Bhattacharya and S. Debnath, 1994. The suitability of tropical marine bivalves as biomonitors of heavy metals in deltaic Sundarbans, north-east India. Chemosphere **29**, 759–770.

Sarvala, J., 1971. Ecology of *Harmothoe sarsi* (Malmgren) (Polychaeta, Polynoidae) in the northern Baltic area. Ann. Zool. Fenn. **8**, 231–309.

Savari, A., A.P.M. Lockwood and M. Sheader, 1991. Effects of season and size (age) on heavy metal concentrations of the common cockle (*Cerastoderma edule* (L.)) from Southampton water. J. Molluscan Studies **57**, 45–57.

Savenko, V.S., 1988. Elemental composition of plankton. Geokhimiya **8**, 1084–1089 (in Russian).

Savvaitova, K.A. and B.M. Miednikov, 1983. Podotrjad Łososievidnyie (*Salmonoidei*), in: Zhizn' Zhivotnyh - Lancetniki - Krugłorotnyie - Chrashchebyie Ryby - Kostnyie Ryby (The Life of Animals - Fish), ed. T.S. Rass (Prosveshchenie, Izdatelstvo, Moscow), **Vol. 4**, 135–158 (in Russian).

Sawidis, Th., and A.N. Voulgaropoulos, 1986. Seasonal bioaccumulation of iron, cobalt and copper in marine algae from Thermaikos Gulf of the Northern Aegean Sea, Greece. Mar. Environ. Res. **19**, 39–47.

Sazonov, Ju.N., 1983. Makrurusovyie (*Macrouridae*), in: Zhizn' Zhivotnyh - Lancetniki - Krugłorotnyie - Chrashchebyie Ryby - Kostnyie Ryby (The Life of Animals - Fish), ed. T.S. Rass (Prosveshchenie, Izdatelstvo, Moscow), **Vol. 4**, 326–330 (in Russian).

Scanlon, P.F., V.D. Stotts, R.G. Oderwald, T.J. Dietric and R.J. Kendall, 1980. Lead concentrations in livers of Maryland waterfowl with and without ingested lead shot present in gizzards. Bull. Environ. Contam. Toxicol. **25**, 855–860.

Schellenberg, A., 1928. Krebstiere oder Crustacea. II: Decapoda, in: Die Tierwelt Deutschlands und der angrenzenden Meeresteile, ed. F. Dah (Verlag von Gustav Fischer, Jena).

Scheuhammer, A.M., 1987. The chronic toxicity of aluminium, cadmium, mercury, and lead in birds: A review. Environ. Pollut. **46**, 263–295.

Scheuhammer, A.M., 1991. Effects od acidification on the availability of toxic metals and calcium to wild birds and mammals. Environ. Pollut. **71**, 329–375.

Schladot J.D., F. Backhaus, P. Ostapczuk and H. Emons, 1997. Eel-pout (*Zoarces viviparus* L.) as a marine bioindicator. Chemosphere **34**, 2133–2142.

Schnier, C., L. Karbe and R. Niedergesäss, 1978. Trace elements in mussels and sediments from the German Bight (North Sea). A comparison of multielement patterns. GKKS 77/E/51 (Gesellschaft fr Kernenergieverwertung in Schiffbau und Schiffahrt mbH. Geesthacht), 1–16.

Scott-Fordsmand, J.J., and M.H. Depledge, 1997. Changes in the tissue concentrations and contents of calcium, copper and zinc in the shore crab *Carcinus maenas* (L.) (Crustacea: Decapoda) during the moult cycle and following copper exposure during ecdysis. Mar. Environ. Res. **44**, 397–414.

Sears, J.R., K.J. Pecci and R.A. Cooper, 1985. Trace metal concentrations in offshore, deep-water seaweeds in the Western North Atlantic Ocean. Mar. Pollut. Bull. **16**, 325–328.

Secor, D.H., J.M. Dean and S.E. Campana (Eds.), 1995. Recent Developments in Fish Otolith Research (University of South Carolina Press, Columbia).

Seeliger, U., and P. Edwards, 1977. Correlation coefficients and concentration factors of copper and lead in sea water and benthic algae. Mar. Pollut. Bull. **8**, 16–18.

Segar, D.A., J.D. Collins and J.P. Riley, 1971. The distribution of the major and some minor elements in marine animals. Part. II. molluscs. J. Mar. Biol. Assoc. UK **51**, 131–136.

Segerstrâle, S.G, 1957. Baltic Sea. Mern. Geol. Soc. America **67**, 751–800.

Seisuma, Z., M. Legzdina and I. Kulikova, 1995. Comparison of heavy metal load in the Gulf of Riga and open Baltic. Archiwum Ochrony Środowiska **3–4**, 129–139.

Senthilkumar, K., C.A. Duda, D.L. Villeneuve, K. Kannan, J. Falandysz and J.P. Giesy, 1999. Butyltin compounds in sediment and fish from the Polish coast of the Baltic Sea. Environ. Sci. Pollut. Res. **6**, 200–206.

Sfriso, A., A. Marcomini and M. Zanette, 1995. Heavy metals in sediments, SPM and phytozoobenthos of the Logoon of Venice. Mar. Pollut. Bull. **30**, 116–124.

Sharp, G.J., H.S. Samant and O.C. Vaidya, 1988. Selected metal levels of commercial valuable seaweed adjacent to and distant from point sources of contamination in Nova Scotia and New Brunswick. Bull. Environ. Contam. Toxicol. **40**, 724–730.

Shiber, J., and E. Washburn, 1978. Lead, mercury and certain nutrient elements in *Ulva lactuca* (Linnaeus) from Ras Beirut, Lebanon. Hydrobiologia **61**, 187–192.

Singer, D.A., 1995. World class base and precious metal deposits – a quantitative analysis. Econ. Geol. **90**, 88–104.

Siudzinski, K., 1977. The zooplankton of the Gdańsk Bay. Stud. Mater., A, 18 (Morski Instytut Rybacki Gdynia) in Polish.

Sivalingam, P.M., 1978. Biodeposited trace metals and mineral content studies of some tropical marine algae. Bot. Mar. **21**, 327–330.

Skaare, J.U., N.H. Markussen, G. Norheim, S. Haugen and G. Holt, 1990. Levels of polychlorinated biphenyls, organochlorine pesticides, mercury, cadmium, copper, selenium, arsenic, and zinc in the harbour seal, *Phoca vitulina*, in Norwegian waters. Environ. Pollut. **66**, 309–324.

Skaare, J.U., E. Degre, P.E. Aspholm and K.I. Ugland, 1994. Mercury and selenium in Arctic and coastal seals off the coast of Norway. Environ. Pollut. **85**, 153–160.

Skóra, K.E., 1991. Notes on cetacea observed in the Polish Baltic sea: 1979–1990. Aquat. Mammals **17.2**, 67–70.

Skóra, K.E., I. Pawliczka and M. Klinowska, 1988. Observations of the harbour porpoise (*Phocoena phocoena*) on the Polish Baltic coast. Aquat. Mammals **14.3**, 113–119.

Skwarzec, B., 1988. Accumulation of ^{210}Po in selected species of Baltic fish. J. Environ. Radioactivity **8**, 111–118.

Skwarzec, B., 1995. Polonium, uranium and plutonium in ecosystem of the southern Baltic. Rozprawy i monografie (Polska Akademia Nauk, Instytut Oceanologii w Sopocie), 183pp. (in Polish, with English abstract).

Skwarzec, B., 1997. Polonium, uranium and plutonium in the southern Baltic Sea. Ambio **26**, 113–117.

Skwarzec, B., 1999. Polonium, uranium and plutonium in the southern Baltic ecosystem. Czech. J. Phys. **49/S1**, 461–466

Skwarzec, B., and R. Bojanowski, 1988. ^{210}Po content in sea water and its accumulation in southern Baltic plankton. Mar. Biol. **97**, 301–307.

Skwarzec, B., and L. Falkowski, 1988. Accumulation of ^{210}Po in Baltic invertebrates. J. Environ. Radioactivity **8**, 99–109.

Skwarzec, B., and R. Bojanowski, 1992. Distribution of plutonium in selected components of the Baltic ecosystem within the Polish economic zone. J. Environ. Radioactivity **15**, 249–263.

Skwarzec, B., A. Kentzer-Baczewska, E. Styczyńska-Jurewicz and E. Neugebauer, 1984. The concentration of trace metals in some crustacea of the Black Sea and the Baltic Sea. Proc. of 121 Symp. of Atomic Spectroscopy on Application of the Emission and Absorption Atomic Spectroscopy in Fishing and Agricultural Sciences (Akademia Rolnicza, Szczecin), 13–15 February 1984, pp. 92–93.

Skwarzec, B., E. Holm, P. Roos and J. Pempkowiak, 1994. Nickel-63 in Baltic fish and sediments. Appl. Radiat. Isot. **45**, 609–611.

Skwarzec, B., D.I. Strumińska and A. Boryło, 2000. Bioaccumulation and distribution of plutonium in fish from Gdańsk Bay. J. Environ. Radioactivity (in press).

Sokolowski, A., M. Wolowicz, H. Hummel and R. Bogaards, 1999. Physiological responses of *Macoma balthica* to copper pollution in the Baltic. Oceanol. Acta **22**, 431–439.

Sokołowski, J., 1958. Birds of Polish Territory (Państwowe Wydawnictwo Naukowe, Warszawa), **Vol. 2**, 569 pp. (in Polish).

Sokołowski, J., 1965. Birds of Poland – Atlas (Państwowe Zakłady Wydawnictw Szkolnych, Warszawa), 272 pp. (in Polish).

Soto, M., M. Kortabitarte and I. Marigómez, 1995. Bioavailable heavy metals in estuarine waters as assessed by metal/shell-weight indices in sentinel mussels *Mytilus galloprovincialis*. Mar. Ecol. Prog. Ser. **125**, 127–136.

Soto, M., M.P. Ireland and I. Marigómez, 2000. Changes in mussel biometry on exposure to metals: implications in estimation of metal bioavailabilty in 'Mussel-Watch' programmes. Sci. Total Environ. **247**, 175–187.

Söderlund, S., Å. Forsberg and M. Pedersén, 1988. Concentrations of cadmium and other metals in *Fucus vesiculosus* L. and *Fontinalis dalecarlica* Br. Eur. from the northern Baltic Sea and the southern Bothnian Sea. Environ. Pollut. **51**, 197–212.

Spanovskaya, W.D., 1983. Smiejstvo Okunievyie (*Percidae*), in: Zhizn' Zhivotnyh – Lancetniki – Krugłorotnyie – Chrashchebyie Ryby – Kostnyie Ryby (The Life of Animals – Fish), ed. T.S. Rass (Prosveshchenie, Izdatelstvo, Moscow), **Vol. 4**, 370–378 (in Russian).

Steimle, F.W., V.S. Zdanowicz and D.F. Gadbois, 1990. Metal and organic contaminants in Northwest Atlantic deep-sea tilefish tissues. Mar. Pollut. Bull. **21**, 530–535.

Stenner, R.D., and G. Nickless, 1974. Distribution of some heavy metals in organisms in Hardangerfjord and Skjerstadfjord, Norway. Water Air Soil Pollut. **3**, 279–291.

Stepnowski, P., and B. Skwarzec, 1999. The bioaccumulation of ^{210}Po and ^{210}Pb in the southern Baltic organisms. Czech. J. Phys. **49/S1**, 205–209.

Stepnowski, P., and B. Skwarzec, 2000a. Tissue and subcellular distribution of ^{210}Po in the crustacean *Saduria entomon* inhabiting the southern Baltic Sea. J. Environ. Radioactivity **49**, 195–199.

Stepnowski, P., and B. Skwarzec, 2000b. A comparison of ^{210}Po accumulation in molluscs from the southern Baltic, the coast of Spitsbergen and Sasek Wielki Lake in Poland. J. Environ. Radioactivity **49**, 201–208.

Stoeppler, M., and H.W. Nürnberg, 1979. Comparative studies on trace metal levels in marine biota. III. Typical levels and accumulation of toxic trace metals in muscle tissue and organs of marine organisms from different European seas. Ecotoxicol. Environ. Safety **3**, 335–351.

Stoeppler, M., M. Burow, F. Backhaus, W. Schramm and H.W. Nürnberg, 1986. Arsenic in seawater and brown algae of the Baltic and the North Sea. Mar. Chem. **18**, 321–334.

Stokes, D., and L. Stokes, 1996a. Stokes Field Guide to Birds – Eastern Region (Little, Brown & Company Limited, Canada), 473 pp.

Stokes, D., and L. Stokes, 1996b. Stokes Field Guide to Birds – Western Region (Little, Brown & Company Limited, Canada), 520 pp.

Stoneburner, D.L., 1978. Heavy metals in tissues of stranded short-finned pilot whales. Sci. Total Environ. **9**, 293–297.

Stronkhorst J., 1992. Trends in pollutants in blue mussel *Mytilus edulis* and flounder *Platichthys flesus* from two Dutch estuaries, 1985–1990. Mar. Pollut. Bull. **24**, 250–258.

Strömgren, T., 1979. The effect of zinc on the increase in length of five species of intertidal Fucales. J. Experiment. Mar. Biol. Ecol. **40**, 95–102.

Struck, B.D., R. Pelzer, P. Ostapczuk, H. Emons and C. Mohl, 1997. Statistical evaluation of ecosystem properties influencing the uptake of As, Cd, Co, Cu, Hg, Mn, Ni, Pb, Zn in seaweed (*Fucus vesiculosus*) and common mussel (*Mytilus edulis*). Sci. Total Environ. **207**, 29–42.

Stureson U., 1976. Lead enrichment in shells of *Mytilus edulis*. Ambio **5**, 253–256.

Stureson U., 1978. Cadmium enrichment in shells of *Mytilus edulis*. Ambio **7**, 122–125.

Stureson U., and R.A. Reyment, 1971. Some minor chemical constituents in the shell of *Macoma balthica*. Oikos **22**, 414–416.

Sudaryanto, A., S. Takahashi, S. Tanabe, M. Muchtar and H. Razak, 2000. Butyltins residues in green mussels (*Perna viridis*) from Indonesian coastal waters and other Asian countries, in: Mussel Watch – Marine Pollution Monitoring in Asian Waters, ed. S. Tanabe (Center for Marine Environmental Studies, Ehime University, Japan), pp. 137–144.

Summers, R.W., J. Cooper and J.S. Pringle, 1977. Distribution and numbers of coastal waders (Charadrii) in the south-western Cape, South Africa, summer 1975–76. Ostrich **48**, 85–97.

Sunila, I, and R. Lindström, 1985. Survival, growth and shell deformations of cooper- and cadmium-exposed mussels (*Mytilus edulis* L.) in brackish water. Estuar. Coast. Shelf Sci. **21**, 555–565.

Sures, B., H. Taraschewski and J. Rokicki, 1997. Lead and cadmium content of two cestodes, *Monobothrium wageneri* and *Bothriocephalus scorpii*, and their fish hosts. Parasitol. Res. **83**, 618–623.

Svieshnikov, V.A., 1987. Klass Piervichnyie kolchetsy (Polychaeta), in: Zhizn' Zhivotnyh – Prostiejshie – Kishechnopolostnyie – Chervy (The Life of Animals – Worms), ed. Yu.I. Polyanskij (Prosveshchenie, Izdatelstvo, Moscow), **Vol. 1**, 363–376 (in Russian).

Swaileh, K.M., 1996. Seasonal variations in the concentrations of Cu, Cd, Pb and Zn in *Arctica islandica* L. (Mollusca: Bivalvia) from Kiel Bay, Western Baltic Sea. Mar. Pollut. Bull. **32**, 631–635.

Swaileh, K.M., and D. Adelung, 1994. Levels of trace metals and effect of body size on metal content and concentration in *Arctica islandica* L. (Mollusca: Bivalvia) from Kiel Bay, Western Baltic. Mar. Pollut. Bull. **28**, 500–505.

Swaileh, K.M., and D. Adelung, 1995. Effect of body size and season on the concentrations of Cu, Cd, Pb and Zn in *Diastylis rathkei* (Kröyer) (Crustacea: Cumacea) from Kiel Bay, Western Baltic. Mar. Pollut. Bull. **31**, 103–107.

Sydeman, W.J., and W.M. Jarman, 1998. Trace metals in seabirds, Steller sea lion, and forage fish and zooplankton from central California. Mar. Pollut. Bull. **36**, 828–832.

Szabo, B.J., 1968. Trace element content of plankton population from the Bahamas. Caribb. J. Sci. **8**, 185–186.

Szefer, P, 1986. Some metals in benthic invertebrates in Gdansk Bay. Mar. Pollut. Bull. **17**, 503–507.

Szefer, P., 1987. Concentration of uranium and thorium by some Baltic seaweeds and lake plants. Mar. Pollut. Bull. **18**, 439–442.

Szefer, P., 1991. Interphase and trophic relationships of metals in a southern Baltic ecosystem. Sci. Total Environ. **101**, 201–215.

Szefer, P., 1992. Uranium and its isotopes in marine biota. Stud. Mater. Oceanol. **62**, 39–85 (in Polish, with English abstract).

Szefer, P., 2000. Possible priorities for future research in the field of marine environmental pollution. Sci. Total Environ. **249**, 385–394.

Szefer, P., and J. Falandysz J, 1983a. Investigations of trace metals in long-tailed duck (*Clangula hyemalis* L.) from the Gdańsk Bay. Sci. Total Environ. **29**, 269–276.

Szefer, P., and J. Falandysz, 1983b. Uranium and thorium content in long-tailed duck (*Clangula hyemalis* L.). Sci. Total Environ. **29**, 277–280.

Szefer, P., and J. Falandysz, 1985. Trace metals in muscle tissue of fish taken from the southern Baltic. Z. Lebensm. Unters. Forsh. **181**, 217–220.

Szefer, P., and K. Szefer, 1985. Occurrence of ten metals in *Mytilus edulis* L. and *Cardium glaucum* from the Gdansk Bay. Mar. Pollut. Bull. **16**, 446–450.

Szefer, P., and J. Falandysz, 1986. Trace metals in the bones of scaup-ducks (*Aythya marila* L.) wintering in the Gdańsk Bay, Baltic Sea, 1982–83 and 1983–84. Sci. Total Environ. **53**, 193–199.

Szefer, P., J. and J. Falandysz, 1987. Trace metals in the soft tissues of scaup-ducks (*Aythya marila* L.) wintering in the Gdańsk Bay, Baltic Sea. Sci. Total Environ. **65**, 203–213.

Szefer, P., and R. Wenne, 1987. Concentration of uranium and thorium in molluscs inhabiting Gdansk Bay, Baltic Sea. Sci. Total Environ. **65**, 191–202.

Szefer, P., and B. Skwarzec 1988. Concentration of elements in some seaweeds from coastal region of the southern Baltic and Zarnowiec Lake. Oceanologia **25**, 87–98.

Szefer, P., and K. Szefer, 1990. Metals in molluscs and associated bottom sediments of the southern Baltic. Helgoländer Meeresunters. **44**, 411–424.

Szefer , and K. Szefer, 1991. Concentration and discrimination factors for Cd, Pb, Zn and Cu in benthos of Puck Bay, Baltic Sea. Sci. Total Environ. **105**, 127–133.

Szefer, P., and M. Wołowicz, 1993. Occurrence of metals in the cockle *Cerastoderma glaucum* from different geographical regions in view of principal component analysis. SIMO-Mar. Pollut. **64**, 253–264.

Szefer, P., and A. Kusak, 2000. Distribution and relationships of trace metals in zoobenthos and associated sediments of the southern Baltic (in preparation).

Szefer, P., S. Nicholson, 2000. Distribution of selected trace elements in soft tissue, shell and byssus of *Perna viridis* from Hong Kong waters (in preparation).

Szefer P., R. Gajewska, R. Bojanowski, Z. Ganowiak and M. Nabrzyski, 1982. Współczynniki koncentracji wybranych pierwiastków w rybach morskich, słodkowodnych i kalmarach (Concentration factors for selected elements in marine and freshwater fish and squids). Biul. MIR **5–6**, 44–48 (in Polish).

Szefer, P., B. Skwarzec and J. Koszteyn, 1985. The occurrence of some metals in mesozooplankton taken from the southern Baltic. Mar. Chem. **17**, 237–253.

Szefer, P., K. Szefer and B. Skwarzec, 1990a. Distribution of trace metals in some representative fauna of the southern Baltic. Mar. Pollut. Bull. **21**, 60–62.

Szefer, P., K. Szefer and J. Falandysz J., 1990b. Uranium and thorium in muscle tissue of fish taken from the southern Baltic. Helgoländer Meeresuntersuch. **44**, 31–38.

Szefer, P., W. Czarnowski, J. Pempkowiak and E. Holm, 1993a. Mercury and major essential elements in seals, penguins, and other representative fauna of the Antarctic. Arch. Environ. Contam. Toxicol. **25**, 422–427.

Szefer, P., J. Pempkowiak, B. Skwarzec, R. Bojanowski and E. Holm, 1993b. Concentration of selected metals in penguins and other representative fauna of the Antarctic area. Sci. Total Environ. **138**, 281–288.

Szefer P., K. Szefer and E. Styczyńska-Jurewicz, 1994a. Metale ciężkie w faunie i florze Wewnętrznej Zatoki Puckiej, in: Zatoka Pucka – Możliwości Rewaloryzacji, eds. L. Dowgiałło-Kruk and P. Ciszewski (Instytut Ochrony Środowiska, Warszawa), pp. 131–143 (in Polish).

Szefer, P., M. Malinga, K. Skóra and J. Pempkowiak, 1994b. Heavy metals in harbour porpoises from Puck Bay in the Baltic Sea. Mar. Pollut. Bull. **28**, 570–571.

Szefer, P., K. Szefer, J. Pempkowiak, B. Skwarzec, R. Bojanowski and E. Holm, 1994c. Distribution and coassociations of selected metals in seals of the Antarctic. Environ. Pollut. **83**, 341–349.

Szefer, P., G.P. Glasby, J. Pempkowiak and R. Kaliszan, 1995a. Extraction studies of heavy-metal pollutants in surficial sediments from the southern Baltic Sea off Poland. Chem. Geol. **120**, 111–126.

Szefer, P., M. Malinga, W. Czarnowski and K. Skóra, 1995b. Toxic, essential and non-essential metals in harbour porpoises of the Polish Baltic Sea, in: Whales, Seals, Fish and Man, eds. A.S. Blix, L. Waloe, O. Utang (Elsevier Science BV), 617–622.

Szefer, P., J. Geldon, A.A. Ali, A. Bawazir and M. Sa'd, 1997a. Distribution and association of trace metals in soft tissue and byssus of mollusc *Perna perna* from the Gulf of Aden, Yemen. Environ. Intern. **23**, 53–61.

Szefer, P., K. Ikuta, S. Kushiyama, K. Szefer, K. Frelek and J. Geldon, 1997b. Distribution and association of trace metals in soft tissue and byssus of *Mytilus edulis* from the east coast of Kyushu Island, Japan. Arch. Environ. Contam. Toxicol. **32**, 184–190.

Szefer, P., K. Ikuta, S. Kushiyama, K. Frelek and J. Geldon, 1997c. Distribution of trace metals in the Pacific oyster, *Crassostrea gigas*, and crabs from the east coast of Kyushu Island, Japan. Bull. Environ. Contam. Toxicol. **58**, 108–114.

Szefer, P., H.M. Fernandes, M.-J. Belzunce, B. Guterstam, J.M. Deslous-Paoli and M. Wolowicz, 1998a. Distribution of metallic pollutants in molluscs *Mytilidae* from the temperate, tropical and subtropical marine environments. First International Symposium, IEP '98 Issues in Environmental Pollution, The State and Use of Science and Predictive Models (Elsevier Science Ltd., Denver, Colorado, U.S.A.), 23–26.08.1998, Section of Abstract Book 4.04.

Szefer, P., J. Geldon J., A.A. Ali, F. Páez-Osuna, A.C. Ruiz-Fernandes and S.R. Guerrero Galvan, 1998b. Distribution and association of trace metals in soft tissue and byssus of *Mytella strigata* and other benthal organisms from Mazatlan harbour, mangrove lagoon of the northwest coast of Mexico. Environ. Intern. **24**, 359–374.

Szefer, P., G.P. Glasby, A. Kusak, K. Szefer, H. Jankowska, M. Wolowicz and A.A. Ali, 1998c. Evaluation of anthropogenic influx of metallic pollutants into Puck Bay, southern Baltic, in: Geochemical Investigations of the Baltic Sea and Surrounding Areas, eds. P. Szefer and G.P. Glasby (Elsevier Science Ltd, Great Britain) Appl. Geochem. (Spec. Issue) **13**, 293–304.

Szefer, P., J. Rokicki, K. Frelek, K. Skóra and M. Malinga, 1998d. Bioaccumulation of selected trace metals in lung nematodes, *Pseudalius inflexus*, of harbour porpoise (*Phocoena phocoena*) in a Polish Zone of the Baltic Sea. Sci. Total Environ. **220**, 19–24.

Szefer, P., K. Ikuta, K. Frelek, I. Zdrojewska and M. Nabrzyski, 1999a. Mercury and other trace metals (Ag, Cr, Co, and Ni) in soft tissue and byssus of *Mytilus edulis* from the east coast of Kyushu Island, Japan. Sci. Total Environ. **229**, 227–234.

Szefer, P., J. Jensen, C. Lockyer, A. Łomża, I. Zdrojewska, K. Frelek, K. Skóra, I. Kuklik and M. Malinga, 1999b. Intercomparison studies on distribution of Cu and Cd in liver and kidney of harbour porpoise (*Phocoena phocoena*) from a Polish sector of the Baltic Sea, coastal waters of Denmark and Greenland, in: Proceedings of the Twelfth Annual Conference of the European Cetaceans Society Monaco 20–24.01.1998 (eds. P.G.H. Evans and E.C.M. Parsons), 1998 (Published in Valencia, February 1999), 358.

Szefer, P., M. Wolowicz, A. Kusak, J.-M. Deslous-Paoli, W. Czarnowski, K. Frelek and M.-J. Belzunce-Segarra, 1999c. Distribution of mercury and other trace metals in the cockle *Cerastoderma glaucum* from the Mediterranean Lagoon, Etang de Thau. Arch. Environ. Contam. Toxicol. **36**, 56–63.

Szefer, P., A.A. Ali, A.A. Ba-Haroon, A.A. Rajeh, J. Geldon and M. Nabrzyski, 1999d. Distribution and relationships of selected trace metals in molluscs and associated sediments from the Gulf of Aden, Yemen. Environ. Pollut. **106**, 299–314.

Szefer, P., M. Domagała-Wieloszewska, J. Warzocha, A. Garbacik-Wesołowska and J. Geldon, 2000a. Distribution and relationships of mercury, lead, cadmium, copper and zinc in perch (*Perca fluviatilis*) from the Pomeranian Bay and Szczecin Lagoon, southern Baltic (submitted).

Szefer, P, M. Wolowicz and P.S. Rainbow, 2000b. Distribution of trace metals in barnacles (*Balanus improvisus*) in the Gulf of Gdańsk, Baltic Sea (in preparation)

Szefer P., T. Ciesielski, I. Zdrojewska, J. Falandysz and K. Szefer, 2000c. Distribution of heavy metals in seabirds wintering in the Gulf of Gdańsk (in preparation).

Szefer, P., T. Ciesielski, I. Zdrojewska, K. Skóra and I. Kuklik, 2000d. Distribution of heavy metals in seals, porpoise and dolphins in the Baltic Sea (in preparation).

Szefer, P., R. Bojanowski, B. Skwarzec, T. Ciesielski, K. Skóra and I. Kuklik, 2000e. Distribution of radionuclides in harbour porpoise, seals and dolphins from the coastal waters of the Baltic Sea (in preparation).

Szefer, P., I. Zdrojewska, J. Jensen, C. Lockyer, K. Skóra K., I. Kuklik and M. Malinga, 2000f. Intercomparison studies on distribution of heavy metals in liver, kidney and muscle of harbour porpoise, *Phocoena phocoena*, from a Polish Sector of the Baltic Sea and coastal waters of Denmark and Greenland (submitted).

Szefer, P., K. Frelek, K. Szefer, Ch.-B. Lee, B.-S. Kim, J. Warzocha and I. Zdrojewska, 2000g. Distribution of mercury and other trace elements in soft tissue, byssus and shells of *Mytilus edulis trossulus* from the southern Baltic (submitted).

Szefer, P., I. Zdrojewska, J. Jensen, C. Lockyer, T. Ciesielski, K. Skóra and I. Kuklik, 2000h. Mercury and selenium in liver, kidney and muscle of harbour porpoise, *Phocoena phocoena*, from the southern Baltic Sea, coastal waters of Denmark and Greenland. Sci. Total Environ. (submitted).

Szefer, P., S.W. Fowler, K. Ikuta, F. Páez Osuna F., Ch.-B. Lee, H.M. Fernandes, A.A. Ali, M.J. Belzunce Segarra, Nicholson C., H. Hummel, B. Guterstam, M. Wolowicz and M. Deslous-Paoli, 2000i. Metallic pollutants in soft tissue, shell and byssus of mollusc Mytilidae from temperate, subtropical and tropical geographical zones (in preparation).

Szefer, P., K. Ikuta, F. Páez Osuna, Ch.-B. Lee, H.M. Fernandes, A.A. Ali, M.J. Belzunce, S.W. Fowler, K. Frelek, H. Hummel, B. Guterstam and M. Deslous-Paoli, 2000j. Toxic metals in soft tissue and byssus of mollusc Mytilidae from different marine ecosystems. III Conference on Trace Metals "Effects on Organisms and Environment" (Polish Academy of Sciences, Sopot) Sopot, 6–8.06.2000. 186/P2-14.

Tamura, Y., T. Maki, H. Yamada, Y. Shimamura, S. Ochiai, S. Nishigaki and Y. Kimura, 1975. Studies on behaviour of accumulation of trace elements in fishes (III) Accumulation of selenium and mercury in various tissues of tuna. Ann. Rep. Tokyo Metr. Res. Lab. P.H. **26**, 200–204.

Tanabe, S., 1999. Butyltin contamination in marine mammals – A review. Mar. Pollut. Bull. **39**, 62–72.

Tanabe, S., M. Prudente, T. Mizuno, J. Hasegawa, H. Iwata and N. Miyazaki, 1998. Butyltin contamination in marine mammals from North Pacific and Asian coastal waters. Environ. Sci. Techn. **32**, 193–198.

Tanabe, S., S. Kan-atireklap, M. S. Prudente and A. Subramanian, 2000. Mussel watch: marine pollution monitoring of butyltins and organochlorines in coastal waters of Thailand, Philippines and India, in: Mussel Watch – Marine Pollution Monitoring in Asian Waters, ed. S. Tanabe (Center for Marine Environmental Studies, Ehime University, Japan), pp. 114–128.

Taylor, J.D., and J.A. Miller, 1989. The morphology of the osphradium in relation to feeding habits in meso- and neogastropods. J. Mollusc. Studies **55**, 227–237.

Taylor, D.L., S. Schliebe and H. Metsker, 1989. Contaminants in blubber, liver and kidney tissue of Pacific walruses. Mar. Pollut. Bull. **20**, 465–468.

Tedengren, M., B. Olsson, B. Bradley and L. Zhou, 1999. Heavy metal uptake, physiological response and survival of the blue mussel (*Mytilus edulis*) from marine and brackish waters in relation to the induction of heat-shock protein 70. Hydrobiologia **393**, 261–269.

Teigen, S.W., J.U. Skaare, A. Bjorge, E. Degre and S. Sand, 1993. Mercury and selenium in harbour porpoise (*Phocoena phocoena*) in Norwegian waters. Environ. Toxicol. Chem. **12**, 1251–1259.

Temara, A., M. Warnau, M. Jangoux and P. Dubois, 1997. Factors influencing the concentrations of heavy metals in the asteroid *Asterias rubens* L. (Echinodermata). Sci. Total Environ. **203**, 51–63.

Tervo, V., K. Erkomaa, H. Sandler, V. Miettinen, R. Parmanne and E. Aro, 1980. Contents of metals and chlorinated hydrocarbons in fish and benthic invertebrates in the Gulf of Bothnia and in the Gulf of Finland in 1979. Aqua Fennica **10**, 42–57.

The Illustrated Encyklopaedia of Birds. Consultant Dr. Christopher M. Perrins in co-operation with ICBP (Copyright© 1990 Marshall Editions Developments Limited for Polish edition by MUZA S.A. Warszawa), 1993, 420 pp.

Theede, H., I. Andersson and W. Lehnberg, 1979. Cadmium in *Mytilus edulis* from German coastal waters. Meeresforsch. **27**, 147–155.

Thompson, D.R., and R.W. Furness, 1989. The chemical form of mercury stored in South Atlantic seabirds. Environ. Pollut. **60**, 305–317.

Thompson, D.R., F.M. Stewart and R.W. Furness, 1990. Using seabirds to monitor mercury in marine environments. The validity of conversion ratios for tissue comparison. Mar. Pollut. Bull. **21**, 339–342.

Thorrold, S.R., C.M. Jones and S.E. Campana, 1997. Response of otolith microchemistry to environmental variations experienced by larval and juvenile Atlantic croaker (*Micropogonias undulatus*). Limnol. Oceanogr. **42**, 102–111.

Thresher, R.E., C.H. Proctor, J.S. Gunn and I.R. Harrowfield, 1994. An evaluation of electron-probe microanalysis of otoliths for stock identification and identification of nursery areas in a southern temperate groundfish, *Nemadactylus macropterus* (Cheilodactulidae). Fish. Bull. **92**, 817–840.

Thyen, S., P.H. Becker and H. Behmann, 2000. Organochlorine and mercury contamination of little terns (*Sterna albifrons*) breeding at the western Baltic Sea, 1978–96. Environ. Pollut. **108**, 225–238.

Tibury, K.L., J.E. Stein, J.P. Meador, C.A. Krone and S.-L. Chan, 1997. Chemical contaminants in harbour porpoise (*Phocoena phocoena*) from the north Atlantic coast: tissue concentrations and intra- and inter-organ distribution. Chemosphere **34**, 2159–2181.

Tohyama, C., S.-J. Himeno, C. Watanabe, T. Suzuki and M. Morita, 1986. The relationship of the increased level of metallothionein with heavy metal levels in the tissue of the harbor seal (*Phoca vitulina*). Ecotoxicol. Environ. Safety **12**, 85–94.

Tomilin, A.G., 1989. Otriad Kitoobraznyje (*Cetacea*), in: Zhizn' Zhivotnyh – Mlekopitayushchie (The Life of Animals – Mammals), ed. W.E. Sokolow (Prosveshchenie, Moscow), **Vol. 7**, 349–393 (in Russian).

Townsend, D.W., R.L. Radtke, D.P. Malone and J.P. Wallinga, 1995. Use of otolith strontium:calcium ratios for hindcasting larval cod distributions relative to water masses on Georges Bank. Mar. Ecol. Prog. Ser. **119**, 37–44.

Tuomainen, K., E. Ilus and T.K. Taipale, 1986. Accumulation of certain long-lived radionuclides by litoral algae and bottom animals, in: Study of Radioactive materials in the Baltic Sea. (International Atomic Energy Agency, Vienna). Report (IAEA-TECDOC-362) of the Final Research Coordination Meeting on the Study of Radioactive Materials in the Baltic Sea organized by the IAEA and held in Helsinki, Finland 24–28 September, 1984, pp. 69–78.

Turekian, K.K., A. Katz and L. Chan, 1973. Trace element trapping in pteropod tests. Limnol. Oceanogr. **18**, 240–249.

Van Netten, C., S.A. Hoption Cann, D.R. Morley and J.P. van Netten, 2000. Elemental and radioactive analysis of commercially available seaweed. Sci. Total Environ. **255**, 169–175.

Vermeer, K., and D.B. Peakall, 1979. Trace metals in seaducks of the Fraser River delta intertidal area, British Columbia. Mar. Pollut. Bull. **10**, 189–193.

Viale, D., 1994. Cetaceans as indicators of a progressive degradation of Mediterranean water quality. Intern. J. Environ. Studies **45**, 183–198.

Viarengo, A., 1989. Heavy metals in marine invertebrates: Mechanisms of regulation and toxicity at the cellular levels. Aquat. Sci. **1**, 295, 317.

Viarengo, A., and L. Canesi, 1991. Mussels as biological indicators of pollution. Aquaculture **94**, 225–243.

Vlastov, B.V., and P.V. Matekin, 1988. Klass Gastropoda, in: Zhizn' Zhivotnyh (The life of Animals – Molluscs – Crustaceans), ed. R.K. Pasternak (Prosveshchenie, Izdatelstvo, Moscow), **Vol. 2**, 14–65 (in Russian).

Voipio, A, K. Erkomaa, E. Karppanen, I. Mäkinen and V. Tervo, 1977. Eräiden raskaiden metallien ja organo klooriyhdisteiden pitoisuudet Itämeren kaloissa ja pohjaeläimissä. Ympäristö ja Terveys **2**, 127–143 (in Finnish).

Voloz, J., M. Simm, H. Jankovski and I. Kotta, 1990. Cadmium, lead, copper and zinc concentrations in *Mesidothea entomon* in the Gulf of Finland (southern part). Proc. Estonian Acad. Biol. **39**, 141–147.

Von Westernhagen, H., H. Rosenthal, V. Dethlefsen, W. Ernst, U. Harms and P.-D. Hansen, 1981. Bioaccumulating substances and reproductive success in Baltic flounder *Platichthys flesus*. Aquatic Toxicol. **1**, 85–99.

Vuorinen, P.J., T. Rantio, A. Witick and M. Vuorinen, 1994. Organochlorines and heavy metals in sea trout (*Salmo trutta m. trutta*) in the Gulf of Bothnia off the coast of Finland. Aqua Fennica **24**, 29–35.

Vuorinen, P.J., H. Haahti, M. Leivuori and V. Miettinen, 1998. Comparison and temporal trends of organochlorines and heavy metals in fish from the Gulf of Bothnia. Mar. Pollut. Bull. **36**, 236–240.

Wagemann, R., 1989. Comparison of heavy metals in two groups of ringed seals (*Phoca hispida*) from the Canadian Arctic. Can. J. Fish. Aquat. Sci. **46**, 1558–1563.

Wagemann, R., and D.C.G. Muir, 1984. Concentrations of heavy metals and organochlorines in marine mammals of northern waters: Overview and evaluation. Can. Techn. Rep. Fish. Aquat. Sci. (Western Region, Department of Fishereies and Oceans Winnipeg, Manitoba) No 1279, p. 97.

Wagemann, R, and B. Hobden, 1986. Low-molecular weight metalloproteins in tissues of the narwhal (*Monodon monoceros*). Comp. Biochem. Physiol. **84C**, 325–344.

Wagemann, R., N.B. Snow, A. Lutz and D.P. Scott, 1983. Heavy metals in tissues and organs of the narwahl (*Monodon monoceros*). Can. J. Fish. Aquat. Sci. **40**, 206–214.

Wagemann, R., R.E.A. Stewart, W.L. Lockhart, B.E. Stewart and M. Povoledo, 1988. Trace metals and methyl mercury: Associations and transfer in harp seals (*Phoca groenlandica*) mothers and their pups. Mar. Mammal Sci. **4**, 339–355.

Wagemann, R., R.E.A. Stewart, P. Beland and C. Desjardins. 1991. Heavy metals and selenium in tissues of beluga whales (*Delphinapterus leucas*) from the Canadian Arctic and the St. Lawrence estuary. Can. Bull. Fish. Aquat. Sci. **224**, 191–206.

Wagemann, R., S. Innes and P.R. Richard, 1996. Overview and regional temporal differences of heavy metals in Arctic whales and ringed seals in the Canadian Arctic. Sci. Total Environ. **186**, 41–66.

Wahbeh, M.I., 1984. Levels of zinc, manganese, magnesium, iron and cadmium in three species of seagrass from Aquaba (Jordan). Aquat. Bot. **20**, 179–183.

Wahbeh, M.I., D.M. Mahasneh and I. Mahasneh, 1985. Concentrations of zinc, manganese, copper, cadmium, magnesium and iron in ten species of algae and sea water from Aqaba, Jordan. Mar. Environ. Res. **16**, 95–102.

Walker, G., and P. Foster, 1979. Seasonal variation of zinc in the barnacle *Balanus balanoides* (L.) maintained on a raft in the Menai Strait. Mar. Environ. Res. **2**, 209–221.

Walker, G., P.S. Rainbow, P. Foster and D.J. Crisp, 1975a. Barnacles: possible indicators of zinc pollution? Mar. Biol. **30**, 57–75.

Walker, G., P.S. Rainbow, P. Foster and D.L. Holland, 1975b. Zinc phosphate granules in tissues surrounding the midgut of the barnacle *Balanus balanoides*. Mar. Biol. **33**, 162–166.

Warnau, M., S.W. Fowler and J.-L. Teyssié, 1996. Biokinetics of selected heavy metals and radionuclides in two marine macrophytes: the seagrass *Posidonia oceanica* and the alga *Caulerpa taxifolia*. Mar. Environ. Res. 41, 343–362.

Watanabe, I., S. Tanabe, M. Amano, N. Miyazaki, E.A. Petrov and R. Tatsukawa, 1998. Age-dependent accumulation of heavy metals in Baikal seal (*Phoca sibirica*) from Lake Baikal. Arch. Environ. Contam. Toxicol. **35**, 518–526.

Watson, D., P. Foster and G. Walker, 1995. Barnacle shells as biomonitoring material. Mar. Pollut. Bull. **31**, 111–115.

Watson, W.S., D.J. Summer, J.R. Baker, S. Kennedy, R. Reid and I. Robinson, 1999. Radionuclides in seals and porpoises in the coastal waters around the UK. Sci. Total Environ. **234**, 1–13.

Weber, A.,M. Krause, H. Marencic and R. Kopp, 1992. Schadstoffumsatz im Zooplankton, in: Prozesse im Schadstoffkreislauf Meer-Atmosphäre: Ökosystem Deutsche Bucht (PRISMA). BMFT-Projekt MFU 0620/6 2. Zwischenbericht, 01.01.–31.12.1991, 179–188.

Weeks, J.M., and P.S. Rainbow, 1991. The uptake and accumulation of zinc and copper from solution by two species of talitrid amphipods (Crustacea). J. Mar. Biol. Assoc. UK **71**, 811–826.

Weeks, J.M., and P.S. Rainbow, 1993. The relative importance of food and seawater as sources of copper and zinc to talitrid amphipods (Crustacea; Amphipoda; Talitridae). J. Appl. Ecol. **30**, 722–735.

Weeks, J.M., P.S. Rainbow and M.H. Depledge, 1995. Barnacles (*Chthamalus stallatus*) as biomonitors of trace metal bioavailability in the waters of Sao Miguel (Azores), in: Proc. Int. Workshop Marine Fauna and Flora of the Azores, August 1991 (Acoreana, Suppl. 1995), 103–111.

Weiss, D., and F. Moldenhawer, 1986. Results of the radiological Baltic-Monitoring Programme of the GDR during 1975-1983, in: Study of Radioactive materials in the Baltic Sea. (International Atomic Energy Agency, Vienna). Report (IAEA-TECDOC-362) of the Final Research Coordination Meeting on the Study of Radioactive Materials in the Baltic Sea organized by the IAEA and held in Helsinki, Finland 24–28 September, 1984, pp. 89–109.

Wenzel, Ch., D. Adelung and H. Theede, 1996. Distribution and age-related changes of trace elements in kittiwake *Rissa tidactyla* nestlings from an isolated colony in the German Bight, North Sea. Sci. Total Environ. **193**, 13–26.

White, D.H., and R.C. Stendell, 1977. Waterfowl exposure to lead and steel shot on selected hunting areas. J. Wildl. Manage. **41**, 469–475.

White, K.N., and Walker, 1981. Uptake, accumulation, and excretion of zinc by the barnacle, *Balanus balanoides* (L.). J. Experiment. Mar. Biol. 51, 285–298.

White, S.L., and P.S. Rainbow, 1984. Regulation of zinc concentration by *Palaemon elegans* (Crustacea: Decapoda): zinc flux and effects of temperature, zinc concentration and moulting. Mar. Ecol. Prog. Ser. **16**, 135–146.

Wiemeyer, S.N., T.G. Lamont and L.N. Locke, 1980. Residues of environmental pollutants and necropsy data for Eastern United States ospreys, 1964–1973. Estuaries **3**, 155–167.

Wiemeyer, S.N., T.G. Lamont, C.M. Bunck, C.R. Sindelar, F.J. Gramlich, J.D. Fraser and M.A. Byrd, 1984. Organochlorine pesticide, polychlorobiphenyl, and mercury residues in Bald Eagles eggs – 1969–79 -and their relationships to shell thinning and reproduction. Arch. Environ. Comtam. Toxicol. **13**, 529–549.

Wiktor, J., 1969. Biologia *Dreissena polymorpha* (Pall.) i jej ekologiczne znaczenie w Zalewie Szczecińskim (The biology of *Dreissena polymorpha* (Pall.) and its biological importance in the Firth of Szczecin). Studia i Materiały (Morski Instytut Rybacki, Gdynia), **Ser. A** (No. 5), 88 pp. (in Polish with English summary).

Wiktor, K., 1985. An attempt to determine trophic structure of the bottom fauna in coastal waters of the Gulf of Gdańsk. Oceanologia **21**, 109–121.

Wiktor, K., 1990. The role of common mussel *Mytilus edulis* L. in biocenosis of the Gulf of Gdańsk. Limnologica (Berlin) **20**, 197–190.

Wilson, J.G., 1983. The uptake and accumulation of Ni by *Cerastoderma edule* and its effect on mortality, body condition and respiration rate. Mar. Environ. Res. **8**, 129–148.

Wilson, J.G., and B. Elkaim, 1992. Estuarine bioindicators – a case for caution. Acta Ecologica **13**, 345–358.

Wilson, J.R., M.A. Czajkowski and M.W. Pienkowski, 1980. The migration through Europe and wintering in West Africa of curlew sandpiper. Wildfowl **31**, 107–122.

Windom, H.L., 1972. Arsenic, cadmium, copper, lead, mercury and zinc in marine biota – North Atlantic Ocean. IDOE Baseline Conf. and Workshop (Brookhaven), 121–147.

Windom, H.L., and D.R. Kendall, 1979. Accumulation and biotransformatiom of mercury in coastal and marine biota, in: The biogeochemistry of mercury in the environment, ed. J.O. Nriagu (Elsevier/North Holland Biomedical Press, Amsterdam), **Chapt. 13**, 303–323.

Windom, H., R. Stickney, R. Smith, D. White and F. Taylor, 1973. Arsenic, cadmium, copper, mercury, and zinc in some species of North Atlantic finfish. J. Fish. Res. Board Can. **30**, 275–279.

Windom, H., D. Stein, R. Sheldon and Jr. R. Smith, 1987. Comparison of trace metal concentrations in muscle tissue of a benthopelagic fish (*Coryphaenoides armatus*) from the Atlantic and Pacific Oceans. Deep-Sea Res. **34**, 213–220.

Witzel, B., 1989. Schwermetallkonzentrationen in Copepoden aus verschiedenen Wassermassen der Deutschen Bucht. Z. Angew. Zool. **3**, 303–332.

Woodhead, D.S., 1984. Contamination due to radioactive materials, in: Marine Ecology, ed. O. Kinne (John Wiley & Sons, New York), pp. 1111–1287.

Wong, C.K.C., R.Y.H. Cheung and M.H. Wong, 2000. Heavy metal concentrations in green-lipped mussels collected from Tolo Harbour and markets in Hong Kong and Shenzhen. Environ. Pollut. **109**, 165–171.

Wright, P., and C.F. Mason, 1999. Spatial and seasonal variation in heavy metals in the sediments and biota of two adjacent estuaries, the Orwell and the Stour, in eastern England. Sci. Total Environ. **226**, 139–156.

Yamamoto, I., and Y. Takizawa, 1982. Contents of heavy metals in fishes and shellfishes from the shore of Hokkaido. Akita J. Med. **9**, 207–214.

Yamamoto, Y., K. Honda, H. Hidala and R. Tatsukawa, 1987. Tissue distribution of heavy metals in Weddell seals (*Leptonychotes weddelli*). Mar. Pollut. Bull. **18**, 164–169.

Yamada, M., T. Aono and S. Hirano, 1999. $^{239+240}Pu$ and ^{137}Cs concentrations in fish, cephalopods, crustaceans, shellfish, and algae collected around the Japanese coast in the early 1990s. Sci. Total Environ. **239**, 131–142.

Yeats, P., G. Stenson and J. Hellou, 1999. Essential elements and priority contaminants in liver, kidney, muscle and blubber of harp seal beaters. Sci. Total Environ. **243/244**, 157–167.

Zafiropoulos, D., and A.P. Grimanis, 1977. Trace elements in *Acartia clausi* from Elefsis Bay of the Upper Saronikos Gulf, Greece Mar. Pollut. Bull. **8**, 79–81.

Zatsepin, V.I., Z.A. Filatova and A.A. Shilejko, 1988. Klass Dwustworchatyie molljuski (*Bivalvia*), in: Zhizn' Zhivotnyh – Molljuski – Igłokożie – Chlenistonogie (The Life of Animals – Molluscs), ed. R.K. Pasternak (Prosveshchenie, Izdatelstvo, Moscow), **Vol. 2**, 65–112 (in Russian).

Zaouali, J., 1977. Contribution a la connaissance de la faune malacologigue du lac Ichkeul (Tunisie septentrionale), etude du Bivalve *Cerastoderma glaucum*. Arch. Inst. Pasteur de Tunis, **1–2**, 119–126.

Zauke, G.-P., M. Krause and A. Weber, 1996. Trace metals in mesozooplankton of the North Sea: Concentrations in different taxa and preliminary results on bioaccumulation in copepod collectives (*Calanus finmarchicus/C. helgolandicus*). Int. Revue Ges. Hydrobiol. **81**, 141–160.

Zauke, G.-P., J. Bohlke, R. Żytowicz, P. Napiórkowski and A. Giziński, 1998. Trace metals in tripton, zooplankton, zoobenthos, reeds and sediments of selected lakes in north-central Poland. Int. Rev. Hydrobiol. **83**, 501–526.

Zauke, G.-P., V.M. Savinov, J. Ritterhoff and T. Savinova, 1999. Heavy metals in fish from the Barents Sea (summer 1994). Sci. Total Environ. **227**, 161–173.

Ziegelmeier, E., 1957. Die Muscheln (Bivalvia) der deutschen Meeresgebiete. Helg. Wiss. Meeresunt. **6**, 51 pp.

Zingde, M.D., S.Y.S. Singbal, C.F. Moraes and C.V.G. Reddy, 1976. Arsenic, copper, zinc and manganese in the marine flora and fauna of coastal and estuarine waters around Goa. Indian J. Mar. Sci. **5**, 212–217.

Chapter 4

Deposits as a Medium for Chemical Elements

A. BOTTOM SEDIMENTS

(i) Introduction

General characteristics

The sediments reflect events taking place in the water column and at the bottom and they may play a fundamental role as a tape-recorder of the historical development. The grain size and the composition of Baltic sediment material are often used as a basic criteria for distinguishing its various types. A great attention is paid to the prevailing bottom dynamic conditions resulting in one of the most commonly used classification system, i.e. erosion, transportation and accumulation (Winterhalter, 1972; Håkanson and Jansson, 1983; Danielsson, 1998).

- Areas of erosion prevail where there is no apparent deposition of fine particles but rather a removal of this materials. These bottoms have hard structure and consist of sand, gravel, consolidated clays and/or rocky material.
- Areas of transportation occur where fine materials are deposited periodically leading to the formation of mixed bottom sediments; such bottom type dominates the open areas of the Baltic Sea. However, it is sometimes difficult to distinguish areas of erosion from areas of transportation.
- Areas of accumulation appear where the fine materials are deposited continuously resulting in the formation of soft bottoms; such areas are generally characterised by high concentration of pollutants.

Hard and sandy sediments deposited within the areas of erosion and transportation have mostly small content of water, nutrients and pollutants. The conditions in the area of transportation are variable, especially for the most mobile substances, e.g. Fe, Mn and P which may be sensitive to any changes such as the redox potential in the sediments. Fine material may be deposited for a long time during stagnant conditions as well as may be resuspended and transported up and away towards the area of accumulation in the deeper parts where continuous deposition takes place (Håkanson and Jansson, 1983). It is estimated that up to 80% of the deposited material at accumulation bottom may be consisted of particles originating from old eroded sediments (Håkanson and Jansson, 1983; Danielsson, 1998). The main sources of sediment material are generally riverine suspended matter or reworked shelf sediments and atmospheric precipitation (Warren, 1981). According to Zöllmer and Irion (1993) the various sources and deposition areas can be identified basing on the spatial distribution pattern of sedimentary clay mineral composition. Several authors (Łomniewski et al., 1975; Blazhchishin, 1982a, 1982b, 1982c; Blazhchishin and Lukashev, 1981; Emelyanov and Pustelnikov, 1982) overviewed distribution of bottom sediments of the Baltic Sea as well as their geochemical composition and litho-stratigraphic characteristics. Iron-bearing minerals in the Baltic Sea have been identified using Mössbauer study (Görlich et al., 1985, 1989).

Baltic Sea

Sedimentation in the Baltic Sea is complex, 83% of the sediments are supplied by erosion, 11% by rivers and only 6% are of biogenic origin reflecting the dynamic nature of ecosystem (Glasby and Szefer, 1998). Average sedimentation rate in the Baltic Sea is ca. 1 mm yr^{-1}. Baltic area can be divided into five major sedimentation zones, i.e. coastal sands, relict clastic deposits, sediment non-deposition, sediment erosion and silts and clays. Muds are detected only below the permanent halocline and the wave-base level (Glasby et al., 1997a). In open part of the Baltic Proper, both areas of transportation and erosion, being well oxidised above and below the halocline, occupy a large bottom areas down to a water depth of 75 m (Kjellin et al., 1987; Carman and Wulff, 1989). Accumulation bottoms generally exist below 75 m where deposited sediments down to 110 m are usually oxic in their surficial layer while those occurred deeper down are anoxic or oxic depending on their spatial distribution. According to Stigebrandt and Wulff (1987), 75% of the total sediment area in the Baltic Proper is located above water depth of 75 m, 12% between 75–100 m and 15 % below 110 m. In some areas abundance of benthic organisms under hypoxic or anoxic conditions is often significantly reduced causing no bioturbation. As a result of this, the area covered by laminated sediments has increased substantially during the last decades. Among parameters responsible for the formation of laminated sediments are probably elevated load of nutrients to the bottom and a decreasing frequency of deep water renewal (Jonsson and Jonsson, 1988; Jonsson et al., 1990; Jonsson, 1992; Persson and Jonsson, 2000).

Kattegat and Skagerrak

In Danish Strait area the grain size and clay mineral composition of sediments do not change remarkably with sediment depth pattern because of their relative uniform distribution for ca. 100 years (Meyenburg and Liebzeit, 1993; Danielsson, 1998). In the open waters of Skagerrak, the muddy and fine grained sediments exist at water depths deeper down than 50–70 m where the surficial sediments ussually contain post-glacial, moderately-sorted silty clays (Paetzel et al., 1994). It should be emphasised that significant quantities, i.e. 40 million tonnes of suspended matter are transported annually from the North Sea to Skagerrak and 50–70% is deposited there (Meyenburg and Liebzeit, 1993).

Overview of Worldwide Literature

The anthropogenic input of trace elements into the coastal and estuarine zones of the southern Baltic Sea has possibly resulted in changes in the ecological equilibrium in this area. Since both natural and anthropogenic material have accumulated together, it is difficult to identify what proportion of the measured concentration of the elements in the sediments is natural and what is anthropogenic. This problem is exacerbated by the variable inputs of elements originating from these two sources, which may vary by several orders of magnitude, as well as the dependence of the element concentrations on factors such as the sediment grain size, mineralogy and organic carbon content (Loring, 1990). Sediments from the Gulf of Gdansk are composed of variable amounts of trace element-poor phases (e.g. quartz sands) and trace element-rich phases (e.g. clay minerals, iron and manganese oxyhydroxides and organic matter). Since the latter phases occur predominantly in the fine-grained fraction of the sediments, increasing concentrations of trace elements with decreasing grain size of the sediments are observed (e.g. Schoer et al., 1982).

To compensate for the influence of grain size on element variability in various size fractions of the sediments, a number of granulometrical and geochemical techniques have been developed (Horowitz, 1991). Some authors have estimated an enrichment of particular elements to identify their sources in sediments and suspended material (Buat-Menard and Chesselet, 1979; Li, 1981a, 1981b, 1982; Borole et al., 1982; Kingston and Greenberg, 1984; Brügmann, 1986a; Szefer and Skwarzec, 1988; Loring 1990; Szefer, 1990a, 1990b). However, the identification of element pollution in sediments is often difficult and therefore determination of additional indices such as correlation coefficients, linear regression characteristics and anthropogenic factor is desirable (Buat-Menard and Chesselet, 1979; Sarin et al., 1979; Li, 1982; Borole et al., 1982; Szefer and Skwarzec, 1988; Windom et al., 1989; Szefer 1990a, 1990b). An approach that is often used is to compare the element concentrations in the surficial layers of sediment with those in the deeper layers (Brügmann and Hennings, 1982; Szefer and Skwarzec, 1988). Bottom sediments have been extensively studied in respect to their geochemical composition (Bonatti et al., 1971; Mo et al., 1973; Trefry and Presley, 1976; Grimanis et al., 1977; Förstner, 1980; Hunt and Smith, 1983; Marchig et al., 1985; Ingri and

Pontér, 1986b; Kersten and Förstner, 1986; Windom et al., 1989; Förstner and Schoer, 1990; Glasby et al., 1990; Förstner and Salomons, 1991; Macdonald et al., 1991; Bricker, 1993; Van Alsenoy et al., 1993; Valette-Silver et al., 1993; Zwolsman et al., 1993; Williamson et al., 1994, 1995, 1996; Magalhães et al., 1995; Stuer-Lauridsen and Dahl, 1995; Szefer et al., 1998a; Baptista Neto et al., 2000; Williams et al., 2000). Evaluation of the geochemistry of interstitial water as well as sediment–water interaction and equilibrium state for trace elements partitioned in the carbonate-interstitial water system have been made by Presley and Trefry (1980) and Kulik and Kersten (1999).

Radiochemical analyses of bottom sediments from the marine environments have been extensively performed since the 70's by Joshi and Ganguly (1972), Koide et al. (1973, 1976), Li and Chan (1979), Cochran and Krishnaswami (1980), Bruland et al. (1981) and others. The distribution and fate of U in marine bottom sediments were discussed on the basis of literature data (Szefer, 1984). Numerous data of radiochemical analyses of bottom sediments are also published recently (e.g. Tkalin et al., 1998; Baskarani et al., 2000).

(ii) Normalisation of Sediment Data

According to Horowitz (1991) the physical and chemical factors which affect sediment chemistry can alter significantly the distribution pattern and concentration of trace elements associated with both suspended and bottom sediments. He discussed comprehensively sediment-trace element manipulations in respect to particular factors such as limitations of analytical data, corrections for grain-size differences, carbonate corrections and normalisation to 'conservative' elements. To compensate for the influence of grain size on element variability in various size fractions of the sediments, a number of granulometrical and geochemical techniques have been developed.

The granulomatric approach involves normalisation of trace element data relative to the different size fractions sieved from the bulk sediment, such as $< 100\ \mu m$ (Luoma and Bryan, 1981; Bryan and Langston, 1992), $< 80\ \mu m$ (Szefer et al., 1995a), $< 63\ \mu m$ (Ackermann et al., 1983; Horowitz et al., 1989, 1990), $60–20\ \mu m$ (Sager et al., 1990), $< 20\ \mu m$ (Glasby et al., 1988), $6\ \mu m$ (De Groot, 1964) and $< 2\ \mu m$ (Irion and Müller, 1987). This approach requires a separation step but the information obtained, although important, is incomplete since it does not reflect the bulk element concentration in the sediment.

The geochemical methods include normalisation of the element concentrations in the sediment relative to a conservative element such as Al (Hirst, 1962a, 1962b; Bruland et al., 1974; Kemp et al., 1976; Borole et al., 1977, 1982; Buat-Menard and Chesselet, 1979; Martin and Meybeck, 1979; Schoer et al., 1982; Brügmann, 1986a; Bernard et al., 1989; Windom et al., 1989; Ingri et al., 1991), Fe (Rule, 1986; Szefer and Skwarzec, 1988; Szefer, 1990a, 1990b), Sc (Ackermann, 1980; Schoer et al., 1982; Norman and De Deckker, 1990), Rb (Allen and Rae, 1986), Li (Loring, 1990), middle rare earth elements (Eu) or the heavy

REE (Yb, Lu) (Olmez et al., 1991; Szefer et al., 1999), Mg (Hilton et al., 1985) and K (Förstner and Salomons, 1980). These conservative elements are dominantly lithogenous in origin. Alternatively, the element concentration in coastal surficial sediments may be compared with that in off-shore sediments which are thought to be unpolluted (Trefry and Presley, 1976; Voutsinou-Taliadouri, 1981; Voutsinou-Taliadouri and Satsmadjis, 1982, 1983; Satsmadjis and Voutsinou-Taliadouri, 1985; Voutsinou-Taliadouri and Georgakopoulou-Grigoriadou, 1989). This approach is viable when the other phases which determine the trace element contents have also been analysed and the system is well understood. An attempt is made to evaluate the relationship between the trace element content of the surficial sediments in the Gulf of Gdansk and the southern Baltic Sea, and the sediment characteristics with the aim of establishing the source of the elements in the sediments (Szefer et al., 1996).

(iii) Chemical Elements in Bottom Sediments

Some preliminary effort has been expended in studying the heavy-metal pollution in the sediments of the Baltic Sea including lagoons by Manheim (1961), Erlenkeuser et al. (1974), Niemistö and Voipio (1974), Suess and Erlenkeuser (1975a, 1975b), Boström et al. (1978, 1983), Niemistö et al. (1978), Hallberg (1979), Brügmann et al. (1980, 1991/1992), Müller et al. (1980), Brügmann (1981), Blazhchishin (1982b), Emelyanov (1982, 1995a, 1995b), Emelyanov et al. (1982), Brzezińska et al. (1984), Irion (1984), Dietrich and Beuge (1986), Emelyanov and Wypych (1987), Brügmann (1988), Belzunce Segarra et al. (1987, 1988, 2000), Skwarzec et al. (1988), Szefer and Skwarzec (1988), Jacobsen and Postma (1989), Leipe et al. (1989, 1994, 1995, 1998), Tervo and Niemistö (1989); Brügmann and Lange (1990), Szefer (1990a, 1990b), Szefer and Szefer (1990), Blomqvist et al. (1992), Brügmann (1992), Helios Rybicka (1992), Belmans et al. (1993), Leivouri and Niemistö (1993, 1995), Szefer and Kaliszan (1993a, 1993b), Szefer et al. (1993a, 1993b, 1995a, 1995b, 1996, 1998b, 1999, 2000), Salonen et al. (1995), Widerlund and Ingri (1995, 1996), Widerlund (1996), Huckriede and Meischner (1996), Huckriede et al. (1996), Lithner et al., (1996), Neumann et al. (1996, 1998), Sohlenius et al. (1996), Brügmann and Matschullat (1997), Neumann et al. (1997), Sternbeck and Sohlenius (1997), Usenkov (1997), Callaway et al. (1998), Emeis et al. (1998), Laima et al. (1998, 2001), Leivouri and Vallius (1998); Osadczuk and Wawrzyniak-Wydrowska (1998), Pempkowiak et al. (1998a, 1999), Pohl et al. (1998), Siegel et al. (1998), Szefer (1998), Szefer et al. (1998b), Uścinowicz et al. (1998), Vallius and Lehto (1998), Virkanen (1998), Falandysz (1999), Lampe (1999), Müller and Heininger (1999), Osadczuk (1999), Vallius (1999a, 1999b), Vallius and Leivouri (1999), Sternbeck et al. (2000), Szefer and Kusak (2000), Glasby et al. (2001), Emelyanov (2001), Heiser et al. (2001) and Neumann et al. (2001). These studies included Skagerrak and/or Kattegat subarea (Kuijpers et al., 1993; Pederstad et al., 1993; Paetzel et al., 1994; Magnusson et al., 1996; Danielsson et al., 1999). Baltic sediments have been analysed recently for concentrations of butyltins (Kannan and Falandysz, 1997; Senthilkumar et al., 1999; Bisseli et al., 2000). Geochemical and geological maps of the southern Baltic Sea and the Vistula Lagoon have been prepared by the Polish Geological Institute (Geological Map of the Baltic Sea Bottom, 1989–1995; Uścinowicz and Zachowicz, 1993; Szczepańska and Uścinowicz, 1994; Geological Atlas of the Southern Baltic, 1995; Kramarska et al., 1999). A more detailed study involving the selective leaching of elements from southern Baltic sediments has been reported by Szefer et al. (1995a, 1995b). A multivariate analysis of the Baltic data has already been attempted by Szefer et al. (1993a, 1993b, 1995a). Belzunce et al. (1988, 2000) presented results of speciation analysis of heavy metals in sediment cores from the Gulf of Gdańsk. Minor- and macroelements of the interstitial waters of Baltic sediments have been studied sporadically (Dietrich and Beuge, 1986; Bolałek, 1992a; Golimowski and Szczepańska, 1996; Brügmann et al., 1998; Neumann et al., 1998).

Horizontal distribution (surficial sediments)

The concentration data of chemical elements in surficial sediments of the Baltic Sea and neighbouring areas are presented in Tables 4.1 and 4.2. The large-scale distribution patterns of selected elements in Baltic sediments have been studied by Borg and Jonsson (1996). These authors observed that the mean concentrations of As and Hg in the surface sediments from the Bothnian Bay were 6- and 4 times higher, respectively than those observed in the Baltic Proper and these decrease southwards (Figs. 4.1 and 4.2). This distribution pattern seems to be related to the historically, very significant anthropogenic input of As into the Bothnian Bay from smelter emission at Rönnskär over several decades. As for Hg, to main its anthropogenic sources belong probably the smelter emissions in the north, and further south in the Bothnia Sea (the Gävle Bight) and earlier large emissions from the chloralkali and cellulose industries (Borg and Jonsson, 1996). The spatial trends for Cd being similar to Cu, Pb and Zn but more distinctly marked, were different from those of As and Hg (Figs. 4.1 and 4.2); it suggests other emission source of these elements in the Baltic Sea. The mean levels of Cd, Cu, Ni, Pb and Zn in the Baltic Proper were higher, even in the case of Cd ca. an order of magnitude as background value, as compared to other subareas, e.g. the Bothnian Sea. Elevated levels of Pb also observed in the Bothnian Bay were probably related to the large smelter emission (Fig. 4.2). This difference seems to be caused by the distribution of organic matter, which highly correlated with Zn and Cu levels in sediments. According to Borg and Jonsson (1996) if these metals levels are normalised to organic matter contents, the distribution pattern is unremarkably changed, i.e. not enough extensively to explain the whole difference between these two Baltic subareas.

According to Emelyanov (1982) concentrations of Au in surficial sediments of the Baltic Sea ranged from 0.1 to 4.5 ng g^{-1} (on the average 3.0 ng g^{-1}). This element occurred mainly in higher levels in clay sediments of central parts of Baltic deeps and in lower levels in coastal areas (Emelyanov, 1982).

TABLE 4.1.

Concentrations of trace elements (in $\mu g\ g^{-1}$ dry wt.), Al, Ca, K, Mg, Na, S, Mn and Fe ($mg\ g^{-1}$ dry wt.) in surface sediments of the Baltic Sea

Region	Sampling date	Sample depth (m)	Fraction (μm)	N	Al	Ag	As	Ba	Be	Bi	References
Baltic Sea											
Southern part	1981–83	60–120	(98% of < 63 μm fraction	5	14.8			130			Belmans et al., 1993
					3.0–22.0			120–140			
Western part	1981–83	25–50		4	13.3			55.7			
					3.0–18.0			73.0–84.0			
North-eastern part	1981–83	110–240		4	15.6			160			Belmans et al., 1993
					13.0–18.0			120–180			
Northern Baltic											
Gulf of Finland	1992–93		< 2 mm	20	76		16				Leivouri, 1998
					59–99		8.0–28				
Gulf of Finland	1992–95	0–10	< 2 mm	42			14				Vallius and Leivouri, 1999
							6.0–28.0				
Bothnian Bay	1991–93	> 60	< 2 mm				278±41				Leivouri and Niemistö, 1995
	1992–93		< 2 mm	8	55±08		239				Leivouri and Niemistö, 1993
					47–66		167–320				
	Pre-1993	< 60	< 2 mm	32	46		91				Leivouri and Niemistö, 1993
					32–54		8–320				
Bothnian Sea	1991–93	> 60	< 2 mm				61±11				Leivouri and Niemistö, 1995
	1992–93			13	62		63				Leivouri and Niemistö, 1993
					44–87		46–85				
	Pre-1993	< 60	< 2 mm	24	49		26				Leivouri and Niemistö, 1993
					33–60		4.0–76 (10)				
Danish Straits		50–600		2	17			75			Belmans et al., 1993
					15.0–19.0			50–100			
Southern Baltic											
Vistula Estuary	1988–89		< 80	4	34.7	3.52					Szefer et al., 1995a
					25.2–48.3	2.0–5.0					
Gulf of Gdansk			< 80	22	26.5	0.78					Szefer et al., 1995a
					9.6–41.1	0.17–1.97					
	1996		< 63	3	55.5		14.8		1.78		Szefer et al., 2000
					31–95		3.9–28.5		0.48–3.03		

Region	Sampling date	Sample depth (m)	Fraction (μm)	N	Al	Ag	As	Ba	Be	Bi	References
Pomeranian Bay	1996		< 63	12	32.8		4.1		0.92		
					21–39		ND–10.9		0.20–2.15		
Central part of Polish EEZ	1988–89		< 80	5	26.5	1.07					Szefer et al., 1995a
					9.5–47.3	0.50–2.56					
	1996		< 63	7	52.1		12.4		1.6		Szefer et al., 2000
					13–94		4.9–23.3		0.32–2.71		
Bornholm Deep	1997		< 63	1	76		29.3		3.8		
Vistula Lagoon	1995	1.0–5.0	< 63	25		0.52	7.32	355		0.31	Szefer et al., 1999
						0.38–0.67	5.0–19.0	320–399		0.24–0.46	
	1996		< 63	27	54.4		7.65		1.87		Szefer et al., 2000
					42–73		4.5–15.1		0.78–2.82		
Szczecin Lagoon	1996		< 63	9	30.3		29.6		1.27		Szefer et al., 2000
					19–43		16.2–54.1		0.75–2.07		
Vistula Estuary	1988–89		< 2 mm	4	11.8	0.55					Szefer et al., 1996
					2.9–22.1	0.14–0.79					
Gulf of Gdansk	1988–89		< 2 mm	22	16.7	0.18					Szefer et al., 1996
					2.5–35.5	0.03–0.46					
Central part of Polish EEZ	1988–89		< 2 mm	5	10.6	0.05					Szefer et al., 1996
					4.4–29.1	0.02–0.07					
Western Baltic											
German coastal waters	1991–95		< 20	96			3.7–25.0				Müller and Heininger, 1999
Szczecin Lagoon			< 20	6			21–30				Müller and Heininger, 1999

TABLE 4.1. – continued

Region	Sampling date	Sample depth (m)	Fraction (μm)	N	Ca	Cd	Co	Cr	Cs	Cu	Fe	References
Baltic Sea												Belmans et al., 1993
Southern part	1981–83	60–120	(98% of < 63 μm fraction	5		1.56	11	47.2		31.6	2.69	
						0.2–4.8	8.0–18.0	13.0–76.0		14.0–50.0	8.0–41.0	
Western part	1981–83	25–50		4		1.13	10	43.8		31.8	18.5	
						0.4–2.0	8.0–13.0	14.0–64.0		19.0–39.0	7.0–27.0	
North-eastern part	1981–83	110–240		4		7.98	18	61.8		83.3	29.8	
						4.2–11.0	9.0–33.0	42.0–77.0		77.0–91.0	19.0–39.0	
Northern Baltic												
Gulf of Finland	1992–93		< 2 mm	20		1.06		85				Leivouri, 1998
						0.34–2.19		53–117				
Gulf of Finland	1992–95	0–10	< 2 mm	43		1.23		87		43		Vallius and Leivouri, 1999
						0.28–2.19		57–138		27–57		
Bothnian Bay	1991–93	> 60	< 2 mm			0.8±0.3				46±15	6.4±1.9	Leivouri and Niemistö, 1995
	1992–93		< 2 mm	8		0.94		73.8		52		Leivouri and Niemistö, 1993
						0.23–1.98		56–81		18–80		
	Pre-1993	< 60	< 2 mm	32		0.53	27	67		29	5.5	Leivouri and Niemistö, 1993
						0.19–1.15 (28)	< 10–67 (27)	39–120		11.0–59	2.6–10.1 (29)	
Bothnian Sea	1991–93	> 60	< 2 mm			0.4±0.2				30±11	5.3±1.8	Leivouri and Niemistö, 1995
	1992–93			13		2.49		80		36		Leivouri and Niemistö, 1993
						0.62–3.28 (9)		48–97		19–45		
	Pre-1993	< 60	< 2 mm	24		0.33	21	60		28	3.9	Leivouri and Niemistö, 1993
						0.04–1.28	< 10–61 (11)	24–110		< 10–42 (20)	1.2–7.5	
Danish Straits		50–600		2		0.4	15	51.5		18.5	26.5	Belmans et al., 1993
						0.30–0.50	12.0–18.0	41.0–62.0		17.0–20.0	23–30.0	
Southern Baltic												
Vistula Estuary	1988–89		< 80	4	6.39	4.77	13.9	47.4	7.28	58.4	34.8	Szefer et al., 1995a
					3.06–7.83	3.39–6.0	11.2–17.0	31.7–68.7	5.1–10.6	49.1–79.0	28.3–46.4	
Gulf of Gdansk	1988–89			22	7.21	1.4	10.7	27.5	5.83	30.2	23.8	Szefer et al., 1995a
					0.35–20.6	0.56–2.63	3.6–20.0	13.3–57.6	3.1–10.4	14.5–71.5	13.3–33.5	
	1996		< 63	3		1.38	8.77	78.2		34.5	4.31	Szefer et al., 2000
						1.0–1.9	3.9–16.5	64–95		18.6–45.1	2.1–7.6	
Pomeranian Bay	1996		< 63	12		3.2	5.08	76.8		21.6	2.63	
						0.8–7.2	3.0–6.6	49–146		11.7–27.9	1.7–3.7	

Region	Sampling date	Sample depth (m)	Fraction (μm)	N	Ca	Cd	Co	Cr	Cs	Cu	Fe	References
Central part	1988–89		< 80	5	29.3	1.42	12.5	30.7	6.34	30.7	21.1	Szefer et al., 1995a
of Polish EEZ					4.53–60.0	0.70–3.18	5.9–28.9	24.4–40.2	4.0–9.3	23.8–40.1	10.7–41.1	
	1996		< 63	7		1.4	11.1	80.3		29	3.21	Szefer et al., 2000
						0.4–4.9	4.5–17.7	44–110		21.9–48	1.6–5.4	
Bornholm Deep	1996		< 63	1		1.2	21.3	85		53.2	5.6	
Vistula Lagoon	1995	1.0–5.0	< 63	25		0.79	11.3	91.4	5.52	22		Szefer et al., 1999
						0.49–1.04	9.1–12.9	68–191	4.1–6.6	17.6–25.8		
	1996		< 63	27		0.95	10.52	88.9		23.2	3.53	Szefer et al., 2000
						0.5–4.3	6.9–14.7	62–113		15.5–30.8	2.4–5.0	
Szczecin Lagoon	1996		< 63	9		4.45	10.26	72.4		59.7	3.73	Szefer et al., 2000
						2.4–8.3	5.8–13.5	38–122		26.0–125	1.8–6.1	
Vistula Estuary	1988–89		< 2 mm	4	0.65	0.95	4.1	34.1	3.48	15.3	10.5	Szefer et al., 1996
					0.09–1.44	0.08–2.19	0.80–8.23	9.2–47.6	1.17–6.14	2.1–30.7	3.5–18.8	
Gulf of Gdansk	1988–89			22	1.48	0.83	5.96	30.4	4.17	13.9	13.7	
					0.09–4.50	0.07–1.98	0.69–13.4	3.2–58.7	0.96–11.1	1.3–34.8	2.5–35.5	
Central part	1988–89			5	2	0.78	4.01	17.1	2.52	4.3	7.26	
of Polish EEZ					0.26–7.0	0.65–1.10	2.15–11.2	6.3–50.3	0.98–6.12	1.6–11.5	1.9–22.1	
Western Baltic												
Western Kieler Bight			< 20			ND				45.0–128		Irion, 1984
Lübeck Bay			< 20			ND–32.8				61.0–344		
German coastal waters	1991–95		< 20	96		0.01–5.1		21–66		18–68		Müller and Heininger, 1999
Szczecin Lagoon	1991–95		< 20	6		2.9–5.3		52–73		52–67		Müller and Heininger, 1999
	1997		< 20							130		Müller and Wessels, 1999

TABLE 4.1. – continued

Region	Sampling date	Sample depth (m)	Fraction (μm)	N	Ga	Hg	In	K	La	Li	Mg	Mn	References
Baltic Sea													
Southern part	1981–83	60–120	(98% of < 63 μm fraction	5		0.162						0.76	Belmans et al., 1993
						0.034–0.40						0.17–2.40	
Western part	1981–83	25–50		4		0.535						0.32	
						0.21–1.3						0.20–0.46	
North-eastern part	1981–83	110–240		4		0.204						7.06	
						0.14–0.28						0.34–26.0	
Northern Baltic													
Gulf of Finland	1992–93		< 2 mm	20		0.13						5.07	Leivouri, 1998
						0.05–0.32						0.56–20.0	
Gulf of Finland	1992–95	0–10	< 2 mm	43		0.18						3.55	Vallius and Leivouri, 1999
						0.05–0.39						2.03–6.48	
Bothnian Bay	1991–93	> 60	< 2 mm			0.25±0.14						8.5±5.3	Leivouri and Niemistö, 1995
	1992–93		< 2 mm	8		0.27						8.96	Leivouri and Niemistö, 1993
						0.06–0.48						1.87–18.0	
	Pre-1993	< 60	< 2 mm	32		0.02				15		3.71	Leivouri and Niemistö, 1993
						0.02–0.97				< 10–28		0.49–18.02	
Bothnian Sea	1991–93	> 60	< 2 mm			0.08±0.04						3.0±1.6	Leivouri and Niemistö, 1995
	1992–93			13		0.09						3.55	Leivouri and Niemistö, 1993
						0.01–0.15						2.03–6.48	
	Pre-1993	< 60	< 2 mm	24		0.05				29		2.29	Leivouri and Niemistö, 1993
						0.01–0.15				12.0–52.0		0.14–9.28	
Danish Straits		50–600		2		0.165						2.26	Belmans et al., 1993
						0.14–0.19						0.81–3.70	
Southern Baltic													
Vistula Estuary	1988–89		< 80	4				7.25		50.6	9.13	0.58	Szefer et al., 1995a
								5.0–8.1		43.4–63.4	7.0–11.1	0.27–1.09	
Gulf of Gdansk	1988–89			22				8.72		44.5	10.2	0.32	
								4.6–17.4		17.5–81.2	7.5–23.7	0.12–0.99	
Dead Vistula River Channel	1993–94		< 2 mm	18		0.5							Falandysz, 1999
						0.04–1.80							

Region	Sampling date	Sample depth (m)	Fraction (μm)	N	Ga	Hg	In	K	La	Li	Mg	Mn	References
Gdańsk Shipyard	1996		< 2 mm	29		0.7							Falandysz, 1999
						0.30–1.30							
Gdynia Seaport	1994		< 2 mm	7		0.25							Kannan and Falandysz, 1998;
						0.08–0.38							Falandysz, 1999
	1996		< 63	3								0.53	Szefer et al., 2000
												0.31–0.83	
Pomeranian Bay	1996		< 63	12								0.74	
												0.47–1.42	
Central part	1988–89		< 80	5				8.72		44.5	22.3	0.56	Szefer et al., 1995a
of Polish EEZ								10.5–17.5		24.0–84.0	9.9–34.4	0.17–0.79	
	1996		< 63	7								1.15	Szefer et al., 2000
												0.21–3.90	
Bornholm Deep	1996		< 63	1									
Vistula Lagoon	1995	1.0–5.0	< 63	25	14		.1–0.3		32.3				Szefer et al., 1999
					12.2–15.5				24.6–36.4				
	1996		< 63	27								0.81	Szefer et al., 2000
												0.47–1.80	
Szczecin Lagoon	1996		< 63	9								2.56	Szefer et al., 2000
												0.99–6.90	
Vistula Estuary	1988–89		< 2 mm	4				8.8		21.7	2.05	0.16	Szefer et al., 1996
								3.7–10.7		7.0–38.4	0.40–3.88	0.06–0.35	
Gulf of Gdansk	1988–89			22				12.9		22.2	6.23	0.12	
								3.1–17.1		3.0–67.2	0.33–9.10	0.036–0.25	
Central part	1988–89			5				10		12.2	2.33	0.16	
of Polish EEZ								3.2–10.2		3.8–35.4	0.46–8.55	0.051–0.34	
Western Baltic													
German coastal waters	1991–95		< 20	96		0.03–1.2							Müller and Heininger, 1999
Szczecin Lagoon			< 20	6		0.5–1.18							Müller and Heininger, 1999

TABLE 4.1. – continued

Region	Sampling date	Sample depth (m)	Fraction (μm)	N	Na	Ni	Pb	Rb	S	Sb	References
Baltic Sea	1981–83		< 100								Belmans et al., 1993
Southern part		60–120	(98% of	5		26.2	54.4		10.5		
			< 63 μm			17.0–38.0	8.0–96.0		3.0–16		
Western part		25–50	fraction	4		22.8	86		9.2		
						17.0–26.0	76.0–100		7.3–11		
North-eastern part		110–240		4		50.3	62.8		23		
						29.0–68.0	54.0–82.0		11.0–40		
Northern Baltic											
Gulf of Finland	1992–93		< 2 mm	20		42	50				Leivouri, 1998
						25–60	26–80				
Gulf of Finland	1992–95	0–10	< 2 mm	43			52				Vallius and Leivouri, 1999
							26–88				
Bothnian Bay	1991–93	> 60	< 2 mm				66±26				Leivouri and Niemistö, 1995
	1992–93		< 2 mm	8		48	79				Leivouri and Niemistö, 1993
						27–58	19–121				
	Pre-1993	< 60	< 2 mm	32		34	27				Leivouri and Niemistö, 1993
						15–58	11.0–98				
Bothnian Sea	1991–93	> 60	< 2 mm				37±12				Leivouri and Niemistö, 1995
	1992–93			13		53	42				Leivouri and Niemistö, 1993
						32–81	17–60				
	Pre-1993	< 60	< 2 mm	24		41	27				Leivouri and Niemistö, 1993
						14–110	14–58				
Danish Straits		50–600		2		25.5	55.5		52.5		Belmans et al., 1993
						25.0–26.0	52.0–59.0		52–53		
Southern Baltic											
Vistula Estuary	1988–89		< 80	4	26.5	48.7	66.5	51.8			Szefer et al., 1995a
					10.5–55.0	38.9–59.7	50.6–80.0	37.5–72.9			
Gulf of Gdansk	1988–89			22	32.7	35.7	43.4	52.2			
					9.7–128	20.7–52.4	21.0–117	17.1–118			
	1996		< 63	3			55.8			1.01	Szefer et al., 2000
							27–75			0.57–1.70	

Region	Sampling date	Sample depth (m)	Fraction (μm)	N	Na	Ni	Pb	Rb	S	Sb	References
Pomeranian Bay	1996		< 63	12			44.1			0.59	
							25–66			0.35–0.90	
Central part	1988–89		< 80	5	147	50.1	45.3	51.6			Szefer et al., 1995a
of Polish EEZ					26.2–220	38.6–57.7	11.2–83.0	15.9–102			
	1996		< 63	7			41.3			1.26	Szefer et al., 2000
							28–62			0.48–2.74	
Bornholm Deep	1996		< 63	1			85			3.49	
Vistula Lagoon	1995	1.0–5.0	< 63	25		38.8	35.7	115		1.08	Szefer et al., 1999
						27.5–47.9	24.0–80.5	92.0–137		0.87–1.97	
	1996		< 63	27			34.6			0.75	Szefer et al., 2000
							23–140			0.50–1.09	
Szczecin Lagoon	1996		< 63	9			109			1.26	Szefer et al., 2000
							60–191			0.43–2.06	Szefer et al., 1996
Vistula Estuary	1988–89		< 2 mm	4	7.03	17.3	25.1	53.3			Szefer et al., 1996
					5.4–12.2	4.7–33.4	10.6–40.3	29.3–74.7			
Gulf of Gdansk				22	9.12	19.9	33.7	63.5			
					5.3–23.5	2.6–51.5	6.2–55.9	18.5–129			
Central part				5	6.16	10.5	27.3	49.5			
of Polish EEZ					5.3–14.7	3.4–28.1	15.0–35.7	20.8–81			
Western Baltic											
Western Kieler Bight			< 20				94.0–173				Irion, 1984
Lübeck Bay			< 20				377–5100				
German coastal waters	1991–95		< 20	96		18–80	22–265				Müller and Heininger, 1999
Szczecin Lagoon	1991–95		< 20	6		36–45	103–161				Müller and Heininger, 1999
	Before flood										
	1997		< 20				180				Müller and Wessels, 1999
	After flood										

TABLE 4.1. – continued

Region	Sampling date	Sample depth (m)	Fraction (μm)	N	Se	Sn	Sr	Ti	V	Yb	Zn	References
Baltic Sea												
Southern part	1981–83	60–120	(98% of < 63 μm fraction	5			54	420			135	Belmans et al., 1993
							25.0–76.0	390–490			40.0–230	
	1994					< 0.001–0.039"						Szpunar et al., 1997
Western part	1981–83	25–50		4			89.7	350			165	Belmans et al., 1993
							74.0–99.0	310–390			100–240	
North-eastern part	1981–83	110–240		4			80.3	515			440	
							60.0–110	330–670			290–560	
Northern Baltic												
Gulf of Finland	1992–93		< 2 mm	20				3.94*	76		175	Leivouri, 1998
								2.42–5.19	57–96		107–243	
	1992–95	0–10	< 2 mm	42							209	Vallius and Leivouri, 1999
											77–513	
Bothnian Bay	1991–93	> 60	< 2 mm								216±92	Leivouri and Niemistö, 1995
	1992–93		< 2 mm	8				3.42*	73		212	Leivouri and Niemistö, 1993
								3.0–4.22	40–93		50–320	
	pre-1993	< 60	< 2 mm	32				3.07*	60		201	Leivouri and Niemistö, 1993
								2.08–4.27	24–120		50–550	
Bothnian Sea	1991–93	> 60	< 2 mm								173±56	Leivouri and Niemistö, 1995
	1992–93			13				3.88*	89		190	Leivouri and Niemistö, 1993
								2.33–5.39	49–114		90–240	
	pre-1993	< 60	< 2 mm	24				2.83*	63		135	Leivouri and Niemistö, 1993
								1.03–4.91	16–130		30–240	
Danish Straits		50–600		2			79.1	190			125	Belmans et al., 1993
							8.1–150	100–280			110–140	
Southern Baltic												
Vistula Estuary	1988–89		< 80	4			164				354	Szefer et al., 1995a
							56.0–350				247–429	
Gulf of Gdansk				22			112				172	
							10.0–277				45.0–639	
	1998			14		45.2"						Senthilkumar et al., 1999
						5.8–130						
Gulf of Gdansk												
Gdynia seaport	1993		< 2 mm			1.8–2.9						Kannan and Falandysz, 1997
				2		1.15–7.32"						Szpunar et al., 1997

Region	Sampling date	Sample depth (m)	Fraction (μm)	N	Se	Sn	Sr	Ti	V	Yb	Zn	References
	1996		< 63	3	ND	ND	103	470	64.9		122	Szefer et al., 2000
							93–110	360–640	27–120		87.2–148	
Pomeranian Bay	1996		< 63	12	ND	ND	170	0.94*	61.7		102	
							93–380	0.40–2.35	18–125		46.1–166	
Central part	1988–89		< 80	5			319				136	Szefer et al., 1996
of Polish EEZ							104–770				80.0–176	
	1996		< 63	7	ND	ND	110	0.56*	73.4		105	Szefer et al., 2000
							98–130	0.39–1.29	25–110		57.6–240	
Bornholm Deep	1996		< 63	1	ND	ND	120	0.43*	144		181	
Vistula Lagoon	1995	1.0–5.0	< 63	25	< 1.0		128		85	2.28	103	Szefer et al., 1999
							111–211		71.0–99.0	1.64–2.91	76.0–137	
	1996		< 63	27	ND	ND	123	0.43*	80.5		94.8	Szefer et al., 2000
							107–188	0.36–0.53	52–120		61.9–200	
Szczecin Lagoon	1996		< 63	9	ND	ND	255	0.26*	34.3		590	Szefer et al., 2000
							163–430	0.11–0.37	ND–68		270–1050	Szefer et al., 1996
	1993–95			2		0.015–0.027"						Szpunar et al., 1997
Vistula Estuary	1988–89		< 2 mm	4			39.8				86.5	Szefer et al., 1996
							24.1–50.9				11.0–156	
Gulf of Gdansk				22			42.3				67.5	
							11.3–102				6.9–190	
Central part				5			67.1				18.3	
of Polish EEZ							35.9–163				6.5–46.0	
Western Baltic												
Western Kieler Bight			< 20								256–538	Irion, 1984
Lübeck Bay			< 20								558–6250	
German coastal waters	1991–95		< 20	96							86–894	Müller and Heininger, 1999
Szczecin Lagoon			< 20	6							534–981	Müller and Heininger, 1999
Western Baltic	1997			19		3.64						Biselli et al., 2000
						0.57–17.0						
North Sea	1997			13		0.41						Biselli et al., 2000
						0.08–0.72						

* – mg g^{-1} dry wt.
" – Value originally reported is converted to tributyltin ion.

TABLE 4.2.

Concentrations of REE ($\mu g\ g^{-1}$dry wt) in surface sediments of the Baltic Sea

Region	Sampling date	Sample depth (m)	Fraction (mm)	N	La	Ce	Pr	Nd	Sm	Eu	Gd	References
Gulf of Gdańsk	1996		< 63	3	44.2±20.9	90.3±44.8	10.8±4.8	40.6±17.3	7.7±2.9	1.2±0.2	7.5±2.3	Szefer et al., 2000
Bornholm Deep	1996		< 63	1	49	98	11.5	43	8.6	1.6	8.1	Szefer et al., 2000
Vistula Lagoon	1995	1.0–5.0	< 63	26	24.6–36.4	53.0–82.7	6.03–9.40	22.7–35.2	4.52–6.87	0.95–1.36	4.17–6.43	Szefer et al., 1999
	1996		< 63	27	33.4±2.4	70.2±5.2	8.1±0.6	31.2±2.4	6.3±0.6	1.2±0.1	5.9±0.5	Szefer et al., 2000
Szczecin Lagoon	1996		< 63	9	26.7±9.4	53.4±19.2	6.3±2.2	23.9±8.6	4.7±1.6	0.9±0.2	4.6±1.5	Szefer et al., 2000
Pomeranian Bay	1996		< 63	12	148±122	304±251	35.6±29.3	132±107	24.3±19.5	2.0±1.1	22.0±17.1	Szefer et al., 2000
Central part of Polish EEZ	1996		< 63	7	69.4±84.9	143±171	16.7±20.2	63.4±75.6	11.7±13.3	1.5±0.7	10.8±11.8	Szefer et al., 2000

TABLE 4.2. – continued

Region	Sampling date	Sample depth (m)	Fraction (mm)	N	Tb	Dy	Ho	Er	Tm	Yb	Lu	References
Gulf of Gdańsk	1996		< 63	3	1.0±0.3	4.9±1.1	0.9±0.2	2.9±0.5	0.4±0.1	2.9±0.6	0.4±0.1	Szefer et al., 2000
Bornholm Deep	1996		< 63	1	1.1	6.2	1.21	3.7	0.5	3.4	0.51	Szefer et al., 2000
Vistula Lagoon	1995	1.0–5.0	< 63	26	0.55–0.84	3.16–4.59	0.57–0.96	1.65–2.46	0.26±0.38	1.64–2.91	0.26–0.38	Szefer et al., 1999
	1996		< 63	27	0.8±0.1	4.3±0.3	0.8±0.1	2.5±0.2	0.3±0.05	2.5±0.5	0.4±0.03	Szefer et al., 2000
Szczecin Lagoon	1996		< 63	9	0.7±0.2	3.3±0.9	0.7±0.2	1.9±0.5	0.3±0.1	1.9±0.5	0.3±0.1	Szefer et al., 2000
Pomeranian Bay	1996		< 63	12	2.8±2.0	13.0±8.5	2.4±1.5	7.4±4.5	1.1±0.7	8.0±4.9	1.3±0.9	Szefer et al., 2000
Central part of Polish EEZ	1996		< 63	7	1.4±1.4	7.0±5.8	1.4±1.0	4.2±3.4	0.6±0.5	4.3±3.9	0.7±0.7	Szefer et al., 2000

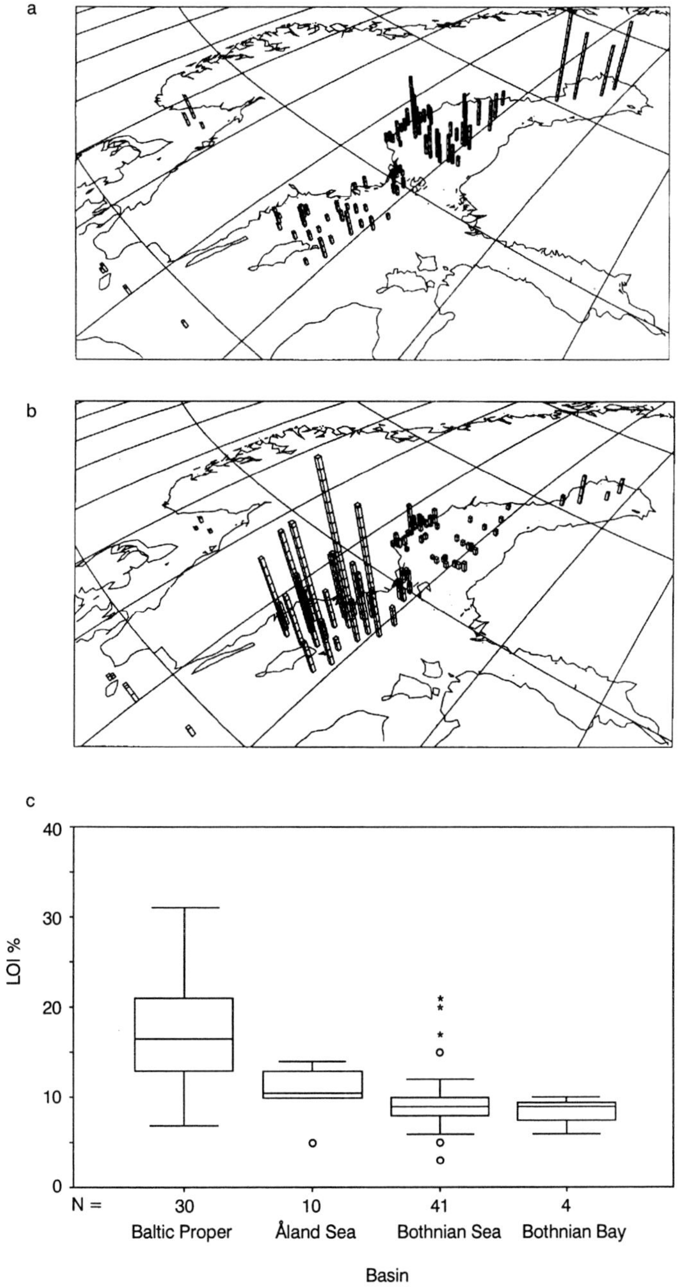

Fig. 4.1. a) The distribution of arsenic in surficial sediment (0–1 cm). One unit of scale on the bars = 10 $\mu g\ g^{-1}$ dry wt. b) The distribution of cadmium in surficial sediment (0–1 cm). One unit of scale on the bars = 0.5 $\mu g\ g^{-1}$ dry wt. c) Organic matter in surficial sediments of the different basins, expressed as loss on ignition (LOI, % dry wt). After Borg and Jonsson (1996); modified.

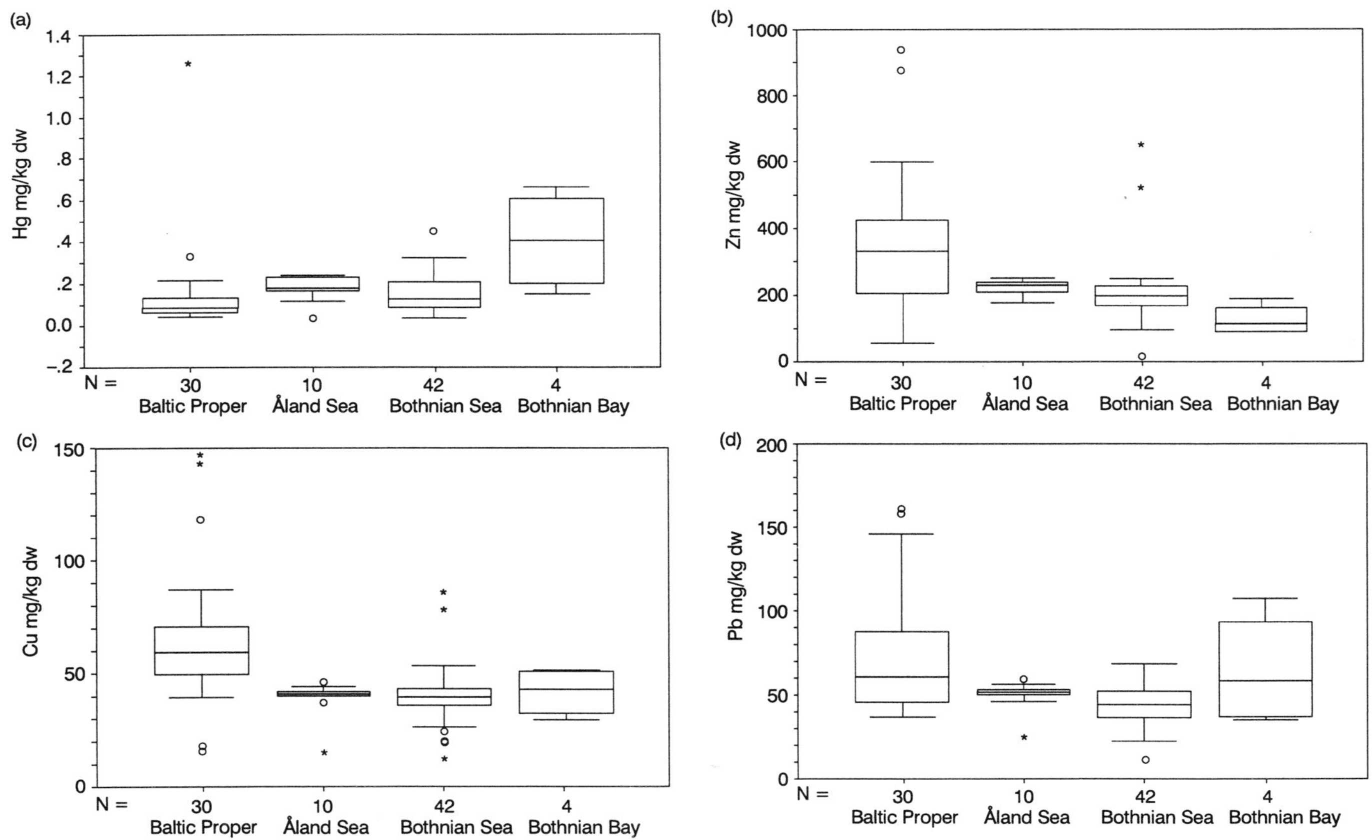

Fig. 4.2. Box-plot of (a) Hg, (b) Zn, (c) Cu and (d) Pb in surficial sediments (0–1 cm). After Borg and Jonsson (1996); modified.

Vertical distribution (sediment cores)

Table 4.3 lists the concentration data of chemical elements in sediment cores from the Baltic Sea. Analyses of sediment cores have reflected the development history of sea, including the anthropogenic impact, since there is a clear enrichment of some trace elements towards the sediment upper layers. Background values of Cd and Hg were generally comparable for different basins of the Baltic Sea, while analogous values of As, Cr, Cu, Zn and especially Pb were lower for the Bothnian Bay (Borg and Jonsson, 1996). To elements showing a general increase towards the sediment surface belong Ag, As, Cd, Cu, Hg, Pb and Zn (Szefer and Skwarzec, 1988; Brügmann and Lange, 1990; Hallberg, 1991; Szefer et al., 1993b, 1995b, 1998b; Paetzel et al., 1994; Borg and Jonsson, 1996; Leivuori and Niemistö, 1995; Neumann et al., 1996, 1998; Leivuori, 1998; Vallius, 1999a, 1999b; Vallius and Leivuori, 1999; Sternbeck et al., 2000). Although the downcore (temporal) trends provide a clear evidence that the upper part of sediment profiles is enriched in some trace elements, the vertical distribution of Cd, Hg, Zn and Pb from the Baltic Proper showed decreasing their concentrations in the uppermost 3–4 cm (Borg and Jonsson, 1996). It is postulated that this pattern reflects a decreased anthropogenic input of the elements, at least Pb, during years corresponding to this depth interval of the sediment core. However, no decrease in the surficial layers was found in the Bothnian Bay in spite of substantially (> 90%) reduction of discharges of trace elements, especially As, from smelters during last ca. 20 years. This expected concentration decreasing towards the surface can not be detected in sediment profiles because of the low rate of sedimentation and bioturbation, i.e. mixing of the surficial layers. Borg and Jonsson (1996) reported scenario for historic input of selected elements to the Baltic Proper based on the dating from varve-counting into a 'mean core' (Fig. 4.3). Up to ca. 1930, no clear changes in levels of trace elements can be observed in the homogenous layer in this core. During 1930–1950 metals such as Cd, Hg, Pb and Zn showed gradual increasing of their concentrations and since 1950 this increase is very evident, coinciding with the appearing of laminae. From ca. 1965–1970 the vertical profile for these elements demonstrates a steep increasing gradient towards the sediment surface.

It is important to note that the vertical distribution of metals in cores from anoxic area, characterised by continuous or periodical lamination along the whole core, is different for the cores from areas with a more recently occurring lamination (Borg and Jonsson, 1996; Persson and Jonsson, 2000). As can be seen in Fig. 4.3 at the areas with continuous lamination, the levels of Cu, Pb, Zn and especially Cd in the surficial sediments are lower, and their increase towards the top layers is more homogenously distributed along the whole cores.

The vertical profiles of selected elements in southern Baltic sediment cores have been investigated by several authors (Szefer and Skwarzec, 1988; Szefer et al., 1993b, 1995b, 1998b). Among 14 cores, three granulometric fractions of < 2, 2–63 and 63–200 μm have been additionally analysed for concentrations of minor

TABLE 4.3.

Concentrations of trace elements ($\mu g\ g^{-1}$ dry wt.), Al, Ca, K, Mg, Na, Mn and Fe ($mg\ g^{-1}$ dry wt.) in sediment cores of the Baltic Sea

Region	Sampling date	Sample depth (m)	Granulometric fraction (μm)	Segment depth (mm)	N	Al	Ag	As	Ca	References
Southern Baltic										
Arkona Basin	1993	45	< 63	0–10	2	4.4**				Neumann et al., 1996, 1998
				< 400		3.6**				
Bornholm Basin	1974	71		0–10	1	13.52*				Suess and Erlenkeuser, 1975
				330–340	1	10.56*				
	1980	88	Silty sediment	0–50	5	56.5			13.2	Szefer and Skwarzec, 1988
						60.5–47.8			17.6–10.9	
				121–152	1	65			8.6	
Oder Lagoon	1993	5	< 63	0–10	2	2.8**				Neumann et al., 1996, 1998
				< 400		2.7**				
Achterwasser	1993	3.7	< 63	0–10	1	2.1**				Neumann et al., 1998
				330–350		3.1**				
Gulf of Gdansk	1978	57.5		0–50	12				4.4	Dietrich and Beuge, 1986
				250–300	12				4.2	
	1980	78	Silty sediment	0–50	5	44.3			17.8	Szefer and Skwarzec, 1988
						48.5–39.0			22.5–14.7	
				250–300	1	50			30	
	1991		2.0–63	0–50	3	130–37.7	3.01–ND		1.59–37.5	Szefer et al., 1998
				200–250	3	126–49.0	4.12–ND		0.40–12.5	
				0–50	1	63.7	0.52			
				50–100	1	62.1	0.19			
				0–50	1	34.2	2.05			
				150–200	1	30.6	4.12			
Gdansk Deep	1980	105	Silty sediment	0–50	5	48.1			5.68	Szefer and Skwarzec, 1988
						56.5–36.3			7.3–6.7	
				200–253	1	63.3			9.1	

	1980	89	0–55	5	60		6.18	
					67.3–49.3		6.0–5.8	
			311–348	1	70.3		9.2	
Western Baltic								
Eckernförder Bucht	1971	28	0–1	1	92.5*			Erlenkeuser et al., 1974
			120–122	1	81.0*			
Southwest of Aero	1971		18–19	1	104*			
			37–38	1	105*			
Baltic Proper	1986–89		0–10	28		15±8		Borg and Jonsson, 1996
			> 10	51		9±3		
Northern Baltic								
Aland Sea	1986–89		0–10	10		14±5		Borg and Jonsson, 1996
			> 10	4		10±1		
Bothnian Bay	1986–89		0–10	4		109±23		Borg and Jonsson, 1996
			> 10	10		6±5		
	1991–93		0–10	32		91±45 (27)		Leivouri and Niemistö, 1995
			15–25	9		7±6		
Bothnian Sea	1986–89		0–10	20		35±12		Borg and Jonsson, 1996
			> 10	5		9±2		
	1991–93		0–10	24		26±25 (10)		Leivouri and Niemistö, 1995
			15–25	5		9±2		

* – Expressed as oxides.
** – Concentration approximated from diagrams.

TABLE 4.3. – continued

Region	Sampling date	Sample depth (m)	Fraction (μm)	Segment depth (mm)	N	Cd	Co	Cr	Cu	Fe	Hg	References
Southern Baltic												
Arkona Basin	1993	45	< 63	0–10	2				30.8–35**	35**		Neumann et al., 1996, 1998
				< 400					20–20.9**	30**		
			< 45	0–10	1				0.16[b]	230[b]		Neumann et al., 1998
				< 400	1				0.2[b]	10[b]		
Bornholm Basin	1974	71		0–10	1	1.88	9.5		45	5.09		Suess and Erlenkeuser, 1975
				330–340	1	0.53	4.8		26	39.5		
	1980	88	Silty sediment	0–50	5	3.28	26.3		42.5	47.7		Szefer and Skwarzec, 1988
						6.8–1.5	29.0–24.0		54.0–39.0	42.0–32.0		
				121–152	1	4.1			45			
	1993	54.1	< 63	0–10	2				44.1			Neumann et al., 1996, 1998
				110–130					30.9			
Oder Lagoon	1993	5	< 63	0–10	2				63–68**	30**		Neumann et al., 1996, 1998
				< 400					13–26.7**	45**		
	1993	5	< 45	0–10	1				0.05[b]	20[b]		Neumann et al., 1998
				< 400	1				0.03[b]	400[b]		
Achterwasser	1993	3.7	< 63	0–10	1				24.9			Neumann et al., 1998
				330–350					21.1			
	1993	3.7	< 63	0–10						2.5[b]		Neumann et al., 1998
				< 400						2.5[b]		
Gulf of Gdansk	1978	57.5		0–50	12						0.776	Dietrich and Beuge, 1986
				250–300	12						0.613	
	1980	78	Silty sediment	0–50	5	4.36	15.6		75.6	32.1		Szefer and Skwarzec, 1988
						6.4–2.5	12.0–19.0		88.0–46.0	46.6–26.3		
				250–300	1	1.8	16		38	36		
Gulf of Gdansk												
Hel	1987			0–50							5.5	Falandysz et al., 1993
				200–250							2.6	
Kuznica	1987			0–50							5.2	
				250–300							1.7	
Gulf of Gdansk												
	1991		2.0–63	0–50	3	3.42–1.36	27.6–15.0	129–76.9	69.6–44.4	46.4–42.8		Szefer et al., 1995, 1998, 1999
				200–250	3	0.65–ND	21.7–16.2	74.0–50.5	36.7–22.5	146–52.2		
				0–50	1	2.97	7.6		21.9	30.4		
				50–100	1	1.19	6.1		12.2	23		
				0–50	1	3.75	9.85	32.9	48	40.5		
				150–200	1	4.76	10	97	56.1	65		
												Belzunce et al., 2000
	1996	104	Bulk sediments	0–25			19	107	71	42.3		
				250–300			21	97	44	42.3		
		78		0–25			11	92	46	24.1		

				250–300			11	67	31	24.4		
		65		0–25			11	93	35	24.5		
				250–300			12	73	24	24.6		
		60		0–25			9	65	22	17.9		
				250–300			13	85	21	21.7		
Gdansk Deep	1980	105	Silty sediment	0–50	5	4.06	22.6		57.8	37.8		Szefer and Skwarzec, 1988
						6.1–2.9	24.0–21.0		69.0–52.0	44.0–30.4		
				200–253	1	2.4	30		54	59		
	1980	89		0–55	5	4.93	22.8		58.8	40.4		
						8.5–2.6	25.0–19.0		62.0–54.0	44.5–35.7		
				311–348	1	5	24		51	56		
Western Baltic												
Eckernförder Bucht	1971	28		0–1	1	1.87	15		71	31.2		Erlenkeuser et al., 1974
				120–122	1	0.28	11		35	34.1		
Southwest of Aero	1971			18–19	1	1.1	12		31	37.5		
				37–38	1	0.3	13		27	40.1		
Baltic Proper	1986–89			0–10	28	2.9±2.5	18±6	39±12	63±28	43±15	0.10±0.05	Borg and Jonsson, 1996
				> 10	51	0.31±0.19	16±4	52±13	45±13	47±17	0.04±0.02	
	1990		< 45	5–7.5	6					0.97–10.7[a]		Carman and Rahm, 1997
				85–95	6					0.97–2.49		
Danish Straits	1991	645		0–20	1				19.6			Paetzel et al., 1994
Skagerrak				250–500	1				15.7			
				0–10	1				17			
				250–500	1				13.4			
Northern Baltic												
Gulf of Finland	1993–95		< 2 mm	top		1.05	18.3	68.6	40.4		0.25	Vallius, 1999a, 1999b
				bottom		0.14	13.3	45.7	22.2		0.07	
Aland Sea	1986–89			0–10	10	0.63±0.22	18±2	38±10	39±9	54±10	0.18±0.06	Borg and Jonsson, 1996
				> 10	4	0.24±0.06	21±1	56±5	40±3	56±15	0.03±0.00	
Bothnian Bay	1986–89			0–10	4	0.94±0.58	21±4	37±4	41±11	48±6	0.40±0.24	Borg and Jonsson, 1996
				> 10	10	0.37±0.27	15±4	37±7	27±7	51±40	0.02±0.01	
	1991–93		< 2 mm	0–10	32	0.53±0.27			29±13		0.20±0.24	Leivouri and Niemistö, 1995
				15–25		0.40±0.30			28±7		0.02±0.01	
Bothnian Sea	1986–89			0–10	20	0.31±0.07	22±4	50±11	39±6	64±14	0.10±0.03	Borg and Jonsson, 1996
				> 10	5	0.10±0.07	17±4	40±7	36±5	47±11	0.03±0.01	
	1991–93		< 2 mm	0–10	24	0.31±0.27			28±14		0.05±0.04	Leivouri and Niemistö, 1995
				15–25		0.10±0.07			36±5		0.03±0.01	

** – Concentration approximated from diagrams.
[a] – Concentration in the interstitial water (mM)
[b] – Concentration in the interstitial water (μM).

TABLE 4.3. – continued

Region	Sampling date	Sample depth (m)	Fraction (μm)	Segment depth (mm)	N	K	Mg	Mn	Na	Ni	References
Southern Baltic											
Arkona Basin	1993	45	< 63	0–10	2			350**			Neumann et al., 1996, 1998
				< 400				450**			
			< 45	0–10	1			80[b]			Neumann et al., 1998
				< 400	1			5[b]			
Bornholm Basin	1974	71		0–10	1					39	Suess and Erlenkeuser, 1975
				330–340	1					61	
	1980	88	Silty sediment	0–50	5	25.4	12.1	4100	23.9	48	Szefer and Skwarzec, 1988
						28.4–21.8	15.1–10.8	6200–3100	35.4–15.9	55.0–44.0	
				121–152	1	28	10.5	1300	32	52	
Oder Lagoon	1993	5	< 63	0–10	2			4700**			Neumann et al., 1996, 1998
				270–290				1700**			
			< 45	0–10				145[b]			
				< 400				45[b]			
Achterwasser	1993	3.7	< 63	0–10	1			880**			Neumann et al., 1998
				330–350	1			420**			
			< 45	0–10	1			48[b]			Neumann et al., 1998
				< 400	1			25[b]			
Gulf of Gdansk	1980	78	Silty sediment	0–50	5	21	11.8	232	28.9	45.4	Szefer and Skwarzec, 1988
						23.3–18.4	14.1–9.3	270–206	34.0–19.0	54.0–39.0	
				250–300	1	25.8	13.6	310	17.8	44	
Gulf of Gdansk											
	1991		2.0–63	0–50	3	39.0–5.15	4.34–0.82	533–232	9.03–7.72	97.8–64	Szefer et al., 1995, 1998, 1999
				200–250	3	47.8–4.39	2.01–1.45	387–260	8.68–3.83	73–45.2	
				0–50	1	37.9	2.16	275	6.6	27.2	
				50–100	1	38.8	1.59	305	7.07	20.6	
				0–50	1	12.75	0.53	509	9.69	50.8	
				150–200	1		0.31	233	7.77	49.4	
											Belzunce et al., 2000
	1996	104	Bulk sediment	0–25				419		46	
				250–300				434		48	
		78		0–25				247		33	
				250–300				273		25	
		65		0–25				255		28	
				250–300				293		24	

		60		0–25				188		20	
				250–300				225		27	
Gdansk Deep	1980	105	Silty sediment	0–50	5	22.9	10.5	368	34.8	48.2	Szefer and Skwarzec, 1988
						24.3–21.0	11.8–9.5	540–276	44.0–30.1	52.0–42.0	
				200–253	1	36.8	16	785	31.7	69	
	1980	89		0-55	5	24.1	11.3	282	31.9	54.8	
						28.1-20.0	13.5-8.9	348-240	40.2-16.2	58.0-50.0	
				311-348	1	30.5	13.3	346	29.8	59	
Western Baltic											
Eckernförder Bucht	1971	28		0-1	1			430		87	Erlenkeuser et al., 1974
				120-122	1			550		72	
Southwest of Aero	1971			18-19	1			580		42	
				37-38	1			790		40	
Baltic Proper	1986-89			0-10	28			0.7±0.4		49±16	Borg and Jonsson, 1996
				> 10	51			0.6±0.3		39±7	
	1990		< 45	5-7.5	6	2.66-3.75[a]	8.4-15.0[a]	5.88-116[a]	101-149[a]		Carman and Rahm, 1997
				85-95	6	3.09-4.88[a]	12.9-15.7[a]	23.3-186[a]	115-167[a]		
Danish Straits	1991	645		0-20	1			880			Paetzel et al., 1994
Skagerrak				250-500				300			
				0-10	1			498			
				250-500				289			
Northern Baltic											
Gulf of Finland	1993-95		< 2 mm	top						36	Vallius, 1999a, 1999b
				bottom						27.6	
Aland Sea	1986-89			0-10	10			2.2±0.7		38±9	Borg and Jonsson, 1996
				> 10	4			0.7±0.1		39±1	
Bothnian Bay	1986-89			0-10	4			5.5±3.6		37±22	Borg and Jonsson, 1996
				> 10	10			1.0±1.0		32±18	
Bothnian Sea	1986-89			0-10	20			2.7±1.2		41±6	Borg and Jonsson, 1996
				> 10	5			3.2±1.8		36±8	

** – Concentration approximated from diagrams.

[a] – Concentration in the interstitial water (mM).

[b] – Concentration in the interstitial water (μM).

TABLE 4.3. – continued

Region	Sampling date	Sample depth (m)	Fraction (μm)	Segment depth (mm)	N	Pb	Sr	Ti	V	Zn	References
Southern Baltic											
Arkona Basin	1993	45	< 63	0–10	2	67.8–75**				127–160**	Neumann et al., 1996, 1998
				240–260		24–29.2**				68–70**	
			< 45	0–10	1	0.4[b]				8[b]	Neumann et al., 1998
				< 400	1	0.4[b]				7.8[b]	
	1974	71		0–10	1	92				204	Suess and Erlenkeuser, 1975
				330–340	1	13				103	
Bornholm Basin	1980	88	Silty sediment	0–50	5	430		3.05		262	Szefer and Skwarzec, 1988
						1260–80		3.26–2.65		310–215	
				121–152	1	70		3.45		184	
	1993	54.1	< 63	0–10	2	76.1				146	Neumann et al., 1996, 1998
				110–130		33				78	
Oder Lagoon	1993	5	< 63	0–10	2	115–118**				750–842**	Neumann et al., 1996, 1998
				270–290		20–34.4**				20–106**	
Achterwasser	1993	3.7	< 63	0–10	1	43.4				173	Neumann et al., 1998
				330–350		21.2				63	
			< 45	0–10	1					0.7[b]	Neumann et al., 1998
				< 400	1					0.8[b]	
Gulf of Gdansk	1978	57.5		0–50	12					106	Dietrich and Beuge, 1986
				250–300	12					76	
	1980	78	Silty sediment	0–50	5	570		2.91		304	Szefer and Skwarzec, 1988
						1760–120		3.33–2.70		400–242	
				250–300	1	69		3.7		162	
Gulf of Gdansk											
	1991		2.0–63	0–50	3	89.7–63.0				505–182	Szefer et al., 1995, 1998, 1999
				200–250	3	48.0–24.0				140–91	
				0–50	1	66.9				100	
				50–100	1	60.4				77.4	
				0–50	1	63.6				445	
				150–200	1	74.1				490	
	1996	104	Bulk sediments	0–25		83				210	Belzunce et al., 2000
				250–300		51				108	
		78		0–25		65				140	
				250–300		60				102	
		65		0–25		44				128	
				250–300		42				75	
		60		0–25		38				102	
				250–300		21				56	

Gdansk Deep	1980	105	Silty sediment	0–50	5	385		3		306	Szefer and Skwarzec, 1988
						1130–108		3.2–2.7		370–240	
				200–253	1	64		4.1		246	
	1980	89		0–55	5	286		3.28		274	
						522–110		3.48–2.76		300–240	
				311–348	1	48		3.9		210	
Western Baltic											
Eckernförder Bucht	1971	28		0–1	1	82				340	Erlenkeuser et al., 1974
				120–122	1	20				125	
Southwest of Aero	1971			18–19	1	57					
				37–38	1	31					
Baltic Proper	1986–89			0–10	28	71±32				360±107	Borg and Jonsson, 1996
				> 10	51	25±15				120±30	
	1990		< 45	5–7.5	6		15–22.6[a]				Carman and Rahm, 1997
				85–95	6		21.2–26.3[a]				
Danish Straits	1991	645		0–20	1	50.3				125	Paetzel et al., 1994
Skagerrak				250–500		5				88.8	
				0–10	1	33.9				101	
				250–500		14.6				73.8	
Northern Baltic											
Gulf of Finland	1993–95		< 2 mm	top					66.6	170.4	Vallius, 1999a, 1999b
				bottom					56.5	86.7	
Aland Sea	1986–89			0–10	10	50±9				226±21	Borg and Jonsson, 1996
				> 10	4	26±8				155±16	
Bothnian Bay	1986–89			0–10	4	65±34				130±46	Borg and Jonsson, 1996
				> 10	10	4.2±2.9				71±21	
	1991–93		< 2 mm	0–10	32	27±21				201±118	Leivouri and Niemistö, 1995
				15–25		4±3				71±22	
Bothnian Sea	1986–89			0–10	20	40±7				193±34	Borg and Jonsson, 1996
				> 10	5	24±6				132±17	
	1991–93		< 2 mm	0–10	24	27±12				135±62	Leivouri and Niemistö, 1995
				15–25		24±6				132±17	

** – Concentration approximated from diagrams.
[a] – Concentration in the interstitial water (mM).
[b] – Concentration in the interstitial water (μM).

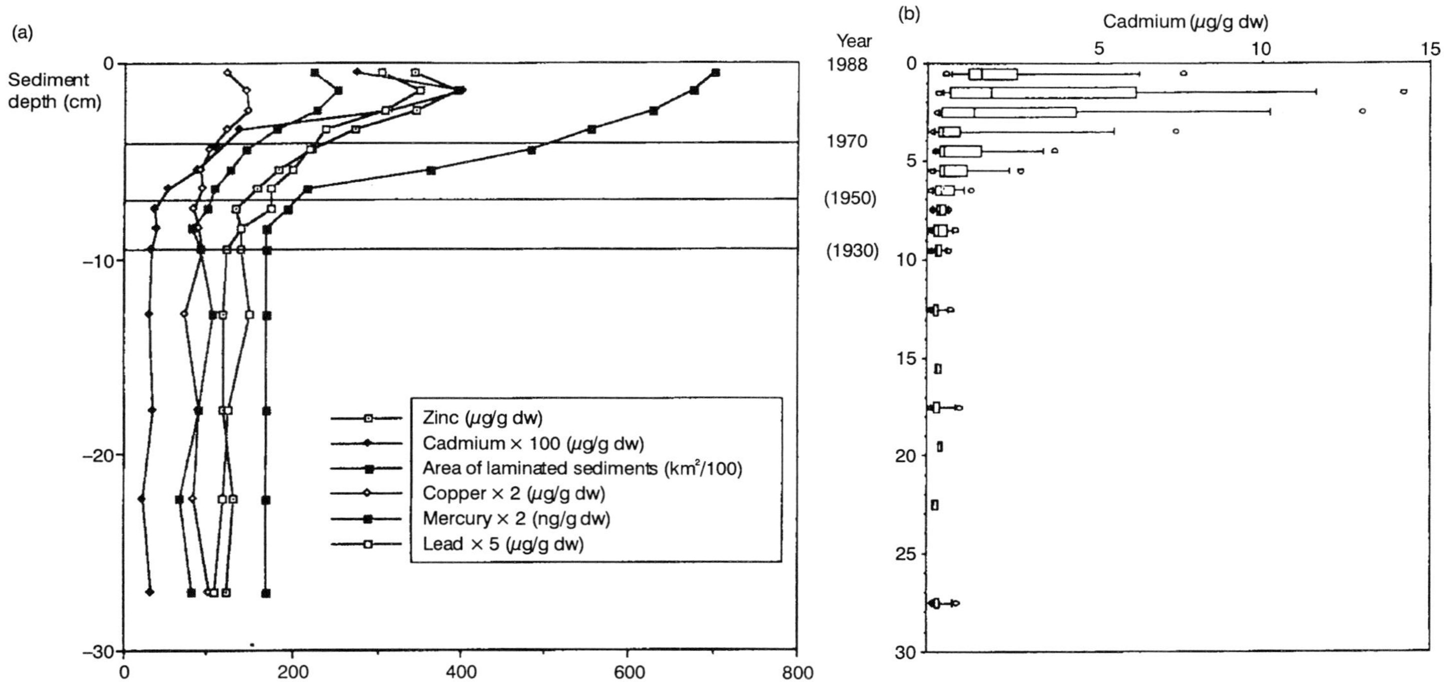

Fig. 4.3. (a) The mean vertical distribution of Cd, Zn, Cu, Pb, Hg in sediment cores (n = 10) from the Baltic Proper. The recent expansion of laminated sediments is also indicated. The dating has been performed by varve-counting down to year 1970 (sediment depth 4.2 cm). The levels for the years 1930 (7.0 cm) and 1950 (9.6 cm) have been estimated from the dry matter curve, assuming a constant mean deposition rate of dry matter. (b) Variation of the Cd concentration at different levels in sediment profiles from the Baltic Proper (n= 10). After Borg and Jonsson (1996); modified.

and major elements. Granulometric and mineralogical characteristics of the sediment cores as well as changes of the organic matter concentration in the particular segments with depth of their location are discussed by Szefer et al. (1993b, 1995b, 1998b). For instance, vertical distribution of heavy metals in core Nos. 8 and 25 from Puck Bay, southern Baltic (Fig. 4.4), is illustrated in Figures 4.5 and 4.6. The data presented graphically show a decrease in the concentrations of Cd, Ag, Pb, Zn and Cu with depth in sediment column in these two cores but not in sediment core taken from the estuarine core No. 38 (Fig. 4.7). The increase in heavy metals in the upper layers of Puck Bay compared to the lower layers reflects the onset of industrialisation, and the resultant increase in heavy-metal pollution, in Poland. By contrast, sediments from estuarine core taken from near the mouth of the Vistula River have a much higher sedimentation rate than those from Puck Bay. Sedimentation rates for the upper layers of nearby sediments have been determined to be in the range 0.91–7.71 mm yr^{-1}. Assuming the average of these two values for the upper layers of sediments for the estuarine core, this implies that the sediments in the upper 20 cm of this core were deposited during the last 45 years or so (although this figure is subject to a wide margin of error). This corresponds to the period of heavy industrialisation in the Vistula Basin (Szefer et al. 1996). In addition, this is a stormy area where extensive sediment resuspension takes place leading to mixing of the sediment. The maximum of many elements in the depth range 5–15 cm in this core may reflect the stagnation of the Polish economy after 1980 when industrial production declined.

As expected, significant variations of metal concentrations in relation to sediment particle size were identified. The fine-grained (sub-colloidal) fraction is mainly enriched in the heavy metals, while the 63–200 μm fraction commonly exhibited the lowest levels of the metals analysed.

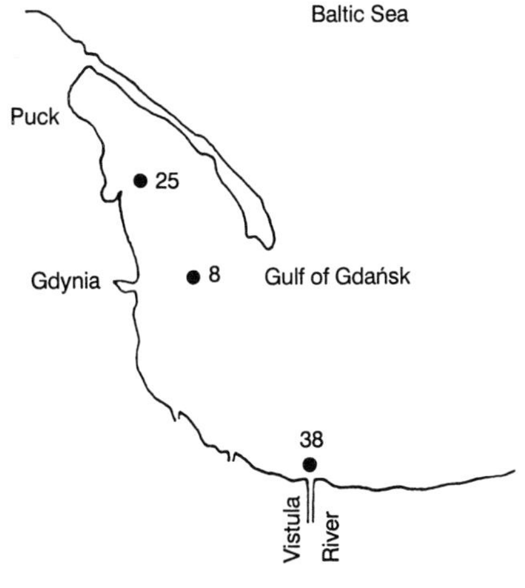

Fig. 4.4. Schematic map showing the locations of sediment cores Nos. 8, 25 (Puck Bay) and 38. After Szefer et al. (1998b).

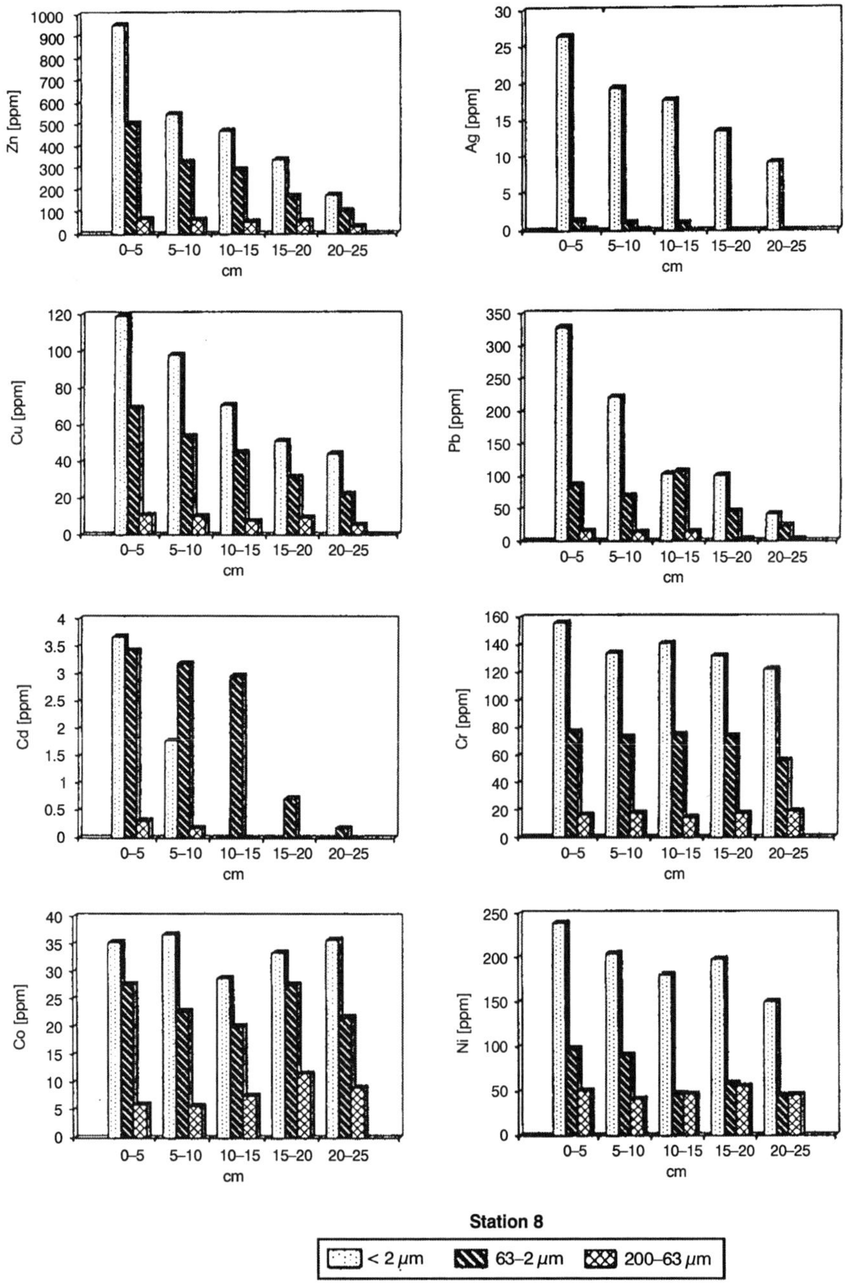

Fig. 4.5. Distribution of trace elements with depth in each of the size fractions (< 2, 2–63 and < 63 μm) in sediment core 8. After Szefer et al. (1998b).

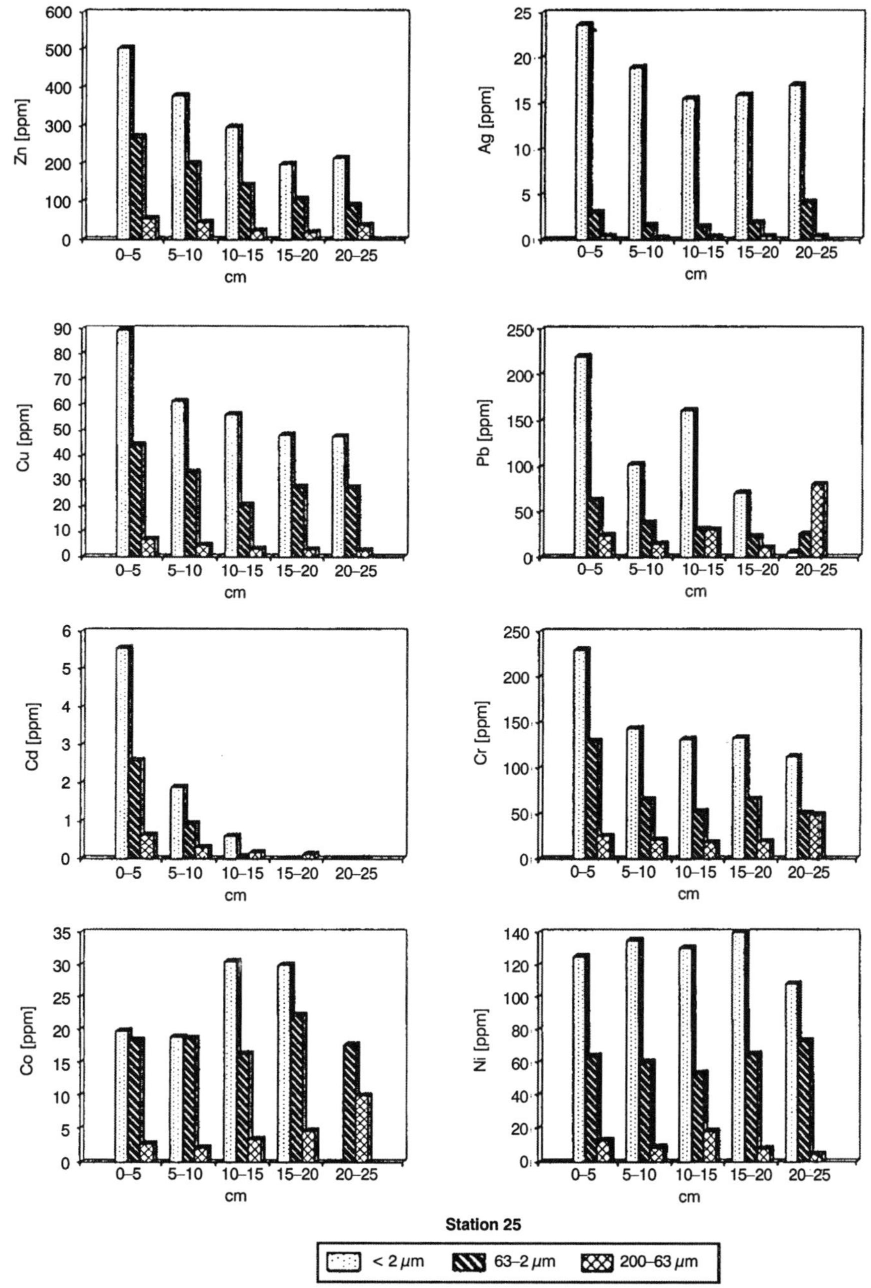

Fig. 4.6. Distribution of trace elements with depth in each of the size fractions (< 2, 2–63 and < 63 μm) in sediment core 25. After Szefer et al. (1998b).

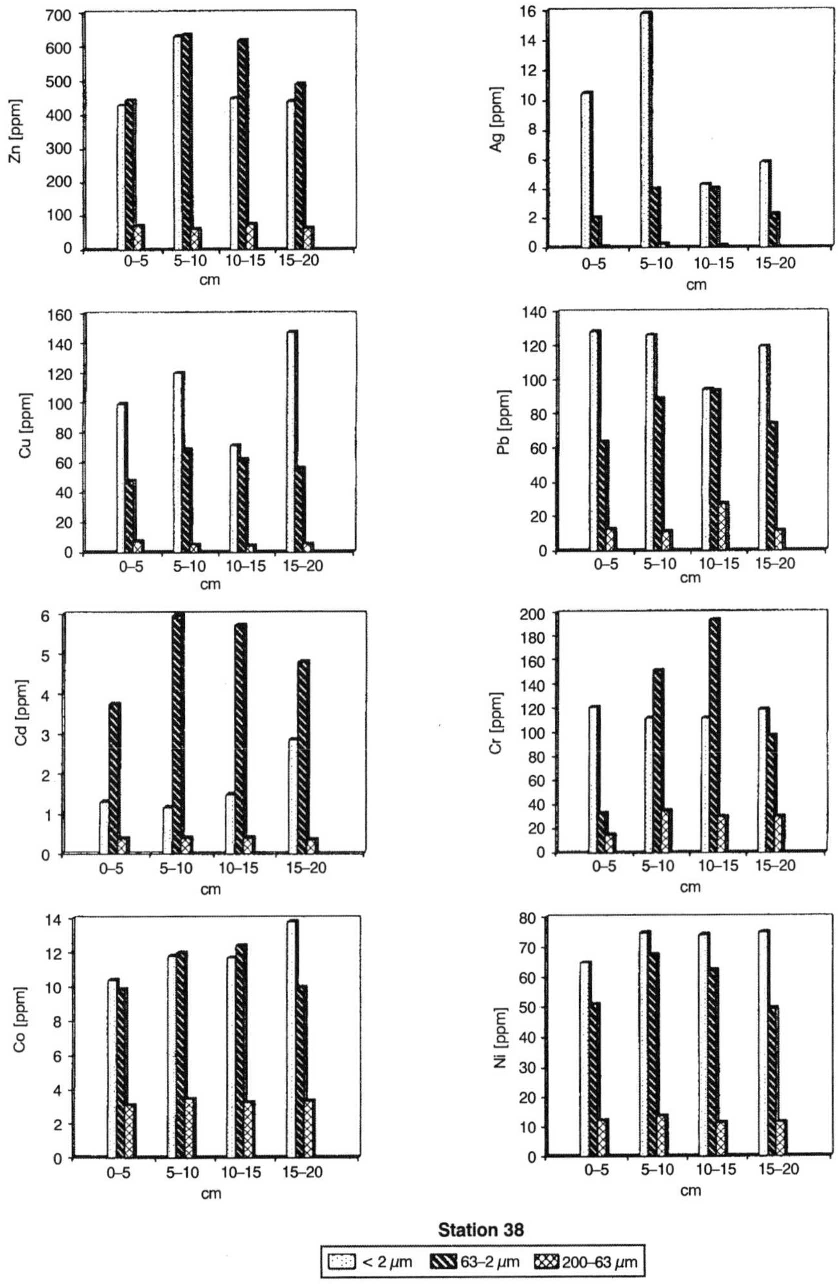

Fig. 4.7. Distribution of trace elements with depth in each of the size fractions (< 2, 2–63 and < 63 μm) in sediment core 38. After Szefer et al. (1998b).

Redox-dependent trends

The changes of redox potential affect the metal distribution in the water column and in the bottom sediments (Kremling, 1983; Borg and Jonsson, 1996; Brügmann et al., 1997, 1998; Kremling, 1983b; Kremling et al., 1987, 1997). The concentrations of dissolved of Cd, Cu and Pb in the water showed decreasing tendency below the redoxcline, while the concentrations of metals such as Co, Fe and Mn indicated increase their contents with water depth under reducing conditions since the reduced forms of these metals are more soluble (Kremling 1983; Kremling et al., 1987, 1997). A mechanism for the Mn deposition in the Baltic sedimentary column has been proposed by many authors (Manheim, 1961; Niemistö and Voipio, 1974; Suess and Djafari, 1977; Blazhchishin, 1982b). It is in agreement with that postulated by Wangersky (1962), Wangersky and Joensuu (1967), Bonatti et al. (1971) and Marchig et al. (1985) for deep sea cores. The core distribution pattern of Mn is probably a result of a resolution of the originally deposited Mn compounds, migration up the sediment column and reprecipitation in the oxidised zone (Wangersky, 1962; Wangersky and Joensuu, 1967). The observed elevated concentrations of H_2S in the waters of the Gotland Deep during a stagnation period in the 1980s (Kremling, 1983) resulted in a further decrease of Cd, Cu and Ni in the anoxic water column. It is postulated that the levels of trace elements in the bottom water were controlled by scavenging with FeS (Kremling, 1983; Dyrssen and Kremling, 1990). An increasingly large volume of the deeper water of the Baltic Proper has been depleted in oxygen because of the continuous input of nutrients and oxygen consuming organic matter (Elmgren, 1989). According to Borg and Jonsson (1996) these anoxic water masses remarkably influence the trapping of Cd, Cu, Pb and Zn in the sediment phase; their concentrations in reduced sediments from the Baltic Proper are greater than those in samples from oxidised sediments from the Åland Sea and the Bothnian Sea (Fig. 4.8). For instance, the median concentration of Cd in sediments from anoxic area is ca. 5 times greater that in those from oxic regions, while an enrichment of Hg is less pronounced in the anoxic conditions (Borg and Jonsson, 1996). This findings is in an agreement with data reported by Hallberg (1991) suggesting the formation of dissolved organic or inorganic complexes of Hg and/or the methylation of this element resulting in its mobility and transport from anoxic to oxic basins (Borg and Jonsson, 1996). It should be stressed that there are differences in enrichment of various elements in black anoxic sediments caused by several competitive mechanisms involving formation of insoluble sulphides as well as inorganic and organic complexes (Kremling, 1983; Dyrssen and Kremling, 1990). The mobility and solubility of these insoluble forms in the Baltic Sea are dependent on a change in the load of oxygen – consuming substances and nutrients (Borg and Jonsson, 1996). Early diagenetic remobilization of As and Cu in near-shore Baltic sediments has been characterised by Widerlund and Ingri (1995) and Widerlund (1996). This process was linked to the aerobic decomposition of organic matter

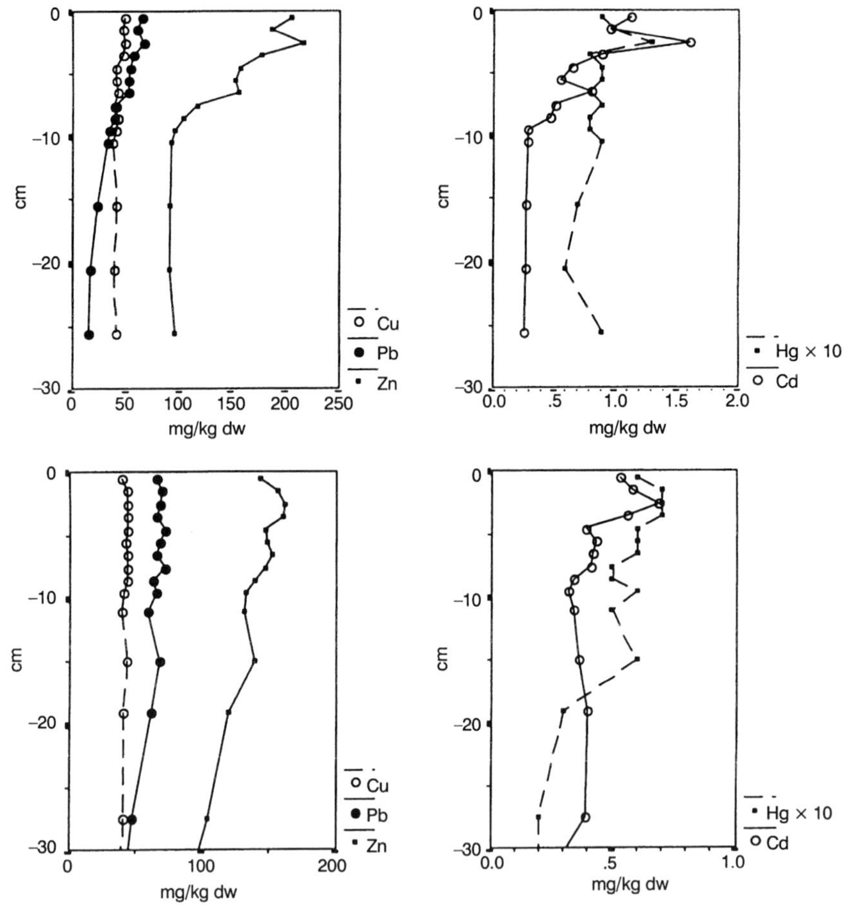

Fig. 4.8. Vertical distribution of Cu, Pb, Zn, Cd and Hg in sediment cores from anoxic sites in the Baltic Proper, showing continuous or periodical lamination through the whole core. To increase clarity, the concentration of Hg is multiplied by 10. Sampling sites SB1, 55°9'N, 15°7'E (above), and NB4, 55°6'N, 18°9'E (below). After Borg and Jonsson (1996); modified.

resulted in releasing back of ca. 3% deposited Cu into the water column of the Kalix River estuary, Gulf of Bothnia. The release of As into the pore water was controlled by the reduction/dissolution of Fe(III)-oxides as a carrier for As down to depths of 10–15 cm in the anoxic zone of the sediment. According to Widerlund and Ingri (1996) upwardly diffusing pore-water Fe and Mn are effectively oxidised and trapped in the oxic surface layer of the sediment, resulting in negligible benthic effluxes of these both elements. In consequence, the concentrations of nondetrital Fe and Mn in permanently deposited and anoxic sediment were comparable to those in settling material in the Kalix River estuary, Gulf of Bothnia.

Metal speciation and mineralogical forms of Fe

For most metals (Cu, Pb and Cd) the silicates and sulphides/organic phases are the dominant substrates in Baltic sediments (Helios Rybicka, 1992). In the clay fraction of sediments the metal-sulphides-organic matter complexes have been sporadically identified. Metals in clay fraction of the sediments are combined in more stable phases as compared with the silty-clayey phase. Metals speciation in the Baltic Sea sediments reflects a complex nature of different processes connected with the precipitation and coprecipitation of trace elements (with carbonates, Mn–Fe–oxyhydroxides) – forming the complexes with organic and inorganic floculated particles – as well as their transport within the crystal lattice of minerals and on exchangeable sites of clay minerals. The last two forms of heavy metals dominate in the clay fraction of sediments (Helios Rybicka, 1992). According to Belzunce Segarra et al. (2000) Mn, Ni, Pb and Zn, are predominantly accumulated in Fe–Mn oxide/hydroxide and organic fractions of Gulf of Gdańsk sediment, especially in carbonate and cation-exchangeable fractions while Cu is mainly associated with the organic fraction. Other elements such as Co, Cr and Fe are mostly found to be associated with the residual mineral component of the sediment, although in samples enriched in Fe there was a significant contribution of these elements in oxidizable fraction, bound with organic matter. According to Pempkowiak et al. (1999) the speciation of selected metals in the four fractions studied differed significantly between sediments from the Baltic Sea and the Norwegian Sea. In contrary to Norwegian Sea sediments, Baltic sediments contained substantial quantities of Cd, Pb and Zn – adsorbed on sediment particles or bound to Fe–Mn oxyhydroxides. Among the metals studied, Cu and secondarily Cd, Pb and Zn exhibited an existence in forms mostly bound to organic matter, especially in the Baltic sediments. This could be explained by high affinity of the metals, particularly Cu to Baltic humic substances which represent some fraction (20–80%) of natural organic matter chemically very active in complexing these metals. It is concluded that atmospheric input is dominantly contributed to the transport of Pb, Zn and Cd to the bottom sediments (Brügmann, 1986b; Nriagu and Pacyna, 1988; Brügmann et al., 1991; Ewers and Schlipköter, 1991; Hallberg, 1991). Fly ash particles and other industrial emissions would be responsible for this input (Schneider, 1987; Förstner et al., 1991; Morgan and Stumm, 1991; Pacyna et al., 1991; Puxbaum, 1991; Stoeppler, 1991; Wedepohl, 1991).

Other major sediment determinants such as amorphic Fe–Mn oxyhydroxides, effectively contribute to accumulate of labile, easily extractable species of Cd, Cu, Pb and Zn in Baltic sediments, especially in estuarine areas (Belzunce Segarra et al., 1987, 1988; Görlich et al., 1989; Szefer et al., 1995a; Danielsson et al., 1999). Surficial sediments from the Gdańsk Basin (Fig. 4.9) have been studied for metal speciation because this area of the deposition for the particulate matter riverine (Fig. 4.10) in origin is found as a highly interested for such studies. As it can be seen in Fig. 4.11 there is a distinctly marked a salinity (hydrological) front ca. 10 km from the Vistula River mouth. Pathways of authigenic Fe-mineral formation

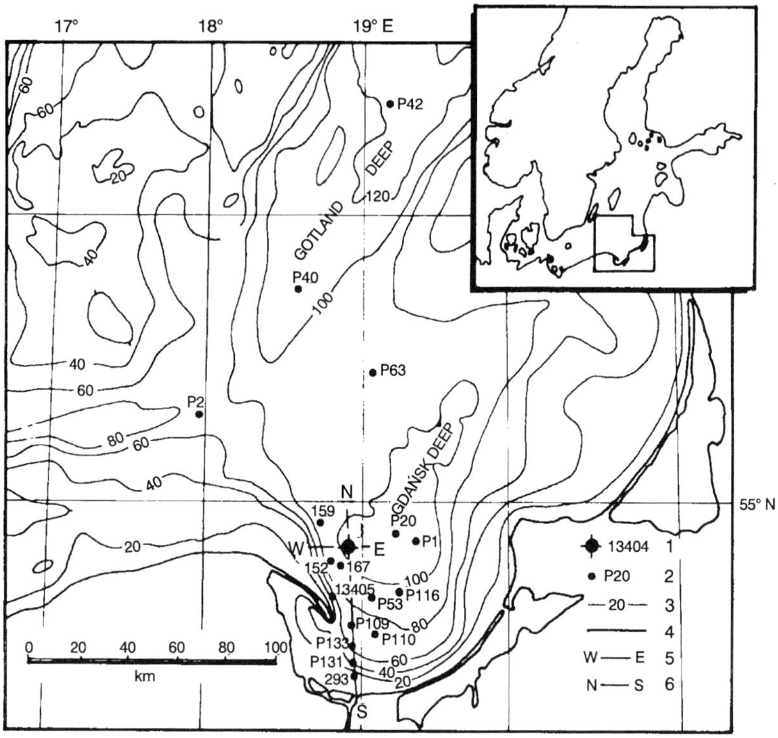

Fig. 4.9. Bathymetric map of the study area (the position within the Baltic Sea is shown in the inset) and location of the sampling stations. 1 = position of the studied core, 13404-1; 2 = grab samples; 3 = isobaths (m); 4 = shoreline; 5 = line of section in Fig. 4.10D; 6 = line of section in Fig. 4.10. The Gdańsk Deep is the deepest part of the Gdańsk Basin. Görlich et al. (1989); modified.

and deposition in the Gdańsk Basin are illustrated in Fig. 4.11. The shape and parameters of the spectra for the samples barely change at 110 K, compared with those at RT (Fig. 4.12). Ferrous ions are located in ferrous hydroxide, an unidentified Fe^{2+}/Fe^{3+} hydroxide, monosulphides and authigenic siderite. Ferric ions are identified in fine-crystalline ferrihydrite, α- and γ-FeOOH. Iron in non-clay magnetic minerals changes its valency and mineral form along the sediment column (Görlich et al., 1989). The presence of lepidocrocite-ferrihydrite- Fe^{2+}/Fe^{3+} hydroxide (and siderite and FeS) parageneses and the Fe^{3+}/Fe^{2+} ratio higher than 2.0 indicate a brackish, stratified estuary-type basin. A ferrous hydroxide-goethite-dominated paragenesis with lower levels FeS and a Fe^{3+}/Fe^{2+} ratio of lower than 1.0 (Fig. 4.13) substantiates identification of a freshwater environment while transitional conditions to or from brackish basins are identified by total Fe and siderite content peaks and enhanced levels of FeS (Görlich et al., 1989).

Speciation studies of Hg in Baltic sediments have been performed by Kannan and Falandysz (1998).

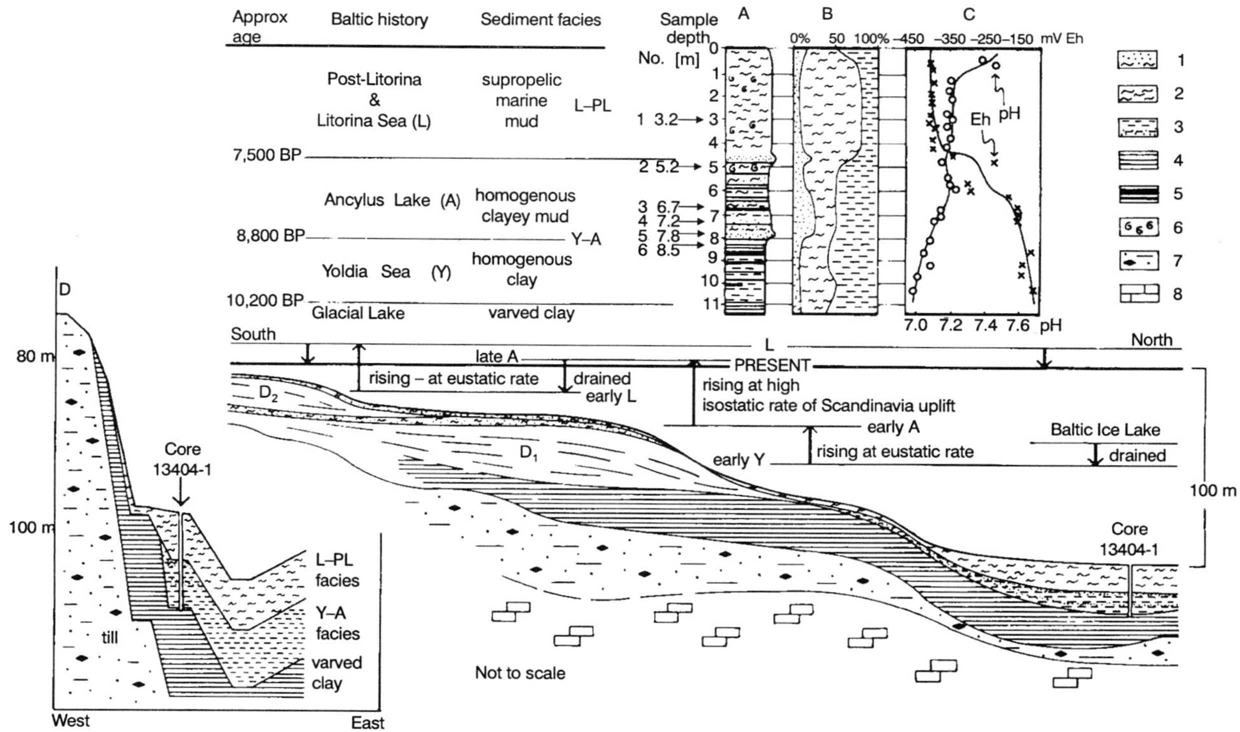

Fig. 4.10. Sediment column in the Gdańsk Basin sampled with core 13404-1 (from R.V. *Meteor)* set against the acoustic data on the sedimentary sequence and a schematic Late Pleistocene to Recent history of sea-level and salinity fluctuations. Core log (A), grain size (B), pH-Eh (C) and the distinguished lithofacies are shown. The locations of the Mössbauer-studied samples are shown adjacent to the sections. The W-E schematic section in the inset (D) is based on 3.5 kHz soundings). Sedimentary sequence in the S-N section is from boomer data, with D_1 and D_2 denoting two bodies of the Vistula River delta foresets. The ages of the lithofacies boundaries are arbitrary (because there is no consensus). Legend: 1 = sand and silty sand; 2 = silt or mud; 3 = clay and clay with sand intercalations; 4 = varved clay; 5 = sulphide bands; 6 = fauna; 7 = diamict; 8 = Cretaceous substratum. Görlich et al. (1989); modified.

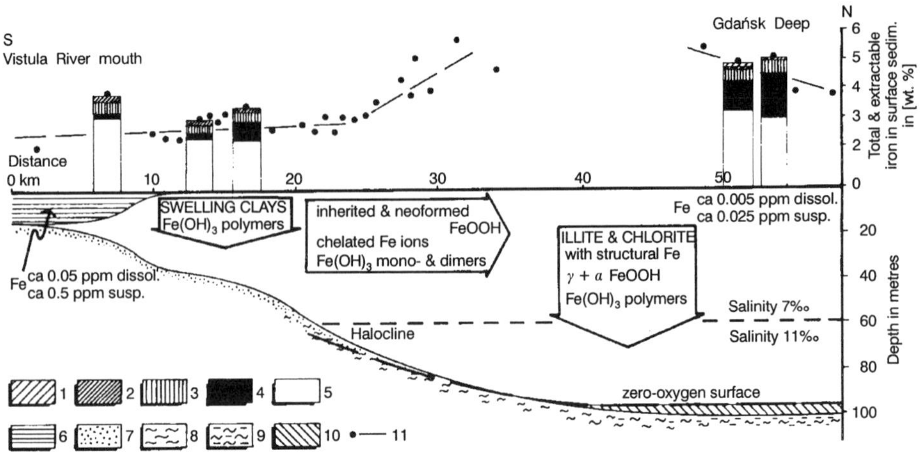

Fig. 4.11 Pathways of authigenic iron-mineral formation in the present-day Gdańsk Basin. Histograms of the sequential extraction of iron are from Belzunce Segarra et al. (1987). 1 = Fe exchangeable at pH 7 (treated with 1*M* NH_4OAc); 2 = Fe bound with carbonates dissolved at pH 5 (treated with 1*M* NaOAc); 3 = Fe in easily reducible oxides (treated with hydroxylamine); 4 = Fe bound with organic matter and sulphides (treated with H_2O_2); 5 = residual fraction dissolved in HNO_3; 6 = freshwater plume; 7 = sand; 8 = mud; 9 = sapropelic mud; 10 = anaerobic zone; 11 = plot of total iron content in grab samples. Görlich et al. (1989); modified.

Surficial sediments from the Baltic Sea have been analysed for concentrations of tri-, di- and monobutyltin (TBT, DBT, MT, respectively) and tri-, di- and monophenyltin compounds (Szpunar et al., 1997; Senthilkumar et al., 1999; Biselli et al., 2000). According to Biselli et al. (2000) the co-toxicants TBT and TPT and their degradation products such as DBT, MBT, DPT and MPT occurred in significantly higher levels in sediments from Baltic Sea marinas than from North Sea marinas. The organotins levels are surprisingly high indicating a long-term contamination of marine sediments although the application of TBT in antifouling paint has been restricted. This suggests that the ecotoxicological risk attributed to organotins has not diminished yet (Biselli et al., 2000).

(iii) Nutrients in Bottom Sediments

Eutrophication and metallic pollutant inputs are features of most estuaries and harbours in industrialised areas. Interactions between eutrophication processes and the cycling of pollutants may be of major importance when an evaluation and prediction of the bioavailability as well as fate of pollutants in the marine ecosystem are required (Hylland et al., 1996; Schaanning et al., 1996; Skei et al., 1996; Virkanen, 1998). Large scale and local variations in the nutrient situation of the Baltic Sea have been reported by several authors (Nehring, 1984a, 1984b, 1985; Grönlund and Leppänen, 1990, 1992; Pitkänen, 1991; Bolałek,

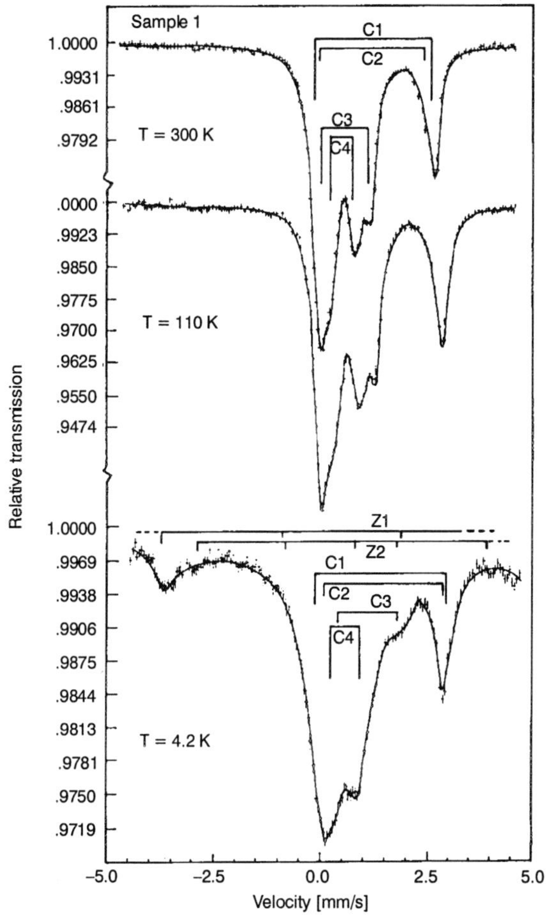

Fig. 4.12. Mössbauer spectra for sample at three temperatures: RT (300 K), 110 K and 4.2 K. As well as quadrupole components (C1 C4), two Zeeman components show up at 4.2 K (denoted Z1 for Fe^{3+} and Z2 for Fe^{2+}). Görlich et al. (1989); modified.

1992b; Falkowska et al., 1993; Wulff et al., 1994a, 1994b, 1996; Pitkänen and Tamminen, 1995; Grönlund et al., 1996; Rahm et al., 1996; Stockenberg and Johnstone, 1997; Voß and Struck, 1997; Danielsson et al., 1998; Struck et al., 1998). Sediments of the Baltic Sea have been studied for the concentration, distribution and transport of selected nutrients e.g. N and P (Rittenberg et al., 1955; Naik and Poutanen, 1984; Carman and Wulff, 1989; Jonsson et al., 1990; Koop et al., 1990; Carman and Jonsson, 1991; Jonsson and Carman, 1994; Conley et al., 1993, 1997; Conley and Johnstone, 1995; Gunnars and Blomqvist, 1997; Carman et al., 1996, 2000; Carman and Aigaras, 1997; Carman and Rahm, 1997; Domanov et al., 1997; Lehtoranta et al., 1997; Graca and Bolalek, 1998; Lehtoranta, 1998; Tuominen et al., 1998; Lampe, 1999; Emeis et al., 2000; Struck et al., 2000). According to Virkanen (1998) eutrophication in the Bay of Töölönlahti, southern Finland, has given rise since

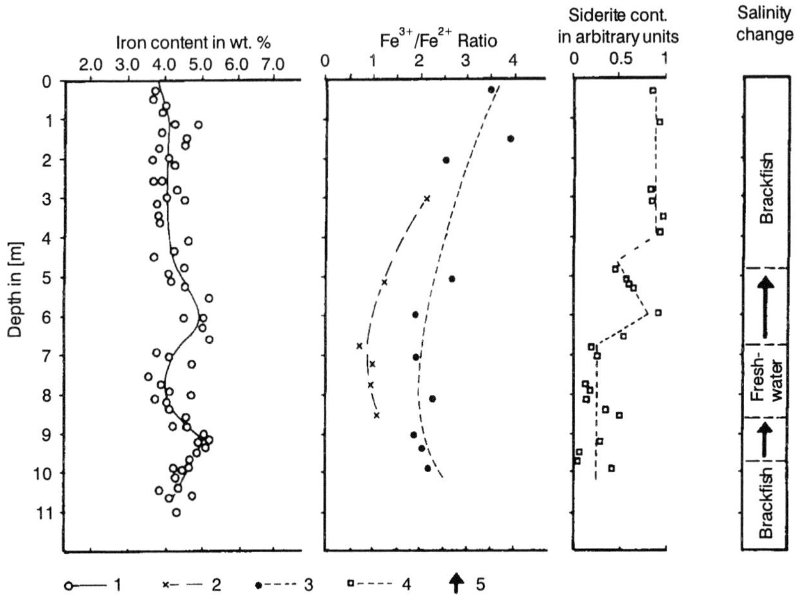

Fig. 4.13. Total Fe content from all available analyses (1) Fe^{3+}/Fe^{2+} ratios (2 = measured with the Mössbauer method; 3 = determined with the wet-chemical method) and siderite content (4) against the sediment depth in core 13404-1. The lines are intended only to lead the eye. On the right-hand side, the environmental changes deduced from all the data from the study are shown (5 = transitional environments). Görlich et al. (1989); modified.

1900 to both temporary and permanent changes in the sediment geochemistry during the course of time. It is reflected in the sediment by a rise of in the levels of P, S as well as organically bound Cu, Fe, Mn, Zn and hydroxide Zn. The total concentrations of Al, Ca and Mn decreased towards the surface, probably as a result of dilution by organic inputs or by biogenic silica. Diagenetic changes of humic substances in Baltic sediments are presented by Pempkowiak et al. (1998b).

The concentrations of particular forms of C, N and P in Baltic sediments (surface and cores) are presented in Tables 4.4 and 4.5.

(iv) Radionuclides in Bottom Sediments

The radioactivity of Baltic sediments has been studied by several authors (Kautsky and Eicke, 1981; Jaworowski et al., 1986; Lazarev et al, 1986; Salo et al., 1986; Tuomainen et al., 1986; Leskinen et al., 1987; Bojanowski et al., 1995a, 1995b; HELCOM, 1995; Panteleev et al., 1995; Suplińska, 1995; Pöllänen, 1997; Pöllänen et al., 1999; Ikäheimonen et al., 2000). Ilus et al. (1998) evaluated a sedimentation rate at two sampling sites at the Gulf of Finland based on ^{210}Pb, ^{137}Cs, and $^{239+240}Pu$ profiles in sediments. The concentration data of radionuclides in surface sediments and core sediments of the Baltic Sea are listed in Tables 4.6 and 4.7. The concentrations of radiocaesium in surficial sediments have been regularly studied since 1984. It is pointed out that in 1986 levels of this radionu-

TABLE 4.4.

Concentrations of various chemical forms of C, N, P and Si (%) in surface sediments of the Baltic Sea and other northern areas

Region	Sampling date	Sample depth (m)	Fraction (μm)	N	C-tot	C-org	C-inorg	N-tot	N-org	N-fix	N-ex	References
Baltic Sea	1981–83											Belmans et al., 1993
Southern part		60–120	(98% of	5		5.12	1.44	0.34				
			< 63 μm			1.7–8.6	0.9–2.2	0.12–0.56				
Western part		25–50	fraction	4		6.13	2.25	0.36				
						5.2–6.8	1.5–3.6	0.29–0.40				
North-eastern part		110–240		4		8.58	5.15	0.56				
						7.9–9.5	1.5–11.0	0.38–0.69				
Gulf of Finland	1992–93		< 2 mm	20		6.61						Leivouri, 1998
						3.48–9.45						
Gulf of Finland	1992–95	0–10	< 2 mm	42		1.23						Vallius and Leivouri, 1999
						0.28–2.19						
Bothnian Bay	1991–93	> 60	< 2 mm			4.0±1.2						Leivouri and Niemistö, 1995
	1992–93		< 2 mm	8		3.96						Leivouri and Niemistö, 1993
						1.65–4.95 (5)						
	1991	< 60	< 2 mm	32		3.7		0.4				Leivouri and Niemistö, 1993
						1.06–8.31		0.12–0.85				
Bothnian Sea	1991–93	> 60	< 2 mm			2.3±1.0						Leivouri and Niemistö, 1995
	1992–93			13		2.49						Leivouri and Niemistö, 1993
						0.62–3.28 (9)						
	1991	< 60	< 2 mm	24		1.9		0.2				Leivouri and Niemistö, 1993
						0.34–7.23		0.03–0.77				
Gulf of Riga	1991/93	27–54		23	4619*	4449*	157*	449*	409*	31.7*	7.9*	Carman et al., 1996
					193–5409	174–5267	12.4–1761	20–521	6.21–481	0.53–42.2	0.29–23.5	
Baltic Proper	1990			6	5460*	6180*	31.6*	603*	584*	27.6*	2.13*	Carman and Rahm, 1997
					4283–8125	4252–8107	16.3–73.3	500–700	476–676	16.7–39.3	1.0–3.1	
Western Baltic												
German coastal waters	1991–95		< 20	102	3.57			3.51				Müller and Heininger, 1999
					0.02–11.5			ND–16				
Szczecin Lagoon			< 20	6	5.2			9.9				Müller and Heininger, 1999
					1.35–8.52			5.9–15				
Danish Straits		50–600		2		3.65	7	0.59				Belmans et al., 1993
						3.4–3.9	4.0–10.0	0.59–0.59				

* – $\mu mol\ g^{-1}$ ds.

TABLE 4.4. – continued

Region	Sampling date	Sample depth (m)	Fraction (μm)	N	P-tot	P-org	P-mob	P-ap	P-inorg	Si-tot	References
Baltic Sea	1981–83										Belmans et al., 1993
Southern part		60–120	(98% of	5	1.1						
			< 63 μm		054–1.60						
Western part		25–50	fraction	4	1.35						
					1.00–1.70						
North-eastern part		110–240		4	1.14						
					0.96–1.20						
Northern Baltic											
Bothnian Bay	1991	< 60	< 2 mm	32	1.48±0.54					29.5±3.8	Leivouri and Niemistö, 1993
					0.80–3.29					24.1–40.8	
Bothnian Sea	1991	< 60	< 2 mm	24	1.27±0.65					33.3±5.8	Leivouri and Niemistö, 1993
					0.45–2.84					24.4–42.4	
Southern Baltic											
Vistula Estuary	1988–89		< 80	4	1.49						Szefer et al., 1995a
					0.60–2.40						
Gulf of Gdansk			< 80	22	0.84						Szefer et al., 1995a
					0.31–1.99						
Central part			< 80	5	1.35						Szefer et al., 1995a
of Polish EEZ					0.75–1.99						
Vistula Estuary	1988–89		< 2 mm	4	0.42						Szefer et al., 1996
					0.16–0.96						
Gulf of Gdansk			< 2 mm	22	0.37						Szefer et al., 1996
					0.10–0.66						
Central part			< 2 mm	5	0.27						Szefer et al., 1996
of Polish EEZ					0.10–0.42						
Gulf of Riga	1991/93	27–54		75	63.4*	17.4*	36.8*	9.2*			Carman et al., 1996
					7.2–90.5	0.1–35.4	0.20–61.8	3.48–18.6			
Baltic Proper	1990			6	37.5*	13.9*	12.6*	11.1*	23.6*		Carman and Rahm, 1997
					30.1–64.9	5.0–19.3	1.3–30.4	7.4–15.2	12.6–45.6		
Western Baltic											
German coastal waters	1991–95		< 20	102	0.43						Müller and Heininger, 1999
					0.1–2.0						
Szczecin Lagoon			< 20	6	1.6						Müller and Heininger, 1999
					0.5–4.1						
Danish Straits		50–600		2	1.05						Belmans et al., 1993
					1.00–1.10						

* – μmol g^{-1}

TABLE 4.5.

Concentrations of C, N, P and Si (%) in sediment cores of the Baltic Sea and other northern areas

Region	Sampling date	Sample depth (m)	Granulometric fraction (μm)	Segment depth (mm)	N	C-tot	C-org	C-inorg	N-tot	N-org	N-fix	N-ex	References
Bothnian Bay	Pre-1984			0–8			4.2		0.5				Naik and Poutanen, 1984
				96–104			2		0.3				
Southern Baltic													
Arkona Basin	1993	45	< 63	0–10	2		5.50**						Neumann et al., 1996, 1998
				< 400			3.25**						
Bornholm Basin	1974	71		0–10	1		5.09						Suess and Erlenkeuser, 1975
				330–340	1		4.7						
	1980	88	Clay	0–7	5		4.94		0.68				Szefer and Skwarzec, 1988
				121–152			4.63		0.53				
Oder Lagoon	1993	5	< 63	0–10	2		11**						Neumann et al., 1996, 1998
				< 400			8**						
Achterwasser	1993	3.7	< 63	0–10	1		13.8**						Neumann et al., 1998
				330–350			10.0**						
Gulf of Gdansk	1980	78	Silty sediment	0–7	1		6.54		1.47				Szefer and Skwarzec, 1988
				250–300	1		3.62		0.44				
Gdańsk Basin	1996		< 2 mm	0–25			4.42		0.56				Belzunce Segarra et al., 2000
							1.61–6.22		0.20–0.76				
				250–300			1.95		0.23				
							1.54–3.52		0.17–0.43				
Gdansk Deep	1980	105	Silty sediment	0–16	1		4.44		1				Szefer and Skwarzec, 1998
				311–348	1		4.45		0.58				
	1980	89		0–10	1		6.38		0.81				
				200–253	1		3.89		0.45				
Gulf of Riga	1991/93	75	< 45	10	23		3340*						Carman and Aigars, 1997
							293–4938						
				50	23		2193*						

Region	Sampling date	Sample depth (m)	Granulometric fraction (μm)	Segment depth (mm)	N	C-tot	C-org	C-inorg	N-tot	N-org	N-fix	N-ex	References
							245–3372						
				100	6		2258*						
							1955–2723						
Western Baltic													
Eckernförder Bucht	1971	28		0–10	1		5.48						Erlenkeuser et al., 1974
				1200–1220	1		6.78						
Southwest of Aero	1971			170–190	1		5.3						
				360–380	1		5.18						
Danish Straits	1991	645		0–20	1		2.24						Paetzel et al., 1994
Skagerrak				200–500	1		1.36						
				0–10	1		2.43						
				200–500	1		1.72						
Baltic Proper	1990		< 45	0–10	6	5460*	6180*	31.6*	603*	584*	27.6*	2.13*	Carman and Rahm, 1997
						4283–8125	4252–8107	16.3–73.3	500–700	476–676	16.7–39.3	1.0–3.1	
				40–50	6	3425*	3375*	49.7*	307*	283*	18.5*	1.45*	
						1925–5162	1880–5138	23.3–75.3	171–488	148–457	6.4–36.4	0.7–2.6	
Northern Baltic	Pre-1984			0–8			7.8		0.5				Naik and Poutanen, 1984
Proper				48–56			4		0.4				

* – μmol g^{-1} dry wt.

** – Concentration approximated from diagrams.

TABLE 4.5. – continued

Region	Sampling date	Sample depth (m)	Granulometric fraction (μm)	Segment depth (mm)	N	P-tot	P-org	P-mob	P-ap	P-inorg	Si-biog	References
	1980	88	Clay	0–7	5	0.186						Szefer and Skwarzec, 1988
				121–152		0.107						
Gulf of Gdansk	1980	78	Silty sediment	0–7	1	0.18						Szefer and Skwarzec, 1988
				250–300	1	0.09						
Gdansk Deep	1980	105	Silty sediment	0–16	1	0.121						Szefer and Skwarzec, 1998
				311–348	1	0.108						
	1980	89		0–10	1	0.141						
				200–253	1	0.099						
Gulf of Riga	1991/93	75	< 45	10	23						846*	Carman and Aigars, 1997
											105–1505	
				50	23						773*	
											154–1258	
				100	6						724*	
											456–1012	
Baltic Proper	1990		< 45	0–10	6	37.5*	13.9*	12.6*	11.1*	23.6*		Carman and Rahm, 1997
						30.1–64.9	5.0–19.3	1.3–30.4	7.4–15.2	12.6–45.6		
				40–50	6	21.6*	6.4*	2.13*	12.7*	15.4*		
						16.9–28.1	0.7–16.6	0.8–5.0	8.3–15.5	11.5–16.7		

* – $\mu mol\ g^{-1}$

TABLE 4.6.

Concentrations of radionuclides (Bq kg^{-1} dry wt.) in surficial sediments of the Baltic Sea

Region	Sampling date	Segment depth (mm)	N	110m-Ag	241-Am	60-Co	134-Cs	137-Cs	40-K	63-Ni	210-Pb	210-Po	239+240-Pu	References
Southern Baltic														
Gulf of Gdansk	1982	0–50						126	720				6.6	Jaworowski et al., 1986
	1994	0–10						303±19					1.50±0.02	Suplińska and Grzybowska, 2000
	1995	0–10						270±13					1.08±0.09	
	1996	0–10						308±17					1.00±0.13	
	1991/92	0–150								0.5±0.1				Skwarzec et al., 1994
Gdańsk Deep	1994	0–10						328±29					2.10±0.24	Suplińska and Grzybowska, 2000
	1995	0–10						269±13					2.15±0.10	
	1996	0–10						371±16					1.93±0.27	
Slupsk Furrow	1982	0–50						27	710				5.9	Jaworowski et al., 1986
	1985	0–100	2									39.2–49.3		
Bornholm Deep	1982	0–50	2					45–73	800–1000				3.4–7.7	Jaworowski et al., 1986
	1994	0–10						129±13						Suplińska and Grzybowska, 2000
	1995	0–10												
	1996	0–10						134±10						
Pomeranian Bay	1993	0–90	7				0.19–0.48	4.0–11.4			9.6–20.2		0.076–0.295	Bojanowski et al., 1995
Arkona Basin	1982	0–100										114±5.2		
Western Baltic														
Mecklenburg Bay	1982	0–100										74.0±6.3		
Lubeck Bight	1994				0.82	ND	1		725		97		1.6	Nielsen, 1996
Northern Baltic														
Loviisa vicinty	1990	0–50	8	5.0–12.0		13.0–44.0	290–580	1800–3500	710–840					Ilus et al., 1992, 1995
Kattegat	1995				0.41	1.2	1.1		815		101		1.1	Nielsen, 1996

TABLE 4.6. – continued

Region	Sampling date	Segment depth (mm)	N	238-Pu	226-Ra	106-Ru	125-Sb	90-Sr	232-Th	234-U	235-U	238-U	References
Southern Baltic													
Gulf of Gdansk	1982	0–50			59								Jaworowski et al., 1986
	1987		2							23.2–25.2	0.5–0.6	22.3–24.4	Skwarzec, 1995
	1995	0–10		29±18*									Suplińska and Grzybowska, 2000
	1996	0–10		50±30*									
Gdansk Deep	1987		2							34.1–51.4	0.8–1.2	33.3–50.6	Skwarzec, 1995
	1995	0–10		53±14*									Suplińska and Grzybowska, 2000
	1996	0–10		91±43*									
Bornholm Deep	1982	0–50			45–73				72				Jaworowski et al., 1986
	1987									41.1±1.3	1.8±0.3	41.8±1.3	Skwarzec, 1995
Słupsk Furrow	1982	0–50			42				66				Jaworowski et al., 1986
Pomeranian Bay	1993	0–90	7	0.003–0.009									Bojanowski et al., 1995
Western Baltic													
Lubeck Bight	1994			0.11	21.4				35.2				Nielsen, 1996
Northern Baltic													
Loviisa vicinty	1990	0–50	8	0.053–0.16		27.0–73.0	13.0–42.0	13–42					Ilus et al., 1992, 1995
Kattegat	1995			0.08	21				32.6				Nielsen, 1996

* – mBq kg^{-1} d.w.

TABLE 4.7.

Concentrations of radionuclides (Bq kg^{-1} dry wt.) in surficial and bottom segments of sediment cores of the Baltic Sea

Region	Sampling date	Segment depth (mm)	N	110m-Ag	241-Am	140-Ba	141-Ce	144-Ce	242-Cm	243+244-Cm	60-Co	References
Gulf of Finland	1986	0–50	1	50		350	110	110			52	Ilus et al., 1987
		50–100	1	ND		ND	ND	ND			5.8	
North Baltic Proper	1986	0–10	1		0.59		32	30	0.27	0.06		Ilus et al., 1987
		40–50	1		1.5		ND	ND	ND	ND		
	1989–90	0–10	2		0.5					0.018	ND–3.4	Ilus et al., 1993
		40–50	2		1.5					ND	ND	
		250–300	2								ND	
	1990	0–50	1								ND	
		150–200	1								ND	

TABLE 4.7. – continued

Region	Sampling date	Segment depth (mm)	N	134-Cs	137-Cs	40-K	95-Nb	63-Ni	210-Pb	210-Po	239+240-Pu	References
Arkona Deep	1983	0–10	1							209±6.4		Bojanowski et al., 1995
		220–240								40.1±3.9		
Gulf of Gdansk	1991/92	0–120						1.3±0.3				Skwarzec et al., 1994
		130–320						1.1±0.2				
		0–100						2.5±0.2				
		230–420						0.1±0.1				
Gdansk Deep	1988	0–60	3	8.6–28.1	140–209						3.52–4.08	Skwarzec and Bojanowski, 1992
		250–300	1	< 1	0.9±0.2						0.02±0.01	
Bornholm Deep	1983	0–30	1							224±5.2		
		210–240	1							45.8±2.1		
	1988	0–60	3	7.0–16.7	55.9–118						1.01–1.89	Skwarzec and Bojanowski, 1992
		250–300	1	< 1	3.1±0.6						0.08–0.01	
Gotland Deep	1988	0–40	2	1.1–6.1	29.8–76.1						7.63–13.2	
		200–250	1	< 1	1.2±0.6						0.01±0.01	
Pomeranian Bay	1993	0–40	2	3.0–4.9	87–117				171–173		1.49–2.08	Bojanowski et al., 1995
		150–200	1	< 0.1	.22				12.0±0.6		0.001±0.001	
		0–40	2	0.5–0.7	4.8–5.1				6.7–7.4		0.056–0.067	
		100–130	1	< 0.21	3.0±0.25				33±0.3		0.011±0.001	
Gulf of Finland	1986	0–50	1	1400	2600	890	2300				3.1	Ilus et al., 1987
		50–100	1	12	150	810	ND				4.3	
	1990	0–50	1	4.7	3100	810					3.3	Ilus et al., 1993
		150–200	1	4.6	130	820					5.1	
	1992	0–20									2.81±0.25	Suplinska, 1995
		180–200									7.88±0.73	
Gulf of Bothnia	1989	0–50	1	5.3	290	750					3.1	Ilus et al., 1993
		50–100	1	3.2	22	750					0.22	
	1990	0–50	1	210	1400	1000					2.7	Ilus et al., 1993
		100–150	1	1.9	32	1100					0.66	
	1992	0–20	1								2.63±0.24	Suplinska, 1995
		80–100									4.36±0.35	
		180–200									0.025±0.015	
North Baltic Proper	1986	0–10	1	19	92	450	72				0.97	Ilus et al., 1987
		40–50	1	ND	73	840	ND				4.5	
	1989–90	0–10	2	46–70	370–410	720–790					1.2–1.3	Ilus et al., 1993
		40–50	2	ND	ND	780–800					4.5–5.6	
		250–300	2	ND	ND	1100					0.01	
	1990	0–50	1	37	310	720					1.2	
		150–200	1	ND	7	1100					0.43	
	1991/92	0–20									7.82±0.49	Suplinska, 1995
		80–100									0.05±0.02	

TABLE 4.7. – continued

Region	Sampling date	Segment depth (mm)	N	238-Pu	103-Ru	106-Ru	125-Sb	89-Sr	90-Sr	129m-Te	Th (tot.) ($\mu g\ g^{-1}$)	U (tot.) ($\mu g\ g^{-1}$)	95-Zr	References
Gulf of Gdansk	1980	0–50	5								7.9–10.0	1.2–1.9		Szefer and Skwarzec, 1988
		250–30	1								12.1	1.2		
Gdansk Deep	1980	0–50	3								7.6–11.2	2.0–2.8		
		20o–253	1								11.9	2.3		
Bornholm Deep	1980	0–50	1								8.8–12.7	1.4–1.9		
		121–152	1								11	1.7		
Pomeranian Bay	1993	0–40	2	0.042–0.06										Bojanowski et al., 1995
		150–200	1	< 0.001										
		0–40	2	0.001–0.003										
		100–130	1	0.001±0.001										
Gulf of Finland	1986	0–50		0.11	1700	750	130	270	24	1400			580	Ilus et al., 1987
		50–100		0.17	8.9	ND	4.4	ND	2.6	ND			ND	
	1990	0–50	1	0.16										
		50–100	1	0.21										
Gulf of Bothnia	1990	0–50		0.08										
		150–200		ND										
		180–200												
North Baltic Proper	1986	0–10	1	0.079	35	48	20						36	Ilus et al., 1987
		40–50	1	0.19	ND	ND	8.7						ND	
	1990	0–50		0.16										
		150–200		0.21										

-clide evidently increase reaching maximum values during 1987–1988 in bottom sediments in several subareas of the Baltic Sea. Thereafter, the radiocaesium concentrations in the surficial layers of sediments remained generally at the same level although its total amount (Bq m^{-2}) attained maximum values later, i.e. during 1990 (HELCOM, 1995). Spatial distribution of radiocaesium in the Baltic Sea indicates that the largest its settling from the water phase into the seabed occurred in northern subareas, the Gulf of Finland and the Bothnian Sea. The highest sediment value of 3 400 Bq kg^{-1} dry wt. was observed during 1987 in the eastern part of the Gulf of Finland while the maximum total amounts of 43 000 Bq m^{-2} at the same sampling site only during 1990 (Figs. 4.14 and 4.15). According to Ilus et al. (1995) for such highest amounts of radiocaesium in sediments are not responsible higher site-specific deposition values but exclusively particle transport and the focusing of sinking radionuclides in the deepest areas of the basins. Although a distinct increase in radiocaesium levels was also noted for other areas such as the Baltic Proper and Bothnian Bay, its amounts in bottom sediments were, however, smaller (Ilus et al., 1995). Accumulation of this radionuclide in the southern Baltic sediments has been studied by Bojanowski et al. (1995b) and Suplińska and Grzybowska (2000). The latter authors concluded that the Chernobyl fallout was the major source of ^{137}Cs in the southern Baltic.

The concentrations and amounts of ^{137}Cs (Fig. 4.15) distinctly increased, like ^{106}Ru, ^{103}Ru, ^{110m}Ag and ^{125}Sb (Fig. 4.16), in surficial sediments of the Baltic Sea

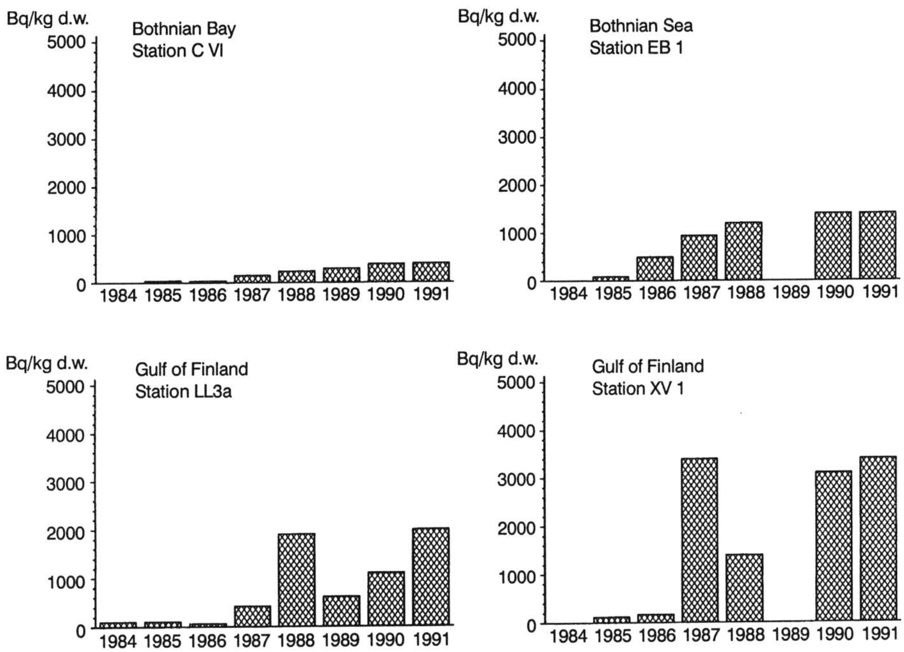

Fig. 4.14. Caesium-137 in surface sediment layer (0–5 cm) at some sampling stations of Gulf of Bothnia and Gulf of Finland in 1984-1991. After Ilus et al. (1995); modified.

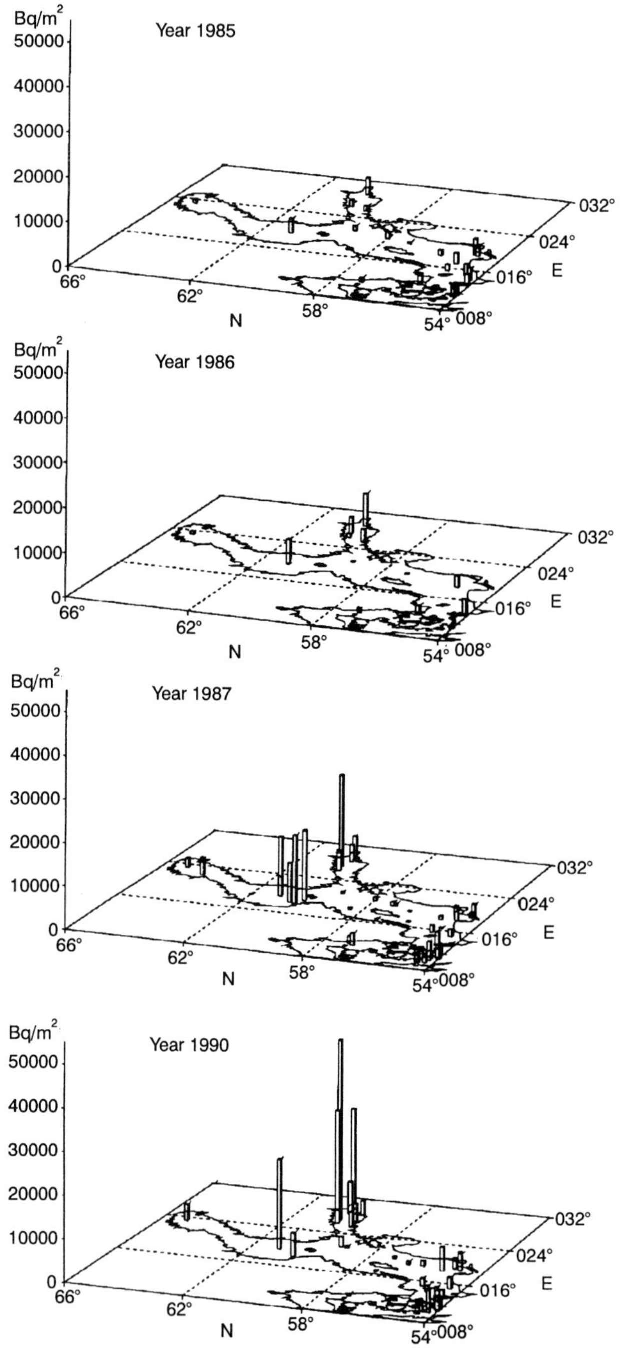

Fig. 4.15. Total amount of caesium-137 at different sampling stations in 1985, 1986, 1987, 1990. After Ilus et al. (1995); modified.

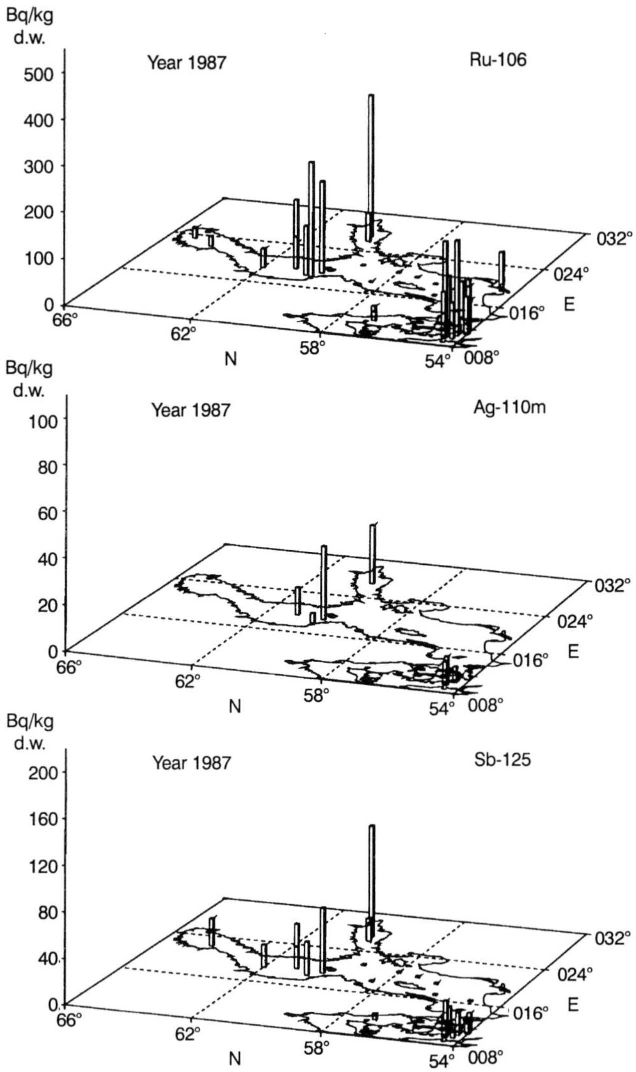

Fig. 4.16. Ruthenium-106, silver-110m, and antimony-125 in surface sediment layer at different sampling stations in 1987. After Ilus et al. (1995); modified.

after the Chernobyl accident. In spite of a rapid removal of plutonium from water phase to bottom sediments (Östlund, 1991; Holm, 1995) the distribution of $^{239+240}$Pu in Baltic sediments does not indicate any pronounced spatial and temporal trends in respect to a Chernobyl influence. Relatively small amounts of $^{239+240}$Pu were present in the Chernobyl fallout, therefore the changes caused by the fallout in their concentrations in surface sediments were also appropriately small (Holm, 1995; Bojanowski et al., 1995a; Ilus et al., 1993, 1995). Holby and Evans (1996) studied the vertical distribution pattern of Chernobyl-derived radionuclides in a Baltic sediment. According to Holm (1995) the Chernobyl accident contributed more significantly to the overall activity concentrations of ^{238}Pu and ^{241}Pu. Analysis of particular segments of Baltic sediment cores supports insignificant contribution of Chernobyl-derived $^{239+240}$Pu in the total its activity concentration in the Baltic Sea ecosystem (Skwarzec and Bojanowski, 1992; Bojanowski et al., 1995a; Supli-ńska, 1995). The major source of plutonium in the Baltic Sea and adjacent areas is the fallout caused by nuclear tests (Holm, 1995; Ilus et al., 1995). Due to different ^{238}Pu/$^{239+240}$Pu ratios estimated for these two kinds of fallout, even very small addition of the Chernobyl-derived plutonium to Baltic sediments can be detected (Skwarzec and Bojanowski, 1992; Holm, 1995; Bojanowski et al., 1995a; Ilus et al., 1995). Relationships between plutonium and some stable elements (As, Cd, Cl, Co, Cu, Fe, Hg, Mn, Mo, Ni, Pb, V) in surface sediment from the Gulf of Bothnia were extensively studied by Östlund (1991). Distribution pattern for ^{60}Co in different subregions of the Baltic Sea indicated, similarly to plutonium, a small contribution of Chernobyl-derived cobalt in its total concentrations in surficial sediments. Because of homogenous horizontal distribution of ^{226}Ra in Baltic sediments its natural origin is postulated by Ilus et al., (1993, 1995). Ikäheimonen et al. (2000) detected a radioactive 'hot' particle in bottom sediments from the Gotland Deep, being an uranium fuel fragment probably originating from the Chernobyl accident.

Skwarzec et al. (1994) reported that ^{63}Ni concentrations in sediment cores from the Gulf of Gdańsk are very low and lay within the values range from 0.5 to 3.8 mBq g^{-1} dry wt. The concentrations of polonium, ^{210}Po, in sediment cores of the southern Baltic ranged from 39.2 to 114.2 Bq kg^{-1} dry wt. and significantly depended on the concentration of organic matter (Skwarzec, 1995).

B. FERROMANGANESE NODULES

(i) Introduction

General Characteristics

The variations in the forms of ferromanganese nodules are closely associated with the structure of bottom (Winterhalter et al., 1981). The morphology of the concretions has been described by Varentsov and Blashchishin (1976) who classified them to three principal types, i.e. spheroidal concretions, discoidal concretions and crusts. The distribution of these groups within the Baltic Sea have been described by several authors (Manheim, 1965; Djafari, 1976; Winterhalter, 1980; Ingri, 1985a; Heuser, 1988; Zhamoida et al., 1996; Szefer et al., 1998c). Spheroidal concretions are found on 'soft bottom', i.e. in areas where a change in sedimentation conditions has lead to the discontinuing of formerly active deposition of clay, silt and mud resulting in the formation of spheroids by their concentric growth. The discoidal concretions are mostly formed around a separate nucleus, e.g. a pebble and are distributed on rather hard, often sandy or silty bottoms. The last group, consisting of concretionary material such as slabs or crust and, however constitutes only a minor part of the total burden of Fe and Mn precipitated in Baltic concretions. The greater fragment of this concretionary type consists of slabs of, e.g. glacial clay encrusted by a layer of oxyhydrates of Fe and Mn only some millimetres thick. Slabs and crusts may be also coalesced into larger aggregates caused by the change in current patterns affecting the environmental conditions (Winterhalter et al., 1981).

These three various groups of ferromanganese concretions generally reflect specific environmental conditions and hence differences in their chemical composition are distinguished by the interelemental ratios of Mn/Fe. This ratio is ca. 0.1 for oxic recent (post-glacial) sediments reflecting the availability of the two elements dispersed originally in the seawater. In temporarily anoxic basins, e.g. in the Gotland Deep the Mn/Fe value is generally greater than unity (Niemistö and Voipio, 1974; Winterhalter et al., 1981).

According to Varentsov (1973) the formation of ferromanganese nodules is most abundant in five regions, i.e. the southern and eastern parts of the Central Baltic Sea, the Gulf of Riga, the eastern part of the Gulf of Finland and the Bothnian Bay. The nature of the deposits is very similar in the various subareas of the Baltic Sea, although the size and form of the nodules may locally vary from spheroidal and discoidal nodules to crusts (Varentsov, 1973; Winterhalter et al., 1981). Based on the abundance, morphology, composition and mode of formation, ferromanganese concretions from the Baltic Sea can be divided into the three main groups, namely those occurred in the Gulfs of Bothnia, Finland and Riga, the Baltic Proper and the western Belt Sea (Glasby et al. 1997a).

Ferromanganese concretions from the Gulf of Bothnia have been studied by several authors (Gripenberg, 1934; Winterhalter, 1966, 1980; Winterhalter and Siivola, 1967, Boström et al., 1982; Ingri, 1985a; Ingri and Pontér, 1986a, 1987; Mellin, 1987; Sanchez et al., 1988; Kimberley, 1989 and Amakawa et al., 1991). Spheroidal nodules in the Gulf of Bothnia are the most abundant at well-oxidised layers of the uppermost water-rich sediments. A maximum abundance is found in Bothnian Bay reaching values of 15–40 kg m^{-2}. They are characterised by the highest Mn/Fe ratios (mean 0.7) for the various concretion groups. Discoidal concretions occur in the least quantities among the three concretion types. They have somewhat lower values of Mn/Fe ratios (mean 0.55) as compared to the spheroidal ones but display negative Ce anomalies suggesting their location at a lower redox level in the surrounding of the redox boundary. Very low Mn/Fe values (mean 0.09) are found for flat slabs and crusts associated with somewhat higher concentrations of P (ca. 3% P_2O_5) which are present in the Fe-rich fraction (Glasby et al. 1997a). The average sedimentation rate of post-glacial sediments in the Bothnian Bay is estimated to be 0.15 mm yr^{-1} which favours the formation of abundant amounts of concretions, but concretions are absent at rate greater than

0.4 mm yr^{-1}. Spheroidal concretions grow at a rate of ca. 0.15 mm yr^{-1} in oxic surficial sediments having high levels of Mn and Fe in the interstitial waters. In the case of concretions which are formed at erosive areas exposing to bottom waters, accretion rates are much lower. According to Glasby et al. (1997a) the maximum age of the ferromanganese concretions is estimated to be ca. 3000 years corresponding to the age of the proposed dissolution event in the early Littorina Sea. In this period the development of a permanent halocline in the Gulf of Bothnia took place. However, most of nodules in this subarea are certainly much younger.

Ferromanganese concretions from the Gulf of Finland have been studied by Gripenberg (1934), Varentsov (1973), Varentsov and Blashchishin (1974, 1976), Butylin et al. (1985), Butylin and Zhamoida (1988), Gorshkov et al. (1992), Zhamoida and Butylin (1992, 1993), Zhamoida et al. (1996) and Glasby (1997a). Concretions from this subarea are abundant in the eastern half of the Gulf with a maximum abundance at level of 18–24 kg m^{-2}. The morphology of concretions in Gulf of Finland is complex, i.e. spheroidal, buckshot, discoidal, rod-like and irregular and also crusts and coatings on shells of mussels and rock fragments are reported. A great number of concretion types estimated to be 18 was however reduced to three major groups for generalising in publication by Zhamoida et al. (1996). The morphology of ferromanganese nodules has been presented by Varentsov and Blashchishin (1976). Among the spheroidal concretions the most abundant buckshot ones occur in the central part of the Gulf of Finland while bean-shaped concretions exist in the central and eastern part of the Gulf. As regards discoidal concretions, concentric horizontal banding around erratic nuclei consisting mostly of granite fragments occur at the surface and within olive-green and brownish silt layers in the eastern part of the Gulf of Finland. The next concretion type includes Fe-rich and low Mn crusts; small crusts with Fe-Mn oxyhydroxide layers on the upper surfaces of rock fragments exist at the sediment-water boundary. Large crusts – flat and irregular in shape, possible coalescence of several crusts, mostly identified to the north of the central part of the Gulf of Finland (Varentsov and Blashchishin, 1976; Glasby et al. 1997a). Mineralogical analysis showed, that the concretions consist mainly of quartz with lesser quantities of alkali feldspar, montmorillonite and traces of chlorite, kaolinite and Ca-rhodochrosite. 7 Å manganite is the principal Mn oxide phase. According to Glasby (1997a) spheroidal concretions have the highest Mn concentration (28.8%) and penny-shaped ones located at greater water depths display the highest Fe concentrations (37.5%). Microbiological studies lead to conclusion that Siderocapsa bacteria predominantly participates in the oxidation of Mn and Fe in the concretions (Glasby et al. 1997a).

Ferromanganese concretions from the Gulf of Riga have been investigated by several authors (Putans et al., 1968; Shterenberg et al., 1968; Shterenberg, 1971, Varentsov, 1973; Varentsov and Blashchishin, 1976 and Varentsov et al., 1977). Concretions in this subarea are the most abundant (up to 17 kg m^{-2}) and occupy a central depression containing muddy sediments. Location of spheroidal concretions is found to be adjacent to the depression and discoidal concretions and crust further away (Glasby et al. 1997a). According to Varentsov and Blashchishin (1976) buckshot concretions having spheroidal form predominant in the Gulf of Riga in contrast to bean-shaped concretions, sometimes ellipsoidal, concentrically layered around a lithogenous, usually clay, nucleus. Discoidal concretions, concentric horizontal banding around erratic nuclei occur rarely in shallower water depths. Fe-rich and low Mn crusts small in size are formed predominantly on the upper surfaces of rock fragments, mostly glacial clay and above the sediment-water boundary. They exist mainly in water depths of 24–48 m; large crusts mostly occurred with numerous deposits of bean-shaped spheroidal concretions. Predominant minerals found in ferromanganese concretions are quartz with feldspar, heavy minerals from the granitic complex and clay minerals from the associated sediment; traces of Ca-rhodochrosite are also identified. Among Mn oxide minerals, the principal minerals are 7 Å manganite and poorly crystalline 10 Å manganite (Glasby et al. 1997a). The trends in chemical composition of concretions have been characterised by Varentsov and Blashchishin (1976). Spheroidal concretions display high Mn concentrations (> 15%) and low Fe concentrations (< 15–20%) and are associated with mudy, organic-rich sediments located near the depression. Discoidal concretions and crusts, being associated with silts and sands, appear at shallower depths away from the depression and have greater Fe concentrations (> 30%) and smaller Mn concentrations (< 5%) (Glasby et al. 1997a).

Concretions from the Baltic Proper are mostly distributed around the margin of the deep basins in a depth range of 48–103 m (Manheim, 1965; Varentsov, 1973; Glasby et al. 1997a). The abundance is generally sporadic, locally their abundance attains values of 10–16 kg m^{-2} (Varentsov and Blashchishin, 1976; Ingri, 1985a). The concretions are formed in the anoxic waters of the deep basins. During major inflows of North Sea waters into the Baltic Sea occurred on the average once every 11 years, the anoxic waters are flushed out of the Basins. In consequence, Mn and Fe oxyhydroxides precipitate out and are incorporated into the concretions (Glasby et al., 1996, 1997a). The ferromanganese concretions are found mostly on lag deposits in surrounding of the halocline where strong bottom currents appear (Glasby et al., 1996, 1997a). The morphological features of the concretions have been reported in detail by Varentsov and Blashchishin (1976). Discoidal concretions are with concretic horizontal banding around erratic nuclei, irregular round to ellipsoidal in shape. Small crusts (Fe-rich and Mn low) are mainly formed on exposed gravel and rocks. Mineralogical analyses of Baltic Proper concretions show that both 10 Å manganite and 7 Å manganite were present in this material. Mössbauer analysis of ferromanganese concretions from the Polish Exclusive Economic Zone showed that they are consisted mainly of poorly crystalline lepidochrocite (Görlich et al., 1989; Szefer et al., 1998c).

The most detailed description of the ferromanganese concretions from Kiel Bay in the western Belt has been undertaken by several authors (Seibold et al., 1971; Djafari, 1976; Suess and Djafari, 1977; Heuser, 1988; Hlawatsch, 1993). The concretions appear in a narrow depth range of 20–28 m at the boundary between sands and mud in areas where active bottom currents occurs. According to Djafari (1976) three types of deposits can be distinguished. The first of them occurs as ferromanganese coatings on molluscs and crabs, the second one appears as spheroidal concretions having sometimes a small nucleus of feldspar, quartz and flint while the third one as discoidal concretions is generally formed around a nucleus of flint or granite. Heuser (1988) performed detailed study of the ferromanganese nodules in the western Kiel Bay observing deposits on shells of *Astarte borealis* similarly to those found in southern Baltic. Lübeck-Mecklenburg Bay has been also recognised, but in less degree, in respect to the distribution and characteristics of concretions. They are generally asymmetrical and discoidal and form on substrates of erratic rock. Their distribution is restricted to the relatively small areas where glacial till is exposed through the mud (Wenk, 1981; Lange, 1987; Moenke-Blankenburg et al., 1989; Nielsen, 1992; Leipe et al., 1994). The concretions are characterised by relatively high Mn/Fe ratios (0.9–3.5). Mineralogically, they consist predominantly of todorokite and quartz, with a lesser quantities of feldspar, kaolonite and montmorillonite (Glasby et al., 1996).

Overview of Worldwide Literature

Since the 1960s numerous studies on the distribution, formation and geochemistry of deep-sea ferromanganese nodules have been performed (Glasby, 1972/73, 1974, 1975, 1977, 1984, 1999; Glasby and Read, 1976; Murray and Brewer, 1977; Johnston and Glasby, 1978; Margolis et al., 1978; Siddiquie et al., 1978; Li, 1982; Pettis and de Forest, 1979; Uchio et al., 1980; Kunzendorf et al., 1983, 1993; Ingri, 1985a, 1985b; Usui et al., 1986, 1987, 1989, 1993; Baturin, 1988; Murad and Schwertmann, 1988; Le Sauve et al., 1989; Takematsu et al., 1990; Neumann and Stüben, 1991; De Carlo and McMurtry, 1992; Usui et al., 1993; Usui and Mita, 1995; Chen and Yao, 1995; Glasby et al. 1997b; Renner et al., 1997; Kasten et al., 1998; Usui and Glasby, 1998; Glasby and Schultz, 1999). A model for the formation of hydrothermal manganese crusts from the Pitcairn Island hotspot has been described by Glasby et al. (1997b). Less attention has been paid to nodules from shallow marine ecosystems which have been analysed by several authors (Winterhalter, 1966, 1972, 1980; Calvert and Price, 1970, 1977; Varentsov, 1973; Varentsov and Blashchishin, 1974, 1976; Varentsov and Sokolova, 1977; Zhamoida et al., 1996). Partitioning of 20 trace metals in Fe-Mn nodules from Sicilian soils, Italy, has been reported by Palumbo et al. (2001).

Radiochemical analyses of manganese nodules from shallow waters of different geographical areas have been extensively performed since 1960/70s (Buchowiecki and Cherry, 1968; Ku and Broecker, 1969; Ku and Glasby, 1972; Andersen and Macdougall, 1977; Nakanishi et al., 1977; Krishnaswami and Cohran, 1978; O'Nions et al., 1978; Ku and Knauss, 1979; Ku et al., 1979; Sharma and Somayajulu, 1979; Lalou et al., 1980; Krishnaswami et al., 1982; Aplin et al., 1986; David et al., 2001 and others). The $^{138}Ce/^{142}Ce$ and $^{143}Nd/^{144}Nd$ ratios were obtained for ferromanganese nodules from the Atlantic and Pacific Oceans (Amakawa et al., 1996). Intercomparison studies of Ce, Nd, Ba, Sr and REE were also performed for ferromanganese nodules from the Baltic and Barents Seas, the Gulf of Bothnia and the Pacific and Atlantic Oceans (Amakawa et al., 1991). Worlwide data concerning the distribution and fate of U in marine ferromanganese nodules have been reviewed by Szefer (1987).

(ii) Chemical Elements in Ferromanganese Nodules

The distribution, formation and geochemistry of ferromanganese nodules in the Baltic Sea have been extensively studied from the mid-1960s to the mid-1980s with a limited attention recently (Manheim, 1961, 1965; Winterhalter and Siivola, 1967; Putans et al., 1968; Shterenberg et al., 1968; Shterenberg, 1971; Djafari, 1976; Varentsov, 1973, 1980; Varentsov and Blashchishin, 1974, 1976; Calvert and Price, 1977; Suess and Djafari, 1977; Varentsov and Sokolova, 1977; Boström et al., 1978; 1982, 1988; Winterhalter, 1980; Kulesza-Owsikowska, 1981; Emelyanov et al., 1982; Varentsov and Blashchishin, 1982; Butylin et al., 1985; Ingri and Pontér, 1986a, 1986b, 1987; Mellin, 1987; Butylin and Zhamoida, 1988; Heuser, 1988; Szefer and Szefer, 1990; Emelyanov, 1992, 1995a; Gorshkov et al., 1992; Zhamoida and Butylin, 1992, 1993; Leipe et al., 1994; Zhamoida et al., 1996). Comparison of geochemical investigations of ferromanganese nodules from the Baltic Sea, the Pacific and Atlantic Oceans has been performed by Varentsov (1980) and Wenk (1981).

Since 1990, there has been an upsurge in interest in southern Baltic Sea ferromanganese nodules (Szefer and Szefer, 1990; Sochan, 1992; Gajewski and Uścinowicz, 1993; Glasby et al., 1996; Szefer et al., 1998c; Hlawatsch et al., 2001) and formation of nodules in the Polish Exclusive Economic Zone (EEZ) is reviewed comprehensively by Glasby et al. (1997a). Data on mass balance of Fe and Mn in the Baltic Sea have been provided by Boström et al. (1983) and Blazhchishin (1984). The latter author estimated the absolute quantities of Mn and Fe in ferromanganese nodules in the different basins of the Baltic. Boström et al. (1983) calculated a riverine input of those two macroelements in the Baltic. Most detailed studies of a supply of Mn and Fe from the Kalix – northern Swedish river have been carried out by Burman (1983), Pontér et al. (1990a, 1990b) and Widerlund (1994) as well as from Finnish area as a result of leaching of till by humic substances (Ingri, 1985a, Carlson, 1982; Hallbach, 1975; Virtanen, 1994).

Normalisation of data

It is reported by Boström et al. (1982) that most analysed fractions contains significant amounts acid-insoluble undesirable component corresponding to an admixed trace element poor, silicate-rich dilutent. These components are released to sample solution during the acid leach treatment because of destroying some sheet-silicates though it is unlikely that significant quantities of trace element derive from acid leaching of the detrital terrigenous matter (Boström et al., 1982).

It usually produces obscure the element interrelations in the acid-soluble fraction and therefore it is recommended to recalculate all the original data to an acid-insoluble free fraction. It is shown that the sum of Fe+Mn clustered ca. 40% (Boström et al., 1978); however this value is not constant resultant from variable amounts of acid-insoluble derived material (Al, Ca, Mg, Na and Ti) as well as variations of the degrees of oxidation and hydratation of Fe-Mn phases (Calvert and Price, 1977; Burns and Burns, 1977). Therefore, to remove the scatter in the sum of Fe+Mn, Boström et al. (1982) normalised all the original data, representing acid-leached fractions of Baltic ferromanganese concretions, to a constant sum of (Fe+Mn) amounting to 40%. These standardised values are denoted as HF fraction (Boström et al., 1982).

Spatial trends

Tables 4.8 and 4.9 present chemical composition of ferromanganese nodules of the Baltic Sea. The primary (uncorrected) concentrations of Mn in ferromanganese nodules (flat, spheroidal and ellipsoidal) and in acid-insoluble residue from the Gulf of Bothnia showed remarkable spatial trends (Boström et al., 1982). The concretions from the Bothnian Bay were richer in Mn than those from the Bothnian Sea, i.e. 67% of the flat nodules contained 2% of Mn; spheroidal and ellipsoidal nodules from the former area were likewise reached in Mn contents. The corrected (normalised) values (HF) showed significant latitudinal variations, similarly to uncorrected values. The hydroxide fractions of the Bothnian Bay concretions are distinctly reached in Mn; 50% of all nodules in this area contain more than 10% of Mn, in some cases even up to 20 % of Mn. The variations of P contents (HF) were much less drastically as compared to those of Mn (HF) and its spatial trend was inversely related to that for Mn (Boström et al., 1982).

Rare earth elements (REE) determinations have previously been carried out on 12 ferromanganese concretions from the Gulf of Bothnia and one composite sample from the Baltic Proper (Ingri and Pontér, 1986a, 1986b, 1987; Amakawa et al., 1991). It was shown that the concretions have REE contents more than 5 times higher than those from the Black Sea and Loch Fyne, Scotland.

Morphology dependent trends

According to Zhamoida et al. (1996), who analysed concretions from the eastern part of the Gulf of Finland, there was a major difference in composition between discoidal concretions and crusts, which are mostly Fe-rich, and spheroidal concretions, which are mostly Mn-rich. Discoidal concretions contained higher levels of Ni, Mo and Ti as compared to discoidal ones. Boström et al. (1982) reported that rounded and ellipsoidal ferromanganese concretions, particularly those from the northern Gulf of Bothnia, were richest in Ba, Cu, Mn and Ni, which probably occured in a Mn oxyhydroxide phase. Flat concretions are enriched in Fe, As, P and REE, probably associated with an Fe oxyhydroxide phase. Other elements such as Al, Cr, Ti and V are present in still another component (Boström et al., 1982).

TABLE 4.8.

Concentrations of trace elements ($\mu g\ g^{-1}$ dry wt.), Al, Fe, Mn, Ti, K, Na, Ca, Mg and Si (% dry wt.) in ferromanganese nodules of the Baltic Sea

Region	Sampling date	Sample depth (m)	Shape	N	Al	Ag	As	B	Ba	Be	Bi	Ca	Cd	References
Baltic Proper	1959	58		24	2.9			60	2500			1.7		Manheim, 1965
	Pre-1961				0.2–2.0	> 0.3	100–300	100–300	640	1.0–3.0		0.9*	3.0–10.0	Manheim, 1961
	Pre-1991	56							3068					Amakawa et al., 1991
Gulf of Bothnia	1976–79		Spheroidal and ellipsoidal		0.56±0.02		410±10		2250±50			1.1±0.01		Boström et al., 1982
			Flat		2.13±0.08		531±7		1170±37			1.44±0.02		
	1980		Discoidal		3.7±1.7		360±64		2345±927			1.8±0.2		Ingri, 1985a
			Flat		8.1±1.7		292±65		792±163			1.9±0.3		
			Spheroidal	20	1.05				2500			1.23		Ingri and Ponter, 1986a
					0.64–2.34				2000–3000			1.02–2.72		
	Pre-1966			8										Winterhalter, 1980
	Pre-1980			15				80						
	Pre-1991	77							1851					Amakawa et al., 1991
		110							3092					
		116							2764					
Gulf of Finland	Pre-1980			25				80						Winterhalter, 1980
	1980–87	30–40	Spheroidal	26	4.49*				3280					Zhamoida et al., 1996
					1.63–6.39				1000–10000					
			Discoidal	29	2.3*				2290					
					0.80–6.95				300–6000					
			Crusts	7	6.01*				2130					
					2.14–11.97				800–4000					
Baltic Sea			Surface, subsurface										3.5–13.5	Suess and Djafari, 1977
			Outer, inner										1.4–1.8	
Southern Baltic	1987		Surface	6		ND							ND	Szefer and Szefer, 1990
Slupsk Furrow			Inner			ND							ND	
			Not described	10					370					Glasby et al., 1996
			Underlying sediments (Clayey till)	12					256					
	1993	70–92	Mainly discoidal	4					1930		0.53		2.1	Szefer et al., 1998
									1580–2400		0.3–0.8		1.2–3.3	
			Crust	1					2290		0.3		3.5	
			Underlying sediments	3					505		0.27		1.1	
			(Sandy mud overlying silty clay)						336–590		0.4–0.8		ND–1.1	

* – Expressed as oxide

TABLE 4.8. – continued

Region	Sampling date	Sample depth (m)	Shape	N	Co	Cr	Cu	Fe	Ga	Ge	Hg	K	Mg	Mn	References
Baltic Proper	1959	58		24	160	10	48	32.1							Manheim, 1965
	Pre-1961				3	37	10	2.5*	< 5	< 10		1.7*	0.99*	18.1	Manheim, 1961
Bornholm Basin			Not described		115–143	10.0–28	19–41	5.9–30.9						5.9–30.9	Glasby et al., 1996
			Underlying sediments (Baltic Ice Lake clay)	17	35	94	35	6						0.05	
Gotland-Gdansk Threshold			Not described	5	30	65	12	11.5						1.5	Glasby et al., 1996
			Underlying sediments (Clayey till)	7	30	82	27	4						0.07	
Kiel Bay	Pre-1976				77		21	10.1						29.3	Glasby et al., 1997
	Pre-1988				96		27	10						26.4	
Gulf of Bothnia	1976–79		Spheroidal and ellipsoidal			26±0.5	74±2	25.9±0.43						14.1±0.43	Boström et al., 1982
			Flat			77±0.3	67±0.3	35.2±0.2						4.8±0.2	
	1980		Discoidal				63±18	29.7±5.9						15.7±6.8	Ingri, 1985a
			Flat			49±19	27±8	29.4±6.5						2.0±1.6	
			Spheroidal	4				18.9						23.6	Ingri and Ponter, 1987
								16.0–24.1						18.4–26.6	
			Discoidal	4				21.1						18.2	
								14.2–27.4						14.3–25.7	
			Flat	4				32.9						4.74	
								20.7–40.9						0.40–16.5	
			Spheroidal	20	242		156	21.1					1.32	18.92	Ingri and Ponter, 1986a
					180–374		55–1084	13.9–27.4					0.97–2.94	12.6–26.1	
	Pre-1966			8	170	30	110	20.7						10.5	Winterhalter, 1980
	Pre-1980			15	140	110	80	16.6						14.6	
Gulf of Finland	Pre-1973			9	100	20	10	19.7						13.3	Varentsov, 1973
	Pre-1980			25	120	100	60	17.7						12.8	Winterhalter, 1980

	1980–87	30–40	Spheroidal	26	67	16	34	20.2*				33.8*	Zhamoida et al., 1996
					30–120	12.0–25	20–80	10.3–40.0				15.5–53.4	
			Discoidal	29	51	13	23	50.4*				8.91*	
					25–100	8.0–30	10.0–80	28.7–70.2				1.75–16.9	
			Crusts	7	44	21	27	32.7*				12.6*	
					30–60	8.0–30	20–40	12.2–51.1				0.59–49.6	
Gulf of Riga	Pre-1973			19	60	20	20	22.8				9.7	Varentsov, 1973
Baltic Sea			Surface, sub-surface				65–270	6.61–25.1				10.5–34.6	Suess and Djafari, 1977
			Outer, inner				27–39	4.30–11.8				30.6–38.7	
Southern Baltic	1987		Surface	6	19.2±0.6		22.7±1.3	17.1±0.7				1.75±0.13	Szefer and Szefer, 1990
Slupsk Furrow			Inner		15.1±2.0		15.2±0.8	15.6±0.5				1.46±0.19	
			Not described	10	50		20	15.5				3.2	Glasby et al., 1996
			Underlying sediments (Clayey till)	12	32		24	1.7				0.05	
	1993	70–92	Mainly discoidal	5	151		40.9	11.8		1.28		6.1	Szefer et al., 1998
					120–185		25.8–58.5	6.3–18.2		0.4–1.9		4.3–7.1	
			Crust	1	156		37.6	37.1		1.8		20.9	
			Underlying sediments	3	7.7		21.9	1.43		1.47		0.05	
			(Sandy mud overlying silty clay)		6.0–9.2		12.6–26.9	1.0–1.8		0.9–2.0		0.03–0.08	

* – Expressed as oxide

TABLE 4.8. – continued

Region	Sampling date	Sample depth (m)	Shape	N	Mo	Na	Ni	P	Pb	Pt	Si	Sn	References
Baltic Proper	1959	58		24	130		750	1.6	40		16.4		Manheim, 1965
	Pre-1961				< 5	1.7*	30–100	0.1–1.0	10.0–30		80*	1.0–3.0	Manheim, 1961
Bornholm Basin			Not described				64–170						Glasby et al., 1996
			Underlying sediments (Baltic Ice Lake clay)	17			66						
Gotland-Gdansk Threshold			Not described	5			50		20				Glasby et al., 1996
			Underlying sediments (Clayey till)	7			55		52				
Kiel Bay	Pre-1976						97						Glasby et al., 1997
	Pre-1988						61						
Gulf of Bothnia	1976–79		Spheroidal and ellipsoidal				327±8	1.55±0.03			1.9±0.04		Boström et al., 1982
			Flat				197±3	2.15±0.00			1.8±0.02		
	1980		Discoidal		269 103		329±94	3.0±1.1			15.2±0.9		Ingri, 1985a
			Flat				70±35	3.1±1.2			33.9±7.4		
			Spheroidal	20	413		557	1.14			0.78		Ingri and Ponter, 1986a
	Pre-1966				313–507		190–1390	0.67–1.46			0.63–1.11		
	Pre-1980			8	370		380	1.77	60				Winterhalter, 1980
				15	330		260	1.35	40				
Gulf of Finland	Pre-1973			9			40	1.24	10				Varentsov, 1973
	Pre-1980			25	290		140	1.65	50				Winterhalter, 1980
	1980–87	30–40	Spheroidal	26	70		134	2.7*	34		17.9		Zhamoida et al., 1996
					20–150		10–500	1.42–4.88	1.0–200		9.91–26.4		
			Discoidal	29	20		34	4.76*	95		11.3		
					4.0–30		5.0–120	3.07–6.48	5.0–400		5.10–37.0		
			Crusts	7	35		42	4.6*	63		23.7		

					12–100	15–80	1.90–7.16	< 5–200		10.6–43.8	
Gulf of Riga	Pre-1973			19		50	0.69	20			Varentsov, 1973
Baltic Sea			Surface, subsurface					53–162			Suess and Djafari, 1977
			Outer, inner					15–57			
Southern Baltic	1987		Surface	6		33.3±2.5		26.7±3.3			Szefer and Szefer, 1990
Slupsk Furrow			Inner			16.2±1.3		ND			
			Not described	10		82		28			Glasby et al., 1996
			Underlying sediments (Clayey till)	12		71		19			
	1993	70–92	Mainly discoidal	5		367		14.1	0.1		Szefer et al., 1998
						71.3–584		5.6–25.7	0.1–0.1		
			Crust	1		550		6.3	0.1		
			Underlying sediments	3		15.3		ND–3.8	0.13		
			(Sandy mud overlying silty clay)			7.9–22.2			0.0–0.2		

* – Expressed as oxide.

TABLE 4.8. – continued

Region	Sampling date	Sample depth (m)	Shape	N	Sr	Ti	Tl	V	W	Zn	Zr	References
Baltic Proper	1959	58		24		2.2		150		80		Manheim, 1965
	Pre-1961				100*	0.49*		3.0–10		100–300		Manheim, 1961
	Pre-1991	56			607							Amakawa et al., 1991
Bornholm Basin			Not described							515–2660		Glasby et al., 1996
			Underlying sediments (Baltic Ice Lake clay)	17						186		
Gotland-Gdansk Threshold			Not described	5	30	65	12	11.5		54		Glasby et al., 1996
			Underlying sediments (Clayey till)	7	30	82	27	4		96		
Kiel Bay	Pre-1976									340		Glasby et al., 1997
	Pre-1988									527		
Gulf of Bothnia	1976–79		Spheroidal and ellipsoidal		454±3	0.047±0. 002		112±1		333±4	31±0.3	Boström et al., 1982
			Flat		472±1	0.23±0.01		166±3		338±3	50±0.2	
	1980		Discoidal		490±97	0.2±0.07		103±24		131±90	71±34	Ingri, 1985a
			Flat		331±55	0.5±0.1		91±44		147±52	115±49	
			Spheroidal	20	439	0.13				540	27.2	Ingri and Ponter, 1986a
	Pre-1966				377–502	0.07–0.64				218–1673	16–51	
	Pre-1980			8				140		310		Winterhalter, 1980
				15		1.7		100		200		
	Pre-1991	77			409							Amakawa et al., 1991
		110			479							
		116			458							
Gulf of Finland	Pre-1973			9		2.8		70		110		Varentsov, 1973
	Pre-1980			25		1.4		100		230		Winterhalter, 1980
	1980–87	30–40	Spheroidal	26	27	0.28*		76		57		Zhamoida et al., 1996

					5.0–50	0.21–0.36		30–120		5.0–150	
			Discoidal	29	28	0.14*		84		36	
					10.0–70	0.04–0.26		40–400		8.0–200	
			Crusts	7	30	0.37*		79		25	
					20–50	0.12–0.63		50–120		12.0–40	
Gulf of Riga	Pre-1973			19		2.9		100		140	Varentsov, 1973
Baltic Sea			Surface, subsurface							420–1650	Suess and Djafari, 1977
			Outer, inner							120–240	
Southern Baltic	1987		Surface	6						165±3	Szefer and Szefer, 1990
Slupsk Furrow			Inner							167±3	
			Not described	10						300	Glasby et al., 1996
			Underlying sediments (Clayey till)	12						54	
	1993	70–92	Mainly discoidal	5			5.6		8.68	279	Szefer et al., 1998
							0.9–8.7		6.6–11.1	132–421	
			Crust	1			6		6.8	285	
			Underlying sediments	3			0.97		1.5	93	
			(Sandy mud overlying silty clay)				0.7–1.2		1.1–1.8	47.6–156	

* – Expressed as oxide

TABLE 4.9.

Concentrations of RRE ($\mu g\ g^{-1}$ dry) in ferromanganese nodules of the Baltic Sea

Region	Sampling date	Sample depth (m)	Shape	N	La	Ce	Pr	Nd	Sm	Eu	Gd	Tb	References
Gulf of Bothnia													
	1976–79		Spheroidal and ellipsoidal*		52±0.1								Boström et al., 1982
			Flat		80±0.6								
	1980		Discoidal		28±12								Ingri, 1985a
			Flat		31±15								
			Spheroidal	20	59.6								Ingri and Ponter, 1986a
					28–153								
			Spheroidal	4	75.7	235	19.2	65.8	12.6	1.9	8.93		Ingri and Ponter, 1987
					39.6–125	81.6–480		37.3–116	7.02–23.2	1.07–3.53	5.33–14.9		
			Discoidal	4	45.3	60.9	10	35.9	6.75	1.08	5.63		
					33.3–65.3	44–80	9.16–10.9	27.5–51.5	4.23–9.53	0.79–1.60	4.03–7.42		
			Flat	4	38.1	67.5	9.12	32.7	6.08	0.91	5.27		
					27.5–48.2	43.8–81.6	7.23–11.0	22.7–42.1	4.04–7.85	0.77–1.05	4.08–6.41		
	Pre-1991	77			101	476		86.9	16.6	2.53	14.4		Amakawa et al., 1991
		110			55.2	112		39.6	7.25	1.23	6.42		
		116			48	93.5		37.7	7.15	1.16	6.48		
Southern Baltic		70–92	Mainly discoidal	5	31.1	52.3	6.88	27	5.71	1.34	5.41	0.81	Szefer et al., 1998
Slupsk Furrow	1993				24.1–36.8	40.8–63.8	5.24–8.18	20.9–31.7	4.58–6.53	1.08–1.66	4.02–6.22	0.63–0.96	
			Crust	1	20.8	32.4	4.55	18	3.87	1.06	3.78	0.58	
			Underlying sediments	3	24.2	57.1	5.96	22.2	4.41	0.98	4.03	0.63	
			(Sandy mud overlying silty clay)		19.5–30.4	46.6–71.0	4.77–7.41	18.1–27.3	3.58–5.33	0.77–1.20	3.19–4.87	0.49–0.79	
Baltic Proper	Pre-1991	56			84.8	145		77.2	15.3	2.81	14.6		Amakawa et al., 1991

TABLE 4.9. – continued

Region	Sampling date	Sample depth (m)	Shape	N	Dy	Ho	Er	Tm	Yb	Lu	Y	Sc	References
Gulf of Bothnia													
	1976–79		Spheroidal and ellipsoidal*						5.5±0.1		29±0.5	1.5±0.1	Boström et al., 1982
			Flat						7.9±0.01		48±1	7.7±0.3	
	1980		Discoidal					792±163	3.0±0.5		25±4	1.0±0.7	Ingri, 1985a
			Flat						3.0±0.8		23±8	4.0±1.3	
			Spheroidal	20					4.17		31.3	5.53	Ingri and Ponter, 1986a
									2.5–9.3		20–84	2.2–19.9	
			Spheroidal	4	8.87	1.57	4.66		3.95	0.56	40.2	5.09	
					5.64–15.3	0.96–2.75	3.05–7.60		2.43–6.93	0.37–0.87	27.4–65.8	4.33–6.30	Ingri and Ponter, 1987
			Discoidal	4	5.24	0.95	3.12		2.52	0.4	28.5	3.21	
					3.43–7.40	0.63–1.33	1.96–4.43		1.57–3.71	0.24–0.57	17.8–40.5	1.80–3.93	
			Flat	4	4.17	0.85	2.59		2.17	0.35	23.3	5.14	
					2.69–5.16	0.48–1.08	1.90–3.11		1.49–2.63	0.22–0.44	16.3–29.3	4.60–5.85	
	Pre-1991	77			10.9		5.83		5.29	0.772			Amakawa et al., 1991
		110			5.09		3		2.79	0.422			
		116			5.18		3.02		2.76	0.411			
Southern Baltic		70–92	Mainly discoidal	5	4.31	0.78	2.37	0.39	2.49	0.37	24.8		Szefer et al., 1998
Slupsk Furrow	1993				3.07–5.04	0.58–0.96	1.92–2.83	0.28–0.51	2.10–2.84	0.29–0.47	19.8–29.2		
			Crust	1	3.07	0.54	1.82	0.27	1.76	0.27	18.7		
			Underlying sediments	3	3.18	0.56	1.68	0.3	1.8	0.31	14.6		
			(Sandy mud overlying		2.54–3.62	0.37–0.69	1.20–2.0	0.21–0.38	1.32–2.16	0.21–0.43	11.5–16.2		
Baltic Proper	Pre-1991	56	silty clay)		12.4		7.3		6.48	0.963			Amakawa et al., 1991

The concretions from this region are characterised by high Mn/Fe values ranging from 3.5 for deposits on mussel shell to 2.5 for discoidal concretions. There is some coassociation between Co, Mn and Sc as well as between Th and Fe in concretions from the Gulf of Finland (Glasby et al., 1997a). It is interesting to note that deposit formed on mussel shells contained one order of magnitude greater concentrations of Zn (mean 1130 $\mu g\ g^{-1}$) as compared to those in the other types of concretions. This is comparable with results obtained for ferromanganese deposit on shells of *Astarte borealis* from the Slupsk Furrow, southern Baltic (Szefer et al., 1998c), containing on the average 970±20.5 $\mu g\ g^{-1}$, i.e. also an order of magnitude more than concretions. Other elements were characterised by similar distribution to Zn, i.e. the levels of Mn, Cu and Pb were higher in the shell deposits in contrast to concretions having higher levels of Fe and Ni (Ikuta and Szefer, 2000). The elevated levels of metals in shell deposits could reflect the very recent deposition of ferromanganese oxyhydroxides in these regions, i.e. in Kiel Bay and Slupsk Furrow, southern Baltic. The ferromanganese concretions, predominantly discoidal in shape, from the Polish Exclusive Economic Zone have relatively high Fe/Mn ratios and high levels of detrital elements (Ca, K, Mg, Na and Ti) but low levels of the transition elements (Co, Cu, Cr, Ni, Pb and Zn) (Glasby et al., 1996).

According to several authors (Ingri and Pontér, 1986a, 1987; Amakawa *et al.*, 1991) the REE contents of the concretions, similarly to some heavy metals, are dependent on the morphology of the concretions. The REE contents of spheroidal concretions were shown to be significantly higher than those of discoidal and flat concretions. The spheroidal concretions have positive cerium anomalies whereas the discoidal and flat concretions have negative cerium anomalies. All three types have negative europium anomalies. The concretions display an enrichment of light REE (LREE) compared with heavy REE (HREE) as might be expected from the results of Sholkovitz (1995). Based on these data, Ingri and Pontér (1987) concluded that the REE in these concretions are ultimately from riverine particulate matter rich in manganese and iron oxyhydroxides. The depth of burial of the concretions in the sediments and their size were considered to be the main factors controlling the REE contents of the concretions.

Element distribution in concretions and associated sediments

The concentrations of Mn, Fe, Cu, Ni, Co, Zn and Pb in the ferromanganese concretions and sediments from the Slupsk Furrow (Szefer et al., 1998c) are in good agreement with data reported by Varentsov (1973) for concretions from the Gulf of Gdańsk. The much higher Mn/Fe ratio in the concretions compared to the underlying sediments reflects the diagenetic remobilisation of Mn from the sediments. In general, Mn, Fe, Ni and Co are greatly enriched in the concretions compared to the underlying sediments whereas Cu and Zn are present in more similar concentrations.

It is seen that the REE contents of concretions from within Slupsk Furrow are somewhat higher than those of the associated sediments but somewhat lower than those of discoidal and flat concretions from the Gulf of Bothnia and much lower than those of spheroidal concretions from the Gulf of Bothnia (Ingri and Pontér, 1987; Szefer et al., 1998c). The ferromanganese crust from Slupsk Furrow has lower REE contents than the ferromanganese concretions there but it is not known if this is significant. The shale-normalised REE distributions for the concretions are relatively flat with a slight negative cerium anomaly but no evidence of a negative europium anomaly (Fig. 4.17a). The sediments also display relatively flat shale-normalised curves but with a slight positive cerium anomaly and again no evidence of a negative europium anomaly (Fig. 4.17c). These flat shale-normalised REE distributions of the concretions and sediments are similar to those of till (Öhlander et al. 1996) from which it is believed the elements in these concretions are ultimately derived (Ingri 1985a).

The data indicate differences in the REE distributions in ferromanganese concretions from within Slupsk Furrow and Gulf of Bothnia (Fig. 4.17a and 4.17b). The positive cerium anomaly of spheroidal concretions from the Gulf of Bothnia reflects the more oxidising conditions in the sediments there. By contrast, the REE distribution in concretions from within Slupsk Furrow is more similar to that of discoidal and flat concretions from the Gulf of Bothnia. Recent sediments in the Bothnian Bay had REE levels very similar to average shale. Small spheroidal nodules from the Gulf of Bothnia showed the highest REE concentrations, with La and Ce most strongly enriched (Fig. 4.18). The lack of positive cerium anomalies in the concretions from within Slupsk Furrow reflects the less oxidising conditions there. Nonetheless, the cerium anomalies in concretions from both areas are much lower than found in Pacific deep-sea nodules (Kunzendorf et al., 1993). Surprisingly, the Ce/La ratios of the concretions from within Slupsk Furrow are somewhat lower than those of the associated sediments suggesting that they are formed under slightly more reducing conditions (Szefer et al., 1998c).

Morphology and mineralogy of iron oxyhydroxide phase

Mössbauer studies of marine sediments from the Baltic Sea have been undertaken by Görlich et al. (1978, 1985, 1989). More recently, Sochan (1992) has used Mössbauer spectroscopy to investigate the mode of occurrence of iron in two ferromanganese concretions from the southern Baltic. He found that 90% of the iron in these concretions was in the trivalent state and 10% in the divalent state. Unfortunately, the spectra were recorded at room temperature which limited their use for the detailed study of the nature of the iron oxyhydroxide phase. In study performed later (Szefer et al., 1998c), ^{57}Fe Mössbauer absorption spectra were recorded with a ^{57}Co (Cr) source for the concretion sample from Slupsk Furrow at several temperatures between 4 and 300K. Basically, the spectra display above 75K a well-resolved quadrupole doublet characteristic of ferric ions in the high-spin (Fe^{3+}, $3d^5$, $^6S_{5/2}$) state (Fig. 4.19). This is the dominant form of iron in the concretions. In addition, there are visible traces of an absorption dip at about 2.5 mm/s which belong to a component with a much larger electric quadrupole interaction characteristic of iron in the high-spin divalent state. The occurrence of

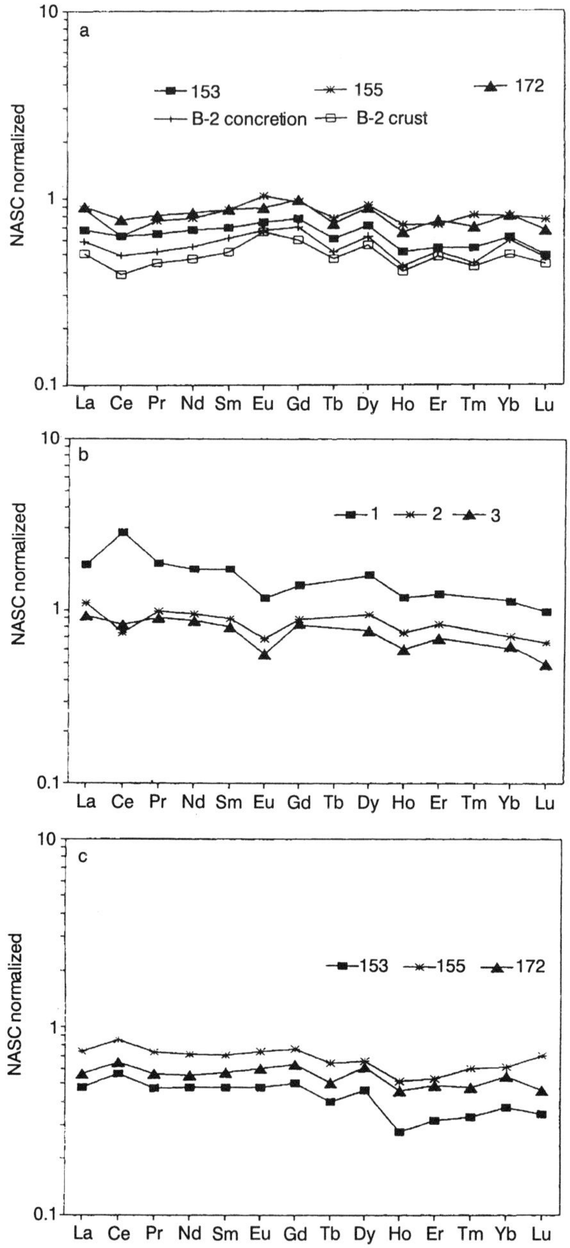

Fig. 4.17. (a) North American shale composite (NASC)-normalized REE data of the ferromanganese crust and concretions from the Polish EEZ. (b) North American shale composite (NASC)-normalised REE data of the ferromanganese concretions from the Gulf of Bothnia. (c) North American shale composite (NASC)-normalised REE data of the associated sediments from the Polish EEZ. After Szefer et al. (1998c).

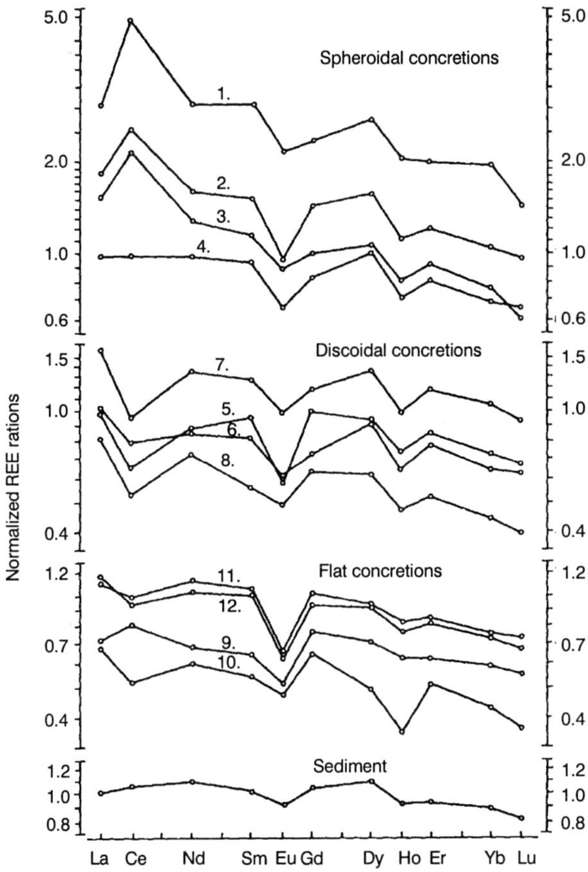

Fig. 4.18. REE ratios normalized to average shale in ferromanganese concretions from the Gulf of Bothnia. After Ingri and Pontér (1987); modified.

ferrous ions within the concretions reflects the incorporation of iron in clay minerals derived from erratic material. Because of the small amounts of the ferrous ion present, it is not possible to identify the clay minerals in which it occurs. Instead, the hyperfine parameters used in the fitting procedure have been constrained close to the values found in previous work on iron in octahedral (Fe^{2+}, Fe^{3+}) and tetrahedral (Fe^{3+}) positions (Görlich et al., 1989) of chlorite and illite. The respective quadrupole doublets were incorporated in the data analysis (dotted lines in Fig. 4.19); the first one accounts for Fe^{2+} with an isomer shift (IS) of 0.96 (3) mm/s (relative to ^{57}Co (Cr) source) and an electric quadrupole coupling constant ($\varepsilon_Q = 1/4e^2qQ$) of 1.38 (2) mm/s and the second one accounting for Fe^{3+} (without further differentiation for various sites) with an IS of 0.39 (5) mm/s and ε_Q of 0.45 (4) mm/s (all values at room temperature). The total iron content in the clay mineral fraction does not exceed 8% (as measured by the subspectra relative area) and the Fe^{3+}/Fe^{2+} ratio of this fraction was calculated to be ca. 2.5.

A quadrupole doublet forms the main feature of the Mössbauer spectrum at higher temperatures (Fig. 4.19). The material contributing to the main component of the spectrum undergoes a magnetic ordering transition at temperatures below 70K which eventually leads to hyperfine splitting at low temperatures (Fig. 4.20). The Mössbauer spectra clearly reveal some characteristic features which may be ascribed to a continuous distribution of hyperfine magnetic fields. In order to fit the spectrum of Fig. 4.20, the method developed by Wivel and Morup (1981) which involves no *a priori* assumptions concerning the specific shape of the distribution of the hyperfine interaction parameters was adopted. The value of the quadrupole coupling constant (ε_Q = 0.40 mm/s) and the full width of an absorption line at the half maximum (FWHM= 0.51(1) mm/s) obtained at 78K were kept constant during the low temperature fits. The resulting hyperfine magnetic field (B_{hf}) distribution was deduced from the spectrum at 10K (Szefer et al., 1998c). The shape of this function, the mean value of the hyperfine magnetic field (from the main maximum in its distribution, $\langle B_{hf} \rangle$ = 46(1)(T) and the values of the remaining parameters (IS = 1.89(1) mm/s, ε_Q = 0.375(5) mm/s) all indicate that the Fe^{3+} occurs mainly in the form of poorly crystalline lepidochrocite (γ–FeOOH) (Görlich et al., 1989; De Grave et al., 1990). This mineral in its natural form is characterised by considerable distortions of Fe sites leading to the broadening of FWHM and the spread of B_{hf} (Carlson and Schwertmann, 1981; Murad and Schwertmann, 1988). It should be noted that this finding differs from

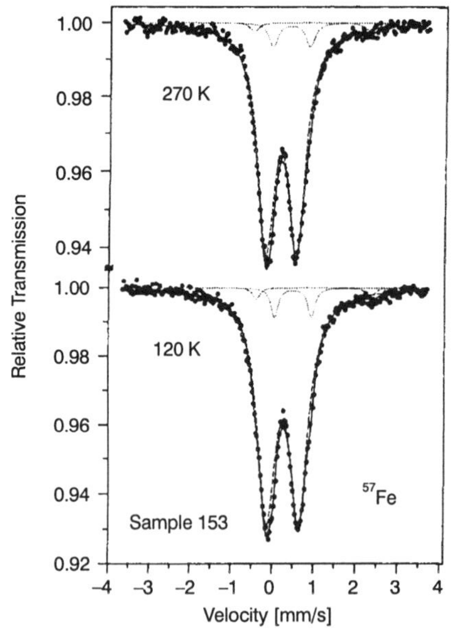

Fig. 4.19. ^{57}Fe Mössbauer effect spectra of the ferromanganese concretion sample 153 at 270 and 120 K. The small contribution of iron in the clay minerals is shown by the dotted lines. The Fe^{3+}/Fe^{2+} ratio in the clay mineral fraction of the sample is about 2.5. After Szefer et al. (1998c).

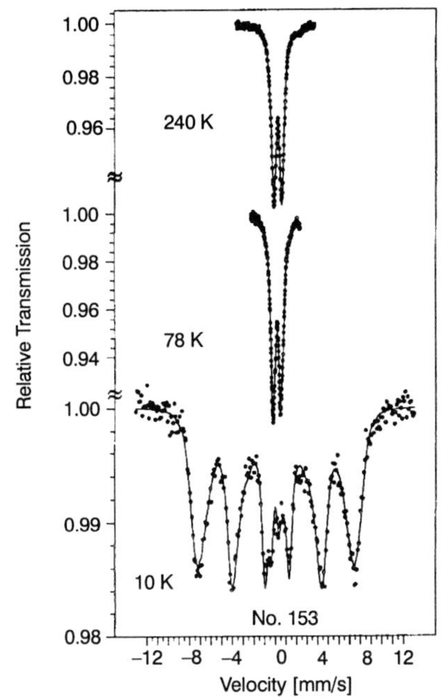

Fig.4.20. ^{57}Fe Mössbauer effect spectra of the ferromanganese concretion sample 153 at 240, 78 and 10 K. Broad absorption lines reflect the spread in hyperfine parameter values. After Szefer et al. (1998c).

the earlier observation of Johnston and Glasby (1978) and Chen and Yao (1995) that akaganéite (β-FeOOH) is the principal iron oxyhydroxide mineral present in deep-sea manganese nodules. This difference may reflect the much lower salinity of Baltic Sea compared to Pacific Ocean seawater since the formation of akaganéite is favoured by the presence of chloride ion (Szefer et al., 1998c).

(iii) Radionuclides in Ferromanganese Nodules

Isotopic composition of ^{138}Ce, ^{142}Ce ^{143}Nd, ^{144}Nd, ^{86}Sr and ^{87}Sr in ferromanganese nodules from the Pacific and Atlantic Oceans, the Baltic and Barents Seas, and the Gulf of Bothnia has been reported by Amakawa et al. (1991). Sanchez et al. (1988) presented data on transuranic radionuclides ^{238}Pu, $^{239+240}$Pu and ^{241}Am in three morphological types of ferromanganese nodules (discoidal, spheroidal and flat) collected from different subregions of the Baltic Sea (Bothnian Bay, Bothnian Sea, Gulf of Finland, Baltic Proper and Gulf of Riga). The data obtained (Table 4.10) show that the levels of these radionuclides are mostly comparable to those in the associated sediments, which are significantly higher than levels detected in the overlying water. It is concluded that the concentrations of ^{238}Pu, $^{239+240}$Pu and ^{241}Am in the nodules may be depended on the redox conditions of the surrounding sediments. More Pu and Am is found in nodules from the Bothnian Sea as relatively less oxidising environment as compared to Both-

TABLE 4.10.

Concentrations of transuranic radionuclides ($\mu Bq\ g^{-1}$ dry wt.) and isotope ratios in ferromanganese nodules of the Baltic Sea; all radionuclides concentrations are expressed in terms of extractable Fe + Mn content (acid-soluble fraction)

Region	Sampling date	Sample depth (m)	Shape	N	^{241}Am	$^{239+240}Pu$	^{238}Pu	$^{238}Pu/^{239+240}Pu$	$^{241}Am/^{239+240}Pu$	Th ($\mu g\ g^{-1}$)	U ($\mu g\ g^{-1}$)	References
Southern Baltic												
Słupsk Furrow	1993	70–92	Mainly discoidal	3						7.03	9.28	Szefer et al., 1998
										4.2–8.3	6.2–11.1	
			Crust	1						3.2	13.1	
			Underlying sediments	3						9	2.1	
										8.0–10.9	1.9–2.5	
Bothnian Bay	Pre-1988	62–80	Discoidal	2	0.12	0.25	ND		0.56			Sanchez et al., 1988
					0.05–0.19	0.21–0.29			0.19–0.92			
		8–108	Flat	4	0.15	0.46			0.32			
					ND–0.24	ND–0.51	ND–0.028		0.17–0.46			
		78–110	Spheroidal	14	0.25	0.68		0.08 (N=7)	0.49			
					ND–0.89	0.03–3.32	ND–0.09	0.021–0.203	0.27–1.26			
		78	Sediment	1	0.79	1.72	ND		0.46			
Bothnian Sea	Pre-1988	60–80	Flat	2	0.25	1.13	0.09		0.22			Sanchez et al., 1988
					0.21–0.28	1.12–1.14	0.09–0.09		0.18–0.25			
		77–116	Spheroidal	4	1.05	2.51	0.05	0.022 (N=3)	0.43			
					0.63–1.46	1.42–3.13	ND–0.07	0.007–0.032	0.26–0.56			
		80	Sediment	1	2.29	10.2	0.46	0.045	0.22			
Gulf of Finland	Pre-1988		Spheroidal	2	0.64	0.21	0.08	0.032	0.25			Sanchez et al., 1988
					0.51–0.76	0.15–0.26	0.07–0.09	0.026–0.037	0.23–0.27			
Baltic Proper	Pre-1988	25–61	Discoidal	3	0.26	0.32	0.09	0.01 (N=1)	0.29			Sanchez et al., 1988
					0.14–0.42	0.36–1.47	0.05–0.13		0.20–0.39			
		25	Flat	1	0.24	0.53	ND		0.45			
Gulf of Riga	Pre-1988	34	Spheroidal	2	0.64	2.61	0.08	0.032	0.25			Sanchez et al., 1988
					0.51–0.76	1.87–3.35	0.07–0.09	0.026–0.037	0.23–0.27			

nian Bay. Therefore, the diagenetic remobilization of the transuranic radionuclides from the bottom sediments to the interstitial waters seems to occur more actively under reducing conditions, resulting in increased their quantities in ferromanganese nodules formed under these conditions. It is possible to use the concentration of the transuranic radionuclides in a given layer of concretion as a dating tool for measuring nodule accretion rate (Sanchez et al., 1988).

An overview of U distribution in ferromanganese concretions has been reported by Szefer (1987). In ferromanganese nodules from the Słupsk Furrow, southern Baltic, were determined concentrations of selected chemical elements including U and Th (Szefer et al., 1998c). The values obtained are typical for those reported for concretions from various geographical regions (Szefer, 1987).

References

Ackermann, F., 1980. A procedure for correcting the grain size effect in heavy metal analysis of estuarine and coastal sediments. Environ. Technol. Letters **1**, 518–527.

Ackermann, F., M. Bergmann and G.U. Schleichert, 1983. Monitoring of heavy metals in coastal and estuarine sediments – a question of grain size: < 20 μm versus < 60 μm. Environ. Technol. Letters **4**, 317–328.

Allen, J.R.L. and J.E. Rae, 1986. Time sequence of metal pollution, Severn Estuary, southwestern UK. Mar. Pollut. Bull. **17**, 427–431.

Amakawa H., J. Ingri, A. Masuda and H. Shimizu, 1991. Isotopic compositions of Ce, Nd and Sr in ferromanganese nodules from the Pacific and Atlantic Oceans, the Baltic and Barents Seas, and the Gulf of Bothnia. Earth Planet. Sci. Letters **105**, 554–565.

Amakawa, H., Y. Nozaki and A. Masuda, 1996. Precise determination of variations in the $^{138}Ce/^{142}Ce$ ratios of marine ferromanganese nodules. Chem. Geol. **131**, 183–195.

Andersen, M.E., and J.D. Macdougall, 1977. Accumulation rates of manganese nodules and sediments: an alpha track method. Geophys. Res. Letters **4**, 351–353.

Andersson, P.S., G.J. Wasserburg and J. Ingri, 1992. The sources and transport of Sr and Nd isotopes in the Baltic Sea. Earth Planet Sci. Letters **113**, 459–472.

Aplin, A., A. Michard and F. Albarède, 1986. $^{143}Nd/^{144}Nd$ in Pacific ferromanganese encrustations and nodules. Earth Planet. Sci. Letters **81**, 7–14.

Baptista Neto, J.A., B.J. Smith and J.J. McAllister, 2000. Heavy metal concentrations in surface sediments in a nearshore environment, Jurujuba Sound, Southeast Brazil. Environ. Pollut. **109**, 1–9.

Baskaran, M., S. Asbill, J. Schwantes, P. Santschi, M.A. Champs, J.M. Brooks, D. Adkison and V. Makeyev, 2000. Concentrations of ^{137}Cs, $^{239,240}Pu$ and ^{210}Pb in sediment samples from the Pechora Sea and biological samples from the Ob, Yenisey Rivers and Kara Sea. Mar. Pollut. Bull. **40**, 830–838.

Baturin, G.N., 1988. The geochemistry of manganese and manganese nodules. D. Reidel. Dordrecht.

Belmans, F., R. van Grieken and L. Brügmann, 1993. Geochemical characterization of recent sediments in the Baltic Sea by bulk and electron microprobe analysis. Mar. Chem. **42**, 223–236.

Belzunce Segarra, M., K. Görlich and E. Helios-Rybicka, 1987. Heavy metals in surface sediments of the Baltic Sea: sequential extraction. Proceedings of 15th Conference of the Baltic Oceanographers (National Agency of Environment Protection, Copenhagen), pp. 85–103.

Belzunce Segarra, M. J., K. Görlich and E. Helios-Rybicka, 1988. Spatial and stratigraphic controls over chemical speciation of heavy metals in the Baltic Sea muds. Proceedings of 16th Conference of the Baltic Oceanographers (Institute of Marine Research, Kiel), **Vol. 1**, 163–176.

Belzunce-Segarra, M.J., M.J. Wilson, J. Bacon, J. Bolalek and P. Szefer, 2000. Chemical forms and vertical distribution of heavy metals in bottom sediments (submitted for publication).

Bernard, P. C., van R.E. Grieken and L. Brügmann, 1989. Geochemistry of suspended matter from the Baltic Sea. 1. Results of individual particle characterization by automated electron microprobe. Mar. Chem. **26**, 155–177.

Biselli, S., K. Bester, H. Hühnerfuss and K. Fent, 2000. Concentrations of the antifouling compound Irgarol 1051 and of organotins in water and sediments of German North and Baltic Sea marinas. Mar. Pollut. Bull. **40**, 233–243.

Blazhchishin, A.I., 1982a. Bottom sediments of the Baltic Sea, in: Geology of the Baltic Sea, eds. V.K. Gudelis and E.M. Emelyanov (Wydawnictwo Geologiczne, Warsaw), 188–214 (in Polish).

Blazhchishin, A.I., 1982b. Main chemical constituents of the sediments of the Baltic Sea, in: Geology of the Baltic Sea, eds. V.K. Gudelis and E.M. Emelyanov (Wydawnictwo Geologiczne, Warsaw), 257–289 (in Polish).

Blazhchishin, A.I., 1982c. Sedimentary environment of the Baltic Sea, in: Geology of the Baltic Sea, eds. V.K. Gudelis and E.M. Emelyanov (Wydawnictwo Geologiczne, Warsaw), 131–141 (in Polish).

Blazhchishin, A.I., 1984. Balance of Fe, Mn and P in the Baltic Sea, ed. A.P. Lisitzin, Geological History and Geochemistry of the Baltic Sea (Nauka, Moscow) 122–129 (in Russian).

Blazhchishin, A.I., and V.K. Lukashev, 1981. Litho-stratigraphic characteristics of the late Quaternary deposits, in: Formation of sediments in the Baltic Sea, eds. A.P. Lisitsyn and E.M. Emelyanov (Izdatelstvo Nauka, Moscow), 232–246 (in Russian).

Blomqvist, S., U. Larsson and H. Borg, 1992. Heavy metal decrease in the sediments of a Baltic Bay following tertiary sewage treatment. Mar. Pollut. Bull. **24**, 258–266.

Bojanowski, R., Z. Radecki and D. Knapińska-Skiba, 1995a. The distribution of ^{137}Cs, $^{239+240}Pu$ and ^{210}Po in the Pomeranian Bay (southern Baltic) ecosystem. Biul. Mors. Inst. Ryb. **3**, 15–24.

Bojanowski, R., D. Knapińska-Skiba, Z. Radecki, J. Tomczak and T. Szczepańska, 1995b. Accumulation of radioactive caesium (^{137}Cs) in Southern Baltic sediments, in: The Baltic, ed. J.E. Mojski. Proceed. Third Mar. Geol. Conference (Polish Geological Institute, Warsaw), 145–150.

Bolałek, J., 1992a. Ionic macroconstituents of the interstitial waters of Puck Bay. Oceanologia **33**, 131–158.

Bolałek, J., 1992b. Phosphate at the water-sediment interface in Puck Bay. Oceanologia **33**, 159–182.

Bonatti, E., D.E. Fisher, O. Joensuu and H.S. Rydell, 1971. Postdepositional mobility of some transition elements, phosphorus, uranium and thorium in deep sea sediments. Geochim. Cosmochim. Acta **35**, 189–201.

Borg, H., and P. Jonsson, 1996. Large-scale metal distribution in Baltic Sea sediments. Mar. Pollut. Bull. **32**, 8–21.

Borole, D.V., S. Krishnaswami and B.L.K. Somayajulu, 1977. Investigations on dissolved uranium, silicon and on particulate trace elements in estuaries. Estuar. Coast. Mar. Sci. **5**, 743–754.

Borole, D.V., S. Krishnaswami and B.L.K. Somayajulu, 1982. Uranium isotopes in rivers, estuaries and adjacent coastal sediments of western India: their weathering, transport and oceanic budget. Geochim. Cosmochim. Acta **46**, 125–137.

Boström, K., J.-O. Burman, B. Boström, C. Pontér, S. Brandlöf and B. Alm, 1978. Geochemistry, mineralogy and origin of the sediments in the Gulf of Bothnia. Finn. Mar. Res. **244**, 8–35.

Boström, K., L. Wiborg and J. Ingri, 1982. Geochemistry and origin of ferromanganese concretions in the Gulf of Bothnia. Mar. Geol. **50**, 1–24.

Boström, K., J.-O. Burman and J. Ingri, 1983. A geochemical massbalance for the Baltic, ed. R.Hallberg, Environmental Biogeochemistry, Proc. 5th Intern. Symp. on Environmental Biogeochemistry (ISEB), Stockholm. Ecol. Bull. **35**, 39–58.

Boström, K., J. Ingri and C. Pontér, 1988. Origin of iron-manganese-rich suspended matter in the Landsort Deep, NW Baltic Sea. Mar. Chem. **24**, 93–98.

Bricker, S.B., 1993. The history Cu, Pb, and Zn inputs to Narragansett Bay, Rhode Island, as recorded by salt-marsh sediments. Estuaries **16**, 589–607.

Bruland, K.W., K. Bertine, M. Koide and E.D. Goldberg, 1974. History of metal pollution in southern California coastal zone. Environ. Sci. Technol. **8**, 425–432.

Bruland, K.W., R.P. Franks, W.M. Landing and A. Soutar, 1981. Southern California inner basin sediment trap calibration. Eart Planet. Sci. Letters **53**, 400–408.

Bruneau, L., 1980. Pollution from industries in the drainage area of the Baltic. Ambio **9**, 145–152.

Brügmann, L., 1981. Heavy metals in the Baltic Sea. Mar. Pollut. Bull. **12**, 214–218.
Brügmann, L., 1986a. Particulate trace metals in waters of the Baltic Sea and parts of the adjacent NE Atlantic. Beiträge zur Meereskunde **55**, 3–18.
Brügmann, L., 1986b. The influence of coastal zone processes on mass balances for trace metals in the Baltic Sea. Rapp. P.-v. Reun. Cons. int. Explor. Mer **186**, 329–342.
Brügmann, L., 1988. Some peculiarities of trace metal distribution in the Baltic waters and sediments. Mar. Chem. **23**, 425–440.
Brügmann, L., 1992. Review of trace element studies in the sediments of the Baltic Sea. Review of Contaminants in Baltic Sea Sediments. International Council for the Exploration of the Sea Report **No. 180**, 69–98.
Brügmann, L., and U. Hennings, 1982. Stratigraphie und Spezifizierung von Metallen in Sedimenten der westlichen Ostsee. Acta Hydrochim. Hydrobiol. **10**, 41–53.
Brügmann, L., and D. Lange, 1990. Metal distribution in sediments of the Baltic Sea. Limnologica **20**, 15–28.
Brügmann, L, and J. Matschullat, 1997. Zur Biogeochemie und Bilanzierung von Schwermetallen in der Ostsee, in: Geochemie und Umwelt-Umweltrelevante Prozesse in Atmo-, Pedo- and Hydrosphäre, eds. J. Matschullat, H.J. Tobschall, H.J. Vogt (Springer Verlag, Berlin) pp. 267–289 (in German).
Brügmann, L., G. Bublitz and U. Hennings, 1980. Der Gehalt von Spurenmetallen in Sedimentkernen der westlichen Ostsee. Z. Angew. Geol. **26**, 398–405.
Brügmann, L., H. Gaul, K.-H. Rohde and U. Ziebarth, 1991/92. Regional distribution and temporal trends of some contaminants in the water of the Baltic Sea. Dr. Hydrogr. Z. **44**, 161–184.
Brügmann, L., P.C. Bernard and R. van Grieken, 1992. Geochemistry of suspended matter from the Baltic Sea. 2. Results of bulk trace metal analysis by AAS. Mar. Chem. **38**, 303–323.
Brügmann, L., R. Hallberg, C. Larsson and A. Löffler, 1997. Changing redox conditions in the Baltic deep basins: Impacts on the concentration and speciation of trace metals. Ambio **26**, 107–112.
Brügmann, L., R. Hallberg, C. Larsson and A. Löffler, 1998. Trace metal speciation in sea and pore water of the Gotland Deep, Baltic Sea, 1994. Appl. Geochem. **13**, 359–368.
Bryan, G.W., and W.J. Langston, 1992. Bioavailability, accumulation and effects of heavy metals in sediments with special reference to United Kingdom estuaries: a review. Environ. Pollut. **76**, 89–131.
Brzezińska, A., A. Trzosińska, W. Żmijewska and L. Wódkiewicz, 1984. Trace metals in suspended matter and surficial bottom sediments from the southern Baltic. Oceanologia **18**, 59–77.
Buat-Menard, P., and R. Chesselet, 1979. Variable influence of the atmospheric flux on the trace metal chemistry of oceanic suspended matter. Earth Planet. Sci. Letters **42**, 399–411.
Buchowiecki, J., and R.D. Cherry, 1968. Thorium, radium and potssium in manganese nodules. Chem. Geol. **3**, 111–117.
Burman, J.-O., 1983. Element transports in suspended and dissolved phases in the Kalix River. Ecol. Bull. (Stockholm) **35**, 99–113.
Burns, R.G., and V.M. Burns, 1977. Mineralogy, in: Marine Manganese Deposits, ed. G.P. Glasby (Elsevier, Amsterdam) 185–248.
Butylin, V.P., and V.A. Zhamoida, 1988. Zonation of modern, shallow water ferromanganese concretions in the Gulf of Finland. Geology and Geochemistry of Ferromanganese Concretions in the World Ocean (Sevmorgeologia, Leningrad) 93–107 (in Russian).
Butylin, V.P., V.A. Zhamoida and M.B. Kosin, 1985. Distribution of chemical elements and concretion formation in the Quaternary deposits of the Gulf of Finland. Glacial Shelves: Problems of Geology and Methods of Investigations (VESEGEI, Ministry of Geology, Leningrad) 43–54 (in Russian).
Callaway, J.C., R.D. Delaune and Jr W.H. Patrick, 1998. Heavy metal chronologies in selected coastal wetland from northern Europe. Mar. Pollut. Bull. **36**, 82–96.
Calvert, S.E., and N.B. Price, 1970. Composition of manganese nodules and manganese carbonates from Loch Fyne, Scotland. Contrib. Miner. Petrol. **29**, 215–233.
Calvert, S.E., and N.B. Price, 1977. Shallow water, continental margin and lacustrine nodules: Distribution and geochemistry, in: Marine Manganese Deposits, ed. G.P. Glasby (Elsevier, Amsterdam), pp. 43–86.

Carlson, L., 1982. Oxide minerals of iron in oxidate accumulation in Finland. Ph.D. thesis (University of Helsinki) 1–24 + 4 papers.

Carlson, L., and U. Schwertmann, 1981. Natural ferrihydrites in surface deposits from Finland and their asssociation with silica. Geochim. Cosmochim. Acta **45,** 421–429.

Carman, R., and F. Wulff, 1989. Adsorption capacity of phosphorus in Baltic Sea sediments. Estuar. Coast. Shelf Sci. **29**, 447–456.

Carman, R., and P. Jonsson, 1991. Distribution pattern of different forms of phosphorus in some surficial sediment types of the Baltic Sea. Chem. Geol. **90**, 91–106.

Carman, R., and J. Aigars, 1997. Vertical and spatial distribution of biogenic silica in the sediment of the Gulf of Riga, Baltic Sea. Toxicol. Environ. Chem. **60**, 245–259.

Carman, R., and L. Rahm, 1997. Early diagenesis and chemical characteristics of interstitial water and sediments in the deep deposition bottoms of the Baltic proper. J. Sea Res. **37**, 25–47.

Carman, R., J. Aigars and B. Larsen, 1996. Carbon and nutrient geochemistry of the surface sediments of the Gulf of Riga, Baltic Sea. Mar. Geol. **134**, 57–76.

Carman, R., G. Edlund and Ch. Damberg, 2000. Distribution of organic and inorganic phosphorus compounds in marine and lacustrine sediments: a ^{31}P NMR study. Chem. Geol. **163**, 101–114.

Chen, J.-C., and Y.-C. Yao, 1995. Geochemistry of manganese nodules from offshore areas of Mariner Islands and Johnston Island. J. SE Asian Earth Sci. **11**, 61–70.

Cochran, J.K., and S. Krishaswami, 1980. Radium, thorium, uranium, and ^{210}Pb in deep-sea sediments and sediment pore waters from the North Equatorial Pacific. American J. Sci. **280**, 849–889.

Conley, D.J., and R.W. Johnstone, 1995. Biogechemistry of N, P, Si in Baltic Sea sediments: response to a simulated deposition of a spring diatom bloom. Mar. Ecol. Prog. Ser. **122**, 265–276.

Conley, D.J., C.L. Schelske and E.F. Stoermer, 1993. Modification of the biogeochemical cycle of silica with eutrophication. Mar. Ecol. Prog. Ser. **101**, 179–192.

Conley, D.J., A. Stockenberg, R. Carman, R.W. Johnstone, L. Rahm and F. Wulff, 1997. Sediment-water nutrient fluxes in the Gulf of Finland, Baltic Sea. Estuar. Coast. Shelf Sci. **45**, 591–598.

Danielsson, Å., 1998. Spatial Modelling in Sediments (Linköping Studies in Arts and Science, Sweden). 89 pp. + Appendices

Danielsson, Å., R. Carman, L. Rahm and J. Aigars, 1998. Spatial estimation of nutrient distributions in the Gulf of Riga sediments using cokriging. Estuar. Coast. Shelf Sci. **46**, 713–722.

Danielsson, Å., I. Cato, R. Carman and L. Rahm, 1999. Spatial clustering of metals in the sediments of the Skagerrak/Kattegat. Appl. Geochem. **14**, 689–706.

David, K., M. Frank, R.K. O'Nions, N.S. Belshaw and J.W. Arden, 2001. The Hf isotope composition of global seawater and the evolution of Hf isotopes in the deep Pacific Ocean from Fe-Mn crusts. Chem. Geol. **178**, 23–42.

De Carlo, E.H., and G.M. McMurtry, 1992. Rare-earth element geochemistry of ferromanganese crusts from the Hawaiian Archipelago, central Pacific. Chem. Geol. **95**, 235–250.

De Grave, E., R.E. Vanderberghe and L.H. Bowen, 1990. Magnetic properties of some selected iron oxides and hydroxides as probed by ^{57}Fe Mössbauer spectroscopy. Proc. XXIV Zakopane School on Physics, Zakopane, Poland, April 1989, ed. A.T. Pędziwiatr, pp. 186–249. World Scientific, Singapore.

De Groot, A.J., 1964. Mud transport studies in coastal waters from the western Scheldt to the Danish frontier, in: Deltaic and Shallow Marine Deposits, ed. L.M.J.U. Van Straaten (Elsevier, Amsterdam), 93–103.

Dietrich, P.G. and P. Beuge, 1986. Quecksilber und Zinc in Schlämmen and Porenwässern der Gdansker Bucht/Ostsee. Z. angewand. Geol. **32**, 240–246.

Djafari, D., 1976. Manganese-iron accumulates in Kiel Bay. Dr. rer. nat. thesis (Universität Kiel) in German.

Domanov, M.M., E.G. Vodyanaya and V.A. Koinov, 1997. Transport of phosphorus and silicon of the water-bottom boundary in the Baltic Sea. Okeanologiya (Moscow) **37**, 60–66.

Dyrssen, D., and K. Kremling, 1990. Increasing hydrogen sulfide concentration and trace metal behavior in the anoxic Baltic water. Mar. Chem. **30**, 193–204.

Elmgren, R., 1989. Man's impact on the ecosystem of the Baltic Sea: energy flows today and at the turn of the century. Ambio **18**, 326–332.

Emeis, K.-C., and T. Neumann, R. Endler, U. Struck, H. Kunzendorf and C. Christiansen, 1998. Geochemical records of sediments in the Eastern Gotland Basin – products of sediment dynamics in a not-so-stagnant anoxic basin? Appl. Geochem. 13, 349–358.

Emeis, K.-C., U. Struck, T. Leipe, F. Pollehne, H. Kunzendorf and C. Christiansen, 2000. Changes in the C:N:P burial rates in some Baltic Sea sediments over the last 150 years – relevance to P regeneration rates and the phosphorus cycle. Mar. Geol. **167**, 43–59.

Emelyanov, E.M., 1982. Trace elements in bottom sediments of the Baltic, in: Geology of the Baltic Sea, eds. V.K. Gudelis and E.M.Emelyanov (Wydawnictwo Geologiczne, Warsaw) 290–306 (in Polish).

Emelyanov, E.M., 1992. The Baltic Proper. In review of contaminants in Baltic sediments. ICES Cooperative Res. Rep. **190**, 16–19.

Emelyanov, E.M., 1995a. Baltic Sea: Geology, Geochemistry, Paleoceanography, Pollution Academy of Natural Sciences, P.P. Shirshov Institute, Kaliningrad).

Emelyanov, E.M., 1995b. Sedimentation in the Gulf of Finland related to anthropogenic load. Okeanologiya (Moscow) **35**, 770–779.

Emelyanov, E.M., 2001. Biogenic components and elements in sediments of the Central Baltic and their redistribution Mar. Geol. **172**, 23–41.

Emelyanov, E.M., and Pustenikov, O.S., 1982. Suspended matter, its composition and the balance of sediments in the Baltic, in: Geology of the Baltic Sea, eds. V.K. Gudelis and E.M. Emelyanov (Wydawnictwo Geologiczne, Warsaw), 161–187 (in Polish).

Emelyanov, E.M., and K. Wypych, 1987. Process of sediment accumulation in Gdansk Basin (Baltic Sea), Akademia Nauk SSSR, Moscow, 273 pp. (in Russian).

Emelyanov, E.M., M.F. Pilipchuk, B.V. Volostnykh, G.S. Khandros and Yu. Shaidurov, 1982. Fe and Mn forms in the geochemical profile of the Baltic Sea. Baltica **7**, 153–171.

Erlenkeuser, H., E. Suess and H. Willkom, 1974. Industrialization affects heavy metal and carbon isotope concentrations in recent Baltic Sea sediments. Geochim. Cosmochim. Acta **38**, 823–842.

Ewers, U., and H.-W. Schlipköter, 1991. Lead, in: Metals and their Compounds in the Environment, ed. E. Merian (VCH, Weinheim), 971–1014.

Falandysz, J., 1999. Mercury in surface sediments associated with shipyard activity in the city of Gdansk. Polish J. Environ. Studies **8**, 265–267.

Falandysz, J., R. Dubrawski and J. Bolałek, 1993. Zawartość rtęci w osadach dennych z Zatoki Puckiej i Gdańskiej. Bromat. Chem. Toksykol. **XXVI**, 29-32 (in Polish with English abstract).

Falkowska, L., J. Bolalek and J. Nowacki, 1993. Nutrients and oxygen in the Gulf of Gdansk. Mar. Pollut. (National Scientific Committee on Oceanic Research, Polish Academy of Sciences, Gdansk) **64** (3), 131–162.

Förstner, U., 1980. Cadmium in polluted sediments, in: Cadmium in Polluted Sediments, ed. J.O. Nriagu (John Wiley and Sons, Inc.) **Part I**, 305–363.

Förstner, U., and W. Salomons, 1980. Trace metal analysis on polluted sediments. Part I: Assessment of sources and intensities. Environ. Technol. Letters **1**, 494–505.

Förstner, U., and J. Schoer, 1990. Metal pollution in the tidal Elbe River. Sci. Total Environ. **97/98**, 347–368.

Förstner, U., and W. Salomons, 1991. Mobilization of metals from sediments, in: Metals and their Compounds in the Environment, ed. E. Merian (VCH, Weinheim) 379–398.

Förstner, U., C. Colombi and R. Kistler, 1991. Dumping of wastes, in: Metals and their Compounds in the Environment, ed. E. Merian (VCH, Weinheim) pp. 333–355.

Gajewski, L.S., and S. Uścinowicz, 1993. Hydrologic and sedimentologic aspects of mining aggregate from the Słupsk Bank (Baltic Sea). Mar. Georesurc. Geotechnol. 11, 229–244.

Geological Map of the Baltic Sea Bottom, 1:200 000, 1989-1995. Mojski, J.E. (ed.). Polish Geological Institute, Warsaw (12 sheets with the legends in Polish and English and explanatory booklets in Polish).

Geological Atlas of the Southern Baltic, 1:500 000, 1995. Mojski, J.E., R. Dadlez, B. Słowańska, Sz. Uścinowicz, J. Zachowicz (eds.). Polish Geological Institute, Sopot – Warsaw (34 Plates and explanatory booklet in English), 63 pp.

Glasby, G.P., 1972–73, Mechanisms of enrichment of the rarer elements in marine manganese nodules. Mar. Chem. **1**, 105–125.

Glasby, G.P., 1974, Mechanisms of incorporation of manganese and associated trace elements in marine manganese nodules. Oceanogr. Mar. Biol. Ann. Rev. **12**, 11–40.

Glasby, G.P., 1975. Minor element enrichment in manganese nodules relative to seawater and marine sediments. Naturwissenschaften **62**, 133–135.

Glasby, G.P. (ed.), 1977. Marine Manganese Deposits (Elsevier Scientific Publishing Company, Amsterdam), 524 pp.

Glasby, G.P., 1984. Manganese nodule research in Federal Republic of Germany: A review. Mar. Mining **4**, 355–402.

Glasby, G.P., 1999. Manganese: predominant role of nodules and crusts, in: Marine Geochemistry, eds. H.D. Schulz and M. Zabel (Springer Verlag).

Glasby, G.P., and A.J. Read, 1976. Deep-sea manganese nodules, in: Handbook of Strata-Bound and Stratiform Ore Deposits, ed. R.H. Wolf (Elsevier Scientific Publishing Company, Amsterdam), 295–340.

Glasby G.P., and P. Szefer, 1998. Marine pollution in Gdańsk Bay, Puck Bay, and the Vistula Lagoon, Poland – An overview. Sci. Total Environ. **212**, 49–57.

Glasby, G.P., and H.D. Schultz, 1999. E_H, pH diagrams for Mn, Fe, Co, Ni, Cu and As under seawater conditions: application of two new types of E_H, pH diagrams to the study of specific problems in marine geochemistry. Aquat. Geochem. 5, 227–248.

Glasby, G.P., P. Stoffers, P. Walter, K.R. Davis and R.M. Renner, 1988. Heavy-metal pollution in Manukau and Waitemata Harbours, New Zealand. N.Z. J. Mar. Freshwater Res. **22**, 595–611.

Glasby, G.P., R.L. Moss and P. Stoffers, 1990. Heavy-metal pollution in Porirua Harbour, New Zealand. N.Z. J. Mar. Freshwater Res. **24**, 233–237.

Glasby, G.P., Sz. Uścinowicz and J.A. Sochan, 1996. Marine ferromanganese concretions from the Polish Exclusive Economic Zone: Influence of major inflows of North Sea water. Mar. Georesourc. Geotechnol. **14**, 335–352.

Glasby, G.P., E.M. Emelyanov, V.A. Zhamoida, G.N. Baturin, T. Leipe, R. Bahlo and P. Bonacker, 1997a Environments of formation of ferromanganese concretions in the Baltic Sea: A critical review, in: Manganese Mineralization: Geochemistry and Mineralogy of Terrestrial and Marine Deposits, eds. K. Nicholson, J.R. Hein, B. Bühn and S. Dasgupta. Geol. Soc. Spec. Publ. **No. 119**, 213–237.

Glasby, G.P., D. Stüben, G. Jeschke, P. Stoffers and C.-D. Garbe-Schönberg, 1997b. A model for the formation of hydrothermal manganese crusts from the Pitcairn Island hotspot. Geochim. Cosmochim. Acta, **61**, 4583–4597.

Glasby, G.P., P. Szefer, J. Gełdon and J. Warzocha, 2001. Heavy-metal pollution of sediments from Szczecin Lagoon and the Gdansk Basin, Poland. Mar. Pollut. Bull. (submitted).

Golimowski, J., and T. Szczepanska, 1996. Voltammetric method for the determination of Zn, Cd, Pb, Cu and Ni in interstitial water. Fresenius J. Anal. Chem. **354**, 735–737.

Görlich, K., E.A. Görlich and L. Stoch, 1978. Iron in the Baltic Sea clay sediments. Mineral. Polon. **9**, 63–79.

Görlich, K., E.A. Görlich and P.Q. Hung, 1985. Iron-bearing minerals in the Baltic Sea surface sediments: A Mössbauer effect study. Mineral. Polon. **15**, 27–35.

Görlich, K., E.A. Görlich, K. Tomala, A.Z. Hrynkiewicz and P.Q. Hung, 1989. ^{57}Fe Mössbauer study of a sediment column in the Gdansk Basin, Baltic Sea: Palaeoenvironmental application. Mar. Geol. **88**, 49–69.

Gorshkov, A.I., G.N. Baturin, V.V. Beresovskaya, G.A. Dubinina and A.V. Sitsov, 1992. Mineralogy and genesis of the buried concretions of the Baltic Sea. Geology of the Oceans and Seas. 10th Intern. Mar. Geol. School, Moscow (Abstracts) **3**, 78–79 (in Russian).

Graca, B., and J. Bolalek, 1998. Forms of phosphorus in sediments from the Gulf of Gdańsk. Appl. Geochem. **13**, 319–327.

Grimanis, A.P., M. Vassilaki-Grimani and G.B. Griggs, 1977. Pollution studies of trace elements in sediments from the upper Saronikos Gulf, Greece. J. Radioanal. Chem. **37**, 761–773.

Gripenberg, S., 1934. A study of the sediments of the North Baltic and adjoining seas. Merentutkimuslaitoksen Julk. Havsforskningsinst. Skr. **No. 96**, 1–231.

Grönlund L., and J.M. Leppänen, 1990. Long-term changes in the nutrient reserves and pelagic production in the western Gulf of Finland. Finnish Mar. Res. **No. 257**, 15–27.

Grönlund L., and J.M. Leppänen, 1992. Variability in the nutrient reserves and pelagic productivity in the western Gulf of Finland. ICES Mar. Sci. Symp. **195**, 499–506.

Grönlund L., K. Kononen, E. Lahdes and K. Mäkelä, 1996. Community development and modes of phosphorus utilization in a late summer ecosystem in the central Gulf of Finland, the Baltic Sea. Hydrobiologia **331**, 97–108.

Gunnars, A. and S. Blomqvist, 1997. Phosphate exchange across the sediment-water interface when shifting from anoxic to oxic conditions – an experimental comparison of freshwater and brackish-marine system. Biogeochem. **37**, 203–226.

Hallbach, P., 1975. Mineralogical and geochemical investigation of Finnish lake ores. Bull. Geol. Soc. Finland **43**, 33–42.

Hallberg, R.O., 1979. Heavy metals in the sediments of the Gulf of Bothnia. Ambio **8**, 265–269.

Hallberg, R.O., 1991. Environmental implications of metal distribution in Baltic Sea sediments. Ambio **20**, 309–316.

Håkanson, L., and M. Jansson, 1983. Lake Sedimentology (Springer Verlag, Berlin), 316 pp.

Heiser, U., T. Neumann, J. Scholten and D. Stüben, 2001. Recycling of manganese from anoxic sediments in stagnant basins by seawater inflow: a study of surface sediments from the Gotland Basin, Baltic Sea. Mar. Geol. **177**, 151–166.

HELCOM, 1995. Radioactivity in the Baltic Sea 1984–1991. Baltic Sea Environment Proceedings (Baltic Marine Environment Protection Commission, HELCOM) **No. 61**.

Helios Rybicka, E., 1992. Heavy metal partitioning in polluted river and sea sediments: clay mineral effects. Miner. Petrogr. Acta, **35A**, 297–305.

Heuser, H., 1988. Observations and investigations on the genesis of shallow-water manganese nodules in Kiel Bay (western Baltic). Berichte-Reports (Geologisch-Palontologisches Institut Universität Kiel) **26** (in German).

Hilton, J., W. Davison, and U. Oschenbein, 1985. A mathematical method for analysis of sediment core data: implications for enrichment factor calculations and trace – metal transport mechanisms. Chem. Geol. **48**, 281–291.

Hirst, D.M., 1962a. The geochemistry of modern sediments from the Gulf of Paria – I. Geochim. Cosmochim. Acta **26**, 309–334.

Hirst, D.M., 1962b. The geochemistry of modern sediments from the Gulf of Paria – II. Geochim. Cosmochim. Acta **26**, 1147–1187.

Hlawatsch, S., 1993. Growth of manganese-iron accumulates in the western Baltic Sea Indicator for environmental change. Diploma Thesis, University of Kiel, (in German).

Hlawatsch, W., F. Lechtenberg, C.D. Garbe-Schönberg, J. Harff and E. Suess, 2001. Ferro-manganese-nodules from the western Baltic Sea: indicator for environmental change in a highly polluted region of Europe. Chem. Geol. (in press).

Holby, O., and S. Evans, 1996. The vertical distribution of Chernobyl-derived radionuclides in a Baltic Sea sediment. J. Environ. Radioact. **33**, 129–145.

Holm, E., 1995. Plutonium in the Baltic Sea. Appl. Radiat. Isot. **46**, 1225–1229.

Horowitz, A.J., 1991. Sediment-trace element data manipulations, in: A Primer on Sediment-Trace Element Chemistry (Lewis Publishers) 80–90.

Horowitz, A.J., F.A. Rinella, P. Lamothe, T.L. Miller, T.K. Edwards, R.L. Roche and D.A. Rickert, 1989. Cross-sectional variability in suspended sediment and associated trace element concentration in selected rivers in the US. Sediment and the Environment. Proceedings of the Baltimore Symposium (IAHS Publication) **No. 184**, pp. 57–66.

Horowitz, A.J., F.A. Rinella, P. Lamothe, T.L. Miller, T.K. Edwards, R.L. Roche and D.A. Rickert, 1990. Variations in suspended sediment and associated trace element concentrations in selected riverine cross sections. Environ. Sci. Technol. **24**, 1313–1320.

Huckriede, H., S., and D. Meischner, 1996. Origin and environment of manganese-rich sediments within black-shake basins. Geochim. Cosmochim. Acta **60**, 1399–1413.

Huckriede, H., S. Clasen and D. Meischner, 1996. Hydrographic and climatic changes recorded in Holocene sediments of the Central Baltic Sea. Baltica **9**, 76–91.

Hunt, C.D., and D.L. Smith, 1983. Remobilization of metals from polluted marine sediments. Can.J. Fish. Aquat. Sci. **40** (Suppl. 2), 132–142.

Hylland, K., M. Sköld, J.S. Gunnarsson and J. Skei, 1996. Interactions between eutrophication and contaminants.IV. Effects on sediment-dwelling organisms. Mar. Pollut Bull. **33**, 90–99.

Ikäheimonen, T.K., R. Pöllänen and S. Klemola, 2000. Radioactive particulate matter in a marine sediment sample, in: The Radiological Exposure of the Population of the European Community to Radioactivity in the Baltic Sea. Marina-Balt Project, ed. S.P. Nielsen. Proceedings of a Seminar held at Hasseludden Conference Centre, Stockholm, 9–11 June 1998, European Commission, Directorate-General Environment, EUR 19200 EN (European Communities, 2000, Belgium), pp. 411–432.

Ikuta, K., and P. Szefer, 2000. Distribution of heavy metals in soft tissue and shell of *Astarte borealis* and *Macoma balthica* from Slupsk Furrow, a southern Baltic (in preparation).

Ilus, E., K.-L. Sjöblom, T.K. Ikäheimonen, R. Saxén and S. Klemola, 1993. Monitoring of radionuclides in the Baltic Sea in 1989–1990. Suppl. 10 to Annual Rep. STUK-A89. STUK-A103, February 1993 (Finnish Centre for Radiation and Nuclear Safety, Helsinki, Finland) 35 pp.

Ilus, E., L. Niemistö and R. Bojanowski, 1995. Radionuclides in sediment and suspended particulate matter, in: Radioactivity in the Baltic Sea 1984–1991. Baltic Sea Environment Proceedings HELCOM **No. 61**, 69–92.

Ilus, E., J. Mattila, S. Klemola and T.K. Ikäheimonen, 1998. Evaluation of sedimentation rate at two sampling stations in the Gulf of Finland based on Pb-210, Cs-137 and Pu-239,240 profiles in sediments. Sateilyturvakeskus, Rapp. STUK-A145, 136–147.

Ingri, J., 1985a. Geochemistry of ferromanganese concretions and associated sediments in the Gulf of Bothnia. Ph.D. thesis (University of Luleå, Sweden).

Ingri, J., 1985b. Geochemistry of ferromanganese concretions in the Barents Sea. Mar. Geol. **67**, 101–119.

Ingri, J., and C. Pontér, 1986a. Scavenging properties of ferromanganese nodules in the Gulf of Bothnia. Rapports et Procès Verbaux des Réunions Condeil International Pour l'Exploration de la Mer **186**, 234–243.

Ingri, J., and C. Pontér, 1986b. Iron and manganese layering in recent sediments in the Gulf of Bothnia. Chem. Geol. **56**, 105–116.

Ingri, J., and C. Pontér, 1987. Rare earth abundance patterns in ferromanganese concretions from the Gulf of Bothnia and the Barents Sea. Geochim. Cosmochim. Acta **51**, 155–161.

Ingri, J., R. Löfvendahl and K. Boström, 1991. Chemistry of suspended particles in the southern Baltic Sea. Mar. Chem. **32**, 73–87.

Irion, G., 1984. Schwermetallbelastung in Oberflächensedimenten der westlichen Ostsee. Naturwissenschaften **71**, 536–538.

Irion, G., and G. Müller, 1987. Heavy metals in superficial sediments of the North Sea. Proceedings of the International Conference on Heavy Metals in the Environment (New Orleans, LA) **Vol. 2**, 38–41.

Jakobsen, R., and D. Postma, 1989. Formation and solid solution behavior of Ca-rhodochrosites in marine muds of the Baltic deeps. Geochim. Cosmochim. Acta **53**, 2639–2648.

Jaworowski, Z., A. Barański, D. Grzybowska, A. Pawlak, J. Tomczak, A. Pietruszewski, M. Suplińska, D. Skiba and T. Wardaszko, 1986. Monitoring of the radioactive materials in the Baltic Sea, in: Study of Radioactive materials in the Baltic Sea. (International Atomic Energy Agency, Vienna). Report (IAEA-TECDOC-362) of the Final Research Co-ordination Meeting on the Study of Radioactive Materials in the Baltic Sea organized by the IAEA and held in Helsinki, Finland 24–28 September, 1984, pp. 141–151.

Johnston, J.H., and G.P. Glasby, 1978. The secondary iron oxyhydroxide mineralogy of some deep sea and fossil manganese nodules: A Mössbauer and X-ray study. Geochem. J. **12**, 153–164.

Jonsson, P., 1992. Large-scale changes of contaminants in the Baltic Sea sediments during the twentieth century. Acta Universitatis Upsaliensis, Comprehensive Summaries of Uppsala Dissertations from the Faculty of Science 407, 52 pp. + Appendices.

Jonsson, P., and B. Jonsson, 1988. Dramatic changes in Baltic sediments during the last three decades. Ambio **17**, 158–160.

Jonsson, P., and R. Carman, 1994. Changes in deposition of organic matter and nutrients in the Baltic Sea during the twentieth century. Mar. Pollut. Bull. **28**, 417–426.

Jonsson, P., R. Carman and F. Wulff, 1990. Laminated sediments in the Baltic – A tool for evaluating nutrient mass balances. Ambio **19**, 152–158.

Joshi, L.U., and A.K. Ganguly, 1972. Anomalous $^{228}Th/^{232}Th$ and $^{230}Th/^{232}Th$ activity ratios in backwater sediments along the west coast of India. IAEA/SM-158/50, Seatle, Washington, July, 1972, 11 pp.

Kannan, K., and J. Falandysz, 1997. Butyltin residues in sediment, fish, fish-eating birds, harbour porpoise and human tissues from the Polish coast of the Baltic Sea. Mar. Pollut. Bull. **34**, 203–207.

Kannan, K., and J. Falandysz, 1998. Speciation and concentrations of mercury including methyl mercury in certain coastal marine sediments. Water Air Soil Pollut. **103**, 129–136.

Kasten, S., G.P. Glasby, H.D. Schulz, G. Friedrich and S.I. Andreev, 1998. Rare earth elements in manganese nodules from the South Atlantic Ocean as indicators of oceanic bottom water flow. Mar. Geol. **146**, 33–52.

Kautsky, H., and H.F. Eicke, 1981. Radiological investigations in the Western Baltic Sea and Baltic Proper, including the Danish Straits, during 1981. Deutsch. Hydrogr. Z. **34**, 125–149.

Kemp, A.L.W., R.L. Thomas, C.J. Deli and J.H. Jaquet, 1976. Cultural impact on the geochemistry of sediments in Lake Erie. J. Fish. Res. Board Can. **33**, 440–462.

Kersten, M., and U. Förstner, 1986. Chemical fractionation of heavy metals in anoxic estuarine and coastal sediments. Water Sci. Technol. **18**, 121–130.

Kimberley, M.M., 1989. Exhalative origins of iron-formations. Ore Geol. Rev. **5**, 13–145.

Kingston, H.M., and R.R. Greenberg, 1984. An elemental ratioing technique for assessing concentration data from a complex water system. Environ. Intern. **10**, 153–161.

Kjellin, B., S. Axberg and A. Elhammer, 1987. Maringeologiska kartbladet 089/480 Gotska Sandön. Geological Survey of Sweden. Ser. AM1. 45 pp.

Koide, M., K.W. Bruland and E.D. Goldberg, 1973. Th-228/Th-232 and Pb-210 geochronologies in marine and lake sediments. Geochim. Cosmochim. Acta **37**, 1171–1187.

Koide, M., K.W. Bruland and E.D. Goldberg, 1976. ^{226}Ra chronology of a coastal marine sediments. Earth Planet. Sci. Letters **31**, 31–36.

Koop, K., W.R. Boynton, F. Wulff and R. Carman, 1990. Sediment-water oxygen and nutrient exchanges along a depth gradient in the Baltic Sea. Mar. Ecol. Prog. Ser. **63**, 65–77.

Kramarska, R., P. Krzywiec and R. Dadlez, 1999. Geological Map of the Baltic Sea Bottom without Quaternary deposits, 1:500 000. Polish Geological Institute, Warsaw (Map + geological cross-sections + explanatory booklet in English).

Kremling, K., 1983. The behaviour of Zn, Cd, Cu, Ni, Co, Fe and Mn in anoxic Baltic waters. Mar. Chem. **13**, 87–108.

Kremling, K., L. Brügmann and A. Jensen, 1987. Trace metals in the Baltic, in: First Periodic Assessment of the State of the Marine Environment of the Baltic Sea Area, 1980–1985; Background Documents. Baltic Sea Environment Proceedings **17B**, pp. 82–11.

Kremling, K., J.J.S. Tokos, L. Brügmann and H.-P. Hansen, 1997. Variability of dissolved and particulate trace metals in the Kiel and Mecklenburg Bights of the Baltic Sea, !990–1992. Mar. Pollut. Bull. **34**, 112–122.

Krishnaswami, S., and J.K. Cochran, 1978. Uranium and thorium series nuclides in oriented ferromanganese nodules: growth rates, turnover times and nuclide behavior. Earth Planet. Sci. Letters **40**, 45–62.

Krishnaswami, S., A. Mangini, J.H. Thomas, P. Sharma, J.K. Cochran, K.K. Turekian and P.D. Parker, 1982. ^{210}Be and Th isotopes in manganese nodules and adjacent sediments: nodule growth histories and nuclide behavior. Earth Planet. Sci. Letters **59**, 217–234.

Ku, T.L., and W.S. Broecker, 1969. Radiochemical studies on manganese nodules of deep-sea origin. Deep-Sea Res. **16**, 625–637.

Ku, T.L., and G.P. Glasby, 1972. Radiometric evidence for the rapid growth rate of shallow-water, continental margin manganese nodules. Geochim. Cosmochim Acta **36**, 699–703.

Ku, T.-L., and K.G. Knauss, 1979. Radioactive disequilibrium in fissure-filling material and its implication in dating of manganese nodules, in La Genèse des Nodules de Manganèse, ed. C. Lalou (Èditions du Centre National de la Recherche Scientifique, Paris). Colloq. Int. C.N.R.S. **No. 289**, 289–293.

Ku, T.L., A. Omura and P.S. Chen, 1979. Be^{10} and U-series isotopes in manganese nodules from central North Pacific, in: Marine Geology and Oceanography of the Pacific Manganese Nodule Province, eds. J.L. Bischoff and D.Z. Piper (Plenum Publishing Corporation), pp. 791–814.

Kuijpers, A., B. Dennegård, Y. Albinsson and A. Jensen, 1993. Sediment transport pathways in the Skagerrak and Kattegat as indicated by sediment Chernobyl radioactivity and heavy metal concentrations. Mar. Geol. **111**, 231–244.

Kulesza-Owsikowska, G., 1981. Mineralogical and geochemical studies of iron-manganese nodules from the southern Baltic Sea. Arch. Mineral. **37**, 149–216 (in Polish).

Kulik, D., and M. Kersten, 1999. Evaluation of the solid solution-aqueous solution equilibrium state for trace metals partitioned in the carbonate-porewater system of a natural sediment. Wiss. Ber.-Forschungszent. Karlsruhe, FZKA 6291, 77–81.

Kunzendorf, H., W.L. Plüger and G.H. Friedrich, 1983. Uranium in Pcific deep-sea sediments and manganese nodules. J. Geochem. Explor. **19**, 147–162.

Kunzendorf, H., G.P. Glasby, P. Stoffers P. and W.L. Plüger, 1993. The distribution of rare earth and minor elements in manganese nodules along an east-west transect in the southern Pacific. Lithos **30**, 45–56.

Laima, M.J.C., H. Matthiesen, L.C. Lund-Hansen and C. Christiansen, 1998. Aspects of phosphorus geochemistry in Baltic Sea sediments: Methods, fluxes, formation of Fe/P particles and resuspension studies., 1st International Symposium on Estuarine Processes, Algarve, Portugal, pp. 111–117.

Laima, M.J., H. Matthiesen, L.C. Lund-Hansen, C. Christiansen, S. Jähmlich, and K.-C. Emeis, 2001. Transport of P, Fe and Mn along a depth gradient in the SW Baltic. Boreal Environ. Res. (in press).

Lalou, C., E. Brichet and P. Bonte, 1980. Some new data on the genesis of manganese nodules, in: Geology and Geochemistry of Manganese. Intern. Congr. Geol., Sydney 1976 (Akadémiai Kiadó Budapest), **Vol. III**, 31–90.

Lampe, R., 1999. The Odra Estuary as a filter and transformation area. Acta Hydrochim. Hydrobiol. **27**, 292–297.

Lange, D., 1987. Mecklenburg Bay Sediments 1:100 000 (Hydrographic Office, GDR).

Lazarev, L.N., Yu.V. Kuznetsov, L.I. Gedeonov, D.I. Gusev, V.V. Anisimov, V.A. Blinov, L.M. Ivanova, T.E. Orlova, A.V. Stepanov, N.A. Tishkova, V.P. Tishkov, V.M. Flegontov, S.M. Vakulovskiy, A.I. Nikitin, I.Yu. Katrich, V.B. Chumichev, O.A. Pavlovskiy and V.D. Stepanova, 1986. Investigation of the radioactive contamination of the Baltic Sea in 1978–83, in: Study of Radioactive materials in the Baltic Sea. (International Atomic Energy Agency, Vienna). Report (IAEA-TECDOC-362) of the Final Research Co-ordination Meeting on the Study of Radioactive Materials in the Baltic Sea organized by the IAEA and held in Helsinki, Finland 24–28 September, 1984, pp. 162–175.

Le Sauve, R., C. Pichocki, G. Pautot, M. Hoffert, Y. Morel, M. Voisset, S. Monti, J. Amosse and A. Kosakevitch, 1989. Geological and mineralogical study of Co-rich ferromanganese crusts from a submerged atoll in the Tuamotu Archipelago (French Polynesia). Mar. Geol., **87**, 227–247.

Lehtoranta, J., 1998. Net sedimention and sediment-water nutrient fluxes in the eastern Gulf of Finland (Baltic Sea). Vie Milieu **48**, 341–352.

Lehtoranta, J., H. Pitkanen and O. Sandman, 1997. Sediment accumulation of nutrients (N, P) in the eastern Gulf of Finland (Baltic Sea). Water Air Soil Pollut. **99**, 477–486.

Leipe, T., L. Brügmann and U. Bittner, 1989. Zur Verteilung von Schwermetallen in rezenten Brackwassersedimenten der Bodengewässer. Chem. Erde **49**, 21–38.

Leipe, T., R. Bahlo, A. Dahmke and J. Harff, 1994. Iron-manganese crusts from Mecklenburg Bay, western Baltic microstructure, chemical compilation and element distribution. Final Rep. to BMFT (BMFT-FK 03 F 0077A) in German.

Leipe, T., T. Neumann and K.-C. Emeis, 1995. Schwermetallverteilung in holozänen Ostseesedimenten – Untersuchungen in Einflußbereich der Oder. Geowissenschaften **13**, 470–478.

Leipe, T., J. Eidam, R. Lampe, H. Meyer, T. Neumann, A. Osadczuk, W. Janke, T. Puff, T. Blanz, F.X. Gingele, D. Dannenberger and G. Witt, 1998. Das Oderhaff – Beiträge zur Rekonstruction der holozänen geologischen Entwicklung und anthropogenen Beeinflussung des Oder-Ästuars. Meereswissenschaftliche Berichte (Mar. Sci. Rep.) No. 28 (Institut für Ostseeforschung, Warnemünde), 61 pp. + 23 appendices.

Leivuori, M., 1998. Heavy metal contamination in surface sediments in the Gulf of Finland and comparison with the Gulf of Bothnia. Chemosphere **36**, 43–59.

Leivuori, M., and L. Niemistö, 1993. Trace metals in sediments of the Gulf of Bothnia. Aqua Fenn. **23**, 89–100.

Leivuori, M., and L. Niemistö, 1995. Sedimentation of trace metals in the Gulf of Bothnia. Chemosphere **31**, 3839–3856.

Leivuori, M., and H. Vallius, 1998. A case study of seasonal variation in the chemical composition of accumulating suspended sediments in the central Gulf of Finland. Chemosphere **36**, 2417–2435.

Leskinen, S., J.K. Miettinen and T. Jaakkola, 1987. Behaviour of $^{239}Pu+^{240}Pu$ and ^{241}Am in the Baltic Sea; measurements and interpretation in 1980–1984. J. Radioanal. Nucl. Chem. **115**, 289–298.

Li, Y.H., 1981a. Ultimate removal mechanisms of elements from the ocean. Geochim. Cosmochim. Acta **45**, 1659–1664.

Li, Y.H., 1981b. Geochemical cycles of elements and human perturbation. Geochim. Cosmochim. Acta **45**, 2073–2084.

Li, Y.H., 1982. Interelement relationship in abyssal Pacific ferromanganese nodules and associated pelagic sediments. Geochim. Cosmochim. Acta **46**, 1053–1060.

Li, Y.-H., and L.-H. Chan, 1979. Desorption of Ba and ^{226}Ra from river-borne sediments in the Hudson Estuary. Earth Planet. Sci. Letters **43**, 343–350.

Lithner, G., D. Broman, C. Naef, H. Borg, A.-M. Johansson, P. Kaerrhage and M.-B. Larsson, 1996. Metals in settling particles and surficial sediments of the Swedish Baltic coast 1988–1989, in: Dev. Prog. Sediment Qual. Assess, eds. M. Munawar and G. Dave. Int. Symp. Sediment Qual. Assess. (SPB Academic Publishing: Amsterdam, Netherlands), 27–48.

Loring, D.H., 1990. Lithium-a new approach for the granulometric normalization of trace metal data. Mar. Chem. **29**, 155–168.

Luoma, S.N., and G.W. Bryan, 1981. A statistical assessment of the form of trace metals in oxidized estuarine sediments employing chemical extractants. Sci. Total Environ. **17**, 165–196.

Łomniewski, K., W. Mańkowski and J. Zalewski, 1975. The Baltic Sea (Państwowe Wydawnictwo Naukowe, Warsaw), (in Polish).

Macdonald, R.W., D.M. Macdonald, M.C. O'Brian and C. Gobeil, 1991. Accumulation of heavy metals (Pb, Zn, Cu, Cd), carbon and nitrogen in sediments from Strait of Georgia B.C., Canada. Mar. Chem. **34**, 109–135.

Magalhães, V.F. and W.C. Pfeiffer, 1995. Arsenic concentration in sediments near a metallurgical plant (Sepetiba Bay, Rio de Janeiro, Brazil). J. Geochem. Explor. **52**, 175–181.

Magnusson, K., R. Ekelund, R. Dave, Å. Granmo, L. Förlin, L. Wennberg, M.-O. Samuelsson, M. Berggren and E. Brorström-Lundén, 1996. Contamination and correlation with toxicity of sediment samples from the Skagerrak and Kattegat. J. Sea Res. **35**, 223–234.

Manheim, F.T., 1961. A geochemical profile in the Baltic Sea. Geochim. Cosmochim. Acta **25**, 52–70.

Manheim, F.T., 1965. Manganese-iron accumulation in the shallow marine environment, eds. D.R. Schink and J.T. Corless, Sympos. Mar. Geochem. Occasional Publications of the Narragansett Marine Laboratory, University of Rhode Island **3**, 217–276.

Marchig, V., P. Möller, H. Bäcker and P. Dulski, 1985. Foraminiferal ooze from the Galapagos Rift area – hydrothermal impact and diagenetic mobilization of elements. Mar. Geol. **62**, 85–104.

Margolis, S.V., T.L. Ku, G.P. Glasby, C.D. Fein and M.G. Audley-Charles, 1978. Fossil manganese nodules from Timor: geochemical and radiochemical evidence for deep-sea origin. Chem. Geol. **21**, 185–198.

Martin, J.M., and M. Meybeck, 1979. Elemental mass-balance of material carried by major world rivers, Mar. Chem. **7**, 173–206.

Mellin, T., 1987. Investigations of manganese nodule occurrences in the Gulf of Bothnia. Cruise Rep. Res. Rep. 1, Havsmineralgruppen. R. Inst. Technol., Sweden (in Swedish).

Meyenburg, G., and G. Liebzeit, 1993. Mineralogy and geochemistry of the core from the Skagerrak/Kattegat boundary. Mar. Geol. **111**, 337–344.

Mo, T., A.D. Suttle and W.M. Sackett, 1973. Uranium concentrations in marine sediments. Geochim. Cosmochim. Acta **37**, 35–51.

Moenke-Blankenburg, L., K. Jahn and L. Brügmann, 1989. Laser-micro-analytical studies on distribution patterns of manganese, iron and barium in Fe/Mn-accumulates of the western Baltic Sea. Chem. Erde **49**, 39–46.

Morgan, J.J., and W. Stumm, 1991. Chemical Processes in the environment, relevance of chemical speciation, in: Metals and their Compounds in the Environment, ed. E. Merian (VCH, Weinheim), 67–103.

Murad, E., and U. Schwertmann, 1988. Iron oxide mineralogy of some deep-sea ferromanganese crusts. Am. Mineral. **73,** 1395–1400.

Murray, J.W., and P.G. Brewer, 1977. Mechanisms of removal of manganese, iron and other trace metals from sea water, in: Marine Manganese Deposits, ed. G.P. Glasby (Elsevier, Scientific Publishing Company, Amsterdam), 291–325 pp.

Müller, A., and P. Heininger, 1999. On sediment pollution in selected German coastal waters of the Baltic Sea. Limnologica **29**, 255–261.

Müller, A., and M. Wessels, 1999. The flood in the Odra River 1997 – Impact of suspended solids on water quality. Acta Hydrochim. Hydrobiol. **27**, 316–320.

Müller, G., J. Dominik, R. Reuther, R. Malisch, E. Schulte, L. Acker and G. Iron, 1980. Sedimentary record of environmental pollution in the western Baltic Sea. Naturwissenschaften **67**, 595–600.

Naik, S., and E.-L. Poutanen, 1984. Humic substances in Baltic Sea sediments. Oceanol. Acta **7**, 431–439.

Nakanishi, T., M. Oohashi, M. Higashi and M. Sakanoue, 1977. Radiochemical studies of deep-sea manganese nodules. Sci. Rep. Kanazawa Univ. **XXII**, 31–43.

Nehring, D., 1984a. Variations in the nutrient situation of the Baltic proper. Limnologica (Berlin) **15**, 277–287.

Nehring, D., 1984b. The further development of the nutrient situation in the Baltic proper. Ophelia Suppl. **3**, 167–179.

Nehring, D., 1985. Langzeitveränderungen essentieller Nährstoffe in der zentralen Ostsee. Acta Hydrochim. Hydrobiol. **13**, 591–609.

Neumann, Th., and D. Stüben, 1991. Detailed geochemical study and growth history of some ferromanganese crusts from the Tuamotu Archipelgo. Mar. Mining **10**, 29–48.

Neumann, T., T. Leipe, T. Brand and G. Shimmield, 1996. Accumulation of heavy metals in the Oder es[illegible]y and its off-shore basins. Chem. Erde **56**, 207–222.

Neumann, T., C. Christiansen, C. Clasen, K.-C. Emeis and H. Kunzendorf, 1997. Geochemical records of salt-water inflows into the deep basins of the Baltic Sea. Cont. Shelf Res. **7**, 95–115.

Neumann, T., T. Leipe and G. Shimmield, 1998. Heavy-metal enrichment in surficial sediments in the Oder River discharge area: source or sink for heavy metals? Appl. Geochem. **13**, 329–337.

Neumann, T., U. Heiser, M.A. Leosson and M. Kersten, 2001. Early diagenetic processes during Mn-carbonate formation: evidence from the isotopic composition of authigenic Ca-rhodochrosites of the Baltic Sea (submitted).

Nielsen, P.E., 1992. Bottom sediments around Denmark and western Sweden. Geol. Survs. Denmark and Sweden.

Nielsen, P.E., 1996. An intercomparison exercise on radionuclides in sediment samples. Rep. Risø-R-914(EN) (Risø National Laboratory, Roskilde, Denmark), 40 pp.

Niemistö, L., and A. Voipio, 1974. Studies on the recent sediments in the Gotland Deep. Merentutkimuslaitoksen Julk. Havsforskningsinst. Skr. **238**, 17–32.

Niemistö, L., T. Vappu and A. Voipio, 1978. Storage of iron and phosphorus in the sediments of the Bothnian Sea. Finnish Mar. Res. **No. 244**, 36–41.

Norman, M.D., and P. De Deckker, 1990. Trace metals in lacustrine and marine sediments: A case study from the Gulf of Carpentaria northern Australia. Chem. Geol. **82**, 299–318.

Nriagu, J.O., and J.M. Pacyna, 1988. Quantitative assessment of worldwide contamination of air, water and soils by trace metals. Nature **333**, 134–139.

Olmez, J., E.R. Sholkovitz, D. Hermann and R.P. Eganhouse, 1991. Rare earth elements in sediments off southern California: A new anthropogenic indicator. Environ. Sci. Technol. **25**, 310–316.

O'Nions, R.K., S.R. Carter, R.S. Cohen, N.M. Evensen and P.J. Hamilton, 1978. Pb, Nd and Sr isotopes in oceanic ferromanganese deposits and ocean floor basalts. Nature **273**, 435–438.

Öhlander, B., M. Land, J. Ingri and A. Widerlund, 1996. Mobility of rare earth elements during weathering of till in northern Sweden. Appl. Geochem. **11**, 93–99.

Östlund, P., 1991. Interrelations between plutonium and some transition elements in Gulf of Bothnian sediments. Ambio **20**, 38–42.

Osadczuk, A., 1999. An estuary or a lagoon? The Zalew Szczeciński (Oderhaff) in the light of analysis of the sedimentary morphology and the character of present-day sedimentation. Quaternary Studies in Poland, Spec. Issue 175–186.

Osadczuk, A., and B.Wawrzyniak-Wydrowska, 1998. Sediments in the Szczecin Lagoon: selected elements and macrozoobenthos. Greifswalder Geographische Arbeiten **16**, 169–193.

Pacyna, J. M., J. Meunch and F. Axenfeld, 1991. European inventory of trace metal emissions to the atmosphere, in: Heavy Metals in the Environment, ed. J.-P. Vernet (Elsevier, Amsterdam), pp. 1–20.

Paetzel, M., H. Schrader and K. Bjerkli, 1994. Do decreased trace metal concentrations in surficial Skagerrak sediments over the last 15–30 years indicate decreased pollution? Environ. Pollut. **84**, 213–226.

Palumbo, B., A. Bellanca, R. Neri and M.J. Roe, 2001. Trace metal partitioning in Fe-Mn nodules from Sicilian soils, Italy. Chem. Geol. **173**, 257–269.

Panteleev, Y., H. Nies, T.K. Ikäheimonen and V. Tishkov, 1995. Radionuclides in seawater, in: Radioactivity in the Baltic Sea 1984–1991. Baltic Sea Environment Proceedings HELCOM **No. 61**, 59–68.

Pederstad, K., E. Roaldset and T. Rönningsland, 1993. Sedimentation and environmental conditions in the inner Skagerrak – outer Oslofjord. Mar. Geol. **111**, 245–268.

Pempkowiak, J., D. Cossa, A. Sikora and J. Sanjuan, 1998a. Mercury in water and sediments of the southern Baltic Sea. Sci. Total Environ. **213**, 185–192.

Pempkowiak, J., M. Piotrowska-Szypryt and J. Kozuch, 1998b. Rates of diagenetic changes of humic substances in Baltic surface sediments. Environ. Intern. **24**, 589–594.

Pempkowiak, J., A. Sikora and E. Biernacka, 1999. Speciation of heavy metals in marine sediments vs their bioaccumulation by mussels. Chemosphere **39**, 313–321.

Persson, J., and P. Jonsson, 2000. Historical development of laminated sediments – an approach to detect soft sediment ecosystem changes in the Baltic Sea. Mar. Pollut. Bull. **40**, 122–134.

Pettis, R.W., and A. de Forest, 1979. Chemical composition of ferromanganese nodules from the Southern Ocean. Aust. J. Mar. Freshwater Res. **30**, 535–539.

Pitkänen, H., 1991. Nutrient dynamics and trophic conditions in the eastern Gulf of Finland: The regulatory role of the Neva Estuary. Aqua Fennica **21**, 105–115.

Pitkänen, H., and T. Tamminen, 1995. Nitrogen and phosphorus as production limiting factors in the estuarine waters of the eastern Gulf of Finland. Mar. Ecol. Prog. Ser. **129**, 283–294.

Pohl, C., U. Hennings, I. Petersohn and H. Siegel, 1998. Trace metal budget, transport, modification and sink in the transition area between the Oder and Peene Rivers and the southern Pomeranian Bight. Mar. Pollut. Bull. **36**, 598–616.

Pontér, C., J. Ingri and K. Boström, 1990a. Geochemistry of manganese in the Kalix River, northern Sweden. Geochim. Cosmochim. Acta 56, 1485–1494.

Pontér, C., J.-O. Burman, J. Ingri and K. Boström, 1990b. Temporal variations in dissolved and suspended iron and manganese in the Kalix River, northern Sweden. Chem. Geol. **81**, 121–131.

Pöllänen, R., 1997. Highly radioactive ruthenium particles released from the Chernobyl accident: particle characteristics and radiological hazard. Radiat. Protect. Dosimetry **7**, 23–32.

Pöllänen, R., T.K. Ikäheimonen, S. Klemola and J. Juhanoja, 1999. Identification and analysis of a radioactive particle in a marine sediment sample. J. Environ. Radioactivity **45**, 149–160.

Presley, B.J., and J.H. Trefry, 1980. Sediment-water interactions and the geochemistry of interstitial water, in: Chemistry and Biogeochemistry of Estuaries, eds. E. Olausson and I. Cato, (John Wiley & Sons Ltd.) pp. 187–232.

Putans, B.D., V.G. Ul'st and V.B. Emss, 1968. Ferromanganese concretions in the sediments of the Gulf of Riga, Lithology, Geochemistry and Economic Minerals of Byelorussia and Baltic States. Minsk, 249–255 (in Russian).

Puxbaum, H., 1991. Metal compounds in the atmosphere, in: Metals and their Compounds in the Environment, ed. E. Merian (VCH, Weinheim), 257–286.

Rahm, L, D. Conley, P. Sandén, F. Wulff and P. Stålnacke, 1996. Time series analysis of nutrient inputs to the Baltic Sea and changing DSi:DIN ratios. Mar. Ecol. Prog. Ser. **130**, 221–228.

Renner, R.M., G.P. Glasby and P. Walter, 1997. Endmember analysis of metalliferous sediments from the Galapagos Rift and East Pacific Rise between 2°N and 42°S. Appl. Geochem. **12**, 383–395.

Rittenberg, S.C., K.O. Emery and W.L. Orr, 1955. Regeneration of nutrients in sediments of marine basins. Deep Sea Rres. **3**, 32–45.

Rule, J.H., 1986. Assessment of trace element geochemistry of Hampton Roads harbour and lower Chesapeake Bay area sediments. Environ. Geol. Water Sci. **8**, 209–219.

Sager, M., R. Pucsko and R. Belocky, 1990. Evaluation of the specification of inorganic constituents in sediments of reservoir at Altenworth of the River Danube. Archiv. Hydrobiol. Suppl. **84**, 37–72.

Salo, A., K. Tuomainen and A. Voipio, 1986. Inventories of some long-lived radionuclides in the Baltic Sea. Sci. Total Environ. **54**, 247–260.

Salonen, V.-P., T. Grönlund, A. Itkonen, M. Sturm and I. Vuorinen, 1995. Geochemical record on early diagenesis of recent Baltic Sea sediments. Mar. Geol. **129**, 101–109.

Sanchez, A.L., J. Gastaud, E. Holm, J. Ingri and R. Löfendahl, 1988. Fallout Pu and Am in Baltic Sea ferromanganese concretions: implications regarding their interactions in sediments. Radiochimica **44/45**, 153–157.

Sarin, M.M., D.V. Borole and S. Krishaswami, 1979. Geochemistry and geochronology of sediments from the Bay of Bengal and the equatorial Indian Ocean. Proceedings of the Indian Academy of Science, **88A**, 131–154.

Satsmadjis, J., and F. Voutsinou-Taliadouri, 1985. An index of metal pollution in marine sediments. Oceanol. Acta **8**, 277–284.

Schaanning, M.T., K. Hylland, D.Ø. Eriksen, T.D. Bergan, J.S. Gunnarsson and J. Skei, 1996. Interactions between eutrophication and contaminants. II. Mobilization and bioaccumulation of Hg and Cd from marine sediments. Mar. Pollut. Bull. **33**, 71–79.

Schneider, B., 1987. Some characterization for atmospheric trace metals over Kiel Bay. Atmos. Environ. **21**, 1275–1283.

Schoer, J., U. Nagel, D. Eggersgluess and U. Förstner, 1982. Metal contents in sediments from the Elbe, Weser and Ems estuaries and from the German Bight (southeastern North Sea): grain size effects. Mitt. Geol.-Paleont. Inst. Univ. Hamburg, SCOPE/UNEP Special, **Vol. 52**, 687–702.

Seibold, E., N. Exon, M. Hartmann, F.-C. Kögler, H. Krumm, G.F. Lutze, R.S. Newton and F. Werner, 1971. Marine Geology of Kiel Bay. Sedimentology of parts of central Europe Guidebook VIII. Intern. Sediment. Congress, 209–235.

Senthilkumar, K., C.A. Duda, D.L. Villeneuve, K. Kannan, J. Falandysz and J.P. Giesy, 1999. Butyltin compounds in sediment and fish from the Polish coast of the Baltic Sea. Environ. Sci. Pollut. Res. **6**, 200–206.

Sharma, P., and B.L.K. Somayajulu, 1979. Growth rates and composition of two ferromanganese nodules from the Central North Pacific, in La Genèse des Nodules de Manganèse, ed. C. Lalou (Èditions du Centre National de la Recherche Scientifique, Paris). Colloq. Int. C.N.R.S. **No. 289**, 281–288.

Sholkovitz, E., 1995. The aquatic chemistry of rare earth elements in rivers and estuaries. Aquat. Geochem. **1**, 1–34.

Shterenberg, L.Y., 1971. Some aspects of the genesis of iron-manganese concretions in the Gulf of Riga. Dokl. Akad. Nauk SSSR **201**, 252–255 (in Russian).

Shterenberg, L.Y., T.I. Gorshkova and E.M. Naktinas, 1968. Manganese carbonates in ferromanganese concretions in the Gulf of Riga. Lithol. Econ. Resourc. **4**, 63–69.

Siddiquie, H.N., D.R. Das Gupta, N.R. Sengupta, P.C. Shrivastava and T.K. Mallik, 1978. Manganese-iron nodules from the Indian Ocean. Ind. J. Mar. Sci. **7**, 239–253.

Siegel, H., W. Matthäus, R. Bruhn, M. Gerth, G. Nausch, T. Neumann and C. Pohl, 1998. The exceptional Oder flood in summer 1997 – Distribution patterns of the Oder discharge in the Pomeranian Bight. Dtsch Hydrogr. Z. **50**, 145–167.

Skei, J., K. Hylland, M.T. Schaanning, J.A. Berge, J.S. Gunnarsson, M. Sköld and D.Ø. Eriksen, 1996. Interactions between eutrophication and contaminants. I. Principles, experimental design and synthesis. Mar. Pollut. Bull. **33**, 64–70.

Skwarzec, B., 1995. Polonium, uranium and plutonium in ecosystem of the southern Baltic Sea. Rozprawy i monografie (Polska Akademia Nauk, Instytut Oceanologii w Sopocie), 183 pp. (in Polish with English summary).

Skwarzec, B., and R. Bojanowski, 1992. Distribution of plutonium in selected components of the Baltic ecosystem within the Polish economic zone. J. Environ. Radioactivity **15**, 249–263.

Skwarzec, B., R. Bojanowski and J. Bolałek, 1988. The determination of Cu, Pb, Cd and Zn in the southern Baltic water, suspension and sediments. Oceanologia **25**, 75–85.

Skwarzec, B., E. Holm, P. Roos and J. Pempkowiak, 1994. Nickel-63 in Baltic fish and sediments. Appl. Radiat. Isot. **45**, 609–611.

Sochan, J., 1992. Mineralogical and geochemical investigations of iron-manganese nodules from the southern Baltic Sea. Unpubl. M.Sc. thesis (Academy of Mining and Metallurgy, Krakow), 89 pp. (In Polish).

Sohlenius, G., J. Sternbeck, E. Andren and P. Westman, 1996. Holocene history of the Baltic Sea as recorded in a sediment core from the Gotland Deep. Mar. Geol. **134**, 183–201.

Sternbeck, J., and G. Sohlenius, 1997. Authigenic sulfide and carbonate mineral formation in Holocene sediments of the Baltic Sea. Chem. Geol. **135**, 55–73.

Sternbeck, J., G. Sohlenius and R.O. Hallberg, 2000. Sedimentary trace elements as proxies to depositional changes induced by a Holocene fresh-brackish water transition. Aquatic Geochem. **6**, 325–345.

Stigebrandt, A., and F. Wulff, 1987. A model for dynamics of nutrients and oxygen in the Baltic Proper. J. Mar. Res. **45**, 729–759.

Stockenberg, A., and R.W. Johnstone, 1997. Benthic denitrification in the Gulf of Bothnia. Estuar. Coast. Shelf Sci. **45**, 835–843.

Stoeppler M., 1991. Cadmium, in: Metals and their Compounds in the Environment, ed. E. Merian (VCH, Weinheim), 803–851.

Struck, U., M. Voss, B. von Bodungen and N. Mumm, 1998. Stable isotopes of nitrogen in fossil Cladoceran exoskeletons: Implications for nitrogen sources in the central Baltic Sea during the past Century. Naturwissenschaften **85**, 597–603.

Struck, U., K.-C. Emeis, M. Voss, C. Christiansen and H. Kunzendorf, 2000. Records of southern and central Baltic Sea eutrophication in $\delta^{13}C$ and $\delta^{15}N$ of sedimentary organic matter. Mar. Geol. **164**, 157–171.

Stuer-Lauridsen, F, and B. Dahl, 1995. Source of organotin at a marine water/sediment interface – A field study. Chemosphere **30**, 831–845.

Suess, E., and H. Erlenkeuser, 1975a. Anthropogenic metal and carbon input into Baltic Sea sediment basins, Geoscientific Studies and the Potential of the Natural Environment. Deutsche UNESCO-Kommission, Köln, pp. 203–214.

Suess, E., and H. Erlenkeuser, 1975b. History of metal pollution and carbon input in the Baltic Sea sediments. Meyniana **27**, 63–75.

Suess, E., and D. Djafari, 1977. Trace metal distribution in Baltic Sea ferromanganese concretions: inferences on accretion rates. Earth Planet. Sci. Letters **35**, 49–54.

Suplińska, M.M., 1995. Plutonium in sediments of the Baltic Sea in the period of 1991–1993. Nukleonika **40**, 33–44.

Suplińska, M., and D. Grzybowska, 2000. Distribution of ^{137}Cs plutonium in Baltic Sea bottom sediments, in: The Radiological Exposure of the Population of the European Community to Radioactivity in the Baltic Sea. Marina-Balt Project, ed. S.P. Nielsen. Proceedings of a Seminar held at Hasseludden Conference Centre, Stockholm, 9–11 June 1998, European Commission, Directorate-General Environment, EUR 19200 EN (European Communities, 2000, Belgium), pp. 421–432.

Szczepańska, T., and S. Uścinowicz, 1994. Geochemical atlas of sediments of the southern Baltic Sea 1 : 500 000, Polish Geological Institute, Warsaw.

Szefer, P., 1984. Uran i jego migracja w morskich i oceanicznych osadach dennych (Uranium and its migration in the marine bottom sediments). Stud. Mater. Oceanol. **44**, 171–212 (in Polish with an English abstract).

Szefer, P., 1987. Rozmieszczenie i migracja uranu w konkrecjach morskich (Distribution and migration of uranium in marine concretions). Stud. Mater. Oceanol. **51**, 225–268 (in Polish with an English abstract).

Szefer, P., 1990a. Interelemental relationships in organisms and bottom sediments of the southern Baltic. Sci. Total Environ. **95**, 119–130.

Szefer, P., 1990b. Mass-balance of metals and identification of their sources in both river and fallout fluxes near Gdansk Bay, Baltic Sea. Sci. Total Environ. **95**, 131–139.

Szefer, P., 1998. Distribution and behaviour of selected heavy metals in various components of the southern Baltic ecosystem, in: Geochemical Investigations of the Baltic Sea and Surrounding Areas, eds. P. Szefer, P. and G.P. Glasby (Elsevier Science Ltd, Great Britain). Appl. Geochem. (Spec. Issue) **13**, 287–292.

Szefer, P., and B. Skwarzec, 1988. Distribution and possible sources of some elements in the sediment cores of the southern Baltic. Mar. Chem. **23**, 109–129.

Szefer, P., and K. Szefer, 1990. Metals in molluscs and associated bottom sediments of the southern Baltic. Helgoländer Meeresunters. **44**, 411–424.

Szefer, P., and R. Kaliszan, 1993a. Inter-element relationships in sediment cores from southern Baltic. International Symposium on Emerging Technologies in Hazardous Waste Management V. (Industrial and Engineering Chemistry Division of the American Chemical Society). Atlanta, U.S.A., 27.09–29.09.1993.

Szefer, P., and R. Kaliszan, 1993b. Distribution of elements in sediment cores of the southern Baltic from the point of view of principal component analysis. SIMO-Mar. Pollut. **64**, 95–102.

Szefer, P., and A. Kusak, 2000. Distribution and relationships of trace metals in zoobenthos and associated sediments of the southern Baltic (in preparation).

Szefer P., J. Pempkowiak J. and R. Kaliszan, 1993a. Distribution of elements in surficial sediments of the southern Baltic determined by factor analysis. SIMO-Mar. Pollut. **64**, 85–93.

Szefer, P., H. Jankowska, M. Wolowicz, A. Kusak and J. Pempkowiak, 1993b. Vertical and horizontal distribution of heavy metals in bottom sediments from Gdańsk Bay, Baltic Sea. International Symposium on Emerging Technologies in Hazardous waste Management V. (Industrial and Engineering Chemistry Division of the American Chemical Society). Atlanta, U.S.A., 27.09–29.09.1993.

Szefer, P., G.P. Glasby, J. Pempkowiak and R. Kaliszan, 1995a. Extraction studies of heavy-metal pollutants in surficial sediments from the southern Baltic Sea off Poland. Chem. Geol. **120**, 111–126.

Szefer, P., A. Kusak, K. Szefer, H. Jankowska, M. Wolowicz and A.A. Ali, 1995b. Distribution of selected metals in sediment cores of Puck Bay, Baltic Sea. Mar. Pollut. Bull. **30**, 615–618.

Szefer, P., G.P. Glasby, K. Szefer, J. Pempkowiak and R. Kaliszan, 1996. Heavy-metal pollution in surficial marine sediments from the southern Baltic Sea off Poland. J. Environ. Sci. Health **31A**, 2723–2754.

Szefer, P., J. Geldon, A.A. Ali, F. Paez-Osuna, A.C. Ruiz-Fernandes and S.R. Guerrero Galvan, 1998a. Distribution and association of trace metals in soft tissue and byssus of *Mytella strigata* and other benthal organisms from Mazatlan harbour, mangrove lagoon of the northwest coast of Mexico. Environ. Intern. **24**, 359–374.

Szefer, P., G.P. Glasby, A. Kusak, K. Szefer, H. Jankowska, M. Wolowicz and A.A. Ali, 1998b. Evaluation of anthropogenic influx of metallic pollutants into Puck Bay, southern Baltic, in: Geochemical Investigations of the Baltic Sea and Surrounding Areas, eds. P. Szefer and G.P. Glasby (Elsevier Science Ltd, Great Britain) Applied Geochem. (Spec. Issue) **13**, 293–304.

Szefer, P., G.P. Glasby, H. Kunzendorf, E.A. Görlich, K. Latka, K. Ikuta and A.A. Ali, 1998c. The distribution of rare earth and other elements and the mineralogy of the iron oxyhydroxide phase in marine ferromanganese concretions from within Slupsk Furrow in the southern Baltic, in: Geochemical Investigations of the Baltic Sea and Surrounding Areas, eds. P. Szefer, P. and G.P. Glasby (Elsevier Science Ltd, Great Britain) Applied Geochem. (Spec. Issue) **13**, 305–312.

Szefer, P., G.P. Glasby, D. Stüben, A. Kusak, J. Geldon, Z. Berner, T. Neumann and J. Warzocha, 1999. Distribution of selected heavy metals and rare earth elements in surficial sediments from the Polish sector of the Vistula Lagoon. Chemosphere **39**, 2785–2798.

Szefer, P., G.P. Glasby, J. Geldon, R.M. Renner, E. Björn' J. Snell and W. Frech, 2000. Distribution and relationships between heavy metals and rare earth elements in surficial sediments from the southern Baltic (in preparation).

Szpunar, J., J. Falandysz, V.O. Schmitt and E. Obrębska, 1997. Butyltins in marine and freshwater sediments of Poland. Bull. Environ. Contamin.Toxicol. **58**, 859–864.

Takematsu, N., Y. Sato, S. Okabe and A. Usui, 1990. Uptake of selenium and other oxyanionic elements in marine ferromanganese concretions of different origins. Mar. Chem. **31**, 271–283.

Tervo, V., and L. Niemistö, 1989. Concentrations of trace metals, carbon, nitrogen and phosphorus in sediments from the northern parts of the Baltic Sea. ICES C.M./E:7, 14 p.

Tkalin, A.V., T.S. Lishavskaya and V.M. Shulkin, 1998. Radionuclides and trace metals in mussels and bottom sediments around Vladivostok, Russia. Mar. Pollut. Bull. **36**, 551–554.

Trefry, J.H., and B.J. Presley, 1976. Heavy metals in sediments from San Antonio Bay and northwest Gulf of Mexico. Environ. Geol. **1**, 283–294.

Tuomainen, K., E. Ilus and T.K. Taipale, 1986. Accumulation of certain long-lived radionuclides by litoral algae and bottom animals, in: Study of Radioactive materials in the Baltic Sea. (International Atomic Energy Agency, Vienna). Report (IAEA-TECDOC-362) of the Final Research Coordination Meeting on the Study of Radioactive Materials in the Baltic Sea organized by the IAEA and held in Helsinki, Finland 24–28 September, 1984, pp. 69–78.

Tuominen, L., A. Heinänen, J. Kuparinen and L.P. Nielsen, 1998. Spatial and temporal variability of denitrification in the sediments of the northern Baltic Proper. Mar. Ecol. Prog. Ser. **172**, 13–24.

Uchio, T., A. Usui, H. Hishida and T. Shoji, 1980. Terrestrial ferromanganese nodules fromOkinawa (Japan) and Bohol (Philippines) Islands. J. Fac. Enginer., Univ. Tokyo (B) **XXXV**, 493–512.

Usenkov, S.M., 1997. Contamination of harbor sediments in the eastern Gulf of Finland (Neva Bay), Baltic Sea. Environ. Geol. (Berlin) **32**, 274–280.

Usui, A., and N. Mita, 1995. Geochemistry and mineralogy of a modern buserite deposit from a hot spring in Hokkaido, Japan. Clays Clay Miner. **43**, 116–127.

Usui, A., and G.P. Glasby, 1998. Submarine hydrothermal manganese deposits in the Izu-Bonin-Mariana arc: An overview. Isl. Arc **7**, 422–431.

Usui, A., M. Yuasa, S. Yokota, M. Nohara, A. Nishimura and F. Murakami, 1986. Submarine hydrothermal manganese deposits from the Ogasawara (Bonin) Arc, off the Japan Islands. Mar. Geol. **73**, 311–322.

Usui, A., A. Nishimura, M. Tanahashi and S. Terashima, 1987. Local variability of manganese nodule facies on small abyssal hills of the Central Pacific Basin. Mar. Geol. **74**, 237–275.

Usui, A., T.A. Mellin, M. Nohara and M. Yuasa, 1989. Structural stability of marine 10 Å manganates from the Ogasawara (Bonin) Arc: Implications for low-temperature hydrothermal activity. Mar. Geol. **86**, 41–56.

Usui, A., A. Nishimura and N. Mita, 1993. Composition and growth history of surficial and buried manganese nodules in the Penrhyn Basin, Southwestern Pacific. Mar. Geol. **114**, 133–153.

Uścinowicz, S., and J. Zachowicz, 1993. Geochemical Atlas of the Vistula Lagoon 1 : 150 000. Polish Geological Institute, Warsaw. 11 pp. + 38 maps (in Polish and English).

Uścinowicz, S., J. Ebbing, C. Laban and J. Zachowicz, 1998. Recent muds of the Gulf of Gdańsk. Baltica **11**, 25–32.

Valette-Silver, N.J., S.B. Bricker and W. Salomons, 1993. Historical trends in contamination of estuarine and coastal sediments: and introduction to the dedicated issue. Estuaries **16**, 575–576.

Vallius, H., 1999a. Anthropogenically derived heavy metals in recent sediments of the Gulf of Finland, Baltic Sea. Chemosphere **38**, 945–962.

Vallius, H., 1999b. Heavy metal deposition and variation in sedimentation rate within a sedimentary basin in central Gulf of Finland. Chemosphere **38**, 1959–1972.

Vallius, H., and O. Lehto, 1998. The distribution of some heavy metals and arsenic in recent sediments from the eastern Gulf of Finland. Appl. Geochem. **13**, 369–377.

Vallius, H., and M. Leivuori, 1999. The distribution of heavy metals and arsenic in recent sediments of the Gulf of Finland. Boreal Environ.Res. **4**, 19–29.

Van Alsenoy, V., P. Bernard and R. van Grieken, 1993. Element concentrations and heavy-metal pollution in sediments and suspended matter from the Belgian North Sea and the Scheldt Estuary. Sci. Total Environ. **133**, 153–181.

Varentsov, I.M., 1973. Geochemical aspect of formation of ferromanganese ores in shelf regions of Recent seas. Acta Mineral.-Petrogr. Szeged **21**, 141–153.

Varentsov, I.M., 1980. Geochemistry of transition metals in the processes of ferromanganese ore formation in recent basins, in: Geology and Geochemistry of Manganese, eds. I.M. Varentsov and Gy. Grasselly (Hungarian Academy of Sciences, Budapest), **Vol. 1**, pp. 367–387.

Varentsov, I.M., and A.I. Blashchishin, 1974. Ferromanganese nodules from the Gulf of Finland. Acta Miner.-Petrogr., Szeged **21**, 303–304.

Varentsov, I.M., and A.I. Blashchishin, 1976. Iron and manganese concretions, in: Geologija Baltijskogo Morja (Geology of the Baltic Sea), eds. V. Gudelis and E. Emelyanov (Mokslas Publishers, Vilnius), pp. 307–348 (In Russian).

Varentsov, I.M., and G.B. Sokolova, 1977. Original variations of the mineral composition of the ferromanganese concretions and crusts, Baltic Sea. Concretions and Concretion Analysis (Nauka, Moscow) 180–187 (in Russian).

Varentsov, I., and A. Blashchishin, 1982. Ferromanganese concretions on the bottom of the Baltic Sea, in: Geology of the Baltic Sea, eds. V.K. Gudelis and E.M. Emelyanov (Wydawnictwo Geologiczne, Warsaw), 307–346 (in Polish).

Virkanen, J., 1998. Effect of urbanization on metal deposition in the Bay of Töölönlahti, southern Finland. Mar. Pollut. Bull. **36**, 729–738.

Virtanen, K., 1994. Geological control of iron and phosphorus precipitation in mines of Ruukk-Vihanti area., Central Finland. Geol. Surv. Finland Bull. **375**, 1–69.

Voß, M. and U. Struck, 1997. Stable nitrogen and carbon isotopes as indicator of eutrophication of the Oder river (Baltic Sea). Mar. Chem. **59**, 35–49.

Voutsinou-Taliadouri, F., 1981. Metal pollution in the Saronikos Gulf. Mar. Pollut. Bull. **12**, 163–168.

Voutsinou-Taliadouri, F., and J. Satsmadjis, 1982. Trace metals in the Pagassitikos Gulf, Greece. Estuar. Coast. Shelf Sci. **15**, 221–228.

Voutsinou-Taliadouri, F., and J. Satsmadjis, 1983. Metals in polluted sediments from the Thermaikos Gulf, Greece. Mar. Pollut. Bull. **14**, 234–236.

Voutsinou-Taliadouri, F., and E. Georgakopoulou-Grigoriadou, 1989. Heavy metal concentrations in surface sediments from Pagassitikos Gulf, Greece. Toxicol. Environ. Chem. **20–21**, 53–58.

Wangersky, P.J., 1962. Sedimentation in three carbonate cores. J. Geol. **70**, 364–375.

Wangersky, P.J., and O.I. Joensuu, 1967. The fractionation of carbonate deep-sea cores. J. Geol. **75**, 148–177.

Warren, L.J., 1981. Contamination of sediments by lead, zinc and cadmium, a review. Environ. Pollut. **2B**, 401–436.

Wedepohl, K.H., 1991. The composition of the upper earth's crust and the natural cycles of selected metals. Metals in natural raw materials. Natural resources, in: Metals and their Compounds in the Environment, ed. E. Merian (VCH, Weinheim), pp. 3–17.

Wenk, B., 1981. Comparison of geochemical investigations of iron-manganese concretions from the Baltic Sea, the Pacific and Atlantic Oceans. Final Rep. (Institut für Meereskunde, Warnemünde) in German.

Widerlund, A., 1994. Geochemistry of iron and arsenic of the Kalix River estuary, northern Bothnian Bay. 3rd Intern. Symp. Environ. Geochem., Kraków (abstract) 438–439.

Widerlund, A., 1996. Early diagenetic remobilization of copper in near-shore marine sediments: a quantitative pore-water model. Mar. Chem. **54**, 41–53.

Widerlund, A., and J. Ingri, 1995. Early diagenesis of arsenic in sediments of the Kalix River estuary, Northern Sweden. Chem. Geol. **125**, 185–196.

Widerlund, A., and J. Ingri, 1996. Redox cycling of iron and manganese in sediments of the Kalix River estuary, Northern Sweden. Aquat. Geochem. **2**, 185–201.

Williams, T.M., J.G. Rees and D. Setiapermana, 2000. Metals and trace organic compounds in sediments and waters of Jakarta Bay and the Pulau Seribu Complex, Indonesia. Mar. Pollut. Bull. **40**, 277–285.

Williamson, R.B., T.M. Hume and J. Mol-Krijnen, 1994. A comparison of the early diagenetic environment in intertidal sands and muds of the Manukau Harbour, New Zealand. Environ. Geol. **24**, 254–266.

Williamson, R.B., J. Mol-Krijnen and L. van Dam, 1995. Trace metal partitioning in bioturbated, contaminated, surficial sediments from Mangere Inlet, New Zealand. N.Z. J. Mar. Freshwater Res. **29**, 117–130.

Williamson, R.B., L.F. van Dam, R.G. Bell, M.O. Green and J.P. Kim, 1996. Heavy metal and suspended sediment fluxes from a contaminated, intertidal inlet (Manukau Harbour, New Zealand). Mar. Pollut. Bull. **32**, 812–822.

Windom, H.M., S.J. Schropp, F.D. Calder, J.D. Ryan, R.G. Smith, Jr., L.C. Burney, F.G. Lewis and C.H. Rowlinson, 1989. Natural trace metal concentrations in estuarine and coastal marine sediments of the Southeastern United States. Environ. Sci. Technol. **23**, 314–320.

Winterhalter, B., 1966. Iron-manganese concretions from the Gulf of Bothnia and the Gulf of Finland. Geoteknillisi Julka. No. 69, 1–77 (in Finnish, with an English abstract).

Winterhalter, B., 1972. On the geology of the Bothnian Sea. Geol. Surv. Fin. Bull. **258**, 66 pp.

Winterhalter, B., 1980. Ferromanganese concretions in the Baltic Sea, in: Geology and Geochemistry of Manganese, eds. I.M. Varentsov and Gy. Grasselly (Hungarian Academy of Sciences, Budapest), **Vol. 3**, pp. 227–254.

Winterhalter, B., and J. Siivola, 1967. An electron microprobe study of the distribution of iron, manganese, and phosphorus in concretions from the Gulf of Bothnia, northern Baltic Sea. Comptes Rendus de la Societéte Geologique de la Finlande **39**, 161–172.

Winterhalter, B., T. Flodén, H. Ignatius, S. Axberg and L. Niemistö, 1981. Geology of the Baltic Sea, in: The Baltic Sea, ed. A. Voipio (Elsevier Scientific Publishing Company), pp. 1–122.

Wivel, C., and S. Morup, 1981. Improved computational procedure for evaluation of overlapping hyperfine parameter distributions in Mössbauer spectra. J. Phys. **E14**, 605–610.

Wulff, F., L. Rahm and M. Rodriguez-Medina, 1994a. Long-term and regional variations of nutrients in the Baltic Sea; 1972–1991. Finnish Mar. Res. **262**, 35–50.

Wulff, F., M. Perttilä and L. Rahm, 1994b. Mass-balance calculations of nutrients and hydrochemical conditions in the Gulf of Bothnia, 1991. Aqua Fennica **24**, 121–140.

Wulff, F., M. Perttilä and L. Rahm, 1996. Monitoring, mass balance calculation of nutrients and the future of the Gulf of Bothnia. Ambio Spec. Rep. **8**, 28–35.

Zhamoida, V.A., and W.P. Butylin, 1992. Distribution of Fe-Mn nodules fields at the bottom of the Gulf of Finland, and the variety of nodules. Institut für Ostseeforschung Warnemünde Mar. Sci. Rep. (abstract) **4**, 169.

Zhamoida, V.A., and W.P. Butylin, 1993. Differentiation of the ferromanganese nodules from the eastern part of the Gulf of Finland by their chemical composition. Third Mar. Geol. Conference, Sopot, 21–24 September, 1993, (abstract), p. 63.

Zhamoida, V.A., W.P. Butylin, G.P. Glasby and I.A. Popova, 1996. The nature of ferromanganese concretions from the eastern Gulf of Finland, Baltic Sea. Mar. Georesourc. Geotechnol. **14**, 161–176.

Zöllmer, V., and G. Irion, 1993. Clay mineral and heavy metal distribution in the northeastern North Sea. Mar. Geol. **111**, 223–230.

Zwolsman, J.J.G., G.W. Berger and G.T.M. VanEck, 1993. Sediment accumulation rates, historical input, post-depositional mobility and retention of major elements and trace metals in salt marsh sediments of the Scheldt estuary SW Netherlands. Mar. Chem. **44**, 73–94.

Chapter 5

Bioavailability and Biomagnification of Chemical Elements and Radionuclides

A. CONCENTRATION AND DISCRIMINATION OF CHEMICAL ELEMENTS

(i) Introduction

It is well known that many species of marine organisms accumulate trace elements, often to surprisingly high values (Phillips, 1980, Bryan et al., 1985, Phillips and Rainbow, 1993). Therefore, it is of interest to investigate the ability of marine fauna and flora to bioaccumulate trace elements from their surroundings, i.e. from seawater or surface sediments as substrata. The quantitative measure of trace element affinity to biota and their biomagnification along the successive trophic levels of the food chain are the concentration and transfer factors, respectively. Bioaccumulation consists of the net retention of chemical substance taken from water and/or diet by an organism over time.

Vertical fluxes of $^{239+240}Pu$ and ^{241}Am and temporal changes in their inventories in the northwestern Mediterranean Sea have been examined through high-resolution water column sampling coupled with direct measurements of the vertical flux of particle-bound transuranics using time-series sediment traps (Fowler et al., 2000). A kinetic model for trace metals as well as uptake routes and biokinetics of plutonium and radiocobalt have been developed to describe metal accumulation in marine zooplankton-copepods and the asteroid *Asterias rubens*, respectively (Guary et al., 1982; Warnau et al., 1999; Fisher et al., 2000).

Analysis of seawater is a most difficult method since the levels of dissolved elements, even in polluted areas, are low; hence probability of sample contamination is very high. Therefore for routine monitoring survey the analysis of marine organisms, usually enriched in heavy metals appears to be the best option. The chemical composition of seawater plays an important role in the bioaccumulation and discrimination of trace elements in marine biota as whole life integrators. Mean values of concentration factor (CF) of a numerous chemical elements for particular marine organisms are listed in Technical Reports Series (1985). The CFs of radionuclides, i.e. ^{90}Sr, ^{137}Cs and $^{239+240}$Pu were computed for different groups of organisms from the Kara and Barents Seas, i.e. brown macroalgae, bivalve molluscs, crustaceans, echinoderms, fish muscle, seabirds muscle and liver as well as seal muscle and liver (Fisher et al., 1999). The analysis showed that coefficients for polar and temperate waters are comparable except 10-fold greater CF for $^{239+240}$Pu in Arctic brown macroalgae, 10-fold smaller partition coefficients (K_d) values for ^{90}Sr in Kara Sea sediment than in typical temperate coastal sediments, and 100-fold greater K_d values for ^{106}Ru in Kara sediment.

The response of various marine organisms, particularly nekton, to changes in trace elements concentrations in the ambient water is relatively low (Fowler, 1982, 1985). Variations in several metal levels in seawater can sometimes be followed by changes in their concentrations in seaweed after a time-lag which may be from one month to three years (Eide et al., 1980). The relationship between the concentration of given metal in the organism and that in the overlying water is dynamic. The concentration and discrimination processes of trace elements in marine biota are related both to the uptake and excretion rates, which are affected by many factors, i.e. size (age) of the organism, physiological status of specimen when caught, residence time of migrant organisms, temperature, light, salinity, the concentration of the element and its physical and chemical forms in seawater, metal interactions and changes in its concentrations in coastal water over short periods of time (Bryan, 1980, 1983; Fowler, 1982; Bryan et al., 1985; Technical Reports Series, 1985).

(ii) Operational Definitions

In order to obtain a better understanding of the relationships between metals in the various phases studied, the concentration factor, CF, and discrimination factor, DF, in the representative biota in the southern Baltic ecosystem were computed according to the following formulae:

$$CF = C_x/C_s$$

where C_x and C_s are the average concentrations of metal in the organism and in seawater or surface sediment, respectively.

To estimate the degree of discrimination between metals in the organism relative to levels of their mobile forms in the ambient water or associated sediments, the discrimination factor, DF, is computed:

$$DF = [(C_a/C_b)_x] / [(C_a/C_b)_s]$$

where $(C_a/C_b)_x$ and $(C_a/C_b)_s$ are ratios of average concentrations of metals in the organism and in the ambient seawater or associated sediment, respectively.

Knowledge of the CF values permits recognition of the relative ability of biota representatives to adsorb and/or take up several trace elements from the medium in which they live. DFs inform us about the affinity of organism to one element compared to the other reference element. If CF values amount to approximate unity this indicates that there is no enrichment of each element in the organism in relation to the reference material as substrata (ambient water, associated sediment, host organ). When DF of ca. 1 it means that there is no discrimination in preference of one element vs. another.

CF and DF values may be estimated on the basis of experimentally derived data, but they are sometimes not comparable with those based on environmental data since short-term laboratory experiments are of insufficient duration for steady state to be reached (Bryan, 1983). It is potentially misleading to use laboratory derived data uncritically and, wherever possible, environmentally derived data have been used in this respect (Technical Reports Series, 1985).

Field-derived concentration factors, CF, for different trace elements in marine biota (relative to ambient water and/or associated sediments) have been computed by several authors (Goldberg, 1965; Cherry and Shannon, 1974; Bryan, 1983; Knauss and Ku, 1983; Spaargaren and Ceccaldi, 1984; Spaargaren, 1985; Szefer et al., 1998a, 1999). For modelling purposes the values are usually represented in terms of a metal concentration relative to that of the overlaying water. It should be stressed that, except for plankton and algae, both CF and DF values do not imply that given element in the biota is concentrated by direct uptake from the surrounding water (Technical Reports Series, 1985). Simply the values relate to the concentration of each of the elements in the organism, which may have been derived by uptake from water, suspended matter, bottom sediments and from food.

CF and/or DF values have been calculated for particulates, seaweeds, plankton, soft tissue and shells of molluscs, whole body of crustaceans, polychaete and other zoobenthal organisms, whole fish and their muscle from the southern Baltic (Bojanowski, 1972; Brzezińska et al., 1984b; Szefer et al., 1985; Szefer, 1991, 1998; Szefer and Szefer, 1991; Szefer and Kusak, 2000). Moreover, CFs and DFs were computed for trace elements in parasites hosted in lung of marine mammal (Szefer et al., 1998b).

(iii) Interphase relations of Chemical Elements and Radionuclides

PLANKTON

Chemical Elements

The average CF values for several elements in mesozooplankton from the Słupsk Furrow are arranged according to the following order: Zn $\geq$ Pb > Cd > Cu, while for mesozooplankton inhabiting the Gdańsk Deep the elements can be ordered in the following sequence: Pb > Cd > Cu > Zn (Szefer et al., 1985). The average DFs indicate spatial variations of trace metals actually either adsorbed or taken in. For instance, the mesozooplankton collected at the Słupsk Furrow exhibited a stronger ability to adsorb and/or take up Zn in comparison with Cd, Cu and Pb, while for samples from the Gdańsk Basin an opposite trend was observed. These variations may be a result of the differences in the rate of supply of metal ions to a given area and/or differences in the species structure of mesozooplankton inhabiting the Słupsk Furrow and the Gdańsk Basin. The species composition is closely associated with the conditions of the environment and the ecological requirements of the mesozooplankton components (see Chapt. 3B).

An estimation of the degree of trace element association with plankton from the southern Baltic was reported by Szefer (1991). According to Fowler (1982), the biological transport of some trace metals by plankton is generally more important than that by physical processes. Changes in the amount of biomass may be expected to influence the chemical composition of seawater. Therefore, it is of interest to determine the relation between the elemental composition of Baltic water and that of plankton from given region. Knowledge of parameters such as the annual elemental removal rate of plankton and the degree of association of each metal with plankton renders it possible to estimate the affinity for metals of this component of the southern Baltic biocenosis. The concentrations of metals in mixed plankton and phytoplankton generally approximate those in zooplankton of the southern Baltic. Primary production in this region amounted to ~80 g C m^{-2} $year^{-1}$ (Łomniewski et al., 1975; Renk, 1978), and phytoplankton contained ~ 30% C_{org} (Łomniewski et al., 1975); hence, the annual removal rate (P) of this biological material was (Szefer, 1991):

$$P = (80 \text{ g C m}^{-2} \text{ year}^{-1} \times 100\%)/(30\% \text{ C}_{org}) = 267 \text{ g m}^{-2} \text{ year}^{-1}$$

Assuming that the annual primary production rate equals the removal rate (Knauss and Ku, 1983) an estimate can be made for the planktonic flux (F) of each metal from the photic layer to bottom sediment down a 100-m water column with a base area of 1 m^2. The following formula proposed by Knauss and Ku (1983) was used:

$$F \text{ (mg m}^{-2} \text{ year}^{-1}) = C_p(\text{mg g}^{-1} \text{ dry weight}) \times P(\text{g m}^{-2} \text{ year}^{-1})$$

where C_p is the concentration of metal in plankton, and P the annual primary production rate.

Bearing in mind that the total biomass of phyto- and zooplankton in a 100-m water column with a base area of 1 m^2 amounted to 54 and 27 g m^{-2}, respectively (Łomniewski et al., 1975; Mańkowski, 1978a; Pliński and Florczyk, 1987) and to quantify the biomass transport, the degree of association, DA, of the metals analysed with Baltic phyto- and zooplankton was estimated according to the formula:

$$DA(\%) = [C_B \text{ (mg m}^{-2})/C_S(\text{mg m}^{-2})] \times 100\%$$

where C_B and C_S are the total metal contents of plankton and water down a column 100 m high and 1 m^2 base area.

Table 5.1 lists values for C_S, F, C_B and DA with average metal concentrations in zooplankton and mixed plankton and phytoplankton of the southern Baltic. The CF values (Tables 5.2) show the highest affinity of plankton for Fe, Pb and Cu and the lowest affinity for U; hence plankton, to a significant extent, participate in the transfer of Fe, Pb and Cu down the water column from surficial photic layers to bottom. Minimum DA values for U may be attributed to the limited bioavailability of the complex anion $UO_2(CO_3)_3^{4-}$ to Baltic organisms (Szefer, 1991).

Radionuclides

According to Skwarzec and Bojanowski (1988) two of four analysed Baltic phytoplankton samples were monocultures contained > 95% of the total population of the centric diatom *Coscinodiscus granii* (Gough) and the chrysophycean *Dinobryon balticum* (Schütt) while the other two samples contained predominantly blue-green alga. The mean CF values of ^{210}Po in phytoplankton from southern Baltic attained a mean level of 4.9×10^3; this value is slightly lower than those reported for mixed-species phytoplankton samples from other areas (Heyraud and Cherry, 1979). This difference can be a result of interspecies and spatial variations in concentrations of the trace elements. The rate and degree to which ^{210}Po is adsorbed from the ambient water depend on the physiological condition of the plant cells as well as the hydrochemical conditions in the medium of the alga (Skwarzec and Bojanowski, 1988).

The mean CFs of ^{210}Po estimated for macro- and mesozooplankton from southern Baltic were 1.83×10^4 and 4.2×10^4, respectively (Skwarzec and Bojanowski, 1988). The analysis of the species composition of mesozooplankton indicated that observed spatial variations in the ^{210}Po content may have been caused by the blooming of blue-green alga observed at the Gdańsk Basin (Skwarzec and Bojanowski, 1988). The mean CF for plankton from the Pomeranian Bay was estimated to be 23×10^3 (Bojanowski et al., 1995).

The concentrations factors increased according to the following order: phytoplankton < macrozooplankton < mesozooplankton. This means that ^{210}Po is absorbed more significantly by zooplankton as consumers than phytoplankton – its producers.

There are a few available data on CFs of ^{137}Cs in Baltic plankton. According to Bojanowski et al. (1995) plankton from the Pomeranian Bay exhibit a weak affinity to this radionuclide resulting in mean CF value amounted to 16.

PHYTO- AND ZOOBENTHOS

Chemical Elements

CF and DF values showed a difference between trace elements either adsorbed or those biologically incorporated in Baltic benthos, but it is impossible to

TABLE 5.1.

Relationships between the chemical composition of plankton and Baltic water. After Szefer (1991)

Metal	C_p (μg g^{-1}dry wt)		C_s ($mg\ m^{-2}$)	F ($mg\ m^{-2}\ year^{-1}$)	C_B ($mg\ m^{-2}$)		DA (%)	
	Zooplankton	Phytoplankton and mixed plankton		Phytoplankton	Zooplankton	Phytoplankton	Zooplankton	Phytoplankton
Fe	1600	2300[a]	750	610	43	124	5.7	16.5
Mn	28	18[b]	1000	4.8	0.76	0.97	0.08	0.1
Ni	8	13[b]	70	3.5	0.22	0.70	0.31	1.0
Ph	14	22[c]	12	5.9	0.38	1.2	3.2	10
Cu	28	157[c]	51	42	0.76	8.5	1.5	17
Co	0.38	1.3[a]	7	0.35	0.01	0.07	0.14	1.0
Zn	160 (800)[d]	514[a]	1100	140	4.3 (22)[d]	28	0.4 (2)[d]	2.5
Cd	1.5	1.6[a]	3	0.43	0.04	0.09	1.3	3.0
U	0.14	0.63	75	0.17	0.004	0.03	0.005	0.04
Th	0.19	0.67	6	0.18	0.005	0.04	0.08	0.7

[a]Concentration of metal in phytoplankton (Brzezińska et al., 1984b).
[b]Concentration of metal in mixed plankton (Boström et al., 1981).
[c]Concentration of metal in phytoplankton (Brügmann, 1978).
[d]Maximum values for Zn were included in the calculation of the mean.

TABLE 5.2.

Concentration factors (× 10^3) expressed on a wet weight basis for metals in various representative organisms and particulates from the southern Baltic[a]. After Szefer (1991)

Biocenosis component	Water (%)	Fe	Zn	Mn	Cu	Pb	Cd	Co	Ni	U	Th
Macroalgae											
Chlorophyta	90.0	44	1.1	7.5	1.2	12	1.4	1.1	0.5	0.03	0.30
Phaeophyta	86.0	43	2.6	9.8	1.8	12	2.4	0.8	1.0	0.05	0.47
Spermatophyta	87.5	21	0.3	20	0.8	24	2.0	1.1	0.7	0.04	0.57[a]
Average	87.8	36	1.4	12	1.3	16	1.9	1.0	0.7	0.04	0.45
Zooplankton dominant species											
Pseudocalanus elongatus, Oithona similis, Evadne nordmanni	82.6	38	3.0 13[b]	0.5	10	15 25[b]	7.0	0.7	2.0	0.03	0.60
Molluscs, soft tissue											
Mytilus edulis	86.6	23	4.0	1.1	3.5	7.9	24	1.6	1.8	0.04	0.35
Cardium glaucum	82.4	18	3.4	0.6	8.5	13	7.6	4.5	15	0.07	0.39
Mya arenaria	83.6	100	2.0	10	7.9	17	9.4	4.7	2.5	0.03	1.30
Macoma balthica	77.1	14	11	0.3	28	9	2.6	3.6	1.3	0.11	0.35
Average	82.4	39	5.1	3.0	12	12	11	3.6	5.2	0.06	0.60
Crustacea, whole											
Saduria entomon	75.7	170	1.5	5.5	23	7.2	2.5	4.0	2.4	0.01	1.2
Fish[c]											
Gadus morhua	80.5	0.5 (1.0)	0.3 (0.4)	0.02 (0.08)	0.36 (0.7)	0.35 (2.8)	0.06 (2.1)	0.07 (0.6)	0.39 (0.1)	0.0004	0.006
Clupea harengus	74.0	1.4 (3.3)	0.7 (2.2)	0.03 (0.15)	0.78 (1.9)	0.53 (3.4)	0.37 (1.4)	0.09 (1.9)	0.56 (1.0)	0.0004	0.006
Sprattus sprattus	67.1	1.8 (4.0)	1.2 (2.7)	0.04 (0.17)	0.80 (2.0)	0.53 (4.4)	0.54 (6.4)	0.09 (1.4)	0.47 (0.9)	0.0006	0.010
Ammodytes tobianus		(1.8)	(0.7)	(0.05)	(1.7)	(5.0)	(2.1)	(0.8)	(0.5)		
Cyclopterus lumpus		(1.7)	(1.0)	(0.05)	(1.6)	(5.2)	(2.9)	(1.1)	(0.4)		
Hyperoplus lanceolatus										(0.001)	(0.02)
Average	73.9	1.2 (2.4)	0.7 (1.4)	0.03 (0.10)	0.65 (1.6)	0.47 (4.2)	0.32 (3.0)	0.09 (1.1)	0.47 (0.6)	0.0005	0.007
Particulates[d]	80.0	115	22		300	140	86	15		< 0.25	47

[a]Baltic water data for Zn, Cd, Pb, Cu, U and Th were taken from Szefer (1977), Bojanowski and Szefer (1979) and Szefer et al. (1985); other Baltic water data were derived from Bojanowski (1972) and Brügmann (1981, 1982).

[b]Maximum values were included in the calculation of the mean.

[c]Calculated for fish muscle; values for whole fish in parentheses; values for whole *Gadus morhua* were estimated from data presented by Szefer et al. (1990b).

[d]Estimated, assuming 80% water, on the basis of data of Brzezińska et al. (1984a), except for U and Th (Szefer, unpublished). Particulates taken from various water layers (seston).

distinguish between these two routes. However the factors, especially DF values, provide information in the selective and specific bioavailability of trace elements to phyto- and zoobenthos relative to concentrations in surrounding water (for algae) and/or associated sediments (for zoobenthos and rooted plants Spermatophyta). CFs and DFs showed that the Baltic organisms had the greatest affinity for Fe and Pb, and the lowest for U (Szefer, 1987; Szefer and Wenne, 1987; Szefer, 1991). The CF values for Cd, Cu, Ni and Zn, and those for Co and Mn were generally within the same order of magnitude. Significant interspecies variations in CFs, especially for soft tissue of bivalvia, were detected. Elements such as Fe, Mn and Th were accumulated to the greatest extent by *Mya arenaria*, Cu, Zn, Ag and U by *Macoma balthica*, and Cd and Ni by *Mytilus edulis* and *Cerastoderma glaucum*, respectively (Szefer, 1991; Szefer and Kusak, 2000)). Brzezińska et al. (1984b) found also that *M. balthica* from the southern Baltic accumulated Cu and Zn to a greater extent and Cd to a lesser extent as compared to *M. edulis*.

Baltic values are comparable to data reported by Bryan (1980) for bivalvia from East Looe estuary: *M. balthica* was characterised by a maximum affinity to Cu and Zn, *C. edule* to Ni and *M. edulis* to Cd. These CF values are also in an agreement with those compiled for world-wide marine biota (Technical Reports Series, 1985). Such similarity may be explained by biotic and abiotic factors modifying metal uptake in spite of natural environmental fluctuations. For instance, both metal speciation and metabolism of the element in the organism have an important influence on the concentrations ultimately achieved; both factors are influenced by environmental parameters such as salinity, temperature, pH and organic ligands in water (Phillips, 1980; Bryan, 1983). If the organism is able to increase the efficiency of excretion in response to increased absorption, then the concentration in the body does not increase in proportion to environmental availability (Bryan, 1980). It means that organisms biologically regulate absorbed metals over a wide range of concentrations in the overlying water. However, a few exceptions are observed, namely for Pb in macroalgae and soft tissue of bivalvia of the southern Baltic (Szefer, 1991). This discrepancy may result from specific interspecies differences and local environmental conditions. DF values amounted to approximately unity were found for some metal pairs, e.g. for ratio of Ni to Zn in macroalgae, soft tissue of bivalvia, whole body of *Saduria entomon* and muscle of fish. The Ni/Zn ratio computed for these organisms is comparable with the same ratio estimated for Baltic water; it means that Ni and Zn are concentrated in similar proportions to those occurred in the surrounding water. These phyto- and zoobenthal organisms from the Gulf of Gdańsk exhibit insignificant discrimination for each of these two metals. All elements with DF values of ~ 1 are not discriminated in Baltic biota (Szefer, 1991). Significant inter-species differences in CFs and DFs, especially for soft tissue of molluscs and macroalgae, were found. For instance, *M. edulis* is characterised by a greater affinity for Cd than for Pb and may therefore be less able to regulate the uptake of Cd compared with Pb from the ambient water and/or surficial sediment (Szefer and Szefer, 1991). By contrast, eel grass, *Zostera marina*, and brown algae *Pilayella littoralis* were less able to regulate Pb uptake from their surroundings. It would therefore appear that *M. edulis* has the greatest potential as a monitor of Cd contamination in the southern Baltic ecosystem (Szefer and Szefer, 1991).

According to Bojanowski (1972), DF values suggested that Baltic seaweeds are characterised by the highest ability to accumulate Fe from water compared to the other elements and reveal lower affinity to Zn than to Pb and Mn. They absorb Zn together with Cd, Co and Ni in proportion approximately equal to those occurring in the Baltic water. Average discrimination factors, DF, for Cu were surprisingly low, similarly to the CF values (Bojanowski, 1972; Szefer and Skwarzec, 1988). It can be assumed that availability of Cu for plants is limited by competitive processes of complexing this metal by organic substances dissolved in water which results in a decrease of the effective concentration of Cu. The mean ratio of Ni/Co concentration for the Baltic seaweeds is very similar to the ratio calculated for seawater which results in the $DF_{Ni/Co}$ being ~ 1 (Szefer, 1991). This means that the plants do not reveal a significant selectivity either to these two metals.

Relatively low CFs and DFs were obtained for Cu in spite of high affinity of this element for organic ligands (Mantoura et al., 1978; Bryan, 1983). It is reported (Bryan, 1983) that competition from ligands present in seawater diminishes the available levels of Cu. Higher CF and DF values for Cu were observed in *M. balthica* and the crustacean *S. entomon* (Szefer, 1991). Inter-tissue analysis of the latter species indicated a maximum concentrations of this metal in the hepatopancreas (Szefer et al., 1990a). It is reported (Kulikova et al., 1985) that in some zoobenthal organisms, e.g. in *S. entomon*, Cu is bound to haemocyanin, the respiratory blue pigment responsible for the transport of oxygen in the blood.

Radionuclides

The CF's of ^{137}Cs for *F. vesiculosus* reached a maximum value of 2.0×10^3 and were distinctly higher in the eastern Gulf of Finland as compared to the Archipelago Sea, the Gulf of Bothnia, Bothnian Sea and the Gulf of Finland (Tuomainen et al., 1986; Ilus et al., 1988). It has been found that CF values of ^{137}Cs in *F. vesiculosus* were negatively correlated with water salinity (Ilus et al., 1988), The brown alga from the Finnish coastal area of the Baltic Sea was characterised by similar values of CF (dry) for ^{137}Cs and ^{90}Sr amounting to 0.67×10^3 and 0.84×10^3, respectively (Tuomainen et al., 1986). According to Holm (1995) before the Chernobyl accident the CF_{Pu}s (dry)/water for Baltic *F. vesiculosus* were 4.2×10^4 on the Swedish east coast and 1.25×10^4 on the west coast. The CF values (dry)/water for *F. vesiculosus* from the Finnish coastal regions ranged from 1.9×10^4 to 10×10^4 (average 3.9×10^4) (Tuomainen et al., 1986). Like for ^{137}Cs, the CF's for $^{239,240}Pu$ increase at lower salinity but not as pronounced a manner (Holm et al., 1995). The CF dry values of ^{241}Am (0.92×10^4 to 4.4×10^4) were somewhat lower than those of $^{239,240}Pu$ in *F. vesiculosus* from the Finnish coastal areas of the Baltic Sea (Tuomainen et al., 1986). Among several species of aquatic plants and seaweeds from the Baltic Sea, only tang, especially *Fucus*, showed higher CF values corresponded to relatively small concentration ranges. According to Weiss and Moldenhaver (1986) in some Baltic seaweeds of 1981, besides ^{137}Cs and ^{90}Sr, also the radionuclides such as ^{95}Zr and ^{95}Nb, presumably originating from the Chinese nuclear weapon test of October 1980, were detected.

According to Stepnowski and Skwarzec (2000) the CF value of ^{210}Po for the total soft tissue of *Mytilus trossulus* from the Baltic Sea was 2.4×10^4 and exceeded the recommended factor for mollusc (1×10^4) for the total soft parts (Technical Reports Series, 1985). Whole body, shell and flesh of *M. edulis* from the Finnish coastal waters exhibited different bioaccumulative abilities in respect to ^{137}Cs and ^{90}Sr (Tuomainen et al., 1986). The average CF's (dry) of ^{137}Cs for whole body, shell and flesh amounted to < 32, < 7.6 and 130, respectively. The CF's of ^{90}Sr were as follows: 0.092×10^4 (whole) and 0.11×10^4 (shell). The average CF values (dry) of $^{239,240}Pu$ calculated for whole, shells and flesh of *M. edulis* from the Finnish coastal region were 0.5×10^4, 15×10^4 and 27×10^4, respectively. The CF's of ^{90}Sr for Baltic mussel shells changed with different water salinity, due to chemical similarity of Sr to Ca, the former is the more accumulated in mussel shell, the less the Ca concentration in the ambient water. Sr is incorporated into the shell instead of Ca. For example, the CF's for ^{90}Sr in shells of mussels in the Greifswald Bodden are within ca. twice as high as in the same species inhabited the offshore, more saline waters of the Baltic Sea (Tuomainen et al., 1986). The CF value of ^{241}Am for whole body of the same species was estimated as 19×10^4 (Tuomainen et al., 1986).

The mean CF values for $^{239,240}Pu$ for seaweeds, molluscs and crustaceans from the southern Baltic were the same order of magnitude, but highly ranged from 0.06×10^4 to 2.71×10^4 (Skwarzec and Bojanowski, 1992; Skwarzec, 1995). The CF values of ^{238}U for phyto- and zoobenthos ranged from 30–55 (Skwarzec, 1995). Since the average $^{234}U/^{238}U$ activity ratio (1.12–1.15) in these Baltic organisms is similar to that in sea water (1.17) it can be supposed that the dissolved species of the radionuclide in the ambient water are the predominant its source in Baltic zoobenthos (Skwarzec, 1995, 1997). This conclusion highly substantiates the suitableness of calculate relative bioaccumulation of U isotope in respect to its concentration in seawater as a substrata. According to several authors (Szefer, 1987; Szefer and Wenne, 1987; Szefer et al., 1990b) the mean CF values were higher for Th than U in Baltic seaweed, soft tissue of mollusc and especially for whole body of crustacean. Minimum CFs obtained for U in Baltic biota may be explained by aquatic chemistry of U. In the sea, U exists primarily as the tricarbonato-uranylate anion $UO_2(CO_3)_3^{4-}$ (Li, 1981), which is not readily available for the formation of organic complexes. Therefore, recovery of such anionic species by marine organisms is expected to be significantly hindered. According to

Nakajima et al. (1979), the negatively charged complexes $UO_2(CO_3)_2^{2-}$ and $UO_2(CO_3)_3^{4-}$ are not taken up by the alga *Chlorella vulgaris*.

The DF values showed that Baltic seaweeds and mollusc soft tissue preferentially incorporate U and Th over Ca in their tissues, whilst for mollusc shells an inverse tendency was observed. Figures from 5.1 to 5.4 illustrate the significant relationships between the logarithm of the DF of U and Th in respect to Ca (y-axis) and the logarithm of the concentrations of Ca (x-axis). The regression lines show that the log $DF_{U/Ca}$ and $DF_{Th/Ca}$ increase with decreasing the log C_{Ca} either for the Baltic seaweed or molluscs, i.e. affinity of the benthos to U and Th grows with the decreasing of their degree of calcification (Szefer, 1987; Szefer and Wenne, 1987).

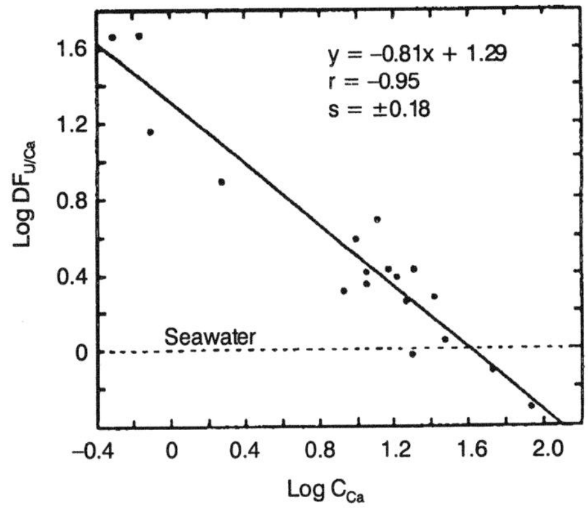

Fig. 5.1. The relationship between the log $DF_{U/Ca}$ and log C_{Ca} for the Baltic seaweeds. After Szefer (1987).

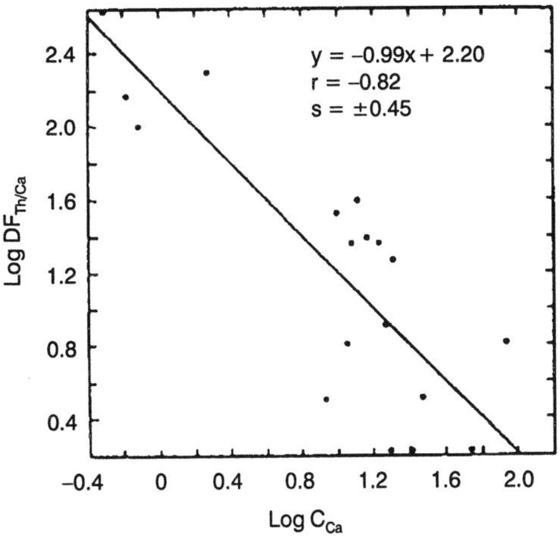

Fig. 5.2. The relationship between the log $DF_{Th/Ca}$ and log C_{Ca} for the Baltic seaweeds. After Szefer (1987).

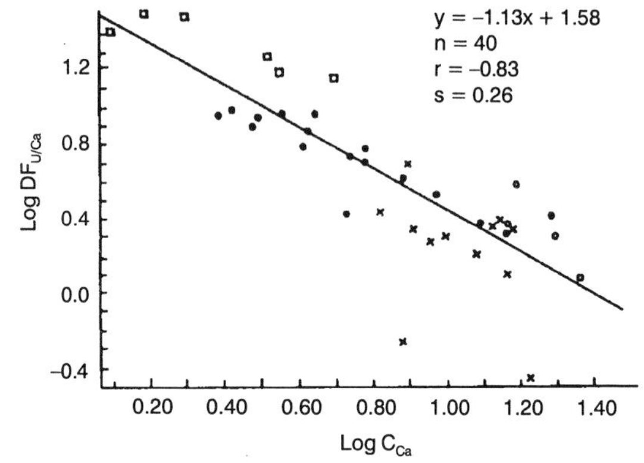

Fig. 5.3. The relationship between log $DF_{U/Ca}$ and log C_{Ca} for the soft tissue of *Mytilus edulis* (●); *Mya arenaria* (x); *Macoma balthica* (□) and *Cardium glaucum* (O). After Szefer and Wenne (1987).

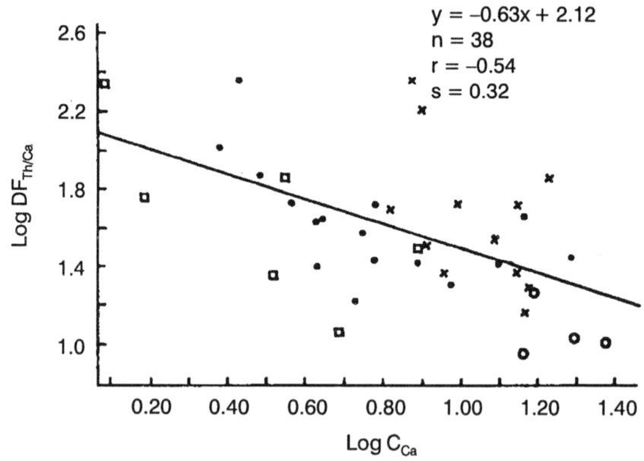

Fig. 5.4. The relationship between the log $DF_{Th/Ca}$ and log C_{Ca} for the soft tissue of *Mytilus edulis* (●); *Mya arenaria* (x); *Macoma balthica* (□) and *Cardium glaucum* (O). After Szefer and Wenne (1987).

FISH

Chemical elements

As can be seen in Table 5.2, CFs of trace elements for fish muscle are generally significantly smaller as compared to those calculated for other representatives of Baltic fauna and flora. It can be explained by more efficient regulation mechanism (homeostasis) in fish, incomplete absorption of metals across the gut, rapid excretion and dilution in muscle. The mean DFs for Fe in respect to Cd, Co, Cu, Ni, Pb, Th, U and Zn are higher than unity, indicating that Fe is more extensively bioaccumulated in fish muscle (Szefer, 1991).

Weiss and Moldenhaver (1986) reported significant differences in CF values (fish flesh or whole fish/water) for chemical elements and radionuclides (Co, Cs,

Fe, Mn, Sr, Zn) in fish from western part of the Baltic Sea. Due to the high CF variability, the authors recommended the consideration of site-specific factors for predicting the radiation exposure resulting from liquid discharges of NPP's along the water-fish-man pathway.

Radionuclides

Weiss and Moldenhaver (1986) computed the water/fish CF's (fish flesh and whole fish) of ^{137}Cs and ^{90}Sr for commercial fish from western part of the Baltic Sea. An inverse correlation between potassium (K) concentrations in water and CF values of ^{137}Cs was observed. The data also demonstrate that the CF's of ^{137}Cs for freshwater fish differ, depending on their trophic level. The following relationship was obtained (Weiss and Moldenhaver, 1986):

$$CF_{(137\text{-}Cs)} \text{ for piscivorous fish} = 4070 \times K\ (\text{mg dm}^{-3})^{-0.62},\ r = -0.995;$$

$$CF_{(137\text{-}Cs)} \text{ for non-piscivorous fish} = 2135 \times K\ (\text{mg dm}^{-3})^{-0.72},\ r = -0.994.$$

The CF's for some fish species, e.g. eel, were within the values of piscivorous and non-piscivorous fish.

The overall average $CF_{(137\text{-}Cs)}$ is expressed as:

$$CF_{(137\text{-}Cs)} = 3070 \times K\ (\text{mg dm}^{-3})^{-0.64},\ r = -0.996.$$

The CF values of $^{239+240}Pu$ for Baltic whole fish depend on species of this nekton and ranged from 100–650 (Skwarzec, 1995, 1997; Skwarzec et al., 2000). It is interesting to note that maximum CF values were reported for cod's intestine and herring's gills amounting to 2.02×10^4 and 1.0×10^4, respectively (Skwarzec et al., 2000).

According to Bojanowski et al. (1995) muscles of fish from the Pomeranian Bay were characterised by significantly higher values of CF for ^{210}Po (4.3×10^3) than for ^{137}Cs (0.166×10^3).

CFs of ^{137}Cs and ^{90}Sr for Baltic cod, herring, place, flounder and the fresh water pike have been graphically presented and extensively discussed in HELCOM (1995). For example, CF values of ^{90}Sr for edible parts of Baltic herring from the northern parts were an order of magnitude higher (ca. 4) as compared with those for the western and southern parts of the Baltic Sea (HELCOM, 1995).

The mean CF_U and CF_{Th} values for fish muscle were 0.5 and 10, respectively; the Baltic values are the same order of magnitude as those reported for several species of marine fish from other remote areas (Ichikawa and Ohno, 1981; Technical Reports Series, 1985). The $DF_{U/Th}$ values confirm the findings that fish exhibit a stronger ability to take up Th than U with respect to seawater (Szefer et al., 1990b).

FISH PARASITES

From the relative bioconcentration found for host tissues and parasite it is clear that the concentration of Cd is similar in organs and cestode *Bothrio-*

cephalus scorpii of Baltic fish turbot (*Scopthalamus maximus*) and freshwater fish (Sures et al., 1997). The bioconcentration of Pb in the cestodes appears to be lower in the intestine of the Baltic fish as opposed to freshwater fish. Similar relative bioconcentrations were detected for parasites *Schistocephalus solidus* and *Thersitina gasterostei* and their host tissues (muscle, gills) of stickleback, *Gasterosteus aculeatus*, from the Gulf of Gdansk, southern Baltic. Concentrations of Cd, Co, Cu, Ni, Pb and Zn were similar in host tissues and the both parasites, i.e. Cestoda and Copepoda (Morozińska-Gogol et al., 1998). The CF and DF values estimated from concentration data for the parasites and their hosts indicate that there is no significant bioaccumulation of the metals in stickleback parasites. i.e. Cestoda and Copepoda.

PORPOISE PARASITES

CF and DF values are reported also for parasites hosted in lung of harbour porpoise (*Phocoena phocoena*) from southern Baltic. It is found (Szefer et al., 1998b) that the CF values for trace elements in lung nematodes (*Pseudalius inflexus*) were generally changed from one to another specimen. The highest factors were obtained for Mn and the lowest for Pb and Ni. The DF values of Cd/Zn, Pb/Cu, Pb/Cr, Pb/Fe, Cu/Cr, Cu/Fe and Cr/Fe in the nematodes compared to those in the host lung, were close to unity; it means that in spite of the interaction between the metals there was no discrimination between Cr, Cu, Pb and Fe as well as between Cd and Zn. Therefore, these trace metals were accumulated in *P. inflexus* in the proportions approximately equal to those observed in the host organ. The mean DFs for the ratios of Mn to metals such as Cd, Cr, Cu, Fe, Ni, Pb and Zn were significantly higher than unity, indicating that Mn is more extensively bioaccumulated in *P. inflexus* with respect to these metals.

The ratios of Pb to other metals especially to Mn, Cd and Zn were significantly less than unity; it suggests that capacity of *P. inflexus* for bioaccumulation of a Pb is insufficient in relation to its abundance in the host lung.

Although DFs can reach very high values for some metals, e.g Mn but it doesn't mean that a parasite is simultaneously able to bioaccumulate this metal proportionally to its level in the host organ. Strictly speaking, nematode levels of highly accumulated metal must not reflect actual its concentration in the lung host as a substrata. Therefore, it is useful to determine relationships between concentrations of selected metal in the parasite and its concentration in the host organ. It is well documented that trace elements, especially Fe, Mn and Zn are bioaccumulated in *P. inflexus* with respect to the host lung of *Phocoena phocoena* (Szefer et al. 1998b). As for mechanisms of such uptake, including the third stage larvae, the cuticle's role appears to be greater in absorbing substances from the environment. The cuticles of larvae are less complex than those of adult nematodes (Bird and Bird, 1991). The lung nematodes, especially of the fourth stage, and the adult individuals have a fully developed digestive tract. Thus, elements can be bioaccumulated in the adult lung nematodes predominantly by the digestive system. A greater bioaccumulation of elements in nematodes might result from a better functioning of trace element elimination process in the host than in the parasite (Bird and Bird, 1991). Zinc, Mn and Fe are distributed throughout the body rather than associated with the surface of the nematodes. Trace elements are suspected to be mainly accumulated from food in the digestive system of *Phocoena phocoena*. However, particulate matter within the orifices of the lung derived from the global atmospheric particulate and/or sea surface debris most probably plays a marginal role as a potential source of the trace elements in the parasites. It is postulated that enhanced levels of Mn, Fe and Zn in the parasite may be dominantly attributed to their active biological uptake from the porpoise digestive tract rather than to scavenging of the particulate metals from within the lung tubules (Szefer et al., 1998b).

B. BIOMAGNIFICATION OF ELEMENTS

(i) Introduction

The inter-relationships between air, water and land result in different habitats for particulate representatives of marine flora and fauna. Bioaccumulation is the transfer of chemical element from water and from diet into given organism. Differences in bioaccumulation due to metal speciation and a metabolic rate of the organism, affect its transferring across biological membranes (Muir et al., 1999). The low molecular weight protein metallothioneine plays an important role in controlling the uptake of elements in the liver of marine vertebrates such as fish and mammals as well as it is responsible for the elimination of elements. In the aquatic ecosystems, chemical elements are taken up by fish and invertebrates via three main routes, i.e. gills, surface part of body (skin) and digestive system. Dissolved species (ionic) enter organism by diffusion across the gill membrane or other respiratory surfaces into the blood (Muir et al., 1999). Trace elements adsorbed onto particle surfaces, e.g. phytoplankton, are accumulated by grazing organisms as predators, e.g. zooplankton. Bioaccumulation of heavy metals, e.g. Me–Hg in larger sea organisms by means of food is a predominant route over uptake from the ambient water. This is due to low levels in water and low elimination rate for these elements by the animal. Absorption from the food involves diffuse transfer across the gut membrane for both trace elements and radionuclides. Transfer along the food chain can lead to biomagnification (bioamplification), i.e. the successive increase in element content with increasing trophic level. Both metabolic rate and production control the food requirements of each organism; hence the metabolic rate is strictly connected with the rate of element uptake. According to Muir et al. (1999), mammals and waterfowls as homeotherms maintain effectively their body temperature and owing to this ability have greater energy requirements for the same body weight as compared to fish as poikilotherms.

For instance, biomagnification of Hg in predator in respective to its prey was observed in the case of fish and marine mammals in Arctic trophic chain (Muir et al., 1999). In the Greenland part of the Arctic elements such as Cd, Hg and Se increased in concentration towards higher trophic levels while Pb in marine biota did not show the same clear increase towards higher trophic levels (Dietz et al., 2000). Potential for secondary poisoning and biomagnification in marine organisms have been evaluated by Nendza et al. (1997).

According to Fowler (1982) and Luoma (1983) the food of marine organisms is often an important source of trace metals in their tissues. This is observed when pollutants are taken in from food and successfully assimilated and retained in tissues of the predator. Predictions of the magnitude of bioaccumulation require knowledge of dominant species, relative biomass and their food consumption.

(ii) Operational Definitions

Because chemical elements move along the food chain from prey to predator, it is important to know whether or not consumers regulate element uptake and whether their food consists of organisms which do or do not regulate the elements. The transfer factor, TF, was therefore calculated as follows (Amiard et al., 1980):

$$TF = C_c/C_p$$

where C_c and C_p are the average chemical element concentrations in the predator (consumer) and prey (diet), respectively.

TF values near unity mean that there is no biomagnification of the chemical elements along the successive levels of the food chain in marine ecosystem. TFs higher than unity show that given element is biomagnified in higher in respect to lower level of trophic web.

Other measures of biomagnification of chemicals in predator in respect to its potential prey as well as identification of diet are stable nitrogen and carbon isotope ratios found for these two trophic levels (DeNiro and Epstein, 1978, 1981). Stable isotope ratios are expressed in conventional δ notation (Voß and Struck, 1997; Thompson et al., 1998; Bearhop et al., 2000; Das et al., 2000) as:

$$\delta^{15}N \text{ and } \delta^{13}C = [(R_{sample}/R_{standard}) - 1] \times 1000$$

where R is the ratios of $^{15}N/^{14}N$ and $^{13}C/^{12}C$ of the sample and the reference gas, respectively.

Isotope values of $\delta^{15}N$ and $\delta^{13}C$ are different for animals and their diets because of a slight selective retention of the heavier isotope and more effective excretion of its lighter counterparts. In consequence, ^{15}N shows a stepwise increase with trophic level along the food chain (Hobson and Welch, 1992; Cabana and Rasmussen, 1994) while a consumer and its diet are characterised by similar values of $\delta^{13}C$ (DeNiro and Epstein, 1978; Hobson and Clark, 1992; Thompson et al., 1995). Therefore this index is generally used to identify relative contributions of various potential primary sources to food intake, e.g. the aquatic vs. terrestrial, inshore vs. offshore or pelagic vs. benthic (Rau et al., 1992; Hobson et al., 1995; Smith et al., 1996; Havelange et al., 1997; Dauby et al., 1998; Das et al., 2000). According to Das et al. (2000), the tissue of consumer reflects dietary input integrated over time, not just the last food intake because the stable isotopes ratios of consumers' tissue are derived from assimilated food. The metabolic rates and the subsequent turn-over of chemical elements are different in liver and muscle and hence it is possible to evaluate of diet integrated between different periods of time measuring off days, weeks or months (Tieszen et al., 1983; Hobson et al., 1996, 1997). For instance, several authors (Kidd et al., 1995a, 1995b, 1998; Das et al., 2000) using these stable isotope indices identified alimentary relations within high trophic levels for lake trout and burbot, and for tuna and dolphin as the top predators in Canadian Arctic freshwater ecosystem and the North-east Atlantic, respectively.

(iii) Trophic Relations of Chemical Elements and Radionuclides

The phytobenthos biomass in the Gulf of Gdańsk is variable and ranged from 0 to 533 g m^{-2}. The average phytoplankton and zooplankton biomass was estimated as 54 and 27 g m^{-2}, respectively (Łomniewski et al., 1975; Pliński and Florczyk, 1987). Considerable variations were observed in the distribution and composition of bottom fauna; the average biomass was ~ 150 g m^{-2}. The average percentage of molluscs accounted for 93.7 % of the total zoobenthic biomass (Wenne and Wiktor, 1982). The global and Polish fish catches (cod, herring, sprat and flat fish) in the Baltic area were esimated as ~ 900,000 and 250,000 t $year^{-1}$, respectively (Mańkowski, 1978b). According to various authors (Sokołowski, 1958, 1965; Brehm, 1962; Mikheev, 1986), ducks *Aythya marila* and *Clangula hyemalis* feed mainly on molluscs and crustaceans in winter and, in summer, they also feed on seaweed. Marti (1983), Rutkowicz (1982) and Cięglewicz et al. (1972) demonstrated that the main food of cod (*Gadus morhua*) is benthic invertebrates such as mollusc and crustaceans. Specimens of *G. morhua* (body length > 45 cm) caught in spring in the southern Baltic were found to have mainly specimens of herring (*Clupea harengus*), sprat (*Sprattus sprattus*) (together 59.5% by weight) and the isopod *Saduria entomon* (4.7%) in the alimentary tracts (Cięglewicz et al., 1972). A greater mass contribution of *S. entomon* (13 and 57.7%, respectively) was found in the food of small specimens of *G. morhua* (15–25 cm) caught in spring and in large specimens (> 45 cm) caught in autumn. The food of *C. harengus* and *S. sprattus* consisted mainly of the zooplankton *Temora longicornis and Pseudocalanus elongatus* (Cięglewicz et al., 1972). The main food of herring caught in spring consisted of plankton, whilst in other seasons its dominant component was nektobenthos. Molluscs are at a low level in the food chain and they mainly feed on zooplankton (small invertebrates) (Zatsepin et al., 1988) and particulate debris. As for higher trophic levels, the food content of Baltic harbour porpoises consisted mainly of fish such as cod, herring and Gobiidae (Szefer et al., 1995a). Sea eagles prey on waterfowl (ducks, geese, loons etc), fish and some mammals (Sokołowski, 1958, 1965).

Biomagnification of metals along the Baltic trophic levels has been studied by Szefer (1991). TF values of selected trace elements calculated for this food chain were generally near or less than 1 (Table 5.3). This means that, in spite of the elevated renal and hepatic levels of some metals in predators, e.g. marine mammals (Szefer, 1991; Szefer et al., 1995a), there is no biomagnification of the trace metals analysed along the successive trophic levels from potential prey to predators in the southern Baltic. This finding is in agreement with that of Fowler (1982) who concluded that preys are generally characterised by higher concentrations of trace metals than predators because food is not the sole source for element input. Factors such as incomplete absorption of metals across the gut, rapid excretion, and dilution in muscle which represents a large mass of the total body weight are responsible for lower contents of metallic pollutants in the predator relative to its prey.

The ratios of stable nitrogen isotopes were analysed in zooplankton exoskeletons extracted from Baltic sediments and in bulk sediments. Combined with results on C_{org} concentrations and abundances of exoskeletons of *Bosmina maritima* in the sediment, the data obtained were used to evaluate major sources of N in the food web over the past century (Struck et al., 1998).

According to Skwarzec et al. (2000) the trophic way and indirect intake from seawater (via gills) are probably the main route of $^{239+240}Pu$ accumulation by Baltic fish. The TF's computed for trophic relations: zooplankton (zoobenthos) – plankton-eating fish – predatory were not constant. Values higher than unity (7.1) for relation cod-sprat let us to conclude that Pu is biomagnified along trophic sector sprat-cod. Values lower than unity (0.09) were obtained for flounder-zoobenthos relation (Skwarzec et al., 2000). According to Bojanowski et al. (1995) bioaccumulation of ^{210}Po in Baltic fish is lower than in crustaceans (*Saduria entomon*, *Crangon crangon*) and mollusc (*Mytilus edulis*). Comparing the distribution of the measured radionuclides along the food chain it is concluded that ^{210}Po concentrations decrease, in contrast to ^{137}Cs which values exhibit increasing trends as the trophic level rises (Bojanowki et al., 1995).

C. AFFINITY OF ELEMENTS IN RESPECT TO BIOTA AND SUBSTRATA

(i) Introduction

In addition to cross correlation data, results of regression analysis are much helpful in interpretation of the established relationship, if any exists, between trace element concentration in biota and the ambient sediment as substrata.

TABLE 5.3.

Transfer factors for metals calculated for trophic relationships of the southern Baltic ecosystem. After Szefer (1991)

Trophic relation Consumer (predator) – food (potential prey)	Fe	Zn	Mn	Cu	Pb	Cd	Co	Ni	U	Th
Aythya marila – *Saduria entomon*	0.10	1.6	0.03	0.3	0.42	0.8	0.2	0.08		
Aythya marila – molluscs (soft tissue)	0.29	0.3	0.04	0.4	0.18	0.1	0.2	0.03		
Aythya marila – macroalgae	0.22	0.9 (3.6)[a]	0.01	2.3 (3.7)'	0.09	0.7	0.4	0.14		
Gadus morhua – *Clupea harengus*	0.30	0.17	0.56	0.37	0.83 (0.12)[c]	1. 5 (0.2)[c]	0.3	0.11	1.30[d]	1.40[d]
Gadus morhua – *Saduria entomon*	0.01	0.27	0.01	0.03	0.38	0.8	0.1	0.04	0.05[d]	0.01[d]
Gadus morhua – molluscs (soft tissue)	0.02	0.06	0.02	0.05	0.16	0.1	0.1	0.01	0.01[d]	0.01[d]
Clupea harengus – zooplankton	0.06	0.61 (0.12)[b]	0.21	0.14	0.16(1.10)[c]	0. 1 (1.0)[c]	1.8	0.34	0.01[d]	0.01[d]
Sprattus sprattus – zooplankton	0.08	0.75 (0.15)[b]	0.25	0.14	0.21 (0.52)[c]	0.6	1.3	0.31	0.01[d]	0.01[d]
Hyperoplus lanceolatus – zooplankton									0.02	0.03
Ammodytes tobianus – zooplankton	0.03	0. 19 (0.04)[b]	0.07	0.12	0.24	0.2	0.8	0.18		
Molluscs (soft tissue) – particulates	0.40	0.25		0.04	0.10	0.2	0.3		> 0.27	0.01

[a] Calculated only for Spermatophyta.
[b] Maximum levels of Zn in zooplankton were included in the calculation of the mean.
[c] Maximum levels of Ph and Cd in *Clupea harengus* were included in the calculation of the mean.
[d] Data for U and Th in muscle of *Gadus morhua, Clupea harengus* and *Sprattus sprattus* were used in calculations.

Based on the regression analysis approach, Bryan and Langston (1992) in their extensive overview using mainly UK estuaries as examples, considered various dominant factors governing the bioavailability, bioaccumulation and biological effects of trace elements in sediments. It is reported that under field conditions, identification of key processes can be achieved by observing a linear relationship between metal concentrations in marine infaunal organisms and various types of sediment leaches (normalised in respect to < 63 μm fraction). Factors of more local importance are also identified based on the marked deviation of some points from a regression line on the scatter-plot. Using regression analysis, several authors reported statistically significant correlations (with or without normalisation to sediment-bound Fe, organics) between concentrations of trace elements in seaweeds, e.g. *Fucus vesiculosus* (Ag, Cd, Hg, Zn) and deposit-feeding organisms, e.g. *Macoma balthica* (Ag, As, Hg, Zn?) and *Nereis diversicolor* (Ag, As, Cd, Cr, Cu) in different coastal and estuarine areas (Bryan and Hummerstone, 1973; Ray et al., 1980; Langston, 1982, 1984, 1985; Bryan et al., 1985; Bryan and Langston, 1992). For instance, uptake of Ag and Pb from sediment substrata by worm *Nereis diversicolor* from the Gannel Estuary was reduced in sediment having high levels of humic substances and tended to be enhanced for Ag in those characterised by elevated levels of Mn and Cr (Bryan, 1985; Luoma and Bryan, 1982). Concentrations of Pb in *M. balthica* were linearly related to the Pb/Fe ratios in sediments (Luoma and Bryan, 1978; Bryan, 1985). Apparently ingested Fe oxides simply compete for adsorbed Pb with the digestive system of the zoobenthal organism. Furthermore, from observations of a wide range of estuaries, it is concluded that increasing concentration of sediment-bound Fe oxides also hinders the availability of Pb in *N. diversicolor* as well as As in *M. balthica* and *N. diversicolor* (Luoma and Bryan, 1982; Langston, 1984, 1985). It means that increasing the level of organic matter or Fe oxides in sediments reduces the bioavailabilty of Ag, As or Pb (Bryan, 1985; Bryan and Langston, 1992). However, in other estuary, i.e. in the Severn Estuary and Bristol Channel, concentrations of Ag in *N. diversicolor* are not only associated with those of the sediment but also with those of the overlying water, as identified by concentration in alga *Fucus vesiculosus*, reflecting dissolved Ag (Bryan and Langston, 1992). This suggests that dissolved and sediment–bound species of Ag are both accumulated by the ragworm. The slopes of the regression line exceed unity suggesting that bioavailabilty of Ag is disproportionally increased as Ag pollution rises (Bryan and Langson, 1992). Similar relationships have been observed for Ag and Zn in the deposit-feeding bivalvia *Macoma balthica* (Bryan, 1985; Harvey and Luoma, 1985; Bryan and Langston, 1992). Normalising sediment trace element concentrations, e.g. Hg or As, with respect to the major binding substrate (% organic matter) highly improves correlations with tissue levels in estuarine infaunal organisms, e.g. deposit-feeding bivalves *M. balthica* and polychaetes *N. diversicolor* (Langston, 1982, 1986; Bryan et al., 1985; Bryan and Langston, 1992). Having established such relationship, based on partitioning in sediments (Luoma and Bryan, 1981), prediction availability of As and Hg is possible, even in the lack of suitable indicator species, using sediment measurements alone and indices of pollution produced as reported for estuarine areas of UK (Bryan and Langston, 1992).

Luoma and Bryan (1981) compared the extraction of metals and substrates from surficial estuarine sediments and determined statistical relationships between extractable metals and extractable substrates. Extractions may help in determining the abundance of the operationally-defined forms of substrates which are the most active in binding metals in sediments as well as in providing information necessary for determining the bioavailabilty of sediment-bound metals. Such statistical approach may aid to identify the important variables in complex ecosystems (Luoma and Bryan, 1981).

(ii) Operational Definitions

In order to establish a relationship, if any exists, between concentration of given element/s in the same or other compartment/s (atmosphere, water, biota, sediment) besides computing a correlation coefficient, r, also regression line is drown described e.g. by the following equation:

$$y = ax + b$$

where y and x are concentrations of given element(s) in the marine component(s) studied and a and b are slope constant and intercept, respectively.

The intercept amounting to ca. 0 suggests natural, terrigenic sources of selected elements (with respect to e.g. Al) in the geological material (sediments). The positive intercept for considered relationships between particular elements indicated that other sources, not only aluminosilicates, are responsible to some extent for the presence of the elements in the sediment studied. The slope constant being ca. 1 means that the concentrations of given element proportionally increase (for positive a value) or decrease (for negative a value) with concentration of other element in the object studied. The slope constant higher or lower than unity suggests that such concentration variations are more or less evident in respect to variations corresponded to the proportional relationship between two compared elements.

(iii) Availability of Chemical Elements to Biota and Sediments

Biota

Evidence of the bioavailability of Zn in sediments comes from studies of marine plants for heavy metal pollutants. Lyngby and Brix (1987) observed significant relationship between Cu and Zn levels in sediments and those in the root-rhizome and leaves of eelgrass *Zostera marina* from the Limfjord, Denmark. Translocation of Zn between these parts was quite low (Lyngby et al., 1982) suggesting that the sediment-bound Zn also controlled the availabilty of dissolved Zn to the leaves. It is found that Cd levels in *Z. marina* reflected its concentration in the ambient water rather than those of the sediment associated to the root-rhizome (Lyngby and Brix, 1982). Apparently this metal is absorbed by leaves and next translocated to the root-rhizome thus providing a route from the overlying water to the sediment (Brinkhuis et al., 1980; Bryan and Langston, 1992).

According to Sandler (1984), the Zn and Cu concentrations of the two crustacean species, i.e. *Pontoporeia affinis* and *Saduria entomon* (*Mesidothea entomon*) from different sampling sites of the Bothnian Sea showed similar patterns, although these organisms have different food habits as detritivorous deposit feeder (*P. affinis*) and carnivore (*S. entomon*). The latter feeds on *P. affinis* and the ostracod *Paracypreides fennica*. Figure 5.5 shows the relationships between concentrations of Zn and Cu in whole body of these two species and the associated sediments. The highest levels of these trace elements occurred in specimens from the shallowest sites with their lowest values in the adjacent sediments. An inverse trend is observed for specimens from the deepest sites. These seemingly contradictory results can be explained by different bioavailabilty of these sediment-bound elements at the shallow and deep areas (Sandler, 1984). Due to different chemical speciation of metals in sediments, they may be less available to biota inhabited sediments with high their total concentrations as compared to those feeding on sediments with low total concentration. It could be a result of greater abundance of Fe oxides in deeper than shallow areas of the Gulf of Bothnia (Boström et al., 1978) and the Bothnian Sea (Niemistö et al., 1978).

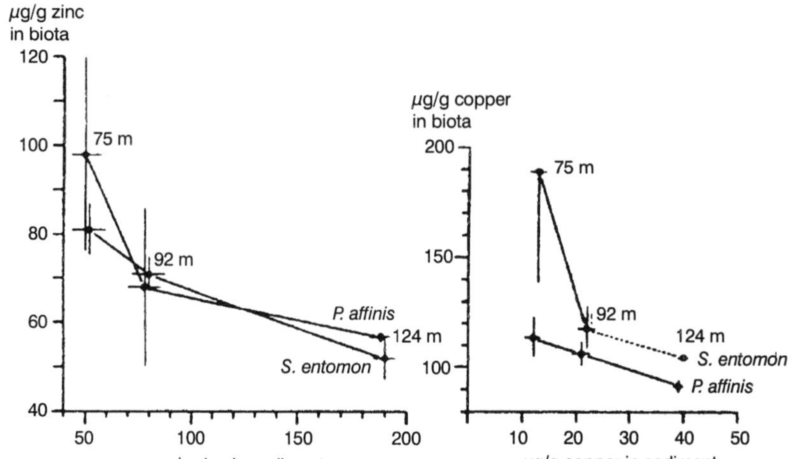

Fig. 5.5. Correlation between zinc and copper contents in biota (*P. affinis* and *S. entomon*) and sediments at the different sites. The vertical and horizontal lines indicate the standard deviations. After Sandler (1984); modified.

As can be seen in Fig. 5.6, the slopes (b) in the equation Metal content (μg) = a (weight)^b for Baltic *Arctica islandica* are lower than unity for Cu and Zn and higher than unity for Cd and Pb. It means that represented on a linear scale, the relationship between metal content and dry weight is curved with larger speci-

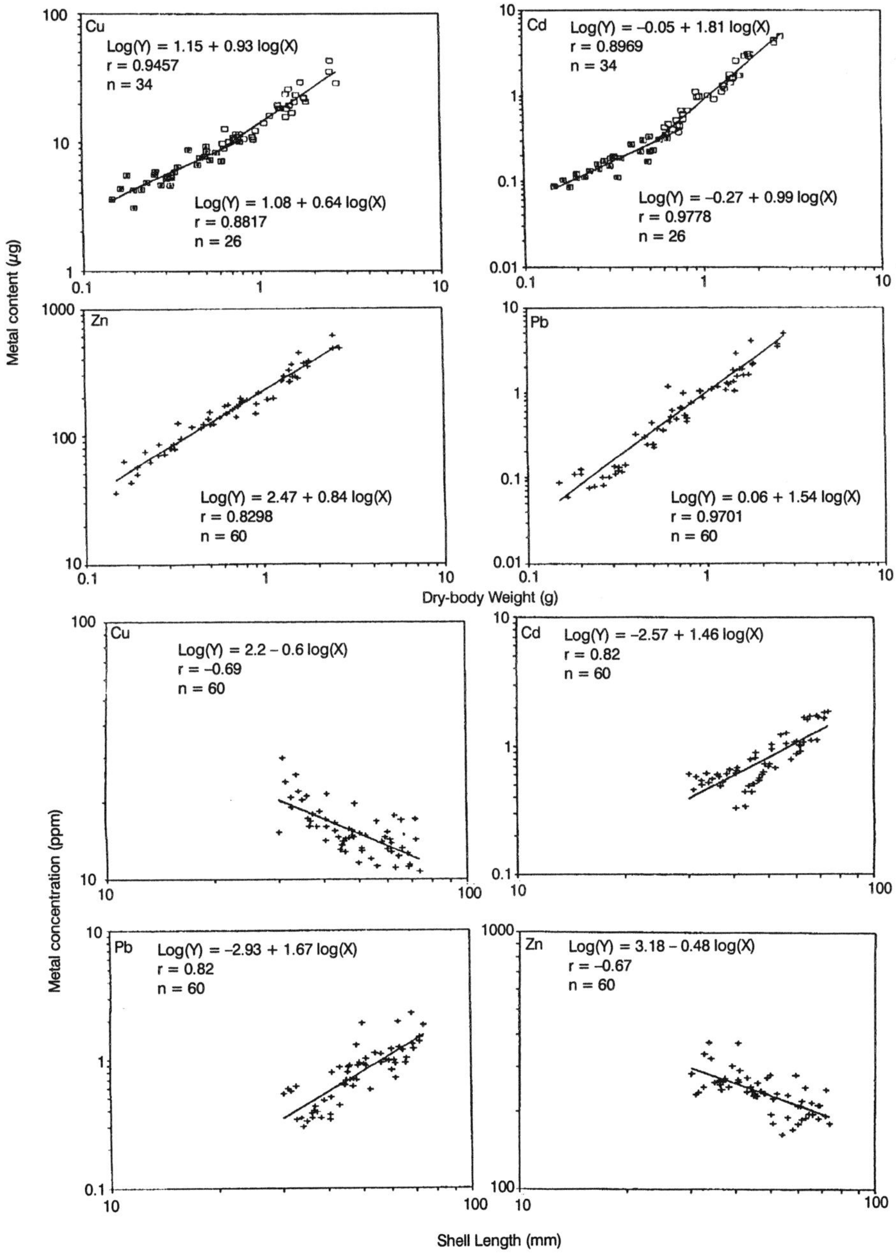

Fig. 5.6. The relationships between Cu, Zn, Cd and Pb contents and dry weight of *Arctica islandica* from Süderfahrt in Kiel Bay in July 1992. ⊡: animals of shell length 30.1–45 mm, □: animals of shell length 45.2–73.7 mm and +: animals of shell length 30.1–73.7 mm. The relationships between the four trace metal concentrations and shell length of *A. islandica* for animals of shell length 30.1–73.7 mm (+) are also presented for comparison. After Swaileh and Adelung (1994); modified.

mens having smaller absolute quantities of Cu and Zn and greater quantities of Cd and Pb than would be expected if content of given metal was directly related to weight. The calculated regression slopes (b) of metal contents against dry body weight (Fig. 5.6) reflect that Cd and Cu appear to be affected by maturation. Since two different regression slopes are obtained for Cd and Cu before and after maturation it is concluded that the both metals are affected by maturation. This does not concerns Pb and Zn (Swaileh and Adelung, 1994). As it results from calculations there is one final slope (1.53) for Cd and Pb from both content (μg)/weight (g) and concentration ($\mu g\ g^{-1}$)/shell length (mm) scatter plots. This means that *A. islandica* from the Western Baltic Sea coast may have one metabolic strategy for the essential elements (Cu, Zn) and another for the non-essential elements (Cd, Pb) (Swaileh and Adelung, 1994).

Suspended Matter and Bottom Sediments

Correlation coefficients were calculated as well as regression analysis was performed for concentration data matrix of minor and major elements in particulate matter and surficial sediments of the Baltic Sea (Boström et al., 1981; Kremling and Petersen, 1984; Szefer, 1990a; Ingri et al., 1991; Brügmann et al., 1992; Szefer et al., 1995b, 1996). Results of regression analysis have been much helpful in identification of the provenience of selected elements in particular subregions of the Baltic Sea. For instance, a linear correlation was found for particulate Ca, Mg and salinity in pelagic samples. It has been concluded that particulate matter from the Bothnian Bay had the lowest and that from the Kattegat-Skagerrak the highest Ca and Mg concentrations (Brügmann et al., 1992). According to Brügmann et al (1982) Cd and Ni showed negative correlation with the salinity; hence exponential $Cd_{susp/0.2m}$ (ng dm^{-3}) = exp {2.0 – [0.14 × Sal. (PSU)]}; r = –0.69 and linear regression equation $Ni_{diss./pelagic}$ ($\mu g\ dm^{-3}$) = 0.72 – [0.013 × Sal. (PSU)]; r = –0.77 show this relationship corresponded to a major input of trace elements by runoff from the land occurring in the northernmost part of the Baltc Sea where lower salinity < 5 PSU is observed. During mixing these water masses with more saline waters, a diluting effect has place, especially for dissolved species of metals, e.g. Ni but also for particulate metal species, e.g. Cd. Elements such as Cd which readily form dissolved chlorocomplexes are increasingly released from particulate matter by desorption during moving from less to more saline waters (Brügmann et al., 1992).

According to Ingri et al. (1991), suspended, non-detrital Fe was linearly correlated with P in subsurface and bottom Baltic water (Fig. 5.7). The intercepts on the Fe–Al axis were 0.57 for subsurface water and 0.67 for bottom water, being closed to the average crust Fe/Al ratio. It is concluded therefore that the intercepts represent the detrital fraction while non-detrital Fe in Baltic particulate matter has been suggested to be present as oxyhydroxide (Emelyanov and Pustelnikov, 1975). The P/Al ratio in suspended material in the Baltic was one to two orders of magnitude higher than in average Earth's crust indicating that the contribution of detrital P was small (Ingri at al., 1991). The Fe–P relation (Fig. 5.7) may be a result of scavenging of P by Fe-oxyhydroxide or the presence of P with

Fe in the organic fraction or a combination of both processes. Three subsurface samples are outside the regression line illustrating strict relationships between non-detrital Fe and P. These three samples being the four northwesternmost ones in the Belt Sea-Kattegat area seemed to be little influenced by Baltic subsurface water and therefore atypical for the Fe–P relation observed in the Belt Sea, the Baltic Proper and the Åland Sea (Ingri et al., 1991). The intercept on the Ba–Al axis (Fig. 5.8) was 0.031. The value is ca. five times higher as compared to the analogous ratio estimated for average crust. It suggests that an additional non-detrital phase must be present being responsible for the enhanced Ba levels. Particulate-bound Ba appears to be partitioned between a detrital, a Mn-rich and an authigenic phase. Ba–S rich particles are commonly distributed in the Baltic Proper and Belt Sea-Kattegat in contrast to poor in this component the Gulf of Bothnia and the Gulf of Finland (Bernard et al., 1989). Boström et al. (1981) reported significant relationships of metal pairs Ti–Al, Si–Al and Fe–Al for suspended matter from the Bothnian Bay, Bothnian Sea and the main part of the Baltic Proper (Fig. 5.9).

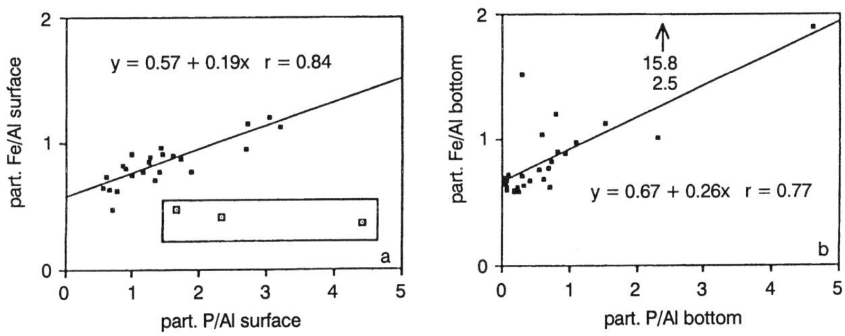

Fig. 5.7. Correlation between suspended Fe and P in water samples from the Baltic proper, Belt Sea and Aland Seas. (a) Bottom water (5 m above bottom), (b) subsurface water (5 m below water surface). Three samples in (a) (unfilled squares) and one sample in (b) (Fe/Al ratio 15.8, P/Al ratio 2.5) have been excluded from the correlation calculations. After Ingri et al. (1991); modified.

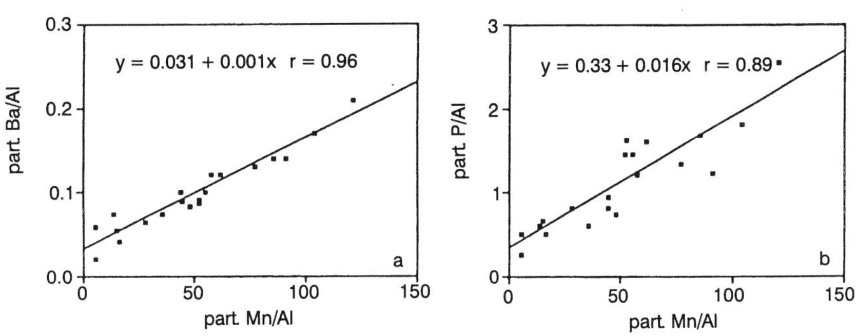

Fig. 5.8. (a) Correlation between suspended non-detrital Ba and suspended non-detrital Mn. (b) Correlation between non-detrital P and non-detrital Mn. The samples were taken at three occasions, August 14, 1984, June 18, 1985 and September 9, 1985 in the Landsort Deep. After Ingri et al. (1991); modified.

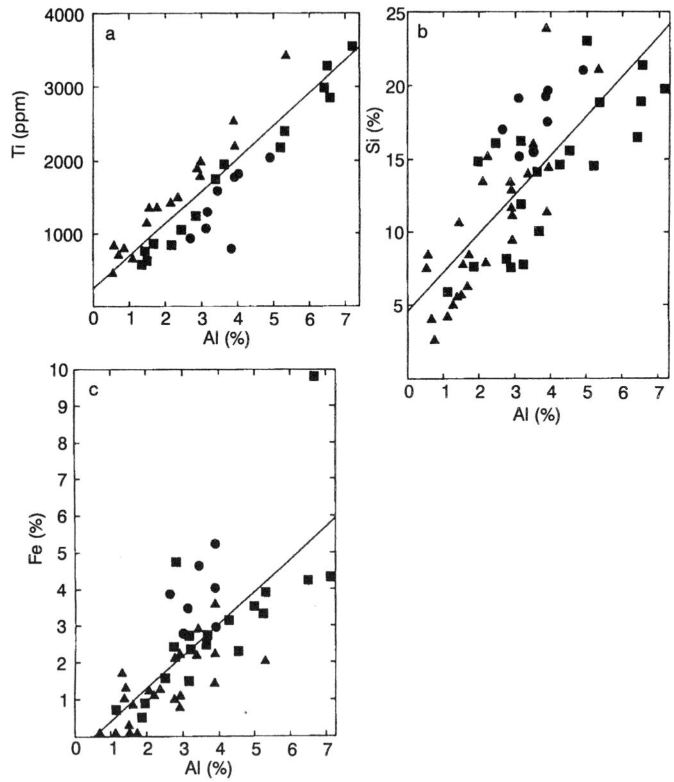

Fig. 5.9. Inter-element correlations for (a) Al and Ti, (b) Al and Si and (c) for Al and Fe. The correlation coefficients are significant at the 1% level. The solid lines represent the best-fit regression line, calculated with Al as independent variable: (●) samples from Bothnian Bay; (■) samples from Bothnian Sea and the main part of Baltic Proper; and (▲) samples from the straits area. After Boström et al. (1981); modified.

The significant correlations were obtained for Al and Fe with most elements in the bottom sediments collected in the southern Baltic (Fig. 5.10). An intercept of ca. 0 was observed for several pairs of elements studied, e.g. Fe–Al, Ni–Al and Co–Al (Fig. 5.11). The positive intercepts of Zn, Pb and P in their plots against Al (Fig. 5.12) show that sources other than aluminosilicates are, to some extent, responsible for the occurrence of these elements in sediments of the southern Baltic. It should be emphasised that points corresponding to the samples denoted as 6 and 16 deviate by more than twice the standard deviation from the linear relationships between Al and trace elements, i.e. Ag, Cd, Cu, P and Zn. Figure 5.12 illustrates the distribution patterns for P–Al and Zn–Al. An explanation for it is that these two untypical samples came from the vicinity of the Vistula River, close to densely populated and heavily industrialised area characterised by high rates of discharge of metallic pollutants (Szefer et al., 1995b, 1996; Glasby and Szefer, 1998). They are consisted of great quantities of finer material enriched in heavy metals, partly anthropogenic in origin. Another two samples (7 and 17)

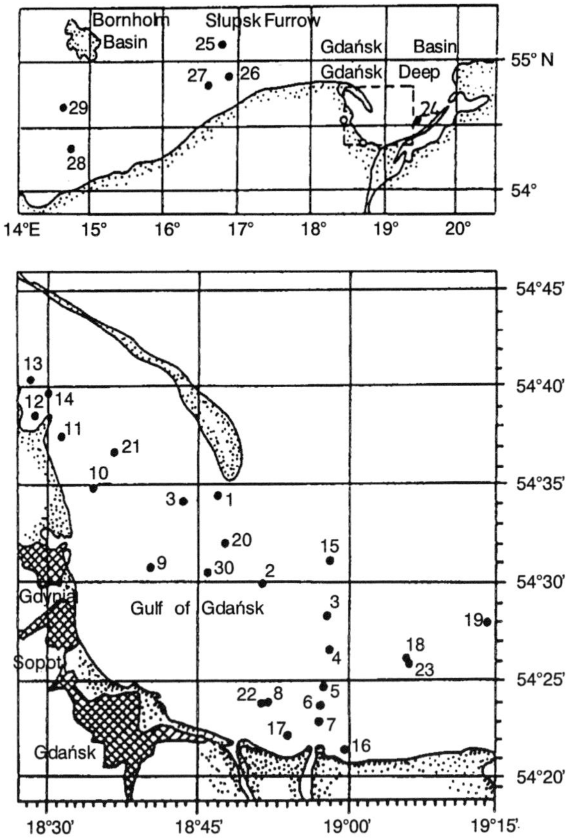

Fig. 5.10. Location of sediment sampling sites in the Southern Baltic. After Szefer et al. (1996).

taken also near the mouth of the Vistula River are dominantly sandy and heavy metal concentrations are therefore indistinguishable from those taken from less contaminated areas because of dilution by coarse-grained material. The distribution pattern of Pb from different sampling sites (Fig. 5.12) suggests the widespread pollution of this metal in the Gulf of Gdansk being more likely the result of atmospheric fallout rather than direct riverine input (Szefer, 1990a, 1990b; Szefer et al., 1996; Glasby and Szefer, 1998; Renner et al., 1998). This Pb input is probably dominantly derived from gasoline (Pacyna et al., 1992; HELCOM, 1996).

Ferromanganese nodules

Boström et al. (1982) estimated statistically the relationships between several elements in ferromanganese nodules from the Gulf of Bothnia. There was a negative correlation between Mn in hydroxide fraction (HF) and the acid insoluble fraction. It is suggested that the concentration of Mn in the HF increases with increasing the total amount of the hydroxide phase.

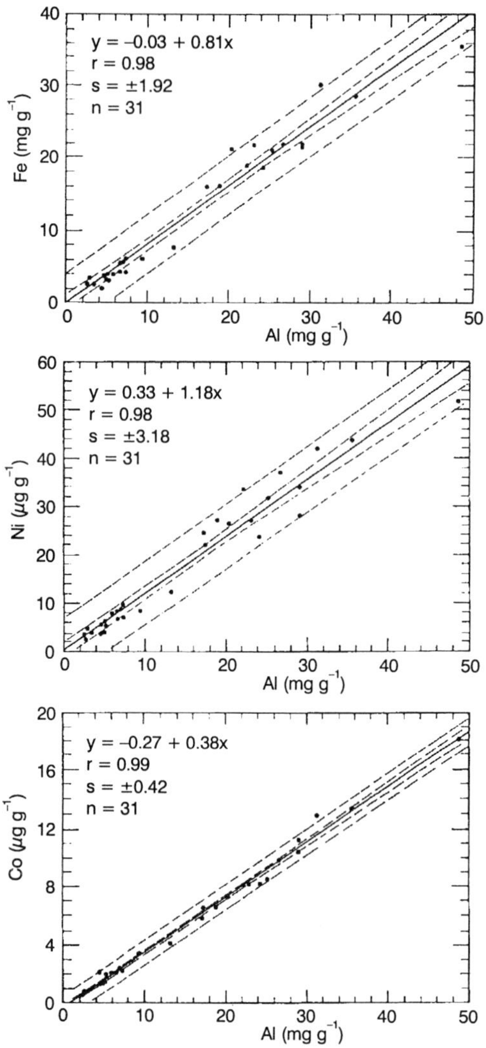

Fig. 5.11. Plot showing the relationship between Fe and Al, Ni and Al and Co and Al in the surficial sediments. After Szefer et al. (1996).

Iron and Mn concentrations indicated a negative correlation which is implicit by the definition of the HF-phase; hence a positive correlation between given element and Mn corresponds to negative correlation for this element with Fe according to equation Fe=40-Mn (Boström et al., 1982). The correlations for metal pairs Al–Fe and Al–Mn were less than 0.70; it means that less than 50% of the covariations were explained by Al when the influence from insoluble fraction (IF), Si or P is removed. The relationship of uncorrected Al concentrations with acid-insoluble residue was the most significant of all tested correlations and indicated that some Al is dissolved from micas, etc., during the acid leaching. This ex-

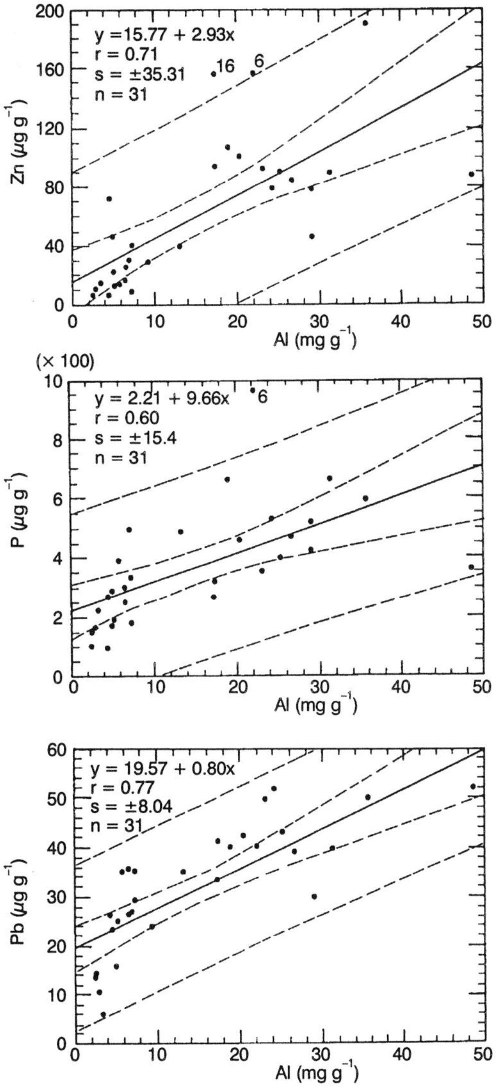

Fig. 5.12. Plot showing the relationship between Zn and Al, P and Al and Pb and Al in the surficial sediments. After Szefer et al. (1996).

planation was supported by the Al–Ti relation, which was very constant (Fig. 5.13). Other elements such as Cr, Sc and V formed refractic compounds and all showed significant correlations with Al, but not with IF. The original mean data after normalisation (removing the influence from IF) indicated significant negative correlation of Si with Al, suggesting that Si to a great extent occurred in separate phases such as quartz and diatomaceous materials. Arsenic and La also were characterised by positive correlations with Fe; the correlations between

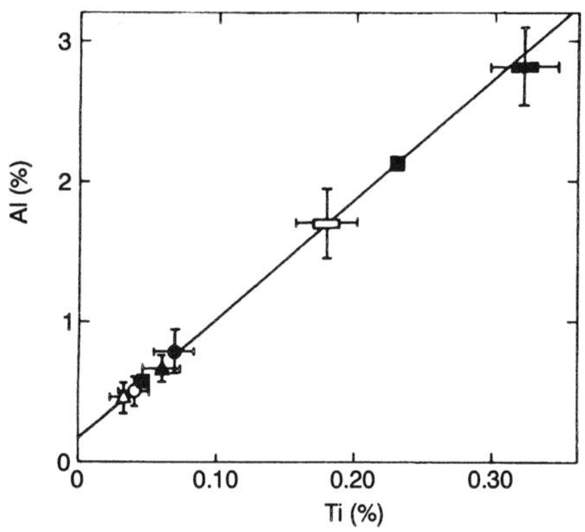

Fig. 5.13. The relations between Al and Ti suggest that terrigenous detritus or authigenic clay minerals are attacked during the leaching. After Boström et al. (1982); modified.

Fe–P and Fe–As are well known (Boström and Valdés, 1969; Calvert and Price, 1977). Apatite is not considered to be a minor source for P, however, because significantly smaller quantities of Ca were present in this phase than this would be expected. Furthermore, La was strongly correlated with Fe (Boström et al., 1982). Barium, Ni and Cu showed pronounced correlations with Mn; the correlations of Mn–Ni and Mn–Cu were obtained in studies of Baltic and deep-sea ferromanganese nodules (Glasby, 1977; Boström et al., 1982). The data obtained for Baltic nodules (Boström et al., 1982) are in an agreement with the high Ni-values reported by other authors (Manheim, 1965) for Baltic nodules. Interrelationship between Mn and Ba is difficult to explain, although Ba is often present as barite in marine deposits (Arrhenius and Bonatti, 1965). Bearing in mind that the brackish waters of the Gulf of Bothnia are characterised by the small concentration of sulfate this is unlikely that this mineral is main source of Ba in the nodules (Boström et al., 1982).

According to Boström et al. (1982) the ferromanganese concretions consist of the following components:

- an acid-insoluble fraction, mostly contained silicates and other rock forming constituents;
- silica, representing leached quartz and diatoms;
- Al, Ca, Cr, Sc, Ti, V and probably Na derived from small amounts of silicate rock which did not belong to the hydroxide phase;
- a Mn-hydroxide component which included Ba, Cu and Ni;
- a Fe-hydroxide component, which included As, La and P.

(iv) Remarks and Recommendations

According to Chapman (1997) there are six basic reasons for measuring bioaccumulation, three of which are possibly known, i.e. bioaccumulation is quantified to:

- determine contaminant-specific bioavailability (for non-metabolised contaminants),
- assist in identification possible causitive agent(s) of toxicity,
- relate body burdens to trophic chain accumulation values in respect to secondary poisoning or bioaccumulation.

Chapman (1997) thinks that in future it would be possible to measure bioaccumulation to:

- develop realistic tissue-residue values,
- identify causitive agent(s) of toxicity,
- assess/predict effects of chronic, low-level exposures.

There is need to study biomagnification of chemical elements in the Baltic food web using stable nitrogen isotopes, namely ^{15}N and ^{14}N. Ratio $\delta^{15}N$ being useful tool to study of bioaccumulation and bioamplification of trace elements, would be also very helpful in identification relative contributions of various potential primary sources to food intake. Bearing in mind that the consumer tissues reflect dietary inputs integrated over time, it would be possible to evaluate of diet integrated between different periods of time. According to Thompson et al. (1998) analysis of stable isotope ratios (^{15}N) revealed that trophic status was not the principal factor in determining of Hg level in marine food chains and that the presence of mesopelagic prey in the diet at least had some influence on Hg levels over and above that from trophic status. The relatively large increase in Hg concentrations in seabirds feeding on mesopelagic prey has clear implications for future human exploitation of mesopelagic marine resources. It appears, therefore, that increase in Hg pollution in mesopelagic food chains requires further research (Thompson et al., 1998). Analyses of the trophic transfer potentials of trace elements in zooplankton, molluscs and fish suggest that slight variations in assimilation efficiency or elimination rate constant may determine whether or not some trace elements are biomagnified. According to Reinfelder et al. (1998) a linear, single model may not be appropriate for fish which, in contrast to many aquatic invertebrates have a large mass of tissue in which the contents of most trace elements show tendency to feedback regulation. It is also recommended to extend studies on biomagnification of radionuclides, especially anthropogenic in origin, along sequential levels of the Baltic trophic web.

References

Amiard, J.-C., C. Amiard-Triquet, C. Metayer, J. Marchand and R. Ferre, 1980. Etude du transfert de Cd, Pb, Cu et Zn dans les chaines trophiques neritiques et estuariennes – I. Etat dans l'estuaire interne de la Loire (France) au cours de l'ete 1978. Water Res. **14**, 665–673.

Arrhenius, G.O.S., and E. Bonatti, 1965. Neptunism and volcanism in the ocean, in: Progress in Oceanography, ed. M. Sears (Pergamon Press, New York, N.Y.) 3, pp. 7–22.

Bearhop, S., S. Waldron, D. Thompson and R. Furness, 2000. Bioamplification of mercury in great skua *Catharacta skua* chicks: the influence of trophic status as determined by stable isotope signatures of blood and feathers. Mar. Pollut. Bull. **40**, 181–185.

Bernard, P. C., van R.E. Grieken and L. Brügmann, 1989. Geochemistry of suspended matter from the Baltic Sea. 1. Results of individual particle characterization by automated electron microprobe. Mar. Chem. **26**, 155–177.

Bird, A.F., and J. Bird, 1991. The Structure of Nematodes. 2nd ed. (New York, Academic Press).

Bojanowski, R. 1976. The occurrence of major and minor chemical elements in the more common Baltic seaweed. Oceanologia **2**, 81–152.

Bojanowski, R., and P. Szefer, 1979. Uranium in water of the Baltic Sea. Stud. Mater. Oceanol. **26**, 253–274 (in Polish with English abstract).

Bojanowski, R., Z. Radecki and D. Knapinska-Skiba, 1995. The distribution of ^{137}Cs, $^{239+240}Pu$ and ^{210}Po in the Pomeranian Bay (southern Baltic) ecosystem. Biul. Mors. Inst. Ryb. **3**, 15–24.

Boström, K., and S. Valdés, 1969. Arsenic in ocean floor. Lithos **2**, 351–360.

Boström, K., J.-O. Burman, B. Boström, C. Ponter, S. Brandlöf and B. Alm, 1978. Geochemistry, mineralogy and origin of the sediments in the Gulf of Bothnia. Finn. Mar. Res. **244**, 8–35.

Boström, K., J.-O. Burman, C. Pontér and J. Ingri, 1981. Selective removal of trace elements from the Baltic by suspended matter. Mar. Chem. **10**, 335–354.

Boström, K., L. Wiborg and J. Ingri, 1982. Geochemistry and origin of ferromanganese concretions in the Gulf of Bothnia. Mar. Geol. **50**, 1–24.

Brehm, A., 1962. The Life of Animals - Birds; title of original - Brehms Tierleben, Dritter Band, Vögel, Urania - Verlag 1956. (Państwowe Wydawnictwo Naukowe, Warsaw) 568 pp. (in Polish).

Brinkhuis, B.H., W.F. Penello and A.C. Churchill, 1980. Cadmium and manganese flux in the eelgrass *Zostera marina*. Mar. Biol. **58**, 187–196.

Brügmann, L., 1978. Zum Gehalt eniger Schwermetalle in marinen Organismen aus dem Auftriebsgebiet vor NW-Afrika. Fisherei–Forschung **16**, 53–58.

Brügmann L., 1981. Heavy metals in the Baltic Sea. Mar. Pollut. Bull. **12**, 214–218.

Brügmann, L., 1982. Trace metals in the Baltic Sea 1980 and 1981. Proc. XIII Conference of Baltic Oceanographers, Helsinki, 24–27 August, Vol. 1, 182–197.

Brügmann, L., P.C. Bernard and R. van Grieken, 1992. Geochemistry of suspended matter from the Baltic Sea. 2. Results of bulk trace metal analysis by AAS. Mar. Chem. **38**, 303–323.

Bryan, G.W., 1980, Recent trends in research on heavy-metal contamination in the sea. Helgoländer Meeresunters. **33**, 6–25.

Bryan, G.W., 1983. Brown seaweed, *Fucus vesiculosus*, and the gastropod, *Littorina littoralis*, as indicators of trace-metal availability in estuaries. Sci. Total Environ. **28**, 91–104.

Bryan, G.W., 1985. Bioavailability and effects of heavy metals in marine deposits. Wastes in the oceans, in: Nearshore Waste Disposal, eds. B.H. Ketchum, J.M. Capuzzo, W.V. Burt, I.W. Duedall, P.K. Park and D.R. Kester (John Wiley & Sons Ltd, New York) **Vol. 6**, 42–79.

Bryan, G.W., and L.G. Hummerstone, 1973. Adaptation of the polychaete *Nereis diversicolor* to estuarine sediments containing high concentrations of zinc and cadmium. J. Mar. Biol. Assoc. UK **53**, 839–857.

Bryan, G.W., and W.J. Langston, 1992. Bioavailability, accumulation and effects of heavy metals in sediments with special reference to United Kingdom estuaries: a review. Environ. Pollut. **76**, 89–131.

Bryan, G.W., W.J. Langston, L.G. Hummerstone and G.R. Burt, 1985. A guide to the assessment of heavy-metal contamination in estuaries using biological indicators. Mar. Biol. Assoc. U.K., Occasional Publication (The Laboratory, Citadel Hill, Plymouth, Devon, England) **No. 4**, 92 pp.

Brzezińska, A., A. Trzosińska, W. Żmijewska and L. Wódkiewicz, 1984a. Trace metals in suspended matter and surficial bottom sediments from the southern Baltic. Oceanologia **18**, 59–77.

Brzezińska, A., A. Trzosińska, W. Żmijewska and L. Wódkiewicz, 1984b. Trace metals in some organisms from the southern Baltic. Oceanologia **18**, 79–94.

Cabana, G., and J.B. Rasmussen, 1994. Modelling food chain structure and contaminant bioaccumulation using stable nitrogen isotopes. Nature **372**, 255–257.

Calvert, S.E., and N.B. Price, 1977. Shallow water, continental margin and lacustrine nodules: Distribution and geochemistry, in: Marine Manganese Deposits, ed. G.P. Glasby (Elsevier, Amsterdam), pp. 43–86.

Chapman, P.M., 1997. Is bioaccumulation useful for predicting impacts. Mar. Pollut. Bull. **34**, 282–283.

Cherry, R.D., and L.V. Shannon, 1974. The alpha radioactivity of marine organisms. At. Energy Rev. IAEA, Venna, **12**, 3–45.

Cięglewicz, W., W. Załachowski, J. Szypuła, S. Krzykawski and I. Krzykawska, 1972. Food habits of commercial fish of the southern Baltic, in: Marine Ecosystems – Biology, ed. K. Siudziński (Morski Instytut Rybacki, Gdynia), **Vol. 2**, pp. 363–450 (in Polish).

Das, K., G. Lepoint, V. Loizeau, V. Debacker, P. Dauby and J.M. Bouquegnau, 2000. Tuna and dolphin associations in the North-east Atlantic; Evidence of different ecological niches from stable isotope and heavy metal measurements. Mar. Pollut. Bull. **40**, 102–109.

Dauby, P., A. Khomsi and J.M. Bouquegneau, 1998. Trophic relationships within intertidal communities of the Britanny coasts: a stable carbon isotope analysis. J. Coast. Res. **14**, 1202–1212.

DeNiro, M.J., and S. Epstein, 1978. Influence of the diet on the distribution of carbon isotopes in animals. Geochim. Cosmochim. Acta **42**, 495–506.

DeNiro, M.J., and S. Epstein, 1981. Influence of the diet on the distribution of nitrogen isotopes in animals. Geochim. Cosmochim. Acta **45**, 341–351.

Dietz, R., F. Riget, M. Cleemann, A. Aarkrog, P. Johansen and J.C. Hansen, 2000. Comparison of contaminants from different trophic levels and ecosystems. Sci. Total Environ. **245**, 221–231.

Eide, I., S. Myklestad and S. Melsom, 1980. Long-term uptake and release of heavy metals by *Ascophyllum nodosum* (L.) Le Jol. (Phaeophyceae) *in situ*. Environ. Pollut. Ser. A **23**, 19–28.

Emelyanov, E.M., and O.S. Pustelnikov, 1975b. Compositions of river and marine forms of suspended matter in the Baltic. Geochem. Int. **12**, 195–208.

Fisher, N.S., S.W. Fowler, F. Boisson, J. Carroll, K. Rissanen, B. Salbu, T.G. Sazykina and K.-L. Sjoeblom, 1999. Radionuclide bioconcentration factors and sediment partition coefficients in Arctic seas subject to contamination from dumped nuclear wastes. Environ. Sci. Technol. **33**, 1979–1982.

Fisher, N.S., I. Stupakoff, S. Sañudo-Wilhelmy, W.X. Wang, J.-L. Teyssié, S.W. Fowler and J. Crusius, 2000. Trace metals in marine copepods: a field test of a bioaccumulation model coupled to laboratory uptake kinetics data. Mar. Ecol. Prog. Ser. **194**, 211–218.

Fowler, S.W., 1982. Biological transfer and transport processes, in: Pollutant Transfer and Transport in the Sea, ed. G. Kullenberg (CRC Press, Boca Raton, Florida), **Vol. II**, 65 pp.

Fowler, S.W., 1985. Heavy metal and radionuclide transfer and transport by marine organisms, in: Heavy Metals in Water Organisms. Symposia Biologica Hungarica (Akadémiai Kiadó, Budapest), **Vol. 29**, pp. 191–205.

Fowler, S.W., V.E. Noshkin, J. La Rosa and J. Gastaud, 2000. Temporal variations in plutonium and americium inventories and their relation to vertical transport in the northwestern Mediterranean Sea. Limnol. Oceanogr. **45**, 446–458.

Glasby, G.P. (ed.), 1977. Marine Manganese Deposits (Elsevier Scientific Publishing Company, Amsterdam), 524 pp.

Glasby G.P., and P. Szefer, 1998. Marine pollution in Gdańsk Bay, Puck Bay, and the Vistula Lagoon, Poland – An overview. Sci. Total Environ. **212**, 49–57.

Goldberg, E.D., 1965. Minor elements in seawater, in: Chemical Oceanography, eds. J.P. Riley and G. Skirrow (Academic Press, London and New York), **Vol. 1**, pp. 163–196.

Guary, J.C., S.W. Fowler and T.M. Beasley, 1982. Routes of plutonium uptake and their relation to biomagnification in starfish. Mar. Pollut. Bull. **13**, 99–102.

Harvey, R.W., and S.N. Luoma, 1985. Effect of adherent bacteria and bacterial extracellular polymers upon assimilation by *Macoma balthica* of sediment-bound Cd, Zn and Ag. Mar. Ecol. Prog. Ser. **22**, 281–289.

Havelange, S., G. Lepoint, P. Dauby and J.M. Bouquegneau, 1997. Feeding of the sparid fish *Sarpa salpa* in a seagrass ecosystem: diet and carbon flux. P.S.Z.N.I. Mar. Ecol. **18**, 289–297.

HELCOM, 1995. Radioactivity in the Baltic Sea 1984-1991. Baltic Sea Environment Proceedings (Baltic Marine Environment Protection Commission, HELCOM) **No. 61**.

HELCOM, 1996. Third periodic assessment of the state of the marine enviroment of the Baltic Sea, 1989–1993. Baltic Sea Environment Proceedings (Baltic Marine Environment Protection Commission, HELCOM), No. **64 B**.

Heyraud, M., and R.D. Cherry, 1979. Polonium-210 and lead-210 in marine food chain. Mar. Biol. **52**, 227–236.

Hobson, K.A., and H.E. Welch, 1992. Determination of trophic relationships within a high Arctic food web using $\delta^{13}C$ and $\delta^{15}N$ analysis. Mar. Ecol. Prog. Ser. **84**, 9–18.

Hobson, K.A., and R.G. Clark, 1992. Assessing avian diets using stable isotopes. II: Factors influencing diet-tissue fractionation. Condor **94**, 189–197.

Hobson, K.A., W.G. Ambrose and P.E. Renaud, 1995. Sources of primary production, benthic-pelagic coupling, and trophic relationships within the North-east water polynia: insights from $\delta^{13}C$ and $\delta^{15}N$ analysis. Mar. Ecol. Prog. Ser. **128**, 1–10.

Hobson, K.A., D. Schell, D. Renouf and E. Noseworthy, 1996. Stable-carbon and nitrogen isotopic fractionation between diet and tissues of captive seals: implications for dietary reconstructions involving marine mammals. Can. J. Fish. Aquat. Sci. **53**, 528–533.

Hobson, K.A, J.L. Sease, R.L. Merrick and J.F. Piatt, 1997. Investigating trophic relationships of pinnipeds in Alaska and Washington using stable isotopes ratios of nitrogen and carbon. Mar. Mammal Sci. **13**, 114–132.

Holm, E., 1995. Plutonium in the Baltic Sea. Appl. Radiat. Isot. **46**, 1225–1229.

Ichikawa, R., and S. Ohno, 1981. Uranium content in marine fish. Bull. Jap. Soc. Scient. Fish. **47**, 289.

Ilus, E., S. Klemola, K.-L. Sjöblom and T. Ikäheimonen, 1988. Radioactivity of *Fucus vesiculosus* along the Finnish coast in 1987. Suppl. 9 to Annual Rep. 1987 (STUK-A74). STUK-A83, May 1988 (Finnish Centre for Radiation and Nuclear Safety, Helsinki, Finland) 36 pp.

Ingri, J., R. Löfvendahl and K. Boström, 1991. Chemistry of suspended particles in the southern Baltic Sea. Mar. Chem. **32**, 73–87.

Kidd, K.A., D.W. Schindler, R.H. Hesslein and D.C.G. Muir, 1995a. Correlation between stable nitrogen isotope ratios and concentrations of organochlorines in biota from a freshwater food web. Sci. Total Environ. **160/161**, 381–390.

Kidd, K.A., D.W. Schindler, D.C.G. Muir, W.L. Lockhart and R.H. Hesslein, 1995b. High concentrations of toxaphene in fishes from a sub-Arctic lake. Science **269**, 240–242.

Kidd, K.A., D.W. Schindler, R.H. Hesslein and D.C.G. Muir, 1998. Effects of trophic position and lipid on organochlorine concentrations in fishes from subarctic lakes in Yukon Territory. Can. J. Fish. Aquat. Sci. **55**, 869–881.

Knauss, K., and T.-L. Ku, 1983. The chemical composition and decay-series radionuclide content of plankton from the East Pacific. Chem. Geol. **39**, 125–145.

Kremling, K., and H. Petersen, 1984. Synoptic survey on dissolved trace metals levels in Baltic surface waters. Mar. Pollut. Bull. **15**, 329–334.

Kulikova, I. Z. Seisuma and M. Legzdina, 1985. Heavy metals in marine organisms, in: Heavy Metals in Water Organisms, ed. J. Salánki (Akadémiai Kiadó Budapest). Symposia Biologia Hungarica **29**, 141–151.

Langston, W.J., 1982. The distribution of mercury in British estuarine sediments and its bioavailability to deposit-feeding bivalves.J. Mar. Biol. Assoc. UK **62**, 667–684.

Langston, W.J., 1984. Availability of arsenic to estuarine and marine organisms: a field and laboratory investigation. Mar. Biol. **80**, 143–154.

Langston, W.J., 1985. Assessment of the distribution and availability of arsenic and mercury in estuaries, in: Estuarine Management and Quality Assessment, ed. J.G. Wilson and W. Halcrow (Plenum Press, New York) pp. 131–146.

Langston, W.J., 1986. Metals in sediments and benthic organisms in the Mersey estuary. Estuar. Coast. Shelf Sci. **23**, 239–261.

Li, Y.H., 1981. Ultimate removal mechanisms of elements from the ocean. Geochim. Cosmochim. Acta **45**, 1659–1664.

Luoma, S.N., 1983. Bioavailability of trace metals to aquatic organisms – A review. Sci. Total Environ. **28**, 1–22.

Luoma, S.N., and G.W. Bryan, 1978. Factors controlling the availability of sediment-bound lead to the estuarine bivalve *Scrobicularia plana*. J. Mar. Biol. Assoc. U.K. **58**, 793–802.

Luoma, S.N., and G.W. Bryan, 1981. A statistical assessment of the form of trace metals in oxidized estuarine sediments employing chemical extractants. Sci. Total Environ. **17**, 165–196.

Luoma, S.N., and G.W. Bryan, 1982. A statistical study of environmental factors controlling concentrations of heavy metals in the burrowing bivalve *Scrobicularia plana* and the polychaete *Nereis diversicolor.* Estuar. Coast. Shelf Sci. **15**, 95–108.

Lyngby, J.E., and H. Brix, 1982. Seasonal and environmental variation in cadmium, copper, lead and zinc concentrations in eelgrass (*Zostera marina* L.) in the Limfjord, Denmark. Aquat. Bot. **14**, 59–74.

Lyngby, J.E., and H. Brix, 1987. Monitoring of heavy metal contamination in the Limfjord, Denmark, using biological indicators and sediment. Sci. Total Environ. **64**, 239–252.

Lyngby, J.E., H. Brix and H.-H. Schierup, 1982. Absorption and translocation of zinc in eelgrass (*Zostera marina* L.). J. Exp. Mar. Biol. Ecol. **58**, 259–270.

Łomniewski, K., W. Mańkowski and J. Zaleski, 1975. Morze Bałtyckie (Państwowe Wydawnictwo Naukowe, Warsaw) 508 pp.

Manheim, F.T., 1965. Manganese-iron accumulation in the shallow marine environment, eds. D.R. Schink and J.T. Corless, Sympos. Mar. Geochem. Occasional Publications of the Narragansett Marine Laboratory, University of Rhode Island **3**, 217–276.

Mantoura, R.F.C., A. Dickson and J.P. Riley, 1978. The complexation of metals with humic materials in natural waters. Estuar. Coast. Mar. Sci. **6**, 387–408.

Mańkowski, W., 1978a. Baltic zooplankton and its productivity, in: Productivity of the Baltic Ecosystem, ed. W. Mańkowski (PAN-KBM, Ossolineum, Wrocław) pp. 113–134 (in Polish, with English abstract).

Mańkowski, W., 1978b. The Baltic as an environment able to support life and a source of food, in: Productivity of the Baltic Ecosystem, ed. W. Mańkowski (PAN-KBM, Ossolineum, Wrocław) pp. 315–326 (in Polish, with English abstract).

Marti, Yu.Yu., 1983. Smiejstvo Treskovyie (Gadidae), in: Zhizn' Zhivotnyh – Lancetniki – Krugłorotnyie – Chrashchebyie Ryby – Kostnyie Ryby (The Life of Animals – Fish), ed. T.S. Rass (Prosveshchenie, Izdatelstvo, Moscow), **Vol. 4**, 310–321 (in Russian).

Mikheev, A.V., 1986. Order Arseniformes, in: The Life of Animals – Birds, eds. V.D. Ilichev and A.V. Mikheev (Prosveshchenie, Izdatelstvo, Moscow) **Vol. 6**, 65–112 (in Russian).

Morozińska-Gogol, J., J. Rokicki, K. Frelek and P. Szefer, 1998. Trace metals in *Gasterosteus aculeatus* L. and its parasites from the Gulf of Gdańsk. Oceanol. Studies **27**, 71–75.

Muir, D., B. Braune, B. DeMarch, R. Norstrom, R. Wagemann, L. Lockhart, B. Hargrave, D. Bright, R. Addison, J. Payne and K. Reimer, 1999. Spatial and temporal trends and effects of contaminants in the Canadian Arctic marine ecosystem: a review. Sci. Total Environ. **230**, 83–144.

Nakajima, A., T. Horikoshi and T. Sakaguchi, 1979. Ion effects on the uptake of uranium by *Chlorella regularis*. Agric. Biol. Chem. **43**, 625–629.

Nendza, M., T. Herbst, C. Kussatz and A. A. Gies, 1997. Potential for secondary poisoning and biomagnification in marine organisms. Chemosphere **35**, 1875–1885.

Niemistö, L., T. Vappu and A. Voipio, 1978. Storage of iron and phosphorus in the sediments of the Bothnian Sea. Finnish Mar. Res. **No. 244**, 36–41.

Pacyna, J.M., J. Meunch and F. Axenfeld, 1992. European inventory of trace metal emissions to the atmosphere, in: Heavy Metals in the Environment, ed. J.P. Vernet (Elsevier, Amsterdam), pp. 1–20

Phillips, D.J.H., 1980. Quantitative Aquatic Biological Indicators (Applied Science Publishers Ltd, London) 488 pp.

Phillips, D.J.H., and P.S. Rainbow, 1993. Biomonitoring of Trace Aquatic Contaminants (Elsevier Science Publishers Ltd., London), 371 pp.

Pliński, M., and I. Florczyk, 1987. Phytoplankton in Gdańsk Bay during 1985–1986. Institute of Oceanography, Gdańsk University.

Rau, G.H., D.G. Ainley, J.L. Bengtson, J.J. Torres and T.L. Hopkins, 1992. $^{15}N/^{14}N$ and $^{13}C/^{12}C$ in Weddell seabirds, seals, and fish: implications for diet and trophic structure. Mar. Ecol. Prog. Ser. **84**, 1–8.

Ray, S., D. McLeese and D. Pezzack, 1980. Accumulation of cadmium by *Nereis virens*. Arch. Environ. Contam. Toxicol. **9**, 1–8.

Reinfelder, J.R., N.S. Fisher, S.N. Luoma, J.W. Nichols and W.-X. Wang, 1998. Trace element trophic transfer in aquatic organisms: a critique of the kinetic model approach. Sci. Total Environ. **219**, 117-135.

Renk, H., 1978. Estimation of the primary production of the southern Baltic, in: Productivity of the Baltic Ecosystem, ed. W. Mańkowski (PAN-KBM, Ossolineum, Wrocław) pp. 93–104 (in Polish, with English abstract).

Renner, R.M., G.P. Glasby and P. Szefer, 1998. Endmember analysis of heavy-metal pollution in surficial marine sediments from the Gulf of Gdańsk and Southern Baltic Sea, in: Geochemical Investigations of the Baltic Sea and Surrounding Areas, eds. P. Szefer, P. and G.P. Glasby (Elsevier Science Ltd, Great Britain) Applied Geochem. (Spec. Issue) **13**, 293–304.

Rutkowicz, S., 1982. Encyklopedia Ryb Morskich (Encyclopaedia of Marine Fish). (Wydawnictwo Morskie, Gdańsk), 663 pp.

Sandler, H., 1984. Zinc and copper concentrations in benthic invertebrates considered in relation to concentrations in sediments and water in the Bothnian Sea (northern Baltic). Finn. Mar. Res. **No. 250**, 19–32.

Skwarzec, B., 1995. Polon, uran i pluton w ekosystemie południowego Bałtyku. Rozprawy i monografie (Polska Akademia Nauk, Instytut Oceanologii w Sopocie), 183pp. (in Polish, with English abstract).

Skwarzec, B., 1997. Polonium, uranium and plutonium in the southern Baltic Sea. Ambio **26**, 113–117.

Skwarzec, B., and R. Bojanowski, 1988. ^{210}Po content in sea water and its accumulation in southern Baltic plankton. Mar. Biol. **97**, 301–307.

Skwarzec, B., and R. Bojanowski, 1992. Distribution of plutonium in selected components of the Baltic ecosystem within the Polish economic zone. J. Environ. Radioactivity **15**, 249–263.

Skwarzec, B., D.I. Strumińska and A. Boryło, 2000. Bioaccumulation and distribution of plutonium in fish from Gdańsk Bay. J. Environ. Radioactivity (in press).

Smith, R.J., K.A. Hobson, H.N. Koopman and D.M. Lavigne, 1996. Distinguishing between populations of fresh- and salt-water harbour seals (*Phoca vitulina*) using stable-isotope ratios and fatty acid profiles. Can. J. Fish. Aquat. Sci. **53**, 272–279.

Sokołowski, J., 1958. Ptaki Ziem Polskich (Państwowe Wydawnictwo Naukowe, Warszawa) **Vol. 2**, 569 pp. (in Polish).

Sokołowski, J., 1965. Ptaki Polski – Atlas (Państwowe Zakłady Wydawnictw Szkolnych, Warszawa) 272 pp. (in Polish).

Spaargaren, D.H., 1985. Origin of life: Oceanic genesis, panspermia or Darwin's "warm little pond"? Experientia **41**, 719–727.

Spaargaren, D.H., and H.J. Ceccaldi, 1984. Some relations between the elementary chemical composition of marine organisms and that of sea water. Oceanol. Acta **7**, 63–76.

Stepnowski, P., and B. Skwarzec, 2000. A comparison of ^{210}Po accumulation in molluscs from the southern Baltic, the coast of Spitsbergen and Sasek Wielki Lake in Poland. J. Environ. Radioactivity **49**, 201–208.

Struck, U., M. Voss, B. von Bodungen and N. Mumm, 1998. Stable isotopes of nitrogen in fossil Cladoceran exoskeletons: Implications for nitrogen sources in the central Baltic Sea during the past Century. Naturwissenschaften **85**, 597–603.

Sures, B., H. Taraschewski and J. Rokicki, 1997. Lead and cadmium content of two cestodes, *Monobothrium wageneri* and *Bothriocephalus scorpii*, and their fish hosts. Parasitol. Res. **83**, 618–623.

Swaileh, K.M., and D. Adelung, 1994. Levels of trace metals and effect of body size on metal content and concentration in *Arctica islandica* L. (Mollusca: Bivalvia) from Kiel Bay, Western Baltic. Mar. Pollut. Bull. **28**, 500–505.

Szefer, P., 1977. Investigations on the uranium and thorium contents of Vistula estuary water, also the inshore water of the Gulf of Gdańsk. Stud. Mater. Oceanol. **17**, 379–392. (in Polish with English abstract).

Szefer, P., 1987. Concentration of uranium and thorium by some Baltic seaweeds and lake plants. Mar. Pollut. Bull. **18**, 439–442.

Szefer, P., 1990a. Interelemental relationships in organisms and bottom sediments of the southern Baltic. Sci. Total Environ. **95**, 119–130.

Szefer, P., 1990b. Mass-balance of metals and identification of their sources in both river and fallout fluxes near Gdansk Bay, Baltic Sea. Sci. Total Environ. **95**, 131–139.

Szefer, P., 1991. Interphase and trophic relationships of metals in a southern Baltic ecosystem. Sci. Total Environ. **101**, 201–215.

Szefer, P., 1998. Distribution and behaviour of selected heavy metals in various components of the southern Baltic ecosystem, in: Geochemical Investigations of the Baltic Sea and Surrounding Areas, eds. P. Szefer, P. and G.P. Glasby (Elsevier Science Ltd, Great Britain). Appl. Geochem. (Spec. Issue) **13**, 287–292.

Szefer, P., and R. Wenne, 1987. Concentration of uranium and thorium in molluscs inhabiting Gdansk Bay, Baltic Sea. Sci. Total Environ. **65**, 191–202.

Szefer, P., and B. Skwarzec 1988. Concentration of elements in some seaweeds from coastal region of the southern Baltic and Zarnowiec Lake. Oceanologia **25**, 87–98.

Szefer, P., and K. Szefer, 1991. Concentration and discrimination factors for Cd, Pb, Zn and Cu in benthos of Puck Bay, Baltic Sea. Sci. Total Environ. **105**, 127–133.

Szefer, P., and A. Kusak, 2000. Distribution and relationships of trace metals in zoobenthos and associated sediments of the southern Baltic (in preparation).

Szefer, P., B. Skwarzec and J. Koszteyn, 1985. The occurrence of some metals in mesozooplankton taken from the southern Baltic. Mar. Chem. **17**, 237–253.

Szefer, P., K. Szefer and B. Skwarzec, 1990a. Distribution of trace metals in some representative fauna of the southern Baltic. Mar. Pollut. Bull. **21**, 60–62.

Szefer, P., K. Szefer and J. Falandysz J., 1990b. Uranium and thorium in muscle tissue of fish taken from the southern Baltic. Helgoländer Meeresuntersuch. **44**, 31–38.

Szefer, P., M. Malinga, W. Czarnowski and K. Skóra, 1995a. Toxic, essential and non-essential metals in harbour porpoises of the Polish Baltic Sea, in: Whales, Seals, Fish and Man, eds. A.S. Blix, L. Waloe, O. Utang (Elsevier Science BV), pp. 617–622.

Szefer, P., G.P. Glasby, J. Pempkowiak and R. Kaliszan, 1995b. Extraction studies of heavy-metal pollutants in surficial sediments from the southern Baltic Sea off Poland. Chem. Geol. **120**, 111–126.

Szefer, P., G.P. Glasby, K. Szefer, J. Pempkowiak and R. Kaliszan, 1996. Heavy-metal pollution in surficial marine sediments from the southern Baltic Sea off Poland. J. Environ. Sci. Health **31A**, 2723–2754.

Szefer, P., J. Geldon J., A.A. Ali, F. Páez-Osuna, A.C. Ruiz-Fernandes and S.R. Guerrero Galvan, 1998a. Distribution and association of trace metals in soft tissue and byssus of *Mytella strigata* and other benthal organisms from Mazatlan harbour, mangrove lagoon of the northwest coast of Mexico. Environ. Intern. **24**, 359–374.

Szefer, P., J. Rokicki, K. Frelek, K. Skóra and M. Malinga, 1998b. Bioaccumulation of selected trace metals in lung nematodes, *Pseudalius inflexus*, of harbour porpoise (*Phocoena phocoena*) in a Polish Zone of the Baltic Sea. Sci. Total Environ. **220**, 19–24.

Szefer, P., A.A. Ali, A.A. Ba-Haroon, A.A. Rajeh, J. Geldon and M. Nabrzyski, 1999. Distribution and relationships of selected trace metals in molluscs and associated sediments from the Gulf of Aden, Yemen. Environ. Pollut. **106**, 299–314.

Technical Reports Series, 1985. Sediment K_ds and concentration factors for radionuclides in the marine environment. International Atomic Energy Agency, Vienna, **No 247**, 73 pp.

Thompson, D.R., R.W. Furness and S.A. Lewis, 1995. Diets and long-term changes in $\delta^{15}N$ and $\delta^{13}C$ values in northern fulmars *Fulmarus glacialis* from two North-east Atlantic colonies. Mar. Ecol. Prog. Ser. **123**, 3–11.

Thompson, D.R., R.W. Furness and L.R. Monteiro, 1998. Seabirds as biomonitors of mercury inputs to epipelagic and mesopelagic marine food chains. Sci. Total Environ. **213**, 299-305.

Tieszen, L.L., T.W. Boutton, K.G. Tesdahl and N.A. Slade, 1983. Fractionation and turn-over of stable carbon isotopes in animal tissues: implication for $\delta^{13}C$ analysis of diet. Oecologia (Berlin) **57**, 32–37.

Tuomainen, K., E. Ilus and T.K. Taipale, 1986. Accumulation of certain long-lived radionuclides by litoral algae and bottom animals, in: Study of Radioactive materials in the Baltic Sea. (International Atomic Energy Agency, Vienna). Report (IAEA-TECDOC-362) of the Final Research Coordination Meeting on the Study of Radioactive Materials in the Baltic Sea organized by the IAEA and held in Helsinki, Finland 24–28 September, 1984, pp. 69–78.

Voß, M., and U. Struck, 1997. Stable nitrogen and carbon isotopes as indicator of eutrophication of the Oder river (Baltic Sea). Mar. Chem. **59**, 35–49.

Warnau, M., S.W. Fowler and J.-L. Teyssié, 1999. Biokinetics of radiocobalt in the asteroid *Asterias rubens* (Echinodermata): sea water and food exposures. Mar. Pollut. Bull. **39**, 159–164.

Weiss, D., and F. Moldenhawer, 1986. Results of the radiological Baltic-Monitoring Programme of the GDR during 1975–1983, in: Study of Radioactive materials in the Baltic Sea. (International Atomic Energy Agency, Vienna). Report (IAEA-TECDOC-362) of the Final Research Co-ordination Meeting on the Study of Radioactive Materials in the Baltic Sea organized by the IAEA and held in Helsinki, Finland 24–28 September, 1984, pp. 89–109.

Wenne, R., and K. Wiktor, 1982. Benthic fauna of the inshore waters of Gdańsk Bay. Stud. Mater. Oceanol. **39**, 137–171 (in Polish, with English abstract).

Zatsepin, V.I., Z.A. Filatova and A.A. Shileyko, 1988. Klass Dwustworchatyie molljuski (Bivalvia), in: Zhizn' Zhivotnyh – Molljuski – Igłokożie – Chlenistonogie (The Life of Animals – Molluscs), ed. R.K. Pasternak (Prosveshchenie, Izdatelstvo, Moscow), **Vol. 2**, 65–112 (in Russian).

Chapter 6

Sources of Chemical Elements

A. ENRICHMENT OF CHEMICAL ELEMENTS

(i) Introduction

Some authors have used the enrichment factors (EF) to identify the sources of elements in sediments and suspended material (Buat-Menard and Chesselet, 1979; Li, 1981a, 1981b, 1982; Borole et al., 1982; Kingston and Greenberg, 1984; Brügmann, 1986; Szefer and Skwarzec, 1988a, 1988b; Loring 1990; Szefer, 1990a, 1990b). However, the identification and differentiation of element pollution in sediments are often difficult and therefore determination of additional indices such as correlation coefficients and linear regression is desirable (Buat-Menard and Chesselet, 1979; Sarin et al., 1979; Li, 1982; Hilton et al., 1985; Szefer and Skwarzec, 1988a; Szefer 1990a, 1990b). An approach that is often used to compare the element concentrations in the surficial layers of sediment with those in the deeper layers is anthropogenic factor, AF. Its values have been reported for Baltic sediments by several authors (Brügmann and Hennings, 1982; Szefer and Skwarzec, 1988a; Neumann et al., 1996; Szefer et al., 1995b, 1998b; Vallius, 1999). In order to identify the provenience of selected minor and major elements in Baltic environmental compartments, the EF is computed. To avoid any misleading results (Wangersky, 1962; Chayes, 1967; Szefer, 1990a; Szefer and Skwarzec, 1988a) other statistical approaches, e.g. the cross-correlation analysis are recommended to be additionally carried out to confirm the classification of elements based on EF criteria. Pempkowiak (1992) reported enrichment factors of heavy metals in the Southern Baltic surface sediments dated with Pb-210 and Cs-134 techniques.

(ii) Operational Definitions

In order to estimate "excess" content of the elements analysed in particular compartments (geological and biological) an enrichment factor (EF) has been calculated according to the formula (Li, 1982; Szefer and Skwarzec, 1988a, 1988b; Knaus and Ku, 1983):

$$EF_{Al}^{M} = \frac{C_X^M - C_X^{Na}(C^M/C^{Na})_S}{C_X^{Al} - C_X^{Na}(C^{Al}/C^{Na})_S} : \frac{C_C^M - C_C^{Na}(C^M/C^{Na})_S}{C_C^{Al} - C_C^{Na}(C^{Al}/C^{Na})_S}$$

where the subscripts x, s and c refer to specific compartment, seawater and the Earth's crust, respectively. In the calculations of EF_{Al}^{M} values, the average Earth's crust composition is taken from Taylor (1964), Riley (1971) and Martin and Meybeck (1979). Aluminium, Yb, Th and Fe were used as a reference elements; Fe can be chosen as a normalizer due to the lack of Al data for mesozooplankton and molluscs (Szefer and Skwarzec, 1988a, Szefer, 1990b; Glasby and Szefer, 1998). It is admissible since $EF_{Al}^{Fe} \sim 1$ (for plankton, suspended matter), therefore $EF_{Al}^{M} \sim EF_{Fe}^{M}$ (Buat-Menard and Chesselet, 1979; Li, 1981b; Szefer and Skwarzec, 1988a). The sea salt components in both sample and standard are subtracted by assuming Na is purely of seawater origin (Bogen, 1974; Boutron, 1979; Li, 1981b; Knauss and Ku, 1983). Baltic seawater data used for the calculation of the EF_{Al}^{M} values were taken from Brügmann (1981, 1982), Kremling and Petersen (1984), Szefer et al. (1985) and Ingri et al. (1991). The EF_{Al}^{M} values for Ag, Cr, Cs, Sr, Rb and Li were not seawater-corrected due to the lack of available Baltic seawater data. However, these uncorrected EF_{Al}^{M} values should be the same as seawater corrected values because of very low levels of these elements in seawater (Szefer and Skwarzec, 1998a; Szefer et al., 1996).

EF's of unity or thereabouts indicate that the element is incorporated in the sample dominantly as lithogenous material whereas EF's much greater than unity indicate that the element is enriched in the sample relative to its concentration in Earth's crust, e.g. it is incorporated as an anthropogenic material.

In order to evaluate quantitatively historical changes in an enrichment of selected elements along sediment core, anthropogenic factor, AF, is calculated as follows:

$$AF = C_s/C_d$$

where C_s and C_d refer to the concentrations of the element in the surface sediments and sediments at depth in the sediment column.

Higher than unity values of AF mean that there is an enrichment of given element in a top (recent) segment in respect to deep one of sediment core (background), i.e. deposited in precivilisation era.

(iii) Major Sources of Chemical Elements

Fallout and riverine input

The origin of various elements in the Vistula River and rain fallout near the coastal areas of the southern Baltic was estimated by Szefer and Szefer (1986) and Szefer (1990b).

Vistula River

Knowledge of the chemical composition of the Vistula River allows to estimate of the river flux F (t year $^{-1}$) for given metal introduced into the Gulf of Gdańsk according to the formula:

$$F = (x)_s \cdot Q + (x)_{pm} \cdot S_{pm}$$

where $(x)_s$ is the average dissolved metal concentration in the Vistula River water (μg dm^{-3}), Q – the river water discharge to Gulf of Gdańsk ($32 \cdot 10^{12}$ dm^{-3} year^{-1}), $(x)_{pm}$ – the average concentration of metal in Vistula particulate matter (μg g^{-1} dry wt) and S_{pm} – the river particulate discharge to Gulf of Gdańsk ($0.51 \cdot 10^6$ t year^{-1}).

The F value represents the total metal load in the Vistula River flux supplied to water as a result of hydro-geochemical processes. It is interesting to know whether the rate of natural production of weathered terrigenous material is equal to the rate of the total load transported by the river, i.e. if continental erosion is the sole source of all mineral components (soluble and insoluble) in the river flux. If this is that case, then riverine transported material originates only from continental denudation. However, when we observed a greater total river load as compared with that resulting from natural weathering processes, then the following possibilities exist: a non-steady state between weathering and transport processes, additional natural sources other than continental erosion (sea salt particles, volcanic dust, plant low-temperature emission) and finally anthropogenic sources. Hence, the theoretical flux, F_T (t year^{-1}), may be related solely to metals introduced to river water as a result of present chemical denundation of the weathered land material (Martin and Meybeck, 1979):

$$F_T = (x)_c \cdot S_{pm} \cdot (Fe)_{pm}/(Fe)_c$$

where $(x)_c$ is the average concentration of metal in the surficial Earth's crust (μg g^{-1} dry wt), $(Fe)_{pm}/(Fe)_c$ – the average ratio of Fe content to the Vistula River particulate matter to that of surficial Earth's crust.

The dissolved transport of Fe in the Vistula is estimated to be ~ 5% (Szefer, 1989), so this metal may be used as a reference element. This is justified since Martin and Meybeck (1979) recommended not only Al as a normalising element but also those elements for which river dissolved transport is negligible, i.e. < 10%. After an estimation of the ratio $(Fe)_{pm}/(Fe)_c$ (1.37) and the total suspended load in the Vistula ($0.51 \cdot 10^6$ t year^{-1}), it may be concluded that the total amount of surficial fresh rock (S_T) from which the Vistula dissolved and particulate products are derived, should amount to ~ $0.7 \cdot 10^6$ t year^{-1}. The difference

(S_D) between S_T and S_{pm} ($\sim 0.2 \cdot 10^6$ t year^{-1}) should be related to the theoretical total dissolved transport of material originating from chemical denudation. The observed Vistula flux obtained by direct measurements of the transport of Ca, Mg, K and Na is, however, significantly greater than the theoretical value, S_D, which is derived from the Fe-content ratio as computed above (Szefer, 1990b). Such a discrepancy may be due to additional sources other than weathering processes, a non-steady state system and/or differential weathering of various types of rocks processes (Martin and Meybeck, 1979). It is important to note that no correction has been made here for the Ca, Mg, K and Na content to eliminate marine salt particles, dust fallout and pollution loads. According to Martin and Meybeck (1979) the correction for surplus of macroconstituents in the world river dissolved load is ~ 10, 10 and 30% for Ca, Mg and Na, respectively. To identify the metal sources in the Vistula River, the flux ratio, FR, instead of the absolute fluxes is considered for each metal (Martin and Meybeck, 1979):

$$FR = [(x)_s \cdot Q + (x)_{pm} \cdot S_{pm}]/[\,(x)_c \cdot S_{pm} \cdot (Fe)_{pm}/(Fe)_c]$$

This is the ratio of observed total flux (F) to theoretical flux (F_T) for the Vistula.

Rain fallout near Gulf of Gdansk

The rain fallout flux, F (t year^{-1}), for each metal to the Gulf of Gdańsk was calculated according to the formula:

$$F' = (x)'_s \cdot Q' + (x)'_{pm} \cdot S'_{pm}$$

where $(x)'_s$ is the average dissolved metal concentration in rain fallout (μg dm^{-3}), Q' – the rain water discharge to the Gulf of Gdańsk ($2.2 \cdot 10^{12}$ dm^{-3} year^{-1}), $(x)'_{pm}$ – the average concentration of metal in the rain particulate matter (μg g^{-1} dry wt) and S'_{pm} – the rain fallout particulate discharge to the Gulf of Gdańsk ($0.081 \cdot 10^6$ t year^{-1}).

The calculations involved data for average annual wet precipitation (440 mm) measured during 1978–1980 on the Hel peninsula (Szefer and Szefer, 1986), the bay surface ($5 \cdot 10^3$ km^2) (Brzezińska and Garbalewski, 1980) and average particulate matter concentration in rain fallout (0.0366 g dm^{-3}) (Szefer and Szefer, 1986).

The F'_T values were also computed for each metal assuming that its source in the marine atmosphere is solely from particles of weathered continental material:

$$F'_T = (x)_c \cdot S'_{pm} \cdot (Fe)'_{pm}/(Fe)_c$$

where $(x)_c$ is the average concentration of metal in the surficial Earth's crust (μg g^{-1} dry wt), $(Fe)'_{pm}/(Fe)_c$ – the average ratio of Fe content to the rain fallout particulate matter to that in the surficial Earth's crust (1.16).

To identify the source of each metal, the flux ratio FR' (quotient of the F' and F'_T values) was computed like in the case of the Vistula flux. Besides the F and FR values, enrichment factor, EF, was also computed. There was a great similarity between these values. According to Szefer (1990b, 1998) Cd, Pb, Zn and Cu in both atmospheric and river influxes had EF values significantly greater than unity (log EF > 0); these metals could therefore be classified as "biophile" or "anomalously enriched elements" (Li, 1981b). The EF values were an order of magnitude greater for Pb in the atmospheric fallout than in the Vistula River runoff which suggest widespread atmospheric contamination of Pb in the southern Baltic (Fig. 6.1). Lead was therefore transported to the southern Baltic mainly by air masses rather than by the Vistula River. Other elements, namely Co, Th, Ca, Mg, Ti, U and, to a lesser extent, Mn and Ni, were generally characterised by EF's close to unity (log EF~0). This means that these metals had insignificant contribution to Baltic biota.

However, based on concentration data from the 1970's, the Vistula River was the major source of the other metallic pollutants which were characterised by greater EF's for the Vistula River runoff than for atmospheric fallout. According to Renner et al. (1998) Cu, Zn and Ag are introduced into the Gulf of Gdańsk principally from the Vistula River whereas Cd and Pb are introduced, in part, by atmospheric transport. Mass transport calculations for Cd and Pb also indicate that a significant proportion of these elements in the southern Baltic Proper has a riverine source (HELCOM, 1998). The dual source of Cd and Pb (atmospheric and riverine) into the southern Baltic may therefore explain the complex interelement relationships displayed by these elements in the sediments of this region. Silver, on the other hand, is introduced into the marine environment mainly associated with sewage sludge (Ravizza and Bothner, 1996). It has been indicated that Cu, Zn, Ag, Cd and Pb in the sediments of the Gulf of Gdańsk, southern Baltic, are of anthropogenic origin (Szefer et al., 1993a, 1993b, 1995a, 1996).

Schneider (1987) reported EFs of selected trace elements and macroconstituents for the atmosphere over Kiel Bight, western Baltic, identifying anthropogenic sources for elements such as As, Cu, Ni, Pb, V and Zn.

Biota

It is pointed out that suspended matter may be rich in biogenic matter or in some elements which have been enriched in particulate matter by biological activities (Krishnaswami and Sarin, 1976; Buat-Menard and Chesselet, 1979; Boström et al., 1981). Therefore an enrichment factor EF was calculated also for body of organisms exposed directly to water body such as plankton, seaweeds, molluscs (Li, 1981b; Knauss and Ku, 1983) as well as for their inner parts, i.e. the liver of marine mammals (Mackey et al., 1995).

Szefer and Skwarzec (1988a, 1988b) reported that Zn, Cd, Pb and partly Cu in seaweeds, plankton and molluscs were significantly concentrated as compared to Earth's crust and seawater; so these metals formed a group of typical 'enriched' metals whose EF^{M}_{Fe} values were higher than 10. However, Ni in seaweeds, mesozooplankton and molluscs might have sources other than the Earth's crust and sea salt ($10 > EF^{M}_{Fe} > 2$) or its biological affinity was a weaker than the typical 'enriched' elements. It is noteworthy that EF^{M}_{Fe} values of Zn, Cd, Pb and Cu for biota (seaweeds, mesozooplankton, molluscs) were generally higher than those calculated for the surface sediments (Figs. 6.1 and 6.2). This can be explained by simple dilution of the pollutants and biogenic fraction with 'natural' material low in heavy metal content in sediment as well as the release of the metal to upper-laying water layers during organic diagenesis (Szefer, 1990a, 1998).

The relative ability of accumulation was assessed for the liver of Baltic mammals (*Phocoena phocoena*) by computing EF values, i.e. the ratio of given element in the liver tissue and its concentration in seawater, normalised to Na (Szefer et al., 2000c). According to Mackey et al. (1995) classification, the EF's corre-

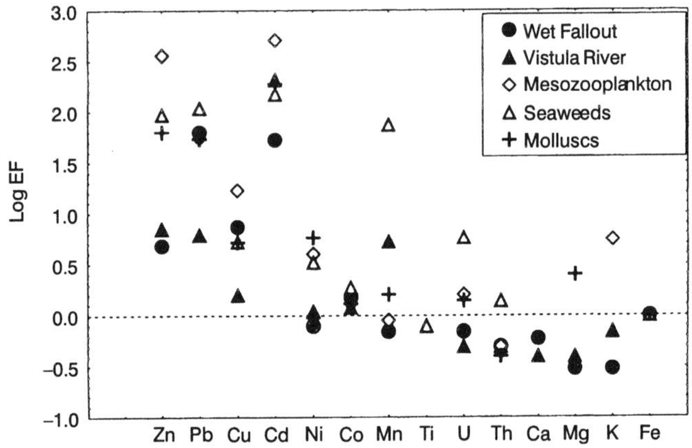

Fig. 6.1. Enrichment factors of various elements in particular biological compartments of the southern Baltic including atmospheric and river input. After Szefer (1998).

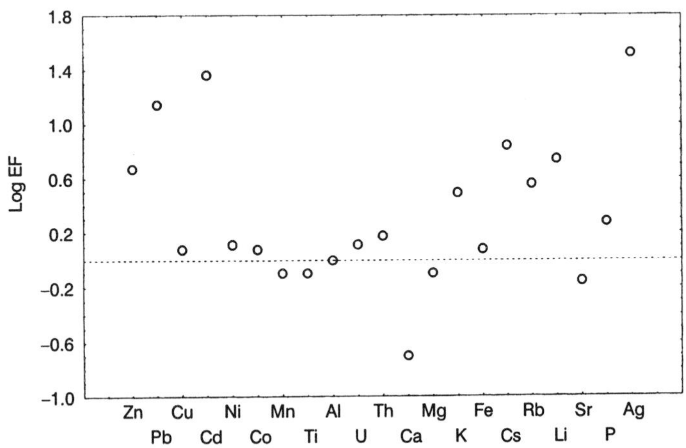

Fig. 6.2. Enrichment factors of elements in bottom sediments of the southern Baltic. After Szefer (1998).

sponded to Baltic mammals fall roughly to three categories: electrolytes, essential trace elements and toxic (or potentially toxic) elements.

Suspended matter

In order to evaluate the distribution of elements in Baltic suspended matter, their concentration data are usually normalised in respect to e.g. Al as an element of terrigenic origin (Boström et al., 1981, 1988; Bernard et al., 1989; Ingri et al., 1991; Brügmann et al., 1992). Most of the elements, i.e. Cd, Ba, Cu, Mn, Pb, V and Zn except Ca, Fe and Mg exhibit significant abilities to enrich in Baltic suspended matter in respect to geochemical background (from 14% for V to 97% for Pb). According to Brügmann et al. (1992), vertical distribution of the excess values appears to show that significant contribution of the metal excess is represented by the easily mobilizable fraction and is imported from the atmosphere (Pb, V and Ba). The remaining elements are enriched in the bottom samples because of remobilization from the sediments and new precipitation in oxic conditions (Mn). In the case of Cd, Cu and Zn, inputs from the atmosphere as well as from the sediments may be responsible for their excess in the particulate material of the Baltic Sea (Brügmann et al., 1992). According to Bernard et al. (1989) nearly particulate Mn is present in authigenic phase of Baltic suspended matter, i.e. as oxides/hydroxides or resuspended carbonates. The extractable Zn fraction constituted the major abundance of this element in particulate material (Bernard et al., 1989). For extremely high levels of Zn is expected the presence of $ZnCO_3$ enriched particles which were identified by a manual electron microprobe (see Fig. 2.19 in Chapter 2D). Baltic particulate matter is also abundant in Ba, which could be precipitated as $BaSO_4$, especially in region where the reiverine influx of dissolved Ba meets more saline water masses with higher sulphate concentrations (Bernard et al., 1989). Although no positive excess of Fe concentration was detected, highly positive excess values were sporadically observed for single particles in samples contained prevailing the leachable fraction (Fig. 2.19).

Pohl et al. (1998) performed intercomparison study of suspended matter and surficial sediments ($< 63\ \mu m$) for metal concentrations in the Pomeranian Bay. From EF values clearly results that accumulation of Zn, Cd and Pb in suspended matter is 4-, 2- and 1.5 times greater, respectively than that in the surficial sediments. These elements are suspected to be anthropogenic in origin.

Surficial bottom sediment

Data reported for southern Baltic sediments (< 2 mm fraction) showed (Szefer et al., 1996) that EF values for Co, Ni, Fe, U, Th, Sr, Cu, Cr and Al were, on average, near unity (log EF $\sim$ 0). These elements were therefore dominantly lithogenous in origin. Zinc, Cd, Pb and Ag were characterised by $EF^{M}_{Al} \gg 1$. This indicated that these elements were significantly enriched relative to crustal material and that their concentrations were not directly controlled by continental weathering. This assumption was supported by the weaker correlations of Zn, Cd, Pb and Ag with Al; P also had EF^{M}_{Al} slightly greater than unity and displayed a weak correlation with Al and might therefore be classed with these elements. The elevated levels of Ag, Cd, Pb, Zn and P in Gulf of Gdansk sediments were therefore classified as anthropogenic in origin (Szefer and Skwarzec, 1988a; Szefer, 1990a, 1990b). These elements were all known to display toxic effects when incorporated into marine organisms (Bryan and Langston, 1992). EF values were < 1 (log EF < 0) for Ca which indicated that this element is associated with phase other than aluminosilicates (e.g. $CaCO_3$). EF values were > 1 for Cs, Rb, Li and K. These alkali elements were apparently adsorbed on clay minerals during their transport to the Baltic ecosystem (Szefer et al., 1996). Manganese, Ca and Sr are weakly correlated with Al and show average $EF_{Al}^{M} < 1$. This implied that a significant proportion of these elements was associated with phases other than clay minerals. Fractionation of Mn from Al was most probably the result of the diagenetic remobilization of Mn in the sediments (Boström, et al. 1983). According to Blazhchishin (1982), Mn is 3–4 times more mobile than Fe in the Baltic watershed. The average EF_{Al}^{Na} is equal to 1.9. This enrichment may be an artefact resulting from the fact that the sediments were not washed with distilled water prior to drying.

The "excess" concentration of particular element relative to its average concentration in the earth's crust was estimated as follows (Bernard et al., 1989; Brügmann et al., 1992; Szefer et al., 1996):

$$M_{exc.} = M_{tot.} - [Al_{tot.} \times (M/Al)_c]$$

where $M_{tot.}$ and Al_{tot} denote the average bulk concentrations of the element and Al in surficial sediments and $(M/Al)_c$ the average element to Al ratio in the earth's crust. Average "excess" concentrations for Ag, Cd, Pb, Zn, Cs, Li, Rb and K were $> 70\%$, for Na and P $\sim 50\%$ and for Cu, Cr, Co, Ni and Fe 10–40%. Enrichments of Mn, Sr, Ca and Mg in the surficial sediments were not in an agreement with the results of the enrichment factors (Szefer et al., 1996).

It is important to note that there was decrease in the EF values from the Oder Lagoon through the Pomeranian Bay to the Arkona Basin; this distribution pattern is attributed to remobilization and solution processes in surficial layers of the sediment and in the mobile nepheloid layer (Pohl et al., 1998).

Lagoonal sediments

Using EF may be problematic, particularly for the heavy metals, since it may reflect primarily the influence of enhanced concentrations of the metal in the overlying waters or diagenetic processes taking place within the sediment column as a result of redox-induced element remobilization or sulphide deposition. Brügmann and Matschullat (1997) have pointed out that, in anoxic basins of the Baltic Sea, Cu, Cd and Hg are fixed in the sediments as sulphides whereas Fe, Mn and Co are mobilised into the water column. However, the surface sediments of the Vistula Lagoon appear to be oxidising suggesting that fixation of heavy metals in these sediments as sulphides is not significant and that the EF's do reflect the concentrations of the heavy metals in the brackish water of the lagoon (Szefer et al., 1999a). Among 22 elements analysed in the Vistula Lagoon sediments, Ag, Sb, As, Cd and Pb were characterised by average EF's > 1 which means that the surficial sediments were significantly enriched in these elements. The distribution patterns of these anthropogenically-derived elements demonstrate the relatively great variability of their concentrations from one site to other. The concentrations of these elements do not vary systematically along the lagoon or with each other. Maximum values of Ag, Sb, As and Cd were observed in the minor Szkarpawa River (Szefer et al., 1999a).

Sediment cores

Since all samples were fine grained clays or silts, they were directly analysed for element concentrations without previous sieving (Szefer and Skwarzec, 1988a). The concentration data concerning top segments of sediments cores collected in the southern Baltic were utilised to compute EF values as well as to AF values (Szefer and Skwarzec, 1988a). Since metals such as Ti, Ni, K, Co, Th in all cores, and partly Mg (core G-2), were positively correlated with Al and Fe ($p < 0.01$) and showed both $EF^{M}_{Fe(Al)}$ and AF values close to unity, crustal weathering may be the main source of their concentration in the sediment analysed. Calcium, and U with Mn were also not enriched relative to the crust ($EF^{M}_{Fe(Al)} \sim 1$, AF ~ 1); however, the lack of positive correlation between U, Mn, Ca and the independent major matrices (Al, C_{org}) suggests the importance of different phases as contributors of these elements in the cores analysed. According to various authors, the marine sediments may be composed of detrital and authigenic or biogenic fractions of U (Ku, 1965; Mo et al., 1973), Mn (Suess, 1979; Brügmann and Hennings, 1982; Marchig et al., 1985) and Ca (Sarin et al., 1979). Copper, Zn and especially Pb and Cd were enriched relative to the crust ($EF^{M}_{Fe(Al)} > 1$); since these metals did not correlate significantly with Al and Fe, their concentrations in sediments were not directly associated with continentally derived aluminosilicate minerals. Based on the AF values (higher than unity), it may be suggested that an anthropogenic source was mainly responsible for the presence of these elements in recent sediments of the Gulf of Gdańsk. It is not surprising because Poland is a major mining country (southern district) with major Pb–Zn deposits in Upper Silesia and major Cu deposits in Lower Silesia. For instance, primary non-ferrous metal production in Poland in 1979 was the major emission source for Zn (3,780 t yr^{-1}), Cd (178 t yr^{-1}), Cu (877 t yr^{-1}) and Pb (2,140 t yr^{-1}), i.e. 80, 86, 66 and 53% of the total emission, respectively. Another important source of Pb emission was gasoline combustion, since nearly 30% of the total anthropogenic Poland emission of Pb (4,568 t yr^{-1}) came from this source. Krüger (1996) has demonstrated the atmospheric input of both Cd and Pb into the Baltic Sea. For Pb, it was shown that 1388 tonnes were introduced into the Baltic in 1985 and this decreased to 627 tonnes in 1990. Using the best available technology in the non-ferrous metals industry and using only unleaded gasoline in Europe, it was estimated that the atmospheric Pb input could be reduced to 190 t yr^{-1}. For Cd, the atmospheric input into the Baltic is about 19 t yr^{-1}.

B. CONCENTRATION RATIO

(i) Introduction

Concentration ratio of one element to another could be useful index of the element origin in the marine environment. It concerns organic compounds, e.g. organic carbon to nitrogen ratio, as well as radionuclides, e.g. plutonium, uranium. Many authors calculated concentration ratio of chemical element in top and deep layers of sediment cores. This index named anthropogenic factor (see Chapter 6A) is quantitative measure of an enrichment of element in respect to its background concentration attributed to precivilisation era.

(ii) Operational Definitions

The origin of radioisotopes can be identified by estimate their appropriate isotopic ratio, e.g. the $^{238}Pu/^{239+240}Pu$ in respect to reference values corresponded to weapon grade plutonium, nuclear test fallout, releases from nuclear fuel reprocessing plants and Chernobyl fallout.

The $^{234}U/^{238}U$ ratio is helpful in identification of sources of U in the marine environments when it is compared to the typical values for e.g. seawater or surficial soil. The $^{235}U/^{238}U$ can be also suitable tool in determining anthropogenic sources of U in the marine ecosystems (Szefer, 1981, 1987).

To recognise possible sources of C_{org} in recent sediments in 1980 (topmost 5 cm), the C_{org} to N ratio (by weight) has been used as an indicator of terrigenous addition to sediments (McMahon and Patching, 1984; Naik and Poutanen, 1984). Banse (1974) interpreted data for the C to N ratio of phytoplankton.

Measure of nutrient sources, nutrient utilisation and changes in the rates of denitrification in the marine environments is the isotopic composition of N in suspended matter and bottom sediments (Altabet and Francois, 1994; Struck et al., 1998). The same ratio is used also as a tool to trace diet because the isotope ratios of a consumer are strictly related to those of their preys. According to Voß and Struck (1997) and Das et al. (2000) stable N and C isotope ratios are expressed in conventional δ notation (see formula in Chapter 5B).

(iii) Major Sources of Nutrients and Radionuclides

The C/N values ranged from 4.5 to 7.1 at station P-2, from 4.4 to 6.2 at station P-10, from 7.8 to 8.4 at station G-2 and from 7.3 to 7.6 at station P-38 (Fig. 6.22).

The mean values calculated for the southern Baltic surface sediments analysed varied from 5.3 to 8.1. It should be mentioned that small variations in the C/N values can also be a result of analytical inaccuracies. Keeping in mind that low C/N ratios (from 3 to 7) are characteristics of phytoplankton (Antia et al., 1963; Banse, 1974; Slawyk et al., 1978) and higher values (more than 10) are attributed to an influx off terrigenous material (Flemer and Biggs, 1971), it is concluded that an important source of C_{org} in southern Baltic sediments is humic substance of planktonic origin (autochthonous). For comparison, the C/N values for sediments from the Ryga Bay are significantly higher (up to 91) which suggest dominant influence of terrigenous humus (allochthonous) on the organic composition of these sediments (Blazhchishin, 1982). According to this author the mean C/N ratio was 10–11.5 for Bothnian Bay, 8.9 for the western Baltic and 6–8 for the central and the southern Baltic. Such a sequence of C/N values may be attributed to a decreasing percentage of terrigenous humus from the north to the south, according to an increase of the living biomass production in the Baltic. Lassig et al. (1978) and Renk (1978) have reported annual primary production rates of phytoplankton carbon of ~ 15–30 g m^{-2} in Bothnian Bay, ~ 60 g m^{-2} in the Bothnian Sea and ~ 80–100 g m^{-2} in the Baltic Proper and the southern Baltic.

The relatively high increase of N and P in the topmost layer (representing the time after about 1960) may reflect both the degradation in top sediment layers and an increasing eutrophication of the southern Baltic during the last 35 years. The latter process is caused by an increasing discharge of P and N compounds (fertilisers, detergents) into the Baltic Sea. According to Pawlak (1980) large amounts of nutrients, i.e. 308,890 t of N and 25,825 t of P enter the Baltic Sea annually. The nutrient situation in the Baltic water has been studied by Nehring (1984a, 1984b, 1985). According to the author, not only pollution but also long-term hydrographic variations are responsible for the increasing PO_4^{3-} and NO_3^- concentrations in the winter surface layer of the Baltic Proper. Moreover, long-term PO_4^{3-} and NO_3^- accumulation has also been found in the deep water of the Gdańsk Deep and other deeps (Nehring, 1984a). So, it may be said that europhication in the Baltic Sea continues unabated. Consequences of this process manifest in the increase of the biomass of zooplankton and zoobenthos and in the yields of the Baltic Sea fisheries as well as in longer anoxic periods in the bottom water (Nehring, 1985).

As has been noted previously, besides eutrophication degradation-diagenetic processes may also be responsible for elevated concentrations of N and P in the surface sediments. The N distribution in the cores studied is probably associated with decomposition of some organic constituents, mainly by bacterial action, giving rise to NH_3, which next may be oxidised to NO_3^- in the oxic layers of sediments. Consequently, the decrease of N with depth in core is observed since in these reactions N is removed from sediment to the interstitial water (Rittenberg et al., 1955).

The $^{238}Pu/^{239+240}Pu$ activity ratios amounting to 0.016, 0.025, 0.25 and 0.47 correspond to weapon grade plutonium, nuclear test fallout, releases from nuclear fuel reprocessing plants and Chernobyl fallout, respectively. The corresponding activity ratios for $^{241}Pu/^{239+240}Pu$ are 4, 16, 25 and 86, respectively (Holm, 1988, 1995). According to Holm (1995) the main source of Pu in the Baltic Sea is nuclear test fallout besides other sources have been also identified.

C. DISTRIBUTION PATTERN OF ELEMENTS IN VIEW OF MULTIVARIATE APPROACH

(i) Introduction

In order to reduce relatively large number of variables to a smaller number of orthogonal factors, the original data are treated by multivariate statistical techniques, e.g. principal component analysis (PCA) or factor analysis (FA). Multivariate data analysis has been presented extensively by Cooley and Lohnes (1971). The statistical analysis of compositional data sets is complicated by the non-negativity and constant-sum constraints, as has been thoroughly documented by Aitchison (1986) and others. Ehrlich and Full (1987) discussed use of statistical methods in the earth sciences. Q-mode factor analysis of compositional data, especially geochemical and petrologic has been also presented (Miesch, 1976a, 1976b; Zhou et al., 1983). In environmental analysis PCA or FA have been used to identify sources of chemical pollutants (Li, 1981b, 1982; Favretto and Favretto, 1984a, 1984b, 1988; Esbensen et al., 1987; Armanino et al., 1996; Zhu et al., 1997; Feng et al., 1998). This or similar multivariate approaches have been successfully used for processing concentrations data concerning, biota, e.g. plankton (Li, 1981b), phyto- and zoobenthos (Julshamn and Grahl-Nielsen, 1996; Szefer and Wołowicz, 1993; Astley et al., 1999; Szefer et al., 1998a, 1999b; Szefer et al., 2000b, 2000d, 2000e), fish (Julshamn and Grahl-Nielsen, 1996; Andres et al., 2000; Szefer et al., 2000a), marine mammals (Julshamn and Grahl-Nielsen, 2000; Szefer et al., 2000c) as well as atmospheric fallout and marine aerosols (Hopke, 1976; Heidam, 1981; Li, 1981b; Piña et al., 2000), suspended matter (Li, 1981b; Bernard et al., 1989; Yeats and Loring, 1991; Jambers et al., 1999; Zwolsman and van Eck, 1999), soils (Davies and Wixson, 1987; stream and marine sediments (Li, 1982; Loring, 1984; Mantovan et al., 1985; Zhou, 1985, 1987; Garrett, 1989; Vogt, 1989; Brügmann and Lange, 1990; Hallberg, 1991; Szefer and Kaliszan, 1993; Szefer et al., 1995a; Emmerson et al., 1997; Szefer, 1998; Danielsson, 1998; Virkanen, 1998; Danielsson et al., 1999; Maurer et al., 1999; Shin and Fong, 1999; Szefer et al., 1999a, 2000f), metalliferous sediments (Renner et al., 1997) and ferromanganese nodules (Li, 1982; Renner et al., 1998).

The statistical multivariate analyses used in environmental data processing are factor analysis (FA), principal component analysis (PCA), end-member analysis, cluster analysis and canonical discriminant analysis (DA). Spatial, interspecies, inter-size and seasonal and other environmental variations in elemental concentrations are tested by analysis of variance (ANOVA) and the multiple comparison test of Tukey (Van Hattum et al., 1991; Zar, 1996). Malinowski (1991) and Beebe et al. (1998) in theirs books showed how to solve different problems using the most widely available chemometric methods.

(ii) Operational Definitions

PCA creates "new" dimensions of the data (Flury and Riedwyl, 1988) and evaluates a reduced number of independent factors or principal components describing the information included in a system of characteristic. It aims at finding a few components or factors that explain the major variations within the data matrix. Each component or factor in PCA or FA, respectively is a weighted linear combination of the original variables. Components or factors only with eigenvalues higher than unity should be preferably considered (Beebe et al., 1998; Danielsson et al., 1999). The factor loading quantities the individual variables' contribution to the respective factor. The ranking of the factors is characterised by the amount of variance which they explain (Struck et al., 1997). The main criticism towards PCA is associated with the difficulties in interpreting the components because of sometimes the lack of information about their meaning in either physical or chemical sense. Moreover in reduction of all the original variables to only a few factors, a relatively small number of components are used to describe a large part of the variation; hence some information is omitted (Danielsson, 1998). However, according to Kuik et al. (1993) also this unexplained variance can be taken into account resulting in improve the reliability of this approach.

Cluster analysis consists of a number of various techniques (Sharma, 1996). In clustering the objects are grouped so that 'similar' objects fall into the same class. Objects in one cluster should be homogenous in relation to some characteristics explaining within cluster properties; they also should be well separated from other the elemental groupings (Danielsson et al., 1999). Cluster analysis assigns particular variables with similar courses to clusters of variables (Struck et al., 1997). Clustering techniques are divided into two basic groups, namely hierarchic and non-hierarchic methods. It is important to decide which clustering procedure is the most suitable. According to Sharma (1996), Wards's minimum variance technique was superior because of giving a larger amount of correct classified observations as compared to most other methods, although it is not always better than average linkage clustering. This finding was supported by Massart and Kaufman (1983). One of major difficulties and criticisms of the technique is defining of objectivity (Danielsson, 1998; Danielsson et al., 1999). It should be noted that clustering technique always produces some clusters, even if the results are completely random and that most methods are biased towards finding spherical and elliptical shaped clusters. When another shape of cluster is obtained, these clusters are not always found causing a loss of information and sometimes even misleading data (Mardia et al., 1989; Everitt and Dunn, 1991; Danielsson, 1998).

Discriminant analysis determines variations between groups of nominal 'elements' which are characterised by numerical variables. Discriminant functions depending linearly on the element concentration studied are formed. The numerical values of the discriminant functions are the coordinates of the locations in a plane described by the two discriminant functions (Struck et al., 1997).

A description of the particular endmember analysis undertaken on the sediment dataset is reported by Renner et al. (1998). Objectives definition of external endmembers in the analysis of mixtures was given by Full et al. (1981). In general, there are indefinitely many sets of extreme points for a particular set of exact mixtures. However, since associations between elements of a geochemical dataset are not arbitrary, a conservative strategy is to seek extreme compositions (datapoints) that are geometrically close to the data and therefore close to observed reality (Full et al., 1981; Ehrlich and Full, 1987; Renner, 1995). A detailed examination of the multivariate analysis was performed by Renner (1988, 1991, 1993a, 1993b, 1995) and Renner et al. (1989, 1997). The abundances of the endmember estimates for any sample are non-negative mixture proportions and therefore also sum to one. Endmember compositions include extreme values for all the elements studied. Depending on the number of endmembers, they are represented geometrically by extreme points or vertices of simplexes (line segments, triangles, tetrahedra etc.). All the datapoints must lie within such a simplex (Renner et al., 1998).

The above mentioned statistical analysed were all applied to estimate data obtained for samples collected in the Baltic Sea and adjacent areas in respect to spatial, species, age or seasonal trends. This approach concerned element concentrations in invertebrates, i.e. *Cerastoderma glaucum* from the Baltic Sea and other regions (Szefer and Wolowicz 1993), *Mytilus edulis*, *Fucus vesiculosus* and *Balanus improvisus* from the Baltic Sea, North Sea and the Hardangerfjord, Norway (Julshamn and Grahl-Nielsen, 1996; Struck et al., 1997; Szefer et al., 2000b, 2000d). Representative vertebrates of three species of fish, i.e. *Perca fluvialitis*, *Gadus virens* and *Platichthys flesus* from the Baltic Sea and the coasts of Norway were also analysed in this respect (Szefer et al., 2000a, Julshamn and Grahl-Nielsen, 1996). The distribution patterns of trace metals in marine mammals, i.e. *Phocoena phocoena* from coastal areas of the Polish, Danish and Greenland, harp seals (*Pagophilus groenlandicus*) and hooded seal (*Cystophora cristata*) from the Greenland Sea were also studied in view of FA or PCA (Szefer et al., 2000c; Julshamn and Grahl-Nielsen, 2000).

Analyses of the effect of size (age), spatial and temporal trends for selected elements in Baltic organisms and their substrata (bottom sediments) were performed and the data obtained for e.g. *Talitrus saltator*, *Mytilus trossulus*, *Balanus improvisus* and *Perca fluviatilis* were processed by ANOVA or ANCOVA multivariate analysis (Rainbow et al., 1998, 2000; Szefer et al., 2000a).

Several authors utilised FA or PCA for quantitative evaluation of both the horizontal and vertical distributions of different elements in geological material, e.g. bottom sediments from the southern Baltic (Szefer and Kaliszan, 1993; Szefer et al., 1993a, 1995a; Szefer, 1998, Szefer et al., 1999a, 2000f).

(iii) Multivariate Distribution Patterns of Elements

BIOTA

Seaweeds

In order to verify the regional influences of seawater on the biochemical composition of *Fucus vesiculosus* from the Baltic Sea and North Sea, which are independent of the presence of trace elements, DA was utilised for evaluation of macroelement concentrations in the seaweed as variables (Struck et al., 1997). Since macroalgae accumulate elements from surrounding solution, groups of locations were formed according to the course of salinity. DA analysis indicated that Baltic and North Seas locations are clearly separated like in the case of cluster analysis. Discriminant analysis was also performed for trace element concentrations (Struck et al., 1997) resulting in reduced number of location groups in comparison with the DA of the concentration patterns of macroelements for the seaweed. This distribution pattern was in an agreement with cluster analysis and indicated the reduced influence of trace-element-independent ecosystem parameters on the uptake of trace elements as compared to the uptake of macroelements (Struck et al., 1997). DA of trace-element concentrations in seaweeds collected at Eckwarderhörne made possible the detection of the pollutants emitted by industrial activity in Wilhelmshaven (Struck et al., 1997).

Molluscs and crustaceans

The first three factors described 44.0% (for the soft tissue) and 46.1% (for the byssus) of the total variance with corresponding eigenvalues amounted to 2.03–1.05 and 2.22–1.32, respectively (Szefer et al., 2000d). As can be seen in Figure 6.3 the objects corresponding to the soft tissue of molluscs inhabited the Pomeranian Bay and the Słupsk Bank region display the highest values of F2 and form a group which is clearly separated from that consisted of tissue samples coming from the Gulf of Gdańsk, characterised by the lowest values of F2. A plot for metals displays loading of Ag, Fe, Co, and partly Pb, Cr and Hg which corresponds to the Gulf of Gdańsk samples described also by the lowest values of F2. It is well isolated from loadings of other metals, especially Ni, Zn, Cd and Cu referring to the Pomeranian Bay and the Słupsk Bank specimens (described by the highest values of F2).

A plot of the samples based on their factor scores shows a clustering of the byssi samples also into two main areas, each corresponding to a geographically distinct zone (Fig. 6.4). Samples from the Pomeranian Bay region have the high-

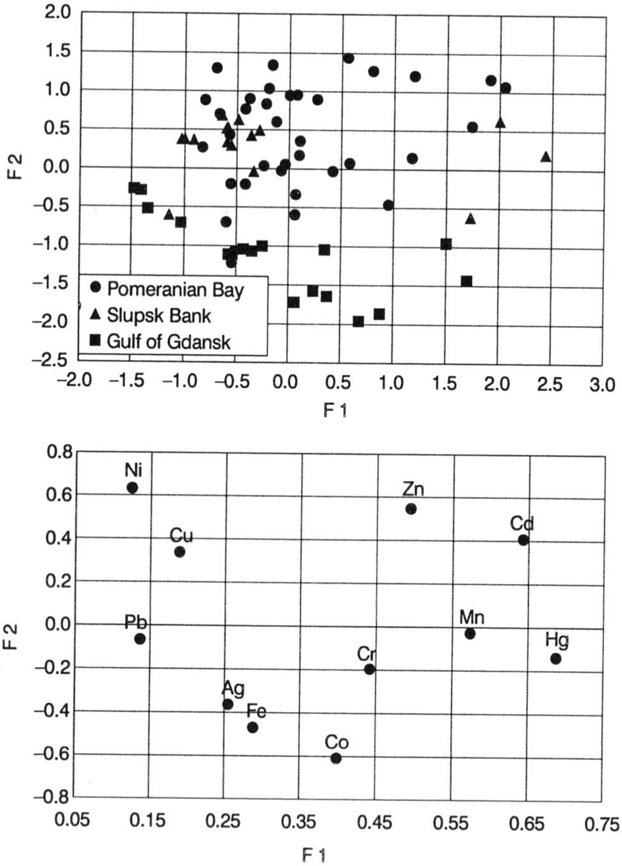

Fig. 6.3. Biplot of scores and loadings (metals) corresponding to *Mytilus* soft tissue from the three areas of the southern Baltic. After Szefer et al. (2000d).

est scores of values of F1 and are the most influenced by input of the fluvial material (Oder River). As can be seen in loading distribution pattern these byssi samples generally have the highest contents of Cd, Mn, Cu, Ni and Zn described by the highest values of F1. Samples from the Gulf of Gdańsk low in F1 may reflect, in part, the high levels of Hg, Cr, Ag, Co, Pb and Fe. It means that these metals are preferably accumulated in byssus of specimens from area adjacent to the Vistula River estuary. Factors 1 and 2 show clear separation of both the byssi and tissue samples, respectively based on their geographic distribution, possibly reflecting a different rate of deposition of clay minerals at the head of the Pomeranian Bay and the Gulf of Gdańsk. Such differentiation between these two groups could be explained by the differences in environmental parameters in the geographical sectors. The Pomeranian Bay, similarly to the Słupsk Bank region, is located in open part of the southern Baltic in contrast to the Gulf of Gdańsk which is partly isolated from open sea by the Hel Peninsula. It is assumed that in

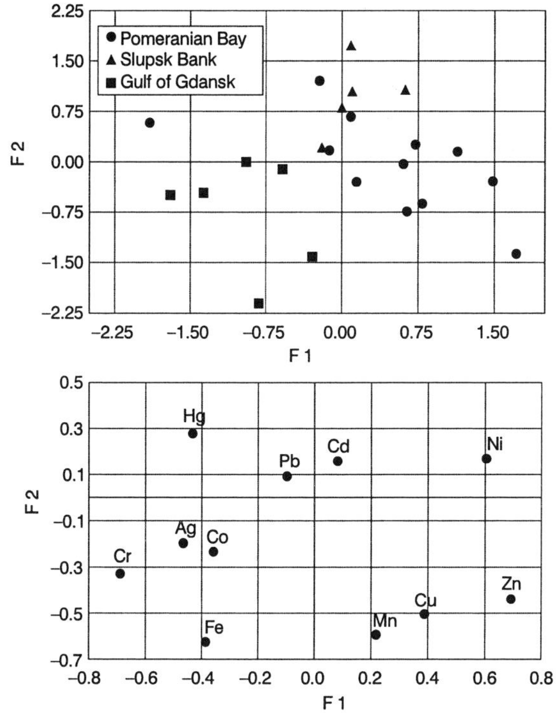

Fig. 6.4. Biplot of scores and loadings (metals) corresponding to *Mytilus* byssus from the three areas of the southern Baltic. After Szefer et al. (2000d).

these neighbouring areas water mixing processes have place; the water coming from the Pomeranian Bay mixes, especially during seasonal storms, with water mass in the Słupsk Bank region. This phenomenon could be responsible mainly for the similarity in distribution of both tissue and byssi objects in two-dimensional scatter-plot (F1/F2). The Pomeranian Bay differs from the Gulf of Gdańsk in respect to geological structure of bottom sediments as a substrata for the *M. edulis trossulus*. Moreover, various sources of metallic pollutants, as mentioned above, are specific for each sector. The Vistula River enters directly the Gulf of Gdańsk, while the Oder River flows directly to the Szczecin Lagoon which is connected with the Pomeranian Bay by means of narrow channel (Szefer et al., 2000d).

In order to study the regional influences of seawater on the biochemical composition of *Mytilus edulis* from the Baltic Sea and North Sea DA was performed for macroelement concentrations in the mussel as variables (Struck et al., 1997). This distribution pattern allows to distinguish Baltic and North Sea locations such as in the case of *F. vesiculosus* in spite of different food habits between these two zoobental organisms. Location groups based on the trace-element concentration patterns showed a less distinctive geographical arrangement in comparison of the location clusters based on macroelement concentration pattern. This picture suggests modified conditions for the accumulation of trace elements in *M. edulis* like in *F. vesiculosus* as compared to the uptake of macroelements (Struck et al., 1997).

Szefer and Wołowicz (1993) processed statistically the concentration data (Cd, Cu, Fe, Mn, Ni, Zn) for the soft tissue of *Cerastoderma glaucum* from four geographical regions, i.e. the Gulf of Gdańsk (Baltic Sea), Marennes-Oleron Bay, Arcachon Bay (French Atlantic coast) and Embiez Islands (Mediterranean Sea) (Fig. 6.5). About 74% of the total variance is explained by the first three factors. The both score and loading data are presented on the first two principal vectors by means of a biplot (Fig. 6.6). Three-dimensional scatter-plot in space determined by PC1, PC2 and PC3 is shown in Fig. 6.7. It follows from comparison between the distribution of the object scores and the loading (variable) vector direction (Fig. 6.6a) that mainly Mn and Fe concentrations in the cockles analysed are responsible for differentiation between populations from Marennes-Oleron Bay and Arcachon Bay. Zn, Cd and partly Ni have a main contribution in distinguishing the Gulf of Gdańsk cluster from the others (Figs. 6.6 and 6.7). Bearing in mind that *Cerastoderma* seems to be appropriate biomonitor for Cd, Cu, Zn and particularly Ni (see Chapter 7A), such distribution pattern implies that anthropogenic sources may be responsible for higher levels of Cd and Zn in *C. glaucum* inhabited the coastal and industrialised zone of the Gulf of Gdańsk. The PCA data display that both inter-regional and seasonal factors have an important influence on the distribution of the metals studied in the cockle tissues (Szefer and Wołowicz, 1993).

The concentration data for the soft tissue and byssus obtained from ca. 10 000 specimens of Mytilidae collected in the Baltic Sea and other geographical areas

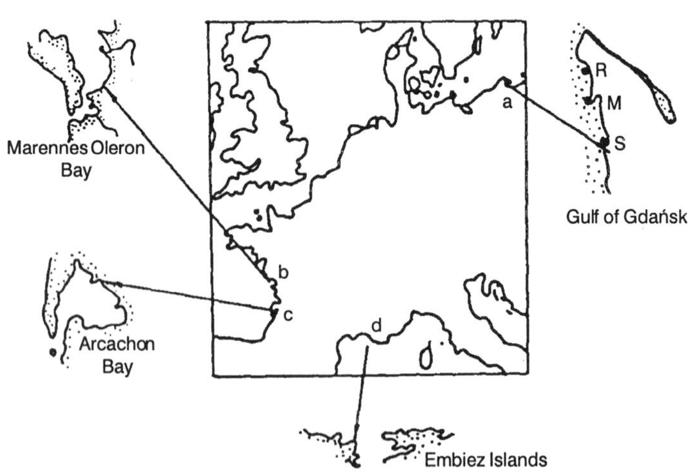

Fig. 6.5. Sampling sites of *Cerastoderma glaucum* populations; a – the Gulf of Gdańsk (Baltic Sea), b – Marennes-Oleron Bay (Atlantic), c – Arcachon Bay (Atlantic), d – Embiez Islands (Mediterranean Sea), R – Rzucewo, M – Mechelinki, S – Sopot. After Szefer and Wołowicz (1993).

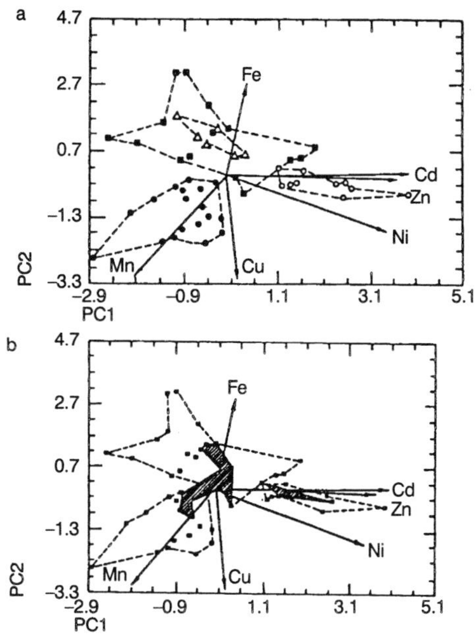

Fig. 6.6. Bi-plot for object scores of the first two principal vectors of 50 mollusc samples: a - regional differences are illustrated by clusters of points corresponding to samples from the Gulf of Gdańsk (○), Marennes-Oleron Bay (●), Arcachon Bay (■) and Embiez Islands (△). Association between principal components (PCI x PC2) and variable (metals) vectors are also indicated; b - season dependent variations are illustrated by clusters of points corresponding to samples collected during January-May. These groupings are indicated by shaded areas. After Szefer and Wołowicz (1993).

were processed by FA. As can be seen in Figs. 6.8 and 6.9, after removing extreme values (corresponded to extremely contaminated samples in highly industrialised areas of Saganoseki, Japan and Öxelosund, Sweden) it is possible to distinguish Baltic population of *M. edulis* from other clusters based on byssus data (Fig. 6.9). The grouping of object samples corresponded to the soft tissue is overlapping with other clusters and hence is inappropriate in identification of Baltic population (Fig. 6.8). Mn is element which allows us to identify Baltic specimens of *Mytilus* among others (Fig. 6.9) and may be used as specific determinant in this respect.

The results of trace element levels in *Balanus improvisus* from the Gulf of Gdańsk, Baltic Sea, were processed using factor analysis. According to Szefer et al. (2000b) the first two factors for the whole body distribution of metals described totally 77.55% of the total variance. Eigenvalues amounted to 3.04 and 0.83. Spatial differences in heavy metal concentrations in this crustacean were well identified.

Fish

The first two factors accounted for 69.8% (for the liver in the Pomeranian Bay) and 61.9% (for the muscle in the Pomeranian Bay) of the total variance

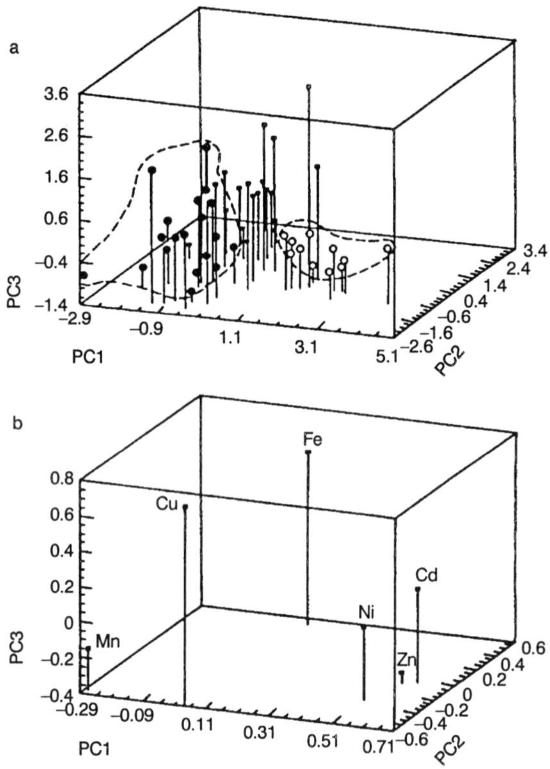

Fig. 6.7. Three-dimensional scatterplot in the space determined by PC1, PC2 and PC3: a – for object scores; samples from the Gulf of Gdańsk (O) and Marennes-Oleron Bay (●) are encircled; b - for loadings (metals). After Szefer and Wołowicz (1993).

with eigenvalues higher than unity (Szefer et al., 2000a). In order to clear graphical presentation of the seasonal and age-dependent differences in several metal clustering, besides overall distribution selected plot, scores corresponding to winter and summer catches as well as age-groups 1 and 3 of fish are specified. As it can be seen in Figure 6.10a the object scores corresponding to the liver of perch inhabited the Pomeranian Bay display the highest values of F1 for summer season and the lowest ones for winter season. Grouping of points represented by old specimens of age-group 3 (characterised by the highest values of F2 and the lowest values of F1) is well isolated (Fig. 6.10b) from that identified as young fish of age-group 1 (described by the lowest values of F2 and the highest values of F1). In order to demonstrate which metals control the grouping of the samples described by object scores, a plot for loadings (metals) is presented in Figure 6.10c. F1 displays loading of hepatic Pb and Cd and corresponds to the winter samples described by the lowest values of F1. It is well isolated from loadings of hepatic Cu and especially Zn, and refers to the summer samples (described by the highest values of F1). Hepatic Cd lets us to identify the oldest specimens, while hepatic

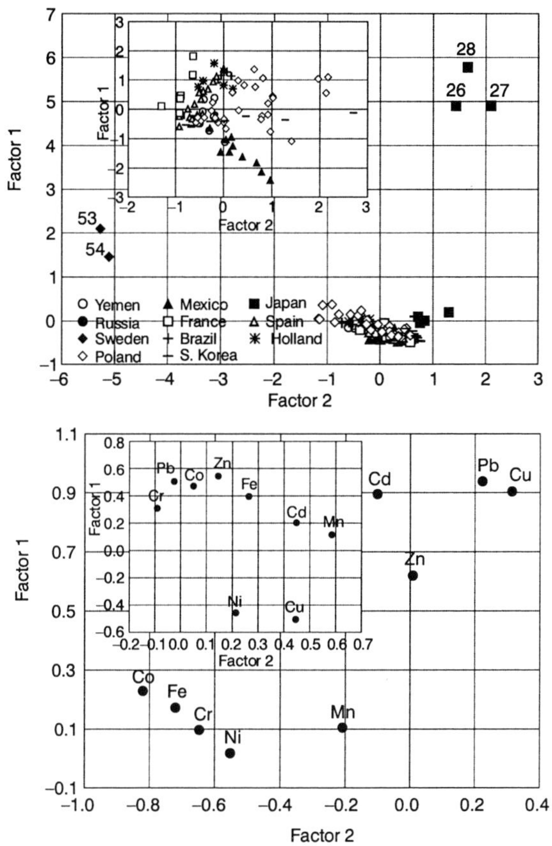

Fig. 6.8. Biplot of scores and loadings (metals) corresponding to Mytilidae soft tissue from the Baltic Sea and other geographical regions. After Szefer et al. (1998a, 2000e).

Cu and mostly Zn are responsible for selection of points corresponding to younger ones.

A biplot of the samples based on their factor scores shows a clustering of the muscle samples also belonging to the Pomeranian Bay. (Fig. 6.11). Seasonal differences, similarly to hepatic objects, are also well marked. Summer muscle samples are clearly separated from winter ones (Fig. 6.11a); however pattern of age-dependent variations (Fig. 6.11b) is not such regular as in the case of hepatic samples. As can be seen in loading distribution pattern (Fig. 6.11c) these muscle samples corresponding to winter season are generally loaded with Cd, Pb and Hg while both muscle Zn and Cu are mainly determinants of summer objects.

There is no regular distribution of both the hepatic and muscle object samples in respect to their sex features.

The observed seasonal variations in selected metals in perch (Szefer et al., 2000a) are reflected by different metal bioavailability depending on the ligands present in the biotopes and the chemical speciations between two the dissolved

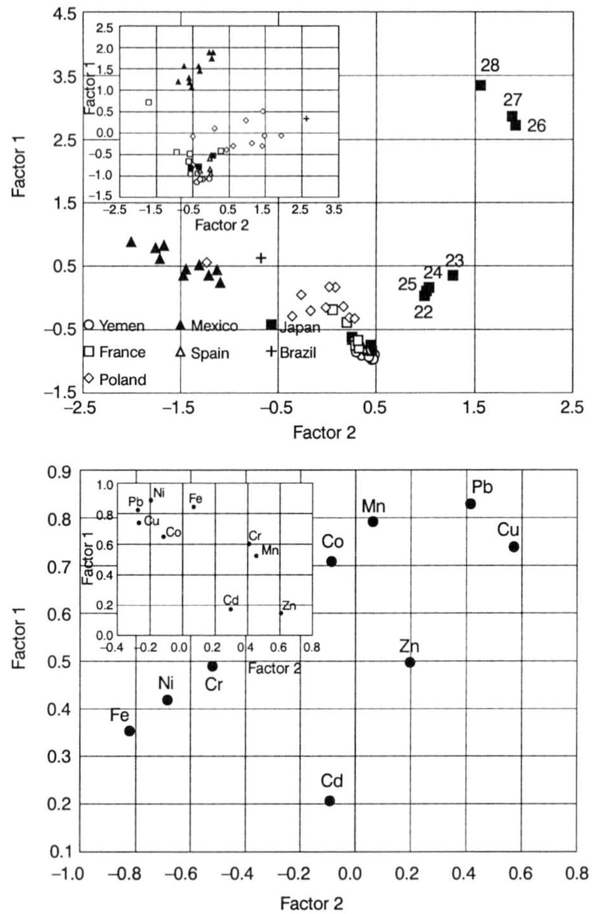

Fig. 6.9. Biplot of scores and loadings (metals) corresponding to Mytilidae byssus from the Baltic Sea and other geographical regions. After Szefer et al. (1998a, 2000e).

and particulate phases (Andres, 2000). Moreover fish metabolism may be dependent on the abiotic conditions, food supply and the stage of the cycle reproduction (Kock et al., 1996; Olsson et al., 1996; Andres et al., 2000).

Marine mammals

The data of trace metal levels in *Phocoena phocoena* from the Baltic and Danish waters and other northern area such as the Greenland were processed using factor analysis. According to Szefer et al. (2000c) the first three factors for hepatic and renal distributions of metals described totally 67.86 and 72.81% of the total variance, respectively. Eigenvalues amounted to 2.56, 1.77 and 1.11 (for liver) and 3.17, 1.56 and 1.10 (for kidney). As it can be seen in Figures 6.12. and 6.13 the hepatic and renal samples corresponding to old specimens of harbour porpoises display the highest values of F1 and form a groups which are clearly

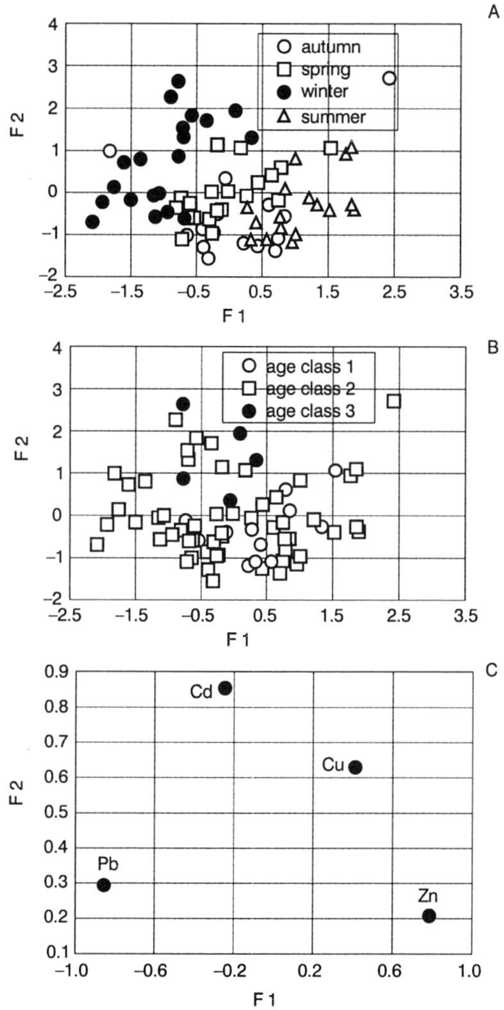

Fig. 6.10. Biplot of scores reflecting seasonal (A) and age (B) differences of metals (C) in the liver of *Perca fluviatilis* from the southern Baltic. After Szefer et al. (2000a).

separated from those consisted of young specimens (characterised by the lowest values of F1). Factor 2 describes spatial differentiation between harbour porpoise populations; specimens inhabited a southern Baltic are identified by object scores (liver and kidney) in the left part of the scatter-plot (lowest values of F2) while Greenlandic group is described by higher values of F2. The third group of object scores corresponding to Danish specimens is overlapped with these two extremely situated clusters. The Danish group confirms the close association of samples corresponding to Greenland and Baltic populations. In order to demonstrate which metals control the grouping of the samples described by object scores, corresponding plots for loadings (metals) are also presented in Figures 6.12 and 6.13.

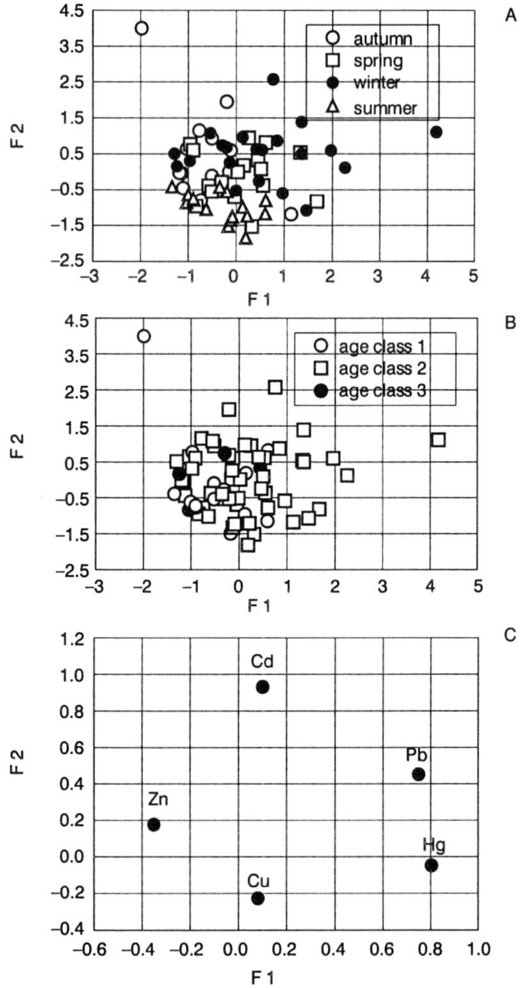

Fig. 6.11. Biplot of scores reflecting seasonal (A) and age (B) differences of metals (C) in the muscle of *Perca fluviatilis* from the Southern Baltic. After Szefer et al. (2000a).

From this distribution pattern clearly results that loadings of hepatic Cd (Fig. 6.12) as well as renal Cd and Zn (Fig. 6.13) accompanied by age and weight are characterised by the highest values of F1 and are well isolated from loadings of other metals, especially hepatic Cr, Cu and Fe (Fig. 6.12) and renal Mn and Fe (Fig. 6.13) described by the lowest values of F1. For geographical differentiation of the object sample distribution are mainly responsible both hepatic and renal metals; Fe and Cr allows identification of samples represented by Baltic specimens described by lower values of F2. Other metals found in the right of the plot (characterised by higher values of F2), especially Cd, Mn, Zn and Cu, make possible recognition of samples represented by Greenlandic specimens (Figs. 6.12 and 6.13).

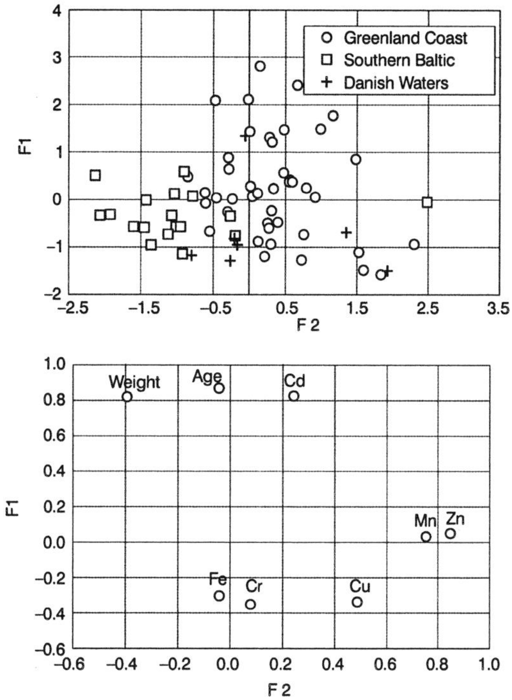

Fig. 6.12. Biplot of scores (a) and selected metals (b) corresponding to the liver of *Phocoena phocoena* from 3 geographical regions. After Szefer et al. (2000c).

Geographical variations in hepatic and renal metals support the above suggestion that the differences in metal bioaccumulation are mainly caused by specific feeding habits of the porpoises inhabited a southern Baltic and the Greenland.

SUSPENDED MATTER AND SEDIMENTS

Suspended matter

A PCA was performed using a data matrix which included the hydrographical data and the relative abundance of the particle types (Bernard et al., 1989). The first four PC represented 70% of the total variance. The first PC described differences between oxygenated surface samples, relatively rich in aluminosilicates, and poor in the oxygen deep water samples containing higher levels of Fe- and Mn-particle concentrations. The second component is mainly loaded by salinity, Ca-rich particle type and temperature, i.e. it possibly describes the differences occurred in the mixing area between the Baltic Sea and the North Sea waters. The third component distinguished the barite particle type and temperature from the depth, suspension content, nitrate concentration and the Fe-rich aluminosilicate particle type (Bernard et al., 1989).

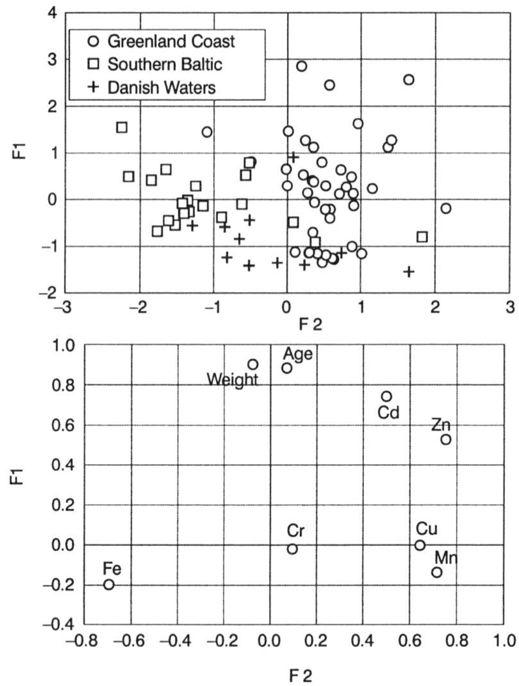

Fig. 6.13. Biplot of scores (a) and selected metals (b) corresponding to the kidney of *Phocoena phocoena* from 3 geographical regions. After Szefer et al. (2000c).

Surface sediments

In order to identify factors governing over geochemical fate of minor- and major elements in southern Baltic their horizontal distribution was characterised using FA, PCA and cluster analysis (Szefer et al., 1993a, 1995a; 1996, 1999a, 2000f; Danielsson et al., 1999). In these studies different processing procedures for sediment subsamples were used as follows:

(a) bulk sediments (< 2 mm) digested using mixture concentrated HNO_3, $HClO_4$ and HF;
(b) sediment fraction (< 80 μm) digested using mixture concentrated HNO_3 and $HClO_4$;
(c) sediment fraction (< 80 μm) leached using 1 M HCl;
(d) sediment fraction (< 63 μm) digested using mixture concentrated HNO_3, $HClO_4$ and HF;
(e) fusing with $LiBO_2$ and dissolution in HNO_3.

Southern Baltic

Approach (a).

The first three factors with eigenvalues > 1.0 were extracted from the data set studied. These accounted for 79.5% of the total variance with F1 contributed to 62.2% of the total data variance. The first factor is mainly influenced by sediment grain size characteristics (i.e. negative loadings are attributed to coarse-grained sediments), i.e. it describes different granulometric structure of geological material. This is undesirable arrangement because coarse-grained structure (sandy) pattern significantly masks the geochemical composition of elements concentrated in clay material (< 80 μm). Therefore concentration data corresponded to these bulk sediments (< 2 mm) were processed by endmember analysis. Use of endmember analysis has enabled to refine the results of previous studies to show different pathways for the introduction of Cu, Zn, and Ag and Cd, and Pb into the Gulf of Gdansk. It is supposed that this difference reflects the fact that Cu, Zn and Ag are introduced into the sediments of the Gulf of Gdansk principally from the Vistula River whereas Cd and Pb are introduced, in part, by atmospheric transport. Renner et al. (1998) identified origin of selected heavy metals in bulk sediments of the southern Baltic using endmember analysis.

Approach (b)

The first three factors with eigenvalues > 1.0 accounted for 73.3% of the total variance. F1 explaining 28.6% of the total data variance was associated with mineralogical composition of the samples studied. It is postulated that F2 (23.0%) corresponded to terrigenic and biogenic phases while F3 is related to (21.7%) geochemical composition of estuarine and open sea sediments considered (see localisation of sampling sites in Fig. 5.10 in Chapter 5C).

Fig. 6.14a illustrates factorial distribution of object scores in the three-dimensional scatter plot. It can be seen that open-sea samples (Nos. 25–29) form a separate cluster which is distinguished from the grouping of points represented by typical estuarine samples (Nos. 6, 7, 16 and 17). The remaining samples are located in mid-distance between the estuarine and open-sea clusters. It means that this region is less exposed to the influx of material of the riverine origin. Fig. 6.14b shows distribution of loadings (variables) in the three-dimensional scatter plot which is similar to that of the object scores presented in Fig 6.14a. Elements such as K, Mg, Ca, Na, Sr as well as physical parameters like salinity and depth of water form a distinct cluster which is isolated from that – down located, consisted of Zn, Cd, Ag, Cu, Cr, Pb, chlorophyll-a and Fe. The localisation of the latter (described by lower values of F3) substantiates identification of samples originating from the Vistula River's mouth (characterised also by lower values of F3). The upper cluster (described by higher values of F3) identifies samples of typical open-sea provenience (having also higher values of F3).

Approach (c)

In order to recognise labile species of metals in sediments, chemical analyses of both the acid and basic extracts have been performed. The first three factors extracted 41, 16.7 and 9.2% of the total variance, respectively (Szefer et al., 1993a, 1995a). Some results are presented in two-dimensional scatter plot (Fig. 6.15). Since Fe and Mn are associated with the first cluster while Al is connected with the second, it is suggested that 1 M HCl leached elements which are split

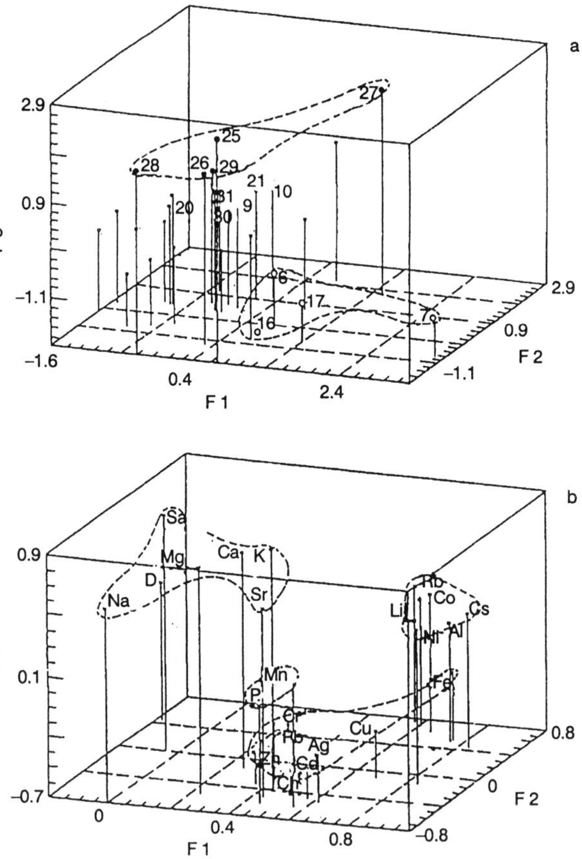

Fig. 6.14. Three-dimensional scatterplot of object sample scores (a) and loadings by individual variables (b) obtained for acid (concentrated HNO_3-$HClO_4$) leachates of sediment (fraction < 80 μm) sample data. Samples are numbered as in Fig. 5.10, Chapt. 5C; Sa = salinity; D = depth of water; Ch = chlorophyll-a. Samples originating close to the subarea of the Vistula estuary *(open circles)* and from the open-sea region *(filled circles)* are indicated. After Szefer et al. (1995a).

into major phase groups, i.e. Fe and Mn hydroxide/carbonate group (Ag, Cd, Cu, P, Pb, Zn and Fe with Mn) and the aluminosilicate group (Co, Cr, Cs, K, Mg, Ni, Rb with Al). The latter bounds the group of elements accompanied Al in Puck Bay area while Fe–Mn phase is responsible mainly for the deposition of labile, easily extractable forms of Zn, Cd, Pb, Cu, Ag and P in the Vistula estuary. These elements, suspected to be anthropogenic in origin, are most probably scavenged by Fe- and Mn-oxyhydroxides at the hydrological front where mixing of the Vistula river water with the brackish Baltic Sea water takes place.

Sequential extraction analysis of heavy metals in the sediments samples taken also from the mouth of the Vistula river in a seawards direction showed considerable enrichment in Zn, Cu and Pb. Intercomparison of surficial and subsurficial metal distributions as well as the preponderance of these metals in a more labile,

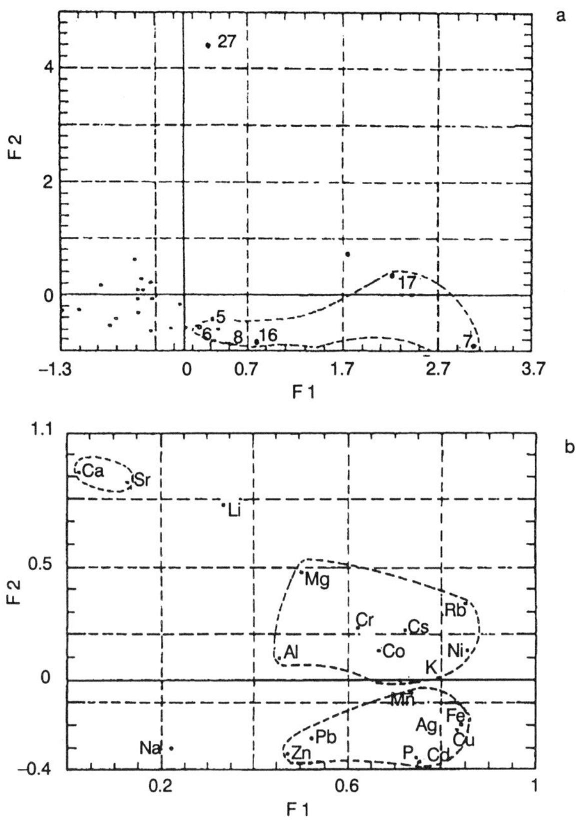

Fig. 6.15. Scatterplots of object sample scores (a) and loadings by individual variables (b) in space spanned by axes F1 and F2 obtained for acid (1 M HCl) leachates of sediment (fraction < 80 μm) sample data. Samples are numbered as in Fig. 5.10, Chapt. 5C; samples originating close to the subarea of the Vistula estuary *(open circles)* and from the open-sea region *(filled circles)* are indicated. After Szefer et al. (1995a).

non-residual sediment fraction suggests anthropogenic inputs of these metals to the Gulf of Gdańsk (Görlich et al., 1989, Belzunce et al., 2000).

Approach (d)

FA was also applied for evaluation the distribution of As, Cd, Co, Cr, Cs, Cu, Ba, Bi, Ga, In, Ni, Pb, Rb, Sb, Se, Sr, Th, Ti, Tl, U, V, Zn, RRE (Ce, Dy, Er, Eu, Gd, Ho, La, Lu, Nd, Pr, Sm, Tb, Tm, Yb) and Al, As, Be, Cd, Co, Cr, Cu, Fe, Mn, Mo, Pb, Sb, Se, Sn, Sr, Ti, V, Zn, REE in surface sediments of the Vistula Lagoon (Polish sector) as well as of three sectors of a southern Baltic Sea, respectively (Szefer et al., 1999a, 2000f). Figure 6.16a illustrates factorial distribution of object scores in three dimensional scatter plot. We can see that the Szczecin Lagoon samples form a separate cluster which is closely fitted to the Pomeranian Bay samples. On the other hand, the Vistula Lagoon object scores are neighbouring to those represented by the Gulf of Gdańsk (properly Puck Bay) sediments. This scatter-plot clearly illustrates a great dissimilarity between

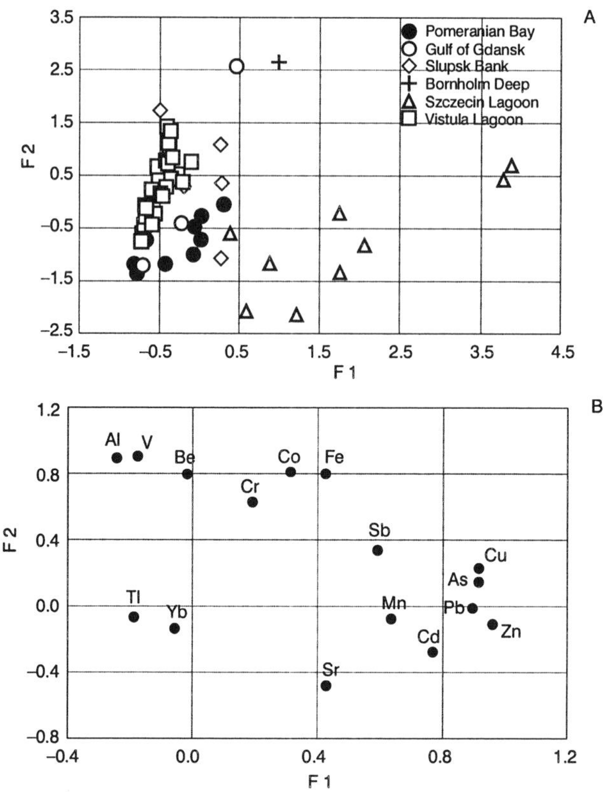

Fig. 6.16. Scatterplots of object sample scores (A) and loadings by individual variables (B) in space spanned by axes F1 and F2 obtained for sediments (fraction < 63 μm) sample data from different areas of the southern Baltic. After Szefer et al. (2000f).

geochemical composition of sediments of the Szczecin and Gdańsk Lagoons. The Pomeranian Bay with the Szczecin Lagoon are elements of the Odra River estuary, while the Gulf of Gdańsk with the Vistula Lagoon, excluding its western inner part named the Puck Bay, are supplied with the Vistula River. Bereft of topographical barriers, the Pomeranian Bay is exposed to permanent, intensive water exchange between itself and the neighbouring Arkona and Bornholm Basins and its exchange with the central part of southern Baltic represented by the Słupsk Bank takes only about 3 weeks. Such long-distance water exchange is possibly reflected by overlapping of the object scores corresponding to the Pomeranian Bay and the Słupsk Bank. Fig. 6.16b shows distribution of loadings (variables) in two-dimensional scatter plot which is similar to that of the object scores presented in Fig. 6.16a. Elements such Zn, Cu, As, Pb, Cd, Sb, Mn, Fe and Sr (described by high values of F1) are isolated from the groupings of points represented by Al, V, Tl, Be and Yb (characterised by lower values of F1). Since Yb and Al are typically terrigenic elements in origin, and Cd and Pb belong to anthropogenically derived metals, it means that the Pomeranian Bay with especially the Szczecin Lagoon are the most polluted areas of the southern Baltic.

Factor analysis was also performed on the 26 samples from the Vistula Lagoon (sampling sites in Fig. 6.17) and additionally a river sample (Szkarpawa River sample 26, uncontaminated locally). The eigenvalue was set to 1.0 as a threshold in order to limit the number of extracted and rotated factors. Four factors (FI-FIV) were obtained which explained 64.3% of the total variance and accounted for 31.1%, 12.4%, 11.8% and 9.0% of the total variance, respectively (Szefer et al., 1999a). The most relevant factors with regard to the distribution of the heavy metals in the sediments are FIII and FI. A plot of the samples based on their factor scores shows a clustering of the samples into three main areas, each corresponding to a geographically distinct zone (Fig. 6.18a). Samples from the western part of the Vistula Lagoon (samples 1–11) and the Szkarpawa River (sample 26) have the highest scores of both these factors and are the most influenced by the riverine input of anthropogenic material. These samples generally have the highest contents of Zn, Pb and Cd (F3) and Ni, Co, Cu, Ag (F1) (Fig. 6.19). Samples from the eastern part of the area (samples 21–25) are moderately high in FI but low in FIII. The high loading of F I in these samples may reflect, in part, the high concentrations of Pb in sample 23. The central area is situated between these two areas and has the lowest scores of both these factors (samples 12–20). The high scores for factor FII in this area indicate that these sediments have higher detrital but lower carbonate contents than in the other two areas. A plot of Factor II v. Factor I shows an even clearer separation of the samples based on their geographic distribution, possibly reflecting a higher rate of deposition of clay minerals at the head of the lagoon (Fig. 6.17). All the samples studied here (with the possible exception of samples 22 and 25) vary from clayey silts to silts and were taken in a limited range of water depths (2–4 m) (Uścinowicz and Zachowicz, 1996).

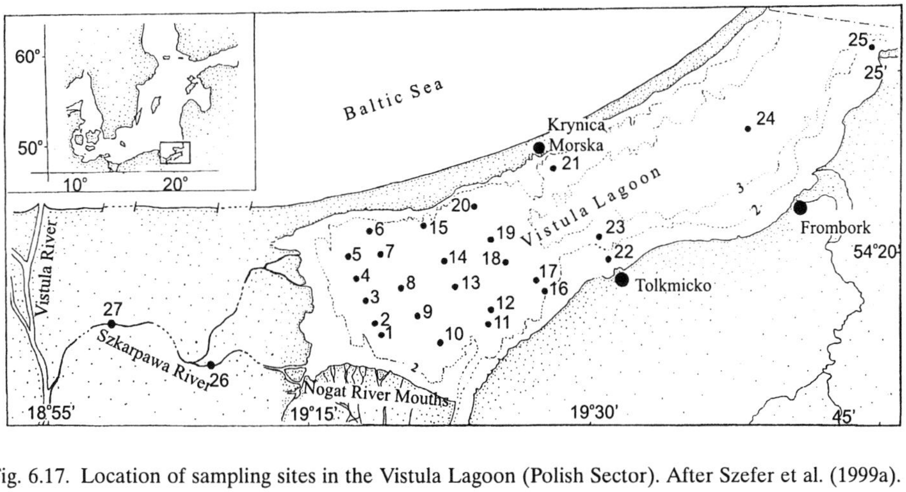

Fig. 6.17. Location of sampling sites in the Vistula Lagoon (Polish Sector). After Szefer et al. (1999a).

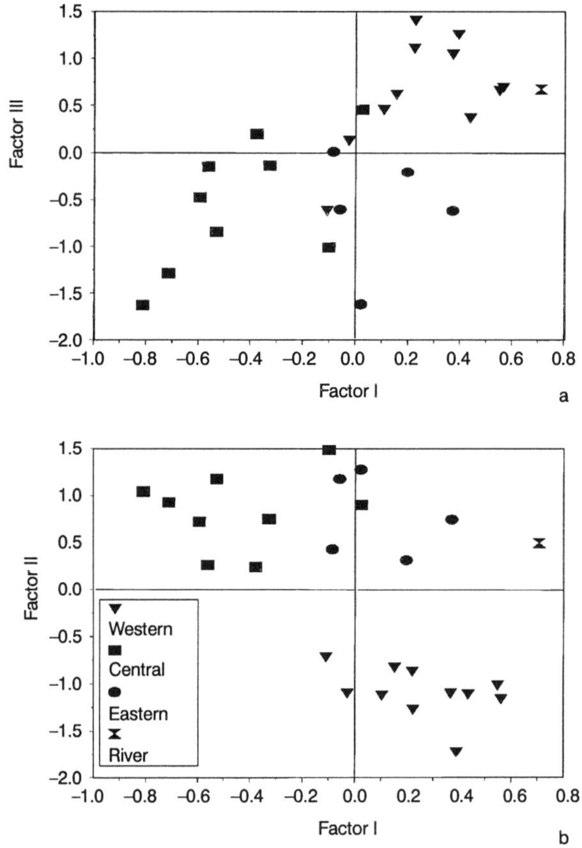

Fig. 6.18. Scatterplots of factor loadings by individual variables (concentrations of the elements analysed and water). Cumulative data for all the cores are displayed: a – loadings in space spanned by F1 and F2, b – loadings in space spanned by F1 and F3. After Szefer et al. (1999a).

Skagerrak/Kattegat

Approach (e)

Danielsson et al. (1999) processed statistically concentration data obtained by analysis of the sediments collected at sampling sites shown in Fig. 6.20. The first component PC1 indicated high positive weigthts for Cd, Co, Fe, Pb and Zn (Danielsson et al., 1999). It means that these elements are co-precipitated with the Fe oxyhydroxides, since according to several authors (Görlich et al., 1989; Szefer et al., 1995a; Drever, 1997) a main mechanism for selected trace elements such as Cd, Pb and Zn is the co-precipitation with the Fe amorphic compound. PC2 exhibited high factor loadings for Cu, Ni and Mo which may correspond to variations in biogenic productivity (Danielsson et al., 1999). This interpretation is in an agreement with data reported by several authors suggesting that these elements are related to the settling and dissolution of biomass (Sclater et al., 1976; Bru-

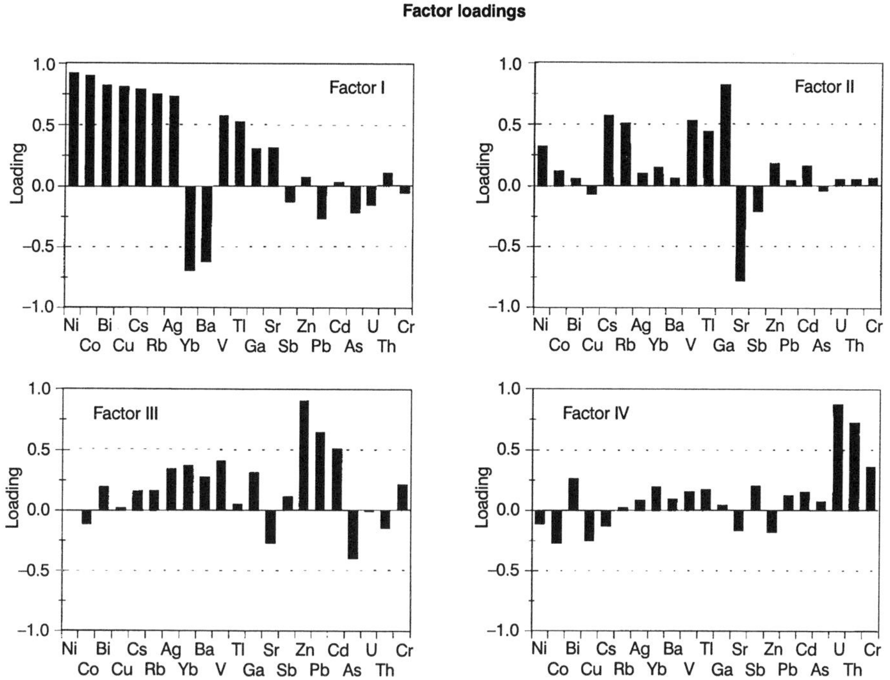

Fig. 6.19. Plots of factor loadings of the four principal factors (FI-F IV) obtained by factor analysis based on compositional data of sediments for samples 1–26. After Szefer et al. (1999a).

land, 1980; Coveney et al., 1991). PC3 demonstrated a high factor loading of Mn; it means that none of the trace elements are associated to the Mn distribution pattern. This can be explained by fact that Mn is continuously mobilised from the deeper reduced part of the sediments into the interstitial waters (Danielsson et al., 1999) and next it migrates to the oxic surficial water where its enrichment takes place (Förstner and Patchinelam, 1976). The PCA data are comparable to those of cluster analysis. Cluster 1 reflects, like PCA, co-precipitation of the trace elements such as Cd, Co, Pb and Zn with Fe oxyhydroxides in the coastal area and near their sources (Goethenburg and Laholm regions). From comparison the metal data from PCA and those from cluster analysis (Fig. 6.21) clearly results that cluster results almost correspond to positive factor loadings. Cluster 2 is related to negative factor loadings; hence is dominated by biophile elements such as Cu, Mo and Ni having also negative or low weights. The separation between these two clusters is connected with differences in the affinity of these two metals' groups to Fe oxyhydroxides phase and organic matter. Cluster 3 is dominated by remarkable greater concentrations of Cr; it comprises all sampling sites of the southwestern part of the Skagerrak and coincides with the sediment transport along the Danish NW coast from the southern North Sea (Danielsson et al., 1999). It is pointed out that this transport is very important source of several trace elements in the Skagerrak (Bengtsson and Stevens, 1996). According to

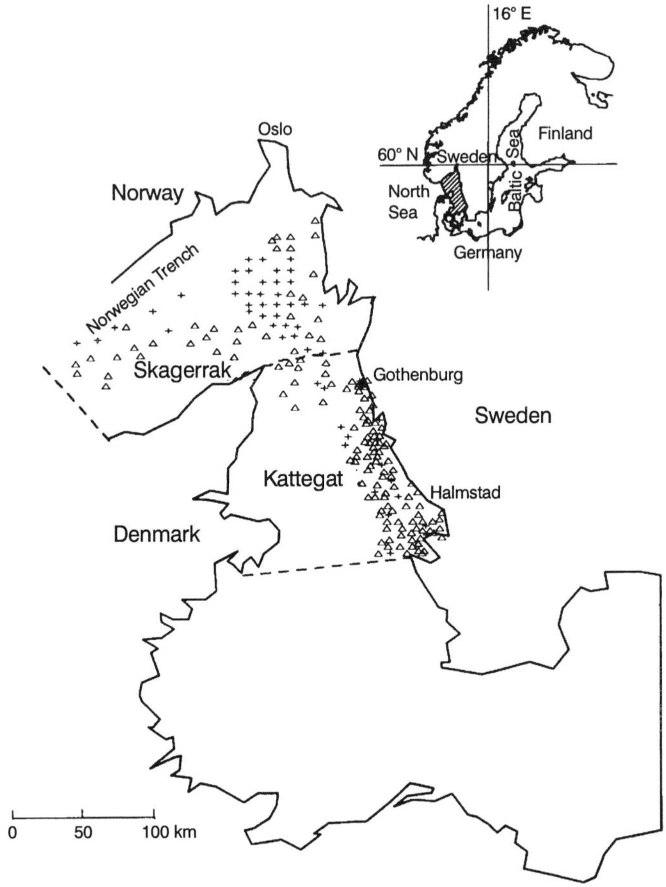

Fig. 6.20. The study area with stations (+ denotes stations also included in the deep sediment comparison). After Danielsson et al. (1999); modified.

Danielsson et al. (1999) it is much probable that Cr, potentially a southern North Sea origin, is accumulated in these sediments having relatively high contents of minerals such as garnet, tourmaline and rutile. Since Mn has insignificant contribution to the clusters, it can be a result of its high mobility as mentioned above.

This multivariate analysis was also performed for particular elements in surface and subsurface sediment layers (Danielsson et al., 1999). Only two trace metals, i.e. Pb and Zn showed an increase their concentrations over time while in the case of Al inverse temporal trend was observed. Enhanced levels of metals in top layers of sediments, resulting from increased pollution, have been also identified in the archipelago of the Bohus coast and in the fjords (Cato, 1997).

Sediment cores

The horizontal distribution of selected metals and their vertical profiles in southern Baltic sediment cores have been investigated widely using dozen sedi-

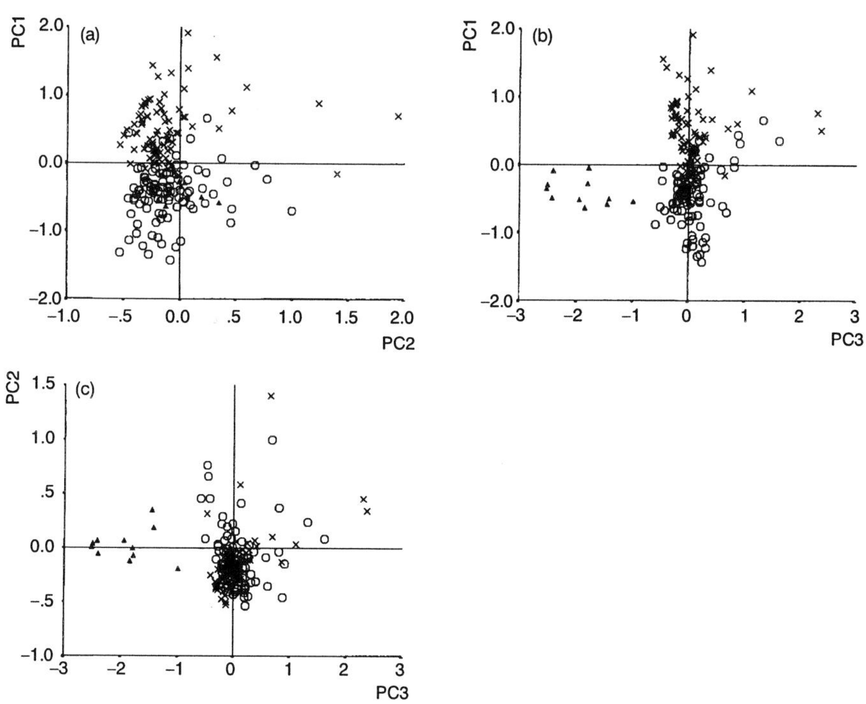

Fig. 6.21. Principal components, with cluster identification as markers (x = 1, o = 2, ▲ = 3). a) component 1 versus 2; b) component 1 versus 3; c) component 2 versus 3. After Danielsson et al. (1999); modified.

ment cores. Only selected results would be presented here. The first three factors extracted 75% of the total variance (Szefer et al., 1993a, 1998b). The sample numbers and depths of samples taken from four sediment cores (collected at sampling sites shown in Fig. 6.22) are listed in Table 6.1. The concentration data obtained were processed by PCA. To illustrate the inter-sample relationships, the object scores and loadings are presented graphically in scatter plots (Figs. 6.23 and 6.24).

The distribution of principal component scores is similar to that of the principal component loadings. Since Al is a typical element of crustal origin and C_{org} represents organic matter molecules, localisation of the two antipathetic PC1 clusters (Fig. 6.24) indicates that Al, Fe, Ti, K, Mg, Th, Co and Ni (as positive values of PC1) in the sediment cores are terrigenous, whereas C_{org}, N, Cu, Pb, Zn, Cd and possibly P (as negative values) are biogenic. These two main groupings of elements let us to identify elements anthropogenic in origin (Pb–Cd) accumulated in recently formed top layers of sediments as well as elements terrigenic in origin (Al) deposited in deeper "background" segments corresponding to precivilisation era.

The concentration data of As, Cd, Co, Cu, Fe, Hg, Mn, Mo, Ni, Pb, V and Zn in sediment cores from 59 stations of the Baltic Proper have been processed using

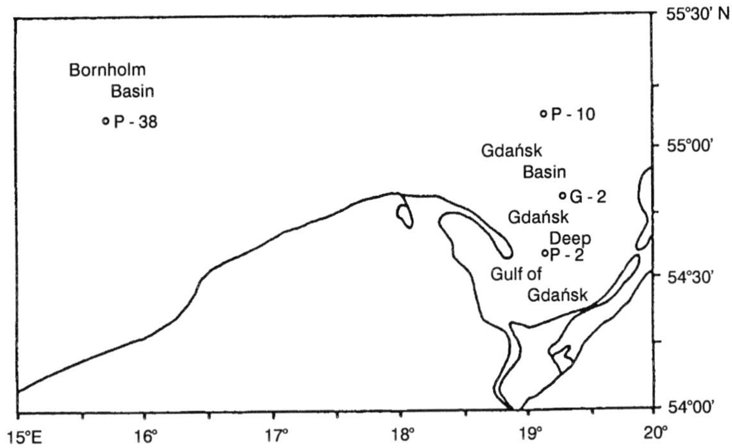

Fig. 6.22. Map of the Southern Baltic region indicating the locations of sampling stations of the cores studied. After Szefer (1998).

TABLE 6.1.
The object numbers and depths of corresponding core segments

Object number*	Sample depth in core [cm]	Object	Sample depth in core [cm]
	Core P-2		Core P-10
2	0.7– 2.0	1	0.0– 1.6
4	3.0– 4.0	3	2.6– 3.9
6	5.0– 6.0	5	5.5– 7.4
7	6.0– 7.0	6	7.4– 9.2
8	7.0– 8.0	7	9.2–11.4
10	10.0–12.0	8	11.4–14.1
11	12.0–15.0	9	14.1–17.8
12	15.0–20.0	10	17.8–21.5
13	20.0–25.0	11	21.5–25.8
14	25.0–30.0	12	25.8–31.3
	Core G-2	13	31.3–34.8
1	0.0– 1.0		Core P-38
2	1.0– 2.0	1	0.0– 0.7
4	3.0– 4.0	2	0.7– 1.8
5	4.0– 5.0	5	5.0– 6.2
6	5.0– 6.0	6	6.2– 8.1
10	9.0–10.0	7	8.1– 9.7
11	10.0–12.0	8	9.7–12.1
12	12.0–15.0		
13	15.0–20.0		
14	20.0–25.3		

* corresponds to the number in Fig. 6.22.

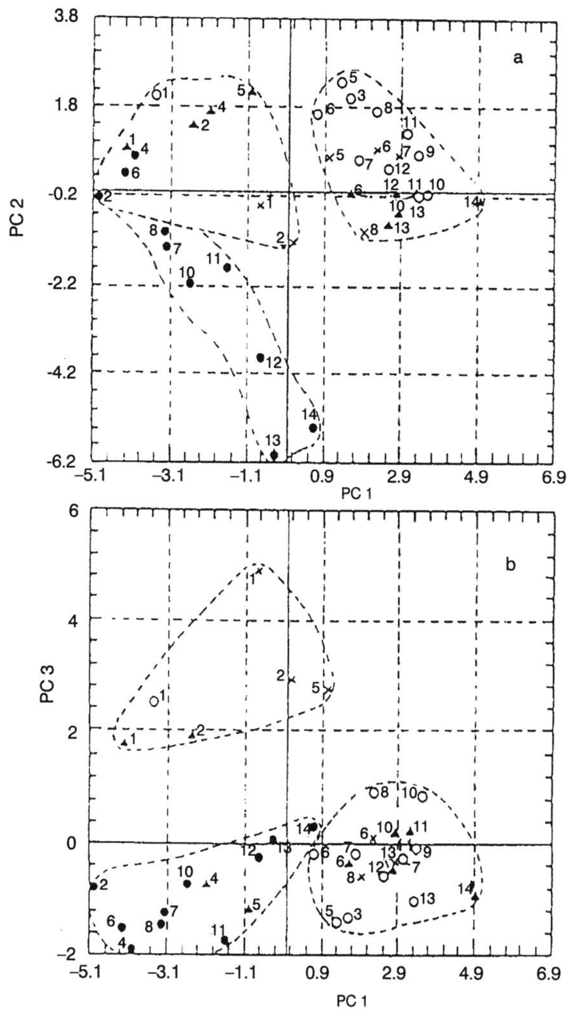

Fig. 6.23. Scatterplots of principal component segment scores. Samples originating from the stations P-2 (●), P-10 (O), P-38 (x), and G-2 (△) are numbered like in Tab. 6.1; a – scores in space spanned by PC1 and PC2, b – scores in space spanned by PC1 and PC3. After Szefer (1998).

claster analysis (Hallberg, 1991). The sampling sites represented different geochemical facies which were divided to three clusters, the first one corresponds to anoxic facies, where the bottom water is generally anoxic over a long period and H_2S is commonly occurred there. The second facies represents areas where bottom water is predominantly oxic while third facies is related to the margin of basin areas representing a mixture of the first and second facies. Both the spatial and downcore (temporal) variations of metals were well explained in light of factor analysis structured as a two-way multivariate ANOVA model (Hallberg, 1991). PC1 is explained by 60% of the total variance while PC2 has less significance

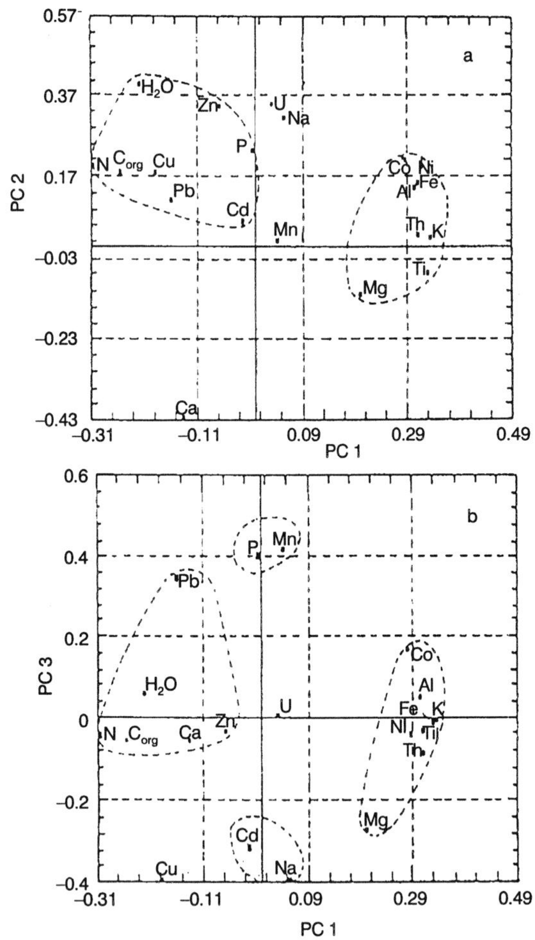

Fig. 6.24. Scatterplots of principal component loadings by individual variables (concentrations of the elements and water). Cumulative data for all the cores are displayed; a – loadings in space spanned by PC1 and PC2, b – loadings in space spanned by PC1 and PC3. After Szefer (1998).

(15%) in all three facies. It means that spatial trends can be explained mainly by only PC1. All the metals studied apart from Hg showed positive covariance between the cores in oxic facies; in anoxic facies exceptions are Fe and V. Possibly under anoxic conditions Fe is transformed into ferrous ion in the reduced bottom water of the basin and will escape the deposition to a significant extent (Hallberg, 1991). This suggestion is supported by occurrence of high levels of ferrous ion in the bottom water in this area (Fonselius, 1969, 1970). The transition facies, in contrast to other two facies, is described by strong negative affinity of As, Hg and Pb to PC1. Perhaps this area between oxidising and reducing conditions have mobilising effect on these elements. Factorial distribution pattern of metals with sediment depth is recognised from their correlation with PC1 which explained

81% of the stratum effect (Hallberg, 1991). Cobalt showed no correlation with PC1 while Fe indicates negative the downcore trend. Further analysis with Model II taking into account besides linear trend also a curvature made possible distinguishing two metal groups (Hallberg, 1991). Elements with an increasing of their concentrations towards the surficial layers of the core exhibit a significant correlation with curvature, i.e. Cd, Cu, Mo, Ni, Pb, V and Zn. This is either can be due to a diagenetic processes taking places in that depth or anthropogenic impact of these metals. Additional statistical test provided strong evidence that the spatial variations in metal concentrations are mainly dependent on the distribution of organic matter while their downcore (temporal) profile is mostly affected by dominating factor such as atmospheric pollution.

The geochemical fluctuations in deposition in the most polluted part of the sediment core of the Bay of Töölönlahti, southern Finland, were summarised in view of PCA (Virkanen, 1998). The first two components accounted for 65.4% of the variance. A cluster containing organic, carbonate and hydroxide Ca together with total P was located in the middle of the sediment sequences, representing a period around the turn of the century. It corresponds to an increase in the nutrient concentrations in the beginning of the eutrophication, i.e. to an expansion in diatom populations indicative of eutrophic conditions (Virkanen, 1998). These ecological conditions have favoured the deposition of Ca–P rich compounds but towards the surficial sediment layers, the total Ca decreases perhaps as a result of dilution by organic matter or biogenic silica. Other grouping dominated by Al, Fe and Mn (bound to organic matter), Cu (carbonates, bound to Fe–Mn oxides and organic material), Zn (exchangeable, carbonates, bound to Fe–Mn oxides and organic material), LOI, and total S was attributed to a depth of 30–40 cm, representing segments deposited during 1930–1950. The presence of total S in this cluster suggests that sulphides besides organic matter may also serve as a sink for Fe, Cu and Zn (Virkanen, 1998).

References

Altabet, M.A., and R. Francois, 1994. Sedimentary nitrogen isotopic ratio as a recorder for surface ocean nitrate utilization. Global Biogeochem. Cycles **8**, 103–116.

Aitchison, J., 1986. The Statistical Analysis of Compositional Data. (Chapman and Hall, London).

Andres, S., F. Ribeyre, J.-N. Tourencq and A. Boudou, 2000. Interspecific comparison of cadmium and zinc contamination in the organs of four fish species along a polymetallic polution gradient (Lot River, France). Sci. Total Environ. **248**, 11–25.

Antia, N.J., C.D. McAllister, T.R. Parsons, K. Stephens and J.D.H. Strickland, 1963. Further measurements of primary productivity using a large volume plastic sphere. Limnol. Oceanogr. **8**, 166–183.

Armanino, C., A. Roda, I. Adriano, M.C. Casolino and M.A. Bacigalupo, 1996. Pattern recognition of particulate-bound pollutants sampled during a long term urban air monitoring scheme. Envi-romertrics **7**, 537–550.

Astley, K.N., H.C. Meigh, G.A. Glegg, J. Braven and M.H. Depledge, 1999. Multi-variate analysis of biomarker responses in *Mytilus edulis* and *Carcinus maenas* from the Tees Estuary (UK). Mar. Pollut. Bull. **39**, 145–154.

Banse, K., 1974. On the interpretation of data for the carbon-to-nitrogen ratio of phytoplankton. Limnol. Oceanogr. **19**, 695–699.

Beebe, K.R., R.J. Pell and M.B. Seasholtz, 1998. Chemometrics: A Practical Guide (John Wiley & Sons, New York).

Belzunce-Segarra, M.J., M.J. Wilson, J. Bacon, J. Bolalek and P. Szefer, 2000. Chemical forms and vertical distribution of heavy metals in bottom sediments (submitted for publication).

Bengtsson, H., and R.L. Stevens, 1996. Heavy-mineral provinces in the southern Skagerrak and northern Kattegat. Norges Geologiske Undersökelse Bulletin **430**, 47–55.

Bernard, P. C., van R.E. Grieken and L. Brügmann, 1989. Geochemistry of suspended matter from the Baltic Sea. 1. Results of individual particle characterization by automated electron microprobe. Mar. Chem. **26**, 155–177.

Blazhchishin, A.I., 1982. Main chemical constituents of the sediments of the Baltic Sea, in: Geology of the Baltic Sea, eds. V.K. Gudelis and E.M. Emelyanov (Wydawnictwo Geologiczne, Warsaw), 257–289 (in Polish).

Bogen, J., 1974. Trace elements in precipitation and cloud water in the area of Heidelberg, measured by instrumental neutron activation analysis. Atmos. Environ. **8**, 835–844.

Borole, D.V., M.M. Sarin and B.L.K. Somayajulu, 1982. Composition of Narbada and Tapti estuarine particles and adjacent Arabian Sea sediments. Indian J. Mar. Sci. **11**, 51–62.

Boström, K., J.-O. Burman, C. Pontér and J. Ingri, 1981. Selective removal of trace elements from the Baltic by suspended matter. Mar. Chem. **10**, 335–354.

Boström, K., J.-O. Burman and J. Ingri, 1983. A geochemical massbalance for the Baltic, ed. R.Hallberg, Environmental Biogeochemistry, Proc. 5th Intern. Symp. on Environmental Biogeochemistry (ISEB), Stockholm. Ecol. Bull. **35**, 39–50.

Boström, K., J. Ingri and C. Pontér, 1988. Origin of iron-manganese rich suspended matter in the Landsort Deep, N.W. Baltic Sea. Mar. Chem. **24**, 93–98.

Boutron, C., 1979. Alkali and alkaline earth enrichments in aerosols deposited in Antarcic snows. Atmos. Environ. **13**, 919–924.

Bruland, K.W., 1980. Oceanic distributions of cadmium, zinc, nickel and copper in the North Pacific. Earth Planet. Sci. Letters **47**, 178–198.

Brügmann, L., 1981. Heavy metals in the Baltic Sea. Mar. Pollut. Bull. **12**, 214–218.

Brügmann, L., 1982. Trace metals in the Baltic Sea 1980 and 1981. Proceedings of the XIII Conference of Baltic Oceanographers, Helsinki, 24–27 August, **Vol. 1**, 182–197.

Brügmann, L., 1986. Particulate trace metals in waters of the Baltic Sea and parts of the adjacent NE Atlantic. Beiträge zur Meereskunde **55**, 3–18.

Brügmann, L., and U. Hennings, 1982. Stratigraphie und Spezifizierung von Metallen in Sedimenten der westlichen Ostsee. Acta Hydrochim. Hydrobiol. **10**, 41–53.

Brügmann, L., and D. Lange, 1990. Metal distribution in sediments of the Baltic Sea. Limnologica **20**, 15–28.

Brügmann, L, and J. Matschullat, 1997. Zur Biogeochemie und Bilanzierung von Schwermetallen in der Ostsee, in: Geochemie und Umwelt-Umweltrelevante Prozesse in Atmo-, Pedo- and Hydrosphäre, eds. J. Matschullat, H.J. Tobschall, H.J. Vogt (Springer Verlag, Berlin) pp. 267–289 (in German).

Brügmann, L., P.C. Bernard and R. van Grieken, 1992. Geochemistry of suspended matter from the Baltic Sea. 2. Results of bulk trace metal analysis by AAS. Mar. Chem. **38**, 303–323.

Bryan, G.W. and W.J. Langston, 1992. Bioavailability, accumulation and effects of heavy metals in sediments with special reference to United Kingdom estuaries: a review. Environ. Pollut. **76**, 89–131.

Brzezińska, A. and C, Garbalewski, 1980. Atmosphere as a source of trace metals input to the Gdańsk Basin. Oceanologia **12**, 43–58.

Buat-Menard, P., and R. Chesselet, 1979. Variable influence of the atmospheric flux on the trace metal chemistry of oceanic suspended matter. Earth Planet. Sci. Letters **42**, 399–411.

Cato, I., 1997. Sedimentological investigations of the Bohus Coast 1995 and recent trends in coastal environmental sediment quality – a report from five trend-monitoring programmes (Geological Survey of Sweden), Report **No. 95**.

Chayes, F., 1967. On the graphical appraisal of the strength of association in petrographic variation diagrams, in: Researches in Geochemistry, ed. P. Abelson (Wiley, New York) **Vol. 2**, 322–339.
Cooley, W.W., and P.R. Lohnes, 1971. Multivariate Data Analysis (John Wiley & Sons, New York) 364 pp.
Coveney, R.M.J., W.L. Watney and C. Maples, 1991. Contrasting depositional models for Pennsylvanian black shale discerned from molybdenum abundances. Geology **19**, 147–150.
Danielsson, Å., 1998. Spatial Modelling in Sediments (Linköping Studies in Arts and Science, Sweden). 89 pp. + Appendices
Danielsson, Å., I. Cato, R. Carman and L. Rahm, 1999. Spatial clustering of metals in the sediments of the Skagerrak/Kattegat. Appl. Geochem. **14**, 689–706.
Das, K., G. Lepoint, V. Loizeau, V. Debacker, P. Dauby and J.M. Bouquegnau, 2000. Tuna and dolphin associations in the North-east Atlantic; Evidence of different ecological niches from stable isotope and heavy metal measurements. Mar. Pollut. Bull. **40**, 102–109.
Davies, B.E., and B.G. Wixson, 1987. Use of factor analysis to differentiate pollutants from other trace metals in surface soils of the mineralized area of Madison county, Missouri, U.S.A. Water Air Soil Pollut. **33**, 339–348.
Drever, J.I., 1997. The Geochemistry of Natural Waters: Surface and Groundwater Environments (Practise Hall, Upper Saddle River), 436 pp.
Ehrlich, R., and W.E. Full, 1987. Sorting out geology – unmixing mixtures, in: Use and Abuse of Statistical Methods in the Earth Sciences, ed. W.B. Size. International Association for Mathematical Geology Studies in Mathematical Geology (Oxford University Press, Oxford), **No. 1**, pp. 33–46.
Emmerson, R.H.C., S.B. O'Reilly-Wiese, C.L. MacLeod and J.N. Lester, 1997. A multivariate assessment of metal distribution in inter-tidal sediments of the Blackwater Estuary, UK. Mar. Pollut. Bull. **34**, 960–968.
Esbensen, K., L. Lindquist, I. Lundholm, D. Nisca and S. Wold, 1987. Multivariate modelling of geochemical and geophysical exploration data. Chemometr. Intellig. Lab. System **2**, 161–175.
Everitt, B.S., and G. Dunn, 1991. Applied Multivariate Data Analysis (Edward Arnold, London).
Favretto, L., and L.G. Favretto, 1984a. Multivariate data analysis of some xenobiotic trace metals in mussels from the Gulf of Trieste. Z. Lebensm. Unters. Forsch. **179**, 201–204.
Favretto, L., and L.G. Favretto, 1984b. Principal component analysis as a tool for studying interedependences among trace metals in edible mussels from the Gulf of Trieste. Z. Lebensm. Unters. Forsch. **179**, 377–380.
Favretto, L., and L.G. Favretto, 1988. Principal component analysis and pollution by trace elements in a mussel survey. J. Chemometr. **3**, 301–308.
Feng, H., J. Cochran, H. Lwiza, B. Brownawell and D. Hirschberg, 1998. Distribution of heavy metal and PBC contaminants in the sediments of an urban estuary: the Hudson River. Mar. Environ. Res. **45**, 69–88.
Flemer, D.A., and R.B. Biggs, 1971. Particulate carbon: nitrogen relations in northern Cheasapeake Bay. J. Fish. Res. Board Can. **28**, 911–918.
Flury, B., and H. Riedwyl, 1988. Multivariate Statistics – A Practical Approach (Chapman & Hall, London), 296 pp.
Fonselius, S.H., 1969. Hydrography of the Baltic deep basins III. Fishery Board of Sweden. Report **No. 23**, 97 pp.
Fonselius, S.H., 1970. Stagnant sea. Environment **12**, 2–11.
Förstner, U., and S.R. Patchinelam, 1976. Bindung und Mobilisation von Schwermetallen in fluviatilen Sedimenten. Chem. Zeitung **100**, 49–57.
Full, W.E., R. Ehrlich and J.E. Klovan, 1981. EXTENDED Q MODEL – Objective definition of external endmembers in the analysis of mixtures. Math. Geol. **13**, 331–344.
Garrett, R.G., 1989. The chi-square plot: a tool for multivariate outlier recognition. J. Geochem. Explor. **32**, 319–341.
Glasby G.P., and P. Szefer, 1998. Marine pollution in Gdańsk Bay, Puck Bay, and the Vistula Lagoon, Poland – An overview. Sci. Total Environ. **212**, 49–57.
Görlich, K., E.A. Görlich, K. Tomala, A.Z. Hrynkiewicz and P.Q. Hung, 1989. ^{57}Fe Mössbauer study of a sediment column in the Gdańsk Basin, Baltic Sea: Palaeoenvironmental application. Mar. Geol. **88**, 49–69.

Hallberg, R.O., 1991. Environmental implications of metal distribution in Baltic Sea sediments. Ambio **20**, 309–316.

Heidam, N.Z., 1981. On the origin of the Arctic aerosol: a statistical approach. Atmos. Environ. **15**, 1421–1427.

HELCOM, 1998. The third Baltic Sea Pollution Load Compilation (PLC-3) (Baltic Marine Environment Protection Commission, Helsinki), Baltic Sea Environment Proceedings **No. 70**.

Hilton, J., W. Davison and U. Ochsenbein, 1985. A mathematical model for analysis of sediment core data: implications for enrichment factor calculations and trace-metal transport mechanisms. Chem. Geol. **48**, 281–291.

Holm, E., 1988. Determination of ^{241}Pu in environmental samples by a radiochemical procedure. Environ. Intern. **14**, 363–365.

Holm, E., 1995. Plutonium in the Baltic Sea. Appl. Radiat. Isot. **46**, 1225–1229.

Hopke, P.K., E.S. Gladney, G.E. Gordon, W.H. Zoller and A.G. Jones, 1976. The use of multivariate analysis to identify sources of selected elements in the Boston urban aerosol. Atmos. Environ. **10**, 1015–1025.

Ingri, J., R. Löfvendahl and K. Boström, 1991. Chemistry of suspended particles in the southern Baltic Sea. Mar. Chem. **32**, 73–87.

Jambers, W., V. Dekov and R. Van Grieken, 1999. Single particle characterisation of inorganic and organic North Sea suspension. Mar. Chem. **67**, 17–32.

Julshamn, K., and O. Grahl-Nielsen, 1996. Distribution of trace elements from industrial discharges in the Hardangerfjord, Norway: A mulivariate data analysis of saithe, flounder and blue mussel as sentinel organisms. Mar. Pollut. Bull. **32**, 564–571.

Julshamn, K., and O. Grahl-Nielsen, 2000. Trace element levels in harp seal (*Pagophilus groenlandicus*) nd hooded seal (*Cystophora cristata*) from the Greenland Sea. A multivariate approach. Sci. Total Environ. **250**, 123–133.

Kingston, H.M., and R.R. Greenberg, 1984. An elemental ratioing technique for assessing concentration data from a complex water system. Environ. Intern. **10**, 153–161.

Knauss, K., and T.-L. Ku, 1983. The chemical composition and decay-series radionuclide content of plankton from the East Pacific. Chem. Geol. **39**, 125–145.

Kock, G., M. Triendl and R. Hofer, 1996. Seasonal pattern of metal accumulation in Arctic char (Salvelinus alpinus) from an oligotrophic Alpine like related to temperature. Can. J. Fish. Aquat. Sci. **53**, 780–786.

Kremling, K., and H. Petersen, 1984. Synoptic survey on dissolved trace metals levels in Baltic surface waters. Mar. Pollut. Bull. **15**, 329–334.

Krishnaswami, S., and M.M. Sarin, 1976. Atlantic surface particulates: composition, settling rates and dissolution in the deep sea. Earth Planet. Sci. Letters **32**, 430–440.

Krüger, O., 1996. Atmospheric deposition of heavy metals into North European marginal seas: Scenarios and trends for lead. GeoJournal **39**, 117–131.

Ku, T.-L., 1965. An evaluation of the U^{234}/U^{238} method as a tool for dating pelagic sediments. J. Geophys. Res. **70**, 3457–3474.

Kuik, P., M. Blaauw, J.E. Sloof and H. Wolterbeek, 1993. The use of Monte Carlo methods in factor analysis. Atmos. Environ. **27A**, 1967–1974.

Lassig, J., J.M. Leppnen, A. Niemi and G. Tamelander, 1978. Phytoplankton primary production in the Gulf of Bothnia 1972–1975 as compared with other parts of the Baltic Sea. Finn. Mar. Res. **244**, 101–115.

Li, Y.H., 1981a. Ultimate removal mechanisms of elements from the ocean. Geochim. Cosmochim. Acta **45**, 1659–1664.

Li, Y.H., 1981b. Geochemical cycles of elements and human perturbation. Geochim. Cosmochim. Acta **45**, 2073–2084.

Li, Y.H., 1982. Interelement relationship in abyssal Pacific ferromanganese nodules and associated pelagic sediments. Geochim. Cosmochim. Acta **46**, 1053–1060.

Loring, D.H., 1984. Trace-metal geochemistry of sediments from Baffin Bay. Can. J. Earth Sci. **21**, 1368–1378.

Loring, D.H., 1990. Lithium – a new approach for the granulometric normalization of trace metal data. Mar. Chem. **29**, 155–168.

Mackey, E.A., R. Demiralp, P.R. Becker, R.R. Greenberg, B.J. Koster and S.A. Wise, 1995. Trace element concentrations in cetacean liver tissues archived in the National Marine Mammal Tissue Bank. Sci. Total Environ. **175**, 25–41.

Malinowski, E.R., 1991. Factor Analysis in Chemistry (A Wiley-Interscience Publication, John Wiley & Sons, Inc.), 350 pp.

Mantovan, P., L. Menegazzo, B. Pavoni and S. Rabitti, 1985. Sediments and pollution in the Northern Adriatic Sea: a statistical analysis. Cont. Shelf Res. **4**, 321–340.

Mardia, K.V., J.T. Kent and J. Bibby, 1989. Multivariate Analysis (Academic Press, London).

McMahon, T.G., and J.W. Patching, 1984. Fluxes of organic carbon in Fjord on the West coast of Ireland. Estuar. Coast. Shelf Sci. **19**, 205–215.

Marchig, V., P. Möller, H. Bcker and P. Dulski, 1985. Foraminiferal ooze from the Galapagos Rift area – hydrothermal impact and diagenetic mobilization of elements. Mar. Geol. **62**, 85–104.

Martin, J.M., and M. Meybeck, 1979. Elemental mass-balance of material carried by major world rivers, Mar. Chem. **7**, 173–206.

Massart, D.L., and L. Kaufman, 1983. The Interpretation of Chemical Data by the Use of Cluster Analysis (John Wiley & Sons, New York).

Maurer, D., M. Mengel, G. Robertson, T. Gerlinger and A. Lissner, 1999. Statistical process control in sediment pollutant analysis. Environ. Pollut. **104**, 21–29.

Miesch, A.T., 1976a. Q-mode factor analysis of compositional data. Comp. Geosci. **1**, 147–159.

Miesch, A.T., 1976b. Q-mode factor analysis of geochemical and petrologic data matrices with constant row-sums. U.S. Geological Survey Professional Paper 574-G, 47 pp.

Mo, T., A.D. Suttle and W.M. Sackett, 1973. Uranium concentrations in marine sediments. Geochim. Cosmochim. Acta **37**, 35–51.

Naik, S., and E.-L. Poutanen, 1984. Humic substances in Baltic Sea sediments. Oceanol. Acta **7**, 431–439.

Nehring, D., 1984a. Variations in the nutrient situation of the Baltic proper. Limnologica (Berlin) **15**, 277–287.

Nehring, D., 1984b. The further development of the nutrient situation in the Baltic proper. Ophelia Suppl. **3**, 167–179.

Nehring, D., 1985. Langzeitveränderungen essentieller Nährstoffe in der zentralen Ostsee. Acta Hydrochim. Hydrobiol. **13**, 591–609.

Neumann, T., T. Leipe, T. Brand and G. Shimmield, 1996. Accumulation of heavy metals in the Oder Estuary and its off-shore basins. Chem. Erde **56**, 207–222.

Olsson, P.E., A. Larson and C. Haux, 1996. Influence of seasonal changes in water temperature on cadmium inducibility of hepatic of hepatic and renal metallothionein in rainbow trout. Mar. Environ. Res. **42**, 41–44.

Pawlak, J., 1980. Land-based inputs of some major pollutants to the Baltic Sea. Ambio Spec. Rep. **9**, 163–167.

Pempkowiak, J., 1992. Enrichment factors of heavy metals in the Southern Baltic surface sediments dated with Pb-210 and Cs-134. Environ. Intern. **17**, 421–428.

Piña, A.A., G.T. Villaseñor, M.M. Fernández, A. Luszczewski Kudra and R.L. Ramos, 2000. Scanning electron microscope and statistical analysis of suspended heavy metal particles in San Luis Potosi, Mexico. Atmos. Environ. **34**, 4103–4112.

Pohl, C., U. Hennings, I. Petersohn and H. Siegel, 1998. Trace metal budget, transport, modification and sink in the transition area between the Oder and Peene Rivers and the southern Pomeranian Bight. Mar. Pollut. Bull. **36**, 598–616.

Rainbow, P.S., W. Fialkowski and B.D. Smith, 1998. The sandhopper *Talitrus saltator* as a trace metal biomonitor in the Gulf of Gdansk, Poland. Mar. Pollut. Bull. **36**, 193–200.

Rainbow, P.S., M. Wolowicz, W. Fialkowski, B.D. Smith and A. Sokolowski, 2000. Biomonitoring of trace metals in the Gulf of Gdansk using mussels (*Mytilus trossulus*) and barnacles (*Balanus improvisus*). Water Res. **34**, 1823–1829.

Ravizza, G.E., and M.H. Bothner, 1996. Osmium isotopes and silver as tracers of anthropogenic metals in sediments from Massachusetts and Cape Cod bays. Geochim. Cosmochim. Acta **60**, 2753–2763.

Renk, H., 1978. Estimation of the primary production of the southern Baltic. Produktywność Ekosystemu Morza Bałtyckiego. PAN-KBM, 93–104 (in Polish, with English abstract).

Renner, R.M., 1988. On the resolution of compositional datasets into convex combinations of extreme vectors, in: Technical Report No. 88/02. (Institute of Statistics and Operations Research, Victoria University of Wellington, Wellington, New Zealand).

Renner, R.M., 1991. An examination of the use of the logratio transformation for the testing of end-member hypotheses. Math. Geol. **23**, 549–563.

Renner, R.M., 1993a. The resolution of a compositional data set into mixtures of fixed source compositions. Appl. Statistics, J. R. Stat. Soc. **42**, 615–631.

Renner, R.M., 1993b. A constrained least-squares subroutine for adjusting negative estimated element concentrations to zero. Comp. Geosci. **19**, 1351–1360.

Renner, R.M., 1995. The construction of extreme compositions. Math. Geol. **27**, 485–497.

Renner, R.M., G.P. Glasby, F.T. Manheim and C.M. Lane-Bostwick, 1989. A partitioning process for geochemical datsets, in: Statistical Applications in the Earth Sciences, ed. F.P. Agterberg and G.F. Bonham-Carter. Geological Survey of Canada Paper 89–9, pp. 319–328.

Renner, R.M., G.P. Glasby and P. Walter, 1997. Endmember analysis of metalliferous sediments from the Galapagos Rift and East Pacific Rise between 2°N and 42°S. Appl. Geochem. **12**, 383–395.

Renner, R.M., G.P. Glasby and P. Szefer, 1998. Endmember analysis of heavy-metal pollution in surficial marine sediments from the Gulf of Gdańsk and Southern Baltic Sea, in: Geochemical Investigations of the Baltic Sea and Surrounding Areas, eds. P. Szefer, P. and G.P. Glasby (Elsevier Science Ltd, Great Britain) Applied Geochem. (Spec. Issue) **13**, 293–304.

Riley, J.P., 1971. The major and minor elements in seawater, in: Introduction to Marine Chemistry, eds. J.P. Riley and R. Chester (Academic Press, London), 64–67.

Rittenberg, S.C., K.O. Emery and W.L. Orr, 1955. Regeneration of nutrients in sediments of marine basins. Deep Sea Res. **3**, 32–45.

Sarin, M.M., D.V. Borole and S. Krishaswami, 1979. Geochemistry and geochronology of sediments from the Bay of Bengal and the equatorial Indian Ocean. Proceedings of the Indian Academy of Science, **88A**, 131–154.

Schneider, B., 1987. Some characterization for atmospheric trace metals over Kiel Bay. Atmos. Environ. **21**, 1275–1283.

Sclater, F.F., E.A. Boyle and J.M. Edmond, 1976. On the marine geochemistry of nickel. Earth Planet. Sci. Letters **31**, 119–132.

Sharma, S., 1996. Applied Multivariate Techniques (John Wiley & Sons, New York).

Shin, P.K.S., and K.Y.S. Fong, 1999. Multiple discriminant analysis of marine sediment data. Mar. Pollut. Bull. **39**, 285–294.

Slawyk, G., Y. Collos, M. Minas and J.R. Grall, 1978. On the relationship between carbon-to-nitrogen composition ratios of the particulate matter and growth rate of marine phytoplankton from the northwest African upwelling. J. Exp. Mar. Biol. Ecol. **23**, 119–131.

Struck, B.D., R. Pelzer, P. Ostapczuk, H. Emons and C. Mohl, 1997. Statistical evaluation of ecosystem properties influencing the uptake of As, Cd, Co, Cu, Hg, Mn, Ni, Pb, Zn in seaweed (*Fucus vesiculosus*) and common mussel (*Mytilus edulis*). Sci. Total Environ. **207**, 29–42.

Struck, U., M. Voss, B. von Bodungen and N. Mumm, 1998. Stable isotopes of nitrogen in fossil Cladoceran exoskeletons: Implications for nitrogen sources in the central Baltic Sea during the past Century. Naturwissenschaften **85**, 597–603.

Suess, E., 1979. Mineral phases formed in anoxic sediments by microbial decomposition of organic matter. Geochim. Cosmochim. Acta **43**, 339–352.

Szefer, P., 1981. Distribution and migration of ^{234}U i ^{238}U isotopes in seawater. Stud. Mater. Oceanol. **34**, 183–228 (in Polish with English abstract).

Szefer, P., 1987. Distribution and migration of uranium isotopes in continental and estuarine waters. Stud. Mater. Oceanol. **51**, 133–193 (in Polish with English abstract).

Szefer, P., 1989. Distribution and possible sources of selected chemical elements in the Baltic Sea ecosystem. (Medical University of Gdańsk), 170 pp. (in Polish).

Szefer, P., 1990a. Interelemental relationships in organisms and bottom sediments of the southern Baltic. Sci. Total Environ. **95**, 119–130.

Szefer, P., 1990b. Mass-balance of metals and identification of their sources in both river and fallout fluxes near Gdansk Bay, Baltic Sea. Sci. Total Environ. **95**, 131–139.

Szefer, P., 1998. Distribution and behaviour of selected heavy metals in various components of the southern Baltic ecosystem, in: Geochemical Investigations of the Baltic Sea and Surrounding Areas, eds. P. Szefer, P. and G.P. Glasby (Elsevier Science Ltd, Great Britain). Appl. Geochem. (Spec. Issue) **13**, 287–292.

Szefer, P., and K. Szefer, 1986. Some metals and their possible sources in rain water of the southern Baltic coast, 1976 and 1978–80. Sci. Total Environ. **57**, 79–89.

Szefer, P., and B. Skwarzec, 1988a. Distribution and possible sources of some elements in the sediment cores of the southern Baltic. Mar. Chem. **23**, 109–129.

Szefer, P., and B. Skwarzec 1988b. Concentration of elements in some seaweeds from coastal region of the southern Baltic and Zarnowiec Lake. Oceanologia **25**, 87–98.

Szefer, P., and M. Wołowicz, 1993. Occurrence of metals in the cockle *Cerastoderma glaucum* from different geographical regions in view of principal component analysis. SIMO-Mar. Pollut. **64**, 253–264.

Szefer, P., and R. Kaliszan, 1993. Inter-element relationships in sediment cores from southern Baltic. International Symposium on Emerging Technologies in Hazardous Waste Management V. (Industrial and Engineering Chemistry Division of the American Chemical Society). Atlanta, U.S.A., 27.09–29.09.1993.

Szefer, P., B. Skwarzec and J. Koszteyn, 1985. The occurrence of some metals in mesozooplankton taken from the southern Baltic. Mar. Chem. **17**, 237–253.

Szefer P., J. Pempkowiak J. and R. Kaliszan, 1993a. Distribution of elements in surficial sediments of the southern Baltic determined by factor analysis. SIMO-Mar. Pollut. **64**, 85–93.

Szefer, P., H. Jankowska, M. Wolowicz, A. Kusak and J. Pempkowiak, 1993b. Vertical and horizontal distribution of heavy metals in bottom sediments from Gdańsk Bay, Baltic Sea. International Symposium on Emerging Technologies in Hazardous waste Management V. (Industrial and Engineering Chemistry Division of the American Chemical Society). Atlanta, U.S.A., 27.09–29.09.1993.

Szefer, P., G.P. Glasby, J. Pempkowiak and R. Kaliszan, 1995a. Extraction studies of heavy-metal pollutants in surficial sediments from the southern Baltic Sea off Poland. Chem. Geol. **120**, 111–126.

Szefer, P., A. Kusak, K. Szefer, H. Jankowska, M. Wolowicz and A.A. Ali, 1995b. Distribution of selected metals in sediment cores of Puck Bay, Baltic Sea. Mar. Pollut. Bull. **30**, 615–618.

Szefer, P., G.P. Glasby, K. Szefer, J. Pempkowiak and R. Kaliszan, 1996. Heavy-metal pollution in surficial marine sediments from the southern Baltic Sea off Poland. J. Environ. Sci. Health **31A**, 2723–2754.

Szefer, P., H.M. Fernandes, M.-J. Belzunce, B. Guterstam, J.M. Deslous-Paoli and M. Wolowicz, 1998a. Distribution of metallic pollutants in molluscs Mytilidae from the temperate, tropical and subtropical marine environments. First International Symposium, IEP '98 Issues in Environmental Pollution, The State and Use of Science and Predictive Models (Elsevier Science Ltd., Denver, Colorado, U.S.A.), 23–26.08.1998, Section of Abstract Book 4.04.

Szefer, P., G.P. Glasby, A. Kusak, K. Szefer, H. Jankowska, M. Wolowicz and A.A. Ali, 1998b. Evaluation of anthropogenic influx of metallic pollutants into Puck Bay, southern Baltic, in: Geochemical Investigations of the Baltic Sea and Surrounding Areas, eds. P. Szefer, P. and G.P. Glasby (Elsevier Science Ltd, Great Britain) Applied Geochem. (Spec. Issue) **13**, 293–304.

Szefer, P., G.P. Glasby, D. Stüben, A. Kusak, J. Geldon, Z. Berner, T. Neumann and J. Warzocha, 1999a. Distribution of selected heavy metals and rare earth elements in surficial sediments from the Polish sector of the Vistula Lagoon. Chemosphere **39**, 2785–2798.

Szefer, P., A.A. Ali, A.A. Ba-Haroon, A.A. Rajeh, J. Geldon and M. Nabrzyski, 1999b. Distribution and relationships of selected trace metals in molluscs and associated sediments from the Gulf of Aden, Yemen. Environ. Pollut. **106**, 299–314.

Szefer, P., M. Domagała-Wieloszewska, J. Warzocha, A. Garbacik-Wesołowska and J. Geldon, 2000a. Distribution and relationships of mercury, lead, cadmium, copper and zinc in perch (*Perca fluviatilis*) from the Pomeranian Bay and Szczecin Lagoon, southern Baltic (submitted).

Szefer, P, M. Wolowicz, and P.S. Rainbow, 2000b. Distribution of trace metals in barnacles (*Balanus improvisus*) in the Gulf of Gdańsk, Baltic Sea (in preparation)

Szefer, P., I. Zdrojewska, J. Jensen, C. Lockyer, A. Łomża, K. Skóra K., I. Kuklik and M. Malinga, 2000c. Intercomparison studies on distribution of heavy metals in liver, kidney and muscle of harbour porpoise, *Phocoena phocoena*, from a Polish Sector of the Baltic Sea and coastal waters of Denmark and Greenland (submitted).

Szefer, P., K. Frelek, K. Szefer, Ch.-B. Lee, B.-S. Kim, J. Warzocha and I. Zdrojewska, 2000d. Distribution of mercury and other trace elements in soft tissue, byssus and shells of *Mytilus edulis trossulus* from the southern Baltic (submitted).

Szefer, P., K. Ikuta, F. Paez Osuna, Ch.-B. Lee, H.M. Fernandes, A.A. Ali, M.J. Belzunce, S.W. Fowler, K. Frelek, H. Hummel, B. Guterstam, and M. Deslous-Paoli, 2000e. Toxic metals in soft tissue and byssus of mollusc Mytilidae from different marine ecosystems. III Conference on Trace Metals "Effects on Organisms and Environment" (Polish Academy of Sciences, Sopot) Sopot, 6–8.06.2000. 186/P2–14.

Szefer, P., G.P. Glasby, J. Geldon, R.M. Renner, E. Björn, J. Snell and W. Frech, 2000f. Distribution and relationships between heavy metals and rare earth elements in surficial sediments from the southern Baltic (in preparation).

Taylor, S.R., 1964. Abundance of chemical elements in the continental crust: a new table. Geochim. Cosmochim. Acta **28**, 1273–1285.

Uścinowicz, S., and J. Zachowicz, 1993. Geochemical Atlas of the Vistula Lagoon. Polish Geological Institute, Warsaw. 11 pp. + 38 maps (in Polish and English).

Vallius, H., 1999. Anthropogenically derived heavy metals in recent sediments of the Gulf of Finland, Baltic Sea. Chemosphere **38**, 945–962.

Van Hattum, B., K.R. Timmermans and H.A. Govers, 1991. Abiotic and biotic factors influencing in situ trace metals levels in macroinvertebrates in freshwater ecosystems. Environ. Toxicol. Chem. **10**, 275–292.

Virkanen, J., 1998. Effect of urbanization on metal deposition in the Bay of Töölönlahti, southern Finland. Mar. Pollut. Bull. **36**, 729–738.

Vogt, N.B., 1989. Polynomial principal component regression: an approach to analysis and interpretation of complex mixture relationships of multivariate environmental data. Chemometr. Intellig. Lab. Systems **7**, 119–130.

Voß, M., and U. Struck, 1997. Stable nitrogen and carbon isotopes as indicator of eutrophication of the Oder river (Baltic Sea). Mar. Chem. **59**, 35–49.

Wangersky, P.J., 1962. Sedimentation in three carbonate cores. J. Geol. **70**, 364–375.

Yeats, P.A., and D.H. Loring, 1991. Dissolved and particulate metal distributions in the St. Lawrence estuary. Can. J. Earth Sci. **28**, 729–742.

Zar, J.H., 1996. Biostatistical Analysis, 3rd edition (Prentice-Hall, Upper Saddle River, NJ).

Zhou, D., 1985. Adjustment of geochemical background by robust multivariate statistics. J. Geochem. Explor. **24**, 207–222.

Zhou, D., 1987. Robust statistics and geochemical data analysis. Math. Geol. **19**, 207–218.

Zhou, D., T. Chang and J.C. Davis, 1983. Dual extraction of R-mode and Q-mode factor solutions. Math. Geol. **15**, 581–606.

Zhu, W., M. Kennedy, E.W. de Leer, H. Zhou and G.J. Alaerts, 1997. Distribution and modelling of rare earth elements in Chinese river sediments. Sci. Total Environ. **204**, 233–243.

Zwolsman, J.J.G., and G.T.M. van Eck, 1999. Geochemistry of major elements and trace metals in suspended matter of the Scheldt estuary, southwest Netherlands. Mar. Chem. **66**, 91–111.

Chapter 7

Monitors of Baltic Sea Pollution

A. TRACE ELEMENTS

(i) Introduction

Coastal degradation, climate changes and growing industrialisation will probably increase the risk of mobilisation of anthropogenically derived and natural toxic agents contributing to the increased potential of their transfer to the marine environments and finally to humans (Knap, 2000). There are numerous articles presenting different points of view on bioaccumulative abilities of selected marine organisms as potential biomonitors of metallic pollutants, also considering their selection and criteria as well as quality assurance in environmental biomonitoring (Bryan, 1976; Phillips, 1980, 1995; Bryan et al., 1985; Phillips and Rainbow, 1993; Chan, 1995; Hansen et al., 1995; Rainbow, 1995; Shulkin and Kavun, 1995; Watson et al., 1995; Wright, 1995; Burgeot et al., 1996; Elliott and Jonge, 1996; Goksøyr et al., 1996; Chapman, 1997; Haynes et al., 1997; Khlebovich, 1997; Nicholson et al., 1997; Blackmore, 1998; Cantillo, 1998; O'Connor, 1998; Lauenstein and Daskalakis, 1998; Batley, 1999; Jeng et al., 2000; Rayment and Barry, 2000; Tanabe, 2000; Wedderburn et al., 2000). Different key issues in ecotoxicology including terms such as contamination, pollution, biomarkers and bioindicators, 'acceptable' variability, 'validation' vs proactivity etc. are explained by Chapman (1995). Many studies on marine biota showing abilities to monitor selected pollutants have shown that some of them caused several problems attributed to variability since various organisms inhabit different substrates ranging from rocky shores to muddy estuaries and adsorb chemical elements from different sources (Phillips, 1980, 1995; Bryan et al., 1985). Such difficulties can be overcome if e.g.

the sampling would be appropriately performed with regard to site and time of collection as well as to sufficient number of relatively standard-sized organisms, etc. Monitoring survey should therefore include the analyses of several species characterised by different food habits (e.g. phytobenthos, filter feeder, deposit feeder, carnivore) in order to evaluate different chemical species of pollutants and their biomagnification along sequential levels of the trophic web (Bryan et al., 1985).

(ii) Abiotic Components

Among abiotic compartments seawater, suspended matter, bottom sediments and ferromanganese nodules have been analysed for concentrations of trace elements. However, sediments, especially their cores have been studied most frequently.

Seawater

Analysis of Baltic seawater is the direct way of assessing pollution status of the environment. However such procedure requires special analytical approach because levels of dissolved species of trace elements are usually very low and hence the possibilities of contamination a sample during collection and analysis are perceptible (Phillips, 1977b; Brügmann, 1981; Bryan et al., 1985). Thus, because of significant elimination of analytical contamination, the seawater concentration data for particular trace elements have become sometimes from one to three orders of magnitude lower than the values obtained prior to 1975 (Bruland, 1983). Perhaps the greatest disadvantage of inter-regional analysis of water samples is the large variation in metal levels attributed to differences in season, time of day, the extent of freshwater influx, depth of sampling, the intermittent flow of industrial effluent as well as hydrological factors, e.g. currents. The interacting effects of these variables may cause as high as an order of magnitude variations in the concentrations of given trace element existed at any one location, especially in estuarine areas. In order to avoid these inconveniences, the use of time-integrated biomonitors is recommended (Phillips, 1977b, 1980).

Trace elements in seawater occur in solution and also in suspended matter being adsorbed to organic and inorganic particulate matter. Additional quantities of metals and metalloids exist in colloidal or chelated forms which are generally difficult to allot to either soluble or particulate phases. Therefore this assignment is in any case somewhat arbitrary because it based on whether the element passes through, or not, a filter of certain pore diameter. Although, a pore size of 0.45 μm is mostly used as a standard, sometimes filters may differ in size from one author to another making the comparison of data difficult or sometimes even impossible (Phillips, 1977b). It is important to note that in estuarine and polluted areas trace elements may be lost from soluble fraction to the sediments by precipitation, or to plankton by adsorption. In consequence of estuarine freshwater-saltwater mixing is generally a decrease in trace element levels in soluble fraction at the cost of increase its levels in particulate fraction. An example of such dissolved metal lost is the deposition of trace elements with amorphic Fe- and Mn oxyhydroxide phase at the hydrological front of the Gulf of Gdańsk (southern Baltic) where estuarine mixing of the brackish water with Vistula river water has place (Szefer et al., 1995).

Kremling and Streu (2000) proved recently that analysis of the dissolved trace element fractions (in addition to their measurements in biota) is an suitable way to monitor metal pollution of Baltic waters. According to these authors there has been significant decreases of the Cd, Cu, Ni, Zn and Pb levels in Baltic Proper surficial water between 1982 and 1995 (Fig. 7.1). This decline possibly reflects reduced loads originated mainly from rivers, waste waters and atmospheric depositions. This negative temporal trend pattern is especially clearly marked for Cd and Pb because of their reduced use in industry and agriculture during the last decades, i.e. for Cd by replacements in electroplating, pigments and stabilisers and by the decreased application of fertilisers and for Pb by limiting of leaded gasoline (Kremling and Streu, 2000).

Suspended matter

Settling suspended particles in the Baltic Sea have potential ability to monitor elemental loads during short periods of time in contrast to surficial sediments (0–10 mm) which may integrate over several years (Jonsson et al., 1990; Lithner et al., 1996). According to Lithner et al. (1996) one way to the applicability of sediment trap for monitoring survey of the present elemental load would be its use in the case of substantial temporal changing of the load. For instance, such situation has taken place off the Swedish Baltic coast, where the load of Pb and As decreased since 1975; i.e. Pb by 50–70%, as a result of reducing atmospheric fallout, and As by more than 90% owing to remedial actions at the Rönnskär smelters (Anon, 1991; Rühling et al., 1992; Notter, 1994; Lithner et al., 1996). However, abundance of these both elements in surficial sediments have not yet reflected significantly the changing loads, e.g. in the case of As, which concentrations in water has decreased an order of magnitude in the Bothnian Bay (Anon, 1991; Borg and Jonsson, 1996; Lithner et al., 1996).

Bottom sediments

In contrast to water samples, the analyses of sediments or of suspended matter are relatively easy (Bryan et al., 1985). In general, better agreement is found between published Baltic data for sediments than for seawater because in the case of the latter samples the measurements need to be carried out near the detection limit of the method used and hence contamination risk significantly grows up (Brügmann, 1981). Moreover, by comparison with water, the undisturbed deposited material may reflect the development history of a sea including the anthropogenic impact from analysis of dated cores (Clifton and Hamilton, 1979; Brügmann, 1981; Bryan et al., 1985). The sediments may therefore serve as a better

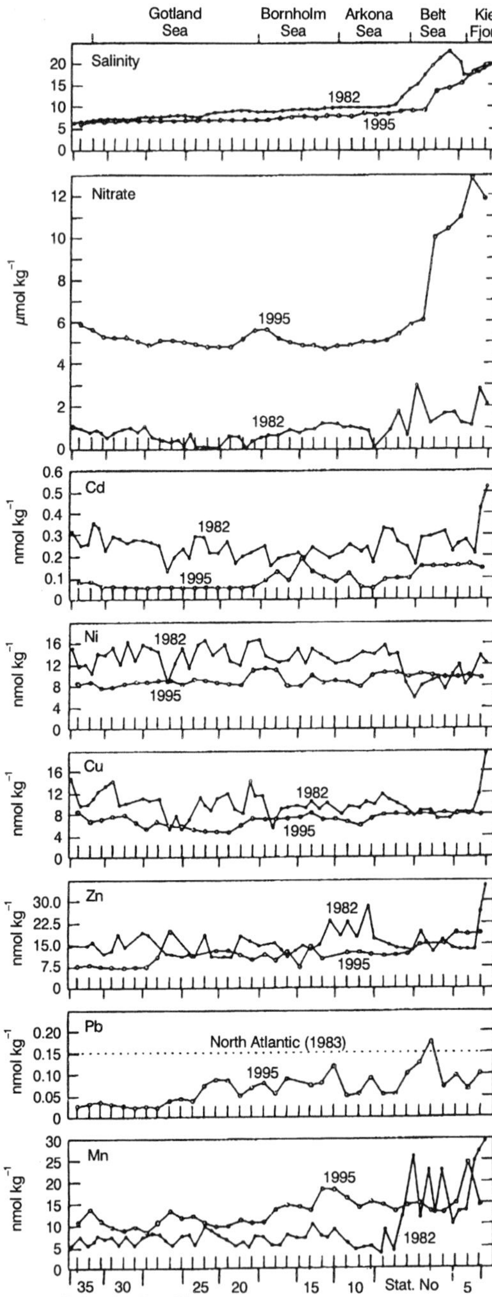

Fig. 7.1. Salinity, nitrate, and trace metal data from 1982 (●) and 1995 (○). In the Pb plot, the dotted line at 0.15 nmol kg^{-1} indicates the suggested average concentration in 1983 (taken from North Atlantic data of Wu and Boyle (1997) due to missing reference values in Baltic Sea waters). After Kremling and Streu (2000); modified.

and integrating monitor of long-term and medium-term metal loads (Brügmann, 1981; Szefer and Skwarzec, 1988a; Szefer, 1998; Szefer et al., 1998b). Since in seawater and sediments, trace elements occur in various chemical forms it is important to know which of them are biologicaly available and capable of having any environmental lability (Phillips, 1977b, 1980; Bryan et al., 1985). An attempt has been therefore made to find correlation, if any exists, between concentrations of metals or metalloids in biota and the ambient environment, i.e. water or bottom sediments (Bryan and Hummerstone, 1973b, 1973c; Luoma and Bryan, 1978; Langston, 1980, 1982; Bryan, 1985; Bryan et al., 1985; Bryan and Langston, 1992; Szefer and Kusak, 2000). One of the examples of such approach is that the concentrations of Pb and As in zoobenthal organisms are sometimes more closely related to easily (1 M HCl) extractable the elements normalised to sediment Fe oxyhydroxide as compared to their total sediment concentrations. An explanation for this is that Fe is the predominant binding substrate for Pb and As in sediments which effectively inhibits the uptake of these both elements in the clams *Scrobicularia plana*. Normalising sediment trace elements, e.g. Pb and As concentrations with respect to the major binding substrate, e.g. Fe concentration, highly improves correlations with tissue burdens in estuarine zoobenthos (Luoma and Bryan, 1978, 1982; Langston, 1980, 1982, 1986; Bryan, 1985; Bryan and Langston, 1992) therefore such intelligent approach is recommended in monitoring survey of pollutants in estuarine waters (Bryan and Langstone, 1992).

In order to quantitative evaluate of metal pollution in the Baltic environment the three main approaches have been used. Firstly, unsieved sediments were standarised geochemically, i.e. in respect to the concentration of Al as a terrigenic normaliser (Szefer et al., 1996). Secondly, both the surficial sediments and sediment cores from the southern Baltic were sieved and normalised granulometrically to $< 80\ \mu m$ or $< 63\ \mu m$ surficial sediment fractions (Szefer et al., 1995; Szefer et al., 1999a) as well as to $< 2\ \mu m$, and for comparison in relation to 63–200 and $> 200\ \mu m$ sediment core fractions (Szefer et al., 1998b). Thirdly, analysis of Baltic sediments for the concentrations of easily extractable metals were performed in order to recognise their bioavailable forms (Szefer et al., 1995). Surprisingly good agreement between data from these three approaches has been obtained indicating anthropogenic origin of Cu, Zn and especially of Pb, Cd and Ag in the coastal, estuarine and lagoonal areas of the southern Baltic. Owing to such approaches, it was possible to eliminate, according to Phillips' recommendation, the variations in trace levels caused by variations in sediment character (sand, mud, silt) at different locations (Phillips, 1977b). The levels of trace elements detected in bottom sediments are associated with both the rates of element deposition and particle sedimentation, the size and nature of particles as well as the concentration of organic matter or other major sediment phases, e.g. Fe- and Mn-oxyhydroxides (Phillips, 1977b; Szefer et al., 1995; Szefer, 1998). In the case of sediments enriched in organic matter a great attention has been paid to eliminate the naturally bound metal concentration with organic matter (reflected by increase in an approximately linear fashion with increased organic matter concentration) by considering levels of pollutants in sediments (falling above the natural pattern) only in relation to the percentage of total C present (Phillips, 1977b).

Ferromanganese nodules

The use of ferromanganese nodules to monitor the metallic pollution in the Baltic Sea was at first suggested ca. 25 years ago. It is found (Djafari, 1976; Suess and Djafari, 1977) that the outer layers of ferromanganese nodules from the Kiel Bay, Baltic Sea, contain anomalously great concentrations of Zn, Pb, Cd and Cu (Figs. 7.2 and 7.3) which seem to be anthropogenic in origin. This finding is in an agreement with reported higher levels of Zn in ferromanganese oxides growing on artificial substrates in the same region (Heuser, 1988). It is considered that the deep-water zone is the ultimately repository for many types of pollutants in the Baltic Sea (Hakansson, 1990). Detailed study in this respect, supporting the increased Zn contents in outer layers of the nodules has been also performed by Hlawatsch (1993) and Hlawatsch et al. (2001) by means laser ablation ICP MS and Scanning Electron Microscopy. The use of ferromanganese nodules as pollution indicators has been investigated by Ingri and Pontér (1986). These authors suggested that specific surface area and the redox conditions govern the scavenging Fe–Mn surface and the enrichment of these elements. Since the enrichment patterns for La, Y and Yb were similar to those of Cu, Ni and Zn, it is postulated that natural processes, e.g. the redox level, play a dominant role in accumulation

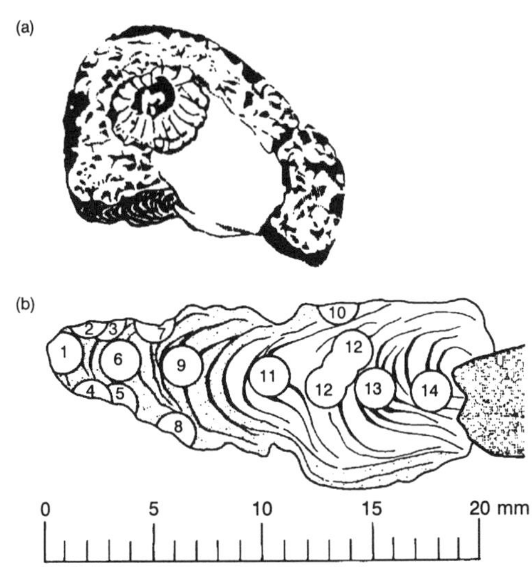

Fig. 7.2. (a) Ferromanganese concretion from the western Baltic Sea with chert core and barnacle growth on the top surface; concentric interior sampled for trace metals. For size see scale Fig. 7.2b. Location: 10°05'E, 54°46'N, southeast slope of Breitgrund flat at 25 m of water depth. (b) Internal structure of saucer-shaped nodule and sites of samples for metal analyses; center of nodule is to the right. The shaded portion is enriched in trace metals. For orientation see Fig. 7.2a. After Suess and Djafari (1977); modified.

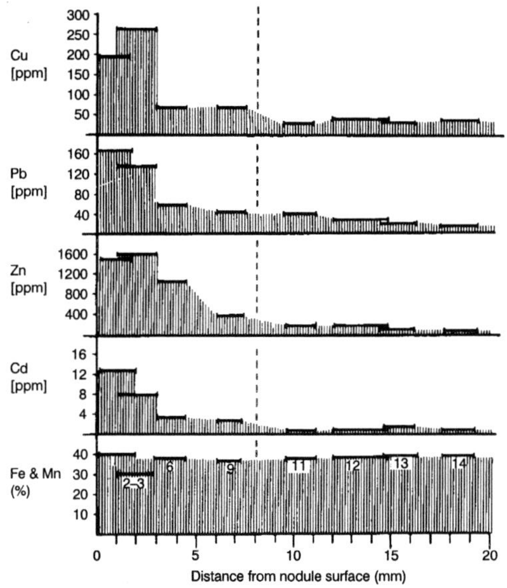

Fig. 7.3. Increasing trace metal concentrations at constant (Fe & Mn) contents with increasing distance from nodule center; the high Cd, Pb, Zn and Cu contents in the outer portion are thought to reflect anthropogenic metal input. Nos. 1-14 and horizontal bars refer to sampling intervals as indicated in Fig. 7.2b. After Suess and Djafari (1977); modified.

of the elements at surficial layers of ferromanganese nodules. According to Ingri and Pontér (1986) it is therefore much questionable the use of ferromanganese concretions for pollution monitoring. Furthermore, the presence of ferromanganese micronodules in surface sediments obscures the interpretation of trace element pollution. Elements terrigenic in origin such as La, Y and Yb are recommended to be used as normalising elements because of their similar enrichment patterns to those of Cu, Ni and Zn. Having this in mind it can be concluded that ferromanganese concretions can only be used under controlled circumstances in monitoring of metallic pollutants in the Baltic Sea (Glasby et al., 1997; Szefer et al., 1998d).

(iii) Biota

Besides suspended matter and bottom sediments, selected representatives of flora and fauna have been also studied in view of their potential use in biomonitoring survey of trace element pollution of the marine environment. Monitor organisms should be good accumulators of metals and their body concentrations must reflect differences in metal bioavailability (Phillips, 1980, 1985; Phillips and Rainbow, 1993). For this reason, the abilities of various representatives of fauna and flora for accumulation of radioactive metals have been assayed at first by several authors (Folsom et al., 1963; Mauchline et al., 1964; Bryan, 1966; Seymour, 1966). First studies of marine organisms, especially molluscs for metallic pollutants have been initiated by Goldberg (1962), Phillips (1976a, 1976b, 1977a, 1978, 1979), Fowler and Oregioni (1976), Bryan and Hummerstone (1977), Goldberg et al. (1978, 1983) and Luoma and Bryan (1978). A sufficient critical overviews on bioaccumulative abilities of benthic organisms were extensively presented by several authors (Phillips, 1977b; Bryan, 1980, 1984; Bryan et al., 1985; Cossa, 1988, 1989; Phillips and Rainbow, 1989; Fowler, 1990; Rainbow et al., 1990; Bryan and Langston, 1992; Wilson and Elkaïm, 1992; Rainbow and Phillips, 1993; Rainbow, 1995). Phillips (1980) and Phillips and Rainbow (1993) in their fundamental books have provided a particularly useful and excellent information on biomonitoring of trace aquatic pollutants and contaminants.

According to Gray (1982) bioindicators can be classified into k-selective species and r-selective species. The former species are characterised by a low reproduction rate, slow growth and a selective advantage in a crowded environment; they are usually located at the end of the food chain (marine mammals, waterfowls, fish). The r-selective species are opportunists, with a selective advantage in an uncrowded environment and they grow fast and rapid.

Various species of organisms have been studied in aspect of their potential use in biomonitoring of trace-element pollutants in the Baltic Sea and adjacent areas, i.e. seaweeds (Bojanowski, 1972; Brix et al., 1983; Caines et al., 1985; Kangas and Autio, 1986; Forsberg et al., 1988; Söderlund et al., 1988; Szefer and Skwarzec, 1988b; Ronnberg et al., 1990; Ostapczuk et al., 1997; Struck et al., 1997), plankton (Szefer et al., 1985; Brügmann and Hennings, 1994), molluscs (Karbe et al., 1977; Phillips, 1977a, 1978, 1979; Möller et al., 1983; Szefer and Szefer, 1985, 1990, 1991; Brix and Lyngby, 1985; Szefer, 1986; Broman et al., 1991; Szefer and Wołowicz, 1993; Ostapczuk et al., 1997; Struck et al., 1997; Rainbow et al., 2000; Szefer and Kusak, 2000; Szefer et al., 2000a), crustaceans (Rainbow et al., 1998; Szefer et al., 2000b), seastar (Brügmann and Lange, 1988), fish (Perttilä et al., 1982; Schladot et al., 1997; Szefer et al., 2000c), waterfowl (Goede et al., 1989); marine mammals (Szefer et al., 2000d).

Phytobenthos

Marine algae would be expected to be the most suitable indicators of dissolved species of metals because, in contrast to animals, the dietary route for trace-element uptake is not involved (Phillips, 1979, 1980, 1990; Bryan et al., 1985).

The evidence for use of bladderwrack, *Fucus vesiculosus*, as an indicator is based on both laboratory and field observations. According to Bryan (1971) and Bryan et al. (1985) occurrence of usually lowest levels of metals in the growing tips of *F. vesiculosus* and their higher, a more constant values in the older tissues can be explained by probably relatively slow accumulation of trace elements as well as the synthesis of more binding sites with age. It means that analyses of the younger parts of the alga at the tips will provide more recent information while analyses of the older fragments will allow to know a value integrated over several months (Bryan et al., 1985). Since *F. vesiculosus*, especially in estuaries, can be contaminated by fine particles of sediment adhering to its body surface then standardised procedure should be used. Owing to use of the standardised procedure, analysis of this brown alga gave good results for biomonitoring of Ag, Cd, Cu, Cr, Hg, Ni, Pb, Zn (Bryan and Hummerstone, 1973c; Morris and Bale, 1975; Phillips, 1977b; Melhuus et al., 1978; Bryan, 1983; Bryan and Langstone, 1992; Phillips, 1980).

According to Phillips (1979) metal concentrations in growing tips of the alga *F. vesiculosus* from the region of the Sound (Öresund) between Sweden and Denmark agree well with available data on the concentrations of dissolved trace elements in waters of the Sound. The alga therefore appears to be responding exclusively to metals in the ambient water, as postulated by other authors (Bryan, 1983; Bryan et al., 1985). Forsberg et al. (1988) and Söderlund et al. (1988) based on concentration data for trace elements in *F. vesiculosus* from the northern Baltic Sea and southern Bothnian Sea recommended the brown seaweed as excellent biomonitor of metal pollution. Elevated concentrations of metals, e.g. Zn, were found in samples taken close to densely populated and heavily industrialised areas (Söderlund et al., 1988). These bioindicative abilities have been also demonstrated by a significant or tendentious increase in concentrations of Al, Co, Cr, Cu, Fe, Mn, Ni, Pb, V and Zn, except Cd, in transplanted *F. vesiculosus* near the city of Stockholm, one of the most densely populated areas around the Baltic Sea (Forsberg et al., 1988). The data for Cd were rather surprising since lower salinity, expected higher pollution with Cd at this area should be reflected by elevated its levels in the *Fucus* biomass. It might be explained by competition from Mn and Zn, suppressed probably Cd uptake (Bryan, 1983; Forsberg et al., 1988). Surprising results have been also obtained for this monitored area of the Archipelago of Stockholm using herbarium species collected in 1933 and 1984. The seaweeds from 1933 contained higher levels of Pb, V and Cu, probably due to mining industry of that time (Forsberg et al., 1988).

On the basis of long-term studies Ostapczuk et al. (1997) demonstrated that *F. vesiculosus* from the North Sea and the Baltic Sea can be useful tool for trend monitoring, depending on the objective. In some cases, however, more detailed information on the chemical form in which the element is present in tissue of the alga is necessary for proper data interpretation. It is recommended (Struck et al., 1997) to consider the concentrations of the macroelements such as Ca, Fe, K, Mg, Na, P and S in the biomatrices to identify and separate independent ecosystem effects, e.g. salinity, temperature. Therefore, the trace element levels in the *F. vesiculosus* do not necessary reflect their total quantities in the ambient water of the Baltic Sea (Kangas and Autio, 1986).

Based on data of analyses of *F. vesiculosus* from Swedish and Finnish coasts (Kangas and Autio, 1986; Forsberg et al., 1988; Söderlund et al., 1988) it is recommended to use of *F. vesiculosus* as biomonitor of metallic pollutants in the Baltic Sea, if the following precautions are taken into account:

- samples should be cut from a fixed part of the *Fucus* thallus and should be free from epiphytes,
- parts of the plants of the same age should be used when comparing spatial distribution,
- samples should be collected at the same or similar time (within a few days) to avoid seasonal variations,
- depth, salinity and water temperature should not be much fluctuated,
- samples should be taken from sites with the same degree of wave-exposure.

Brown seaweed *Pilayella littoralis* from the Gulf of Gdańsk is, as compared to *M. edulis*, less able to regulate Pb uptake from their surroundings (water, sediment); hence it would appear that this Baltic seaweed has a great potential as a biomonitor of Pb in the Baltic environment (Szefer and Szefer, 1991).

Green alga *Enteromorpha* sp. has been used as a biomonitor of trace-element contamination in the marine ecosystems (Bojanowski, 1972; Hägerhäll, 1973; Stenner and Nickless, 1974; Seeliger and Edwards, 1977; Melhuus et al., 1978; Szefer and Skwarzec, 1988b). From these field data clearly results that the seaweed responds to variations of concentrations of dissolved species of As, Cd, Cu, Hg, Pb and Zn and therefore it can be use as effective their biomonitor. Bearing in mind that *E. intestinalis* absorbed higher levels of trace elements, e.g. Co, Mn and Zn at lower salinity (Munda, 1984), advantages of this green alga over *F. vesiculosus* are that *E. intestinalis* often penetrates farther upstream, into regions of very low salinity. Moreover it may also reflect changes in ambient element concentrations more rapidly than *F. vesiculosus* (Bryan et al., 1985).

The concentrations of the trace metals were significantly elevated near the cities of Aalborg (Pb, Cu) and Struer (Cd) at the Limfjord, Denmark. The application of eelgrass as a monitoring organism is highly recommended (Brix et al., 1983). According to Szefer and Szefer (1991) *Z. marina* can be appropriate plant for biomonitoring of Pb pollution in the Gulf of Gdańsk, Poland.

The results strongly suggest (Lyngby and Brix, 1982; Brix and Lyngby, 1982, 1983; Brix et al., 1983) that *Z. marina* can be used as an monitor organism of trace metal contamination and bioavailability in coastal areas; among the properties required of this plant are the following:

- the concentration of some trace metals in above- and belowground parts of *Zostera marina* should be used as a measure of the bioavailable fraction of trace metals in ambient and interstitial water (sediment) in this area,
- in order to get information on the dynamics of chemical elements in coastal Baltic waters, data on the distribution of the elements in the individual plants are needed,
- because of significant seasonal variations in trace elements in eelgrass *Z. marina*, its parts of the same age should be taken at the sampling site in the same or similar time.

Plankton

According to Phillips (1980) phytoplankton have been rarely used as appropriate biomonitors for the comparison of elemental pollutant abundance at more than one sampling site. The main reasons for this limitation seem to be the difficulty in obtaining reasonable size of sample free from other strange particles or other organisms, e.g. zooplankton, and the knowledge of ability of particular species in the accumulation of pollutants by phytoplankton. The use of these organisms as biomonitors affords little time-integration although in the case of single species studied, the concentrations of pollutants detected will be a complex composite of the quantities of available trace elements in the water column as well as the species succession in the phytoplankton community (Phillips, 1980). In spite of questionable abilities of phytoplankton as biomonitors, the uptake of pollutants from Baltic seawater column by these organisms plays an important role in the transferring these trace elements along the successive levels of the trophic chain to its higher organisms (Szefer, 1991). Phillips (1977a, 1978) found that the variations in trace element levels in soft tissue of blue mussel *M. edulis* from the east and west Swedish coasts were attributed to different species composition of phytoplankton populations inhabited the two water areas. Low saline waters of the Baltic Sea were dominated by well adopted blue-green algae while Danish Strait waters hosted other phytoplanktonic species preferred more marine conditions.

The use of zooplankton like phytoplankton as a biomonitoring tool to detect spatial and temporal trends in the Baltic Sea is not recommended. According to several authors (Martin and Knauer, 1973; Boström et al., 1974; Szefer et al., 1985; Diaz and Fernandez-Puelles, 1988; Pohl, 1992; Weber et al., 1992; Brügmann and Hennings, 1994) this is because:

- metal concentrations in different species may vary over a rather broad range,
- some zooplankton species my accumulate the metals depending on their life stage and age,
- some metals seem to be well regulated by the zooplankton,
- non-biogenic material adheres strongly to phytoplankton biomass or becomes incorporated into the zooplankton (e.g. rust particles, paint chips, clay particles) and may contaminate zooplankton samples,
- there is no possible to separate phyto- and zooplankton using different mesh sizes of the nets, i.e. a higher percentage of phytoplankton in the samples may result in higher metal contents.

Nonetheless, zooplankton have already been used frequently to study metal contamination in the marine environment (Phillips, 1980). It may be at least a valuable tool for identification of pollution hot spots (Balogh, 1988).

Molluscs

Various species of molluscs are recommended to be used as biomonitors of trace-element pollution in the marine ecosystems (Phillips, 1980, 1990; Bryan et al., 1985). For instance, a deposit-feeding clam, *Macoma balthica*, generally lies within a few cm of the sediment surface and occurs in the majority of estuaries. It appears to act as biomonitor for Ag, Cd, Cr, Hg, Ni and especially for Zn. There is some doubt about its use for Cu (Bryan and Hummerstone, 1977; Bryan, 1980). Because of their world-wide distribution and potential as indicators, species of *Mytilus* as a filter feeder have become the subject of various monitoring programmes of the 'Mussel Watch" type (Goldberg et al., 1978, 1983; Koide et al., 1982; Cossa, 1989; Fabris et al., 1994). In *Mytilus*, metals are probably adsorbed both from solution and from ingested phytoplankton and other suspended particles (George, 1980). It is evaluated that soft tissue of this mussel appears to be a good bioindicator for Cd, Cr, Hg, Ni, Pb and Zn but not for Cu (Boyden, 1975, 1977; Bryan, 1980, Bryan et al., 1985). Julshamn (1981) concluded that *M. edulis* from polluted waters of Sorfjorden (Norway) was acceptable for monitoring of Pb and probably Hg but appeared to be useless for Cd, Cu and Zn in this respect. It is concluded that *M. edulis* is an unreliable biomonitor for Ag and As (Bryan and Hummerstone, 1977; Langston, 1984). According to Roesijadi et al. (1984) in contrast to Cd, trace elements such as Ag, Cu, Hg and Zn can be successfully biomonitored using the soft tissue of *M. edulis*.

The common cockle *Cerastoderma edule* usually inhabited relatively saline waters of estuaries, it is, as a filter feeder, most likely be able to absorb trace metals from solution and particulate matter (Bryan et al., 1985). Several authors (Boyden, 1975; Bryan and Hummerstone, 1977; Bryan, 1980) postulated that *C. edule* seems to be appropriate accumulator of Ag, Cd, Cu and Zn, and particularly good one for Ni. Based on inter-comparison studies of *C. edule* and seaweed *F. vesiculosus*, it should be emphasised that Ni levels in this bivalve can underestimate the degree of dissolved Ni pollution at its greater concentrations (Bryan et al., 1985). However, changes of Cd levels in this bivalvia are approximately proportional to those in *F. vesiculosus* and levels of Ag increase more rapidly than those in the brown alga in response to pollution of the surrounding area. Szefer et al. (1999b) reported significant spatial and seasonal variations in concentrations of trace elements in *C. glaucum* from Thau Lagoon, the Mediterranean Sea. It is concluded that *Cerastoderma* is not particularly useful as indicator, although it reflects environmental pollution with Ag, As, Cd and Ni. It also responds to high levels of Cu and Zn but, probably as a result of regulation, underestimate moderate levels of pollution. Soft tissue of *C. glaucum* from the Gulf of Gdańsk, southern Baltic, contained higher levels of Zn, Cd and Ni as compared to that from other geographical regions (Szefer and Wołowicz, 1993). Particulate contamination of *Cerastoderma* specimens often makes difficulties in their use as indicators of Cr and Pb (Bryan et al., 1985).

The concentration levels of trace metals in *M. edulis* from the Limfjord, Denmark, were significantly greater in the soft tissue than in the shells. The results suggest, that shells of this species are of no practical use in the monitoring of the metals investigated (Brix and Lyngby, 1985). According to Szefer (1991) and Szefer and Szefer (1991) *M. edulis* has a great potential as a biomonitor of Cd contamination in the southern Baltic ecosystem. Several authors (Rainbow et al., 2000; Szefer et al., 2000a) concluded that *M. trossulus* is suitable biomonitor to employ in programmes designed to trace changes in trace element pollution in the Gulf of Gdańsk, Baltic Sea. It has been reported that in comparison to soft tissue, byssus of *M. trossulus* is more effective bioaccumulator of trace elements except Cd, in the southern Baltic and other geographical regions (Szefer et al., 1998c, 2000a). Study of metals in *Mytilus edulis* along the Swedish coasts disclosed a tendency towards increasing concentrations of Cd and Zn at some locations in the open coastal archipelagos of Stockholm and land compared to the other coastal parts of the Baltic. This increase in concentration at locations not directly affected by industrial metal discharge was argued to be a result of the influence of low salinity on the forms metals and on their bioavailability (Phillips, 1976a, 1976b, 1977a, 1978, 1980; Struck et al. 1997). According to Möller et al. (1983), *M. edulis* from near the Kiel sewage outlet, southwestern Baltic, accumulated higher levels of Ag, Au, Cd, Cr, Hg and Ni reflecting elevated levels of these elements in the ambient water. Therefore in biomonitoring survey, especially concerning areas with a great salinity gradient, a special attention should be paid to more complex interpretation of the data matrix considering besides trace element also macroelement concentrations in *M. edulis* and *F. vesiculosus* from the North Sea and the Baltic Sea (Struck et al. 1997).

It has been found a significant relationship between concentrations of Ag, As, Cd and Pb in perwinkle (*Littorina littorea*) and bladder wrack (*Fucus vesiculosus*) suggesting that, directly or indirectly, concentrations in this gastropod species reflect those of the ambient water (Bryan et al., 1985). Moreover, significant correlation exists between these two benthic species for Cu, Fe, Hg and Zn but slope constants were relatively low, perhaps as a results of regulation by the perwinkle. It seems to be a suitable indicator of pollution with dissolved Ag, Cd and Pb and perhaps As and Hg (Bryan et al., 1985). According to Bauer et al. (1997) malformations in male perwinkles are closely related to the tributyltin (TBT) contamination: the reduction of male mamilliform penial glands showed strong correlations to TBT concentrations in soft tissues. The intersex index (ISI) being the average value of the intersex stages in *L. littorea* is recommended as the most sensitive biological parameter for the assessment of the TBT contamination in the Baltic Sea and the North Sea, i.e. in those regions where the dogwelk *Nucella lapillus*, as the more sensitive species in European surveys, is absent. As it results from numerous literature data, *N. lapillus* was mainly used for TBT biomonitoring in all European programs (Bryan and Langston, 1992; Huet et al., 1996; Evans et al., 1996, 2000; Skarphédinsdóttir et al., 1996; Minchin et al., 1995, 1996, 1997; Morgan et al., 1998; Følsvik et al., 1999; Miller et al., 1999; Santos et al., 2000). This species is well suited in biomonitoring survey, especially in areas of high contamination, like Kiel/Schilksee sampling site, the Baltic Sea, where other alternative monitoring species such as *N. lapillus* are extinct (Bauer et al., 1997). Fig. 7.4 shows that there is highly significant relationship between the ISI values and TBT concentrations as well as the temporal stability of the data. It should be em-

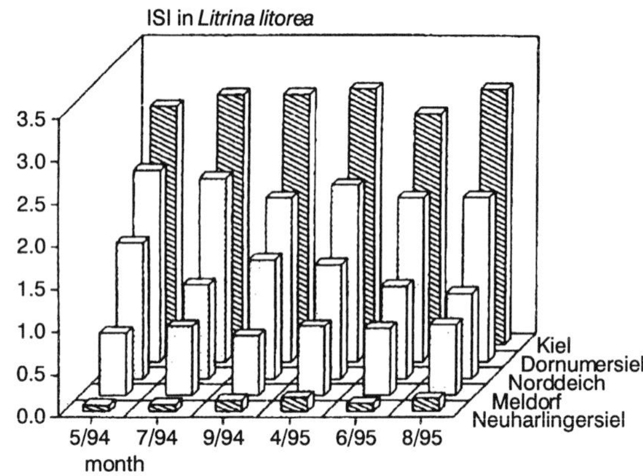

Fig. 7.4. *Littorina littorea*. ISI values for different stations during the German TBT survey 1994-1995. After Bauer et al. (1997); modified.

phasised that the TBT levels in German coastal zone of the Baltic Sea and the North Sea, being much higher than in France, Irish and English waters (Oehlmann et al., 1993; Minchin et al., 1997; Bailey and Davies, 1989, Gibbs et al., 1990, 1991), are too high for the survival of more sensitive species, e.g. *N. lapillus* and *Ocenebra erinacea*. Perwinkle represents, therefore, a monitoring system for many parts of northern Europe and northern America and can be complementary to other recommended TBT-effect monitoring species (Bauer et al., 1997). The field data have been supported by results of laboratory experiments. As can be seen in Fig. 7.5a in all control groups, irrespective of the size-class, no intersex trends can be marked. In all groups exposed to TBT, the third size-class (6–8 mm) displays the lowest ISI's. The most impressive reactions to TBT exposure can be observed in the first size-class (2–4 mm). All analysed females exposed to TBT exhibit intersex stages 3 and 4; stage 3 with incomplete supplanted pallial glands is dominant in group exposed to 100 ng TBT-Sn dm^{-3}. The females exposed to 400 ng dm^{-3} exclusively develop stage 3 with a total replacement of the pallial female glands by a prostate. As already detected in the German TBT survey, and also in laboratory investigations, a reduction of the number of male penial glands can be noted depending on the TBT exposure. From Fig. 7.5b results that after 12 months, males of the control group and third size-group (6–8 mm) develop an average value of ca. 12 glands while specimens of the same size-class exposed to 400 ng TBT-Sn dm^{-3} exhibit only an average of ca. 3 penial glands. Similar negative trend pattern can be observed for the other size-classes. The most advanced diminishing of penial glands is registered in the first size-class (2–4 mm) and highly significant majority of males in this class exposed to 400 ng TBT-Sn dm^{-3} develop no penial glands. Additionally, considering the results of the field TBT survey, the reduction of the number of mamilliform penial glands seems to be a good marker for assessment of TBT contamination in the Baltic Sea, not only in juveniles but also in adult specimens, because the penis is shed and rebuilt every year (Bauer et al., 1997).

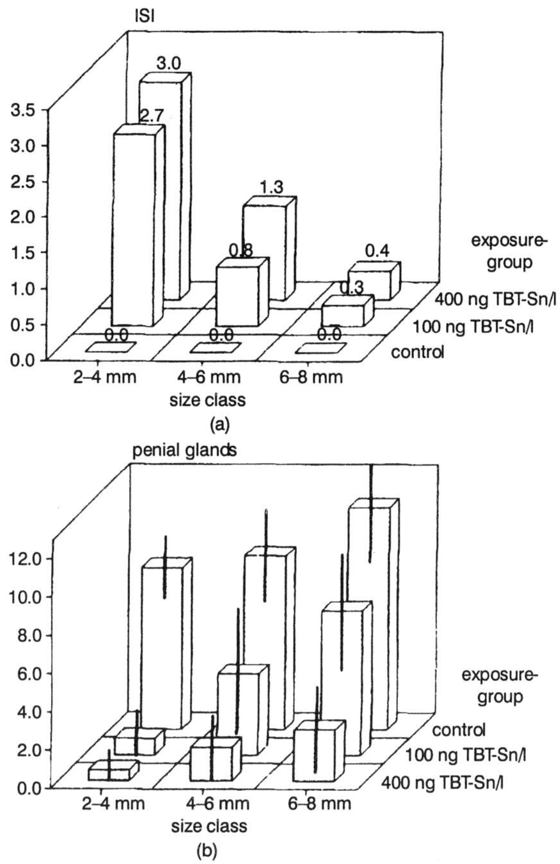

Fig. 7.5. *Littorina littorea*. Results of laboratory experiments with juvenile specimens exposed to different aqueous TBT concentrations. (a) ISI values for different size-classes analysed after 12 months of TBT exposure. (b) Average number of penial glands in males after 12 months of TBT exposure (with standard deviation). After Bauer et al. (1997); modified.

The best way to conduct a monitoring survey using mussels is undoubtedly to collect a large size-range of mussels at each sampling location, and to analyse the mussels individually. This approach, however, demands a very great number of analyses, and will therefore probably not realistic in greater monitoring surveys. Since Cu is regulated in *M. edulis* therefore *Mytilus* can not be used as an indicator organism for Cu pollution (Brix and Lyngby, 1985). This note is in an agreement with data obtained by Julshamn (1981) who reported that *M. edulis* from Sorfjorden, Norway, was useless as biomonitor for Cd, Cu and Zn, but was acceptable for Pb and probably Hg. Theede et al. (1979) found that in numerous places on the German coast of the Baltic the Cd content of the edible common mussel *Mytilus edulis* is higher than in individuals from the North Sea coast. The highest Cd content is found in mussels from the innermost part of the Kiel Fjord. Specimens from the outer parts of the Kiel and Flensburg Fjord contain less Cd.

Polychaete worms

The ragworm, *Nereis diversicolor*, is common in intertidal estuarine systems; it is regarded as a deposit/omnivorous species and depending on the conditions, may deposit feed, filter feed or ingest larger particles (Goerke, 1971; Svieshnikov, 1987). Several authors considered the ability of this polychaete to accumulate trace-elements in marine systems (Bryan and Hummerstone, 1973a, 1973b, 1977, Bryan, 1980, Phillips, 1980; Bryan et al., 1985; Bernds et al., 1998). *N. diversicolor* is considered as indicator of Ag, Cd, Cu, Cr, Hg, Ni and Pb and inter-comparison between heavy-metal concentrations in its body and those of surface sediments showed that in most cases they are significantly related (Bryan et al., 1985). Such relationships were observed for Ag, As, Co, Cr, Cu, Hg and Pb (Langston, 1980; Luoma and Bryan, 1982; Bryan et al., 1980, 1985; Bryan and Langston, 1992). The more reliable correlations between concentrations of Ag, As and Hg in *N. diversicolor* and associated sediments were obtained in the case of use sediment levels of these elements to be normalised to organic matter (Langston, 1986; Bryan and Langston, 1992). In spite of close dependence of trace elements in the polychaete on those in the sediments in some cases tissue concentrations may be additionally influenced by those of the surrounding water. An example of such situation is Ag and Cd in *N. diversicolor* from the Severn Estuary and Bristol Channel because their levels in the body were unrelated to those of the associated sediments but were very closely related to those of *F. vesiculosus* taken up dissolved species of Ag and Cd from the overlying water (Bryan et al., 1985; Bryan and Langston, 1992).

Crustaceans

The significance of trace elements in decapods has been described by Rainbow (1988). The mobility of many species of crustaceans makes impossible to their use as indicators in estuary areas. Extended studies of trace metal uptake rates have been performed by Rainbow et al. (1999) who investigated the possible effect of the life history strategy on the trace element biology of crustaceans such as amphipods and crabs from coastal area of NW Europe. According to Bryan (1968) in the decapod crustaceans, Zn and Cu are regulated against environmental changes. On the other hand, Cd is accumulated in polluted environments by the shrimp *Crangon crangon* (Dethlefsen, 1977). Szefer and Kusak (2000) found a significant relationship between concentrations of Ag, Mn and Fe in *Crangon* sp. and concentrations of these elements in the associated sediments ($< 63\ \mu m$) from the Gulf of Gdańsk, southern Baltic.

As sessile and widely distributed, barnacles are the crustaceans with the greatest potential as indicators (Ireland, 1974; Walker et al., 1975; Walker and Foster 1979; White and Walker, 1981; Phillips and Rainbow, 1988; Powell and White, 1990; Rainbow and Blackmore, 2001). Extremely high levels of Zn, Cu and Cd amounting to 113000, 3230 and 60 $\mu g\ g^{-1}$ dry wt., respectively contained barnacle, *Semibalanus balanoides*, from heavily polluted mining waters of Dulas Bay (Rainbow et al., 1980).

Barnacle (*Balanus improvisus*) from the Gulf of Gdańsk appeared to be suitable biomonitor of trace elements (Rainbow et al., 2000; Szefer et al., 2000b). These authors have shown significant geographical and temporal differences in the local bioavailabilities of trace elements to Baltic barnacles as reflected in the levels of accumulated metallic pollutants (Rainbow et al., 2000). Based on the data obtained (Rainbow et al., 2000; Szefer et al., 2000b) it is concluded that *B. improvisus* is appropriate candidate to be employed in biomonitoring surveys of the Gulf of Gdańsk, Baltic Sea.

Among crustaceans, the amphipods are considered as promising biomonitors of trace metal pollutants in the marine environments (Rainbow and Moore, 1990). Comparative studies of decapod, amphipod and barnacle crustaceans in aspect to their abilities to accumulate of Cd, Cu and Zn have been performed by Rainbow and White (1989).

The talitrid amphipod crustacean *Talitrus saltator* from the strandline of sites around the Gulf of Gdańsk, southern Baltic, indicated spatial trends for trace metal concentrations (Fig. 7.6). These trends have resulted from significant geographical differences in the local bioavailabilities of trace metals, which are variably dependent on outflows from the Vistula River (Cd, Fe, Mn, Zn) or from local sources around the Gulf of Gdańsk (Cu, Pb). According to Rainbow et al. (1998)

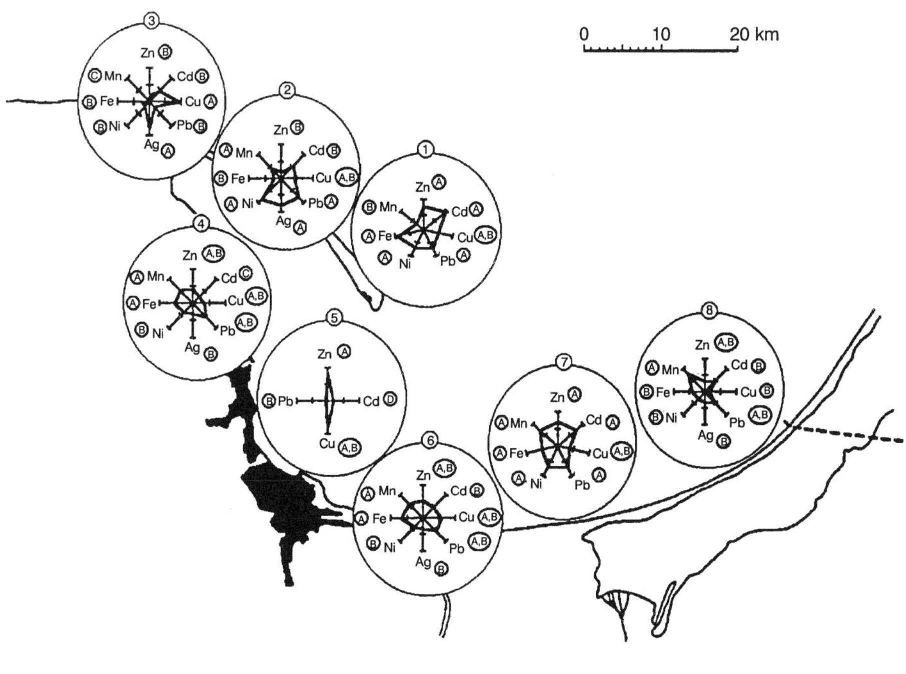

Fig. 7.6. Spatial distributions of metals along the coast of the Gulf of Gdansk, southern Baltic. Standardized deviates of concentrations are used as units on the axes. the mark in the middle of each axis represents the mean concentration of the respective metal in *Talitrus saltator.* The letters in the circles indicate the ANCOVA groups. After Rainbow et al. (1998); modified.

T. saltator is an appropriate biomonitor because it lives on the shore and does not require expensive equipment for its collection. This study has provided baseline concentration data which could be compared with future changes in metal pollution in monitored the Gulf of Gdańsk, Baltic Sea (Rainbow et al., 1998).

There is a lack of available information concerning heavy metal concentrations in Baltic crabs. An example of such studies are concentration data reported for these crustaceans from Belgian, French and English coastal waters (Guns et al., 1999; Rainbow et al., 1999).

Starfish

The asteroid *Asterias rubens* has been proposed by several authors as a valuable sentinel organism for monitoring metallic pollutants in the marine environments (Phillips, 1990; Bjerregaard, 1988; Temara et al., 1996, 1997, 1998). It has been shown that this species efficiently bioconcentrates metals such as Cd, Hg, Mn, Pb, Se and Sr (Binyon, 1978; Bjerregaard, 1988; den Besten et al., 1990; Sorensen and Bjerregaard, 1991; Rouleau et al., 1993; Hansen and Bjerregaard, 1995; Temara et al., 1996, 1998). Among other echinoderms has been studied sea urchin in respect to effects of metals on its development (Lee and Xu, 1984). Distinct differences and strong relationships in trace element concentrations (Fig. 7.7) were found between starfish from different sampling sites of the Baltic Sea, and these are attributed to the composition of the sediments acting as a substrate for their prey such as mussels and snails (Brügmann and Lange, 1988).

Fish

Fish accumulate trace elements from both food and ambient water (Phillips, 1977b). Fish muscle is usually monitored in the interest of public health, but is too much insensitive to most species of metal levels since the metals are either regulated or the tissue concentrations are inconveniently low (Phillips, 1977b, 1980; Bryan, 1984). However, Hg is an important exception and the analysis of fish muscle, in which a high percentage occurs as methyl Hg (Me–Hg); hence muscle is used in monitoring Hg pollutants (Olsson, 1976; Schladot et al., 1997; Sonesten, 2001a). The flounder *Platichthys flesus* is possibly the best indicator since it is often distributed throughout the estuary and is able to enter fresh water (Bryan et al., 1985). An important feature of the flounder is that the larger fish (> 10 cm) occupy home feeding ranges within the estuary (Bryan et al., 1985). Moriarty et al. (1984) recommended miller's thumb, *Cottus gobio*, from the river Ecclesbourne, Derbyshire, for monitoring of heavy-metal pollution. However, the results suggested that there is less profitable for use of concentration data than mass (content) of pollutant in a tissue, e.g. Cd in liver and Pb in gills. According to Olsson (1976) there are significant differences between sexes and ages in respect to Hg concentrations in northern pike, *Esox lucius*, from Lake Marmen, Sweden. Abilities of this species to monitor Hg pollution in Swedish waters have been also studied by Johnels et al. (1968).

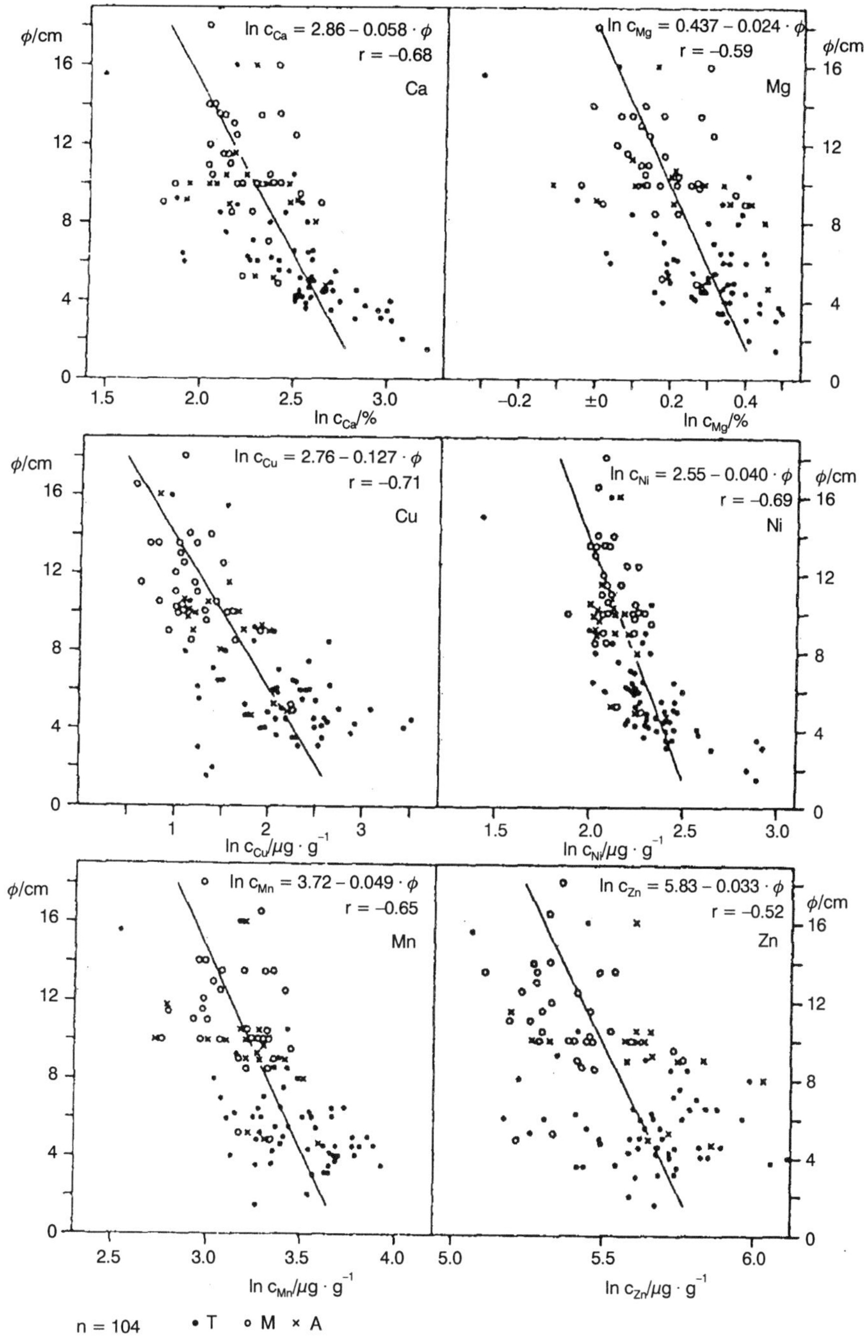

Fig. 7.7. Lines of regression between maximum animal diameter and the contents of Ca, Mg, Cu, Ni, Mn and Zn (logarithmic scale). After Brügmann and Lange (1988); modified.

Several authors studied different species of Baltic fish for concentration of trace elements in aspect to their use as potential biomonitors. According to Schladot et al. (1997), eel-pout (*Zoarces viviparus*) as a sedentary fish of shallow waters can be used as biomonitor for monitoring of Hg, Me–Hg and As in the Baltic ecosystem. In muscle and liver, the concentrations of these elements were enhanced relative to the ambient water and dependent strongly on the sampling site. Figure 7.8 illustrates clearly spatial variations in Hg concentrations in eel-

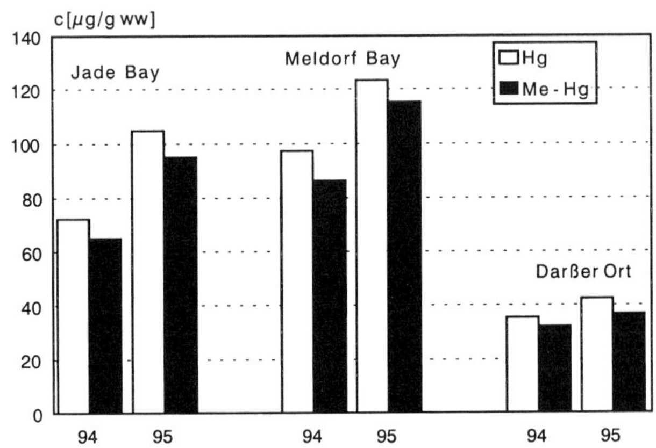

Fig. 7.8. Hg and methyl mercury (Me-Hg) content in muscles of eel-pout samples from the North Sea (Jade Bay and Meldorf Bay) and the Baltic Sea (Darßer Ort). After Schladot et al. (1997); modified.

pout muscle. As results from UBA (1996) due to higher levels, the liver is more useful for monitoring of Pb and Tl than muscle. However, for Cd, Ni and Co no bioaccumulation could be detected.

Perttilä at al. (1982) reported that Baltic cod liver exhibited spatial differences, which, with the exception of Pb, followed the spatial differences of metal concentrations (Hg, Cd, Cu and Zn) in seawater. These differences suggest that, in spite of the extensive migration of cod, it is a better indicator species for aquatic pollution than is herring. Cod feeds mainly on herring and benthic animals, and thus harmful substances are accumulated to higher levels in cod than in herring.

According to Szefer et al. (2000c) perch (*Perca fluviatilis*) indicated significant spatial variations of muscle Cu reaching maximum values in the Pomeranian Bay, southern Baltic.

The use of catalytic converters for automobile exhaust purification is resulted in emission in the platinum-group-metals, i.e. Pt, Pd and Rh. According to Sures et al. (2001) automobile catalyst emitted Pd is bioavailable for European eels (*Anguilla anguilla*).

Waterfowls

Seabirds, as predators located at the top of marine food webs, have a great potential as monitors of metallic pollutants owing to their biomagnification along trophic levels. It is well known about general seabird ecology, the numbers and productivity of many populations what it also makes them particularly appropriate as a choice of biomonitor. The colonial habit of breeding waterfowl has also several advantages. Moreover seabirds can be sometimes used to monitor of fish stocks and fisheries activities (Furness and Camphuysen, 1997). The chronic effects of metallic pollutants as well as effects of acidification may have series of consequences on reproduction, disease, immunosupression and behaviour of waterfowl (Scheuhammer, 1987, 1991). According to Bearhop et al. (2000a), Hg levels in feathers in great scua (*Catharacta skua*) were significantly correlated with those in blood at the time of their growth, suggesting that blood and feathers reflect Hg intake over the same time period. However, blood appeared to be a better biomonitor than feathers (Bearhop et al., 2000b). Using seabirds to monitor Hg pollution has been considered by Thompson et al. (1990). Since the influence of egg contamination on metal burdens in chicks of kittiwake *Rissa tridactyla* from the German Bight decreased with increasing chick age and dietary metal intake gained importance, particularly older chicks ($\geq$ 6 days old) were suitable biomonitors of Hg and Cd pollution around Helgoland Island (Wenzel et al., 1996).

Recent monitoring survey of Hg in seabirds showed spatial variations and the rates of increase in pollution of this element in ecosystem over the last 150 years. This assessment of pollution has been possible owing to analysis of Hg concentrations in feathers of museum specimens. Long-term studies of feathers of Swedish birds since 1840 evidently showed that the rise of Hg content in several bird species to its present values started in the 1940's. The supply of Hg compounds added to Swedish soils as seed dressings was absorbed to birds tissues through the digestive system (Berg et al., 1966). In order to elucidate the feasibility of using feathers as a monitoring object, Appelquist et al. (1984) examined the influence of factors such as ultraviolet light, heating, freezing and weathering on the Hg concentration in feathers of guillemots (*Uria aalge*) and black guillemots (*Cepphus grylle*) from North Baltic as well as from Danish and Greenland waters. According to Furness and Camphuysen (1997) pelagic seabirds indicate higher increases in Hg pollution than most coastal specimens, and such increases have been greatest in seabirds feeding on mesopelagic prey. Apparently this finding is related to patterns of methylation of Hg in low-oxygen, deeper water. Accurate evaluation of long-term trends in Hg pollution assumes that the seabird diet composition has remained unchangeable over decades (Furness and Camphuysen, 1997). Following Goede and de Bruin (1984) either several parts, or the whole feather of *Calidris canutus* and *Limosa lapponica*, can be used in monitoring survey of the Hg pollution and it is emphasised that, with time, contamination may occur via the feather oils. In the case of Zn, only the vane is suitable as a monitoring tissue, sampled just after moult. The shaft reflected the levels of As, Pb and Se deposited in the feather during formation; these elements like Zn should be sampled soon after moult. Figure 7.9 shows changes of mean Se concentrations in dunlin feathers with time. Temporal trends of Se and Hg in the kidney of dunlin caught in Scandinavia and surrounding areas are presented in Figure 7.10. Al-

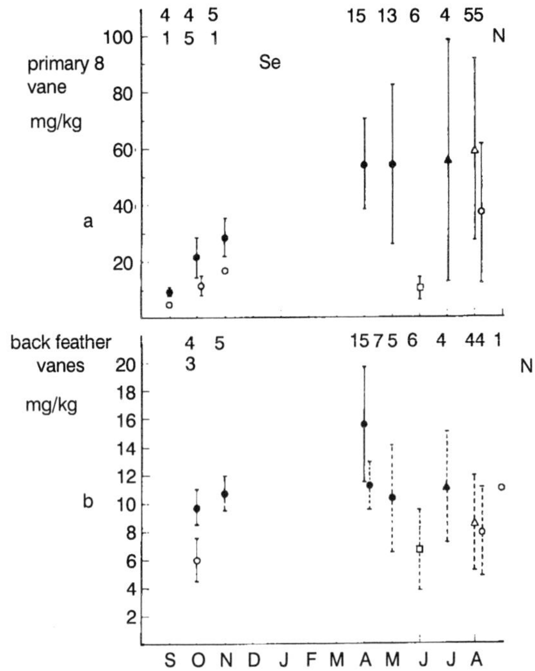

Fig. 7.9. Mean selenium concentrations with standard deviations in Dunlin feathers with time; (●) eastern Wadden Sea; (○) western Wadden Sea; (▲) Finnmark; (△) Ottenby; (□) Vikna. After Goede et al. (1989); modified.

though Se is accumulated significantly in the kidney of Scandinavian *Calidris alpina*, after the waterfowl's departure from the marine to freshwater environments, levels decline rapidly (Goede et al., 1989).

Marine mammals

Changes in the marine environments due to chemical pollution affect sequential trophic levels of food web including its the highest elements, i.e. marine mammals. They are, therefore, doubly injured, directly by pollution and indirectly by the decreasing food stocks (Viale, 1994). Harbour porpoise is rare species in the Baltic Sea (Skóra et al., 1988; Skóra, 1991) constituting a final link in the Baltic food chain. It is interesting organism for pollution studies because of its widespread distribution all over the world. Similar situation concerned also a Dutch coast where a record of the number of dead harbour porpoises was very high (De Wolf, 1983). This decline in numbers of alive specimens since 1960 was related to poisoning because very high levels of Hg and other organic toxicants have been detected in dead animals (De Wolf, 1983). The maximum Baltic value (114 $\mu g\ g^{-1}$) for renal Hg in harbour porpoise, *Phocoena phocoena* (Szefer et al., 2000d), is higher that value of 18 $\mu g\ g^{-1}$ [estimated by Viale (1994) as high] reported for young died striped dolphin, *Stenella coeruleoalba*, from Corsican coasts. It is an

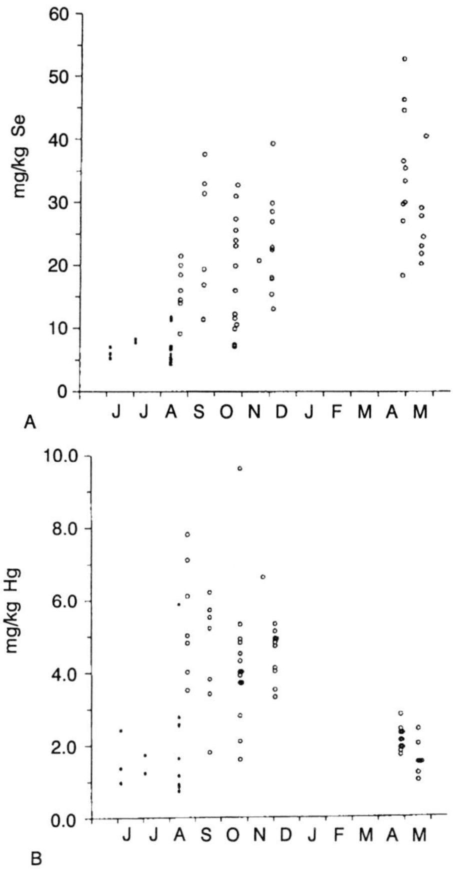

Fig. 7.10. Selenium and mercury kidney concentrations (DW) with time in Dunlin caught in Scandinavia (●), breeding or on post-nuptial migration) and in the Dutch Wadden Sea (O). After Goede and Wolterbeek (1994); modified.

example of significantly elevated levels of renal Hg, most possibly anthropogenic in origin. For some dolphins stranded on Spanish and French coasts, the following diagnoses were discovered: encephalitis and respiratory troubles, a paramyxovirus named delphinoid distemper virus. This collective pathology indicates a strong immunodepressive response of cetaceans to the global water quality changes (Viale, 1994). It is well known that synergic effects of chemical pollutants including Hg cause a viral epizootic damaging some cetaceans and leading to collective pathology (Viale, 1994). Therefore biomonitoring of Hg in marine mammals including Baltic porpoises, is justified bearing in mind that the serious diseases are connected with high levels of chemical toxicants, especially Hg.

Parasites

According to MacKenzie (1999) parasites can be used as an early warning system to monitor the effects of pollutants on marine organisms. The use of parasites as monitors of aquatic pollution has been also reviewed by Möller (1987), Khan and Thulin (1991), MacKenzie et al. (1995) and Lafferty (1997). Some studies indicate higher concentrations of trace elements in some fish parasites, i.e. cestodes and acanthocephalans than in the tissues of their final hosts (Gabrashanska and Nedeva, 1996; Taraschewski and Sures, 1996; Galli et al., 1998). However, there were insignificant differences between metal concentrations in the parasites *Thersitina gasterostei* and cestodes *Schistocephalus solidus* and their host tissues i.e. muscle and gills of stickleback *Gasterosteus aculeatus* from the Gulf of Gdańsk, Baltic Sea (Morozińska-Gogol et al., 1998).

Based on both concentration and discrimination factors it is well documented that Cd, Cr, Cu and especially Fe, Mn and Zn are bioaccumulated in *Pseudalius inflexus* with respect to the host lung of harbour porpoise (*Phocoena phocoena*) of the southern Baltic (Szefer et al., 1998a) (Fig. 3.30). A greater bioaccumulation of these elements in nematodes might result from a better functioning of metal elimination process in the host than in the parasite (Bird and Bird, 1991). Further investigations of metal bioaccumulation in this parasite are needed to evaluate its utility in monitoring of metallic pollutants in the Baltic environment.

B. RADIONUCLIDES

(i) Introduction

Several authors studied zoobenthal organisms such as seaweeds and mussels (Ilus et al., 1987; Carlson and Holm, 1990; Dahlgaard, 1994, 1996; Charmasson et al., 1999) in respect to their abilities to biomonitor of contaminants, i.e. radionuclides in the marine environments. Charmasson et al. (1999) have reported results from a 14-year monitoring (1984–1997) of man-made radionuclide (^{137}Cs and ^{106}Ru) levels in *M. galloprovincialis* collected monthly on the French Mediterranean coast. Long-term variations of radionuclide concentrations in the soft tissue demonstrated seasonal variations which are associated with the reproductive cycle of this mussel as well as to variations in land-based inputs of man-made readionuclides. Studies on biokinetics in benthal fauna have been performed by several authors (Grillo et al., 1981; Fowler and Carvalho, 1985; Warnau et al., 1996a, 1996b, 1999). Application of molluscs for radioecological monitoring of the Chernobyl outburst has been recommended by Frantsevich et al. (1996).

(ii) Biomonitoring Survey

According to several authors (Ilus et al., 1987, 1988; Neumann et al., 1991; Dahlgaard and Boelskifte, 1992; Holm, 1995) *F. vesiculosus* collected from the Baltic Sea is a useful bioindicator in monitoring programs integrating and concentrating low concentrations of radionuclides. It should be emphasised that a dilution effect of radionuclide concentrations caused by growth of *Fucus* is more significant parameter than biological loss of radionuclides (Dahlgaard and Boelskifte, 1992).

It has been reported (Christansen and Strålberg, 2000) that the contribution of ^{137}Cs from Sellafield discharges is now negligible and that the main source of the radionuclide found in the *Fucus* along the Norwegian coast is the Chernobyl fallout being transported to the sea by runoff from land into rivers entering the Baltic Sea.

According to Ilus et al. (1981, 1987, 1988; Carlson, 1990) *F. vesiculosus* from the Finnish coast can be used in monitoring of radioactive substances in the area adjacent to nuclear power stations and as well as in survey concerning the dispersion pattern and fate of radioactive fallout in the marine ecosystem. It is proved evidently that this brown alga is the most sensitive biomonitor of ^{60}Co and ^{65}Zn and hence it is effective to detect of these radionuclides from the Chernobyl fallout. This note is in an agreement with data reported by Neumann et al., (1991) indicating that *F. vesiculosus* is a sensitive indicator for many radionuclides released into receiving water. For instance, it was observed that under conditions of regular maintenance of the nuclear power plants at Ringhals (Swedish west coast) and Simpevarp (the Baltic Proper), activation products, i.e. ^{60}Co and ^{65}Zn in *F. vesiculosus* can be identified at long distances along the coastal line from the discharge point (Neumann et al., 1991).

According to Holm (1995) the Pu concentrations along the Swedish coast, before and after the Chernobyl accident, were comparable reflecting no significant impact on $^{239,240}Pu$ in water on concentration in this brown alga (Holm, 1995).

The ^{129}I levels are strongly dominated by reprocessing discharge from La Hague and Sellafield in the western Norwegian coast and inner Danish water as well as in the Baltic Sea and NW Greenland (Hou et al., 2000).

Mussels appear to be appropriate organisms to monitor radioactive contaminants in the Baltic environment. This zoobenthal organisms collected at the most southern subareas, e.g. Bornholm Sea and especially Kattegat were characterised by lower levels of ^{137}Cs and ^{60}Co than those from Bothnian Sea, reaching maximum values in 1986 and 1988. It means that mussels from these more southern subareas were influenced by a relatively low Chernobyl-derived fallout (HELCOM, 1995).

As regards echinoderms, very few studies have been performed using *A. rubens* as a biomonitor of radioactive contaminants or radiotraces, i.e. Pu, ^{57}Co and ^{203}Hg (Guary et al., 1982; Warnau et al., 1999). However, it has been reported for several echinoderm species, including asteroids, that radioactive elements are willingly bioaccumulated (Grillo et al., 1981; Fowler and Carvalho, 1985; Nakamura et al., 1986; Hutchins et al., 1996a, 1996b; Warnau et al., 1996a, 1996b; Fowler and Teyssié, 1997).

As presented in Chapter 3D elevated levels of Chernobyl radiocaesium (^{137}Cs) in Baltic subareas such as the Bothnian Sea, the Gulf of Finland, the Åland and the Archipelago Seas corresponded to maximum levels of this radioisotope in fish in 1986 and 1987 with tendency to their decrease during the following years. Fish muscle appears to be appropriate biomonitor of radionuclide contamination in the Baltic ecosystem and its drainage area (Ilus, 1987, 1992, 1993; HELCOM, 1995; Sonesten, 2001b). According to Rissanen and Ikäheimonen (2000) flesh of salmon reflects the concentrations of some radionuclides in the ambient waters. The authors detected in 1996–1997 significantly higher flesh levels (36 Bq kg^{-1}) of ^{137}Cs in salmon (*Salmo salar*) from River Tornionjoki (Torneälven) than those (0.37 Bq kg^{-1}) from River Teno. The Tornionjoki salmon originated from the Gulf of Bothnia, the Baltic Sea. It contained ^{134}Cs (< 0.2–1.3 Bq kg^{-1}) originating from the Chernobyl accident. Similar radiocaesium concentrations have been measured in pike (*Esox lucius*) in the Baltic Sea. The elevated concentrations of ^{137}Cs (two orders of magnitude) in the Tornionjoki salmon as compared to the Teno salmon are attributed to several factors, e.g. several Finnish and Swedish rivers have transported radionuclide fallout during the 60's and, particularly after the Chernobyl accident from large catchment areas into the Baltic Sea (Rissanen and Ikäheimonen, 2000).

(iii) Recommendations and future trends

Seabirds can effectively reflect long-term changes in Hg pollution of epipelagic and mesopelagic marine waters, based on inter-specific dietary preferences. Moreover, measured trends in seabirds are in general accordance with model predictions for the surficial marine waters. Based on these findings, Thompson et al. (1998) greatly recommend the use of seabirds as monitors of Hg pollution in the marine environments.

As it has been recommended by Rainbow (1995), Rainbow and Phillips (1993) and Rainbow et al. (2000) whole specimens of barnacle can be effective biomonitors of metallic pollutants in the marine environments. However, it is concluded that barnacle shell can not be considered to be an ideal biomonitoring material (Watson et al., 1995). Potential solutions connected with the use of barnacle shell as biomonitor have been proposed by Watson et al. (1995). The following requirements should be considered:

- use an internal standard which is digested with each batch of samples;
- use large numbers of specimens per sample,
- alternatively, sample barnacles of the same size, or sample a wide range of barnacle sizes from each site and compare a regression lines between metal content and shell weight, or use weight adjusted metal concentrations,
- sample from each different site at the same time.

Parasite nematodes and their Turbot and Trench host organs have been studied to evaluate the relationship, if any exists, between parasitism and pollution in the Baltic and lake environments (Sures et al., 1997). It is suggested to investigate the differences between accumulation of Cd and Pb by the two tapeworm species, i.e. *Bothriocephalus scorpii* and *Monobothrium wageneri*. Further studies dealing with physiological properties of heavy metals accumulation by parasitized marine and freshwater fish are recommended to explain these differences.

Aarkrog (2000) in his millennium article wrote that "Radioecology may briefly be described as the science which studies the interaction between radionuclides and the biogeosphere". This definition is closely related to concept of Dahlgaard and Boelskifte (1992) who recommended study of biological factors such as biomass turnover rates as well as environmental effects on accumulation of radionuclides and their biological loss in the case of use of bioindicator in environmental monitoring. This concept concerns also trace elements. The SENSI model is helpful in evaluation of ability of *Fucus* to monitor pollutants and contaminants by including ecological factors resulting in increase the correlation between expected and measured values. Moreover, the SENSI model may be used successfully to quantify an uncontrolled discharge and to estimate routinely the quality of discharge data (Dahlgaard and Boelskifte, 1992).

References

Aarkrog, A., 2000. Trends in radioecology at the turn of the millennium. J. Environ. Radioactivity **49**, 123–125.

Anon, 1991. Metaller i svenska havsområden (Metals in Swedish sea areas). (The Swedish Environmental Protection Agency), Rep. **No. 3696** (in Swedish).

Appelquist, H., S. Asbirk and I. Drabæk, 1984. Mercury monitoring: mercury stability in bird feathers. Mar. Pollut. Bull. **15**, 22–24.

Bailey, S.K., and I.M. Davies, 1989. The effects of tributyltin on dogwhelsk (*Nucella lapillus*) from Scotisch coastal waters. J. Mar. Biol. Assoc. UK **69**, 335–354.

Balogh, K., 1988. Comparison of mussels and crustacean plankton to monitor heavy metal pollution. Water Air Soil Pollut. **37**, 281–292.

Batley, G.E., 1999. Quality assurance in environmental monitoring. Mar. Pollut. Bull. **39**, 23–31.

Bauer, B., P. Fioroni, U. Schulte-Oehlmann, J. Oehlmann and W. Kalbfus, 1997. The use of *Littorina littorea* for tributyltin (TBT) effect monitoring – results from the German TBT survey 1994/1995 and laboratory experiments. Environ. Pollut. **96**, 299–309.

Bearhop, S., G.D. Ruxton and R. Furness, 2000a. Dynamics of mercury in blood and feathers of great skua. Environ. Toxicol. Chem. **19**, 1638–1643.

Bearhop, S., S. Waldron, D. Thompson and R. Furness, 2000b. Bioamplification of mercury in great skua *Catharacta skua* chicks: the influence of trophic status as determined by stable isotope signatures of blood and feathers. Mar. Pollut. Bull. **40**, 181–185.

Berg, W., A. Johnels, B. Sjöstrand and T. Westermark, 1966. Mercury content in feathers of Swedish birds from the past 100 years. Oikos **17**, 71–83.

Bernds, D., D. Wübben and G.-P. Zauke, 1998. Bioaccumulation of trace metals in polychaetes from the German Wadden Sea: Evaluation and verification of toxicokinetic models. Chemosphere **37**, 2573–2587.

Binyon, J., 1978. Some observations upon the chemical composition of the starfish *Asterias rubens* L with particular reference to strontium uptake. J. Mar. Biol. Assoc. UK **58**, 441–449.

Bird, A.F., and J. Bird, 1991. The Structure of Nematodes. 2nd ed. (New York, Academic Press).

Bjerregaard, P., 1988. Effect of selenium and cadmium uptake in selected benthic invertebrates. Mar. Ecol. Prog. Ser. **48**, 17–20.

Blackmore, G., 1998. An overview of trace metal pollution in the coastal waters of Hong Kong. Sci. Total Environ. **214**, 21–48.

Bojanowski, R. 1972. The occurrence of major and minor chemical elements in the more common Baltic seaweed. Oceanologia **2**, 81–152.

Borg, H., and P. Jonsson, 1996. Large-scale metal distribution in Baltic Sea sediments. Mar. Pollut. Bull. **32**, 8–21.

Boström, K., O. Joensuu and I. Brohm, 1974. Plankton: its chemical composition and its significance as a source of pelagic sediments. Chem. Geol. **14**, 255–271.

Boyden, C.R., 1975. Distribution of some trace metals in Poole Harbour, Dorset. Mar. Pollut. Bull. **6**, 180–187.

Boyden, C.R., 1977. Effect of size upon metal content of shellfish. J. Mar. Biol. Ass. UK **57**, 675–714.

Brix, H., and J.E. Lyngby, 1982. The distribution of cadmium, copper, lead, and zinc in eelgrass (*Zostera marina* L.). Sci. Total Environ. **24**, 51–63.

Brix, H., and J.E. Lyngby, 1983. The distribution of some metallic elements in eelgrass (*Zostera marina* L.) and in sediment in the Limfjord, Denmark. Estuar. Coast. Shelf Sci. **16**, 455–467.

Brix, H., and J.E. Lyngby, 1985. The influence of size upon the concentrations of Cd, Cr, Cu, Hg, Pb and Zn in the common mussel (*Mytilus edulis* L.). Symposia Biologia Hungarica **29**, 253–271.

Brix, H., J.E. Lyngby and H.-H. Schierup, 1983. Eelgrass (*Zostera marina* L.) as an indicator organism of trace metals in the Limfjord, Denmark. Mar. Environ. Res. **8**, 165–181.

Broman, D., L. Lindquist and I. Lundbergh, 1991. Cadmium and zinc in *Mytilus edulis* L. from the Bothnian Sea and the northern Baltic proper. Environ. Pollut. **74**, 227–244.

Bruland, K.W., 1983. Trace elements in sea-water, in: Chemical Oceanography, eds. J.P. Riley and R. Chester (Academic Press, London) 2nd ed., **Vol. 8**, pp. 157–220.

Brügmann, L., 1981. Heavy metals in the Baltic Sea. Mar. Pollut. Bull. **12**, 214–218.

Brügmann, L., and D. Lange, 1988. Trace metal studies on the starfish *Asterias rubens* L. from Western Baltic Sea. Chem. Ecol. **3**, 295–311.

Brügmann, L., and U. Hennings, 1994. Metals in zooplankton from the Baltic Sea, 1980–84. Chem. Ecol. **9**, 87–103.

Bryan, G.W., 1966. The metabolism of zinc and Zn^{65} in crabs, lobsters and freshwater crayfish, in: Radioecological Concentration Processes, eds. B. Åberg and F.P. Hungate. Proc. of the Intern. Symp. (Oxford, Pergamon Press, Stockholm) 1966, 1005–1016.

Bryan, G.W., 1968. Concentrations of zinc and copper in the tissues of decapod crustaceans. J. Mar. Biol. Assoc. UK **48**, 308–321.

Bryan, G.W., 1971. The effects of heavy metals (other than mercury) on marine and estuarine organisms. Proc. of the Royal Society, B **177**, 389–410.

Bryan, G.W., 1976. Heavy metal contamination in the sea, in: Marine Pollution, ed. R. Johnston (Academic Press, London, New York) pp. 185–302.

Bryan, G.W., 1980. Recent trends in research on heavy-metal contamination in the sea. Helgoländer Meeresunters. **33**, 6–25.

Bryan, G.W., 1983. Brown seaweed, *Fucus vesiculosus*, and the gastropod, *Littorina littoralis*, as indicators of trace-metal availability in estuaries. Sci. Total Environ. **28**, 91–104.

Bryan, G.W., 1984. Pollution due to heavy metals and their compounds, in: Marine Ecology, ed. O. Kinne (John Wiley & Sons Ltd, Chichester) **vol. 5**, Part 3, 1289–1431.

Bryan, G.W., 1985. Bioavailability and effects of heavy metals in marine deposits. Wastes in the oceans, in: Disposal Nearshore Waste, eds. B.H. Ketchum, J.M. Capuzzo, W.V. Burt, I.W. Duedall, P.K. Park and D.R. Kester (John Wiley & Sons Ltd, New York) **vol. 6**, 42–79.

Bryan, G.W., and L.G. Hummerstone, 1973a. Adaptation of the polychaete *Nereis diversicolor* to manganese in estuarine sediments. J. Mar. Biol. Assoc. UK **53**, 859–872.

Bryan, G.W. and L.G. Hummerstone, 1973b. Adaptation of the polychaete *Nereis diversicolor* to estuarine sediments containing high concentrations of zinc and cadmium. J. Mar. Biol. Assoc. UK **53**, 839–857.

Bryan, G.W., and L.G. Hummerstone, 1973c. Brown seaweed as an indicator of heavy metals in estuaries in South-West England. J. Mar. Biol. Assoc. U.K. **53**, 705–720.

Bryan, G.W., and L.G. Hummerstone, 1977. Indicators of heavy metal contamination in the Looe Estuary (Cornwall) with particular regard to silver and lead. J. Mar. Biol. Assoc. U.K. **57**, 75–92.

Bryan, G.W., and W. Langston, 1992. Bioavailability, accumulation and effects of heavy metals in sediments with special reference to United Kingdom estuaries: a review. Environ. Pollut. **76**, 89–131.

Bryan, G.W., W.J. Langston and L.G. Hummerstone, 1980. The use of biological indicators of heavy metal contamination in estuaries with special reference to an assessment of the biological availability of metals in estuarine sediments from south-west Britain. Mar. Biol. Assoc. UK. Occasional Publication **No.1**, 73 pp.

Bryan, G.W., W.J. Langston, L.G. Hummerstone and G.R. Burt, 1985. A guide to the assessment of heavy-metal contamination in estuaries using biological indicators. Mar. Biol. Assoc. U.K., Occasional Publication (The Laboratory, Citadel Hill, Plymouth, Devon, England) **No. 4**, 92 pp.

Burgeot, T., S. Woll and F. Galgani, 1996. Evaluation of the micronucleus test on *Mytilus galloprovincialis* for monitoring applications along French coasts. Mar. Pollut. Bull. **32**, 39–46.

Caines, L.A., A.W. Watt and D.E. Wells, 1985. The uptake and release of some trace metals by Aquatic Bryophytes in acidified waters of Scotland. Environ. Pollut. **10B**, 1–18.

Cantillo, A.Y., 1998. Comparison of results of Mussel Watch programs of the United States and France with Worldwide Mussel Watch studies. Mar. Pollut. Bull. **36**, 712–717.

Carlson, L., 1990. Effects of biotic and abiotic factors on the accumulation of radionuclides in *Fucus vesiculosus* L. Dissertation (Lund University, Sweden) 111 pp.

Carlson, L., and E. Holm, 1990. Radioactivity in the Baltic Sea following the Chernobyl accident. SSI project P392-86 (University of Lund, Sweden) 21 pp.

Chan, K.M., 1995. Metallothionein: potential biomarker for monitoring heavy metal pollution in fish around Hong Kong. Mar. Pollut. Bull. **31**, 411–415.

Chapman, P.M., 1995. Ecotoxicology and pollution – key issues. Mar. Pollut. Bull. **31**, 167–177.

Chapman, P.M., 1997. Is bioaccumulation useful for predicting impacts. Mar. Pollut. Bull. **34**, 282–283.

Charmasson, S., E. Barker, D. Calmet, A.-S. Pruchon and H. Thébault, 1999. Long-term variations of man-made radionuclide concentrations in a bio-indicator *Mytilus galloprovincialis* from the French Mediterranean coast. Sci. Total Environ. **237/238**, 93–103.

Christensen, C.Ch., and E. Strålberg, 2000. Can the outflow of radiocaesium from the Baltic Sea be detected in brown algae along the Norwegian coast?, in: The Radiological Exposure of the Population of the European Community to Radioactivity in the Baltic Sea. Marina-Balt Project, ed. S.P. Nielsen. Proceedings of a Seminar held at Hasseludden Conference Centre, Stockholm, 9–11 June 1998, European Commission, Directorate-General Environment, EUR 19200 EN (European Communities, 2000, Belgium), pp. 433–438.

Clifton, R.J., and E.I. Hamilton, 1979. Lead-210 chronology in relation to levels of elements in dated sediment core profiles. Estuar. Coast. Shelf Sci. **8**, 259–269.

Cossa, D., 1988. Cadmium in *Mytilus* spp.: Worldwide survey and relationship between seawater and mussel content. Mar. Environ. Res. **26**, 265–284.

Cossa, D., 1989. A review of the use of *Mytilus* spp. as quantitative indicators of cadmium and mercury contamination in coastal waters. Oceanol. Acta **12**, 417–432.

Dahlgaard, H., 1994. Marine radioecology, in: Radioecology: Lectures in Environmental Radioactivity, ed. E. Holm (World Scientific Singapore, New Jersey, London, Hong Kong)

Dahlgaard, H., 1996. Polonium-210 in mussels and fish from the Baltic-North Sea estuary. J. Environ. Radioactivity **32**, 91–96.

Dahlgaard, H., and S. Boelskifte, 1992. "Sensi": A model describing the accumulation and time-integration of radioactive discharges in the bioindicator *Fucus vesiculosus*. J. Environ. Radioactivity **16**, 49–63.

Den Besten, P.J., H.J. Herwig, D.I. Zandee and P.A. Voogt, 1990. Cadmium accumulation and metallothionein-like proteins in the sea star *Asterias rubens*. Arch. Environ. Contam. Toxicol. **19**, 858–862.

De Wolf, P., 1983. Bio-indicators and the quality of the Wadden Sea. Environ. Monit. Assess. **3**, 355–367.

Dethlefsen, V., 1977. Uptake, retention and loss of cadmium by brown shrimp (*Crangon crangon*). Berichte der Deutschen Wissenschaftlichen Komission für Meeresforschung **26**, 137–152.

Diaz, C., and M.L. Fernandez-Puelles, 1988. Contents of heavy metals in zooplankton of Canary Islands. Biol. Instit. Espanol. Oceanogr. **5**, 57–61 (in Spanish).

Djafari, D., 1976. Manganese-iron accumulates in Kiel Bay. Dr. rer. nat. thesis (Universität Kiel) in German.

Elliott, M., and V.N. De Jonge, 1996. The need for monitoring the monitors and their monitoring. Mar. Pollut. Bull. **32**, 248–249.

Evans, S.M., P.M. Evans and T. Leksono, 1996. Widespread recovery of dogwhelks, *Nucella lapillus* (L.), from tributyltin contamination in the North Sea and Clyde Sea. Mar. Pollut. Bull. **32**, 263–269.

Evans, S.M., E. Kerrigan and N. Palmer, 2000. Causes of imposex in the dogwhelk *Nucella lapillus* (L.) and its use as a biological indicator of tributyltin contamination. Mar. Pollut. Bull. **40**, 212–219.

Fabris, J.G., B.J. Richardson, J.E. O'Sullivan and F.C. Brown, 1994. Estimation of cadmium, lead, and mercury concentrations in estuarine waters using the mussel *Mytilus edulis planulatus* L. Environmental Toxicology and Water Quality: An International Journal **9**, 183–192.

Folsom, T.R., D.R. Young, J.N. Johnson and K.C. Pillai, 1963. Manganese-54 and zinc-65 in coastal organisms of California. Nature, Lond. **200**, 327–329.

Forsberg, A., S. Söderlund, A. Frank, L.R. Petersson and M. Pedersen, 1988. Studies on element content in seaweed, *Fucus vesiculosus*, from Archipelago of Stockholm. Environ. Pollut. **49**, 245–263.

Fowler, S.W., 1990. Critical review of selected heavy metal and chlorinated hydrocarbon concentrations in the marine environment. Mar. Environ. Res. **29**, 1–64.

Fowler, S.W., and B. Oregioni, 1976. Trace metals in mussels from the N. W. Mediterranean. Mar. Pollut. Bull. **7**, 26–29.

Fowler, S.W., and F.P. Carvalho, 1985. Americium biokinetics in benthic organisms as a function of feeding mode. Bull. Environ. Contam. Toxicol. **35**, 826–834.

Fowler, S.W., and J.L. Teyssié, 1997. Assimilation and excretion of selected heavy metals and radionuclides ingested by seastars. Radioprotect. - Colloque **32** **C2**, 317–322.

Følsvik, N., J.A. Berge, E.M. Brevik and M. Walday, 1999. Quantification of organotin compounds and determination of imposex in populations of dogwhelks (*Nucella lapillus*) from Norway. Chemosphere **38**, 681–691.

Frantsevich, L., A. Korniushin, I. Pankov, A. Ermakov and T. Zakharchuk, 1996. Application of molluscs for radioecological monitoring of the Chernobyl outburst. Environ. Pollut. **94**, 91–100.

Furness, R.W. and Kees (C.J.) Camphuysen, 1997. Seabirds as monitors of the marine environment. ICES J. Mar. Sci. **54**, 726–737.

Gabrashanska, I.M., and I. Nedeva, 1996. Content of heavy metals in the system fish-cestodes. VII European Multicolloquium of Parasitology, Parma, Italy, 2–6 September 1996.

Galli, P., G. Crosa and A.O. Ambrogi, 1998. Heavy metals concentrations in Acanthocephalans parasites compared to their fish host. Chemosphere **37**, 2983–2988.

George, S.G., 1980. Correlation of metal accumulation in mussels with the mechanism of uptake, metabolism and detoxification: a review. Thallasia Jugoslavica **16**, 347–365.

Gibbs, P.E., G.W. Bryan, P.L. Pascoe and G.R. Burt, 1990. Reproductive abnormalities in female *Ocenebra erinacea* (Gastropoda) resulting from tributyltin-induced imposex. J. Mar. Biol. Assoc. UK **70**, 639–656.

Gibbs, P.E., G.W. Bryan and P.L. Pascoe, 1991. TBT-induced imposex in the dogwhelk, *Nucella lapillus*: geographical uniformity of the response and effects. Mar. Environ. Res. **32**, 79–87.

Glasby, G.P., E.M. Emelyanov, V.A. Zhamoida, G.N. Baturin, T. Leipe, R. Bahlo and P. Bonacker, 1997. Environments of formation of ferromanganese concretions in the Baltic Sea: A critical review, in: Manganese Mineralization: Geochemistry and Mineralogy of Terrestrial and Marine Deposits, eds. K. Nicholson, J.R. Hein, B. Bühn and S. Dasgupta. Geol. Soc. Spec. Publ. **No. 119**, 213–237.

Goede, A.A., and M. de Bruin, 1984. The use of bird feather parts as a monitor for metal pollution. Environ. Pollut. **8B**, 281–298.

Goede, A.A., T. Nygard, M. de Bruin and E. Steinnes, 1989. Selenium, mercury, arsenic and cadmium in the lifecycle of the dunlin, *Calidris alpina*, a migrant wader. Sci. Total Environ. **78**, 205–218.

Goerke, H., 1971. De Ernährungsweise der *Nereis* – Arten (Polychaeta, Nereidae) der deutschen Küsten (Veroffentlichungen des Instituts für Meeresforschung in Bremerhaven), **13**, pp. 50.

Goksøyr, A., J. Beyer, E. Egaas, B.E. Grøsvik, K. Hylland, M. Sandvik and J.U. Skaare, 1996. Biomarker responses in flounder (*Platichthys flesus*) and their use in pollution monitoring. Mar. Pollut. Bull. **33**, 36–45.

Goldberg, E.D., 1962. Elemental composition of some pelagic fishes. Limnol. Oceanogr. **7**, (Supplement), 1xxii–1xxv.

Goldberg, E.D., V.T. Bowen, J. W. Farrington, G. Harvey J.H. Martin, P.L. Parker, R.W. Risebrough, W. Robertson, E. Schneider and E. Gamble, 1978. The Mussel Watch. Environ. Conserv. **5**, 101–125.

Goldberg, E.D., M. Koide, V. Hodge, A.R. Flegal and J. Martin, 1983. U. S. Mussel Watch: 1977–1978 results on trace metals and radionuclides. Estuar. Coast. Shelf. Sci. **16**, 69–93.

Gray, J.S., 1982. Effects of pollutants on marine ecosystems. Neth. J. Sea Res. **16**, 424–443.

Grillo, M.C., J.C. Guary and S.W. Fowler, 1981. Comparative studies on transuranium nuclide biokinetics in sediment-dwelling invertebrates, in: Impacts of Radionuclide Releases into the Marine Environment (IAEA, Vienna), pp. 273–291.

Guary, J.C., S.W. Fowler and T.M. Beasley, 1982. Routes of plutonium uptake and their relation to biomagnification in starfish. Mar. Pollut. Bull. **13**, 99–102.

Guns, M., P. Van Hoeyweghen, W. Vyncke and H. Hillewaert, 1999. Trace metals in selected benthic invertebrates from Belgian coastal waters (1981-1996). Mar. Pollut. Bull. **38**, 1184–1193.

Hägerhäll, B., 1973. Marine botanical-hydrographical trace element studies in the Öresund area. Botanica Mar. **16**, 53–64.

Hakansson, L., 1990. Baltic research developments. Ambio Special Report, **7**.

Hansen, S.N., and P. Bjerregaard, 1995. Manganese kinetics in the sea star *Asterias rubens* (L.) exposed via food or water. Mar. Pollut. Bull. **31**, 127–132.

Hansen, I.V., J.M. Weeks and M.H. Depledge, 1995. Accumulation of copper, zinc, cadmium and chromium by the marine sponge *Halichondria panicea* Pallas and the implications for biomonitoring. Mar. Pollut. Bull. **31**, 133–138.

Haynes, D., J. Leeder and P. Rayment, 1997. A comparison of the bivalve species *Donax deltoides* and *Mytilus edulis* as monitors of metal exposure from effluent discharges along the Ninety Mile Beach, Victoria, Australia. Mar. Pollut. Bull. **34**, 326–331.

HELCOM, 1995. Radioactivity in the Baltic Sea 1984–1991. Baltic Sea Environment Proceedings (Baltic Marine Environment Protection Commission, HELCOM) **No. 61**.

Heuser, H., 1988. Observations and investigations on the genesis of shallow-water manganese nodules in Kiel Bay (western Baltic). Berichte-Reports (Geologisch-Paläontologisches Institut Universität Kiel) **26** (in German).

Hlawatsch, S., 1993. Growth of manganese-iron accumulates in the western Baltic Sea. Indicator for environmental change. Diplom thesis, Univesität Kiel (in German).

Hlawatsch, W., F. Lechtenberg, C.D. Garbe-Schönberg, J. Harff and E. Suess, 2001. Ferro-manganese-nodules from the western Baltic Sea: indicator for environmental change in a highly polluted region of Europe. Chem. Geol. (in press).

Holm, E., 1995. Plutonium in the Baltic Sea. Appl. Radiat. Isot. **46**, 1225–1229.

Hou, X., H. Dahlgaard and S.P. Nielsen, 2000. Iodine-129 time series in Danish, Norwegian and Northwest Greenland coast and the Baltic Sea by seaweed. Estuar. Coast. Shelf Sci. **51**, 571–584.

Huet, M., Y.M. Paulet and M. Glémarec, 1996. Tributyltin (TBT) pollution in the coastal waters of west Brittany as indicated by imposex in *Nucella lapillus*. Mar. Environ. Res. **41**, 157–167.

Hutchins, D.A., I. Stupakoff and N.S. Fisher, 1996a. Temperature effect on accumulation and retention of radionuclides in the sea star, *Asterias forbesi*: implications for contaminated northern waters. Mar. Biol. **125**, 701–706.

Hutchins, D.A., J.L. Teyssié, F. Boisson, S.W. Fowler and N.S. Fisher, 1996b. Temperature effect on uptake and retention of contaminant radionuclides and trace metals by the brittle star, *Ophiothrix fragilis*. Mar. Environ. Res. **41**, 363–378.

Ilus, E., J. Ojala, K.-L. Sjöblom and K. Toumainen, 1981. *Fucus vesiculosus* as bioindicator of radioactivity in Finnish coastal waters. 1. Gulf of Finland. ISBN 951-46-5408-0 (Inst. Rad. Prot.).

Ilus, E., K.-L. Sjöblom, R. Saxén, H. Aaltonen and T.K. Taipale, 1987. Finnish studies on radioactivity in the Baltic Sea after the Chernobyl accident in 1986. Suppl. 11 to Annual Rep. (STUK-A55). STUK-A66, April 1987 (Finnish Centre for Radiation and Nuclear Safety, Helsinki) 40 + 5 pp.

Ilus, E., S. Klemola, K.-L. Sjöblom and T. Ikäheimonen, 1988. Radioactivity of *Fucus vesiculosus* along the Finnish coast in 1987. Suppl. 9 to Annual Rep. 1987 (STUK-A74). STUK-A83, May 1988 (Finnish Centre for Radiation and Nuclear Safety, Helsinki, Finland) 36 pp.

Ilus, E., K.-L. Sjöblom, S. Klemola, H. Arvela, 1992. Monitoring of radionuclides in the environs of Finnish nuclear power plants in 1989-1990. Suppl. 9 to Annual Rep. STUK-A89. STUK-A102, January 1992 (Finnish Centre for Radiation and Nuclear Safety, Helsinki) 91 pp.

Ilus, E., K.-L. Sjöblom, T. Ikäheimonen, , R. Saxén and S. Klemola, 1993. Monitoring of radionuclides in the Baltic Sea in 1989-1990. Suppl. 10 to Annual Rep. STUK-A89. STUK-A103, February 1993 (Finnish Centre for Radiation and Nuclear Safety, Helsinki, Finland) 35 pp.

Ingri, J., and C. Ponter C., 1986. Scavenging properties of ferromanganese nodules in the Gulf of Bothnia. Rapports et Procès Verbaux des Réunions Condeil International Pour l'Exploration de la Mer **186**, 234–243.

Ireland, M.P., 1974. Variations in the zinc, copper, manganese and lead content of *Balanus balanoides* in Cadigan Bay, Wales. Environ. Pollut. **7**, 65–75.

Jeng, M.-S., W.-L. Jeng, T.-Ch. Hung, Ch.-Y. Yeh, R.-J. Tseng, P.-J. Meng and B.-Ch. Han, 2000. Mussel Watch: a review of Cu and other metals in various marine organisms in Taiwan, 1991–98. Environ. Pollut. **110**, 207–215.

Johnels, A.G., M. Olsson and T. Westermark, 1968. *Esox lucius* and some other organisms as indicators of mercury contamination in Swedish lakes and rivers. Bull. Off. Int. Epiz. **69**, 1439–1452.

Jonsson, P., R. Carman and F. Wulff, 1990. Laminated sediments in the Baltic – A tool for evaluating nutrient mass balances. Ambio **19**, 152–158.

Julshamn, K., 1981. Studies on major and minor elements in molluscs in western Norway. VII. The contents of 12 elements, including copper, zinc, cadmium and lead in common mussel (*Mytilus edulis*) and brown seaweed (*Ascophyllum nodosum*) relative to the distance from the industrial sites in Sorfjorden, inner Hardangerfjord. Fisk. Dir. Skr., Ser. Ernæring **1**, 267–287.

Kangas, P., and H. Autio, 1986. Macroalgae as indicators of heavy metal pollution. Publication of the Water Research Institute, National Board of Waters, Finland **No. 68**, 183–189.

Karbe, L., Ch. Schnier and H.O. Siewers, 1977. Trace elements in mussels (*Mytilus edulis*) from coastal areas of the North Sea and the Baltic. Multielement analyses using instrumental neutron activation analysis. J. Radioanal. Chem. **37**, 927–943.

Khan, R.A. and J. Thulin, 1991. Influence of pollution on parasites of aquatic animals. Advances in Parasitology **30**, 201–238.

Khlebovich, V.V., 1997. Selection and criteria for biological indicator species for Arctic monitoring. Mar. Pollut. Bull. **35**, 381–383.

Knap, A., 2000. NIEHS/UNESCO Meeting on Indicators of Ocean and Human Health. Mar. Pollut. Bull. **40**, 461–462.

Koide, M., D.S. Lee and E.D. Goldberg, 1982. Metals and transuranic records in mussel shells, byssal threads and tissues. Estuar. Coast. Shelf. Sci. **15**, 679–695.

Kremling, K., and P. Streu, 2000. Further evidence for a drastic decline of potentially hazardous trace metals in Baltic Sea surface waters. Mar. Pollut. Bull. **40**, 674–679.

Lafferty, K.D., 1997. Environmental parasitology, what can parasites tell us about human impacts on the environment? Parasitology Today **13**, 251–255.

Langston, W.J., 1980. Arsenic in U.K. estuarine sediments and its availability to benthic organisms. J. Mar. Biol. Assoc. UK **60**, 868–881.

Langston, W.J., 1982. The distribution of mercury in British estuarine sediments and its availability to deposit-feeding bivalves. J. Mar. Biol. Assoc. UK **62**, 667–684.

Langston, W.J., 1984. Availability of arsenic to estuarine and marine organisms: a field and laboratory evaluation. Mar. Biol. **80**, 143–154.

Langston, W.J., 1986. Metals in sediments and benthic organisms in the Mersey estuary. Estuar. Coast. Shelf Sci. **23**, 239–261.

Lauenstein, G.G., and K.D. Daskalakis, 1998. U.S. long-term coastal contaminant temporal trends determined from mollusk monitoring programs, 1965–1993. Mar. Pollut. Bull. **37**, 6–13.

Lee, H.D., and C.H. Xu, 1984. Effects of metals on sea urchin development: a rapid bioassay. Mar. Pollut. Bull. **15**, 18–21.

Lithner, G., D. Broman, C. Naef, H. Borg, A.-M. Johansson, P. Kaerrhage and M.-B. Larsson, 1996. Metals in settling particles and surficial sediments of the Swedish Baltic coast 1988–1989, in: Dev.

Prog. Sediment Qual. Assess, eds. M. Munawar and G. Dave. Int. Symp. Sediment Qual. Assess. (SPB Academic Publishing: Amsterdam, Netherlands), 27–48.

Luoma, S.N., and G.W. Bryan, 1978. Factors controlling the availability of sediment-bound lead to the estuarine bivalve *Scrobicularia plana.* J. Mar. Biol. Assoc. U.K. **58**, 793–802.

Luoma, S.N., and G.W. Bryan, 1982. A statistical study of environmental factors controlling concentrations of heavy metals in the burrowing bivalve *Scrobicularia plana* and the polychaete *Nereis diversicolor.* Estuar. Coast. Shelf Sci. **15**, 95–108.

Lyngby, J.E., and H. Brix, 1982. Seasonal and environmental variation in cadmium, copper, lead and zinc concentrations in eelgrass (*Zostera marina* L.) in the Limfjord, Denmark. Aquat. Bot. **14**, 59–74.

MacKenzie, K., 1999. Parasites as pollution indicators in marine ecosystems: a proposed early warning system. Mar. Pollut. Bull. **38**, 955–959.

MacKenzie, K., H.H. Williams, B. Williams, A.H. McVicar and R. Siddall, 1995. Parasites as indicators of water quality and the potential use of helminth transmission in marine pollution studies. Advances in Parasitology **35**, 85–144.

Martin, J.H., and G.A. Knauer, 1973. The elemental composition of plankton. Geochim. Cosmochim. Acta **37**, 1639–1653.

Mauchline, J., A.M. Taylor and E.B. Ritson, 1964. The radioecology of beach. Limnol. Oceanogr. **9**, 187–194.

Melhuus, A., K.L. Seip, H.M. Seip and S. Myklestad, 1978. A preliminary study of the use of benthic algae as biological indicators of heavy metal pollution in Sorfjorden, Norway. Environ. Pollut. **15**, 101–107.

Miller, K.L., T.F., Fernandes and P.A. Read, 1999. The recovery of populations of dogwhelks suffering from imposex in the Firth of Forth 1987–1997/98. Environ. Pollut. **106**, 183–192.

Minchin, D., J. Oehlmann, C.B. Duggan, E. Stroben and M. Keatinge, 1995. Marine TBT antifouling contamination in Ireland, following legislation in 1987. Mar. Pollut. Bull. **30**, 633–639.

Minchin, D., B. Stroben, J. Oehlmann, B. Bauer, C.B. Duggan and M. Keatinge, 1996. Biological indicators used to map organotin contamination in Cork Harbour, Ireland. Mar. Pollut. Bull. **32**, 188–195.

Minchin, D., B. Bauer, J. Oehlmann, U. Schulte-Oehlmann and C.B. Duggan, 1997. Biological indicators used to map organotin contamination from a fishing port, Killybegs, Ireland. Mar. Pollut. Bull. **34**, 235–243.

Morgan, E., J. Murphy and R. Lyons, 1998. Imposex in *Nucella lapillus* from TBT contamination in south and south-west Wales: a continuing problem around ports. Mar. Pollut. Bull. **36**, 840–843.

Moriarty, F., H.M. Hanson and P. Freestone, 1984. Limitations of body burdens as an index of environmental comtamination: heavy metals in fish *Cottus gobio* L. from the River Ecclesbourne, Derbyshire. Environ. Pollut. Ser. A **34**, 297–320.

Morozińska-Gogol, J., J. Rokicki, K. Frelek and P. Szefer, 1998. Trace metals in *Gasterosteus aculeatus* L. and its parasites from the Gulf of Gdańsk. Oceanol. Studies **27**, 71–75.

Morris, A.W., and A.J. Bale, 1975. The accumulation of cadmium, copper, manganese and zinc by *Fucus vesiculosus* in the Bristol Channel. Estuar. Coast. Mar. Sci. **3**, 153–163.

Möller, H., 1987. Pollution and parasitism in the aquatic environment. Intern. J. Parasit. **17**, 353–361.

Möller, H., R. Schneider and Ch. Schnier, 1983. Trace metal and PCB content of mussels (*Mytilus edulis*) from the southwestern Baltic Sea. Int. Revue ges. Hydrobiol. **68**, 633–647.

Munda, I.M., 1984. Salinity dependent accumulation of Zn, Co and Mn in *Scytosiphon lomentaria* (Lyngb.) and *Enteromorpha intestinalis* (L.) from the Adriatic Sea. Bot. Mar. **27**, 371–376.

Nakamura, R., M. Nakahara, Y. Suzuki and T. Ueda, 1986. Relative importance of food and water in the accumulation of radionuclides by sea urchin *Strongylocentrotus nudus*. Bull. Japan. Soc. Scient. Fish. 52, 703–710.

Neumann, G., M. Notter and H. Dahlgaard, 1991. Bladder-wrack (*Fucus vesiculosus* L.) as an indicator for radionuclides in the environment of Swedish nuclear power plants. Report 3931 (Swedish Environmental Protection Agency, Solna, Sweden), pp. 35.

Nicholson, M.D., R.J. Fryer and C.A. Ross, 1997. Designing monitoring programmes for detecting temporal trends in contaminants in fish and shellfish. Mar. Pollut. Bull. **34**, 821–826.

Notter, M., 1994. Metals and the environment (The Swedish Environmental Protection Agency), Rep. No. **4245**.

O'Connor, T.S., 1998. Mussel Watch results from 1986 to 1996. Mar. Pollut. Bull. **37**, 14–19.

Oehlmann, J., E. Stroben and P. Fioroni, 1993. Fréquence et degré d'expression du pseudohermaphrodisme chez quelques Prosobranches Sténoglosses des côtes françaises (surtout de la baie de Morlaix et de la Manche). 2. Situation jusqu'au printemps de 1992. Cahiers Biologie Marine **34**, 343–362.

Olsson, M., 1976. Mercury level as a function of size and age in northern pike, one and five years after the mercury ban in Sweden. Ambio **5**, 73–76.

Ostapczuk, P., M. Burrow, K. May, C. Mohl, M. Froning, B. Süßenbach, E. Waidmann and H. Emons, 1997. Mussels and algae as bioindicators for long-term tendencies of element pollution in marine ecosystems. Chemosphere **34**, 2049–2058.

Perttilä, M., V. Tervo and R. Parmanne, 1982. Heavy metals in Baltic herring and cod. Mar. Pollut. Bull. **13**, 391–393.

Phillips, D.J.H., 1976a. The common mussel *Mytilus edulis* as an indicator of pollution by zinc, cadmium, lead and copper. I. Effects of environmental variables on uptake of metals. Mar. Biol. **38**, 59–69.

Phillips, D.J.H., 1976b. The common mussel *Mytilus edulis* as an indicator of pollution by zinc, cadmium, lead and copper. II. Relationship of metals in the mussel to those discharged by industry. Mar. Biol. **38**, 71–80.

Phillips, D.J.H., 1977a. The common mussel *Mytilus edulis* as an indicator of trace metals in Scandinavian waters. I. Zinc and cadmium. Mar. Biol. **43**, 283–291.

Phillips D.J.H., 1977b. The use of biological indicator organisms to monitor trace metal pollution in marine and estuarine environments – A review. Environ. Pollut. **13**, 281–317.

Phillips, D.J.H., 1978. The common mussel *Mytilus edulis* as an indicator of trace metals in Scandinavian waters. II. Lead, iron and manganese. Mar. Biol. **46**, 147–156.

Phillips, D.J.H., 1979. Trace metals in the common mussel, *Mytilus edulis* (L.), and in the alga *Fucus vesiculosus* (L.) from the region of the Sound (Öresund). Environ. Pollut. **18**, 31–43.

Phillips, D.J.H., 1980. Quantitative Aquatic Biological Indicators (Applied Science Publishers Ltd, London) 488 pp.

Phillips, D.J.H., 1985. Organochlorines and trace metals in green-lipped mussels *Perna viridis* from Hong Kong waters: a test of indicator ability. Mar. Ecol. Prog. Ser. **21**, 251–258.

Phillips, D.J.H., 1990. Use of macroalgae and invertebrates as monitors of metal levels in estuaries and coastal waters, in: Heavy Metals in the Marine Environment, eds. R.W. Furness and P.S. Rainbow (CRC Press, Boca Raton), pp. 81–99.

Phillips, D.J.H., 1995. The chemistries and environmental fates of trace metals and organochlorines in aquatic ecosystems. Mar. Pollut. Bull. **31**, 193–200.

Phillips, D.J.H., and P.S. Rainbow, 1988. Barnacles and mussels as biomonitors of trace elements: a comparative study. Mar. Ecol. Prog. Ser. **49**, 83–93.

Phillips, D.J.H., and P.S. Rainbow, 1989. Strategies of trace metal sequestration in aquatic organisms. Mar. Environ. Res. **28**, 207–210.

Phillips, D.J.H., and P.S. Rainbow, 1993. Biomonitoring of Trace Aquatic Contaminants. Elsevier Science Publishers Ltd., London, 371 pp.

Pohl, Ch., 1992. Correlation between trace metal concentrations (Cd, Cu, Pb, Zn) in seawater and zooplankton organisms (Copepoda) of the Arctic and Atlantic Ocean. Berichte Polarforsch. **101**, 198 pp.

Powell, M.I., and K.N. White, 1990. Heavy metal accumulation by barnacles and its implications to their use as biological monitors. Mar. Environ. Res. **30**, 91–118.

Rainbow, P.S., 1988. The significance of trace metal concentrations in decapods. Symp. Zool. Soc. Lond. **No. 59**, 291–313.

Rainbow, P.S., 1995. Biomonitoring of heavy metal availability in the marine environment. Mar. Polut. Bull. **31**, 183–192.

Rainbow, P.S., and S.L. White, 1989. Comparative strategies of heavy metal accumulation by crustaceans: Zinc, copper and cadmium in a decapod, an amphipod and a barnacle. Hydrobiologia **174**, 245-262.

Rainbow, P.S., and P.G. Moore, 1990. Seasonal variation in copper and zinc concentrations in three talitrid amphipods (Crustacea). Hydrobiologia **196**, 65–72.

Rainbow, P.S., and D.J.H. Phillips, 1993. Cosmopolitan biomonitors of trace metals. Mar. Polut. Bull. **26**, 593–601.

Rainbow, P.S., and G. Blackmore, 2001. Barnacles as biomonitors of trace metal availabilities in Hong Kong coastal waters: changes in space and time. Mar. Environ. Res. **51**, 441–463.

Rainbow, P.S., A.G. Scott, E.A. Wiggins and R.W. Jackson, 1980. Effect of chelating agents on the accumulation of cadmium by the barnacle *Semibalanus balanoides*, and complexation of soluble Cd, Zn and Cu. Mar. Ecol. Progr. Ser. **2**, 143–152.

Rainbow, P.S., D.J.H. Phillips and M.H. Depledge, 1990. The significance of trace mrtal concentrations in marine invertebrates. A need for laboratory investigation of accumulation strategies. Mar. Polut. Bull. **21**, 321–324.

Rainbow, P.S., W. Fialkowski and B.D. Smith, 1998. The sandhopper *Talitrus saltator* as a trace metal biomonitor in the Gulf of Gdansk, Poland. Mar. Pollut. Bull. **36**, 193–200.

Rainbow, P.S., C. Amiard-Triquet, J.C. Amiard, B.D. Smith, S.L. Best, Y. Nassiri and W.J. Langston, 1999. Trace metal uptake rates in crustaceans (amphipods and crabs) from coastal sites in NW Europe differentially enriched with trace metals. Mar. Ecol. Prog. Ser. **183**, 189–203.

Rainbow, P.S., M. Wolowicz, W. Fialkowski, B.D. Smith and A. Sokolowski, 2000. Biomonitoring of trace metals in the Gulf of Gdansk using mussels (*Mytilus trossulus*) and barnacles (*Balanus imorovisus*). Wat. Res. **34**, 1823–1829.

Rayment, G.E., and G.A. Barry, 2000. Indicator tissues for heavy metal monitoring – additional attributes. Mar. Pollut. Bull. **41**, 353–358.

Rissanen, K., and T.K., Ikäheimonen, 2000. Caesium and plutonium concentrations in salmon caught in River Teno (Norwegian Sea) and in River Tornionjoki (Gulf of Bothnia), in: The Radiological Exposure of the Population of the European Community to Radioactivity in the Baltic Sea. Marina-Balt Project, ed. S.P. Nielsen. Proceedings of a Seminar held at Hasseludden Conference Centre, Stockholm, 9–11 June 1998, European Commission, Directorate-General Environment, EUR 19200 EN (European Communities, 2000, Belgium), pp. 439–448.

Roesijadi, G., J.S. Young, A.S. Drum, and J.M. Gurtisen, 1984. Behaviour of trace metals in *Mytilus edulis* during a reciprocal transplant field experiment. Mar. Ecol. Prog. Ser. **18**, 155–170.

Ronnberg, O., K. Adjers, C. Ruokolahti and M. Bondestam, 1990. *Fucus vesiculosus* as an indicator of heavy metal availabilty in a fish farm recipient in the Northern Baltic Sea. Mar. Pollut. Bull. **21**, 388–392.

Rouleau, C., E. Pelletier and H. Tjälve, 1993. The uptake and distribution of $^{203}HgCl_2$ and CH_3HgCl_2 in the sea star *Asterias rubens* after 24-exposure studied by impulse counting and whole body autoradiography. Aquatic Toxicol. **26**, 103–116.

Rühling, Å., G. Brumelis, N. Goltsova, K. Kvietkus, E. Kubin, S. Liivs, S. Magnusson, A. Mäkinen, K. Pilegaard, L. Rasmusson, E. Sander and E. Steinnes, 1992. Atmospheric heavy metal deposition in northern Europe 1990. Nordic Council of Ministers, NORD, 12 p.

Santos, M.M., N. Vieira and A.M. Santos, 2000. Imposex in the dogwhelk *Nucella lapillus* (L.) along the Portuguese coast. Mar. Pollut. Bull. **40**, 643–646.

Scheuhammer, A.M., 1987. The chronic toxicity of aluminium, cadmium, mercury, and lead in birds: A review. Environ. Pollut. **46**, 263–295.

Scheuhammer, A.M., 1991. Effects od acidification on the availability of toxic metals and calcium to wild birds and mammals. Environ. Pollut. **71**, 329–375.

Schladot J.D., F. Backhaus, P. Ostapczuk and H. Emons, 1997. Eel-pout (*Zoarces viviparus* L.) as a marine bioindicator. Chemosphere **34**, 2133–2142.

Shulkin, V.M., and V.Ia. Kavun, 1995. The use of marine bivalves in heavy metal monitoring near Vladivostok, Russia. Mar. Pollut. Bull. **31**, 330–333.

Seeliger, U., and P. Edwards, 1977. Correlation coefficients and concentration factors of copper and lead in sea water and benthic algae. Mar. Pollut. Bull. **8**, 16–18.

Seymour, A.H., 1966. Accumulation and loss of zinc-65 by oysters in a natural environment, in: Disposal of Radioactive Waters into Seas, Oceans and Surface Waters. (Vienna, International Atomic Energy Agency), pp. 605–619.

Skarphédinsdóttir, H., K. Ólafsdóttir, J. Svavarsson and T. Jóhannesson, 1996. Seasonal fluctuations of tributyltin (TBT) and dibutyltin (DBT) in the dogwhelk, *Nucella lapillus* (L.), and the blue mussel, *Mytilus edulis* L., in Icelandic waters. Mar. Pollut. Bull. **32**, 358–361.

Skóra, K.E., 1991. Notes on cetacea observed in the Polish Baltic sea: 1979–1990. Aquat. Mammals **17.2**, 67–70.

Skóra, K.E., I. Pawliczka and M. Klinowska, 1988.. Observations of the harbour porpoise (*Phocoena phocoena*) on the Polish Baltic coast. Aquat. Mammals **14.3**, 113–119.

Sonesten, L., 2001a. Mercury content in roach (*Rutilus rutilus* L.) in circumneutral lakes effects of catchment area and water chemistry. Environ. Pollut. **112**, 471–481.

Sonesten, L., 2001b. Land use influence on ^{137}Cs levels in perch (*Perca fluviatilis* L.) and roach (*Rutilus rutilus* L.). J. Environ. Radioactivity **55**, 125–143.

Sorensen, M., and P. Bjerregaard, 1991. Interactive accumulation of mercury and selenium in the sea star *Asterias rubens*. Mar. Biol. **108**, 269–270.

Söderlund, S., Å. Forsberg and M. Pedersén, 1988. Concentrations of cadmium and other metals in *Fucus vesiculosus* L. and *Fontinalis dalecarlica* Br. Eur. from the northern Baltic Sea and the southern Bothnian Sea. Environ. Pollut. **51**, 197–212.

Stenner, R.D., and G. Nickless, 1974. Distribution of some heavy metals in organisms in Hardangerfjord and Skjerstadfjord, Norway. Water Air Soil Pollut. **3**, 279–291.

Struck, B.D., R. Pelzer, P. Ostapczuk, H. Emons and C. Mohl, 1997. Statistical evaluation of ecosystem properties influencing the uptake of As, Cd, Co, Cu, Hg, Mn, Ni, Pb, Zn in seaweed (*Fucus vesiculosus*) and common mussel (*Mytilus edulis*). Sci. Total Environ. **207**, 29–42.

Suess, E., and D. Djafari, 1977. Trace metal distribution in Baltic Sea ferromanganese concretions: inferences on accretion rates. Earth Planet. Sci. Letters **35**, 49–54.

Sures, P H. Taraschewski and J. Rokicki, 1997. Lead and cadmium content of two cestodes, *Monobothrium wageneri* and *Bothriocephalus scorpii*, and their fish hosts. Parasitol. Res. **83**, 618-623.

Sures, B., S. Zimmermann, J. Messerschmidt, A. von Bohlen and F. Alt, 2001. First report on the uptake of automobile catalyst emitted palladium by European eels (*Anguilla anguilla*) following experimental exposure to road dust. Environ. Pollut. **113**, 341–345.

Svieshnikov, V.A., 1987. Klass Piervichnyie kolchetsy (Polychaeta), in: Zhizn' Zhivotnyh – Prostiejshie – Kishechnopolostnyie – Chervy (The Life of Animals – Worms), ed. Yu.I. Polyanskij (Prosveshchenie, Izdatelstvo, Moscow), **Vol. 1**, 363–376 (in Russian).

Szefer, P, 1986. Some metals in benthic invertebrates in Gdansk Bay. Mar. Pollut. Bull. **17**, 503–507.

Szefer, P., 1991. Interphase and trophic relationships of metals in a southern Baltic ecosystem. Sci. Total Environ. **101**, 201–215.

Szefer, P., 1998. Distribution and behaviour of selected heavy metals in various components of the southern Baltic ecosystem, in: Geochemical Investigations of the Baltic Sea and Surrounding Areas, eds. P. Szefer, P. and G.P. Glasby (Elsevier Science Ltd, Great Britain). Appl. Geochem. (Spec. Issue) **13**, 287–292.

Szefer, P., and K. Szefer, 1985. Occurrence of ten metals in *Mytilus edulis* L. and *Cardium glaucum* from the Gdansk Bay. Mar. Pollut. Bull. **16**, 446–450.

Szefer, P., and B. Skwarzec, 1988a. Distribution and possible sources of some elements in the sediment cores of the southern Baltic. Mar. Chem. **23**, 109–129.

Szefer, P., and B. Skwarzec 1988b. Concentration of elements in some seaweeds from coastal region of the southern Baltic and Zarnowiec Lake. Oceanologia **25**, 87–98.

Szefer, P., and K. Szefer, 1990. Metals in molluscs and associated bottom sediments of the southern Baltic. Helgoländer Meeresunters. **44**, 411–424.

Szefer, P., and K. Szefer, 1991. Concentration and discrimination factors for Cd, Pb, Zn and Cu in benthos of Puck Bay, Baltic Sea. Sci. Total Environ. **105**, 127–133.

Szefer, P., and M. Wolowicz, 1993. Occurrence of metals in the cockle *Cerastoderma glaucum* from different geographical regions in view of principal component analysis. SIMO-Mar. Pollut. **64**, 253–264.

Szefer, P., and A. Kusak, 2000. Distribution and relationships of trace metals in zoobenthos and associated sediments of the southern Baltic (in preparation).

Szefer, P., B. Skwarzec and J. Koszteyn, 1985. The occurrence of some metals in mesozooplankton taken from the southern Baltic. Mar. Chem. **17**, 237–253.

Szefer, P., G.P. Glasby, J. Pempkowiak and R. Kaliszan, 1995. Extraction studies of heavy-metal pollutants in surficial sediments from the southern Baltic Sea off Poland. Chem. Geol. **120**, 111–126.

Szefer, P., G.P. Glasby, K. Szefer, J. Pempkowiak and R. Kaliszan, 1996. Heavy-metal pollution in surficial marine sediments from the southern Baltic Sea off Poland. J. Environ. Sci. Health **31A**, 2723–2754.

Szefer, P., J. Rokicki, K. Frelek, K. Skóra and M. Malinga, 1998a. Bioaccumulation of selected trace metals in lung nematodes, *Pseudalius inflexus*, of harbour porpoise (*Phocoena phocoena*) in a Polish Zone of the Baltic Sea. Sci. Total Environ. **220**, 19–24.

Szefer, P., G.P. Glasby, A. Kusak, K. Szefer, H. Jankowska, M. Wolowicz and A.A. Ali, 1998b. Evaluation of anthropogenic influx of metallic pollutants into Puck Bay, southern Baltic, in: Geochemical Investigations of the Baltic Sea and Surrounding Areas, eds. P. Szefer and G.P. Glasby (Elsevier Science Ltd, Great Britain) Applied Geochem. (Spec. Issue) **13**, 293–304.

Szefer, P., H.M. Fernandes, M.-J. Belzunce, B. Guterstam, J.M. Deslous-Paoli and M. Wolowicz, 1998c. Distribution of metallic pollutants in molluscs *Mytilidae* from the temperate, tropical and subtropical marine environments. First International Symposium, IEP '98 Issues in Environmental Pollution, The State and Use of Science and Predictive Models (Elsevier Science Ltd., Denver, Colorado, U.S.A.), 23–26.08.1998, Section of Abstract Book 4.04.

Szefer, P., G.P. Glasby, H. Kunzendorf, E.A. Görlich, K. Latka, K. Ikuta and A.A. Ali, 1998d. The distribution of rare earth and other elements and the mineralogy of the iron oxyhydroxide phase in marine ferromanganese concretions from within Slupsk Furrow in the southern Baltic, in: Geochemical Investigations of the Baltic Sea and Surrounding Areas, eds. P. Szefer, P. and G.P. Glasby (Elsevier Science Ltd, Great Britain) Applied Geochem. (Spec. Issue) **13**, 305–312.

Szefer, P., G.P. Glasby, D. Stüben, A. Kusak, J. Geldon, Z. Berner, T. Neumann and J. Warzocha, 1999a. Distribution of selected heavy metals and rare earth elements in surficial sediments from the Polish sector of the Vistula Lagoon. Chemosphere **39**, 2785–2798.

Szefer, P., M. Wolowicz, A. Kusak, J.-M. Deslous-Paoli, W. Czarnowski, K. Frelek and M.-J. Belzunce-Segarra, 1999b. Distribution of mercury and other trace metals in the cockle *Cerastoderma glaucum* from the Mediterranean Lagoon, Etang de Thau. Arch. Environ. Contam. Toxicol. **36**, 56–63.

Szefer, P., K. Frelek, K. Szefer, Ch.-B. Lee, B.-S. Kim, J. Warzocha and I. Zdrojewska, 2000a. Distribution of mercury and other trace elements in soft tissue, byssus and shells of *Mytilus edulis trossulus* from the southern Baltic (submitted).

Szefer, P, M. Wolowicz and P.S. Rainbow, 2000b. Distribution of trace metals in barnacles (*Balanus improvisus*) in the Gulf of Gdańsk, Baltic Sea (in preparation).

Szefer, P., M. Domagała-Wieloszewska, J. Warzocha, A. Garbacik-Wesołowska and J. Geldon, 2000c. Distribution and relationships of mercury, lead, cadmium, copper and zinc in perch (*Perca fluviatilis*) from the Pomeranian Bay and Szczecin Lagoon, southern Baltic (submitted).

Szefer, P., I. Zdrojewska, J. Jensen, C. Lockyer, A. Łomża, K. Skóra K., I. Kuklik, M. Malinga, 2000d. Intercomparison studies on distribution of heavy metals in liver, kidney and muscle of harbour porpoise, *Phocoena phocoena*, from a Polish Sector of the Baltic Sea and coastal waters of Denmark and Greenland (submitted).

Tanabe, S., 2000. Asia-Pacific Mussel Watch Progress Report. Mar. Pollut. Bull. **40**, 651.

Taraschewski, H., and B. Sures, 1996. Heavy metal concentrations in parasites compared to their fish hosts bioconcentration by acanthocephalans and cestodes. VII European Multicolloquium of Parasitology, Parma, Italy, 2–6 September 1996.

Temara, A., G. Ledent, M. Warnau, H. Paucot, M. Jangoux and P. Dubois, 1996. Experimental cadmium contamination of *Asterias rubens* L (Echinodermata). Mar. Ecol. Prog. Ser. **140**, 83–90.

Temara, A., M. Warnau, M. Jangoux and P. Dubois, 1997. Factors influencing the concentrations of heavy metals in the asteroid *Asterias rubens* L (Echinodermata). Sci. Total Environ. **203**, 51–63.

Temara, A., P. Aboutboul, M. Warnau, M. Jangoux and P. Dubois, 1998. Uptake and fate of lead in the common asteroid *Asterias rubens* L (Echinodermata). Water Air Soil Pollut. **102**, 201–208.

Theede, H., I. Andersson and W. Lehnberg, 1979. Cadmium in *Mytilus edulis* from German coastal waters. Meeresforsch. **27**, 147–155.
Thompson, D.R., F.M. Stewart and R.W. Furness, 1990. Using seabirds to monitor mercury in marine environments. The validity of conversion ratios for tissue comparison. Mar. Pollut. Bull. **21**, 339–342.
Thompson, D.R., R.W. Furness and L.R. Monteiro, 1998. Seabirds as biomonitors of mercury inputs to epipelagic and mesopelagic marine food chains. Sci. Total Environ. **213**, 299–305.
UBA (Umweltbundesamt), 1996. Annual Report of the German Environmental Specimen Bank, Berlin 1996.
Viale, D., 1994. Cetaceans as indicators of a progressive degradation of Mediterranean water quality. Intern. J. Environ. Studies **45**, 183–198. Walker, G., P.S. Rainbow, P. Foster and D.J. Crisp, 1975a. Barnacles: possible indicators of zinc pollution? Mar. Biol. **30**, 57–75.
Walker, G., and P. Foster, 1979. Seasonal variation of zinc in the barnacle *Balanus balanoides* (L.) maintained on a raft in the Menai Strait. Mar. Environ. Res. **2**, 209–221.
Walker, G., P.S. Rainbow, P. Foster and D.L. Holland, 1975. Zinc phosphate granules in tissues surrounding the midgut of the barnacle *Balanus balanoides*. Mar. Biol. **33**, 162–166.
Warnau, M., S.W. Fowler and J.L. Teyssié, 1996a. Biokinetics of selected heavy metals and radionuclides in two marine macrophytes: the seagrass *Posidonia oceanica* and the alga *Caulerpa taxifolia*. Mar. Environ. Res. **41**, 343–362.
Warnau, M., J.L. Teyssié and S.W. Fowler, 1996b. Biokinetics of selected heavy metals and radionuclides in the common Mediterranean echinoid *Paracentrotus lividus*: sea water and food exposure. Mar. Ecol. Prog. Ser. **141**, 83–94.
Warnau, M., S.W. Fowler and J.-L. Teyssié, 1999. Biokinetics of radiocobalt in the asteroid *Asterias rubens* (Echinodermata): sea water and food exposures. Mar. Pollut. Bull. **39**, 159–164.
Watson, D., P. Foster and G. Walker, 1995. Barnacle shells as biomonitoring material. Mar. Pollut. Bull. **31**, 111–115.
Weber, A.,M. Krause, H. Marencic and R. Kopp, 1992. Schadstoffumsatz im Zooplankton, in: Prozesse im Schadstoffkreislauf Meer-Atmosphäre: Ökosystem Deutsche Bucht (PRISMA). BMFT-Projekt MFU 0620/6 2. Zwischenbericht, 01.01.–31.12.1991, 179–188.
Wedderburn, J., I. McFadzen, R.C. Sanger, A. Beesley, C. Heath, M. Hornsby and D. Lowe, 2000. The field application of cellular and physiological biomarkers, in the mussel *Mytilus edulis*, in conjunction with early life stage bioassays and adult histopathology. Mar. Pollut. Bull. **40**, 257–267.
Wenzel, Ch., D. Adelung and H. Theede, 1996. Distribution and age-related changes of trace elements in kittiwake *Rissa tidactyla* nestlings from an isolated colony in the German Bight, North Sea. Sci. Total Environ. **193**, 13–26.
White, K.N., and Walker, 1981. Uptake, accumulation, and excretion of zinc by the barnacle, *Balanus balanoides* (L.). J. Experiment. Mar. Biol. 51, 285–298.
Wilson, J.G., and B. Elkaïm, 1992. Estuarine bioindicators – a case for caution. Acta Œcologica **13**, 345–358.
Wright, D.A., 1995. Trace metal and major ion interactions in aquatic animals. Mar. Pollut. Bull. **31**, 8–18.
Wu, J., and E.A. Boyle, 1997. Lead in the western Atlantic Ocean: completed response to leaded gasoline phaseout. Geochim. Cosmochim. Acta **61**, 3279–3283.

Chapter 8

Estimate of Health Risk

A. SEAFOOD TRACE ELEMENTS

(i) Introduction

Pollutants e.g. trace elements and radionuclides, may be magnified in successive levels of the food chain and in consequence pose a risk to consumers. For instance, van Oostdam et al. (1999) assessed the impact on human health of exposure to current concentrations of pollutants and contaminants in the Canadian Arctic. Johansen et al. (2000) estimated that human intake of Cd (1004 μg/person per week) and Hg (846 μg/person per week) from Greenland marine food significantly exceeded limits established by FAO/WHO while the intake of Pb was very low. The first step in risk estimate is to identify a hazard by establishing if a cause-effect relationship exists. When a hazard is identified then the relationship between exposure and the probability of an adverse effect observing is estimated.

(ii) Measures of Health Risk

Hagel (2000) made survey of the quantities and utilisation of sea products to provide a database for the collective dose computations under the Marina-Balt Project. Basing on total annual catches and landings of the various countries bordering the Baltic Sea over 1990–94 (ICES CM, 1995), it has been estimated of their shares for different compartments of the sea as well as the flow of marine products from the Baltic Sea through import and export, gross amounts of fish, crustaceans and molluscs available for human consumption in the EU Member

States and the other countries bordering the Baltic Sea. According to Hagel (2000) import and export data on seafood have been derived from EUROSTAT C data, FAO fishery statistics (FAO, 1993, 1994, 1995) and individual members of the Marina-Balt Working Group 4.

It is shown that proportion of the total annual catches and landings for the surrounding Baltic countries to total amounts of fish is < 1% for the Russian Federation and > 60% for Finland. Corresponding indices for crustaceans and molluscs appear to be of great importance only for Denmark and Sweden and are restricted to the Kattegat and Belt Sea compartment (Hagel, 2000). Such limitations are caused by fact, that in contrast to the North Sea bottom fauna, greater specimens of shrimp *Crangon crangon* with a length over 50 mm are sporadically observed in the western Baltic (Dornheim, 1969). At lower salinity shrimps do not occur at all in enough great amounts interesting from commercial point of view. A similar the abundance pattern is observed for molluscs; although *Mytilus edulis* occurs in very big quantities in the Baltic Sea (Öst and Kilpi, 1997), however its maximum length rarely exceeds value of 30 mm (Kautsky, 1982) which is unsuitable for commercial exploitation (Hagel, 2000). Among fish, *Gadus morrhua*, *Clupea harengus*, *Platichthys flesus* and *Pleuronectes platessa* are commercially exploited although catches and landings of fish from the Baltic Sea have significantly decreased in the beginning of the 1990s. Inverse trend is however noted for crustaceans and mollusc which catches and landings seem to increase in the recent years exploitation (Hagel, 2000).

Biomass of marine products from the Baltic Sea was converted from gross to net values by taking 50% of the gross weight for fish, one third for crustaceans and one sixth for mollusc. The net values obtained are useful as a basis for calculations of the collective doses to man. According to Hagel (2000) an approximation of a critical group consumption rate of seafood can be calculated by multiply the average per capita supply by 5.

Three trace elements, i.e. Pb, Cd and Hg are most important from ecotoxicological point of view and therefore human exposure has been frequently assessed in respect to these elements (Hansen et al., 1990; Dabeka and McKenzie, 1995; van Oostdam et al., 1999). It is difficult to estimate precisely to which extent the anthropogenic activity contributes to the total environmental input of these elements. It should be emphasised that Pb, Cd and Hg accumulate in human tissues and hence they are harmful to human health (van Oostdam et al., 1999). It is known that most of human exposure to Pb is from food. The current WHO TDI (tolerable daily intakes) and WHO PTWI (Provisional Tolerable Weekly Intake) for Pb are estimated to be 3.57 $\mu g\ kg^{-1}$ body wt. day^{-1} and 25 $\mu g\ kg^{-1}$ body wt. $week^{-1}$, respectively (WHO, 1993). It is important to note that seafood such as shellfish and crustaceans contains elevated levels of Cd and therefore is important source of this element in consumer tissues. The PTWI value, as established for Cd by the FAO/WHO (1989) amounts on 7 $\mu g\ kg^{-1}$ body wt. equalling 420 μg Cd $week^{-1}$ for a 60-kg person. Among different species of Hg in the marine environment, MeHg (methylmercury) is distinguished itself by the strongest toxic effect in men. Although the inorganic species of Hg is predominantly released to the environment from natural and anthropogenic sources, several microorganisms in aquatic ecosystems are able to convert inorganic Hg to MeHg; the last one is biomagnified in the food chain. Food containing elevated levels of MeHg, i.e. fish and marine mammals can be a very remarkable source of exposure for human (van Oostdam et al., 1999). For instance, in populations consuming more fish or marine mammals, blood MeHg values are significantly greater than in those consuming marine foods less than once a week (Hansen et al., 1990; van Oostdam et al., 1999). Ponce et al. (2001) based on a case study of the risks and benefits of fish consumption demonstrated that across all considered fish intake rates (0-300 g day^{-1}) and fish methyl-Hg concentrations (0.5-2 $\mu g\ g^{-1}$), fish consumption had a strong net positive health impact in the population consisting of 100,000 individuals of all ages and both genders. However, under the same exposure conditions fish consumption had a strong net negative health impact in women of child-bearing age and their children. They are at very high risks (methyl-Hg induced neurodevelopmental delay during pregnancy) relative to other subgroups (Ponce et al., 2001). The PTWI of Hg is established at level of 5 $\mu g\ kg^{-1}$ body wt. (FAO/WHO, 1972), equalling 300 μg Hg $week^{-1}$ for a 60-kg person. The WHO TDI's for the total and methyl Hg are set at 0.714 and 0.471 $\mu g\ kg^{-1}$ body wt. day^{-1}, respectively (WHO, 1990). Gajewska et al. (2000) reported temporal trends in concentrations of the total Hg in Baltic fish caught in 1971–1997. The Hg levels were within the maximum levels admissible in Poland, although a slight increase in the total Hg content was detected for some samples in 1997.

The suitability of seal meat for human consumption is questionable if the hunting of ringed seals is ever reintroduced in the Baltic Sea (Fant et al., 2001). Although food standard limits for Cd, Hg and Pb are recommended for fish, seafood and vegetables but they are not yet available for meat products and no standard limits exist for Se in food. Levels of Hg in the Baltic ringed seals, in respect to current food standards, exceeded the allowable limit (0.5–1.0 $\mu g\ g^{-1}$ for fish) in muscle, especially in kidney and liver (Fant et al., 2001). The WHO and WHO PTWI's of Hg is 0.3 mg, which corresponds to on average 200 g of Baltic ringed seal meat. The hepatic and renal levels of Cd exceeded mostly the limits (0.1–0.5 $\mu g\ g^{-1}$ in fish and seafood) (Fant et al., 2001).

A great attention is recently focused to pollution of seafood by organotin. It should be emphasised that TBT has ability to accumulate through the food chain resulting in biomagnification of this pollutant as well as its breakdown products in particular trophic levels, e.g. shellfish, squid and fish and in top predators as whales dolphins, seals and fish-eating waterfowls (Kannan and Falandysz, 1997a, 1997b; Senthilkumar et al., 1999; Tanabe, 1999; Belfroid et al., 2000; Hoch, 2001). Belfroid et al. (2000) reported tolerable average levels (TARL) for TBT in seafood products, which were calculated based on the TDI of TBT and the seafood consumption of the average consumer in 24 countries. Among the Baltic states only Germany, Poland and Sweden have been considered because data for the remaining countries were unavailable (Table 8.1). The TARLs for these countries

TABLE 8.1.

Average seafood consumption per country and calculated tolerable average residue level of TBT in seafood products. After Belfroid et al. (2000); modified

Country	Per capita supply		Tolerable average residue level/day in ng g^{-1} seafood product for an average person of 60 kg
	in kg yr^{-1}	in g day^{-1}	
Australia	19.2	52.6	285
Bangladesh	9.4	25.8	582
Canada	22.7	62.2	241
France	27.9	76.4	196
Germany[a]	15.6	42.7	351
Hong Kong	59.6	163	92
India	3.8	10.4	1440
Indonesia	15.2	41.6	360
Italy	23.1	63.3	237
Japan	71	195	77
Korea Republic	50.3	138	109
Malaysia	53.5	147	102
Netherlands	14.6	40	375
Papua N. Guinea	26.2	71.8	209
Poland[a]	16.5	45.2	332
Portugal	58.7	161	93
Singapore[b]	53.5	147	102
Solomon Islands	20	54.8	274
Sweden[a]	30.8	84.4	178
Taiwan[b]	59.6	163	92
Thailand	25.9	71	211
UK	20.1	55.1	272
USA	21.6	59.2	253
Vietnam	12.6	34.5	435

[a] – Baltic country

[b] – Data were unavailable for Singapore and Taiwan, therefore, data for Malaysia and Hong Kong, respectively, were used, that resemble these countries in terms of culture and proximity to the sea.

were estimated as 351, 332 and 178 ng TBT g^{-1} seafood for an average person of 60 kg, respectively. It should be stressed that the TARL is based on the average consumer and that variations in consumer weight and consumption patterns were not taken into account. However, advantage of this approach is that the TARLs can be compared directly with measured residue levels of TBT in seafood and these values can be the basis for governments to derive the maximum limit (MRL) of TBT in seafood for their country. The MRL values are constitutional tools to ensure the health of the population (Belfroid et al., 2000). As can be seen in Table 8.1 the average seafood consumption for Sweden is ca. two times higher than that for Germany and Poland. According to Kannan and Falandysz (1997a) organotins levels in muscle tissue of several fish species from the Baltic Sea devoted to human consumption approached or even exceeded the TDI for human. Taking into account the data obtained, the authors recommended the need for seafood consumption advisory guidelines; however their suggestion was rejected by Robinson et al. (1999) who argued that the TDI is exceeded for only one sample in Poland. The authors view was supported by Keithly et al. (1997) who concluded that commercially marketed seafood caught from traditional fishery areas makes insignificant risk to the average consumer' in eight countries all over the world.

B. SEAFOOD RADIOACTIVE DOSE

(i) Introduction

The health effects from radionuclides, emitting ionising radiation, are known as carcinogenic; these are well documented by assaying of human populations exposed to high levels of radiation (BEIR, 1990). Radionuclides enter the Baltic Sea as fallout from atmosphere, e.g. the Chernobyl accident in 1986 was an additional source of radioactive material to the Baltic via atmospheric trajectory. Radionuclides can be bioconcentrated and biomagnified in sequential food chain levels resulting in contamination of Baltic seafood. This pathway is particularly important for anthropogenic ^{137}Cs and ^{210}Po. According to Aarkrog et al. (2000a) the collective dose from consumption of Greenland foods contaminated by ^{137}Cs and ^{90}Sr was low amounting to 0.6 mSv/average Greenlander. This dose corresponds to the relative high consumption of marine products (fish, shrimps, marine mammals) by Greenlanders, although 10–20 times higher doses were estimated for groups consuming of reindeer, lamb or freshwater fish. It is shown that doses from the shorter-lived radionuclides, e.g. ^{137}Cs and longer-lived radionuclides, e.g. ^{239}Pu are mainly delivered from seafood production in the Barents Sea and further away from the Arctic Ocean, respectively (Nielsen et al., 1997). Intake of ^{226}Ra, ^{210}Pb, ^{238}U, ^{234}U, ^{232}Th, ^{230}Th, ^{228}Th and ^{210}Po with food including sea fish in Poland has been estimated by Pietrzak-Flis et al. (1997, 2001).

(ii) Measures of Health Risk

The knowledge of the exposure rates to man to radionuclides is extremely important from both the radiation protection and hygienic points of view. The radiological dose received by a consumer in an exposure medium consists of three factors (van Oostdam et al., 1999):

- the concentration of the given radionuclide in the exposure medium (Bq kg^{-1});
- the amount of that exposure medium taken in/consumed per year (kg);
- the dose conversion factor (Sieverts/Becquerel) for the given radionuclide.

Nielsen et al. (1999) and Nielsen (2000a) carried out an assessment of the radiological consequences of radioactivity in the Baltic Sea based on data concerning input and observed levels of radionuclides in the sea for the period 1950–1996. The authors considered discharges of radioactivity in the Baltic environment taking into account the following sources: fallout from the Chernobyl accident in 1986, atmospheric nuclear-weapons fallout, discharges of radionuclides from the two European reprocessing plants Sellafield and La Hague transported into the Baltic Sea as well as discharges of radionuclides from nuclear installations bordering the Baltic Sea area (Nielsen et al., 1999). Doses to man – estimated using a computer model – were related to members of public from the ingestion of radionuclides in seafood produced in the Baltic Sea and from exposure to radioactivity in coastal areas (Nielsen et al., 1995; Nielsen, 2000a). Dose rates from man-made radioactivity to individual members of critical groups have been computed taking into account rates of annual intake (90 kg fish, 10 kg crustaceans and 10 kg molluscs) as well as beach occupancy time amounting to 700 h yr^{-1}.

The total collective dose from man-made radioactivity in the Baltic Sea is estimated as 2600 manSv; ca. two thirds of this dose originated from Chernobyl fallout, ca. one quarter of fallout from nuclear weapons testing, ca. 8% from European reprocessing facilities and ca. 0.04% from nuclear installations bordering the Baltic Sea (Nielsen et al., 1999; Nielsen, 2000a). An estimation of radioactivity of the dumpings of low-level radioactive waste in the Baltic Sea in the 1960's by Sweden and the former Soviet Union showed insignificant doses to man. Doses related to naturally occurring radioactivity in seafood, i.e. ^{210}Po, were compared with those corresponding to man-made radioactivity; it is shown that dose rates and doses from natural radioactivity dominate except for the year 1986 when the Chernobyl-derived dose rate exceeded the natural level (Nielsen, 2000a).

Skwarzec (1997) reported that the annual intake of ^{210}Po, ^{239}Pu, ^{240}Pu, ^{234}U and ^{238}U from fish food by Poles is equivalent to values of 10 Bq (Po), 7 mBq (Pu) and 24 mBq (U) per capita. The dose equivalents, DE, estimated using the annual intakes and the fractional absorption values taken from reports (ICRP, 1979, 1986, 1991) and UNSCEAR (1982) for human bone marrow are 4.9 μSv for ^{210}Po, 0.0017 μSv for $^{234+238}$U, and 0.0011 μSv for $^{239+240}$Pu. Higher values of DE (43 μSv yr^{-1}) for Po were obtained for spleen. This estimation indicated that the impact of the consumption of Baltic fish on the annual internal radiation dose for a statistical citizen of Poland is insignificant amounting to ca. 1% (Jagiellak, 1989; Skwarzec, 1997).

An evaluation of the consequences of the 1986 Chernobyl accident is presented in a new report by the UN Scientific Committee on the Effects of Atomic Radiation (UNSCEAR)* to the UN General Assembly. It is concluded that '*there is no evidence of a major public health impact attributable to radiation exposure fourteen years after the accident*', although a high level of thyroid cancers in children is reported. There have been ca. 1800 such cases in children exposed so far and now more is expected (UNSCEAR, 2000). This conclusion is generally in an agreement with that established by the International Conference entitled "One Decade After Chernobyl Summing Up the Consequences of the Accident" organised by the IAEA in Vienna in 1996. Under the Conference it is established that in addition to clinically observed health effects involving hundreds of occupationally affected persons, a very significant increase in thyroid cancer in children among those individuals, who inhabited the affected areas during 1986 is '*the only clear evidence to date of a public health impact of radiation exposure as a result of the Chernobyl accident*'. It is also concluded that reports of increases of malignancies in the general population '*are not consistent – and the reported increases could reflect differences in the follow-up of exposed populations and increased ascertainment following the Chernobyl accident and which require further investigations*' (UNSCEAR, 2000).

(iii) Remarks and Recommendations

Although incidental, but evident exceeding of TARLs in seafood in nine of the 22 countries suggests that there is need for these country-specific maximum residue limits (MRL) for seafood TBT levels and that TBT levels should be monitored in seafood more regularly (Belfroid et al., 2000).

According to Nielsen (1995, 2000b), most important future radiological changes in the Baltic Sea are expected to be continuing decrease of ^{137}Cs concentrations due to the outflow of water through the Kattegat and to a smaller extent the increase of ^{99}Tc concentrations caused by water inflow from the North Sea. Therefore future monitoring programmes should follow these changes in order to receive proper information on the radionuclides exchange between the Baltic Sea and North Sea.

Satellite monitoring of the ionosphere in order to monitor extreme situations caused by natural and man-produced accidents is recommended (Boyarchuk, 1998). Promising results are obtained for box modelling of the radiological consequences of releases of radionuclides into large marine environments such as the Arctic Ocean and the North Atlantic Ocean (Iosjpe and Strand, 1998). Mathematical models of environmental radionuclide distribution and transport have been developed to assess the impact on man of potential and actual releases of radioactivity, both planned and accidental, from various nuclear sources (Thiessen et al. (1999).

As for future radiological studies, according to Aarkrog (1998) the radiological impact of marine radionuclides is generally lower than that of radionuclides in the terrestrial environment. Therefore it appears that scientific studies on terrestrial radioactivity are needed. However, radionuclides in the marine environment can be used as effective tracers for biochemical processes (sedimentation processes) and for sea currents and Aarkrog (1998) recommends that environmental scientists should concentrate on radiological studies of both the marine and terrestrial environments and consider the whole global ecosystem in its entirety.

Trends in radioecology at the turn of the millennium have been presented in detail by Aarkrog (2000b).

Papers dealing with bioconcentration factors of radionuclides for marine fauna and flora as well as transfer factor for particular trophic levels with a special emphasis to man should be continued (Skwarzec and Bojanowski, 1992; Holm, 1995; Skwarzec, 1997).

References

Aarkrog, A., 1998. A retrospect od anthropogenic redioactivity in the global marine environment. Radiat. Protect. Dosimetry **75**, 23–31.

Aarkrog, A., 2000a. A retrospect of earlier EU-studies of the radiological consequences of radioactive discharges to the aquatic environment, in: The Radiological Exposure of the Population of the European Community to Radioactivity in the Baltic Sea. Marina-Balt Project, ed. S.P. Nielsen. Proceedings of a Seminar held at Hasseludden Conference Centre, Stockholm, 9–11 June 1998, European Commission, Directorate-General Environment, EUR 19200 EN (European Communities, 2000, Belgium), pp. 321–332.

Aarkrog, 2000b. Trends in radioecology at the turn of the millennium. J. Environ. Radioactivity **49**, 123–125.

Aarkrog, A., H. Dahlgaard and S.P. Nielsen, 2000. Environment radioactive contamination in Greenland: a 35 years retrospect. Sci. Total Environ. **245**, 233–248.

BEIR, 1990. Health effects of exposure to low levels of ionizing radiation, in: BEIR V Report, Committee on the Biological Effects of Ionizing Radiation (National Academy of Sciences, Washington, National Academic Press).

Belfroid, A.C., M. Purperhart and F. Ariese, 2000. Organotin levels in seafood. Mar. Pollut. Bull. **40**, 226–232.

Boyarchuk, K.A., 1998. New approach to the satellite monitoring of radioactive pollution. First International Symposium, IEP '98 Issues in Environmental Pollution, The State and Use of Science and Predictive Models (Elsevier Science Ltd., Denver, Colorado, U.S.A.) 23–26.08.1998, Section of Abstract Book 5.05.

Dabeka, R.W., and A.D. McKenzie, 1995. Survey of lead, cadmium fluoride, nickel and cobalt in food composites and estimation of dietary intakes of these elements by Canadian in 1986–1988. J. AOAC Int. **78**, 897–909.

Dornheim, H., 1969. Beitrage zur Biologie der Garnele *Crangon crangon* (L.) in der Kieler Bucht, in: Berichte der Deutschen Wissenschaftlich Kommission für Meerforschung. Neue Folge-Band XX, 179–215.

Fant, M.L., M. Nyman, E. Helle and E. Rudbäck, 2001. Mercury, cadmium, lead and selenium in ringed seals (*Phoca hispida*) from the Baltic Sea and from Svalbard. Environ. Pollut. **111**, 493–501.

FAO/WHO, 1972. Evaluation of certain food additives and the contaminants. WHO Technical Report Series **No. 776**.

FAO/WHO, 1989. Evaluation of certain food additives and the contaminants mercury, lead and cadmium. WHO Technical Report Series **No. 505**.

FAO Yearbook, 1993. Fishery Statistics, Commodities, **Vol. 77**.

FAO Yearbook, 1994. Fishery Statistics, Catches and Landings.**Vol. 78**.

FAO Yearbook, 1995. Fishery Statistics, Commodities, **Vol. 81**.

Gajewska, R., E. Malinowska, M. Nabrzyski and Z. Ganowiak, 2000. Porównanie zawartości rtęci w rybach bałtyckich poławianych w latach 1971–1997 (A comparative study on total mercury content of the Baltic fish 1971–1997). Bromat. Chem. Toksykol. **XXXIII**, 233–236 (in Polish, with English summary).

Hagel, P., 2000. Survey of the quantities and utilisation of marine products, in: The Radiological Exposure of the Population of the European Community to Radioactivity in the Baltic Sea. Marina-Balt Project, ed. S.P. Nielsen. Proceedings of a Seminar held at Hasseludden Conference Centre,

Stockholm, 9–11 June 1998, European Commission, Directorate-General Environment, EUR 19200 EN (European Communities, 2000, Belgium), pp. 131–175.

Hansen, J.C., U. Tarp and J. Bohm, 1990. Prenatal exposure to methyl mercury among Greenlandic Polar Inuits. Arch. Environ. Health **45**, 355–358.

Hoch, M., 2001. Organotin compounds in the environment – an overview. Appl. Geochem. **16**, 719–743.

Holm, E., 1995. Plutonium in the Baltic Sea. Appl. Radiat. Isot. **46**, 1225–1229.

ICES CM, 1995/Assess: 13, Report of the Baltic Fisheries Assessment Working Group.

ICES CM, 1995/Assess: 18, Report of the Working group on the Assessment of Demersal and Pelagic Stocks in the Baltic.

ICRP – International Commission on Radiological Protection, 1979. Publication 30, Annales of the ICRP 3 (Pergamon Press, Oxford).

ICRP – International Commission on Radiological Protection, 1986. Publication 48, Annales of the ICRP 16 (Pergamon Press, Oxford).

ICRP – International Commission on Radiological Protection, 1991. Publication 60, Annales of the ICRP 26 (Pergamon Press, Oxford).

Iosjpe, M, and P. Strand, 1998. Some aspects of modelling of radiological consequences from releases into marine environment, in: First International Symposium, IEP '98 Issues in Environmental Pollution, The State and Use of Science and Predictive Models (Elsevier Science Ltd.), Denver, Colorado, U.S.A. 23–26.08.1998, Section of Abstract Book 5.11.

Jagiellak, J., 1989. Źródła promieniowania jonizującego i ocena równoważnika dawki otrzymanej przez ludność Polski. Bezpieczeństwo i ochrona radiologiczna. Biuletyn Informacyjny **2**, 28–31 (in Polish).

Johansen, P., T. Pars and P. Bjerregaard, 2000. Lead, cadmium, mercury and selenium intake by Greenlanders from local marine food. Sci. Total Environ. **245**, 187–194.

Kannan, K., and J. Falandysz, 1997a. Butyltin residues in sediment, fish, fish-eating birds, harbour porpoise and human tissues from the Polish coast of the Baltic Sea. Mar. Pollut. Bull. **34**, 203–207.

Kannan, K., and J. Falandysz, 1997b. Response to the comment on: Butyltin residues in sediment, fish, fish-eating birds, harbour porpoise and human tissues from the Polish coast of the Baltic Sea. Mar. Pollut. Bull. **38**, 61–63.

Kautsky, N., 1982. Growth and size structure in a Baltic *Mytilus edulis* population. Mar. Biol. **68**, 117–133.

Keithly, J.C., R.D. Cardwell and G. Henderson, 1997. Tributyltin in seafood from Asia, Australia, Europe, and North America, in: Harmful Effects of the Use of Antifouling Paints for Ships (Parametrix, Kirkland, Washington), pp. 79–93.

Nielsen, S.P., 1995. A box model for North-East Atlantic coastal waters compared with radioactive tracers. J. Mar. Syst. **6**, 545–560.

Nielsen, S.P., 2000a. Modelling and assessment of doses to man, in: The Radiological Exposure of the Population of the European Community to Radioactivity in the Baltic Sea. Marina-Balt Project, ed. S.P. Nielsen. Proceedings of a Seminar held at Hasseludden Conference Centre, Stockholm, 9–11 June 1998, European Commission, Directorate-General Environment, EUR 19200 EN (European Communities, 2000, Belgium), pp. 177–311.

Nielsen, S.P., 2000b. Conclusions and recommendations, in: The Radiological Exposure of the Population of the European Community to Radioactivity in the Baltic Sea. Marina-Balt Project, ed. S.P. Nielsen. Proceedings of a Seminar held at Hasseludden Conference Centre, Stockholm, 9–11 June 1998, European Commission, Directorate-General Environment, EUR 19200 EN (European Communities, 2000, Belgium), pp. 313–317.

Nielsen, S.P., M. Ohlenschlæger and O. Karlberg, 1995. The radiological exposure of man from ingestion of Cs-137 and Sr-90 in seafood from the Baltic Sea (Risø National Laboratory) Risø-R-819 (EN).

Nielsen, S.P., M. Iosjpe and P. Strand, 1997. Collective doses to man from dumping of radioactive waste in the Arctic Seas. Sci. Total Environ. **202**, 135–146.

Nielsen, S.P., P. Bengtson, R. Bojanowski, P. Hagel, J. Herrmann, E. Ilus, E. Jakobson, S. Motiejunas, Y. Panteleev, A. Skujna and M. Suplinska, 1999. The radiological exposure of man from radioactivity in the Baltic Sea. Sci. Total Environ. **237/238**, 133–141.

Öst, M., and M. Kilpi, 1997. A recent change in size distribution of blue mussels (*Mytilus edulis*) in the western part of the Gulf of Finland. Ann. Zool. Fennici **34**, 31–36.

Pietrzak-Flis, Z., E. Chrzanowski and S. Dembinska, 1997. Intake of ^{226}Ra, ^{210}Pb and ^{210}Po with food in Poland. Sci. Total Environ. **203**, 157–165.

Pietrzak-Flis, Z., L. Rosiak, M.M. Suplinska, E. Chrzanowski and S. Dembinska, 2001. Daily intakes of ^{238}U, ^{234}U, ^{232}Th, ^{230}Th, ^{228}Th and ^{226}Ra in the adult population of central Poland. Sci. Total Environ. **273**, 163–169.

Ponce, R.A., E.Y. Wong and E.M. Faustman, 2001. Quality adjusted life years (QALYs) and dose-response models in environmental health policy analysis – methodological considerations. Sci. Total Environ. **274**, 79–91.

Robinson, S., J. Volosin, J. Keithly and R. Cardwell, 1999. Comment on: Butyltin residues in sediment, fish, fish-eating birds, harbour porpoise and human tissues from the Polish coast of the Baltic Sea (Kannan and Falandysz 1997). Mar. Pollut. Bull. **38**, 57–61.

Senthilkumar, K., C.A. Duda, D.L. Villeneuve, K. Kannan, J. Falandysz and J.P. Giesy, 1999. Butyltin compounds in sediment and fish from the Polish coast of the Baltic Sea. Environ. Sci. Pollut. Res. **6**, 200–206.

Skwarzec, B., 1997. Polonium, uranium and plutonium in the southern Baltic Sea. Ambio **26**, 113–117.

Skwarzec, B., and R. Bojanowski, 1992. Distribution of plutonium in selected components of the Baltic ecosystem within the Polish economic zone. J. Environ. Radioactivity **15**, 249–263.

Tanabe, S., 1999. Butyltin contamination in marine mammals – a review. Mar. Pollut. Bull. **39**, 62–72.

Thiessen, K.M., M.C. Thorne, P.R. Maul, G. Pröhl and H.S. Wheater, 1999. Modelling radionuclide distribution and transport in the environment. Environ. Pollut. **100**, 151–177.

UNSCEAR (United Nation Scientific Committee on the Effects of Atomic Radiation), 1982. Sources and Effect of Ionizing Radiation ((United Nations, New York).

UNSCEAR, 2000. Radiological Consequences of Chernobyl Accident: UN Scientific Committee on Effects of Atomic Radiation confirms earlier IAEA assessment. Sci. Total Environ. **258**, 209.

Van Oostdam, J., A. Gilman, E. Dewailly, P. Usher, B. Wheatley, H. Kuhnlein, S. Neve, J. Walker, B. Tracy, M. Feeley, V. Jerome and B. Kwavnick, 1999. Human health implications of environmental contaminants in Arctic Canada: a review. Sci. Total Environ. **230**, 1–82.

WHO, 1990. Environmental health criteria. Methyl Mercury. International Programme on Chemical Safety, **Vol. 101**.

WHO, 1993. 41st Report of the Joint Expert Committee on Food Additives (JEFCA).

Chapter 9

Global Input of Chemical Elements and Pollution Status of the Baltic Sea

(i) Introduction

The ‘industrial revolution’ began in the eighteenth century in England but in the countries around the Baltic Sea it started later in the 1850–1860s. Industrial production has grown steadily, particularly from the 1950s until the present. In consequence, large quantities of various chemical anthropogenically-derived compounds introduce to the Baltic Sea every day. These substances come from land and marine point sources such as industrial plants, power plants, waste disposal site, waste water treatment plants as well as from diffuse, non-point sources through rivers or land run-off, e.g. agricultural pollution, domestic waste and traffic (Backlund et al., 1993). Riverine and direct point sources of load of nutrients, i.e. N and P as well as heavy metals, i.e. Cd, Cu, Hg, Pb and Zn into the Baltic Sea by particular subregions have been estimated by HELCOM (1998). Moreover, both the point source and diffuse loads of nutrients given for particular Baltic countries have been estimated there. The Baltic drainage basin also receives different pollutants from long-range atmospheric transport from British Isles, Central and Eastern Europe, and even from more remote regions. There are numerous anthropogenic emitters in the countries bordering the Baltic Sea. The structure of industry is in principle different in particular Baltic countries. The metal, pulp and paper industries are the most important branches in Sweden and Finland. Food industry dominates in Denmark while industrial structure in Germany is a very diversified. Industries in these countries have generally advanced and hence direct pollutant emissions have been significantly decreased over last two decades. However, still actual problems are connected with the diffuse sources of toxic substances and they remain to be solved. On the other hand, in countries of the former communist block many industrial plants have outdated technology. In those countries there are problems associated with waste handling and therefore excessive quantities of nutrients and industrial pollutants are transported to the Baltic via rivers (Backlund et al., 1993; HELCOM, 1998). The amounts and type of pollutants are therefore considerably different in particular sub-areas of the Baltic. For instance, the load of air-borne pollutants is higher in the southern than in the northern part because the former part is more densely populated and more heavily industrialised. Moreover, in the south there is more extended atmospheric transport of pollutants from remote areas (Pacyna, 1984; Pacyna et al., 1984). On the other hand, the Bothnian Sea and the Bothnian Bay are mainly supplied with pollutants by sources in Sweden and Finland, although some minor their amounts reach this area by means of water currents and winds from the south. It is supposed that various pollutants of the industrial wastes discharge to Lake Ladoga and next follow the water through the Neva River to the Gulf of Finland and thus affect the Baltic (Bruneau, 1980). The Baltic Proper is heavily polluted by sources located along the eastern and south-eastern coasts. On the Swedish side, the water entering the Baltic originates partly from the central industrial district with numerous old mines and steel mills, refinery and ammonia plants and others. In Russia on the southeast side of the Baltic there are fertiliser plants and paper mills. Especially the Polish rivers (Vistula and Oder), the St. Petersburg area and the northern Estonia, Latvia and Lithuania contribute considerably to the high total emissions of pollutants to the Baltic Proper (Bruneau, 1980; Backlund et al., 1993; Enell, 1996; Tammemäe, 1998).

(ii) Chemical Budget

The first available information on trace element inputs to the Baltic Sea appeared in the 1970s (Suess and Erlenkeuser, 1975; Åkerblom, 1977). However, mass balances for trace elements and nutrients apart further input data for the Baltic Sea were published later (Hallberg, 1979; Pawlak, 1980; Rodhe et al., 1980; Dybern and Fonselius, 1981; Boström et al., 1983; Brügmann, 1986) and in the more recent reports (Lithner et al., 1990; Löfvendahl, 1990; Brügmann and Lange, 1990; Brügmann et al., 1991/1992; Hallberg, 1991; HELCOM, 1991, 1993; Kihlström, 1992; Pacyna, 1992, 1993; Forsberg, 1993; Backlund et al., 1993; Brügmann, 1994; Wulff et al., 1994, 1996; Schneider, 1995; Enell, 1996; Brügmann and Matschullat, 1997; Brügmann et al., 1997; Matschullat, 1997; Danielsson, 1998). Inputs of Fe and Mn to the Baltic have been reported by Blazhchishin (1982). Mass balance for As, Ge and Sb in the Baltic Sea was estimated by Andreae and Froelich (1984). A budget for chemical elements was also calculated for surrounding areas, e.g. the German Bight in the North Sea (Kühn et al., 1992; Beddig et al., 1997; Puls et al., 1997; Radach and Heyer, 1997; Sündermann and Radach, 1997).

Brügmann and Matschullat (1997) have evaluated the mass balances for Cd, Cu, Hg, Pb and Zn in the Baltic Sea and shown that 47% Zn, 34% Cu, 28% Pb,

25% Hg and 20% Cd introduced into the Baltic each year are fixed in the sediments. It is also shown that 65% Pb, 51% Zn, 48% Cd, 13% Cu and 11% Hg are introduced into the Baltic from the atmosphere in respect to the combined atmospheric and fluvial (riverine, industrial and municipal) inputs of these trace elements. Budgets for trace elements and nutrients in the Baltic Sea are presented in Figs. 9.1 and 9.2.

Matschullat (1997) presented in reliable manner the total riverine and atmospheric inputs of selected trace elements into the Baltic Sea. It is pointed out that the atmospheric input of many anthropogenically-derived trace elements. e.g. Cd, Cu and Pb exceeded their riverine input which is in an agreement with the HELCOM (1991) report emphasising atmospheric input as a predominant transport of some heavy metals. Annual atmospheric and riverine inputs of selected trace elements into the Baltic Sea are presented in Fig. 9.3. Elemental inputs with distinguishing between natural river and atmospheric loads as well as the respective anthropogenic contribution are listed in Table 9.1. While Cd, Cu, Pb and Zn are characterised almost by identical inputs, As, Co, Cr, Hg and Ni are generally transported via the rivers. An anthropogenic share of the total load is very high (> 70%) for As, Cd, Cu, Hg, Pb and Zn; less impressive values of anthropogenic input are obtained for Co, Cr and Ni (44–57%) (Table 9.1). According to Andreae and Froelich (1984) between ca. 12 and 26% of the emitted As, Sb and Ge end up in the Baltic Sea, i.e. 281 x 10^6, 75 x 10^6 and 46 x 10^6 g, were deposited annually in the Baltic Sea, respectively. The ratio between the atmospheric and riverine fluxes showed a progression for As, for which the flux to the Baltic Sea is carried mostly by the rivers; for Ge exhibiting the anomalously high molar Ge/Si ratios in the Baltic, the atmospheric transport predominates. As for Sb, the atmospheric component is also indicated to be the most important in transferring of this element to the Baltic Sea. It is concluded that anthropogenic inputs are an important component in the mass balance of As and Sb, and probably are dominant in the case of Ge (Andreae and Froelich, 1984).

(iii) Pollution Status of the Baltic Sea in Respect to other Seas

A comparison between the Baltic and Black Seas as enclosed seas under man-induced changes has been made by Leppäkoski and Mihnea (1996).

Annual loadings of Cd, Cr, Cu, Hg, Mn, Ni, Pb and Zn for the Baltic Sea, Adriatic Sea and Black Sea through wastewater and 'natural' waters have been estimated by Sekulić and Vertačnik (1997). The evaluation of data on dissolved species of trace elements concerning the Black Sea and the North Aegean Sea indicated that in this interrelated system water mass exchanges play an important role in the trace element distribution (Zeri et al., 2000). The Baltic, Adriatic and especially the Black Sea are almost closed basins. The connection with other seas has place by means of 5–15 km wide channels (Øre Sund and Femer Bælt to the North Sea) for the Baltic Sea, 1–5 km-wide channels (Bosphorus and Dardanelles to the Aegean Sea) for the Black Sea and ca. 80 km-wide channel (Strait of

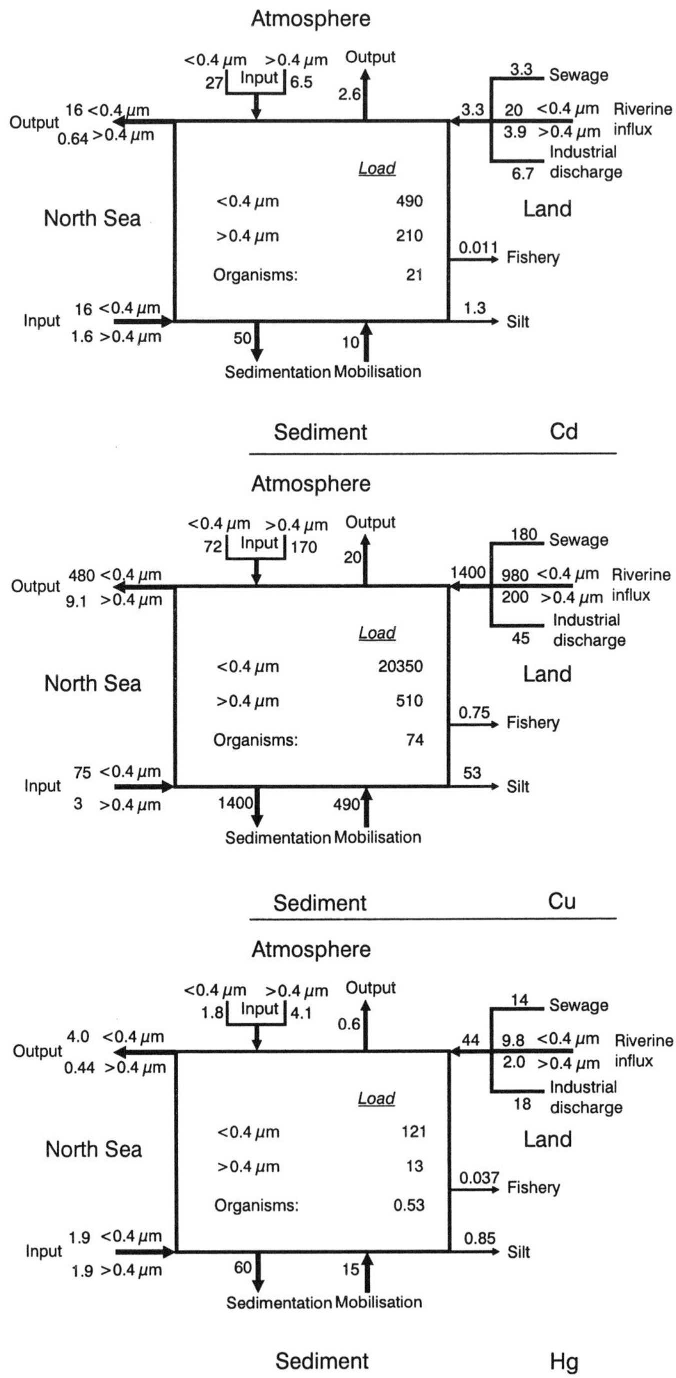

Fig. 9.1. Budgets for trace elements in the Baltic Sea. After Brügmann and Matschullat (1997); modified.

Fig. 9.1. – continued.

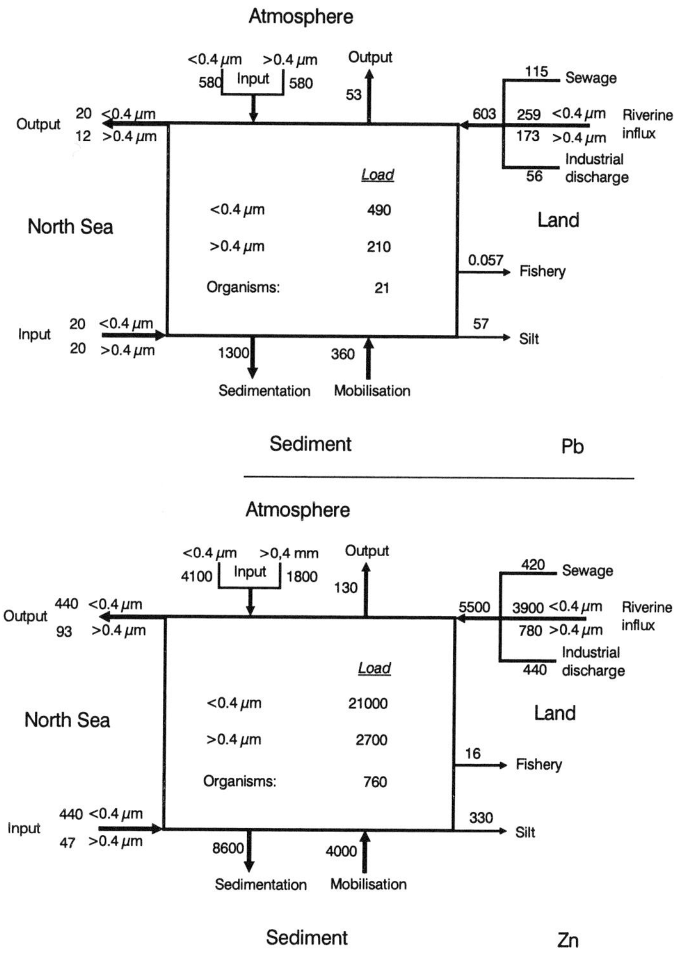

Otranto to the Mediterranean Sea) for the Adriatic Sea (Great Geographical Atlas, 1990). The closed Seas, however, differ significantly in respect to their biological and physical-chemical characteristics.

It should be emphasised that high annual input of suspended matter concerns all the three closed seas; this particulate matter, enriched in heavy metals is settled down in the vicinity of its terrestrial source and hence the concentrations of chemical pollutants are elevated exclusively in the narrowest littoral zones while their low levels are detected in the deep-sea (Sekulić and Vertačnik, 1997). Several 'black spots', e.g. great estuaries and seaport towns, heavily contaminated by chemical elements, are identified in each of the Seas (Sekulić and Vertačnik, 1997). Therefore the present pollution status has ecological implications primarily on the enhanced point-source spots. As can be seen in Fig. 9.4 among these Seas,

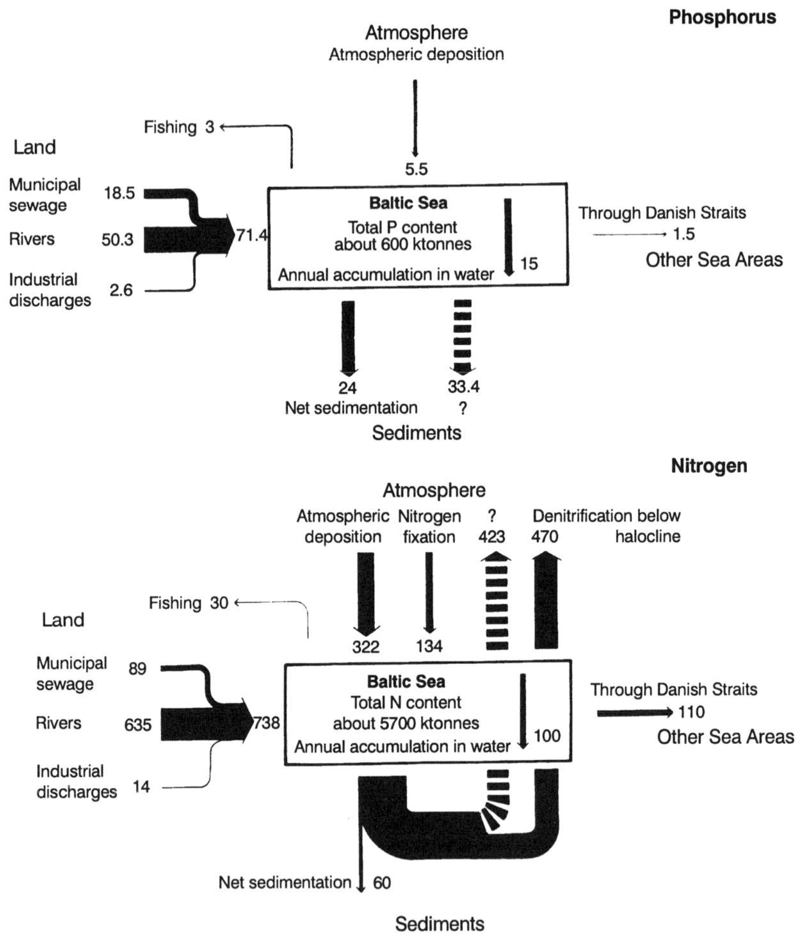

Fig. 9.2. Budget for nutrients in the Baltic Sea, expressed as thousands of tons per year. After Forsberg (1993); modified.

the Baltic is the most heavily loaded with trace elements. In contrast, the Adriatic is characterised by the lowest loading taken absolutely and relatively compared to its volume, while the Black Sea, and especially the Baltic have significantly higher loading. The amounts of 'natural' waters are several orders of magnitude higher than 'anthropological' waters. Owing to the expected 'natural' input of chemical substances this loading highly exceeds the anthropological one. Relatively low levels found in the Seas mean that these great natural systems are very stable, with a great autopurification possibilities.

(iv) Status of the Baltic in Past, Today and in the Future

One hundred years ago the Baltic was a clean oligotrophic sea (Jonsson, 1992). In the 1940s, the Baltic Sea was a basin poor in nutrients, hence was char-

TABLE 9.1.

Element input to the Baltic Sea (t yr^{-1}), split by natural river and atmospheric inputs, total fluvial and airborne inputs, the sum total for the Baltic Sea and the respective anthropogenic share. After Matschullat (1997); modified

	As	Cd	Co	Cr	Cu	Hg	Ni	Pb	Zn
Natural fluv.	58	5.1	120	270	310	4.9	165	140	1700
Natural atm.	10	3	3	20	130	0.1	5	20	350
Σ Natural	68	8.1	123	290	440	5	170	160	2050
Fluv. + diffuse	200	60	200	440	1300	50	300	1500	6000
Atmospheric	50	60	20	100	1200	20	100	1300	5000
Σ Baltic Sea	250	120	220	540	2500	70	400	1800	11000
% Anthropog.	73	93	44	46	82	93	57	91	81

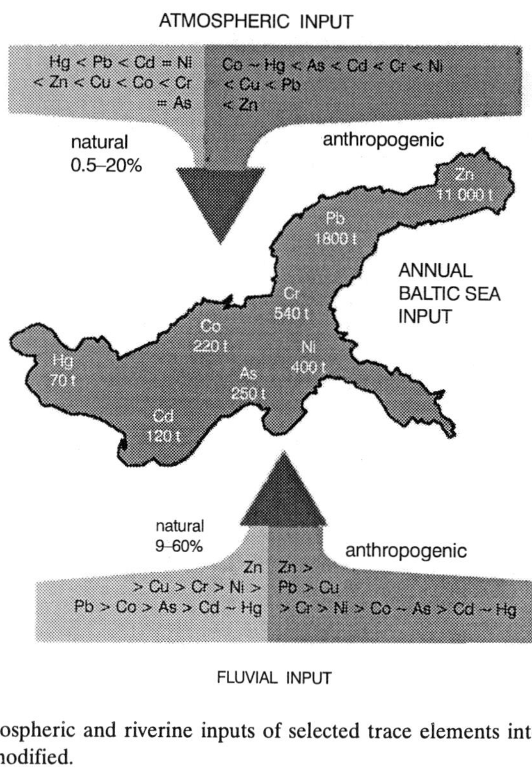

Fig. 9.3. Annual atmospheric and riverine inputs of selected trace elements into the Baltic Sea. After Matschullat (1997); modified.

acterised by low biological production, clear water and rocky shores densely overgrown by the bladderwrack, providing food and shelter for many species. Since 1900, there has been a 4-fold increase in N input and 8-fold increase in P input into the Baltic Sea. Its bottom waters were then sufficiently oxygenated creating favourable conditions for spawning of cod in the deep areas of the Baltic Proper, except of periods of oxygen depletion in the Gotland Deep (Jansson and Dahl-

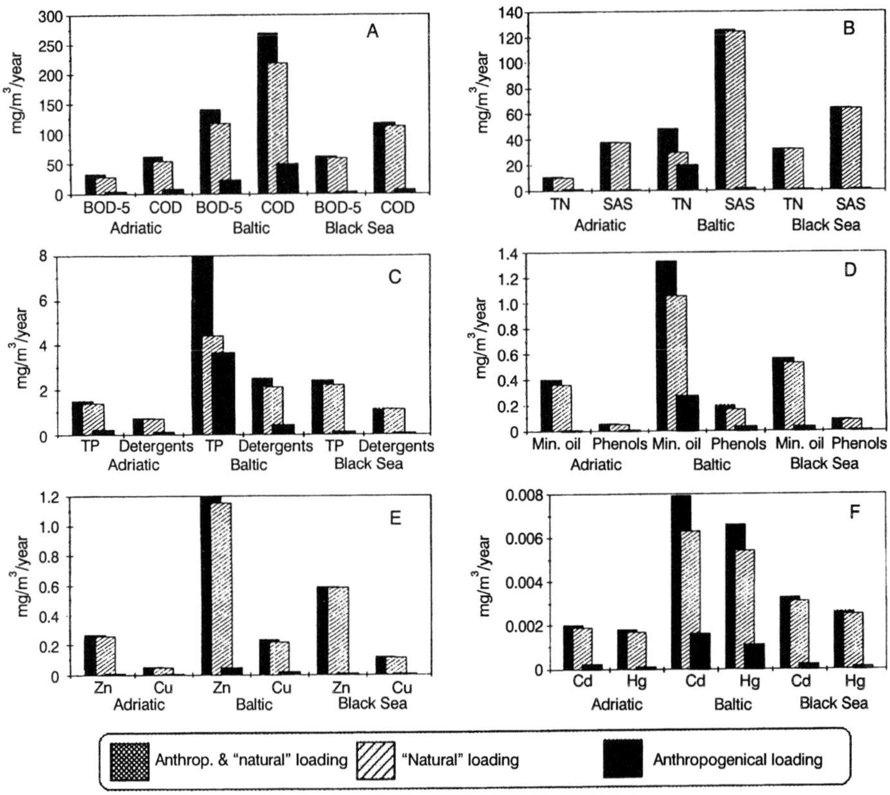

Active volume: Adriatic Sea ~35 000 km^3, Baltic Sea ~22 000 km^3, Black Sea ~50 000 km^3

Fig. 9.4. Comparison of specific loading of sea active volumes mg m^{-3} year^{-1}: (A), BOD_5 and COD; (B), total nitrogen (TN) and surface active substances (SAS); (C), total phosphates (TP) and detergents; (D), mineral oils and phenols; (E), Zn and Cu; and (F), Cd and Hg loading through waters. After Sekulić and Vertačnik (1997); modified.

berg, 1999). Top levels of the food chain like seals, harbour porpoises and sea eagles were common and people living along the Baltic coasts could eat fish without risking their health.

The today's Baltic status is different. The impact of Man on the Baltic ecosystem in the 20th century is summarising by Elmgren (1989). Eutrophication and toxic compounds now affect the whole area of its ecosystem, even open sea subareas. The bladderwrack has been shaded or even totally replaced, e.g. in the Southern Baltic, by filamentous green and brown algae as a result of increased plankton blooms and organic particle production. Moreover, a light penetration is lowered by 3 m; the O_2 content of waters below the halocline has gradually declined over this period and H_2S in these waters sometimes dominated (Jansson and Dahlberg, 1999). Extinction of macrofauna in these anoxic areas has led to the formation of laminated sediments, the area of which has increased 4-times since 1940 (Jonsson and Jonsson, 1988; Jonsson et al., 1990; Jonsson, 1992; Jonsson and Carman, 1994; Persson and Jonsson, 2000). Marine mammals and waterfowls have suffered from declines in reproduction rates caused by increased inputs of organic contaminants although a slow recovering of seals and sea eagles is now observed. The bulk of metals introduced into the Baltic is of anthropogenic in origin, e.g. more than 90% of Cu, Zn and Pb and almost 60% of Ag. According to Kihlström (1992) and Jansson and Dahlberg (1999), anthropogenic loads of Cd, Hg, Pb and Zn to the Baltic Proper are 2.3- to 12.6-times higher than those corresponding to the natural loads. However, there is some evidence that heavy metal concentrations in surficial sediments have not increased significantly in the offshore areas over the last decade (HELCOM, 1996). Recent sediments exhibit metal levels starting to rise during the 1950s and attaining a maximum values during the 1960s and 1970s. However, heavy metal levels are decreasing since the 1980s (probably as a result of economical crisis of the Baltic States belonged to the former Soviet block) but are still higher than in the 1940s. It should be emphasised that sediment concentration is attributed not only to anthropogenic load of metals but also to the quantities of organic matter deposited in the sea floor, and hence they are strictly related to the eutrophication process. Some improvement in the quality of Baltic water over the last few years is also detected (HELCOM, 1996). For instance, Pb decreased in Baltic fish, perhaps as a result of the reduced air emission from car traffic (HELCOM, 1996; Jansson and Dahlberg 1999). Variations in O_2 concentrations may affect the efficiency of binding and holding of trace elements in the sediment particles. The declined heavy metal levels in the 1980s can therefore be explained also by reduced input to the Baltic Sea but also by the above factors.

The future of the Baltic ecosystem depends on the development of all the Baltic countries. A substantial contribution to the improvements achieved to date is the result of economic regress in Baltic countries of the Central and East Europe where the economies are now growing again. It should be stressed that the present ecological gains in the Baltic countries will be lost when western achievements, e.g. intensive agricultural, wetland draining and the intensive using of chemical substances production will be implanted to the Baltic countries of the former communist block (Jansson and Dahlberg 1999). The future use of the Baltic Sea basin should consider the limitations and sensitivity of the whole ecosystem to a much greater extent than it has been done up to day. Remediation of much polluted Baltic areas, i.e. more than 130 ecologically endangered areas (pollution 'hot spots') is highly desired. It would require upgrading of present industrial systems and adoption of modern sewage treatment plants. This is particularly important in the former communist countries (Glasby and Szefer, 1998). However, the remediation should not be focused exclusively on these east countries since the 70 major industrial regions and individual plants in the drainage basin are scattered around the Baltic Sea (Backlund et al., 1993). For instance, in Estonia and the Latvia, the former Soviet military navy bases, there are the largest military objects, most polluted with oil products as well as different chemicals and heavy metals (Tammemäe, 1998). Another example of pollution emitter is the Swedish pulp and paper industry removed annually to the Baltic Sea significant quantities of Pb (4 tonnes) and other metals, nutrients, i.e. 400 and 3100 tonnes of P and N, respectively (Enell, 1996). According to Governmental Regulation No. 294 there are possibilities to provide means for 'environmental investigations and analyses' at the privatised objects as well as for 'remediation and sanitation of soil, ground water bodies within polluted areas' in Estonian and Latvian former Soviet military navy bases in the Baltic (Tammemäe, 1998). HELCOM has defined a goal of 'Forward to 1950' which would bring the Baltic back to the relatively clean state that existed in 1950. It is estimated that this project would cost 7 billion ECU over a period of 60 years. To realise this goal, it should be set up sophisticated programs to monitor environmental quality (Anon, 1990). Brügmann and Matschullat (1997) have recommended the need for an increase in the number of sampling stations and of elements analysed as well as improved analytical precision if accurate mass balances for a larger number of elements in the Baltic Sea to be estimated. Although it seems impossible to restore the Baltic to the exact ecological status that existed in 1950, however it would be possible to restore the water quality of this Sea and improve its trophic situation so more that some symptoms of improvement in the quality of the Baltic waters over the last decade is an encouraging first announcement.

Visions of sustainability in the Baltic Sea region (BSR) have been presented by Raskin et al. (1998). Dreborg et al. (1999) described integrated scenarios of the BSR to the year 2030. The scenario exercise suggests several preconditions, necessary for the BSR to realise a sustainability vision; they include the development and diffusion of clean and efficient technologies, reorientation of consumer demand towards less resource-intensive products, public support for strong sustainability policies and a co-operative climate between nations in the BSR.

(v) General Remarks and Recommendations

The input of selected chemical elements to the Baltic Sea is not well known. It is mainly caused by incomplete data base, uncertainties about how much of the gross input finally arrives in the open sea and the very limited knowledge of the nature and extent of the exchange at the interface with the sea floor and with the atmosphere (Brügmann, 1994). Although the data matrix utilised to mass balance calculations has been recently improved significantly, it is still impossible to present quite precise data on trace element inputs to the Baltic Sea. A main reason for it is that reliable data for the water discharge of the rivers alone have not been available up to date. This problem is particularly connected with incomplete pollution data given by some Baltic countries, because a lot of the figures were only estimated as totals by sub-regions, and figures from the Kaliningrad region have been missing completely. Many uncertainties remain with regard to trace elements due to incomplete load data sets, especially from Russia and partly from all Contracting Parties (HELCOM, 1998). It is well known that the major part of the pollution load is transported by rivers to the Baltic Sea. It is therefore an important task to start investigations on collecting load data for point and diffuse sources comprised the entire Baltic catchment area. On the other hand the EMEP and HELCOM atmospheric networks are still extending and focused on macroelements, e.g. S, N, P in fallout (Matschullat, 1997). National and international research programmes should increase our knowledge of the chemistry, biology, hydrography and meteorology of the Baltic Sea and provide us with indispensable information for effective protective measures (Rheinheimer, 1998). The protective measures brought about by HELCOM have already contributed to an improvement upon ecological situation in the Baltic Sea. For instance, modern, more effective treatment plants for sewage have been constructed. The concentrations of some pollutants, e.g. Pb in water decreased significantly. According to Rheinheimer (1998) the harmful substances which accumulated in the sediments, however, will still pose a threat for the near future. The atmospheric inputs are still on a large scale. The progress in traffic, industrial production and tourism, particularly in the eastern countries, will probably contribute to an increase in pollution at least at this early stage before an achievement of improving in the treatment plants for sewage and exhaust gases (Rheinheimer, 1998).

Successful chemical balancing requires obviously high quality data. In order to obtain relevant and reliable data sets in future estimate approaches it is essential that the laboratories continue the implementation of the quality assurance programme certified by international accreditation. It however needs time and the laboratories in the eastern Baltic countries still require support in terms of training and founding for improvement of analytical equipment development of analytical skills (HELCOM, 1998).

References

Anon, 1990. Status of the Baltic Sea – a sea in transition. Ambio **4**, 24 pp.

Åkerblom, A. (ed.), 1977. 3rd Soviet-Swedish Symposium on the Pollution of the Baltic. Ambio Spec. Rep. (Stockholm, Sweden), **5**, 294 pp.

Andreae, M.O., and P.N. Froelich, Jr, 1984. Arsenic, antimony, and germanium biogeochemistry in the Baltic Sea. Tellus **36B**, 101–117.

Backlund, P., B. Holmbom and E. Leppäkoski, 1993. Industrial Emissions and Toxic Pollutants. The Baltic Sea Environment (Uppsala University, Sweden) **Session 5**, 36 pp.

Beddig, S., U. Brockmann, W. Dannecker, D. Körner, T. Pohlmann, W. Puls, G. Radach, A. Rebers, H.-J. Rick, M. Schatzmann, H. Schlünzen and M. Schulz, 1997. Nitrogen fluxes in the German Bight. Mar. Pollut. Bull. **34**, 382–394.

Blazhchishin, A.I., 1982. Main chemical constituents of the sediments of the Baltic Sea, eds. V.K. Gudelis and E.M. Emelyanov, Geology of the Baltic Sea (Wydawnictwo Geologiczne, Warsaw), 257–289 (in Polish).

Boström, K., J.-O. Burman and J. Ingri, 1983. A geochemical massbalance for the Baltic. Environ. Biogeochem. Ecol. Bull. (Stockholm) **35**, 39–58.

Bruneau, L., 1980. Pollution from industries in the drainage area of the Baltic. Ambio **9**, 145–152.

Brügmann L., 1981. Heavy metals in the Baltic Sea. Mar. Pollut. Bull. **12**, 214–218.

Brügmann, L., 1986. The influence of coastal zone processes on mass balances for trace metals in the Baltic Sea. Rapp. P.-v. Réun. Cons. int. Explor. Mer **186**, 329–342.

Brügmann, L., 1994. Effects of toxic metal pollutants on the ecology of the Baltic Sea, in: Use of Mechanistic Information in Risk Assessment, eds. H.M. Bolt, B. Hellman and L. Dencker. Proc. of the 1993 EUROTOX Congres, Upsalla, Sweden, June 30–July 3, 1993 (Springer-Verlag, Berlin Heidelberg New York), pp. 32–42.

Brügmann, L., and D. Lange, 1990. Metal distribution in sediments of the Baltic Sea. Limnologica **20**, 15–28.

Brügmann, L, and J. Matschullat, 1997. Zur Biogeochemie und Bilanzierung von Schwermetallen in der Ostsee, in: Geochemie und Umwelt-Umweltrelevante Prozesse in Atmo-, Pedo- and Hydrosphäre, eds. J. Matschullat, H.J. Tobschall, H.J. Vogt (Springer Verlag, Berlin) pp. 267–289 (in German).

Brügmann, L., H. Gaul, K.-H. Rohde and U. Ziebarth, 1991/92. Regional distribution and temporal trends of some contaminants in the water of the Baltic Sea. Dr. Hydrogr. Z. **44**, 161–184.

Brügmann, L., R. Hallberg, C. Larsson and A. Löffler, 1997. Changing redox conditions in the Baltic deep basins: Impacts on the concentration and speciation of trace metals. Ambio **26**, 107–112.

Danielsson, Å., 1998. Spatial Modelling in Sediments (Linköping Studies in Arts and Science, Sweden). 89 pp. + Appendices.

Dreborg, K.-H., S. Hunhammar, E. Kemp-Benedict and P. Raskin, 1999. Scenarios for the Baltic Sea region: a vision of sustainability. Int. J. Sustain. Dev. World Ecol. **6**, 34–44.

Dybern, B.I. and S.H. Fonselius, 1981. Pollution, in: The Baltic Sea, ed. A. Voipio (Elsevier Scientific Publishing Company, Amsterdam), pp. 351–382.

Elmgren, R., 1989. Man's impact on the ecosystem of the Baltic Sea: energy flows today and at the turn of the century. Ambio **18**, 326–332.

Enell, M, 1996. Load from the Swedish pulp and paper industry (nutrients, metals and AOX): Quantities and shares of the total load on the Baltic Sea, in: Environmental Fate and Effects of Pulp and Paper Mill Effluents, eds. M.R. Servos, K.R. Munkittrick, J.H. Carey, G.J., Van der Kraak (St. Lucie Press, Delray Beach, Florida) 229–237.

Forsberg, C., 1993. Eutrophication of the Baltic Sea. The Baltic Sea Environment (Uppsala University, Sweden) **Session 3**, 32 pp.

Glasby, G. P., and P. Szefer P., 1998. Marine pollution in Gdansk Bay, Puck Bay, and the Vistula Lagoon, Poland – An overview. Sci. Total Environ. **212**, 49–57.

Great Geographical Atlas, Revised 1990 Edition (Printed in the United States of America by Rand McNally & Company), 304 pp. + 144 pp. of Map Index.

Hallberg, R.O., 1979. Heavy metals in the sediments of the Gulf of Bothnia. Ambio **8**, 265–269.

Hallberg, R.O., 1991. Environmental implications of metal distribution in Baltic Sea sediments. Ambio **20**, 309–316.

HELCOM, 1991. Nitrogen and Agriculture International Workshop. Baltic Sea Environment Proceedings **No. 45**.

HELCOM, 1993. Second Baltic Sea Pollution Load Compilation. Baltic Sea Environmental Proceedings **45**.

HELCOM, 1996. Third Periodic Assessment of the State of the Marine Environment of the Baltic Sea, 1989–93; Background Document. Baltic Sea Environment Proc. (Baltic Marine Environment Protection Commission, Helsinki) **No. 64B**.

HELCOM, 1998. The third Baltic Sea Pollution Load Compilation (PLC-3). Baltic Sea Environment Proceedings **No. 70**.

Jansson, B.-O., and K. Dahlberg, 1999. The environmental status of the Baltic Sea in the 1940s, today, and in the future. Ambio **28**, 312–319.

Jonsson, P., 1992. Large-scale changes of contaminants in the Baltic Sea sediments during the twentieth century. Acta Universitatis Upsaliensis, Comprehensive Summaries of Uppsala Dissertations from the Faculty of Science 407, 52 pp. + Appendices.

Jonsson, P., and B. Jonsson, 1988. Dramatic changes in Baltic sediments during the last three decades. Ambio **17**, 158–160.

Jonsson, P., and R. Carman, 1994. Changes in deposition of organic matter and nutrients in the Baltic Sea during the twentieth century. Mar. Pollut. Bull. **28**, 417–426.

Jonsson, P., R. Carman and F. Wulff, 1990. Laminated sediments in the Baltic – A tool for evaluating nutrient mass balances. Ambio **19**, 152–158.

Kihlström, J.K., 1992. Toxicology – The Environmental Impact of Pollutants. The Baltic Sea Environment (Uppsala University, Sweden) **Session 6**, 30 pp.

Kühn, W., G. Radach and M. Kersten, 1992. Cadmium in the North Sea – a mass balance. J. Mar. Systems **3**, 209–224.

Leppäkoski, E., and P.E. Mihnea, 1996. Enclosed seas under man-induced change: a comparison between the Baltic and Black Seas. Ambio **25**, 380–389.

Lithner, G., H. Borg, U. Grimås, A. Göthberg, G. Neumann and H. Wrådhe, 1990. Estimating the load of metals to the Baltic Sea. Ambio Spec. Rep. **7** Sept., 7–9.

Löfvendahl, R., 1990. Changes in the flux of some major dissolved components in Swedish rivers during the present century. Ambio **19**, 210–219.

Matschullat, J., 1997. Trace element fluxes to the Baltic Sea: problem of input budgets. Ambio **26**, 363–368.

Pacyna, J.M., 1984. Estimation of the atmospheric emissions of trace elements from anthropogenic sources in Europe. Atmos. Environ. **18**, 41–50.

Pacyna, J.M., 1992. The Baltic Sea environmental programme. The topical area study for atmospheric deposition of pollutants. Final Technical Report and Final Synthesis Report. NILU Rep. **No. 46**, 141 pp.

Pacyna, J.M., 1993. Atmospheric deposition of heavy metals to the Baltic Sea, in: Intern. Conf. Heavy Metals in the Environment, eds R.J. Allen and J.O. Nriagu (CEP Consultants, Toronto), **1**, pp. 93–96.

Pacyna, J.M., A. Semb and J.E. Hanssen, 1984. Emission and long-range transport of trace elements in Europe. Tellus, **36B**, 163–178.

Pawlak, J., 1980. Land-based inputs of some major pollutants to the Baltic Sea. Ambio **9**, 163–167.

Persson, J., and P. Jonsson, 2000. Historical development of laminated sediments – an approach to detect soft sediment ecosystem changes in the Baltic Sea. Mar. Pollut. Bull. **40**, 122–134.

Puls, W., W. Gerwinski, M. Haarich, M. Schirmacher and D. Schmidt, 1997. Lead budget for the German Bight. Mar. Pollut. Bull. **34**, 410–418.

Radach, G., and K. Heyer, 1997. A cadmium budget for the German Bight in the North Sea. Mar. Pollut. Bull. **34**, 375–381.

Raskin, P., E. Kemp-Bendict, K.-H. Benedict and S. Hunhammar, 1998. Visions of sustainability in the Baltic Sea region: beyond conventional development. Baltic 21 Research Report (SEI, Stockholm Environment Institute – Boston Center, MA, USA) June 1998, 32 pp.

Rheinheimer, G., 1998. Pollution in the Baltic Sea. Naturwissenschaften **85**, 318–329.

Rodhe, H., R. Söderlund and J. Ekstedt, 1980. Deposition of airborne pollutants on the Baltic. Ambio **9**, 168–173.

Schneider, B., 1995. Bilanzen und Kreisläufe von Spurenmetallen in der Ostsse. Geowissenschaften **13**, 464–469.

Sekulić, B., and A. Vertačnik, 1997. Comparison of anthropological and "natural" input of substances through waters into Adriatic, Baltic and Black Sea. Wat. Res. **31**, 3178–3182.

Suess, E., and H. Erlenkeuser, 1975. History of metal pollution and carbon input in the Baltic Sea sediments. Meyniana **27**, 63–75.

Sündermann, J., and G. Radach, 1997. Fluxes and budgets of contaminants in the German Bight. Mar. Pollut. Bull. **34**, 395–397.

Tammemäe, O., 1998. Remediation of polluted environment at naval of the Baltic Sea, in: Environmental Contamination and Remediation Practises at Former and Present Military Bases, eds. F. Fonnum et al. (Kluwer Academic Publishers, the Netherlands), pp. 305–311.

Wulff, F., M. Perttilä and L. Rahm, 1994. Mass-balance calculations of nutrients and hydrochemical conditions in the Gulf of Bothnia, 1991. Aqua Fennica **24**, 121–140.

Wulff, F., M. Perttilä and L. Rahm, 1996. Monitoring, mass balance calculation of nutrients and the future of the Gulf of Bothnia. Ambio Spec. Rep. **8**, 28–35.

Zeri, C., F. Voutsinou-Taliadouri, A.S. Romanov, E.I. Ovsjany and A. Moriki, 2000. A comparative approach of dissolved trace element exchange in two interconnected basins: Black Sea and Aegean Sea. Mar. Pollut. Bull. **40**, 666–673.

Author Index

Roman type indicates references corresponding to authors within the text.
Italic type indicates references corresponding to authors within the tables.
Asterisk indicates references corresponding to authors within figure caption.

Species Index

Roman type indicates references corresponding to species within the text.
Italic type indicates references corresponding to species within the tables.
Asterisk indicates references corresponding to species within figure caption.

Subject Index